Bapan Adak and Samrat Mukhopadhyay (Eds.)
Smart and Functional Textiles

Also of interest

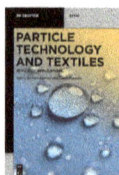

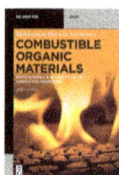

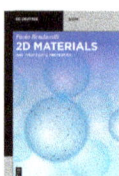

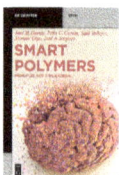

Smart and Functional Textiles

—

Edited by
Bapan Adak and Samrat Mukhopadhyay

DE GRUYTER

Editors

Dr. Bapan Adak
School of Material Science & Engineering (MSE)
Nanyang Technological University (NTU)
50 Nanyang Avenue
Block N4.1, Singapore, 639798
bapan.iitd15@gmail.com

Prof. Samrat Mukhopadhyay
Indian Institute of Technology Delhi
Department of Textile & Fibre Engineering
Hauz Khas
New Delhi 110016, India
samrat@textile.iitd.ac.in

ISBN 978-3-11-075972-3
e-ISBN (PDF) 978-3-11-075974-7
e-ISBN (EPUB) 978-3-11-075993-8

Library of Congress Control Number: 2022950923

Bibliographic information published by the Deutsche Nationalbibliothek
The Deutsche Nationalbibliothek lists this publication in the Deutsche Nationalbibliografie;
detailed bibliographic data are available on the internet at http://dnb.dnb.de.

© 2023 Walter de Gruyter GmbH, Berlin/Boston
Cover image: piranka/iStock/Getty Images Plus
Typesetting: Integra Software Services Pvt. Ltd.
Printing and binding: CPI books GmbH, Leck

www.degruyter.com

Preface

Smart textile – also known as functional fabric or e-textile – is changing the way we view fabrics for both industrial and everyday uses. By virtue of advance fibres and polymers, functional nanomaterials, nanotechnology-based processes, advanced manufacturing techniques, and smart small electronics, the smart fabrics are continually advancing in their capabilities for a wide range of technical applications. This book will cover almost all the application areas of smart and functional textiles focusing on the fundamentals, advancements, current challenges, and future perspectives.

Chapter 1 lays the foundation of the whole book. With a coverage of nano-finished and coated/laminated textiles, electroconductive textiles, stimuli-responsive smart and functional textiles, fundamentals of photochromic textiles, thermochromic textiles, concepts of 3D printing, and wearable energy devices, this chapter will provide a solid foundation for the detailed journey ahead.

Chapter 2 deals with nanotechnology in textile finishing. The use of nanotechnology in various finishes like water and oil repellency, antistatic, flame retardancy, UV protection, antimicrobial, antiviral, anti-odour, anti-crease, super absorbency, and multi-functional properties have been discussed at length. The process technique for application and mechanism of action of various nanomaterials have also been deliberated and a section on biomimicry adds to the depth.

The innovations in coating and lamination processes are covered in detail in Chapter 3, and interesting developments covered include smart coatings based on nanotechnology, electrospun nanofibre coating, and active coatings. Chapter 4 is a natural sequence of the previous chapter and deals with the speciality coating and lamination. It covers a variety of polymers and chemicals used in specialty coating and laminating of textiles, base fabric preparation, and coating and lamination procedures as well as their applications. There are detailed sections on anti-biofouling coating, electrically conductive coating which make this chapter an interesting read.

Chapter 5 attempts to offer a thorough overview and comprehension of how nanocomposites might be synthesized and applied in the rapidly developing field of smart and functional textiles. Additionally, it covers the potential of different nanoparticles for making functional polymer nanocomposites, different forms of the nanocomposites, their structure–property relationships, and the various ways by which polymer nanocomposites can be used in smart fabrics. The strength of the chapter lies in a strong focus on how these polymer nanocomposite-enhanced smart and functional textiles are used in a variety of technical fields.

Chapter 6 provides comprehensive information on electroconductive textiles made from graphene, carbon nanotubes (CNTs), MXene, and conductive polymers. All the conductive polymers have been dealt with in detail including a special section on MXenes. The chapter goes into great detail about numerous sensors such as strain, gas, pH, and humidity sensors.

https://doi.org/10.1515/9783110759747-202

The purpose of Chapter 7 is to provide comprehensive information concerning the development strategies and applications of graphene-based functional textiles. The various methods for creating functional textiles, fundamental properties of graphene, graphene synthesis routes, and graphene application methods on textile substrates have been discussed in detail along with proper schematics for understanding.

Chapter 8 discusses smart textiles for personal thermal comfort. After reviewing the fundamentals of thermal comfort, including the principles of heating and cooling and strategies for fabricating thermoregulation textiles, the different passive heating and cooling textiles have been discussed. There are sections on advanced heating and cooling which makes this chapter a very interesting read.

Chapter 9 details the recent progresses in thermochromic, photochromic, solvatochromic, electrochromic, piezochromic, and mechanochromic textiles.

As part of Chapter 10, the authors have detailed the development of protective textiles designed to resist ballistics and impacts; extreme cold weather; fire; and hazardous nuclear, biological, and chemical warfare agents. In designing and developing protective smart and functional textiles, the importance of nanotechnology has been critically highlighted.

Chapter 11 details the advancements in military and defence textiles. Radar absorbing material (RAM) coating, working mechanism of thermal infrared (TIR) stealth, multispectral camouflage net, design of high-altitude sleeping bags for military, HAPO chambers, use of advanced materials for military shelters, use of electrospun nanofibres and sensors in security and defence, programmable fibre-based military uniform, and wearable motherboard have been discussed at length.

In Chapter 12, the authors present the latest research advances on sensors and actuators for e-textiles along with their significance in a functional system. There are detailed sections on optical and chemical sensors and a very detailed section on the research possibilities with these sensors and actuators.

Chapter 13 is a critical overview of the medical textile sector. The chapter covers very interesting developments such as organic and inorganic antibacterial agents, advances in healthcare monitoring, wound dressings, compression garments, and sutures. Chapter 14 is a sequel to the previous chapter focusing on chitosan-based wound care and haemostatic systems.

Chapter 15 explores the area for energy harvesting. It covers areas like thermoelectric devices based on inorganic materials, photovoltaic energy harvesters, and critically discusses the construction of photovoltaic textiles, updates on the developments of dye-sensitized solar cells, and triboelectric energy harvesters.

In Chapter 16, the behaviour of the filter media and filtration process required for automotive filtration have been discussed in depth from the fundamental scientific concepts. Advanced filtration materials for electronic cars, improved interior air quality, enhanced thermal and sound insulation, advanced structural materials, and airbags are just a few of the topics that have been discussed.

If one is interested to know about the 3D printing processes and their developments in the textile field, Chapter 17 would be really useful. In addition to discussions on fused deposition moulding (FDM), stereolithography, and selective laser sintering (SLS), critical strategies for adhesion improvement between 3D printed objects and fabrics as well as the potential of 4D printing have also been discussed.

Chapter 18 specifically deals with the development of functional textiles during the Covid times. In addition to covering advancements in face masks, it covers electrospun nanofibres and their potential use against COVID-19, durable nanofinishes, coated and laminated liquid-proof textiles, personal protective equipments (PPEs) for single- and multiple-time uses, wearable technologies, and COVID-19-relevant textile testing as well.

Overall, the intention of the editors is to make sure that the readers get updated about the recent technology improvements without a compromise in the treatment of the fundamentals and science that underlies the basis for such intriguing developments. The future research perspectives have been a hallmark of every chapter. Editors are pleased to recognize the significant contributions each author has made to the quality and content of this book. Regardless of how hard we work or what we accomplish, there is always room for improvement. We seek criticism, feedback, and encouraging remarks from readers so that future versions can be improved.

Bapan Adak and Samrat Mukhopadhyay

Contents

Preface —— V

List of authors —— XI

Sagnik Ghosh, Bapan Adak, Samrat Mukhopadhyay
1   Introduction to functional, smart, and intelligent textiles:
    perspectives and potential applications —— 1

S. Wazed Ali, Swagata Banerjee, Mandira Mondal
2   Nanotechnology in textile finishing: an approach towards imparting
    manifold functionalities —— 63

Lelona Pradhan, Saptarshi Maiti, Aranya Mallick, Mohammad Shahid,
Sandeep P. More, Ravindra V. Adivarekar
3   Coating- and lamination-based smart textiles: techniques,
    features, and challenges —— 97

Subhankar Maity, Kunal Singha, Pintu Pandit
4   Speciality coatings and laminations on textiles —— 151

Jagadeshvaran P. L., Sampath Parasuram, Suryasarathi Bose
5   Polymer-based nanocomposites for smart and functional textiles —— 183

Subhankar Maity, Kunal Singha, Pintu Pandit
6   Emerging trends in electroconductive textiles —— 221

Arobindo Chatterjee and Vinit Kumar Jain
7   Graphene-based functional textile materials —— 267

Md Omar Faruk, Md Milon Hossain
8   Smart and functional textiles for personal thermal comfort —— 329

Santanu Basak, Animesh Laha
9   Stimuli-responsive smart and functional textiles —— 355

Unsanhame Mawkhlieng, Abhijit Majumdar
10 Protective smart and functional textiles —— 375

Sudipta Mondal, Bapan Adak, Samrat Mukhopadhyay
11 Functional and smart textiles for military and defence applications —— 397

Akanksha Pragya, Kony Chatterjee, Tushar K. Ghosh
12 Sensors and actuators for textiles: from materials to applications —— 469

R. Rathinamoorthy
13 Medical textiles: materials, applications, features, and recent advancements —— 533

Chetna Verma, Manali Somani, Ankita Sharma, Pratibha Singh, Surabhi Singh, Shamayita Patra, Mukesh Kumar Singh, Samrat Mukhopadhyay, Bhuvanesh Gupta
14 Design and development of chitosan-based textiles for biomedical applications —— 591

Anupam Chowdhury, Srijan Das, Wazed Ali
15 Smart textiles for energy harvesting applications —— 607

Ajay K. Maddineni and Dipayan Das
16 Fibrous materials for automotive applications —— 635

Bapan Adak, Samrat Mukhopadhyay, Shanmugam Kumar
17 Smart/functional textiles and fashion products enabled by 3D and 4D printing —— 683

Andrew J. Hebden, Parikshit Goswami
18 Functional and smart textiles in care, treatment, and diagnosis of COVID-19 —— 721

Index —— 751

List of authors

Dr. Bapan Adak
School of Material Science and Engineering,
Nanyang Technological University,
50 Nanyang Avenue, Block N4.1,
Singapore 639798
bapan.iitd15@gmail.com
Chapters 1, 11, 17

Prof. Ravindra V. Adivarekar
Department of Fibres and Textile Processing
Technology,
Institute of Chemical Technology,
N. P. Marg, Matunga (E),
Mumbai 400019, India
rv.adivarekar@ictmumbai.edu.in
Chapter 3

Dr. S. Wazed Ali
Department of Textile and Fibre Engineering,
Indian Institute of Technology Delhi,
Hauz Khas,
New Delhi 110016, India
wazed@textile.iitd.ac.in
Chapters 2, 15

Swagata Banerjee
Department of Textile and Fibre Engineering,
Indian Institute of Technology Delhi,
Hauz Khas,
New Delhi 110016, India
swagatabanerjee888@gmail.com
Chapter 2

Dr. Santanu Basak
National Institute of Natural Fibre Engineering
and Technology,
Regent Park, Kolkata 700040, India
shantanubasak@gmail.com
Chapter 9

Prof. Suryasarathi Bose
Polymer Processing Group,
Department of Materials Engineering,
Indian Institute of Science, Bangalore,
Karnataka 560012, India
sbose@iisc.ac.in
Chapter 5

Dr. Arobindo Chatterjee
Department of Textile Technology,
Dr. B. R. Ambedkar National Institute of
Technology,
Jalandhar 144 011, Punjab, India
chatterjeea@nitj.ac.in
Chapter 7

Dr. Kony Chatterjee
Fiber and Polymer Science Program,
North Carolina State University,
Raleigh, North Carolina, USA 27695
and
Department of Textile Engineering,
Chemistry and Science,
North Carolina State University,
Raleigh, North Carolina, USA 27695
kchatte@ncsu.edu
Chapter 12

Anupam Chowdhury
Department of Textile and Fibre Engineering,
Indian Institute of Technology Delhi,
Hauz Khas,
New Delhi 110016, India
anupam0429@gmail.com
Chapter 15

Prof. Dipayan Das
Department of Textile and Fibre Engineering,
Indian Institute of Technology Delhi,
Hauz Khas,
New Delhi 110016, India
dipayan@textile.iitd.ac.in
Chapter 16

https://doi.org/10.1515/9783110759747-204

Srijan Das
Department of Textile and Fibre Engineering,
Indian Institute of Technology Delhi,
Hauz Khas,
New Delhi 110016, India
ttc212726@textile.iitd.ac.in
Chapter 15

Md Omar Faruk
Department of Materials Science and
Engineering,
Binghamton University,
State University of New York at Binghamton,
New York, USA 13902
mfaruk2@binghamton.edu
Chapter 8

Sagnik Ghosh
Department of Textile and Fibre Engineering,
Indian Institute of Technology Delhi,
Hauz Khas,
New Delhi 110016, India
email2sagnikghosh@gmail.com
Chapter 1

Prof. Tushar K. Ghosh
Fiber and Polymer Science Program,
North Carolina State University,
Raleigh, North Carolina, USA 27695
and
Department of Textile Engineering,
Chemistry and Science,
North Carolina State University,
Raleigh, North Carolina, USA 27695
tghosh@ncsu.edu
Chapter 12

Prof. Parikshit Goswami
Technical Textile Research Centre,
The University of Huddersfield,
HD1 3DH, UK
P.Goswami@hud.ac.uk
Chapter 18

Prof. Bhuvanesh Gupta
Bioengineering Laboratory,
Department of Textile and Fibre Engineering,
Indian Institute of Technology Delhi
New Delhi 110016, India
bgupta@textile.iitd.ac.in
Chapter 14

Dr. Andrew Hebden
Technical Textile Research Centre,
The University of Huddersfield,
HD1 3DH, UK
a.hebden@hud.ac.uk
Chapter 18

Dr. Md Milon Hossain
Mechanical and Aerospace Engineering,
Cornell University, Ithaca,
New York, USA 14850
mh2276@cornell.edu
Chapter 8

Vinit Kumar Jain
Department of Textile Technology,
Dr. B. R. Ambedkar National Institute of
Technology,
Jalandhar 144 011, Punjab, India
vinitjain475@gmail.com
Chapter 7

Prof. Shanmugam Kumar
James Watt School of Engineering,
The University of Glasgow,
Glasgow G12 8QQ, UK
Msv.Kumar@glasgow.ac.uk
Chapter 17

Dr. Animesh Laha
Welspun India Limited,
Welspun House, 6th Floor,
Kamala City, Senapati Bapat Marg, Lower Parel (W),
Mumbai 400 013, India
asanimesht@gmail.com
Chapter 9

Dr. Ajay Kumar Maddineni
Global Technical Group,
Donaldson Filtration Solutions,
Gurgaon, 122001, Haryana, India
maddineniajay@gmail.com
Chapter 16

Dr. Saptarshi Maiti
Department of Fibres and Textile Processing
Technology,
Institute of Chemical Technology,
N. P. Marg, Matunga (E),
Mumbai 400019, India
maiti.sapta@gmail.com
Chapter 3

Dr. Subhankar Maity
Uttar Pradesh Textile Technology Institute,
Kanpur 208001, Uttar Pradesh, India
maity.textile@gmail.com
smaity@uptti.ac.in
Chapter 4, 6

Prof. Abhijit Majumdar
Department of Textile and Fibre Engineering,
Indian Institute of Technology Delhi,
New Delhi 110016,
India
majumdar@textile.iitd.ac.in
Chapter 10

Dr. Aranya Mallick
Department of Fibres and Textile Processing
Technology,
Institute of Chemical Technology,
N. P. Marg, Matunga (E),
Mumbai 400019, India
as.mallick@ictmumbai.edu.in
Chapter 3

Dr. Unsanhame Mawkhlieng
National Institute of Fashion Technology Shillong,
Shillong 793012, Meghalaya, India
and
Department of Textile and Fibre Engineering,
Indian Institute of Technology Delhi,
New Delhi 110016, India
unsan.hame@gmail.com
Chapter 10

Mandira Mondal
Department of Textile and Fibre Engineering,
Indian Institute of Technology Delhi,
Hauz Khas, New Delhi 110016, India
mandiramandal2012@gmail.com
Chapter 2

Dr. Sandeep P. More
Department of Fibres and Textile Processing
Technology,
Institute of Chemical Technology,
N. P. Marg, Matunga (E),
Mumbai 400019, India
sp.more@ictmumbai.edu.in
Chapter 3

Prof. Samrat Mukhopadhyay
Department of Textile and Fibre Engineering,
Indian Institute of Technology Delhi,
Hauz Khas,
New Delhi 110016, India
samrat@iitd.ac.in
Chapters 1, 11, 14, 17

Sudipta Mondal
Product Development Department,
Kusumgar Corporates Pvt. Ltd.,
Plot No. 1809, GIDC, 3rd Phase,
Vapi, Valsad 396195,
Gujarat, India
sudiptam1990@gmail.com
Chapter 11

Dr. Pintu Pandit
Department of Textile Design,
National Institute of Fashion Technology,
Patna 800001, Bihar, India
pintu.pandit@nift.ac.in
Chapters 4, 6

Sampath Parasuram
Polymer Processing Group,
Department of Materials Engineering,
Indian Institute of Science, Bangalore,
Karnataka 560012, India
parasurams@iisc.ac.in
Chapter 5

Dr. Shamayita Patra
Shri Vaishnav Vidyapeeth Vishwavidyalaya,
Indore 453111, Madhya Pradesh, India
shamayitapatra@yahoo.co.in
Chapter 14

Dr. Jagadeshvaran P. L.
Polymer Processing Group,
Department of Materials Engineering,
Indian Institute of Science,
Bangalore 560012, Karnataka, India
jagadeshvara@iisc.ac.in
Chapter 5

Lelona Pradhan
Department of Fibres and Textile Processing
Technology,
Institute of Chemical Technology,
N. P. Marg, Matunga (E),
Mumbai 400019, India
lelonapradhan@gmail.com
Chapter 3

Dr. Akanksha Pragya
Fiber and Polymer Science Program,
North Carolina State University,
Raleigh, North Carolina, USA 27695
and
Department of Textile Engineering,
Chemistry and Science,
North Carolina State University,
Raleigh, North Carolina, USA 27695
apragya@ncsu.edu
Chapter 12

Dr. R. Rathinamoorthy
Department of Fashion Technology,
PSG College of Technology,
Coimbatore, Tamil Nadu, India
r.rathinamoorthy@gmail.com
Chapter 13

Dr. Mohammad Shahid
Department of Applied Science,
Dr. K. N. Modi University, INS-1,
RIICO Industrial Area Ph-II,
Newai, Tonk 304021, Rajasthan, India
mshahid96@gmail.com
Chapter 3

Ankita Sharma
Bioengineering Laboratory,
Department of Textile and Fibre Engineering,
Indian Institute of Technology Delhi
New Delhi 110016, India
ankitaiit0411@gmail.com
Chapter 14

Dr. Mukesh Kumar Singh
Uttar Pradesh Textile Technology Institute,
Kanpur 208001, Uttar Pradesh, India
mksinghuptti@gmail.com
Chapter 14

Pratibha Singh
Bioengineering Laboratory,
Department of Textile and Fibre Engineering,
Indian Institute of Technology Delhi
New Delhi 110016, India
pratibhas1402@gmail.com
Chapter 14

Dr. Surabhi Singh
Bioengineering Laboratory,
Department of Textile and Fibre Engineering,
Indian Institute of Technology Delhi
New Delhi 110016, India
surabhiwc@gmail.com
Chapter 14

Kunal Singha
Department of Textile Design,
National Institute of Fashion Technology,
Kolkata 700098, West Bengal, India
kunal.singha@nift.ac.in
Chapters 4, 6

Manali Somani
Bioengineering Laboratory,
Department of Textile and Fibre Engineering,
Indian Institute of Technology Delhi
New Delhi 110016, India
manali.100mani@gmail.com
Chapter 14

Dr. Chetna Verma
Bioengineering Laboratory,
Department of Textile and Fibre Engineering,
Indian Institute of Technology Delhi
New Delhi 110016, India
vermachetna18@gmail.com
Chapter 14

Sagnik Ghosh, Bapan Adak, Samrat Mukhopadhyay

1 Introduction to functional, smart, and intelligent textiles: perspectives and potential applications

Abstract: "Smart textiles", also known as functional fabrics or e-textiles, are changing the way of thinking about fabrics. The term "functional textiles" refers to textiles with integrated functions that control or adjust according to the application. Nowadays, electronics and photonics have greatly influenced the evolution of technical textiles. Smart textiles are functional fabrics integrated with a sensor array or functional nanomaterial/polymer or an optical fibre. Various fibres (conductive and high-performance), chemicals/additives (finishing and coating chemicals, smart polymers, nanomaterials, etc.) and technologies (spinning, weaving, knitting, nonwoven, braiding, finishing, coating, lamination, etc.) are combined to create smart fabrics, depending on their use.

This chapter covers all the main areas of applications of smart and functional textiles, including textiles with various functionalities (antimicrobial, UV-resistant, fire-retardant, oil/water-repellent, stain-repellent, wrinkle-resistant, anti-order, antistatic, superabsorbent, etc.), coated/laminated high-performance textiles, conductive textiles, textile-based sensors, energy-harvesting textiles, medical textiles, protective textiles, textiles for military and defence, automotive textiles, and so on. This chapter also discusses the potential of emerging 3D printing technologies for making smart and functional textiles. Finally, the current challenges as well as future perspectives of functional and smart textiles have also been summarized.

Keywords: E-textiles, smart textiles, functional textiles, intelligent textiles, conductive textiles, flexible wearable electronics

1.1 Introduction

Our lives are incomplete without textiles, which serve as clothing, protection, and for aesthetic purposes. Nevertheless, in recent years, as technology has developed and requirements have changed, the demand for smart materials and smart textiles has become increasingly global. In addition to their basic function, functional textiles have in-built features that meet the end-use requirements. As a result, they can be controlled and adjusted according to their intended use. The term "smart textiles" refers to textiles that can detect environmental conditions or stimuli such as mechanical, thermal, chemical, electrical, or magnetic. Smart and intelligent textiles can be used in very diverse fields, ranging from simple to very complicated applications including protection, military and defence,

https://doi.org/10.1515/9783110759747-001

medical and healthcare, sportswear, automotive, and energy harvesting. Therefore, smart and intelligent textiles are considered as next-generation textiles [1, 2].

The developments in the area of functional and smart textiles are largely driven by material and technological advancements in fibres, fabrics, chemicals and additives (polymer, functional nanomaterials, finishing chemicals, adhesives, etc.), and manufacturing processes. Smart or intelligent textiles are produced by collaborating with other branches of science and technology, such as material science, nanotechnology, design, electronics, and computer engineering [3, 4].

This chapter highlights the history of development of smart textiles, different types of smart textiles, technology and approaches for making functional and smart textiles, different materials used, and all the main applications of functional and smart textiles.

1.2 What are functional, smart, and intelligent textiles?

1.2.1 Functional textiles

Functional textiles can be defined as follows: It is nothing, but a kind of technical textiles used mostly for its technical performance and functional aspects, rather than their aesthetic value. A few examples of functional textiles have been summarized:

i) UV-protective textile: It protects the human body by absorbing or reflecting harmful UV radiation.
ii) Antimicrobial textile: A textile on which no microbial growth takes place.
iii) Flame-retardant textile: A textile material which is able to restrict catching and spreading of fire.
iv) Waterproof fabric: It restricts the penetration of water up to a certain pressure.
v) Antistatic textile: It restricts development of any static charge on it.

1.2.2 Smart textiles

According to the European Standard CEN/TR 16,298, "Smart textiles are the textile materials or textile systems that possess supplementary intrinsic and functional properties that are not normally associated with traditional textiles". In other words, smart textiles can be defined as textiles that are able to sense and analyse a signal, in order to respond in an adapted manner. These input stimuli can be mechanical, thermal, or chemical (like change in pressure, temperature, pH, electrical, and magnetic), and the output response can be changed in colour, shape, flow property, conductivity, and so on. In other words, smart textiles are all functional, but not all functional textiles are smart.

Whether it is for pleasure, performance, or safety, smart textiles are designed to give extra value to the user. These include applications for monitoring health, safety, security, and lifestyle. Thus, in contrast to the conventional definition of textiles, it has not only integrated new functionalities but has also introduced alternatives for use in a variety of different sectors and is in the form of either passive smart textiles or active smart textiles [2].

1.2.2.1 Passive smart textiles

Passive smart textiles are the earliest generation of smart textiles that can merely sense ambient conditions or stimuli. Examples are optical fibres, conductive material, insulating materials, antibacterial, self-cleaning, or waterproof breathable textiles. Briefly, a cotton fabric loaded with silver nanoparticles on the surface shows strong antibacterial effect, and TiO_2-loaded textiles exhibit self-cleaning feature, but both come under passive smart textiles, since antibacterial effect of silver and photocatalytic property of TiO_2 make the textile smart but not responsive.

1.2.2.2 Active smart textiles

Active smart textiles, as opposed to passive smart textiles, may perform interactive behaviours on their own, automatically, in response to a specified demand. The interactive behaviours, in this context, are typically connected with receiving stimuli and delivering feedback [5]. For example, chlorhexidine (antibiotic drug)-loaded crosslinked gelatine electrospun mat behaves as an active smart textile, as it releases the loaded drug (due to volume expansion of the mat) that kills gram-negative (*E. coli*) and gram-positive (*S. epidermidis*) bacteria in response to the change in surrounding pH (surrounding pH changes with the presence of bacteria). Here, input stimuli are the changes in pH, and the feedback response is volume expansion of material [6]. Figure 1.1 illustrates the defining features of passive and active smart textiles [5].

1.2.3 Ultra-smart or intelligent textiles

The third generation of smart textiles includes ultra-smart textiles that are capable of sensing, reacting to, and adapting to the surrounding environment or stimuli. Fundamentally, they are made up of a unit that functions similarly to the brain, with cognition, reasoning, and decision-making capabilities. The production of smart textiles is now a reality, owing to the successful amalgamation of customary clothing technology with other branches of science and technology, such as sensor and actuator technology, communication, material science, structural mechanics, biology, artificial intelligence, and advanced processing technologies. Examples are spacesuit and wearable computers.

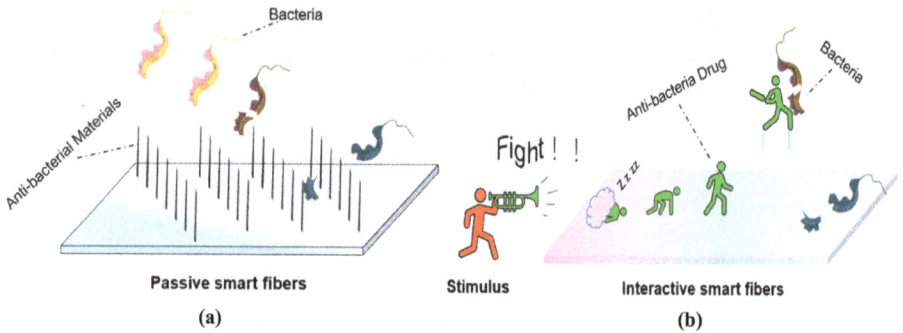

Figure 1.1: The schematic representation of crucial aspects of: (a) passive smart and (b) active smart textiles with antibacterial function (reproduced with kind permission from [5], Copyright 2020, Elsevier).

1.3 History of developments of smart textiles

Textiles have evolved since the Stone Age, when men and women wore clothing made of plant leaves/barks and animal skins to protect themselves from the sun rays, cold, wind, rain, sand, and dust. Although pioneers have been inventing new products within the textile industry for a long time, the terms "smart fabrics", "smart textiles", and "e-textiles" are still sparkling and new to many of us. Makers and designers who develop multiple materials that provide comfort to the wearer have facilitated the convergence of fashion and technology. It is surprising to know that the first conductive gold threads were woven into clothes for a dazzling accent during the Elizabethan age (1600). We now frequently employ silver or nickel threads for conductivity, although the idea of metallic threads for adorning clothes has existed for millennia. In 1989, Japan was the first country to define the idea of a "smart material". Silk thread with shape-memory behaviour was the first textile material to be dubbed "smart textile", in retrospect. Further key progresses in this field are listed in Figure 1.2 [7, 8].

1.3.1 Few recent developments in smart textiles

In July 2021, DuPont bought Laird Performance Materials, a leading provider of high-performance electromagnetic shielding and thermal management systems. In February 2021, they also bought Tex Tech's Core Matrix Technology, a monolithic fabric structure capable of reducing blackface trauma and boosting ballistic and fragmentation performance for defence personnel. This new technology not only improves the comfort of wearing bullet-resistant body armour but also offers enhanced protection. MesoMat, a Canadian start-up patented a technology, where a stretchable copper wire can transmit electricity. The company uses self-assembly physics to create ultra-durable fibre that can be stretched by more than 50%, while maintaining conductivity.

MesoMat's patented soft solder makes it simple to connect the fabric to typical electronic components [9]. Nyoka Technologies, a start-up based in India, produces soft and flexible circuit systems that perceive, actuate, and analyse data. It creates Zeal, a women's safety jacket that senses an attack and uses a moderate electric shock to repel the attacker. It also includes an SOS alarm and delivers location information [10]. Clim8, a French wearable technology start-up, makes smart thermal technologies that may be integrated into temperature-regulating clothing. This technology incorporates small thermal sensors that track the temperature of the user's skin. The system then assesses the surroundings, the user's profile, and other unique requirements, before activating heat and regulating body temperature [11].

1.4 The present market scenario and future perspectives

The global smart textiles market reached a value of USD 2.41 billion in 2020. Looking forward, according to the IMARC group experts' analysis, the worldwide smart textiles market is anticipated to be worth USD 11.92 billion by 2026, at a compound annual growth rate (CAGR) of 33.17 % for the predicted period, as shown in Figure 1.3(a). The leading manufacturers in the smart textile market are Adidas (Germany), Alphabet (USA), AIQ Smart Clothing (Taiwan), DuPont (USA), Gentherm (USA), Interactive Wear (Germany), Jabil (USA), Outlast Technologies (USA), Hexoskin (Canada), and Sensoria (USA). The smart textiles market has developed rapidly in recent years, owing, mostly, to the innovation and application of technologies such as artificial intelligence (AI) and the internet of things (IoT) in smart textiles, as well as the expansion of wearable electronics in these years. Figure 1.3(b) illustrates the total number of global wearable devices shipment in the period 2015–2021. In 2020, the sensing sector had the major market share of the smart textiles market. The unforeseen emergence of the COVID-19 pandemic has affected the market of smart textiles, particularly in 2020 and 2021. As a result of the obtrusive lockdown in practically all major nations to reduce the prevalence of COVID-19 pandemic, organizations have been driven to adopt remote working practices, which disrupted the supply chain and interrupted industrial activities, resulting in manufacturing delays. However, due to social-distancing guidelines, a large portion of the global population has been forced to work from home and perform indoor fitness activities, which has increased the mandate for integrated smart apparels that can keep track of breathing rate, oxygen saturation, and so on, thus propelling the smart textiles industry forward. Apart from that, smart textiles have applications in athletics, defence, and space industries, where sensing and monitoring are now managed through additional devices but would be considerably handier and more effective, if combined with textiles. The military and defence segments are expected to generate a substantial marketplace during the upcoming periods. Smart textiles for military applications should include superior insulating

1990	MIT students started researching smart apparel for military use
1996	The conductive fabric superstore (for EMF blocking purposes)
1998	Sabine Seymour launches Moondial, merging "silicon and style."
2000	MIT researchers developed E-broidery: Design and fabrication of textile-based computing
2003	Georgia Tech Motherboard shirt
2008	Mika Satomi and Hannah Perner-Wilson Launch kobakant
2009	Forster Rohner launches the Climate Dress
2011	Midi Puppet Glove, conductive thread and electronics, by MICA fibre
2013	Machina Launches the Midi Controller Jacket on Kickstarter
2014	Dupont presents their stretchable, conductive ink at printed Electronics
2014	The MIT Biosuit
2014	Bebop sensors launches wearable tech and textile circuits using Dupont ink
2015	Google's Project Jacquard directs tech eyes to e-textiles
2016	No more washing: Introduction of self-cleaning textiles
2016	Smart threads that can change the colour of textile
2016	Aerochromics, smart clothing that monitors pollution in real time
2016	Georgia Tech. researchers developed Energy harvesting textiles from motion and sunshine: Next generation wearable electronics
2017	Kassim Denim x Neue Labs: The World's First Connected Jeans
2017	NanoZoo, the game changing smart diaper bag
2017	HyperFace, A smart textile collection designed for privacy protection
2017	Bluetooth-enabled jacket, Levi's® Commuter X Jacquard by Google
2017	DuPont Intexer connected, intelligent garment
2017	Twistron Yarnology generates power when stretched
2017	Sweater made up of thermosensitive yarn
2017	Hard disk fabric Allen School's Networks & Mobile Systems Lab
2018	Introduction of artificial intelligence within smart textiles
2018	Functional smart jacket that produces warmth on demand
2019	Intelligent sleepwear Phyjama launched
2019	Copenhagen Fashion Summit promises to introduce sustainability within smart textiles
2019	Samsung applied for a patent for a smart shirt that can track symptoms of diseases like pneumonia and bronchitis
2019	Nike introduced the Adapt-branded self-lacing shoes, which fit themselves to the shape of the user's foot
2020	Adidas partnered with Google to use the Jacquard technology in GMR insoles for football players
2020	Xenoma launched smart pajamas that monitor the user's heart rate and sleep patterns and detect falls or trips.
2021	DuPont developed monolithic fabric structure that reduced blackface trauma and boosted ballistic and fragmentation performance for defense personnel.
2026	The global smart clothing market will worth more than $11.92 bn, according to data bridge market research estimates

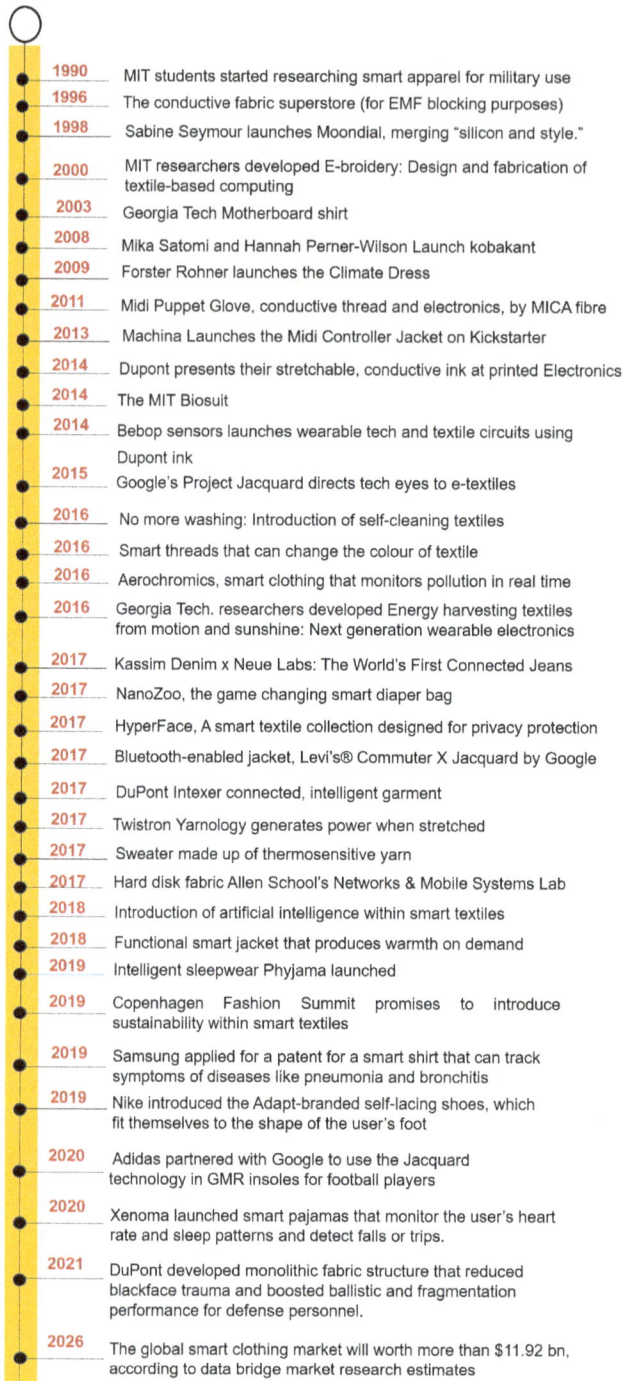

Figure 1.2: History and key developments of smart textiles.

characteristics with ballistics safeguard within waterproof fabric, health monitoring capabilities, and integration with precise GPS systems and motion detectors. It is anticipated that product protection, thermal considerations, lack of standards, and regulations will be major impediments to market expansion, creating difficulties in scaling up and introducing new technologies. In addition, lab prototypes that are more or less successful should be scaled up, made more resilient and dependable, and most importantly, mass-produced on an affordable scale [1].

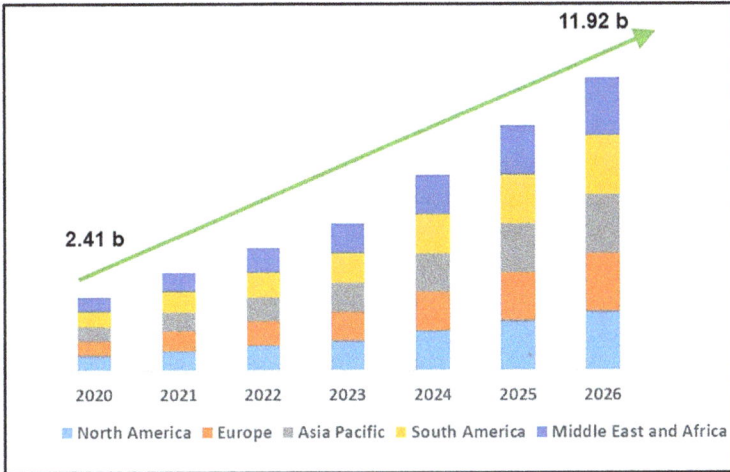

(a)

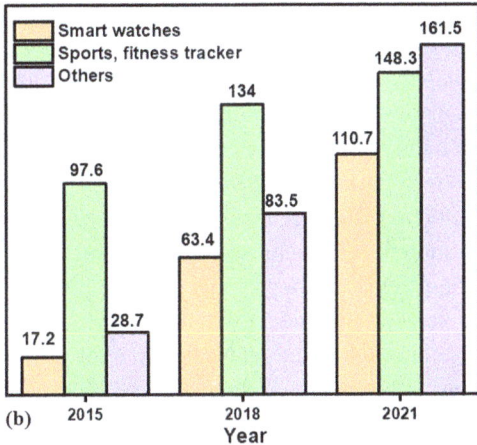

(b)

Figure 1.3: (a) The global market forecast for smart textiles from 2020 to 2026, in billion US dollars [12]; (b) worldwide total number of wearable device shipments from 2015 to 2021, in million units [13].

1.5 Approaches/techniques for making of smart and functional textiles

This book covers all the key areas where smart and functional textiles are being used, till date. The timeline of development of smart fibres and textiles is depicted in Figure 1.4 [14]. The key scientific principles, concerns, and offered answers relating to diverse results, prototypes, and achieved accomplishments in the top labs of both academia and industry, across the world, are presented in a meticulous and scientific manner. In the following sections, a brief overview of potential applications of functional and smart textiles (like UV-protective, water/oil-repellent, fire-retardant, antimicrobial and antiviral, crease-resistant, and so on; chromic textiles, stimuli- responsive textiles, shape-memory textiles, electroconducting textiles, textile sensors, textile actuators, energy-harvesting textiles, energy storage textiles, medical textiles, automotive textiles, protective textiles, etc.) have been provided. Further, in the subsequent chapters, actual solutions and realizations that are meant to be of interest to industry are given in detail and discussed. Researchers and students studying smart textiles as well as companies designing and manufacturing smart textiles-based products will largely benefit from this book.

1.5.1 Nano-finished functional and smart textiles

The growing consumer desire for long-lasting, functional clothing made via sustainable processing has generated an opportunity for nanomaterials to be incorporated into textile substrates. Conventional textile apparels (like cotton, jute, and linen) exhibit high absorbency, softness, and breathability, but their high-end applications are somewhat limited, due to their crease, inferior strength, flammability, degradation, and soiling. Synthetic fibres, on the other hand, have better antibacterial and stain/crease-resistant properties, although they are often uncomfortable. The development of new fibre types that combine the advantages of both natural and synthetic fibres with additional novel functionalities was limited to conventional finishes and was a challenge, until textiles met with nanotechnology [15]. The past and upcoming trends in fibres and demands from textiles are depicted in Figure 1.5 [14].

Nanotechnology refers to the study, manipulation, and control of matter at the nanoscale (1–100 nm), with the goal of altering the physical, chemical, and biological aspects of materials through engineering, to create improved materials and devices for the next generation [16]. A drastic change in overall properties (like optical, physical, chemical, electrical, and biological) takes place within nanodimensions, mainly due to combined effect of quantum confinement and high surface-to-volume ratio [17]. These properties are used to impart desirable textile traits, such as soft hand,

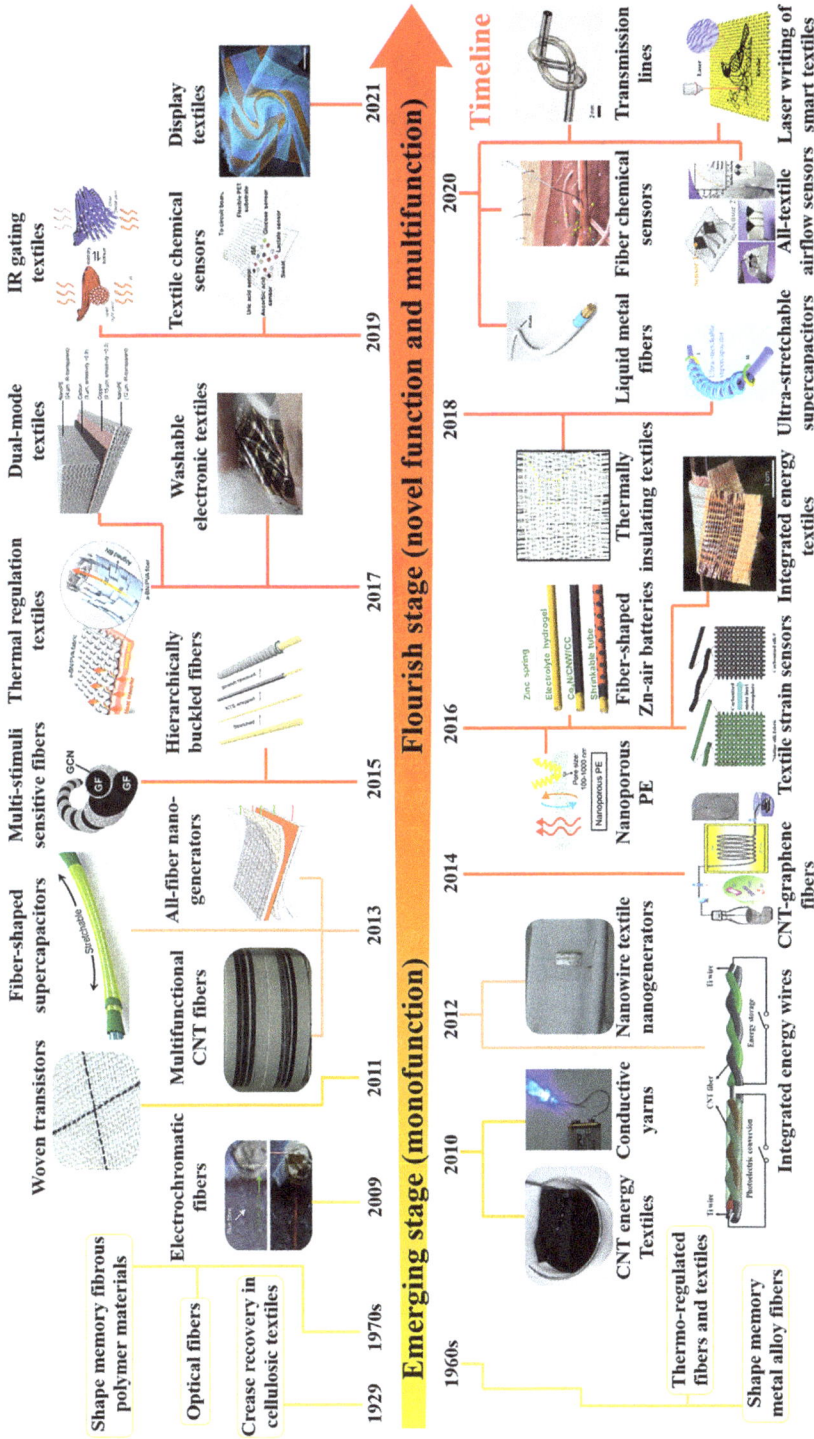

Figure 1.4: Timeline demonstrating the progress of smart fibres and textiles (reproduced with kind permission from [14], Copyright 2021, American Chemical Society).

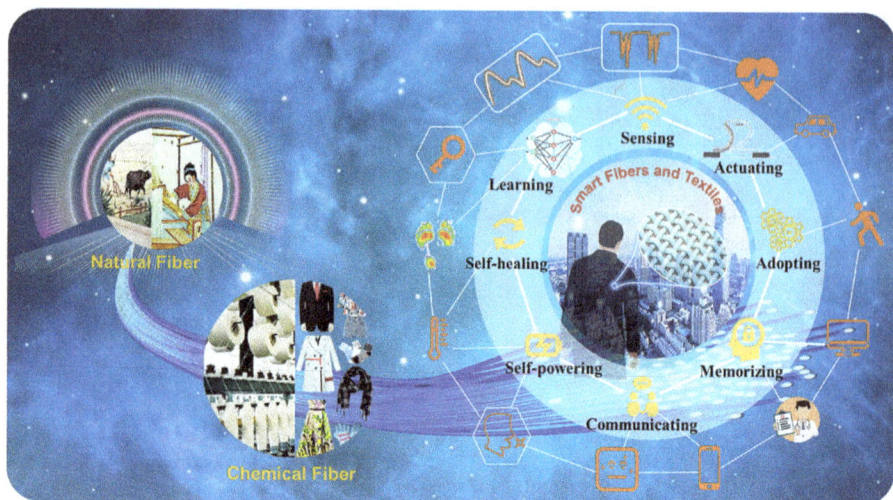

Figure 1.5: The past and upcoming trends in fibres, along with modern demands from textiles (reproduced with kind permission from [14], Copyright 2021, American Chemical Society).

superior tensile properties, durability, water repellence, distinctive surface structure, fire retardance, and antibacterial properties, as depicted in Figure 1.6.

When it comes to the introduction of nanotechnology in textiles, Taiwan-born Dr. David Soane is one of the pioneers of Nano-Tex; he first produced multifunctional nanomolecules in 1998, using nanotechnology and added unusual properties to textiles, without changing a fabric's aesthetics or handle. Almost simultaneously, the early work of Prof. W. Barthlott on the "Lotus Effect" to generate super hydrophobic surfaces for self-cleaning has served as the foundation for stain-repellent and oil/water-repellent textile finishes for NanoSphere®, Schoeller Textiles A. G., Switzerland [18]. Nanotechnology-enhanced textiles have reached the stage of commercialization and are being used in several industries, including sports, cosmetics, space technology, apparel, as well as for material technologies in adverse situations [19].

1.5.1.1 Potential of different nanomaterials in making functional and smart textiles

When it comes to the specific nanomaterial category, semiconductor ceramics and metal oxide nanomaterials are used to obtain properties such as flame retardance, water-oil repellence, and thermal resistance. Oxides nanoparticles of zinc, titanium, and aluminium are applied for oxidative catalysis, antistatic, shielding and UV protection [21–25]. Indium and tin oxide nanoparticles are used for infrared protective clothing. Polypropylene or polyethylene-coated silicon dioxide and aluminium oxide nanoparticles are employed as

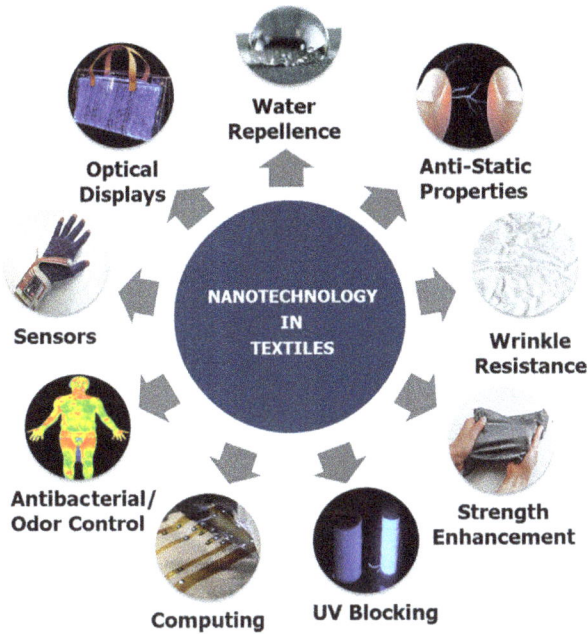

Figure 1.6: Nanotechnology applications in textiles (reproduced with kind permission from [20] Copyright 2021, American Chemical Society).

super water-repellent surfaces [27]. Ceramic nanoparticles are used for enhancing abrasion resistance [28]. Fe, Pd/Pt nanoparticles are used to get enhanced magnetic properties and conductive heating, whereas nanosilver is used for bacterial protection and wound dressings [29]. Table 1.1 summarizes the numerous nanomaterials used in making smart and functional textiles. The operating mechanisms and results for each category of nanomaterials for specified end use have been discussed in the subsequent chapters.

1.5.2 Coated and laminated smart textiles

In addition to new finishes, nanotechnology has also introduced new applications methods such as plasma, layer-by-layer assembly (LBL), coatings, laminations, and voided patterns. All these application methods provide excellent overall design flexibility (one or both sides and bulk application) with very low add-on, superior mechanical and functional properties along with durability. Their large surface area-to-volume ratio, high surface energy, and functional groups provide better affinity for fabrics, leading to increased functional endurance [20]. Additionally, nanoparticle coatings do not affect the breathability or hand feel of fabrics. Nanocoating may be accomplished using a variety of processes, including sol–gel coating, polymer dispersions, electrodeposition,

Table 1.1: Different nanomaterials and their use in making functional and smart textiles.

Nanomaterial class	Nanomaterial	Functionality	References
Metal oxide nanoparticle	SiO_2	Moisture management Controlled delivery of active agents (e.g. medicinal products and fragrances)	[30, 31]
	TiO_2	Self-cleaning, antibacterial, UV protection, wrinkle resistant, strain resistant	[22–25]
	Fe_3O_4	Electromagnetic shielding Microwave absorption	[32]
	ZnO	Energy harvesting Antibacterial UV protection	[21]
	Indium-tin oxide	IR protection	[33]
	Antimony oxide	Flame retardance	[34]
Nanoclay	Montmorillonite	Flame retardance Controlled delivery of active agents (e.g., medicinal products and fragrances)	[35]
Nanowire	Copper nanowire	Conductive materials Heat management Antistatic	[36]
	Polyaniline nanowire	Electrode Antistatic	[37]
	Zinc oxide nanowire	Energy harvesting Antibacterial	[38]
	Silver nanowires	Energy harvesters Antimicrobial	[39]
Conducting polymer nanoparticles	Polypyrrole	Antimicrobial EMI shielding Glucose sensor Ion exchange biosensor Microwave absorption Electroconductive/antistatic	[40]
	Polyaniline	Electromagnetic shielding Microwave absorption	[41]
	PEDOT:PSS	Conductivity Thermoregulating Sensor	[42]

Table 1.1 (continued)

Nanomaterial class	Nanomaterial	Functionality	References
Metal nanoparticles	Silver	Antibacterial Conductive Energy harvesting	[43]
	Gold	Biosensor	[44]
	Copper	Energy harvesters Antibacterial	[45]
	Aluminium	Energy harvesting	[46]
Carbon-based nanomaterials	Carbon nanotube Carbon nanofibre	Strain sensor Optical sensor Thermal sensor Electrical field sensitivity Bio-based glucose sensor Energy harvesters	[47]
Nanocellulose	Cellulose nanocrystals	Bio-based glucose sensor Ion sensor pH sensor Actuator Ion interchange Energy harvesters Energy storing	[48]

atomic-layer deposition (ALD), microwave heating, dip coating, physical vapour deposition (PVD), chemical vapour deposition (CVD), spin coating, and vacuum evaporation. Polymeric coating methods, in contrast to textile printing, include direct coating, foamed and crushed foam coating, transfer coating, hot melt extrusion coating, calendar coating, and rotary screen coating. Each approach has limitations, and its use is determined by the end use. Some useful textile applications of coatings are listed in Table 1.2 [49–51]. Electrostatic, covalent bonding, and hydrogen bonding are the most common interactive forces that hold the coating material to the base textile [52]. On the other hand, lamination is the process of joining numerous substrates by bonding them together to create a stable multilayer structure with features that a single substrate cannot attain. Lamination combines a fabric and the film; it replaces or supplements sewing to obtain laminated fabrics with enhanced functional properties, hand feel, and more consistent qualities [4]. The major difference between lamination and coating is that in lamination, adhesion can be obtained by adhesive, heat, pressure, or mechanical bonding, while in coating, the material is applied to the surface of a substrate and cured [49]. Nowadays, electro-spun nanofibrous mats are also coated onto textile surfaces for making sensors, improving filtration, abrasion, and corrosion resistance [53]. Readers can find out more

Table 1.2: Polymeric coating materials and their uses in making functional/smart textiles [49].

Coating material	End use
Polyvinyl chloride (PVC)	Tarpaulins, tents, seat upholstery, leather protective clothing, aprons, banners, bunting
Polyvinylidene chloride (PVDC)	Improved flame retardance, gas barrier applications
Polyurethane (PU)	Aircraft life jackets, waterproof protective clothing, solvent resistance, abrasion resistance, ageing resistance
Acrylic	Back coating for upholstery including auto seats, binders for nonwovens and glass fibres
Polyolefins (LDPE, HDPE, polypropylene)	Chemical protective, lightweight tarpaulins, stiff/interlining coating
Perfluorochemical, polyacrylates, silicone-based products, and vinyl acetate	Stain release, soil release, water repellent
Poly-hexamethylene biguanide hydrochloride (PHMB), cyclodextrin	Deodorant/antibacterial clean room fabric finishes
Activated carbon-based coatings	Chemical odour absorbing
Butyl rubber, chloro-sulphonated polyethylene/rubber (e.g. Hypalon - DuPont)	Protective clothing – especially for chemicals and acids, thermal insulation, lightweight life jackets, aircraft carpet backing, linear and covers for portable waste reservoir, inflatable boat/life raft

details of coated and laminated textiles (materials, synthesis, process techniques, potential applications, current challenges, and future trends) in Chapters 3 and 4.

1.5.3 Polymer-based nanocomposites for making smart and functional textiles

Besides coated and laminated textiles, polymeric nanocomposites have recently made significant advances and can be used for a wide range of applications, such as sensors, self-cleaning, antibacterial, moisture management, fire protection, actuators, harvesting energy, transportation, aerospace, biomedicine, and security [3, 54–57]. Nanocomposites are a class of nanomaterials that contains a dispersed phase reinforced within the matrix, where at least one dimension of the dispersed phase is within the nanoscale, and the constituent materials are physically/chemically different. Nanoparticles (like lamellar clay), nanorods/nanotubes [like carbon nanotubes (CNTs)], and nanofibres are some examples of non-reinforcement with discrete inorganic units. The composite material properties depend upon the properties of each of its phases, their relative proportions, and their geometry.

Similar to conventional composites, nanocomposites can be divided into three catego-ries, based on their matrix materials: metal, ceramic, and polymer matrix nanocompo-sites, as depicted in Figure 1.7. Polymer matrix composites are based on matrix materials such as polyurethane, polyolefins (polypropylene and polyethylene), polyester, polyam-ide, vinyl, and other special polymers [58]. In addition to being lightweight, easy to pro-cess, and offering various improved functionalities (conductivity, gas barrier, weather resistance, EMI shielding, corrosion resistance, etc.), polymeric nanocomposites have also shown promise for applications in batteries, sensors, microelectronics, and wearables. However, they lack strength and thermal resistance, when compared to metallic or ce-ramic nanocomposites [59]. However, metal (Al, Fe, Cr, Mg, W, etc.) and ceramic (SiN, SiC, Al_2O_3, etc.) matrix nanocomposites have the potential to serve a wide range of industries, from transportation and aerospace to defence and military. Apart from the matrix type, nanocomposites can also be categorized primarily by the nature of their reinforcements as follows: fibre-reinforced composites (e.g. cellulose whiskers, CNTs, and nanofibre-reinforced), particulate composites (e.g. metal nanoparticles, spherical silica, and semi-conductor nanoclusters-reinforced), and laminar composites. The reinforcing material helps carry the load, provide structural support/strength, thermal stability, conducting properties conductivity, and so on to the composite, whereas the matrix binds the rein-forcements together, providing rigidity and shape to the structure [60].

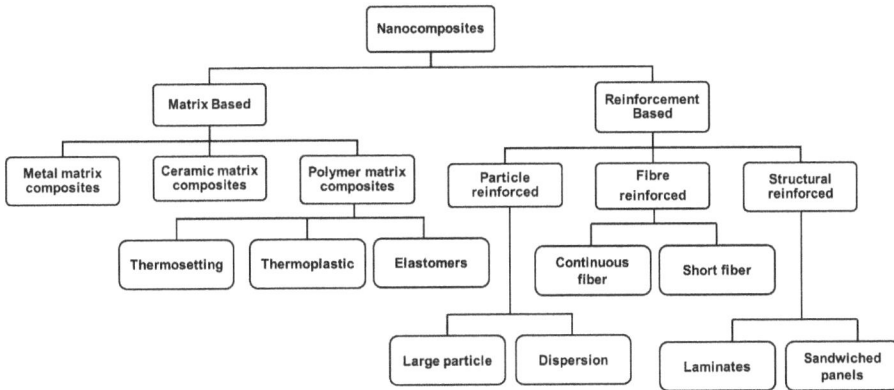

Figure 1.7: Schematic displaying the overall classification of nanocomposites.

This book covers most of the smart textile-based innovations, and the principal focus is on polymeric nanocomposites. Potential applications of polymeric nanocomposites are illustrated in Figure 1.8. They are available in the forms of laminated nanocompo-sites, nanocomposite fibres, nanocomposite coatings, nanocomposite membranes, and nanocomposite hydrogels and can be made through in situ polymerization, melt ho-mogenization, electrodeposition, sol–gel method, solution dispersion, templating, and other advanced processes such as self-assembly and ALD. Laminated nanocomposites consist of a layered structure and are suitable for applications like filtration and moisture management. In contrast, nanocomposite coatings eradicate the wash durability issue associated with the nanoparticle-coated textiles, since this structure embeds nanoparticles within the polymer matrix. Thus, nanocomposite coatings help in achieving effects like UV resistance, antimicrobial, conductivity, fire protection, and self-cleaning with better human, environmental, and ecological safety. Fibre nanocomposites are a different type of nanocomposites, where the fibre itself exhibits a nanocomposite structure; examples of this include composites generated by incorporation of functional nanoparticles within the fibre matrix. Lastly,, nanocomposite hydrogels that kick out our misconception about nanocomposites are all rigid materials, mainly used as stimuli-responsive materials.

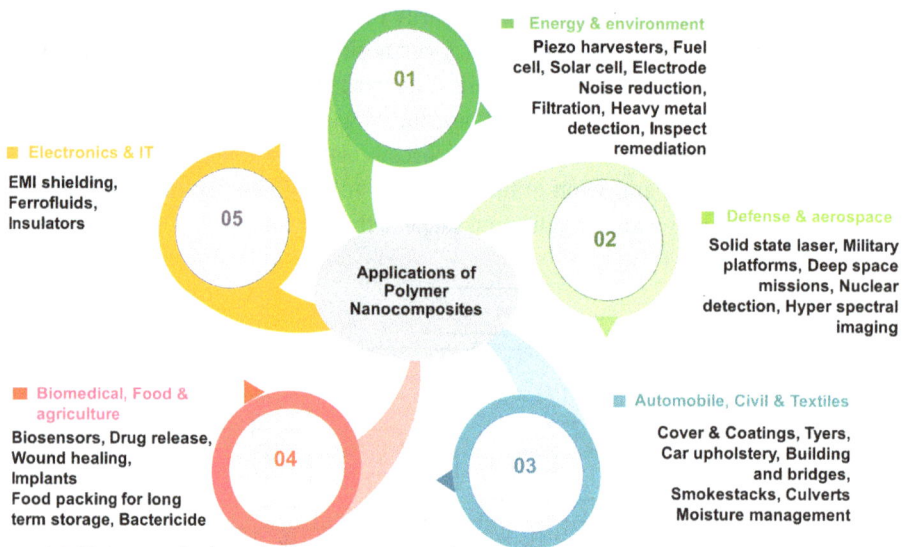

Figure 1.8: Various applications of polymer nanocomposites.

Briefly, till date, research on nanotechnology-based textiles has primarily focused on simplifying techniques to produce multifunctional fabrics; however, results obtained are generally much lower than their theoretical values, may be due to lack of uniformity, orientation, nanoparticle agglomeration, and interfacial adhesion. For modern smart textile manufacturers, optimizing and applying these techniques from lab experiments to industrial platforms and avoiding unintentional cytotoxic release are, therefore, of paramount importance. However, the push to commercialize these methods will become more intense in the near future, since very few coating methods are available for technical textiles, biosensors, and metal surfaces.

1.6 Properties and potential applications of smart and functional textiles

1.6.1 Electroconductive textiles and their applications

The most important performance characteristic requirement of e-textiles is electrical conductivity, which helps it receive, store, and transmit data and signals to serve its primary functions. Textile fibres and polymers are natural insulators, because they do not allow the flow of electric charges [61]. However, for diversified application of textiles beyond their garmenting use, the textiles are modified to impart electrical conductivity. Thus, according to the norms, such conductive textile materials are smart but not intelligent, since they do not have the ability to react to their environment; however, they make many smart textile applications possible, especially those that monitor body functions like sensors, actuators, communication, heating fabrics, and clothing that gets rid of static electricity [62–64]. An overview of the field of electronic textiles is depicted in Figure 1.9 [65].

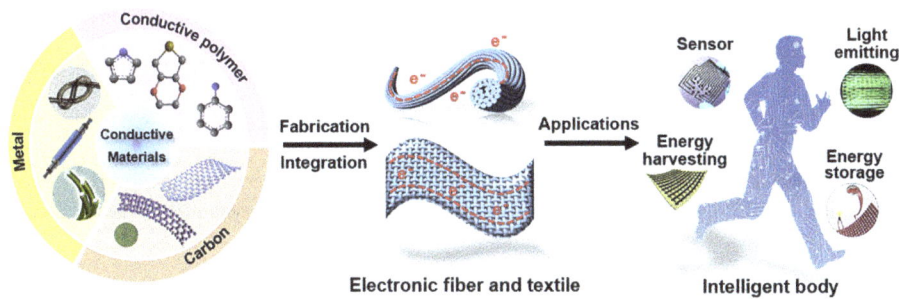

Figure 1.9: An overview of the field of electronic textiles (reproduced with kind permission from [65], Copyright 2021, Elsevier).

There are different types of conventional methods for the preparation of electroconductive textiles like metal coating, inserting metal fibres and metal wires within the structure of textiles, and blending of metal fibre during spinning or weaving. Other than such metalized textiles, various non-metallic electroconductive materials (such as carbon black, graphite, graphene, CNTs, MXene, and conductive polymers) are used for the development of electroconductive textiles, with much better functional and mechanical properties. Figure 1.10 depicts how an insulating textile fibre can be converted into e-textiles [66]. Briefly, electroconductive textiles can be classified as metal-based electroconductive textiles, graphene-based electroconductive textiles, MXene-based electroconductive textiles, CNT-based electroconductive textiles, and conductive polymer-based electroconductive textiles. Figure 1.11 gives an overview of the range of conductivity of metallic conductors, conducting polymers, and conductive polymeric composites [67].

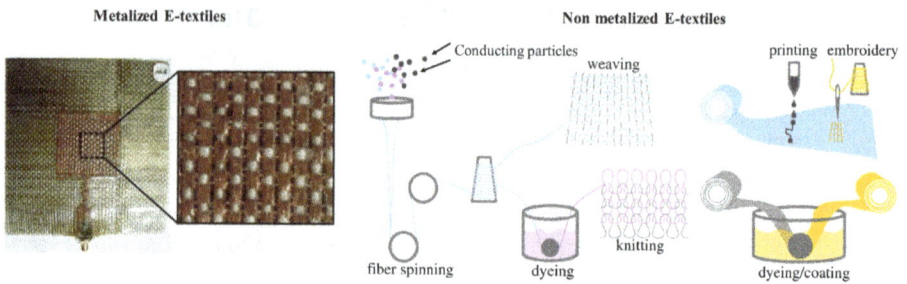

Figure 1.10: Schematic of the e-textile manufacturing process via spinning, weaving, knitting, dyeing, coating, embroidery, and printing (reproduced with kind permission from [66], Copyright 2021, Springer Nature).

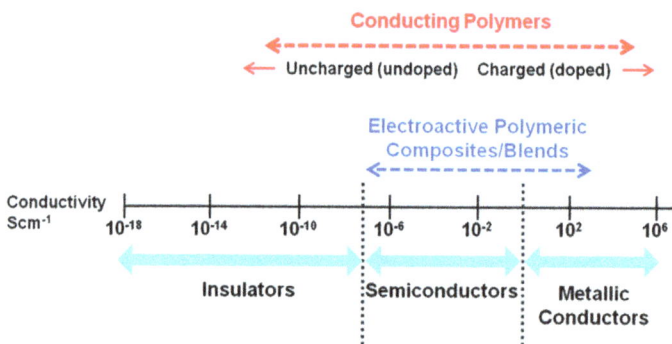

Figure 1.11: Range of conductivity of metallic conductors, conducting polymers, and conductive polymeric composites/blends (reproduced with kind permission from [67], Copyright 2021, Royal society of chemistry).

In addition to electromagnetic interference (EMI) shielding, and electrostatic charge dissipation, metal-based electroconductive textiles can also be prepared by blending metal fibres with common textile fibres during spinning. The shielding efficiency increases at low porosity and high content of conducting material. Although metallic yarn/fabric can be prepared using a weaving or knitting process in a conventional set-up, prepared fabrics lack textile properties such as flexibility, drape, and comfort [68, 69]. These limitations of metallic electroconductive textiles can be successfully overcome by coating conducting materials on textile fibres, yarns, and fabrics. Among them, there are various forms of carbonaceous materials available, such as carbon black, graphite, graphene, and CNTs other than conducting polymers like polyacetylene, polyphenylene, polypyrrole, polythiophene, polyphenylene sulphide, and polyaniline. Due to the lack of affinity to textiles, carbon blacks are not found to be very efficient in the electrification of textiles. In contrast, graphene, in which carbon atoms are arranged in 2D single layer and form honeycomb lattice, possesses good electrical, thermal conductivity, as well as excellent optical and mechanical properties [70–72]. But graphene tends to agglomerate and even restack to form graphite through van der Waals interactions, which makes it difficult to use. Its oxide derivatives such as nonpolar graphene quantum dot, graphene oxide (GO), and reduced GO are currently the most popular for producing scalable quantities of graphene materials through wet chemical processing, owing to their negative surface charge, good colloidal stability, higher yield, and excellent dispersibility in various solvents [73–77]. The negative surface charge allows better interaction with the functional groups of the fibres, helps them bind with the fabric better, and makes them flexible, washable, and long-lasting. It has been reported in many scientific publications that treating textiles with graphene or graphene derivatives in conductive polymeric solutions can improve conductivity (initially) and durability. Different methods have been used to coat textiles with graphene materials, such as dip coating, brush coating, vacuum filtration, electrophoresis, direct electrochemical deposition, kinetic trapping method, wet transfer of monolayer, or screen printing [78–85]. The successful and durable coating of graphene onto the textile substrates by suitable methodology will produce electroconductive textiles and may have many potential applications such as energy storing devices, drug delivery systems, flexible heating pad, EMI shielding,

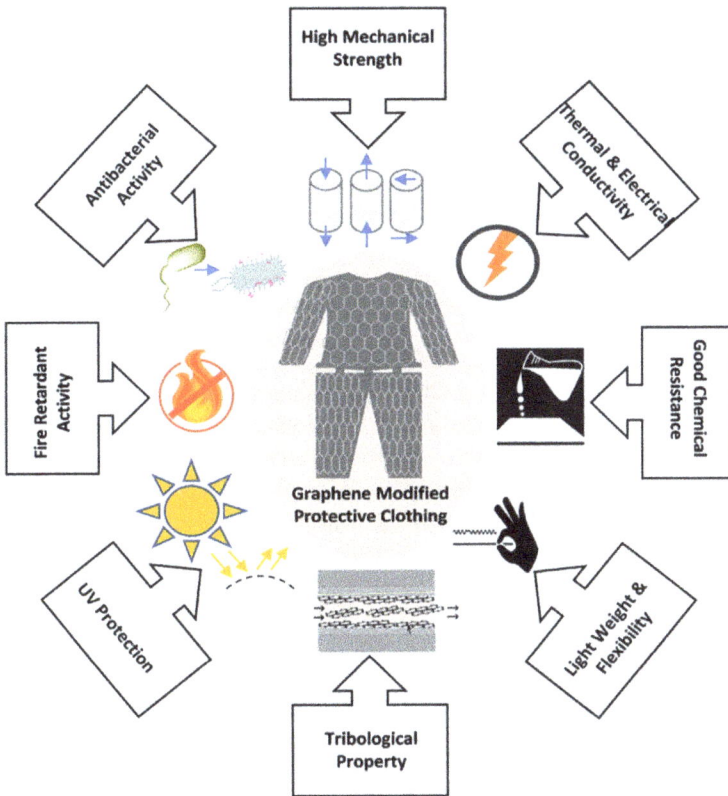

Figure 1.12: Potential of graphene in textiles (reproduced with kind permission from [87], Copyright 2019, Wiley).

cooling garment, heating garment, sensors, actuators, antibacterial clothing, antistatic clothing, and so on [86]. Multifunctional properties of graphene-modified protective clothing are depicted in Figure 1.12 [87], which is further described in detail in Chapter 7.

Other than these, nowadays, MXenes, a category of novel 2D material made of transition metal carbides and nitrides, are also garnering huge research interest, due to their high metallic conductivity, tuneable chemistry, suitable surface termination groups, electrochemical and optoelectronic properties, and surface hydrophilicity and have found many applications, such as wearable electronics, EMI shielding, energy storage, conductive electrodes, and biomedicine, as illustrated in Figure 1.13 [88]. It is possible to load MXenes on textile surfaces by different techniques such as spraying, nip-n-dip, spinning, coating, printing for converting the conventional textiles into high-end smart and functional textiles, since MXenes have good dispersibility in many common solvents such as water, ethanol, dimethylformamide, N-methyl-2-pyrrolidone, and dimethyl sulphoxide, and also in polymeric solutions [89–95].

However, conductive polymers, which combine some of the mechanical properties of plastics with the electrical properties typical of metals, play an important role in the development of modern-day smart textiles [61]. Intrinsically conducting polymers (such as polyacetylene, polypyrrole, polyaniline, polyphenylene, polythiophenes, and polyfuran), (conductivity 10^2–10^4 S/cm Figure 1.14a) and also extrinsically conductive polymer-based composites/nanocomposites (conductivity 10^{-5}–10^3 S/cm) are extensively used nowadays to prepare electroconductive textiles [63, 96–98]. As depicted in Figure 1.14b, these conductive polymers have long π-conjugated chains and hetero atoms in their chemical structure; doping of suitable impurities (like I_2, Br_2, Li, Na, AsF_5, BF_4^-, and ClO_4^-) to this polymer system can generate either holes or electrons in sufficient numbers, and this localized charge creates a local distortion of the crystal lattice. These can travel across the conjugated polymer chain during the application of potential difference and, thus, conduct electricity. The processes are termed as p-doping and n-doping.

In general, there are several ways of preparing conductive polymer-based electroconductive textiles, such as via blending, grafting, and coating. Coating textile techniques include electroless plating, electrodeposition, vapour deposition, sputtering of thin films, knife-over-roll coating, spraying, dip coating, soft lithography, printing, embossing, or imprint. Among them, the most common and familiar method of coating is in situ chemical polymerization, where the textile substrate is first soaked with monomer, followed by treatment in oxidant atmosphere, or vice versa. Applications of such materials are mainly focused on health, sport, automotive, and fitness, as depicted in Figure 1.15. Further, most recent and sophisticated achievements along with difficulties and future prospects in this field will be discussed in subsequent chapters.

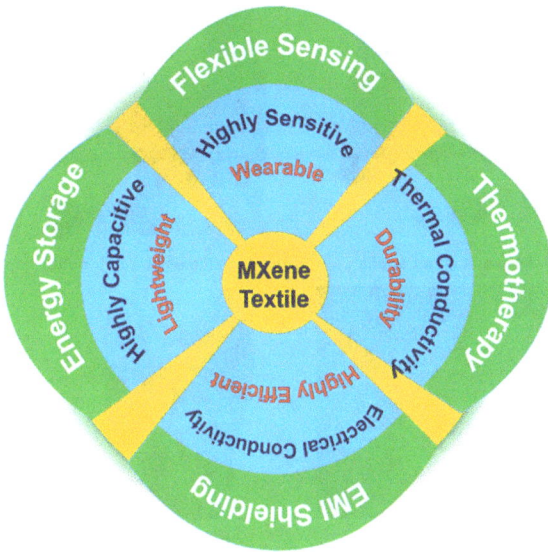

Figure 1.13: Emerging application fields for MXene in textile-based smart systems (reproduced with kind permission from [88], Copyright 2020, American Chemical Society).

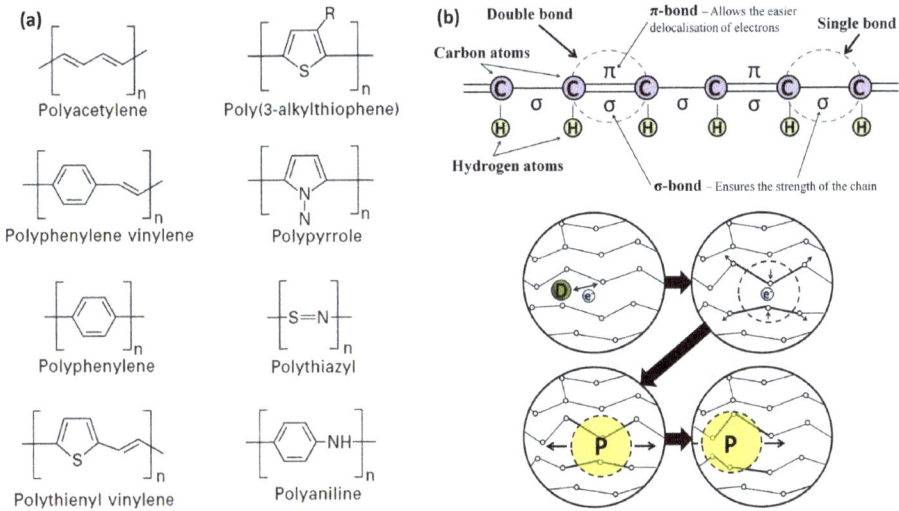

Figure 1.14: (a) Structure of various conducting polymers with their conjugated backbone and (b) mechanism of electrical conductivity in conducting polymers (reproduced with kind permission from [99], Copyright 2014, Elsevier).

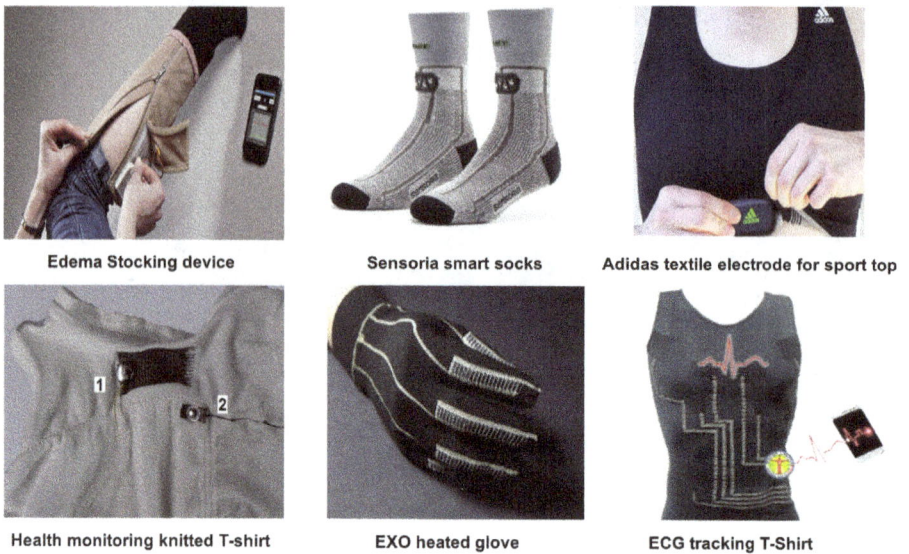

Figure 1.15: Commercialized applications of e-textiles.

1.6.2 Thermoregulating smart textiles and their applications

The American Society of Heating, Refrigeration, and Air-Conditioning Engineers defines thermal comfort as "the state of mind that shows happiness with the surrounding thermal environment" [100]. It is directly influenced by a person's mood, body, mind, physical state, psychological, and other conditions. Sometimes, even people who always live in the same kind of environment still feel thermal discomfort because of interacting and less tangible variables [101]. Thermoregulation is a negative feedback system that helps the human body keep its internal core temperature fairly stable, even when the outside temperature changes a lot [102]. It consists of three steps: thermosensation, central control, and efferent reactions to maintain the balance between generated and dissipated heat [103]. Although the human body is capable of self-thermal control via thermoregulatory processes such as muscular activity, sweat gland/sudoriferous gland functioning, and vasoconstriction and vasodilation, that is, narrowing and widening of blood vessels, with the increase in global warming and predicted climate change, the humanity will confront increasingly frequent and extreme heatwaves, potentially posing serious sustainability concerns [104]. For this, there is an urgent need to develop and promote useful technology for controlling personal thermal comfort in both indoor and outdoor environments, and thermoregulatory textiles has the potential to do so.

Based on heat transfer mechanism and structural characteristics, thermoregulatory textiles can be classified as passive, active, and responsive textiles. A basic summary of the mechanisms involved for thermoregulation in textiles, along with appropriate examples is depicted in Figure 1.16 [105]. Passive thermoregulation is the manipulation of generated and dissipated body heat without using external energy, whereas active thermoregulation employs the heating and cooling of the human body with auxiliary/ external energy inputs [102]. A wide range of new materials and possible innovative solutions, such as encapsulating nanofibres, biomimetic materials, phase change materials, and advanced approaches for producing and utilizing responsive fabrics (temperature and moisture) for improved cooling and higher thermal insulation are broadly discussed in Chapter 8.

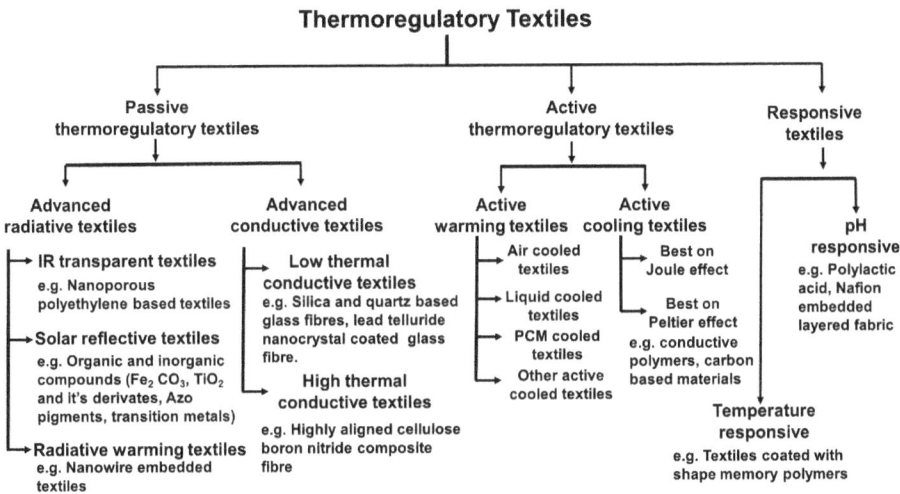

Figure 1.16: Brief classification of thermoregulating smart textiles.

However, achieving thermal comfort under high temperature and humid condition using passive thermoregulatory textiles is tough. To address these issues, active personal thermal garments have been developed. In order to regulate the human body temperature, these textiles require external energy and are of different designs, including water-cooled, air-cooled, and PCM-cooled systems. Among these, PCMs are the most suited candidate for textile applications, since they can be incorporated within textiles by either microencapsulation, into the structure of another material's matrix, fabric coating, hollow fibre spinning, and emulsion electrospinning [105, 106]. Examples of PCMs are hydrated salts, fatty acids, paraffin waxes, polyalcohol, and eutectics of organic and inorganic compounds. Of all these, alcohol and their derivatives are the most popular, owing to their solid–solid phase transition, odourless, small volumetric change, leakage proof and non-corrosive nature, long lifespan, and transition involving reversible breaking of the nearest hydrogen bonds within the molecular crystals [107]. The important properties of PCMs that can be used to store thermal energy are listed in Figure 1.17 [108]. However, the major limitations of organic PCMs include the supercooling during the cooling cycle and the low thermal conductivity. The incorporation of metal matrix, nano- and micro-sized metal fillers, and dispersion of high thermal conductivity materials like CNTs, CNFs, graphite, exfoliated graphite, or graphene in PCMs have been extensively studied, in order to improve the thermal conductivity of PCMs. However, their inherent cytotoxicity limits their commercial applications [109, 110].

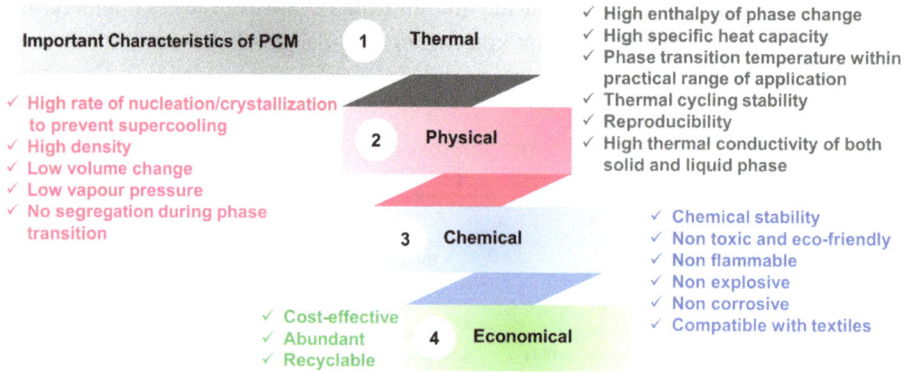

Figure 1.17: Important properties of PCMs for smart textiles application.

In addition, in order to acclimatize people to hot conditions, researchers are investigating the potential of personal thermal wears that are combined with thermometric modules, vapour compression chillers, and the Joule heating effect. But, compared to Joule heating fabrics, thermoelectric modules that follow the Peltier effect are more efficient for warming. However, due to inefficiency, developments in the Peltier effect's cooling applications are restricted. In a nutshell, personal thermal management is a reliable and economical method for achieving personal thermal comfort in difficult situations, while also balancing energy costs. However, to address the low-efficiency constraints of passive personal thermal wears and to improve the wearability of these materials, more new materials and innovation and structural improvements are necessary [105].

1.6.3 Stimuli-responsive smart and functional textiles

Tremendous developments in science and technology, along with rapidly increasing ambitious demands from textiles, have resulted in a number of breakthroughs in the last two decades, one of them being the introduction of stimuli-responsive materials. Stimuli-responsive materials demonstrate noticeable change in their configuration, shape, colour, and so on, upon exposure to an external stimulus such as physical (light, temperature, mechanical, electrical, magnetic) or chemical (electrochemical, pH, ionic strength and biological). Broadly, this class of smart textiles can be divided into chromic textiles (colour-changing textiles) and shape-memory textiles. Their operating mechanisms and application in textiles, along with other potential applications will be discussed in the following sections.

1.6.3.1 Chromic textiles

The suffix, "chromic", denotes reversible change in colour in a compound along with associated physical properties. Colorants that modify, radiate, or erase colour in response to the external stimuli are known as chromic colourants. These external stimuli can be light (photochromic), heat (thermochromic), pressure (piezochromic), electricity (electrochromic), electron beam (carsolchromic), or a liquid (solvatechromic) [111, 112]. The colour-altering effect is mainly caused by physical and chemical changes in dye molecules, and it can be predominantly transitory or, in some cases, permanent. The literature refers to these dyestuffs as intelligent dyestuffs, which have the potential to be used to manufacture smart textile materials, due to their ability to generate new innovative designs that can dynamically adjust looks of the garments to suit people's moods, styles, and so on, despite their poor resistance to visible light, heat, and other environmental stimuli, as well as their high price. Photochromism and thermochromism, both, have substantial commercial uses in ophthalmic purposes, including the familiar spectacles that turn into sunglasses when exposed to sunlight, dye lasers, information recording materials, security printing, and camouflage [113].

1.6.3.2 Photochromic textiles

Photochromic textiles contain photochromic dyes/pigments that induce a reversible change in colour upon irradiation with electromagnetic radiation, notably in the ultraviolet (UV) to visible range and revert to their normal hue, after the illuminated element is removed.

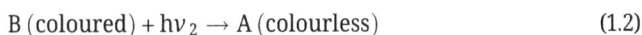

$$A \text{ (colourless)} + h\nu_1 \rightarrow B \text{ (coloured)} \tag{1.1}$$

$$B \text{ (coloured)} + h\nu_2 \rightarrow A \text{ (colourless)} \tag{1.2}$$

The absorption spectra of these two types are distinct from one another. Depending on the nature of reversal mechanism, photochromic textiles can further be classified as T-type (thermally driven) and P-type (photochemically driven) [113]. Moreover, the open ring structures of P-type photochromic molecules are colourless, while closed ring structures are coloured, and vice versa for T-type molecules [114, 115]. T-type photochromic molecules are widely used in memory disks, optical switches, sensors, and dye-sensitized solar cells (SCs), whereas computing, optical circuits, memory technologies, and ultrahigh-density storage systems, all use P-type photochromic materials. Spiropyrans, Spirooxazines, and Naphthopyrans fall under T-type, whereas Diarylethenes and Fulgides are of P-Type. All the above-mentioned photochromic colorants are commercially popular and operate through one of the following mechanisms such as variation in the oxidation state, ring opening, trans-cis isomerization, tautomerism, photodimerization, or homolytic or heterolytic cleavage of the chemical bond, as depicted in Figure 1.18, along with the commercial SolarActive® Colour Changing T-shirt [115]. Solar Active International produces durable photo colourable textured yarns that can be converted to fabric by knitting, weaving, or embroidery [116]. Embroidery threads of polypropylene are manufactured by mass colouration method, where different photochromic components are added within the melt hopper. Indoors (in absence of UV radiation), the fabric appears white, but outdoors (in the presence of UV radiation) as soon as the UV radiation penetrates the thread, photochromic compounds get activated and result in photo coloration to specific colours. These threads show reversibility and return to the original colourless state, as soon as the UV source is removed [117].

The first photochromic dyed fabrics appeared in the 1980s. Kanebo Ltd., a Japanese manufacturer, used microencapsulated spiropyrans that were printed on textile materials. The dyes reversibly changed from light blue to dark blue, when exposed to UV rays with wavelengths of 350–400 nm. As spiropyrans are unstable, possess low thermal bleaching rate, and show poor temperature fastness, they were replaced by more stable spirooxazines that contain oxazine group instead of pyran core [118] by the end of the 1980s. Toray Industries Inc. invented the process of coating fabrics with photochromic dyes, around this time. The fabric, known as Sway UV®, was able to turn blue or violet when exposed to UV rays with wavelengths of 350–380 nm. The colour of the fabric vanished within 30s of removal of UV radiation source. T-shirts, polo shirts, and jumpers were all made with Sway UV® [118]. Spirooxazines are extremely resistant to photodegradation. The extent of fading and colour strength of these compounds are determined by the presence of distinct alkyl groups (-R) with the nitrogen atom [119]. Unlike other dyestuffs, photochromic colourants are applied as disperse dyes, either through mass colouration or via exhaust or continuous method, along with a dispersing agent [115]. A typical end use has been towelling and beachwear in the textile sector, UV-light sensitive curtain, and knapsacks. Other potential applications include chromic sensors that provide a self-contained response (without using any additional electrical circuitry) to a change in environment by a visible colour change, and responsive camouflage fabrics for military applications [113, 120–123].

1.6.3.3 Thermochromic textiles

Thermochromic systems undergo a reversible change in colour, as the temperature changes [124]. Depending on the operating mechanism, thermochromics can be classified into two categories: intrinsic thermochromics, in which heating is the direct cause of colour change, and indirect thermochromics, where heating induces changes in the environment in which the chromophore is placed, resulting in colour change [125].

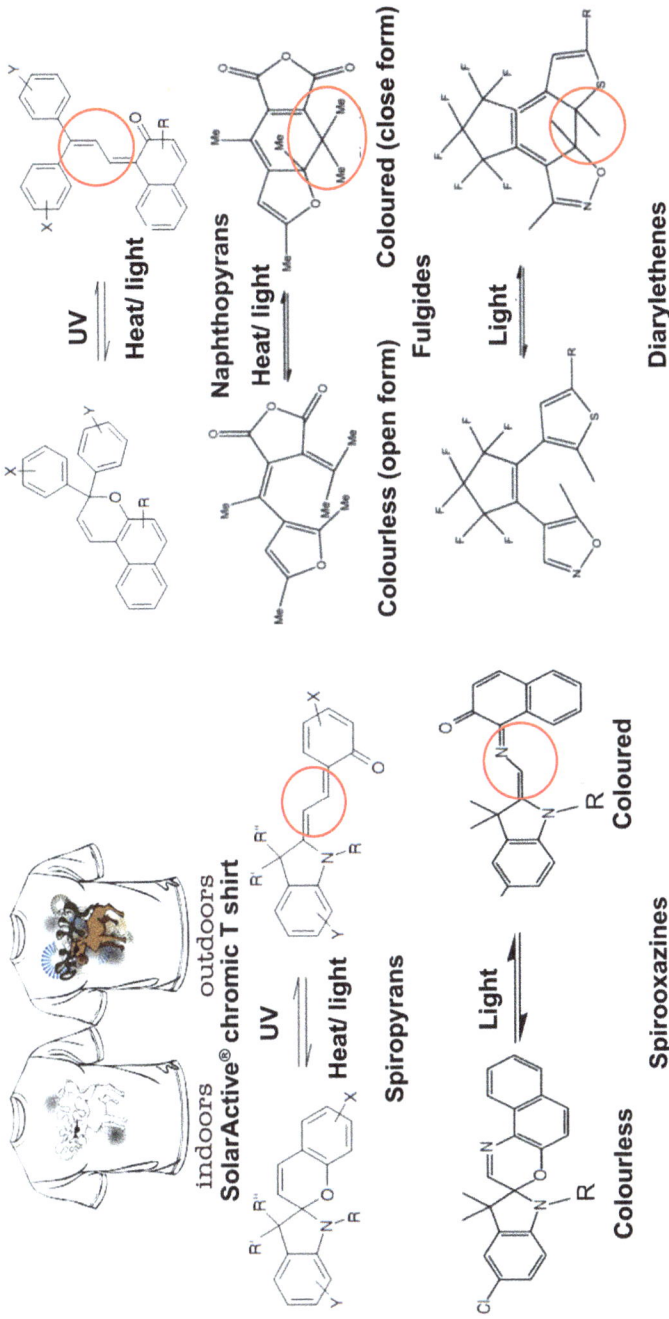

Figure 1.18: SolarActive® Colour Changing T-shirt and working mechanism of different types of photochromic textiles.

Although, many inorganic and transition metal complexes, organic compounds (leuco dye and liquid crystalline polymers (poly(alkoxythiophenes)) are popular as thermochromics, but their textile application is somewhat limited to organic compounds, since the first one operates at high temperatures and, thus, are not suitable for textile applications [113].

Organic thermochromics show their transition through an equilibrium between molecular species such as acid-base, keto-enol, and various crystal structures, whereas inorganic ones operate through phase transition, equilibrium between different molecular structures, change in ligand geometry, and change in the number of solvent molecules in the coordination sphere. Textile application of such material takes place either by microencapsulation followed by printing or pad-dry-cure along with a binder [115]. So far, liquid crystal and molecular rearrangement are the only two types of thermochromic systems that have been successfully used in commercial textiles. Both systems are based on microencapsulation [113]. Microencapsulation helps enclose the active components within a tiny hard shell, ensuring proper protection of the materials against the sensitive environment. Printing ink comprises a colourless electron-donating chromatic organic compound (fluoran dyes, N-acyl leuco-methylene blue derivatives, diphenylmethane compounds, etc.), an electron acceptor, and a single or bicomponent reaction medium (mainly alcohols, ketones, hydrocarbons, thiols, ester, and alcohol-acrylonitrile mixture) [115]. Of these two components one solvates the electron donor above the melting point, rendering the system colourless. When the medium's temperature drops below its solidification point, the electron donor becomes the electron acceptor, resulting in a coloured complex. Solvation stops, once the temperature drops below the medium's solidification point, and the electron donor becomes the electron acceptor, resulting in a coloured complex. The colours are determined by the chemical structures of the electron donors and acceptors. The temperatures at which the colour appears and departs, as well as the sharpness of these transitions and the colour intensities, are all controlled by the reaction medium components [126].

Applications of photochromic materials are limited to non-textile including photochromic glasses, UV sensors, holographic recording media, nonlinear optics and memory devices, medical thermography, or food packaging. Their lack of affinity with the common textile fibres, poor colour strength, inferior fabric handle, poor fastness, and activation temperatures typically in the range of 10–20 °C for a majority of these materials which is well above room temperature in cold countries and well below room temperature in hot countries, prevent thermochromic effects on textiles from being properly produced [113]. However, recently, it has gained attention in textiles including ski outfits, colour-changing tablecloths (in contact with a hot dish), and colour-changing chairs (upon sitting).

1.6.3.4 Solvatochromic textiles

Solvatochromism, in which a fabric changes colour when it gets wet or comes into contact with certain solvents, has been explored, recently, for the personal hygiene industry, along with in novelty textile accessories like handkerchiefs, interior decoration, performance costumes, display textiles, and in other fields like chemical sensors, writing ink, and anti-counterfeiting [127–134] This colour comes under the category of structural colours (Japanese jewel beetles and neon tetra are examples of organisms that display structural colour in nature), which overcome the disadvantages of the chemical colours that have a tendency to fade in air over time, emit high effluent during processing, and barely respond to external stimuli. Thin-film interference, multi-layer interference, and photonic crystals are three common optical methods used to generate structural colour [135]. These structural colours are mostly applied on textiles through nanotechnology-based coating approaches (such as L-b-L and thin-film deposition) in the form of microspheres. For example, silica (SiO_2) nanoparticles with a controlled size distribution of (207–350 nm) can be put together on black woven cotton textile fabrics to make a structural colour fabric [136].

In another study, it was shown that "soft" copolymerized poly (styrene–butyl acrylate–acrylic acid) [P(St–BA–AA)] microspheres can be coated onto commercial polyester, nylon, cotton, spandex, and carbon fibre fabrics. Fabrics made this way are strong, can let steam through, and are washable. When wetted with a polar solvent (water/ethanol), they show brilliant colour within 10 s. This is because the refractive index of the microspheres is different from that of the solvent. When the fabric dries, it goes back to its original colour [137]. These textiles have the potential to be used as smart fabrics, since they possess attributes that traditional dyeing cannot achieve, as well as the ability to satisfy textile requirements.

1.6.3.5 Other chromic textiles

Electrochromism may also be of interest to the apparel business, as the colour is created by applying an electric current to the fabric. Commonly used electrochromic materials include inorganic transition metal oxides, Prussian blue (hexacyanometallate) and organic conductive polymers, and metallo-polymers [138]. These materials have the potential to be employed in fashion and high-performance clothes, but there has been no evidence of successful commercialization owing to restrictions in prices, durability, application, and so on. Piezochromic textiles lead to a shift of colour in material in response to pressure and have been found suitable mainly for fabricating flexible sensors [139]. Readers can find a more detailed discussion on stimuli- responsive textiles in Chapter 9.

1.6.3.6 Shape-memory textiles

Shape-memory materials are defined as those that can "remember" a macroscopic permanent shape and can be manoeuvred so that a temporary shape can be "fixed" under the right conditions of temperature and stress, and then "relax" back to its original, stress-free state when told to do so by external command such as heat, electricity, or the environment, as shown in Figure 1.19 [140].

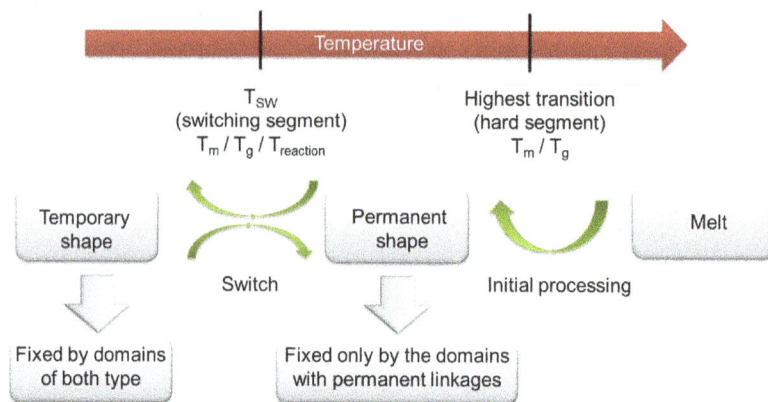

Temperature

T_{SW}
(switching segment)
$T_m / T_g / T_{reaction}$

Highest transition
(hard segment)
T_m / T_g

Temporary shape

Permanent shape

Melt

Switch

Initial processing

Fixed by domains of both type

Fixed only by the domains with permanent linkages

Figure 1.19: Schematic illustration of the basic mechanism of the shape-memory effect (reproduced with kind permission from [140], Copyright 2015, Elsevier).

Briefly, shape-memory polymeric structures (one-way, two-way, or triple-shape-memory) contain at least two contrasting structural "phases": A stable network and another external trigger-controlled phase, as depicted in Figure 1.20. The stable network is achieved by introducing chemical cross links, crystalline phases or interpenetrating networks and it helps to retain the original shape [141]. The second trigger-controlled temporary phase can be fixed by crystallization (for polyolefin, polyether, and polyester), glass transition (for epoxidized natural rubber, polyurethane, and polymethacrylate), reversible non-covalent bonds (for supramolecular interactions like hydrogen bonds, ionic interactions, and metal complexes), redox reactions (for responsive copper-cross-linked hydrogels), and segmental rearrangement [140, 142]. For example, chemically cross-linked polyethylene/polypropylene blends (30–70% polyethylene) demonstrate triple-shape-memory behaviour, owing to their crystallization and crystallization type of transition. The initial permanent shape at 25.0 °C is depicted in Figure 1.21(a); whenever heated at >120 °C, the polyethylene part melts, and it helps the composite attain first temporary shape upon deformation at 130 °C [Figure 1.21(b)]; further at >165 °C, the polypropylene part melts and helps attain second temporary shape upon deformation at 175 °C [Figure 1.21(c)]. When the structure is cooled down below 165 °C but above 120 °C, the polypropylene part gets locked, and below 130 °C, the entire structure gets locked; thus, temporary shapes are stable [Figure 1.21(d)], and further heating under stress-free condition reverts to the permanent shape at 175.0 °C as depicted in Figure 1.21(e) [143].

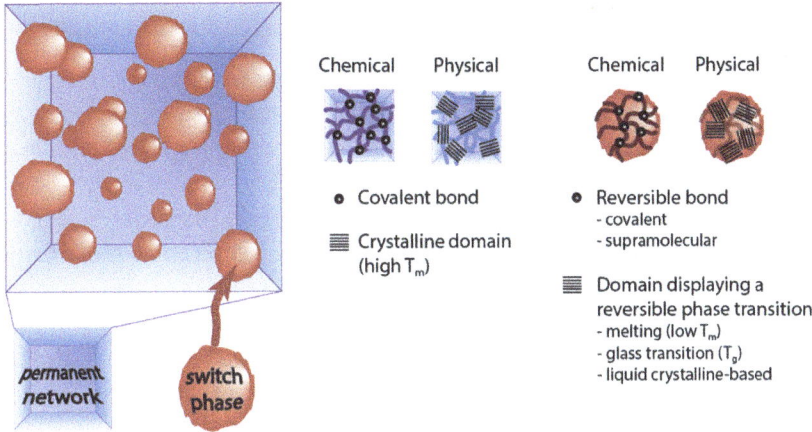

Figure 1.20: General structure of shape-memory materials composed of a stable phase and a soft phase (reproduced with kind permission from [140], Copyright 2015, Elsevier).

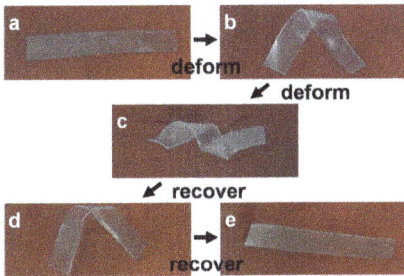

Figure 1.21: Photographs of cross-linked polyethylene/polypropylene composite film (a) permanent shape at 25 °C, (b) first temporary shape at 130 °C, (c) second temporary shape at 175 °C, (d) first temporary shape at 130 °C, and (e) permanent shape at 175 °C (reproduced with kind permission from [143], Copyright 2013, American Chemical Society).

Among all the stimulus-responsive smart and functional materials, shape-memory materials have gained scientific significance in the past decades, owing to their several advantages:

(i)   Abundance of commercially available raw materials.
(ii)   Material development processes are simple, do not require harsh temperature, or physiological conditions.
(iii) Materials are responsive to different stimuli, such as heat [144], water [145], light [146], magnetic field [147], electricity [148], pH value [149], and solvent [150] and can be easily tailor-made.
(iv)   Allow simple and flexible single or multistep programming to obtain the desired shape fixity and recovery responses.

(v) Nanomaterials induced additional functionality that offer a wide range of different properties can be easily incorporated.

(vi) Biodegradable, biocompatible, and comfortable smart biomedical implants can be developed easily [151].

Shape-memory textiles are available in fibre as well as fabric form (upon subsequent weaving) and are made by several manmade fibre spinning methods like wet spinning, melt spinning, dry spinning, reaction spinning, and electrospinning. Fibres with less elongation are suitable for subsequent weaving, whereas fibres with large elongation and elasticity promote knitting ability [152–154]. Characteristic properties of shape-memory materials like elongation, switch temperature, shape recovery, and shape fixity are tailorable and depend on fibre spinning parameters as well as the proportion of switch phase to permanent network. During the manufacturing process, these fibres are subjected to stress-induced orientation, resulting in increased molecular orientation, higher shape recovery, lower shape fixity, and higher recovery stress [155]. Owing to their easy processability, soft handle, switching at body temperature, biocompatibility, and biodegradability, shape-memory polymers are mostly suited for textile applications, along with wide-ranging applications in all aspects of daily life. Smart fabrics for firefighters, Nitinol-reinforced underwire, oricalco shirt (sleeve length shortens as atmospheric temperature increases), diabetic socks, diaper edging, and self-tightening shoelace are some textile uses; heat-shrinkable tubes for electronics, mechanical remover for blood clots, stent, packaging films, screw/fasteners, self-disassembling mobile phones, self-deployable sun sails in spacecraft, implants for minimally invasive surgery, and intelligent medical devices are examples of other uses [155–157]. These are only a few of the many conceivable uses for shape-memory technology, which has a lot of potential in many other areas.

1.6.4 Energy-harvesting textiles for smart applications

The stock of burning fossil fuel, which accounts for up to 89% of total energy consumption, has been steadily depleting over the last few decades and is expected to run out within the next 150 years [158]. It also contributes significantly to the greenhouse effect, air pollution, and particulate matter emissions, which, in turn, contribute to an overall increase in the Earth's surface temperature, resulting in global warming and the ultimate threat to human health [159]. We must, therefore, exercise extreme caution when using active energy sources. Since the world has entered the era of Internet of Things, smart wearable sensors, robotics, as well as all artificial intelligences, we are being forced to use passive energy to sustainably power billions of distributive devices. In the present scenario, portable energy storage devices like batteries seem to be an innate choice to provide a ubiquitous energy solution. However, from the wearable electronics point of view, the adoption of battery is quite difficult, since it has a rigid and bulky structure that makes it unfit at skin interfaces [160], contains toxic electrochemicals that may restrict their bio-integrated application [161], shows irreversible chemical reaction, has low specific power, inferior charge discharge rate capability, and limited lifetime [162–165]. Therefore, a ubiquitous, environmentally acceptable, and sustainable energy solution is very desirable.

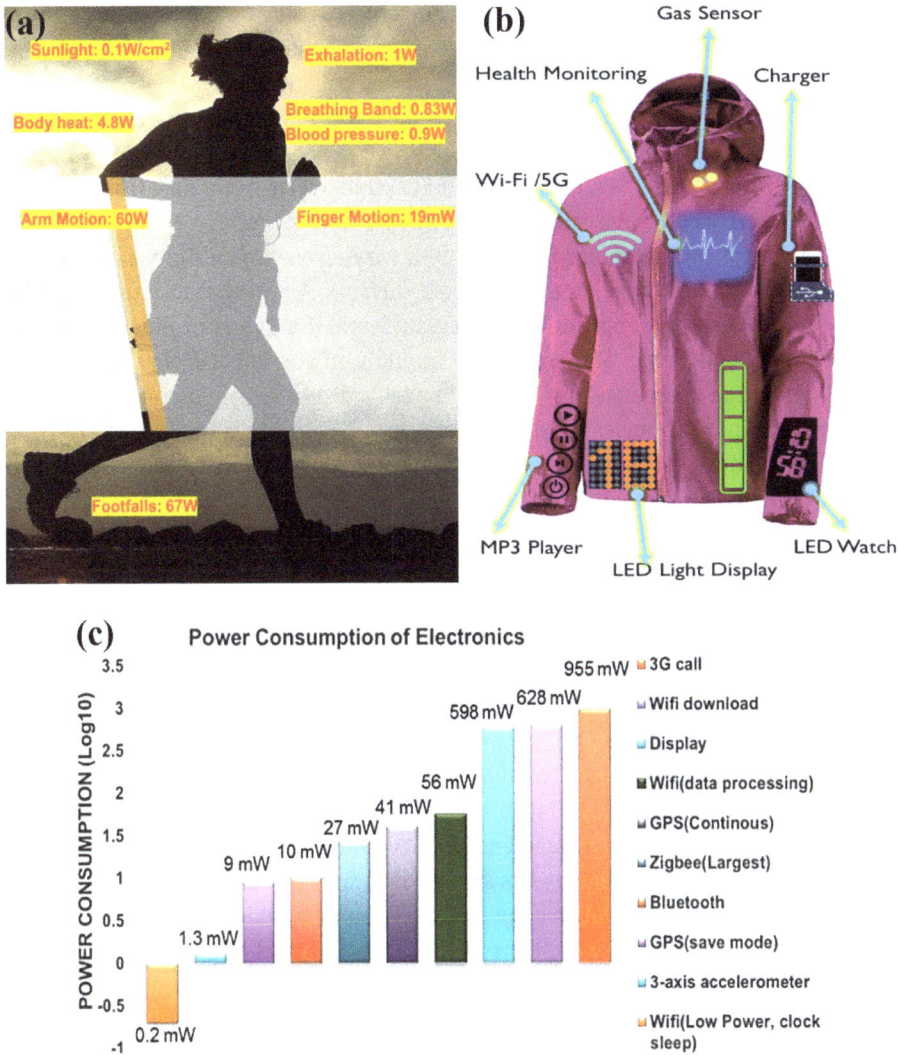

Figure 1.22: (a) Quantitative energy associated with various body movements [159]; (b) smart textile-based energy harvesters integrated within jacket (reproduced with kind permission from [167], Copyright 2019, Royal society of chemistry); and (c) power required to run one body electronics (reproduced with kind permission from [166], Copyright 2019, American chemical society).

The human body and its surroundings, including biomechanical motions, body heat, biochemical, and fluid flow are some rich sources of renewable energy. The energy associated with body heat is 2.4–4.8 W, whereas net biomechanical energy reaches up to 67 W, and fluid flow can result in 100 W of renewable energy, as depicted in Figure 1.22(a) [159]. Whereas, as illustrated in Figure 1.22(c), the power required to run modern on-body electronics is within the range of 200 µW to 1 W [166]. Thus, converting 1% of the power from the human body may be sufficient to power the majority of portable electronics. Targeted wearable electronics, usually used for sensing and remote monitoring applications are to be powered by smart textile-based energy harvesters, as depicted in Figure 1.22(b) [167]. In recent years, many working mechanisms and wearable device designs have emerged to capture this type of energy. These include piezoelectric nanogenerators (PENGs), triboelectric nanogenerators (TENGs), thermoelectric generators (TEGs), SCs, biofuel cells (BFCs), and hybrid generators (HGs).

TENGs are based on the triboelectric effect, a form of contact electrification in which rubbing of two dissimilar materials with varying electron affinity can generate opposite electrostatic charge on the contacted surfaces. A potential difference will be generated between two oppositely charged surfaces, if a gap is produced between them by applying a force, and a polarization-induced current will flow, as a result [168]. This kind of current is also called Maxwell's displacement current, and is the basic idea behind a triboelectric nanogenerator [169]. An ultra-flexible 3D TENG on commercial clothing for biomechanical energy harvesting is depicted schematically in Figure 1.23 [170].

The piezoelectric effect is now widely used for biomechanical energy harvesting, which is based on the generation of electric field under mechanical stress, due to the shifting of the positive and negative charge centres within the piezoelectric material [171]. A photograph of a textile PENG made by using weaving with cotton to form an energy elbow pad is depicted in Figure 1.24 [172]. In contrast, thermoelectric effect uses temperature differences over space, that is, temperature difference between an adult and its surroundings can be used for continuous energy harvesting, and it has the potential to provide heat flow of up to 10 mW/cm^2 [173]. Photograph of TEG-integrated smart T-shirt is depicted in Figure 1.25 [174].

Aside from the previously stated energy harvesting via biomechanical movements and body heat, biochemical energy is a sort of renewable and eco-friendly on-body energy source that is widely available but often overlooked, and it manifests itself in the form of body fluids such as sweat, tears, blood, and saliva and also in form of biofuels such as glucose, lactate, and fructose that can provide up to 100 W in total, for a healthy adult [175, 176]. The above-mentioned biofuels need to be oxidized within a BFC by using suitable biocatalysts (mainly microbe- or enzyme-based) in the anode; this process releases electrons that move towards the cathode via external circuits, reduces oxygen, and subsequently converts the biochemical energy into electricity, as shown in the following equations[159, 177]:

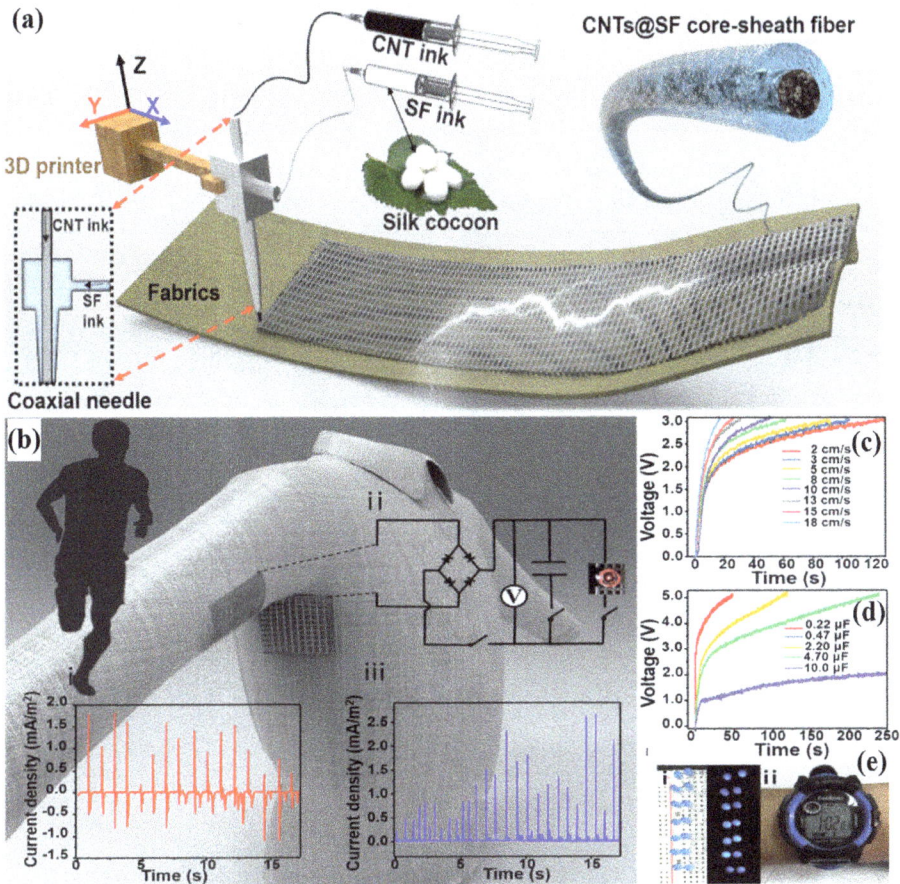

Figure 1.23: Schematic illustration showing smart clothes for (a) biomechanical energy harvesting, using ultra-flexible 3D textile TENG prepared through direct printing of core-shell architecture on commercial clothes, (b) its performance; inset shows (i) the output current density of the printed smart gridline pattern placed underarm sleeve, (ii) circuit diagram, and (iii) output current density of the smart pattern, (c) charging curves of a capacitor, (d) charging curves of different capacitances charged using the prepared pattern displaced at different speeds, and (e) digital image showing LEDs and an electrical watch driven by the power generated by the 3D printed E-textile (reproduced with kind permission from [170], Copyright 2019, Elsevier).

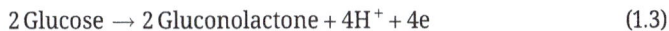

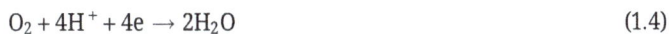

$$2\,\text{Glucose} \rightarrow 2\,\text{Gluconolactone} + 4\text{H}^+ + 4\text{e} \qquad (1.3)$$

$$\text{O}_2 + 4\text{H}^+ + 4\text{e} \rightarrow 2\text{H}_2\text{O} \qquad (1.4)$$

Figure 1.26 depicts a schematic of enzyme/CNT composite fibres used to create a woven textile BFC, in which an LED device attached to enzymatic power fibres starts glowing when a glucose solution is dropped on a fabric.

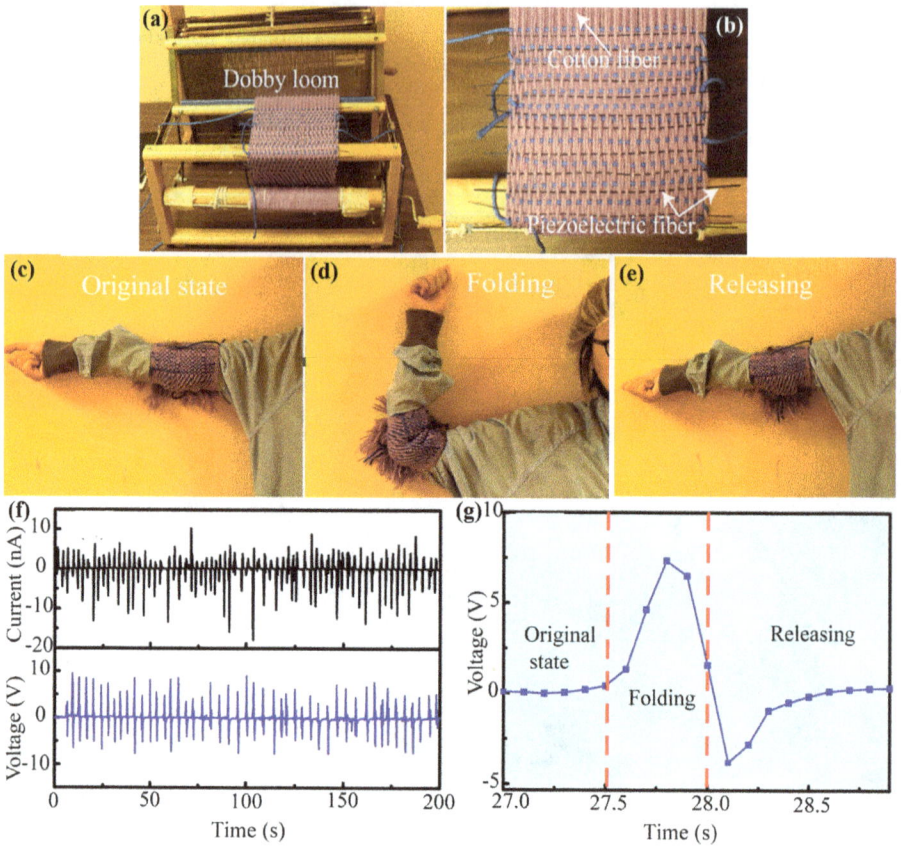

Figure 1.24: (a) Textile-based PENG prepared using dobby loom, by taking cotton yarn as a warp and piezoelectric fibres as weft; (b) and (c) Electrical properties (open-circuit voltages and short-circuit currents) of the textile PENG in a 90° folding-release action of the elbow (reproduced with kind permission from [172], Copyright 2017, American chemical society).

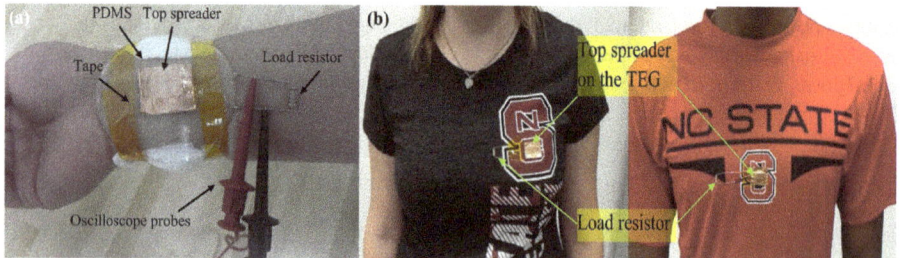

Figure 1.25: (a) TEG device used on the wrist, (b) TEG-integrated T-shirts (reproduced with kind permission from [174], Copyright 2016, Elsevier).

Moving on to environmental sources, the worldwide annual potential of solar energy is 1,575–49,837 exajoules (EJ), which is at least three times higher than global energy consumption of 589.1 EJ in 2020, and more interestingly, it also exceeds the estimated global energy consumption in 2050 [178]. Direct sunlight has a potential to produce about 100 mW/cm^2 of ambient solar energy, which is sufficient and can be easily used for powering on-body electronics based on photovoltaic effect [159]. When an appropriate wavelength of light strikes these cells, the photon's energy is transferred to the electrons within the atom of the semiconducting material in the p-n junction. As a result, electrons jump to a higher energy state known as the conduction band. This creates a "hole" in the valence band where the electron leapt. The movement of the electron caused by increased energy results in the formation of two charge carriers, an electron-hole pair. The motion of the electron in the cell generates an electric current. The hole can move as well, but in the opposite direction, as the p-side thus generates the photocurrent [179]. SCs are commonly classified into three generations. The first generation is wafer-based, for example, crystalline silicon (c-Si). Thin film materials, such as cadmium telluride (Cd-Te), amorphous silicon (a-Si), and copper indium gallium selenide are used in the second generation of SCs (CIGS). Dye-sensitized SCs (DSSCs), organic SCs (OSCs), and perovskite SCs (PSCs) are examples of third-generation SCs. These third-generation SCs have sparked a lot of interest in the textile industry since they offer tempting traits such as flexibility and lightweight, along with an abundance of raw materials, ease of manufacture, and cheap cost [159]. Schematic demonstration of a stitchable textile OSC integrated with clothing is depicted in Figure 1.27 [180].

There are three oft used fabrication techniques to realize textile-based TENG, PENGs, TEG, SC, and hybrid textile triboelectric-piezoelectric nanogenerators (TPENGs) in energy harvesting, namely layer stacking (multilayered structure), yarn intersection (2D interlacing and 3D interlacing), and 3D printing [159]. In addition to these energy harvesters, energy storage devices are also necessary to continuously power the wearable devices. Thus, the ultimate aim is to integrate energy harvesting and energy storage devices within a wearable structure/fabric. Figure 1.28 depicts required properties in a wearable energy-harvesting and storage device [167].

For the development and construction of wearable energy storage devices, it is essential to have small volumes, lightweight, high safety, good electrochemical performance, mechanical durability in terms of stretching and twisting reliability, compression stability, and finally, high energy density [182]. Keeping in mind that batteries and traditional supercapacitors are stiff and bulky, ultracapacitors or electrochemical supercapacitors, based on ion adsorption or redox reaction that combines the energy storage capabilities of batteries with the discharge properties of capacitors, have become a better alternative to batteries for flexible, wearable, and portable electronic devices. They have a high specific energy, power density (up to 10 kW kg^{-1}), long charge-discharge cycling stability (up to 10,000 times), along with increased energy storage capabilities ranging from a hundred to several thousands in the same volume [163, 183–185]. The

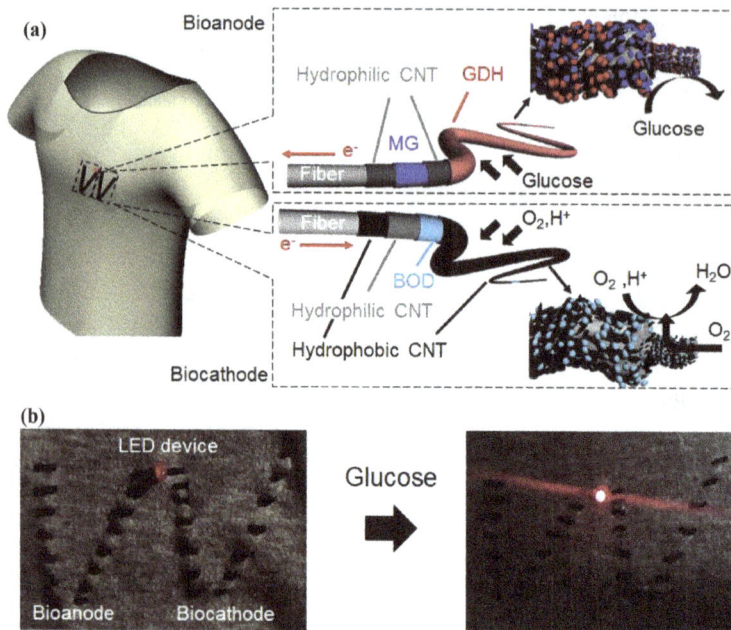

Figure 1.26: Schematic illustration of a woven textile BFC enzyme/CNT composite fibres (reproduced with kind permission from [181], Copyright 2019, Elsevier).

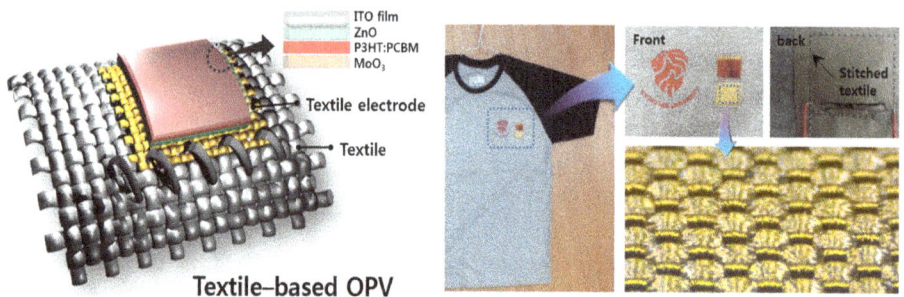

Figure 1.27: Stitchable textile SC integrated with t-shirt (reproduced with kind permission from [180], Copyright 2014, Elsevier).

main features of fibre-based Li-ion, non-Li ion batteries, and supercapacitors are shown in Table 1.3.

The charging/discharging process of a supercapacitor is depicted schematically in Figure 1.29a. The most commonly used electrode materials for supercapacitors include transition-metal oxides, high surface area-based carbon materials, and conducting polymers with carbon and metal oxides. Depending on the configuration of the electrode materials, these supercapacitors can be classified as 1D (yarn/fibre-shaped) and 2D (planner). For the 1D system, two yarn/fibrous electrodes are either arranged in

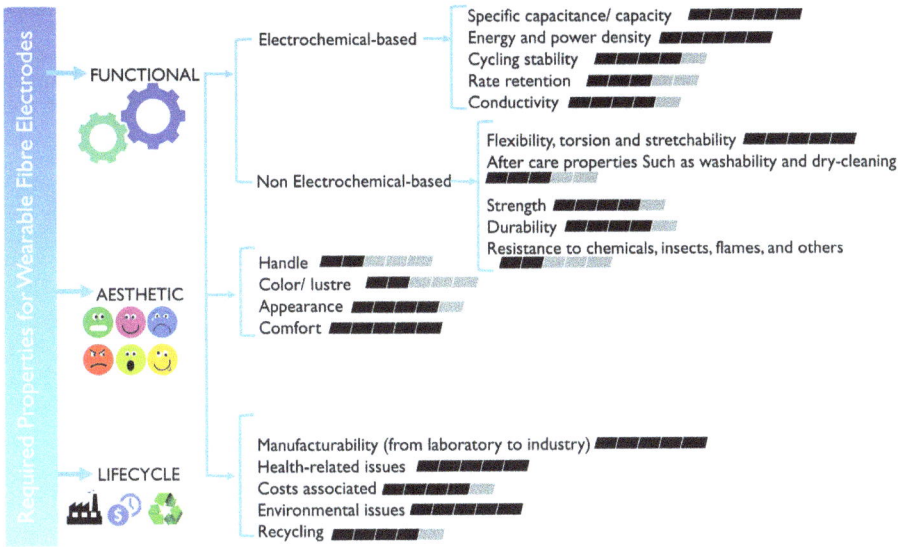

Figure 1.28: Required properties in wearable energy-harvesting and storage device (reproduced with kind permission from [167], Copyright 2019, Royal society of chemistry).

Table 1.3: Main features of fibre-based Li-ion, non-Li-ion batteries, and fibre-based supercapacitors.

Types	Key aspects	Materials employed	References
Flexible Li-ion batteries (fibre-based)			
Aligned multiwalled CNT-based wire-shaped battery	Twisted architecture	Multiwalled CNT fibre, Li wire, MnO_2	[186]
Waste cotton cloth-derived collector-based Lithium sulphur battery	Interlaced woven architecture promoted migration of Li ions	Cotton cloth, lithium foil, lithium bis(trifluoro methane sulfonyl) imide (LiTFSI)	[187]
Aqueous lithium-ion battery	Fibre-shaped, safe, environmentally friendly with good electrochemical performance	Polyimide-CNT fibre, Lithium manganese oxide ($LiMn_2O_4$)- CNT fibre, lithium sulphate aqueous solution	[188]
Flexible, micro sized and shape-versatile single fibre-based lithium-ion battery	Wearable, and shape versatile	Single carbon fibre, $LiFePO_4$, $Li_4Ti_5O_{12}$, polyethylene oxide with a lithium LiTFSI	[189]

Table 1.3 (continued)

Types	Key aspects	Materials employed	References
Non-Li-ion batteries			
Flexible, aqueous solvent-based zinc-air battery	Cable-type, high power/ energy density, safety, eco-friendly due to nonreactive zinc metal	Zn, gel polymer electrolyte	[190]
Woven textile like aluminium-air battery	Abundance, three-electron-redox properties, high volumetric capacity	CNT/Ag–nanoparticle hybrid sheets, Al, PVA-based hydrogel electrolyte	[191]
Ni–Fe battery	Safety, high ionic conductivity, low cost	CoP@Ni(OH)$_2$ NWAs/CNT, S-α-Fe$_2$O$_3$	[192]
Sodium battery	Abundance, environmental sustainability, low cost	Sodium foil, NiS$_2$⊂PCF	[193]
Fibrous supercapacitors			
Carbon-based	Electrochemical double-layer capacitor	GF@3D-G (graphene-based electrode), H$_2$SO$_4$-PVA	[194]
Metal oxide-based	Pseudocapacitor	Elastic fibre, PVA-based hydrogel electrolyte, MnO$_2$	[195]
Thermally drawn supercapacitor	High productivity, the longest functional supercapacitor	Thermal drawable materials (PVDF, CPE, COC), CB, activated carbon	[196]

parallel, twisted form or in core-sheath structure. Fabrication of 1D flexible poly(3,4-ethylenedioxythiophene): poly(styrenesulfonate)–polyacrylonitrile (PDEOT: PSS-PAN)/Ni cotton (PNF/NiC) capacitor yarn is illustrated in Figure 1.29b. In contrast, the 2D system consists of two planner electrodes separated by a solid/gel electrolyte and separator [197, 198]. Based on their energy storage strategy, supercapacitors are classified as electric double-layer capacitors (EDLCs), pseudocapacitors, or hybrid capacitors. EDLCs store energy by forming an electrical double layer of electrolyte ions through ion adsorption-dislodging at the electrode–electrolyte interface, whereas pseudocapacitors store energy through a redox reaction in which charge is largely transferred near the electrode material's surface [184, 199]. Charge storage mechanism diagrams for (a) EDLCs and (b) pseudo-capacitance are depicted in Figure 1.30.

Electrode surface area plays a vital role in EDLC, which makes carbon-based materials such as graphene and CNT suitable for EDLC, whereas transition metal oxides that exhibit redox reactions (e.g. RuO$_2$, NiO, MnO$_2$, and Fe$_2$O$_3$) and conducting polymers (e.g. polypyrrole, poly[3,4-ethylenedioxythiophene] (PEDOT) and polyaniline) are mainly used for pseudocapacitors [182]. Although EDLCs have a high-power density and superior

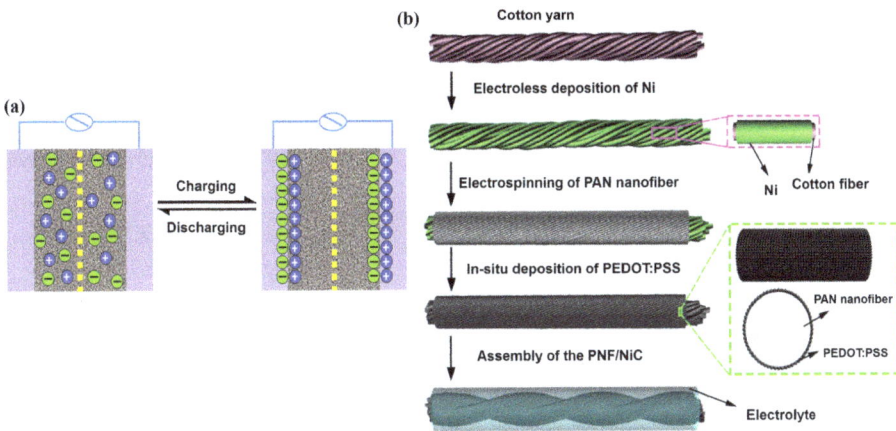

Figure 1.29: Schematic showing (a) how a supercapacitor is charged and how it is drained. Reproduced with kind permission from [200], Copyright 2019, Elsevier. (b) Fabrication of the PNF/NiC capacitor yarn (reproduced with kind permission from [201], Copyright 2019, MDPI).

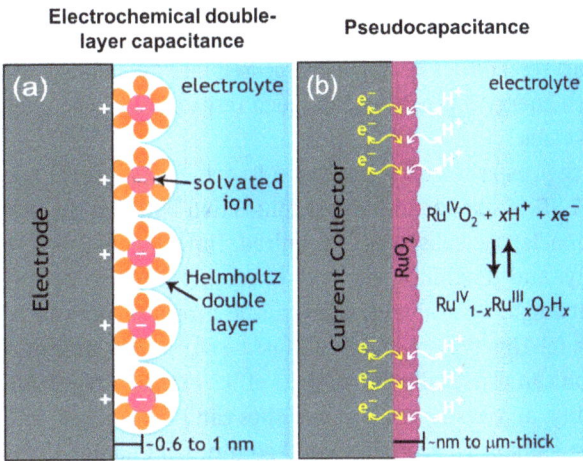

Figure 1.30: Schematic of charge storage mechanism of (a) EDLCs and (b) pseudo-capacitor (reproduced with kind permission from [202], Copyright 2016, Elsevier).

stability, they possess low capacitance, whereas pseudocapacitors show the opposite performance. A comparative analysis on characteristics of electrode materials for supercapacitors is shown in Table 1.4 [203].

In summary, the implementation of biomechanical energy harvesters within textiles provides a wide range of unique benefits. To begin with, the drape of the textile structure assures very high sensitivity to mechanical deformation, which aids in very efficient energy harvesting. Secondly, these kinds of assemblies will keep the wearer comfortable because textiles naturally allow air to pass through them. Taking advantage of industry,

Table 1.4: Characteristics of electrode materials for supercapacitors [203].

Materials used	Strengths	Weaknesses
Metal oxides	High faradaic capacitance, high energy density	High mass density, low stretchability, low flexibility
Carbon-based materials	High conductivity, chemical stability, low mass density, long cycle life, good stretchability, large specific surface area	High faradaic capacitance, high industrial cost, high energy density, low product throughputs
Conductive polymers	High faradaic capacitance, high conductivity, high stretchability, high flexibility	Low power density, low energy density

these kinds of assemblies can be made in huge numbers [159]. However, the use of electromagnetic and electrostatic transduction inside textiles to capture biomechanical energy is limited, because the former requires a magnet and coil setup and the latter, a DC voltage supply [204, 205]. On the other hand, due to their inherent designing flexibility, triboelectric, piezoelectric, and hybrid nanogenerators have the potential to be built into wearable textile structures to power electronics on the body. Even though a lot of preliminary work has been done so far, researchers have mostly focused on how to make the lab grade devices and how much energy it produces. Not much research has been done to make a specific product marketable.

The following are the future prospects in this field – (a) To achieve long-term mechanical durability in textile TENG, a strong mechanical connection between the fabric electrode and the active triboelectric substance is required. However, in order to improve comfort, stretchability, and flexibility, we have to sacrifice the strong mechanical interlocking; (b) Lightweight wearable devices can be produced on a large scale with improved efficiency; (c) This type of material must be shielded from external impurities, as contamination can drastically diminish its efficiency, (d) Breathability, tactile sensations, washability, and aesthetic characteristics can be enhanced; and (e) Determination of evaluation standards. Readers should go through Chapter 15 for more detail discussion on energy-harvesting textiles.

1.6.5 Functional and smart textiles in protective and defence application

Protective textiles are a class of technical textiles that protects the wearer (may be common people or armed force personnel) from extreme environments and various risks. Nowadays, a wide variety of smart protective textiles is being used in military and defence applications, too. A few examples of protective textiles are: firefighting clothes/suits (protect from catching/spreading fires), heat-protective textiles (protect from extreme heat exposure), UV-protective textiles (protect from serious effect of sunrays), antimicrobial textiles (protect the growth of microbes), extended cold weathering clothing systems (ECWCS, protect wearer from the extreme shock of sub-zero temperatures), nuclear biological and chemical (NBC) suits (protect from the serious effects of nuclear, biological, and chemical hazards), space suits (protect astronauts from the impact of extreme temperature, low oxygen level, small bits of space dust, etc.), bulletproof vest (protects the army/police from the impact of bullets), stab resistant body armour (protects from the impact of stabs), camouflage fabrics (help hide amidst nature and protect from enemies), military shelters, military parachutes, aerostat/airship, and so on.

With the advent of nanotechnology, textile products for defence applications have demonstrated significant improvements in functional properties as well as smartness. For soft body armour applications, high performance fibre- (such as Kevlar, ultra-high-molecular-weight polyurethane) based fabrics are generally treated with shear thickening fluids (STFs) for improving impact resistance. Researchers have reported that incorporation of halloysite nanotube, nanosilica, and so on improve the performance of STF significantly [206]. Of late, extensive research has been going on radar/microwave absorptive textiles that have the potential for radar and microwave invisibility, showing a potential use in stealth applications. Conductive nanocomposite coatings containing various conductive nanofillers (such as graphite, CNT, carbon nanofibre, and fullerene) have great potential in making radar-absorptive textiles. Bhattacharyya and Joshi [207] observed a significant enhancement in microwave absorption in radar frequency (8–18 GHz), that is, in the X (8–12 GHz) and Ku (12–18 GHz) bands, by incorporating 10 wt% iron-nickel co-deposited nanographite (FeNiNG) making TPU/FeNiNG nanocomposite for coating on textiles, showing potential for stealth application.

Moreover, to increase protection/safety and survivability of people working in extreme environments and hazardous situations, material scientists, textile researchers, and technologists have come forward with real-time information technology. Hence, sensors and actuators are performing a strong role in protection of military personals. Textile materials and equipment used by soldiers have been increasingly integrated with electronic capabilities and components, during the last two decades. During a war, a soldier wearing a "Smart Shirt" or "wearable motherboard" will be able to transmit information on his wound and his condition, immediately, to nearby medical units. Based on the heartbeat and respiratory rate of the soldier, the physician can determine the extent of his injury, thereby determining the priority for treatment. Nowadays, it has been possible to make clothes that detect toxic gases using smart textiles. When a chemical penetrates the hazardous material gear, sensors in the suit can alert the user. A "smart fabric" uses metal organic frameworks (MOFs) to support electronic sensors.

Coated or laminated multilayered, high helium/hydrogen gas barrier and weather-resistant textiles are potentially used in making hulls/ballonets of aerostats/airships, which are used in military surveillance [208]. In particular, polyurethane nanocomposite-based, highly flexible, high helium gas barrier and weather-resistant films and coatings have huge potential, in this regard [209, 210].

Nowadays, "*n*" number of functional and smart clothing are being explored by military forces across the world to increase safety and effectiveness. Readers can find a detailed discussion on various functional/smart textiles for protective and military/defence applications, as well as their mechanism and technologies, in Chapter 10 and Chapter 11.

1.6.6 Functional and smart textiles for medical uses

Medical textiles are textile structures that have been designed for use in healthcare applications and range from a face mask, wet wipes, diaper, bed covers, surgical attire, and dressings to more complex textiles like spacer fabrics for compression bandages, drug-releasing textiles for wound healing, flight socks, light-emitting fabrics used for photodynamic therapy, textile electrodes for nerve or muscle injury and orthopaedic treatment, hernia meshes, stents, sutures, composite heart valves or blood vessels, and many more wearable devices for detection of COVID-19. Based on the wide variety of product ranges, it has become a very happening and diverse field, located at the intersection of textile engineering, chemistry, microbiology, medicine, and comfort [211]. As a result, innovations in any of these areas may spur advancements in others. For example, new gel-forming, superabsorbent compounds discovered in chemistry led to the creation of novel infant nappy and adult intemperance products used in medicine. In brief, medical textile products are classified according to their usage as extracorporeal devices, healthcare and hygiene products, implantable materials, and non-implantable materials [212]. Healthcare and hygiene products include surgical clothing gowns, masks, beddings, blankets, cover stock, absorbent layer, incontinence diaper sheet, wipes, and surgical hosiery. Extracorporeal devices include artificial liver, artificial kidney, artificial heart, and mechanical lung, whereas implantable materials include sutures and vascular grafts. More details about these medical textile products are summarized in Figure 1.31 as well as discussed with more details in Chapters 13, 14, and 18.

However, in the initial days, to make conventional textiles suitable for surgical clothing, it was thought that alteration of the macroporous structure was essential. The second-generation surgical clothing and bandages were made coated/laminated, keeping in mind that coated/laminated structure can prevent penetration of body fluids or microorganisms. However, wearers realized quickly that the coating or membrane must be breathable, that is, water vapour- permeable, otherwise, the wearer comfort suffers; consequently, they are being phased out in favour of third-generation medical costumes such as three-dimensional spacer fabrics (textile engineering) or bio-functional clothing (chemistry). In recent years, the results of extensive study within the domains of material, colloidal, interfacial, surface science, and engineering have enabled textile materials to achieve better qualities such as filtration, antibacterial and antiviral activity, breathability, and so on, which are crucial in hospital textiles and medical staff apparel and have made textiles a frontline warrior in the current pandemic of COVID-19, for successful prevention of infectious disease [213]. In addition, since it is an airborne disease,

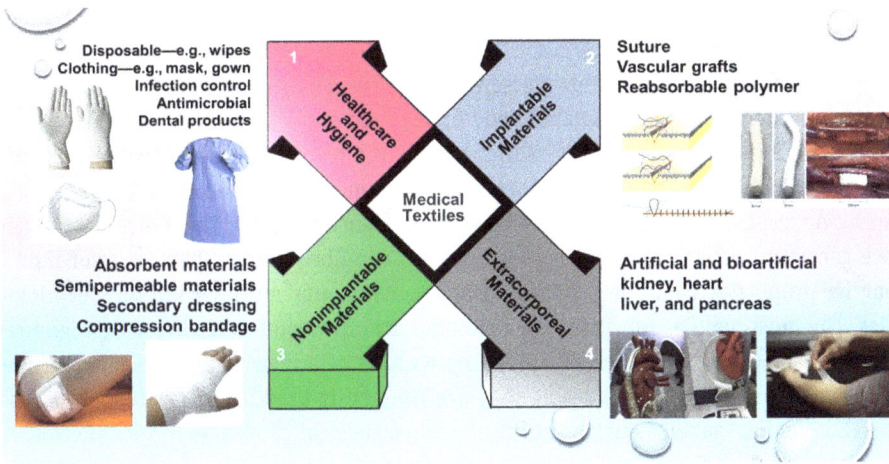

Figure 1.31: Brief classification of medical textile-based products.

it causes an abrupt increase in the number of new patients, which can lead to a massive failure of healthcare. For this reason, the World Health Organization (WHO) recommends that patients with moderate symptoms and no major chronic diseases be cared for at home, while maintaining contact with their healthcare professionals. Under these conditions, nanotechnology- based wearable smart textiles, with their own sensing and actuation functionalities, allow the remote monitoring of a patient's physiological and physical data and signals via non-invasive sensors implanted in garment materials. These data or signals can be used to aid diagnosis and management of COVID-19 as well as chronic illnesses. Figure 1.32 depicts an overview of required wearable assistive technologies for COVID-19 patients [214].

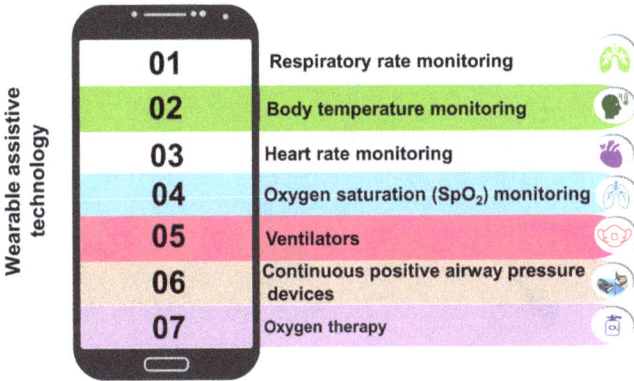

Figure 1.32: Overview of required wearable assistive technologies for COVID-19 patients.

Further, in pandemics of highly contagious diseases like COVID-19, the risk of infection for healthcare workers is much higher than for the general population, because they work directly with patients. Personal Protective Equipment (PPE) is important to protect workers from contaminated body fluids, such as droplets from coughs, sneezes, and aerosols generated from procedures, as well as other contaminated body fluids emanating from infected patients [215]. The key components of PPE like aprons, coveralls, masks or respirators, and goggles are depicted in Figure 1.33 [213]. These medical protective garments, which are typically made of synthetic fibres due to their superior liquid barrier properties, could be produced using weaving, nonwoven, or knitting technologies. The most popular fabric for such clothing is nonwoven fabric, due to its high levels of sterility and infection control, along with its relatively quick and inexpensive manufacturing process. As a result, they are frequently used in the production of disposable medical textiles such as surgical gowns, surgical caps, and surgical masks. A typical nonwoven fabric of this type is made from polypropylene and constructed with a spunbond–meltblown–spunbond (SMS) process [216, 217]. In order to maintain level-4 protection against pathogens, these nonwoven fabrics are given repellent finishes. In addition to repellent finishes, antimicrobial finishes in medical gowns are becoming increasingly popular for controlling, destroying, or suppressing pathogens that cause

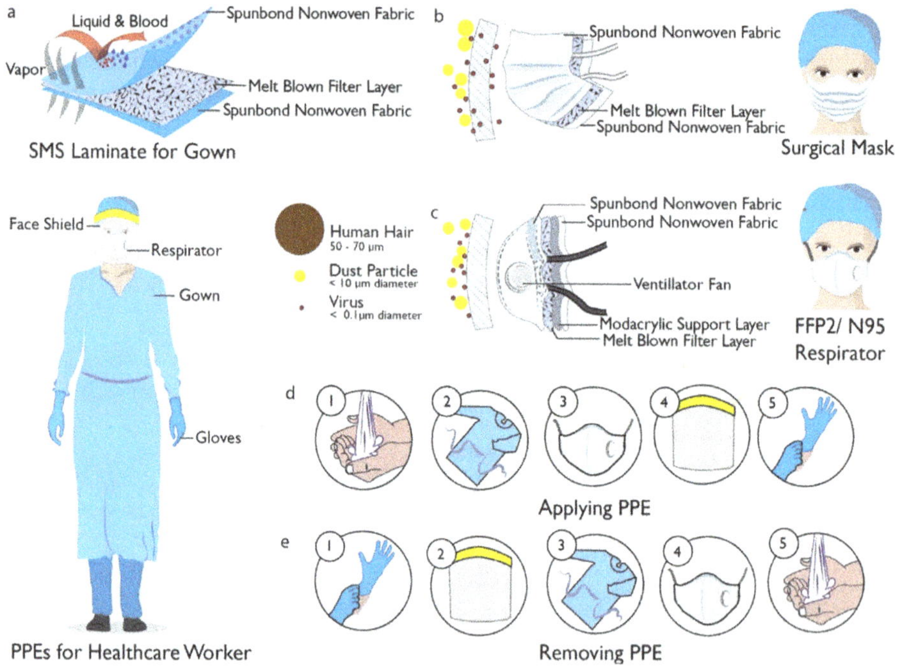

Figure 1.33: PPE for healthcare workers (a) liquid proof SMS laminate for gown, (b) surgical facemask, (c) respirator, and (d and e) methods of safely handling PPE (reproduced with kind permission from [213], Copyright 2020, American chemical society).

odours, staining, and deterioration. Various antimicrobial agents, along with their major pathways to inhibit pathogens, have been discussed broadly in Chapter 2, Chapter 5, and Chapter 18.

These various interconnections between diverse factors make medical textiles a tricky but engrossing area of experimentation that requires repeated clinical trials to validate smart textiles. The goal of this section of the book is to provide the most recent and sophisticated achievements in the field of complex textile constructs used for health purposes.

1.6.7 Smart and functional automotive textiles

The transportation sector, including ground transportation (automotive and railway) and aerospace (airplanes, spaceships, satellites, etc.) contribute the most in global technical textiles market, with a wide range of novel textile structures such as car upholstery and seating, headliners, floor covering, pillar coverings, door and side-panel coverings, sun-visors, as well as airbags, safety belts, thermal and sound insulators, battery separators, filter fabrics, hose/belt products, tyres, and a variety of textile-reinforced flexible and hard composites. These are mostly woven, weft knitted, warp knitted, tufted, nonwovens, and laminated fabrics. Nowadays, fibrous materials have gained enormous popularity in automotive applications due to their desirable properties in meeting the higher level of air quality, better passenger comfort, thermal insulation, noise reduction, fuel efficiency, stringent emission, and safety regulations. Chapter 16 summarizes the recent technical developments and applications of fibrous materials in automotive applications.

1.6.8 3D printed smart textiles and their application

3D printing, a layer-by-layer additive manufacturing technology that combines computer-aided design (CAD) and computer-aided manufacturing (CAM), is becoming popular, day-by-day, within the modern day's textile apparel and fashion accessories, since it has advantages in product design, material selections, and the incorporation of novelty to comply with customers' body movements and ensure their demand and comfort, along with the advantages of cost efficiency, sustainability, and waste remediation. However, it is also possible to scan a person's body and create personalized apparel using 3D printing. Depending on the mechanism of layer formation and deposition, 3D printing can be classified as fused deposition modelling (FDM), selective laser sintering (SLS), stereolithography (SLA), binder jetting, and inkjet printing [218]. The mechanism of operation, materials suitable for each process, and their textile and fashion applications will be elaborated in Chapter 17. Mainly, 3D printing is used for making 3D

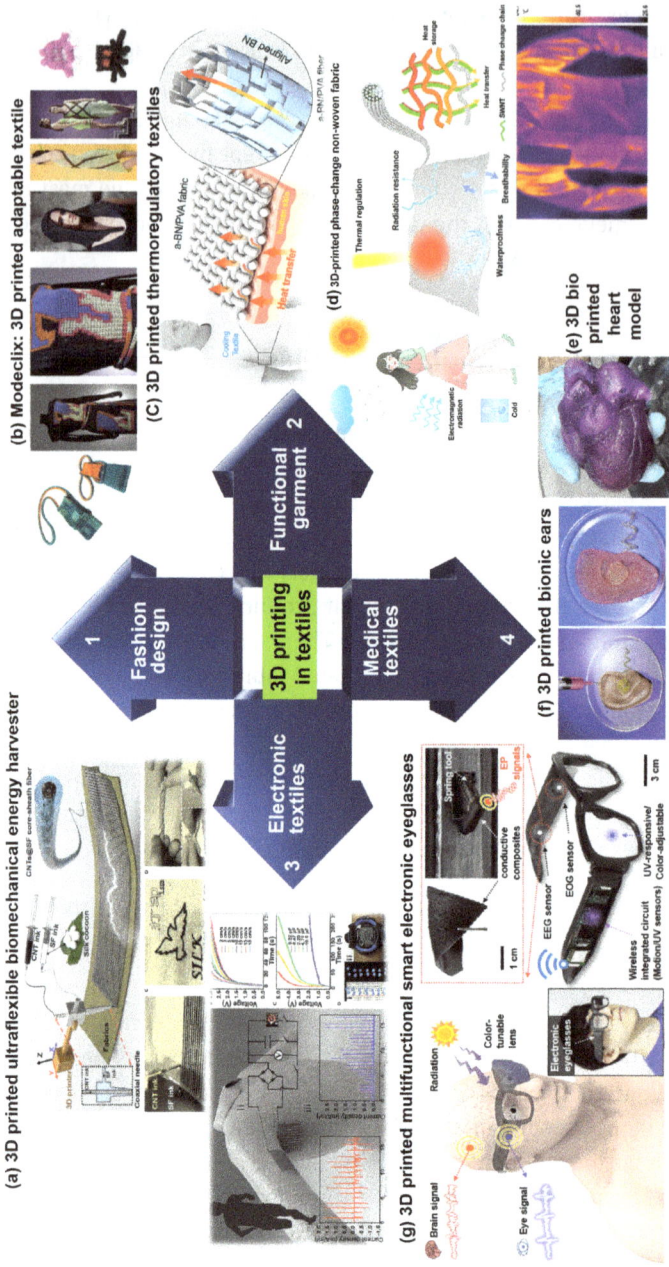

Figure 1.34: Potential of 3D printing technology in manufacturing modern-day textiles: (a) 3D printed ultra-flexible biomechanical energy harvester (reproduced with kind permission from [170], Copyright 2019, Elsevier). (b) Modeclix: 3D printed adaptable textile. (reproduced with kind permission from [220], Copyright 2018, Elsevier). (c) 3D printed thermoregulatory textiles (reproduced with kind permission from [221], Copyright 2017, American Chemical Society). (d) 3D printed phase-change nonwoven fabric (reproduced with kind permission from [222], Copyright 2022, American Chemical Society). (e) 3D bioprinted heart model (reproduced with kind permission from [223], Copyright 2020, American Chemical Society). (f) 3D printed bionic ears (reproduced with kind permission from [224], Copyright 2013, American Chemical Society). (g) 3D printed multifunctional smart electronic eyeglasses (reproduced with kind permission from [225], Copyright 2020, American Chemical Society).

textile-like structures or for nanocoating on conventional textiles to make structural composites, thermoregulating textiles, flexible electronics – especially for applications in electronic textiles, flexible displays, soft actuators and robots – energy-harvesting and storage devices, as well as biomedical devices, as depicted in Figure 1.34 [219].

3D printing technology, on the other hand, has significant drawbacks. Due to the small dimensions of the fibres, high porosity of the textile structure, and the required discontinuities within the textile, printing textiles using any of the current 3D printing processes is a very challenging task taking into account, the shortcoming of poor adhesion. Moreover, the synthetic thread employed in this technology produces stiff and unyielding clothing, which can be uncomfortable for consumers; likewise, constructing complicated woven structures can be difficult. Hence, future research is more inclined towards the making of flexible, wearable, and breathable textile structures using 3D printing.

1.7 Current challenges and outlook

The challenges in developing and commercializing different functional and smart textiles have already been discussed in respective sections, with their features. This section will summarize the major challenges faced by textile/garment industries in developing smart and functional textiles with new features and commercializing them in the market.

Most advanced functional textiles are treated with nanomaterials or polymer nanocomposites. The intrinsic issues of nanotechnology are creating problems in commercializing these nanomaterial-treated functional textiles. Nanotechnology is still one of the most expensive technologies currently available. Hence, application of nanotechnology or the use of expensive nanomaterials can result in increased costs for final products. Although nanofibres offer great potential for improving filtration efficiency, their high production costs and slow production speed have kept them from reaching mass production. This is the main reason why their use in filtration and other applications has not fully taken off. Moreover, a finished or coated textile product may also be more costly with the use of expensive nanomaterials.

Despite this, nanomaterials can achieve similar or better properties than bulk materials even at far lower concentrations, which can compensate the costs marginally. Different polymer nanocomposites have great potential in making functional and smart textiles, but the agglomeration issue of most of the nanoparticles limit their potential in enhancing the functional properties/performance. However, researchers have been successful in overcoming this issue to a great extent, by using some special strategy for improving dispersion in polymer matrix [226]. According to a few recent research reports, nanomaterials have strong negative environmental effects and are toxic to living organisms. Particularly, carbonaceous nanomaterials cause inhalation problems and many other fatal diseases because of their small size. On the other hand, currently, most nanotechnology-based textile research and development are taking place only in the lab, and commercialization of these products in bulk scale is a big challenge for the textile researchers or material scientists. Although there are several challenges at present, considering the potential of nanotechnology, material and textile scientists have high expectations on nanotechnology-based textiles for developing smart and functional textiles for the next generation.

In wearable smart electronics or E-textiles, sensors, actuators, circuits, power storage, and many other electronic components are used, which are extremely rugged during manufacturing as well as during wearing, causing reduction in work efficiency/performance; sometimes, they stop functioning. Moreover, incorporation of such electronic components should not affect comfort while wearing the smart textiles and they should be washable, but both are very challenging to achieve. Since the very early days of smart textiles, the wearer has had to remove all electronic components before washing most commercial smart textiles.

A company, Eleksen, developed a wireless keyboard fabric called as "Eleck Tex" which is electroconductive, flexible, lightweight, wearable, and most importantly, washable, because of waterproof coating on the electronic components [227].

Most smart textiles require power supplies that are lightweight and have a high capacity to run autonomously for hours at a time. A majority of smart textiles, however, use traditional rechargeable batteries that are bulky, heavy, and cannot be integrated fully with the E-textile architecture. Due to the presence of a lot of toxic electrochemicals, they are generally not suitable for skin contact. Hopefully, with the advent of energy-harvesting textiles (that can generate power from body motion, body fluid flow, body heat, etc.) and with more research on flexible/elastic and lightweight batteries/supercapacitors, SCs, and thermo/piezogenerators, this issue may be solved in the near future.

Some other barriers identified include the lack of standardization, the absence of regulations for new products, ethical and social issues, lack of coordination and collaboration among value chain partners, high production and selling cost, and so on.

Another big challenge with smart textiles is commercialization of the developed products. Research on smart textiles has been ongoing for the last three decades, but the number of commercial products in the market are still limited. The reasons may be the same as described earlier.

Based on the above-mentioned discussions, smart textiles still have a long way to go before becoming mainstream technology. However, despite these challenges, the potentials of smart and functional textiles include their ability to communicate with other devices, conductivity, energy-harvesting ability, and protection from environmental hazards. In order to find multidirectional applications and eliminate existing issues, material scientists and textile technologists must conduct extensive research. Particularly, wearable textile-based systems are getting considerable attention in health monitoring, defence, protection, and safety, and in providing a healthy lifestyle. This fact offers engineers and scientists an opportunity to learn new skills and expand their knowledge. During their development, interdisciplinary teams of materials scientists, physicists, chemists, process engineers, and textile and electronics manufacturing professionals will be formed. By combining knowledge from different groups of people, this type of research can ultimately provide new opportunities based on the combination of their knowledge, as well as increased dialogue between people, who, otherwise, would not interact. Research on smart textiles will lead to innovative solutions for integrating electronics in unusual environments, advancing science, and enriching our understanding.

References

[1]   Koncar V. Introduction to smart textiles and their applications. Smart Text. Their Appl., Elsevier; 2016, p. 1–8. https://doi.org/10.1016/B978-0-08-100574-3.00001-1.
[2]   Li L, Cheung TW. Sustainable development of smart textiles: a review of 'self-functioning' abilities which makes textiles alive. J Text Eng Fash Technol 2018;4. https://doi.org/10.15406/jteft.2018.04.00133.
[3]   Kausar A. Sensing Materials: Nanocomposites. Ref. Module Biomed. Sci., Elsevier; 2021, p. B9780128225486000000. https://doi.org/10.1016/B978-0-12-822548-6.00048-0.
[4]   Hu JL. Introduction to active coatings for smart textiles. Act. Coat. Smart Text., Elsevier; 2016, p. 1–7. https://doi.org/10.1016/B978-0-08-100263-6.00001-0.
[5]   Liu L, Xu W, Ding Y, Agarwal S, Greiner A, Duan G. A review of smart electrospun fibers toward textiles. Compos Commun 2020;22:100506. https://doi.org/10.1016/j.coco.2020.100506.
[6]   Nagarajan S, Soussan L, Bechelany M, Teyssier C, Cavaillès V, Pochat-Bohatier C, et al. Novel biocompatible electrospun gelatin fiber mats with antibiotic drug delivery properties. J Mater Chem B 2016;4:1134–41. https://doi.org/10.1039/C5TB01897H.
[7]   Source: The Smart Textiles Timeline, 1900-2021 n.d.
[8]   https://www.retail-insight-network.com/comment/smart-clothing-timeline/ accessed on 06 September 2022 n.d.
[9]   https://mesomat.com/stretchable-circuitry/ n.d. https://mesomat.com/stretchable-circuitry/ (accessed May 29, 2022).
[10]   https://nyokastechnologies.com/civilian.html n.d. https://nyokastechnologies.com/civilian.html (accessed May 29, 2022).
[11]   https://myclim8.com/tech/#textile n.d. https://myclim8.com/tech/#textile (accessed May 29, 2022).
[12]   https://www.databridgemarketresearch.com n.d. https://www.databridgemarketresearch.com (accessed May 29, 2022).
[13]   https://www.mordorintelligence.com n.d. https://www.mordorintelligence.com (accessed May 29, 2022).
[14]   Wang H, Zhang Y, Liang X, Zhang Y. Smart Fibers and Textiles for Personal Health Management. ACS Nano 2021;15:12497–508. https://doi.org/10.1021/acsnano.1c06230.
[15]   Cherenack K, van Pieterson L. Smart textiles: Challenges and opportunities. J Appl Phys 2012;112:091301. https://doi.org/10.1063/1.4742728.
[16]   Bawa R, Bawa SR, Maebius SB, Flynn T, Wei C. Protecting new ideas and inventions in nanomedicine with patents. Nanomedicine Nanotechnol Biol Med 2005;1:150–8. https://doi.org/10.1016/j.nano.2005.03.009.
[17]   Akkerman QA. Lead Halide Perovskite Nanocrystals: A New Age of Semiconductive Nanocrystals 2019. https://doi.org/10.13140/RG.2.2.23651.81442.

[18]   Vigneshwaran N. Modification of textile surfaces using nanoparticles. Surf. Modif. Text., Elsevier; 2009, p. 164–84. https://doi.org/10.1533/9781845696689.164.

[19]   [No title found]. Asian J Nanosci Mater n.d.;2.

[20]   Yetisen AK, Qu H, Manbachi A, Butt H, Dokmeci MR, Hinestroza JP, et al. Nanotechnology in Textiles. ACS Nano 2016;10:3042–68. https://doi.org/10.1021/acsnano.5b08176.

[21]   Bashari A, Shakeri M, Shirvan AR. UV-protective textiles. Impact Prospects Green Chem. Text. Technol., Elsevier; 2019, p. 327–65. https://doi.org/10.1016/B978-0-08-102491-1.00012-5.

[22]   Lam YL, Kan CW, Yuen CWM. Effect of concentration of titanium dioxide acting as catalyst or co-catalyst on the wrinkle-resistant finishing of cotton fabric. Fibers Polym 2010;11:551–8. https://doi.org/10.1007/s12221-010-0551-7.

[23]   Lee HJ, Kim J, Park CH. Fabrication of self-cleaning textiles by TiO_2-carbon nanotube treatment. Text Res J 2014;84:267–78. https://doi.org/10.1177/0040517513494258.

[24]   Dastjerdi R, Mojtahedi MRM, Shoshtari AM, Khosroshahi A. Investigating the production and properties of Ag/TiO_2/PP antibacterial nanocomposite filament yarns. J Text Inst 2010;101:204–13. https://doi.org/10.1080/00405000802346388.

[25]   Morawski AW, Kusiak-Nejman E, Przepiórski J, Kordala R, Pernak J. Cellulose-TiO2 nanocomposite with enhanced UV–Vis light absorption. Cellulose 2013;20:1293–300. https://doi.org/10.1007/s10570-013-9906-6.

[26]   Jeong S-M, Ahn J, Choi YK, Lim T, Seo K, Hong T, et al. Development of a wearable infrared shield based on a polyurethane–antimony tin oxide composite fiber. NPG Asia Mater 2020;12:32. https://doi.org/10.1038/s41427-020-0213-z.

[27]   Hsieh C-T, Wu F-L, Chen W-Y. Super water- and oil-repellences from silica-based nanocoatings. Surf Coat Technol 2009;203:3377–84. https://doi.org/10.1016/j.surfcoat.2009.04.025.

[28]   Çöpoğlu N, Çiçek B. Abrasion resistant glass-ceramic coatings reinforced with WC-nanoparticles. Surf Coat Technol 2021;419:127275. https://doi.org/10.1016/j.surfcoat.2021.127275.

[29]   Yin IX, Zhang J, Zhao IS, Mei ML, Li Q, Chu CH. The Antibacterial Mechanism of Silver Nanoparticles and Its Application in Dentistry. Int J Nanomedicine 2020;Volume 15:2555–62. https://doi.org/10.2147/IJN.S246764.

[30]   Yuan B, Jiang C, Li P, Sun H, Li P, Yuan T, et al. Ultrathin Polyamide Membrane with Decreased Porosity Designed for Outstanding Water-Softening Performance and Superior Antifouling Properties. ACS Appl Mater Interfaces 2018;10:43057–67. https://doi.org/10.1021/acsami.8b15883.

[31]   Ward CJ, DeWitt M, Davis EW. Halloysite Nanoclay for Controlled Release Applications. In: Nagarajan R, editor. ACS Symp. Ser., vol. 1119, Washington, DC: American Chemical Society; 2012, p. 209–38. https://doi.org/10.1021/bk-2012-1119.ch010.

[32]   Nasouri K, Shoushtari AM. Fabrication of magnetite nanoparticles/polyvinyl pyrrolidone composite nanofibers and their application as electromagnetic interference shielding material. J Thermoplast Compos Mater 2018;31:431–46. https://doi.org/10.1177/0892705717704488.

[33]   Kanehara M, Koike H, Yoshinaga T, Teranishi T. Indium Tin Oxide Nanoparticles with Compositionally Tunable Surface Plasmon Resonance Frequencies in the Near-IR Region. J Am Chem Soc 2009;131:17736–7. https://doi.org/10.1021/ja9064415.

[34]   Palve AM, Vani OV, Gupta RK. Metal Oxide-Based Compounds as Flame Retardants for Polyurethanes. In: Gupta RK, editor. ACS Symp. Ser., vol. 1400, Washington, DC: American Chemical Society; 2021, p. 121–36. https://doi.org/10.1021/bk-2021-1400.ch008.

[35]   Zhang J, Hereid J, Hagen M, Bakirtzis D, Delichatsios MA, Fina A, et al. Effects of nanoclay and fire retardants on fire retardance of a polymer blend of EVA and LDPE. Fire Saf J 2009;44:504–13. https://doi.org/10.1016/j.firesaf.2008.10.005.

[36]   Ramachandran T, Vigneswaran C. Design and Development of Copper Core Conductive Fabrics for Smart Textiles. J Ind Text 2009;39:81–93. https://doi.org/10.1177/1528083709103317.

[37]   Rehman A, Houshyar S, Reineck P, Padhye R, Wang X. Multifunctional Smart Fabrics through Nanodiamond-Polyaniline Nanocomposites. ACS Appl Polym Mater 2020;2:4848–55. https://doi.org/10.1021/acsapm.0c00789.

[38]   Zhang Y, Ram MK, Stefanakos EK, Goswami DY. Synthesis, Characterization, and Applications of ZnO Nanowires. J Nanomater 2012;2012:1–22. https://doi.org/10.1155/2012/624520.

[39]   Yao S, Yang J, Poblete FR, Hu X, Zhu Y. Multifunctional Electronic Textiles Using Silver Nanowire Composites. ACS Appl Mater Interfaces 2019;11:31028–37. https://doi.org/10.1021/acsami.9b07520.

[40]   Zahid M, Anwer Rathore H, Tayyab H, Ahmad Rehan Z, Abdul Rashid I, Lodhi M, et al. Recent developments in textile based polymeric smart sensor for human health monitoring: A review. Arab J Chem 2022;15:103480. https://doi.org/10.1016/j.arabjc.2021.103480.

[41]   Maldonado XS. Nanodiamond-polyaniline nanocomposites yield smart textiles. MRS Bull 2021;46:96–96. https://doi.org/10.1557/s43577-021-00037-z.

[42]   Ahmed A, Jalil MA, Hossain MdM, Moniruzzaman Md, Adak B, Islam MT, et al. A PEDOT:PSS and graphene-clad smart textile-based wearable electronic Joule heater with high thermal stability. J Mater Chem C 2020;8:16204–15. https://doi.org/10.1039/D0TC03368E.

[43]   Shahariar H, Kim I, Soewardiman H, Jur JS. Inkjet Printing of Reactive Silver Ink on Textiles. ACS Appl Mater Interfaces 2019;11:6208–16. https://doi.org/10.1021/acsami.8b18231.

[44]   Yoon J, Lee SN, Shin MK, Kim H-W, Choi HK, Lee T, et al. Flexible electrochemical glucose biosensor based on GOx/gold/MoS2/gold nanofilm on the polymer electrode. Biosens Bioelectron 2019;140:111343. https://doi.org/10.1016/j.bios.2019.111343.

[45]   Mahmoodi S, Elmi A, Hallaj Nezhadi S. Copper Nanoparticles as Antibacterial Agents. J Mol Pharm Org Process Res 2018;06. https://doi.org/10.4172/2329-9053.1000140.

[46]   Lee M, Kim JU, Lee KJ, Ahn S, Shin Y-B, Shin J, et al. Aluminum Nanoarrays for Plasmon-Enhanced Light Harvesting. ACS Nano 2015;9:6206–13. https://doi.org/10.1021/acsnano.5b01541.

[47]   Babu VJ, Anusha M, Sireesha M, Sundarrajan S, Abdul Haroon Rashid SSA, Kumar AS, et al. Intelligent Nanomaterials for Wearable and Stretchable Strain Sensor Applications: The Science behind Diverse Mechanisms, Fabrication Methods, and Real-Time Healthcare. Polymers 2022;14:2219. https://doi.org/10.3390/polym14112219.

[48]   Golmohammadi H, Morales-Narváez E, Naghdi T, Merkoçi A. Nanocellulose in Sensing and Biosensing. Chem Mater 2017;29:5426–46. https://doi.org/10.1021/acs.chemmater.7b01170.

[49]   Singha K. A Review on Coating & Lamination in Textiles: Processes and Applications. Am J Polym Sci 2012;2:39–49. https://doi.org/10.5923/j.ajps.20120203.04.

[50]   Bidoki SM, Wittlinger R. Environmental and economical acceptance of polyvinyl chloride (PVC) coating agents. J Clean Prod 2010;18:219–25. https://doi.org/10.1016/j.jclepro.2009.10.006.

[51]   Krebs FC. Fabrication and processing of polymer solar cells: A review of printing and coating techniques. Sol Energy Mater Sol Cells 2009;93:394–412. https://doi.org/10.1016/j.solmat.2008.10.004.

[52]   Madou MJ. Manufacturing Techniques for Microfabrication and Nanotechnology. 0 ed. CRC Press; 2011. https://doi.org/10.1201/9781439895306.

[53]   Parvinzadeh Gashti M, Pakdel E, Alimohammadi F. Nanotechnology-based coating techniques for smart textiles. Act. Coat. Smart Text., Elsevier; 2016, p. 243–68. https://doi.org/10.1016/B978-0-08-100263-6.00011-3.

[54]   Nanomaterials for Biosensors. Elsevier; 2018. https://doi.org/10.1016/C2015-0-04697-4.

[55]   Nag A, Mukhopadhyay SC, Kosel J. Flexible carbon nanotube nanocomposite sensor for multiple physiological parameter monitoring. Sens Actuators Phys 2016;251:148–55. https://doi.org/10.1016/j.sna.2016.10.023.

[56]   Edwards JV, Prevost N, French A, Concha M, DeLucca A, Wu Q. Nanocellulose-Based Biosensors: Design, Preparation, and Activity of Peptide-Linked Cotton Cellulose Nanocrystals Having

Fluorimetric and Colorimetric Elastase Detection Sensitivity. Engineering 2013;05:20–8. https://doi.org/10.4236/eng.2013.59A003.

[57] Babar AA, Wang X, Iqbal N, Yu J, Ding B. Tailoring Differential Moisture Transfer Performance of Nonwoven/Polyacrylonitrile-SiO$_2$ Nanofiber Composite Membranes. Adv Mater Interfaces 2017;4:1700062. https://doi.org/10.1002/admi.201700062.

[58] Camargo PHC, Satyanarayana KG, Wypych F. Nanocomposites: synthesis, structure, properties and new application opportunities. Mater Res 2009;12:1–39. https://doi.org/10.1590/S1516-14392009000100002.

[59] Zaferani SH. Introduction of polymer-based nanocomposites. Polym.-Based Nanocomposites Energy Environ. Appl., Elsevier; 2018, p. 1–25. https://doi.org/10.1016/B978-0-08-102262-7.00001-5.

[60] Rajak DK, Pagar DD, Kumar R, Pruncu CI. Recent progress of reinforcement materials: a comprehensive overview of composite materials. J Mater Res Technol 2019;8:6354–74. https://doi.org/10.1016/j.jmrt.2019.09.068.

[61] Grancarić AM, Jerković I, Koncar V, Cochrane C, Kelly FM, Soulat D, et al. Conductive polymers for smart textile applications. J Ind Text 2018;48:612–42. https://doi.org/10.1177/1528083717699368.

[62] Ala O, Fan Q. Applications of Conducting Polymers in Electronic Textiles. Res J Text Appar 2009;13:51–68. https://doi.org/10.1108/RJTA-13-04-2009-B007.

[63] Bajgar V, Penhaker M, Martinková L, Pavlovič A, Bober P, Trchová M, et al. Cotton Fabric Coated with Conducting Polymers and its Application in Monitoring of Carnivorous Plant Response. Sensors 2016;16:498. https://doi.org/10.3390/s16040498.

[64] Maity S, Chatterjee A. Conductive polymer-based electro-conductive textile composites for electromagnetic interference shielding: A review. J Ind Text 2018;47:2228–52. https://doi.org/10.1177/1528083716670310.

[65] Zhang Y, Wang H, Lu H, Li S, Zhang Y. Electronic fibers and textiles: Recent progress and perspective. IScience 2021;24:102716. https://doi.org/10.1016/j.isci.2021.102716.

[66] Lund A, Wu Y, Fenech-Salerno B, Torrisi F, Carmichael TB, Müller C. Conducting materials as building blocks for electronic textiles. MRS Bull 2021;46:491–501. https://doi.org/10.1557/s43577-021-00117-0.

[67] Kaur G, Adhikari R, Cass P, Bown M, Gunatillake P. Electrically conductive polymers and composites for biomedical applications. RSC Adv 2015;5:37553–67. https://doi.org/10.1039/C5RA01851J.

[68] Cheng KB, Ramakrishna S, Lee KC. Electromagnetic shielding effectiveness of copper/glass fiber knitted fabric reinforced polypropylene composites. Compos Part Appl Sci Manuf 2000;31:1039–45. https://doi.org/10.1016/S1359-835X(00)00071-3.

[69] Alagirusamy R, Eichhoff J, Gries T, Jockenhoevel S. Coating of conductive yarns for electro-textile applications. J Text Inst 2013;104:270–7. https://doi.org/10.1080/00405000.2012.719295.

[70] Novoselov KS, Geim AK, Morozov SV, Jiang D, Zhang Y, Dubonos SV, et al. Electric Field Effect in Atomically Thin Carbon Films. Science 2004;306:666–9. https://doi.org/10.1126/science.1102896.

[71] Geim AK. Graphene: Status and Prospects. Science 2009;324:1530–4. https://doi.org/10.1126/science.1158877.

[72] He Q, Wu S, Yin Z, Zhang H. Graphene-based electronic sensors. Chem Sci 2012;3:1764. https://doi.org/10.1039/c2sc20205k.

[73] Stankovich S, Dikin DA, Dommett GHB, Kohlhaas KM, Zimney EJ, Stach EA, et al. Graphene-based composite materials. Nature 2006;442:282–6. https://doi.org/10.1038/nature04969.

[74] Stankovich S, Dikin DA, Piner RD, Kohlhaas KA, Kleinhammes A, Jia Y, et al. Synthesis of graphene-based nanosheets via chemical reduction of exfoliated graphite oxide. Carbon 2007;45:1558–65. https://doi.org/10.1016/j.carbon.2007.02.034.

[75] Kim J, Cote LJ, Kim F, Yuan W, Shull KR, Huang J. Graphene Oxide Sheets at Interfaces. J Am Chem Soc 2010;132:8180–6. https://doi.org/10.1021/ja102777p.

[76] Li D, Müller MB, Gilje S, Kaner RB, Wallace GG. Processable aqueous dispersions of graphene nanosheets. Nat Nanotechnol 2008;3:101–5. https://doi.org/10.1038/nnano.2007.451.

[77] Paredes JI, Villar-Rodil S, Martínez-Alonso A, Tascón JMD. Graphene Oxide Dispersions in Organic Solvents. Langmuir 2008;24:10560–4. https://doi.org/10.1021/la801744a.

[78] Shateri-Khalilabad M, Yazdanshenas ME. Fabricating electroconductive cotton textiles using graphene. Carbohydr Polym 2013;96:190–5. https://doi.org/10.1016/j.carbpol.2013.03.052.

[79] Tang X, Tian M, Qu L, Zhu S, Guo X, Han G, et al. Functionalization of cotton fabric with graphene oxide nanosheet and polyaniline for conductive and UV blocking properties. Synth Met 2015;202:82–8. https://doi.org/10.1016/j.synthmet.2015.01.017.

[80] Javed K, Galib CMA, Yang F, Chen C-M, Wang C. A new approach to fabricate graphene electroconductive networks on natural fibers by ultraviolet curing method. Synth Met 2014;193:41–7. https://doi.org/10.1016/j.synthmet.2014.03.028.

[81] Cao Y, Zhu M, Li P, Zhang R, Li X, Gong Q, et al. Boosting supercapacitor performance of carbon fibres using electrochemically reduced graphene oxide additives. Phys Chem Chem Phys 2013;15:19550. https://doi.org/10.1039/c3cp54017k.

[82] Maiti UN, Maiti S, Das NS, Chattopadhyay KK. Hierarchical graphene nanocones over 3D platform of carbon fabrics: A route towards fully foldable graphene based electron source. Nanoscale 2011;3:4135. https://doi.org/10.1039/c1nr10383k.

[83] Woltornist SJ, Alamer FA, McDannald A, Jain M, Sotzing GA, Adamson DH. Preparation of conductive graphene/graphite infused fabrics using an interface trapping method. Carbon 2015;81:38–42. https://doi.org/10.1016/j.carbon.2014.09.020.

[84] Neves AIS, Bointon TH, Melo LV, Russo S, de Schrijver I, Craciun MF, et al. Transparent conductive graphene textile fibers. Sci Rep 2015;5:9866. https://doi.org/10.1038/srep09866.

[85] Skrzetuska E, Puchalski M, Krucińska I. Chemically Driven Printed Textile Sensors Based on Graphene and Carbon Nanotubes. Sensors 2014;14:16816–28. https://doi.org/10.3390/s140916816.

[86] Krishnamoorthy K, Navaneethaiyer U, Mohan R, Lee J, Kim S-J. Graphene oxide nanostructures modified multifunctional cotton fabrics. Appl Nanosci 2012;2:119–26. https://doi.org/10.1007/s13204-011-0045-9.

[87] Bhattacharjee S, Joshi R, Chughtai AA, Macintyre CR. Graphene Modified Multifunctional Personal Protective Clothing. Adv Mater Interfaces 2019;6:1900622. https://doi.org/10.1002/admi.201900622.

[88] Ahmed A, Hossain MM, Adak B, Mukhopadhyay S. Recent Advances in 2D MXene Integrated Smart-Textile Interfaces for Multifunctional Applications. Chem Mater 2020;32:10296–320. https://doi.org/10.1021/acs.chemmater.0c03392.

[89] Eom W, Shin H, Ambade RB, Lee SH, Lee KH, Kang DJ, et al. Large-scale wet-spinning of highly electroconductive MXene fibers. Nat Commun 2020;11:2825. https://doi.org/10.1038/s41467-020-16671-1.

[90] Levitt AS, Alhabeb M, Hatter CB, Sarycheva A, Dion G, Gogotsi Y. Electrospun MXene/carbon nanofibers as supercapacitor electrodes. J Mater Chem A 2019;7:269–77. https://doi.org/10.1039/C8TA09810G.

[91] Cheng W, Zhang Y, Tian W, Liu J, Lu J, Wang B, et al. Highly Efficient MXene-Coated Flame Retardant Cotton Fabric for Electromagnetic Interference Shielding. Ind Eng Chem Res 2020;59:14025–36. https://doi.org/10.1021/acs.iecr.0c02618.

[92] Uzun S, Seyedin S, Stoltzfus AL, Levitt AS, Alhabeb M, Anayee M, et al. Knittable and Washable Multifunctional MXene-Coated Cellulose Yarns. Adv Funct Mater 2019;29:1905015. https://doi.org/10.1002/adfm.201905015.

[93] Zhang X, Wang X, Lei Z, Wang L, Tian M, Zhu S, et al. Flexible MXene-Decorated Fabric with Interwoven Conductive Networks for Integrated Joule Heating, Electromagnetic Interference Shielding, and Strain Sensing Performances. ACS Appl Mater Interfaces 2020;12:14459–67. https://doi.org/10.1021/acsami.0c01182.

[94] Zhang C, McKeon L, Kremer MP, Park S-H, Ronan O, Seral-Ascaso A, et al. Additive-free MXene inks and direct printing of micro-supercapacitors. Nat Commun 2019;10:1795. https://doi.org/10.1038/s41467-019-09398-1.

[95] Zhang Y, Wang Y, Jiang Q, El-Demellawi JK, Kim H, Alshareef HN. MXene Printing and Patterned Coating for Device Applications. Adv Mater 2020;32:1908486. https://doi.org/10.1002/adma.201908486.

[96] Ding Y, Invernale MA, Sotzing GA. Conductivity Trends of PEDOT-PSS Impregnated Fabric and the Effect of Conductivity on Electrochromic Textile. ACS Appl Mater Interfaces 2010;2:1588–93. https://doi.org/10.1021/am100036n.

[97] Nardes AM, Kemerink M, Janssen RAJ, Bastiaansen JAM, Kiggen NMM, Langeveld BMW, et al. Microscopic Understanding of the Anisotropic Conductivity of PEDOT:PSS Thin Films. Adv Mater 2007;19:1196–200. https://doi.org/10.1002/adma.200602575.

[98] Nardes AM, Kemerink M, de Kok MM, Vinken E, Maturova K, Janssen RAJ. Conductivity, work function, and environmental stability of PEDOT:PSS thin films treated with sorbitol. Org Electron 2008;9:727–34. https://doi.org/10.1016/j.orgel.2008.05.006.

[99] Balint R, Cassidy NJ, Cartmell SH. Conductive polymers: Towards a smart biomaterial for tissue engineering. Acta Biomater 2014;10:2341–53. https://doi.org/10.1016/j.actbio.2014.02.015.

[100] Therm Environ Cond Hum Occup 2014.

[101] Lin Z, Deng S. A study on the thermal comfort in sleeping environments in the subtropics – Developing a thermal comfort model for sleeping environments. Build Environ 2008;43:70–81. https://doi.org/10.1016/j.buildenv.2006.11.026.

[102] Fang Y, Chen G, Bick M, Chen J. Smart textiles for personalized thermoregulation. Chem Soc Rev 2021;50:9357–74. https://doi.org/10.1039/D1CS00003A.

[103] Vriens J, Nilius B, Voets T. Peripheral thermosensation in mammals. Nat Rev Neurosci 2014;15:573–89. https://doi.org/10.1038/nrn3784.

[104] Zhang X. Heat-storage and thermo-regulated textiles and clothing. Smart Fibres Fabr. Cloth., Elsevier; 2001, p. 34–57. https://doi.org/10.1533/9781855737600.34.

[105] Farooq AS, Zhang P. Fundamentals, materials and strategies for personal thermal management by next-generation textiles. Compos Part Appl Sci Manuf 2021;142:106249. https://doi.org/10.1016/j.compositesa.2020.106249.

[106] Sarier N, Onder E. Organic phase change materials and their textile applications: An overview. Thermochim Acta 2012;540:7–60. https://doi.org/10.1016/j.tca.2012.04.013.

[107] Yang T, King WP, Miljkovic N. Phase change material-based thermal energy storage. Cell Rep Phys Sci 2021;2:100540. https://doi.org/10.1016/j.xcrp.2021.100540.

[108] Sarbu I, Dorca A. Review on heat transfer analysis in thermal energy storage using latent heat storage systems and phase change materials. Int J Energy Res 2019;43:29–64. https://doi.org/10.1002/er.4196.

[109] Rykaczewski K. Rational design of sun and wind shaded evaporative cooling vests for enhanced personal cooling in hot and dry climates. Appl Therm Eng 2020;171:115122. https://doi.org/10.1016/j.applthermaleng.2020.115122.

[110] Reinertsen RE, Færevik H, Holbø K, Nesbakken R, Reitan J, Røyset A, et al. Optimizing the Performance of Phase-Change Materials in Personal Protective Clothing Systems. Int J Occup Saf Ergon 2008;14:43–53. https://doi.org/10.1080/10803548.2008.11076746.

[111] E. Wilusz. Military Textiles. Woodhead Publishing Series in Textiles; 2008.

[112] H Mattila. Intelligent Textiles and Clothing. Woodhead Publishing Series in Textiles; 2006.

[113] Chowdhury MA, Joshi M, Butola BS. Photochromic and Thermochromic Colorants in Textile Applications. J Eng Fibers Fabr 2014;9:155892501400900. https://doi.org/10.1177/155892501400900113.

[114] Ortica F. The role of temperature in the photochromic behaviour. Dyes Pigments 2012;92:807–16. https://doi.org/10.1016/j.dyepig.2011.04.002.

[115] Kumar Gupta V. Photochromic Dyes for Smart Textiles. In: Papadakis R, editor. Dyes Pigments – Nov. Appl. Waste Treat., IntechOpen; 2021. https://doi.org/10.5772/intechopen.96055.

[116] N.d.http://www.solaractiveintl.com/newsrelease.htm/05.08.2005 (accessed May 29, 2022).

[117] Etters, J N D HC. Textile Chemist and Colourist & American Dyestuff Reporter, 2000, 32; 20 2000.

[118] Rijavec T, Bračko S. Smart dyes for medical and other textiles. Smart Text. Med. Healthc., Elsevier; 2007, p. 123–49. https://doi.org/10.1533/9781845692933.1.123.

[119] Oliveira MM, Salvador MA, Delbaere S, Berthet J, Vermeersch G, Micheau JC, et al. Remarkable thermally stable open forms of photochromic new N-substituted benzopyranocarbazoles. J Photochem Photobiol Chem 2008;198:242–9. https://doi.org/10.1016/j.jphotochem.2008.03.019.

[120] U.S Pat. No. 4,681,791; 1987, 1987.

[121] U.S Pat. No. 5,985,381; 1999, 1999.

[122] A Design Research Program for Textiles and Computational Technology. Nord Text J 2002;No. 1:56–63.

[123] U.S Pat. No. 10,095,299; 2002, 2002.

[124] Day JH. Thermochromism. Chem Rev 1963;63:65–80. https://doi.org/10.1021/cr60221a005.

[125] Christie RM. Chromic materials for technical textile applications. Adv. Dye. Finish. Tech. Text., Elsevier; 2013, p. 3–36. https://doi.org/10.1533/9780857097613.1.3.

[126] John E. Berkowitch. Trends in Japanese Textile Technology. DIANE Publishing; 2000.

[127] Xuan R, Ge J. Photonic Printing through the Orientational Tuning of Photonic Structures and Its Application to Anticounterfeiting Labels. Langmuir 2011;27:5694–9. https://doi.org/10.1021/la200571y.

[128] Xuan R, Ge J. Invisible photonic prints shown by water. J Mater Chem 2012;22:367–72. https://doi.org/10.1039/C1JM14082E.

[129] Saito H, Takeoka Y, Watanabe M. Simple and precision design of porous gel as a visible indicator for ionic species and concentration. Chem Commun 2003:2126. https://doi.org/10.1039/b304306a.

[130] Colodrero S, Ocaña M, Míguez H. Nanoparticle-Based One-Dimensional Photonic Crystals. Langmuir 2008;24:4430–4. https://doi.org/10.1021/la703987r.

[131] Mackiewicz M, Karbarz M, Romanski J, Stojek Z. An environmentally sensitive three-component hybrid microgel. RSC Adv 2016;6:83493–500. https://doi.org/10.1039/C6RA15048A.

[132] Xuan R, Wu Q, Yin Y, Ge J. Magnetically assembled photonic crystal film for humidity sensing. J Mater Chem 2011;21:3672. https://doi.org/10.1039/c0jm03790g.

[133] Sharma AC, Jana T, Kesavamoorthy R, Shi L, Virji MA, Finegold DN, et al. A General Photonic Crystal Sensing Motif: Creatinine in Bodily Fluids. J Am Chem Soc 2004;126:2971–7. https://doi.org/10.1021/ja038187s.

[134] Puzzo DP, Arsenault AC, Manners I, Ozin GA. Electroactive Inverse Opal: A Single Material for All Colors. Angew Chem Int Ed 2009;48:943–7. https://doi.org/10.1002/anie.200804391.

[135] Kinoshita S, Yoshioka S, Miyazaki J. Physics of structural colors. Rep Prog Phys 2008;71:076401. https://doi.org/10.1088/0034-4885/71/7/076401.

[136] Gao W, Rigout M, Owens H. The structural coloration of textile materials using self-assembled silica nanoparticles. J Nanoparticle Res 2017;19:303. https://doi.org/10.1007/s11051-017-3991-7.

[137] Gong X, Hou C, Zhang Q, Li Y, Wang H. Solvatochromic structural color fabrics with favorable wearability properties. J Mater Chem C 2019;7:4855–62. https://doi.org/10.1039/C9TC00580C.

[138] Kelly FM, Cochrane C. Color-Changing Textiles and Electrochromism. In: Tao X, editor. Handb. Smart Text., Singapore: Springer Singapore; 2015, p. 859–89. https://doi.org/10.1007/978-981-4451-45-1_16.

[139] Bai L, Bose P, Gao Q, Li Y, Ganguly R, Zhao Y. Halogen-Assisted Piezochromic Supramolecular Assemblies for Versatile Haptic Memory. J Am Chem Soc 2017;139:436–41. https://doi.org/10.1021/jacs.6b11057.

[140] Hager.MD, Bode S, Weber C, Schubert US. Shape memory polymers: Past, present and future developments. Prog Polym Sci 2015;49–50:3–33. https://doi.org/10.1016/j.progpolymsci.2015.04.002.

[141] Hu J, Zhu Y, Huang H, Lu J. Recent advances in shape–memory polymers: Structure, mechanism, functionality, modeling and applications. Prog Polym Sci 2012;37:1720–63. https://doi.org/10.1016/j.progpolymsci.2012.06.001.

[142] Lin T, Ma S, Lu Y, Guo B. New Design of Shape Memory Polymers Based on Natural Rubber Crosslinked via Oxa-Michael Reaction. ACS Appl Mater Interfaces 2014;6:5695–703. https://doi.org/10.1021/am500236w.

[143] Zhao J, Chen M, Wang X, Zhao X, Wang Z, Dang Z-M, et al. Triple Shape Memory Effects of Cross-Linked Polyethylene/Polypropylene Blends with Cocontinuous Architecture. ACS Appl Mater Interfaces 2013;5:5550–6. https://doi.org/10.1021/am400769j.

[144] Chen S, Hu J, Liu Y, Liem H, Zhu Y, Meng Q. Effect of molecular weight on shape memory behavior in polyurethane films. Polym Int 2007;56:1128–34. https://doi.org/10.1002/pi.2248.

[145] Chen S, Hu J, Chen S. Studies of the moisture-sensitive shape memory effect of pyridine-containing polyurethanes. Polym Int 2012;61:314–20. https://doi.org/10.1002/pi.3192.

[146] Koerner H, Price G, Pearce NA, Alexander M, Vaia RA. Remotely actuated polymer nanocomposites – stress-recovery of carbon-nanotube-filled thermoplastic elastomers. Nat Mater 2004;3:115–20. https://doi.org/10.1038/nmat1059.

[147] Weigel T, Mohr R, Lendlein A. Investigation of parameters to achieve temperatures required to initiate the shape-memory effect of magnetic nanocomposites by inductive heating. Smart Mater Struct 2009;18:025011. https://doi.org/10.1088/0964-1726/18/2/025011.

[148] Kim H, Abdala AA, Macosko CW. Graphene/Polymer Nanocomposites. Macromolecules 2010;43:6515–30. https://doi.org/10.1021/ma100572e.

[149] Lendlein A, Kelch S. Shape-Memory Polymers. Angew Chem Int Ed 2002;41:2034. https://doi.org/10.1002/1521-3773(20020617)41:12<2034::AID-ANIE2034>3.0.CO;2-M.

[150] Du H, Zhang J. Solvent induced shape recovery of shape memory polymer based on chemically cross-linked poly(vinyl alcohol). Soft Matter 2010;6:3370. https://doi.org/10.1039/b922220k.

[151] Joshi M, editor. Nanotechnology in Textiles: Advances and Developments in Polymer Nanocomposites. 1st ed. Jenny Stanford Publishing; 2020. https://doi.org/10.1201/9781003055815.

[152] Zhuo H, Hu J, Chen S, Yeung L. Preparation of polyurethane nanofibers by electrospinning. J Appl Polym Sci 2008;109:406–11. https://doi.org/10.1002/app.28067.

[153] Kaursoin J, Agrawal AK. Melt spun thermoresponsive shape memory fibers based on polyurethanes: Effect of drawing and heat-setting on fiber morphology and properties. J Appl Polym Sci 2007;103:2172–82. https://doi.org/10.1002/app.25124.

[154] Hu J, Meng Q, Zhu Y, Lu J, Zhuo H. Shape memory fibers prepared via wet, reaction, dry, melt, and electro spinning., 2009.

[155] Hu J, Lu J. Shape Memory Fibers. In: Tao X, editor. Handb. Smart Text., Singapore: Springer Singapore; 2015, p. 183–207. https://doi.org/10.1007/978-981-4451-45-1_3.

[156] Behl M, Lendlein A. Shape-memory polymers. Mater Today 2007;10:20–8. https://doi.org/10.1016/S1369-7021(07)70047-0.

[157] Gök MO, Bilir MZ, Gürcüm BH. Shape-Memory Applications in Textile Design. Procedia – Soc Behav Sci 2015;195:2160–9. https://doi.org/10.1016/j.sbspro.2015.06.283.

[158] Looney, B. BP. Statistical Review of World Energy: An Unsustainable Path 2021. https://www.bp.com/content/dam/bp/business-sites/en/global/corporate/pdfs/energy-economics/statistical-review/bp-stats-review-2021-full-report.pdf (accessed May 29, 2022).

[159] Chen G, Li Y, Bick M, Chen J. Smart Textiles for Electricity Generation. Chem Rev 2020;120:3668–720. https://doi.org/10.1021/acs.chemrev.9b00821.

[160] Chen R-J, Zhang Y-B, Liu T, Xu B-Q, Lin Y-H, Nan C-W, et al. Addressing the Interface Issues in All-Solid-State Bulk-Type Lithium Ion Battery via an All-Composite Approach. ACS Appl Mater Interfaces 2017;9:9654–61. https://doi.org/10.1021/acsami.6b16304.

[161] Fu J, Cano ZP, Park MG, Yu A, Fowler M, Chen Z. Electrically Rechargeable Zinc-Air Batteries: Progress, Challenges, and Perspectives. Adv Mater 2017;29:1604685. https://doi.org/10.1002/adma.201604685.

[162] Placke T, Kloepsch R, Dühnen S, Winter M. Lithium ion, lithium metal, and alternative rechargeable battery technologies: the odyssey for high energy density. J Solid State Electrochem 2017;21:1939–64. https://doi.org/10.1007/s10008-017-3610-7.

[163] González A, Goikolea E, Barrena JA, Mysyk R. Review on supercapacitors: Technologies and materials. Renew Sustain Energy Rev 2016;58:1189–206. https://doi.org/10.1016/j.rser.2015.12.249.

[164] Muzaffar A, Ahamed MB, Deshmukh K, Thirumalai J. A review on recent advances in hybrid supercapacitors: Design, fabrication and applications. Renew Sustain Energy Rev 2019;101:123–45. https://doi.org/10.1016/j.rser.2018.10.026.

[165] Yu M, Feng X. Thin-Film Electrode-Based Supercapacitors. Joule 2019;3:338–60. https://doi.org/10.1016/j.joule.2018.12.012.

[166] Tao X. Study of Fiber-Based Wearable Energy Systems. Acc Chem Res 2019;52:307–15. https://doi.org/10.1021/acs.accounts.8b00502.

[167] Tebyetekerwa M, Marriam I, Xu Z, Yang S, Zhang H, Zabihi F, et al. Critical insight: challenges and requirements of fibre electrodes for wearable electrochemical energy storage. Energy Environ Sci 2019;12:2148–60. https://doi.org/10.1039/C8EE02607F.

[168] Wang ZL. On the first principle theory of nanogenerators from Maxwell's equations. Nano Energy 2020;68:104272. https://doi.org/10.1016/j.nanoen.2019.104272.

[169] Wang ZL. On Maxwell's displacement current for energy and sensors: the origin of nanogenerators. Mater Today 2017;20:74–82. https://doi.org/10.1016/j.mattod.2016.12.001.

[170] Zhang M, Zhao M, Jian M, Wang C, Yu A, Yin Z, et al. Printable Smart Pattern for Multifunctional Energy-Management E-Textile. Matter 2019;1:168–79. https://doi.org/10.1016/j.matt.2019.02.003.

[171] Elahi H, Eugeni M, Gaudenzi P. A Review on Mechanisms for Piezoelectric-Based Energy Harvesters. Energies 2018;11:1850. https://doi.org/10.3390/en11071850.

[172] Lu X, Qu H, Skorobogatiy M. Piezoelectric Micro- and Nanostructured Fibers Fabricated from Thermoplastic Nanocomposites Using a Fiber Drawing Technique: Comparative Study and Potential Applications. ACS Nano 2017;11:2103–14. https://doi.org/10.1021/acsnano.6b08290.

[173] Bhatnagar V, Owende P. Energy harvesting for assistive and mobile applications. Energy Sci Eng 2015;3:153–73. https://doi.org/10.1002/ese3.63.

[174] Hyland M, Hunter H, Liu J, Veety E, Vashaee D. Wearable thermoelectric generators for human body heat harvesting. Appl Energy 2016;182:518–24. https://doi.org/10.1016/j.apenergy.2016.08.150.

[175] Jeerapan I, Sempionatto JR, Wang J. On-Body Bioelectronics: Wearable Biofuel Cells for Bioenergy Harvesting and Self-Powered Biosensing. Adv Funct Mater 2020;30:1906243. https://doi.org/10.1002/adfm.201906243.

[176] Zebda A, Alcaraz J-P, Vadgama P, Shleev S, Minteer SD, Boucher F, et al. Challenges for successful implantation of biofuel cells. Bioelectrochemistry 2018;124:57–72. https://doi.org/10.1016/j.bioelechem.2018.05.011.

[177] Bandodkar AJ, Wang J. Wearable Biofuel Cells: A Review. Electroanalysis 2016;28:1188–200. https://doi.org/10.1002/elan.201600019.

[178] World Energy Outlook 2021.

[179] J.M.K.C. Donev et al. Energy Education – Photovoltaic effect. 2015.

[180] Lee S, Lee Y, Park J, Choi D. Stitchable organic photovoltaic cells with textile electrodes. Nano Energy 2014;9:88–93. https://doi.org/10.1016/j.nanoen.2014.06.017.

[181] Yin S, Jin Z, Miyake T. Wearable high-powered biofuel cells using enzyme/carbon nanotube composite fibers on textile cloth. Biosens Bioelectron 2019;141:111471. https://doi.org/10.1016/j.bios.2019.111471.

[182] Xue Q, Sun J, Huang Y, Zhu M, Pei Z, Li H, et al. Recent Progress on Flexible and Wearable Supercapacitors. Small 2017;13:1701827. https://doi.org/10.1002/smll.201701827.

[183] Huang Y, Zhu M, Huang Y, Pei Z, Li H, Wang Z, et al. Multifunctional Energy Storage and Conversion Devices. Adv Mater 2016;28:8344–64. https://doi.org/10.1002/adma.201601928.

[184] Shao Y, El-Kady MF, Wang LJ, Zhang Q, Li Y, Wang H, et al. Graphene-based materials for flexible supercapacitors. Chem Soc Rev 2015;44:3639–65. https://doi.org/10.1039/C4CS00316K.

[185] Barik R, Raulo A, Jha S, Nandan B, Ingole PP. Polymer-Derived Electrospun Co_3O_4@C Porous Nanofiber Network for Flexible, High-Performance, and Stable Supercapacitors. ACS Appl Energy Mater 2020;3:11002–14. https://doi.org/10.1021/acsaem.0c01955.

[186] Ren J, Li L, Chen C, Chen X, Cai Z, Qiu L, et al. Twisting Carbon Nanotube Fibers for Both Wire-Shaped Micro-Supercapacitor and Micro-Battery. Adv Mater 2013;25:1155–9. https://doi.org/10.1002/adma.201203445.

[187] Joshi A, Raulo A, Bandyopadhyay S, Gupta A, Srivastava R, Nandan B. Waste cotton cloth derived flexible current collector with optimized electrical properties for high performance lithium–sulfur batteries. Carbon 2022;192:429–37. https://doi.org/10.1016/j.carbon.2022.03.018.

[188] Zhang Y, Wang Y, Wang L, Lo C-M, Zhao Y, Jiao Y, et al. A fiber-shaped aqueous lithium ion battery with high power density. J Mater Chem A 2016;4:9002–8. https://doi.org/10.1039/C6TA03477B.

[189] Yadav A, De B, Singh SK, Sinha P, Kar KK. Facile Development Strategy of a Single Carbon-Fiber-Based All-Solid-State Flexible Lithium-Ion Battery for Wearable Electronics. ACS Appl Mater Interfaces 2019;11:7974–80. https://doi.org/10.1021/acsami.8b20233.

[190] Park J, Park M, Nam G, Lee J, Cho J. All-Solid-State Cable-Type Flexible Zinc-Air Battery. Adv Mater 2015;27:1396–401. https://doi.org/10.1002/adma.201404639.

[191] Xu Y, Zhao Y, Ren J, Zhang Y, Peng H. An All-Solid-State Fiber-Shaped Aluminum-Air Battery with Flexibility, Stretchability, and High Electrochemical Performance. Angew Chem Int Ed 2016;55:7979–82. https://doi.org/10.1002/anie.201601804.

[192] Li Q, Zhang Q, Liu C, Sun J, Guo J, Zhang J, et al. Flexible all-solid-state fiber-shaped Ni–Fe batteries with high electrochemical performance. J Mater Chem A 2019;7:520–30. https://doi.org/10.1039/C8TA09822K.

[193] Chen Q, Sun S, Zhai T, Yang M, Zhao X, Xia H. Yolk-Shell NiS_2 Nanoparticle-Embedded Carbon Fibers for Flexible Fiber-Shaped Sodium Battery. Adv Energy Mater 2018;8:1800054. https://doi.org/10.1002/aenm.201800054.

[194] Meng Y, Zhao Y, Hu C, Cheng H, Hu Y, Zhang Z, et al. All-Graphene Core-Sheath Microfibers for All-Solid-State, Stretchable Fibriform Supercapacitors and Wearable Electronic Textiles. Adv Mater 2013;25:2326–31. https://doi.org/10.1002/adma.201300132.

[195] Zhang Q, Sun J, Pan Z, Zhang J, Zhao J, Wang X, et al. Stretchable fiber-shaped asymmetric supercapacitors with ultrahigh energy density. Nano Energy 2017;39:219–28. https://doi.org/10.1016/j.nanoen.2017.06.052.

[196] Khudiyev T, Lee JT, Cox JR, Argentieri E, Loke G, Yuan R, et al. 100 m Long Thermally Drawn Supercapacitor Fibers with Applications to 3D Printing and Textiles. Adv Mater 2020;32:2004971. https://doi.org/10.1002/adma.202004971.

[197] Vlad A, Singh N, Galande C, Ajayan PM. Design Considerations for Unconventional Electrochemical Energy Storage Architectures. Adv Energy Mater 2015;5:1402115. https://doi.org/10.1002/aenm.201402115.

[198] Kim BC, Hong J-Y, Wallace GG, Park HS. Recent Progress in Flexible Electrochemical Capacitors: Electrode Materials, Device Configuration, and Functions. Adv Energy Mater 2015;5:1500959. https://doi.org/10.1002/aenm.201500959.

[199] Bonaccorso F, Colombo L, Yu G, Stoller M, Tozzini V, Ferrari AC, et al. Graphene, related two-dimensional crystals, and hybrid systems for energy conversion and storage. Science 2015;347:1246501. https://doi.org/10.1126/science.1246501.

[200] Chen T, Dai L. Carbon nanomaterials for high-performance supercapacitors. Mater Today 2013;16:272–80. https://doi.org/10.1016/j.mattod.2013.07.002.

[201] Sun X, He J, Qiang R, Nan N, You X, Zhou Y, et al. Electrospun Conductive Nanofiber Yarn for a Wearable Yarn Supercapacitor with High Volumetric Energy Density. Materials 2019;12:273. https://doi.org/10.3390/ma12020273.

[202] Zhang X, Cheng X, Zhang Q. Nanostructured energy materials for electrochemical energy conversion and storage: A review. J Energy Chem 2016;25:967–84. https://doi.org/10.1016/j.jechem.2016.11.003.

[203] Lee J, Jeon S, Seo H, Lee JT, Park S. Fiber-Based Sensors and Energy Systems for Wearable Electronics. Appl Sci 2021;11:531. https://doi.org/10.3390/app11020531.

[204] Li Z, Zuo L, Luhrs G, Lin L, Qin Y. Electromagnetic Energy-Harvesting Shock Absorbers: Design, Modeling, and Road Tests. IEEE Trans Veh Technol 2013;62:1065–74. https://doi.org/10.1109/TVT.2012.2229308.

[205] Miljkovic N, Preston DJ, Enright R, Wang EN. Jumping-droplet electrostatic energy harvesting. Appl Phys Lett 2014;105:013111. https://doi.org/10.1063/1.4886798.

[206] Joshi M, Adak B. Advances in Nanotechnology Based Functional, Smart and Intelligent Textiles: A Review. Compr. Nanosci. Nanotechnol., Elsevier; 2019, p. 253–90. https://doi.org/10.1016/B978-0-12-803581-8.10471-0.

[207] Bhattacharyya A, Joshi M. Functional properties of microwave-absorbent nanocomposite coatings based on thermoplastic polyurethane-based and hybrid carbon-based nanofillers: Microwave-absorbent nanocomposite coatings. Polym Adv Technol 2012;23:975–83. https://doi.org/10.1002/pat.2000.

[208] Adak B, Joshi M. Coated or Laminated Textiles for Aerostat and Stratospheric Airship. In: ul-Islam S, Butola BS, editors. Adv. Text. Eng. Mater., Hoboken, NJ, USA: John Wiley & Sons, Inc.; 2018, p. 257–87. https://doi.org/10.1002/9781119488101.ch7.

[209] Joshi M, Adak B, Chatterjee U. Polyurethane Nanocomposite-Based Advanced Materials for Aerostat/Airship Envelopes. Coat. Laminated Text. Aerostats Airsh. 1st ed., Boca Raton: CRC Press; 2022, p. 165–97. https://doi.org/10.1201/9780429432996-6.

[210] Adak B, Joshi M. Recent Developments in Gas Barrier Polymer Nanocomposite Coatings. In: Joshi M, editor. Nanotechnol. Text. 1st ed., Jenny Stanford Publishing; 2020, p. 661–94. https://doi.org/10.1201/9781003055815-22.

[211] Saber D, Abd El-Aziz K. Advanced materials used in wearable health care devices and medical textiles in the battle against coronavirus (COVID-19): A review. J Ind Text 2022;51:246S-271S. https://doi.org/10.1177/15280837211041771.

[212] Azam Ali M, Shavandi A. Medical textiles testing and quality assurance. Perform. Test. Text., Elsevier; 2016, p. 129–53. https://doi.org/10.1016/B978-0-08-100570-5.00007-4.

[213] Karim N, Afroj S, Lloyd K, Oaten LC, Andreeva DV, Carr C, et al. Sustainable Personal Protective Clothing for Healthcare Applications: A Review. ACS Nano 2020;14:12313–40. https://doi.org/10.1021/acsnano.0c05537.

[214] Islam MdM, Mahmud S, Muhammad LJ, Islam MdR, Nooruddin S, Ayon SI. Wearable Technology to Assist the Patients Infected with Novel Coronavirus (COVID-19). SN Comput Sci 2020;1:320. https://doi.org/10.1007/s42979-020-00335-4.

[215] Kilinc FS. A Review of Isolation Gowns in Healthcare: Fabric and Gown Properties. J Eng Fibers Fabr 2015;10:155892501501000. https://doi.org/10.1177/155892501501000313.

[216] Virk RK, Ramaswamy GN, Bourham M, Bures BL. Plasma and Antimicrobial Treatment of Nonwoven Fabrics for Surgical Gowns. Text Res J 2004;74:1073–9. https://doi.org/10.1177/004051750407401208.

[217] Rutala WA, Weber DJ. A Review of Single-Use and Reusable Gowns and Drapes in Health Care. Infect Control Hosp Epidemiol 2001;22:248–57. https://doi.org/10.1086/501895.

[218] Chakraborty S, Biswas MC. 3D printing technology of polymer-fiber composites in textile and fashion industry: A potential roadmap of concept to consumer. Compos Struct 2020;248:112562. https://doi.org/10.1016/j.compstruct.2020.112562.

[219] Chatterjee K, Ghosh TK. 3D Printing of Textiles: Potential Roadmap to Printing with Fibers. Adv Mater 2020;32:1902086. https://doi.org/10.1002/adma.201902086.

[220] Bloomfield M, Borstrock S. Modeclix. The additively manufactured adaptable textile. Mater Today Commun 2018;16:212–6. https://doi.org/10.1016/j.mtcomm.2018.04.002.

[221] Gao T, Yang Z, Chen C, Li Y, Fu K, Dai J, et al. Three-Dimensional Printed Thermal Regulation Textiles. ACS Nano 2017;11:11513–20. https://doi.org/10.1021/acsnano.7b06295.

[222] Yang Z, Ma Y, Jia S, Zhang C, Li P, Zhang Y, et al. 3D-Printed Flexible Phase-Change Nonwoven Fabrics toward Multifunctional Clothing. ACS Appl Mater Interfaces 2022;14:7283–91. https://doi.org/10.1021/acsami.1c21778.

[223] Mirdamadi E, Tashman JW, Shiwarski DJ, Palchesko RN, Feinberg AW. FRESH 3D Bioprinting a Full-Size Model of the Human Heart. ACS Biomater Sci Eng 2020;6:6453–9. https://doi.org/10.1021/acsbiomaterials.0c01133.

[224] Mannoor MS, Jiang Z, James T, Kong YL, Malatesta KA, Soboyejo WO, et al. 3D Printed Bionic Ears. Nano Lett 2013;13:2634–9. https://doi.org/10.1021/nl4007744.

[225] Lee JH, Kim H, Hwang J-Y, Chung J, Jang T-M, Seo DG, et al. 3D Printed, Customizable, and Multifunctional Smart Electronic Eyeglasses for Wearable Healthcare Systems and Human–Machine Interfaces. ACS Appl Mater Interfaces 2020;12:21424–32. https://doi.org/10.1021/acsami.0c03110.

[226] Joshi M, Adak B, Butola BS. Polyurethane nanocomposite based gas barrier films, membranes and coatings: A review on synthesis, characterization and potential applications. Prog Mater Sci 2018;97:230–82. https://doi.org/10.1016/j.pmatsci.2018.05.001.

[227] Eleksen offers wireless fabric keyboard n.d. https://www.infoworld.com/article/2674081/eleksen-offers-wireless-fabric-keyboard.html (accessed September 29, 2022).

S. Wazed Ali, Swagata Banerjee, Mandira Mondal

2 Nanotechnology in textile finishing: an approach towards imparting manifold functionalities

Abstract: Nanotechnology relates to the science of nanomaterials and their respective technological applications. In the past few decades, this technology has been explored in various scientific domains that demonstrated its potential as a promising technology. The textile sector has also benefitted with the various functionalities offered by nanotechnology. Nanodimensions of materials exhibit certain exclusive properties that are not seen in the bulk form of the material. These unique features of nanomaterials have been exploited in the textile field through the application of various finishes. Water repellence, antistatic, flame retarding ability, antimicrobial, and anti-crease are some examples of the nanofinishes that are imparted to textiles. The unique features of nanomaterials can be exploited to bestow multiple functionalities in textile substrates, and not just a single property. Detailed discussions of more such finishes have been covered in this chapter. The application of these finishes takes place through different methods such as sol–gel and layer-by-layer that do not affect the feel or the aesthetic appeal of textiles. Use of this technology has broadened the spectrum of textile applications. This chapter provides an overall idea about the various finishes on textiles that can be obtained by adopting nanotechnological approaches to impart value addition and, consequently, their potential fields of applications.

Keywords: Nanotechnology, textiles, nanofinishing, plasma, antimicrobial, aesthetic

2.1 Introduction

With the technological advancements in the textile industry, nanotechnology is getting closely entwined with the science of textiles [1]. Nanotechnology deals with the characteristics and properties of materials in their nanodimensions. The properties of a material in the nanoscale are strikingly different from its properties in the bulk form. This is ascribed to the quantum effect observed at the nanolevel. Imparting such unique properties to textiles through nanomaterials is gaining ground because it helps to widen the spectrum of textile applications [2].

Textile is an ideal substrate for several high-end applications, apart from clothing, owing to its flexibility. Much functionality can be imparted to textiles as per the end use requirements through various established techniques. These conventional techniques usually alter the feel and handling of fabrics. They, in turn, affect the aesthetic appeal of textiles. The durability of the functionalities imparted through conventional procedures

https://doi.org/10.1515/9783110759747-002

is also an issue. This has led to the emergence of nanotechnology-based techniques for the value addition of textiles [3].

Nanotechnology-based approaches work with textiles at the nanoscale. These approaches do not alter the aesthetic feel of the fabrics. The comfort properties of textile are in no way compromised through the nanotechnological approaches. The durability issues of the finishes are also taken care in these non-conventional approaches. Nanotechnology has been used to impart several functionalities such as hydrophobicity, UV resistance, anti-microbial resistance, and so on [4]. The nanomaterials used for these finishes are of different shapes and sizes. They may be applied as nanocoatings, through sol–gel techniques, layer-by-layer methods, and other procedures. They do not affect the breathability of textiles and hence can be preferred over the conventional methods of finishing.

2.2 Nanofinishing methods of textiles

The commonly used methods for nanofinishing are listed below:

2.2.1 Sol–gel method

The sol–gel system allows homogenous synthesis of a multi-component system. Sol refers to the state of colloidally dispersed particles in a fluid medium. The stability of the dispersion is governed by the Brownian motion and the dimensions of the dispersed phase. Gel, in the term "sol–gel", relates to the three-dimensional mesh-like structure formed by the solid wherein the liquid is entrapped. Hydrolysis and polycondensation reactions mainly govern the mechanism of the sol–gel formation. A schematic representation of the sol–gel method is shown in Figure 2.1. For sol–gel formation, a precursor mix is prepared. This mix gradually converts to a more solid-like material following different reactions such as solvent evaporation, dehydration, and crosslinking between the dispersed phase via chemical bonds [5]. Solvent removal from the sol–gel is an important step in the process. Often, the solvent is vaporized. This vaporization induces capillary pressure and the gel shrinks. This is called the xerogel. On the other hand, when the solvent is eliminated at a temperature and pressure higher than the critical point of the solvent, it leads to less shrinkage and, thus, the formed product is called aerogel. For the preparation of cryogels, freeze drying technique is used for solvent removal. Sol–gel technology can be used to impart various finishes such as hydrophobicity, UV protection, self-cleaning, anti-bacterial activity, and even multifunctional properties to textiles [6].

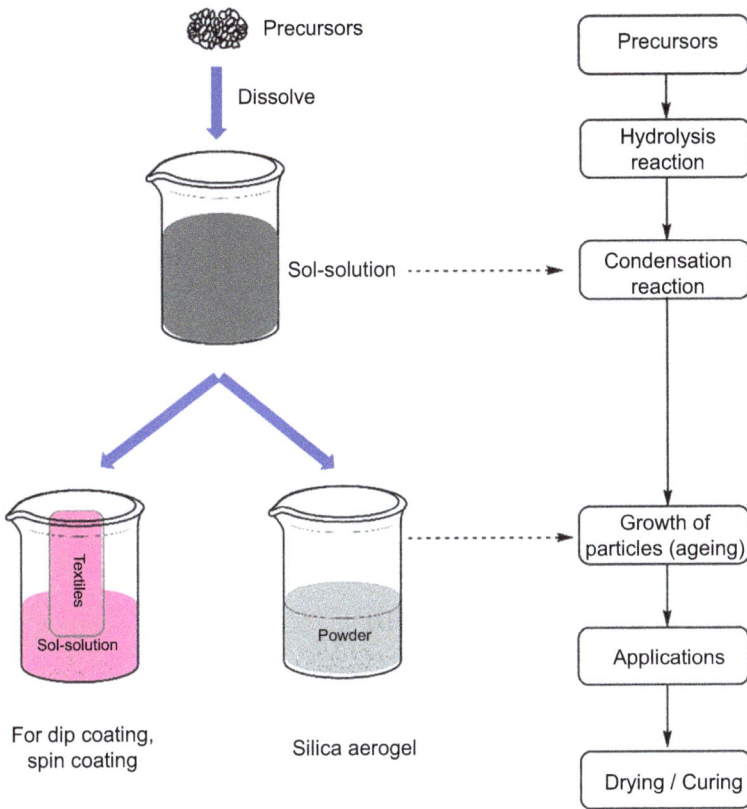

Figure 2.1: Schematic representation of sol–gel method for the nanofinishing of textiles [7].

2.2.2 Layer-by-layer deposition

This is one of the extensively used methods for the application of nanoparticles on a textile substrate. This method is favoured for the preparation of many products, as the thickness of the layer can be controlled up to the nanoscale. There is negligible variation due to the type of material used, and it can be applied on a wide range of substrates (Figure 2.2). In the layer-by-layer (l-b-l) deposition method, oppositely charged polyelectrolytes or finishing agents are utilized alternately and stepwise, and complication such as by-product formation is not observed. The layers are fabricated by the deposition of alternate cationic and anionic functional materials on the textile surface. This method is very economical, simple, and can be controlled as per the requirements [4, 5]. Self-assembled multiple nanolayers can be developed by following this method [10].

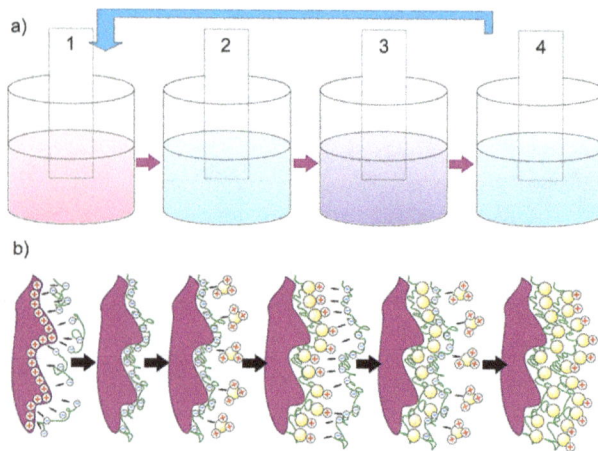

Figure 2.2: Layer-by-layer deposition technique and a schematic representation of the layer formation on the substrate [11].

2.2.3 Chemical vapour deposition (CVD)

This is a widely used method where thin nanofilms or nanocoatings are developed on a heated substrate from vapour precursors. Chemical vapour deposition has several advantages over physical vapour deposition. The main advantage of chemical vapour deposition is the good deposition rate due to the chemical reactions. In addition, the quality achieved from this technique is excellent. This technology is mostly used to prepare hydrophobic textiles, antibacterial textiles, and electronic textiles [8, 9].

2.2.4 Physical vapour deposition (PVD)

It is a type of vacuum deposition method that is generally used for the production of coatings and thin films. Initially, the material transforms from a condensed phase to the vapour phase, and finally again transforms to the condensed phase. Sputtering and the subsequent evaporation is the most commonly used physical vapour deposition method. The other methods are pulsed layer deposition and ion plating. There are several steps involved in the physical vapour deposition process. Firstly, the material is sputtered or evaporated with the help of a high energy source. Then, the material in the vapour phase is transported to the substrate where the material will be deposited. During the transport stage, reaction occurs between the material to be deposited and the selected reactive gas. Finally, the deposition of the material takes place on the surface of the textile substrate [10, 11]. The several advantages of the physical deposition process are: (a) no limitation on the usage of any inorganic material; however, only some types of organic materials are allowed, (b) comparatively eco-friendly than the chemical vapour deposition process, since no chemical process is involved, and (c) this process takes places at a substantially lower temperature, and hence can be used to coat films on temperature-sensitive substrates. But the process is costly, complex, and many difficulties arise when coating complex shapes [14].

2.2.5 Plasma

In the textile industry, the conventional finishing processes lead to several environmental issues that make the plasma technology more attractive in the eyes of technocrats. Plasma treatment of textiles does not alter the bulk properties of the substrate. In obtaining finishing from this technology, no changes are observed in the properties of bulk textiles. New functionalities can be obtained by controlling the deposition up to the nanometer range. The factors on which the efficiency of the plasma treatment depends are the types of plasma, the operating conditions (applied voltage, exposure time, used gas type, ratio of gases in the mixture, etc.), and the properties of the substrate. There are various types of functional textiles that are obtained by this method [16].

2.3 Water- and oil-repellent finish

Water and oil repellence in textiles have become important requirements in the textile market. Such properties in textiles make the handling of the product easy and user-friendly. The main factors that govern the repellence effect in textiles are surface energy and the hierarchical structures present on their surface. Fluorocarbon-based and non-fluorocarbon-based compounds are generally used for this purpose. Nanoparticles enhance the performance of hydrophobicity and oleophobicity. The fibrous structure developed by nanotechnology produces a highly micro rough surface, like a "lotus" surface, which shows the capability to repel water or other liquids. Significant efforts are made to achieve all these properties without any change in the breathability and the handling property of the textile materials [17].

A coating named "OmniBlock" created by the sol–gel process caused a crosslinking between the Si–O–Si groups. This coating contained silica nanoparticles of the nanometer range (say, 200 nm), a perfluorooctanoic acid-free fluoropolymer, 3-glycidoxypropyl trimethoxysilane, and tetraethylorthosilicate. The dip-dry-cure method was adopted for the application of perfluorooctanoic acid-free fluoropolymer-coated silica nanoparticles on the cotton fabric surface. The substrate showed large contact angles of 154 and 121 °C for water and n-dodecane, respectively, indicating good liquid repellence [18]. In order to add water repellent property to the bleached jute fabric, commercial water repellent (NUVA N2114 polymer) and silica nanosol were applied to it via the pad–dry–cure technique. The finished jute fabric with silica nanocomposite attained excellent semi durable water repellence. The water repellence property obtained only by silica nanofinishing results in sustainability of the property for less than 5 cycles of home washing, whereas the water repellence property sustained up to 15 washing cycles by the silica nanocomposite-finished jute fabric [19]. The layer-by-layer technique, with electrostatic self-assembly, was adopted for the coating of a cotton fabric, resulting in the formation of poly (diallyldimethylammonium chloride) (PDDA) and ZnO/SiO_2 colloidal nanocomposite solution. The number of deposition layers on the fibres indicated the deposition of ZnO/SiO_2 nanocomposite. A post treatment of the finished fabric with stearic acid resulted in a good water repellent property. Additionally, the finished fabric, without any usage of stearic acid, exhibited UV resistance properties [20]. The sols were prepared of titanium(IV) butoxide where the functional additive used was boric acid. The pad-dry method renders the application of sols on the cotton fabrics. This treatment enhanced the water repellent property of the finished cotton fabric. Moreover, the treated fabric showed flame-retardant property and good thermal stability [21]. Quaternary ammonium-functionalized MSNs and fluorinated mesoporous silica nanoparticles can get adhered to the textile surfaces using a binder of polydimethylsiloxane. This synergetic coating resulted in a hydrophobic coating, with a sliding angle of 2° and a contact angle of 152°. Moreover, this treatment exhibited antibacterial activity against both the Gram-positive, *Staphylococcus aureus* (*S. aureus*) and Gram-negative, *Escherichia coli* (*E. coli*) bacteria. With the deposition of the coating, the breathability of the fabric was not hampered [22]. The water repellent property is enhanced when the opposite charge attraction is increased by cationized bovine serum albumin (cBSA) and the roughness is increased with the help of an alkali. Micro roughness and generation of functional groups are the results of alkaline etching. Cationized bovine serum albumin was further applied to the etched fabric. A generation of positively charged functional group was observed due to the final treatment of the fabric with cBSA. These positively charged functional groups were responsible for the initiation of strong electrostatic attraction, which propagates the crosslinking of silica nanoparticles and the surface of the fibre. This overall treatment results not only in excellent hydrophobicity but also in oleophilic characteristics. Moreover, the process exhibits mechanical and chemical stability, with a self-cleaning property [23].

Some research studies discussed that both the oleophobic and the hydrophobic properties can be imparted by treating the textile material with special nanofinishes. The combination of ZnO nanoparticles and SiO_2 nanoparticles organically imparts water and oil repellent properties when applied to the cotton fabric. A mixture of nanoparticles was applied to the cotton fabric by the padding method. High water absorption time, low wetting capacity, and a water-repellent degree of 75 on the AATCC photographic scale were observed in the treated fabric, leading to a desirable hydrophobic effect. The treatment on the cotton fabric showed a good oil-repellent behaviour and it resulted in the efficient protection and rejection of oily substances from the cotton fabric surface. This treatment showed no impact on the properties of the cotton fibres [24]. The desired water-repellent and oil-repellent properties can be obtained by spraying an aqueous dispersion on the silk fabric. Silica nanoparticles, silane quaternary ammonium salt, organic fluoropolymer and alkoxy silanes were the constituents of the dispersion. The maximum contact angle of water and of oil drop was more than 150°. Additionally, antimicrobial property was obtained. The vapour permeability of the fabric was reduced slightly due to the coating on the fabric. However, the coating, in triplicates, sustained even after abrasion testing, as per ASTM D3884-09 [25].

Borah et al. [17] imparted water-repellent property on eri silk fabric by applying nano-silica. In a study, rice husk was used as the main source for the extraction of nano-silica. The test methods used for the evaluation of water repellence efficiency were spray test, water absorbency, and water contact angle measurement. The water repellence property was enhanced by applying silicone-based polymer to it, which resulted in the creation of hydrophobic-to-superhydrophobic surface characteristics. The bending length of the treated fabrics increased with the increase in the concentration of nano-silica and polymer, probably due to the crosslinking taking place between the two. This, in turn, affected the handling properties of the fabric. Additionally, the brightness and whiteness were slightly increased due to the incorporation of nano-silica onto the eri silk fabric. However, the tensile strength was slightly reduced with the increase in the concentration of nano-silica [17].

Hydrophobic and oleophobic finishing on a textile substrate can be imparted by modifying the textile surface such that the contact angle between the water/oil and the textile surface is increased. The modification is achieved by increasing the micro roughness of the textile surface with the help of chemicals or nanoparticle finishing. The nanofinish can lead to superhydrophobic and superoleophobic features on the textile substrates. In some research studies, superhydrophobic properties were introduced along with other additional functionalities. In the case of nanocoating, it is necessary to keep the properties of the fabrics intact along with their enhanced functionalities, as there may be reduction in vapour permeability as well as mechanical and chemical stability. Both the hydrophobicity and oleophobicity can be introduced in a textile substrate via a single step – a simple and easy finishing route. Nanoparticle finishing shows enhanced performance than traditional finishing without hampering the original properties of the textile materials.

2.4 Self-cleaning textiles: lotus effect and the use of semiconductors

Self-cleaning in textiles is inspired by nature. There are several plants and animal species that possess natural self-cleaning properties. The natural self-cleaning property is a result of certain characteristic features present on their body surfaces. One of the well-known plants showing self-cleaning property is the lotus leaf. The self-cleaning property imparted to textiles that biomimicks the lotus leaf is called the "lotus effect". This effect is a result of the hierarchical structures (Figure 2.3) in combination with a waxy coating present on the leaf surface. The presence of a waxy coating makes the surface hydrophobic. The micro-scale mounds and the nanoscale roughness make the surface self-cleaning by bead-up and roll-off mechanisms of the dirt particles (Figure 2.4). It was, however, observed that heat treating these surfaces at a temperature of 150 °C adversely affects the nanoscale roughness, thus decreasing the contact angle of the surfaces. Annealing at a temperature of 150 °C basically led to the melting of the nanoscale hair-like structures, without affecting the wax composition. This led to a reduction in the contact angle, hampering the self-cleaning property, thus emphasizing the role of the nanostructures in the self-cleaning activity [26].

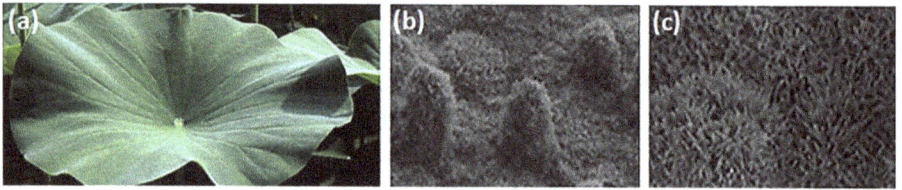

Figure 2.3: (a) Lotus leaf, (b) microstructures on lotus leaf, and (c) structures on the lotus leaf observed at the nanoscale (reproduced with kind permission from [27], Copyright 2016, Elsevier).

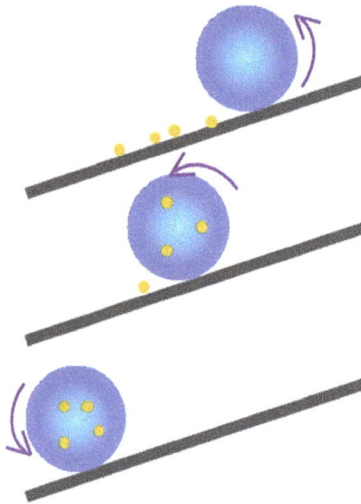

Figure 2.4: Roll-off mechanism of the self-cleaning effect (reproduced with kind permission from [27], Copyright 2016, Elsevier).

A few characteristics were thus thought to be essential for a surface to have self-cleaning property. These are a high contact angle (superhydrophobicity), multi-level roughness, and a very low tilt angle. Keeping these points in mind, self-cleaning textiles have been developed through various approaches. Sometimes, ridges and grooves have been created at the nanoscale using plasma, which helped to increase the contact angle, leading to superhydrophobicity. Nanocoatings have been employed to impart nanoscale roughness at different hierarchical levels. Polymer coatings that lower the surface energy of the substrate are also used to impart water-repellent effects.

In a study by Liu et al. [28], a hydrophilic cotton fabric was successfully turned hydrophobic by mimicking the lotus leaf hierarchy. Carbon nanotubes, both pristine as well as a surface modified with poly(butylacrylate), were deposited on the cotton fabric, which resulted in hierarchical structures, as observed on the lotus leaf. The artificially introduced structures have been found to impart hydrophobicity to the substrate and can be useful as self-cleaning textiles [28]. Silver nanoparticle can be used to produce a highly water-repellent coating, which also provides dirt resistance. Self-cleaning property is a more prominent property in the case of silver nanoparticle-adhered fabric, compared to conventional fabrics [29]. Hydrophobicity and self-cleaning characteristics were imparted to polyester-based textiles by a two-step process of chemical etching, followed by coating. Chemical etching of fibres promoted the Cassie–Baxter state, while a coating with poly(dimethylsiloxane) reduced the surface energy. The functionalities thus imparted were durable to abrasion, laundering, UV radiation, and diverse pH conditions [30]. In another study, modified zirconia was synthesized via the sol–gel technique, which was further coated onto the cotton fabric to impart superhydrophobicity. The result was a covalent interaction between the coating and the cotton substrate. The coated fabrics can possess a high water contact angle and low hysteresis. These properties promote the self-cleaning property on to the textile as well. Moreover, these fabrics demonstrated good durability to laundering at different pH conditions [31]. Self-cleaning, based on the lotus effect, was mimicked on a cotton fabric through a coating of fluoroalkyl-functional siloxane, pre-treated with low-pressure water vapour plasma. The plasma pre-treatment helped to enhance the surface polarity, increasing the surface area as well as roughness. The coating along with the precursor helped to lower the surface energy. The plasma pre-treated fabric showed better results compared to fabrics that are not pre-treated with plasma. The pre-treatment helped to form a network of fluoroalkyl-functional siloxane and imparted a specific hierarchy that improved their self-cleaning ability [32].

Apart from bio mimicking the self-cleaning structures on textiles, a second approach – semiconductor photo catalysis-based self-cleaning is widely utilized. A semiconductor, on being irradiated with light of appropriate wavelength, results in the excitation of the electron from its valence band to the conduction band. This creates electrons and holes in the conduction and the valence band, respectively. In the presence of air, these electrons produced oxygen radicals and the holes produced hydroxyl radicals that promoted the decomposition of contaminants by the reduction and oxidation mechanisms, respectively (Figure 2.5) [29, 30].

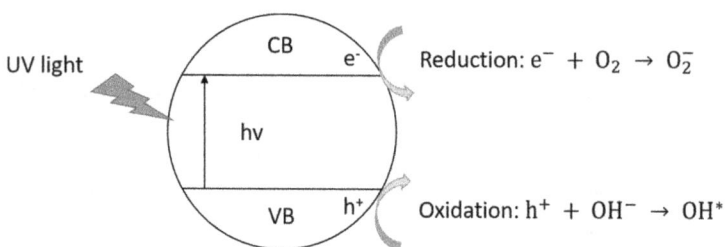

Figure 2.5: Mechanism of self-cleaning using semiconductors [35].

The widely used semiconductors for this purpose include titanium dioxide (TiO_2) and zinc oxide (ZnO) in their nanoform due to the favourable band of 3.37 and 3.2 eV, respectively. Nanocrystalline particles of TiO_2 and ZnO are synthesized and coated onto cotton substrates. The structural analysis of the synthesized TiO_2 nanoparticles revealed that they existed in their anatase form, which is reported for photocatalytic effect. It was observed that the crystallite size significantly affects the self-cleaning activity. The TiO_2 particles of smaller dimensions, with monocrystalline morphology, exhibit superior self-cleaning ability, compared to the bigger polycrystalline particles. A similar trend was also observed in the case of the ZnO nanoparticles. With reduction in size, finer dispersion was produced that could be applied uniformly on the substrate, resulting in better self-cleaning properties. The binder used for the coating of the textiles also affects the self-cleaning performance. A relatively low binder concentration is preferred, as the agglomeration of the nanoparticles is encouraged with higher binder concentrations, leading to low self-cleaning activity [36]. TiO_2 nanocoatings had shown adverse effect on the handling properties of textiles. Better handling properties were observed with a composite nanocoating of TiO_2–SiO_2. A more fuller, softer, and flexible fabric was obtained with the composite TiO_2–SiO_2 coating, compared to the nano-coating of TiO_2-SiO_2 [37]. Cotton fabrics, with superhydrophobic character combined with photocatalytic self-cleaning activity, were prepared by a coating of anatase TiO_2, combined with meso-tetra(4-carboxyphenyl)porphyrin, modified further with trimethoxy(octadecyl)silane. The coated fabrics showed sufficiently high water contact angles with superhydrophobic characteristics. The photocatalytic activity of the treated samples was quantified based on the change in the concentration of the methylene blue absorbed samples when exposed to visible light. These fabrics have shown sufficient photocatalytic activity under visible light, as was observed in the case of photodegradation of methylene blue [38]. In another study, composite TiO_2 coatings were synthesized in order to improve their photocatalytic self-cleaning performance. A nanocomposite of $AgI/AgCl/TiO_2$, with photocatalytic, self-cleaning, and antimicrobial activity, was prepared via the ion exchange method. Polyester fabrics pre-treated with oxygen plasma, when immobilized with $AgI/AgCl/TiO_2$ nanocomposite, showed better visible light absorption. The enhanced self-cleaning and antimicrobial properties of these fabrics were attributed to the inclusion of AgI and AgCl in the composite [39]. Cotton textiles dip-coated with anatase TiO_2 sol and poly(dimethylsilox-ane) had a high water contact angle and showed superhydrophobic feature. These textiles also show the ability to self-clean by the photocatalytic mechanism. These effects were stable under diverse pH conditions as also long-term UV exposure. This simple method of using a fluorine-free compound for self-cleaning could thus be extended for practical applications [40]. In a nanocomposite of reduced graphene oxide and zinc oxide, the quantity of zinc precursor influenced the conversion of graphene oxide to reduced graphene oxide. Cotton fabrics, when coated with this nanocomposite, showed improved efficiency in the degradation of organic stains and dyes. The coated cotton fabrics effectively degraded a complex dye stain of rhodamine B. This nanocomposite had the advantage of utilizing the abundantly available sunlight for its photocatalytic activity [41]. Cotton fabrics were coated with a combination of anatase TiO_2 and meso-tetra(4-carboxyphenyl)porphyrinato copper(II) (CuTCPP), and superior self-cleaning abilities could be exhibited in the case of methylene blue, coffee, and wine stains under visible light irradiation. CuTCPP showed greater photo stability, compared to TCPP; thus CuTCPP/TiO_2-coated cotton fabrics showed better performance compared to TCPP/TiO_2-coated fabrics [42]. The TiO_2-coated textiles often face an issue of wash durability. Therefore, SiO_2 can be used in conjunction with it to resolve this problem. Nano-titania was deposited on one-half of silica particles and applied on a cotton fabric under neutral pH conditions. These structures attach to the cotton substrate through their SiO_2 surface, while the photocatalytic TiO_2 layer points outwards. The unique morphology of these structures prevents a direct contact between TiO_2 and a substrate, explaining the superior performance of fabrics coated with TiO_2. This type of technology shows better performance at lower concentration under neutral pH, without any deterioration in properties even after long term UV exposure [43].

2.5 Flame-retardant textiles

Flame-retardant properties are the key requirement of fabrics as they can protect the wearer and the surroundings from fire hazards. Several chemicals and treatments are used to obtain flame-retardant properties in fabrics. Fire-retardant finishes have been developed actively with the passage of time. In the initial stage of development, certain chemicals were used in fabric for flame-retardant finishes, but they are hazardous for health as well as for environment. Halogen-based flame retardants are very popular as commercial flame-retardant agents. However, they cause environmental toxicity with risk to human health. As a halogen-free flame retardant, a flame-retardant agent containing nitrogen, phosphorus, boron, and so on is generally imparted to textiles. Phosphorus-based flame retardants are most commonly available in the commercial market. They possess significant flame-retardant properties, with good char formation ability. Thus, flame-retardant finishes have been explored in order to improve performance with fewer hazards. This has led to the promotion of using bio macromolecules in textiles [44].

The chemical and physical properties, and thermal stability of textiles can be improved by nanoparticles. In recent years, nanoparticles are most promising alternatives to impart functional finishes to textiles, instead of the traditional flame-retardant compounds.

In the textile industry, the blend that is mostly used is the polyester/cotton blend. Polyester melts but cotton fibre does not melt, causing a scaffolding effect in the polyester/cotton blend. Due to this reason, flammability of polyester/cotton blend increases. Researchers observed that the flammability of polyester/cotton blend was reduced by a layer-by-layer assembly of a nanocoating of nitrogen, phosphorus, and silicon. Self-extinguishing ability and a slight delay in ignition were achieved by a 15 bilayer nanocoating on polyester/cotton blend, during a vertical flammability test. In a cone calorimetry test, heat release was highly decreased. Both the condensed phase and vapour phase flame-retardant activities have been observed by this nanocoating [45]. Bleached jute fabric was prepared as a flame-retardant fabric by applying a finish of nano-zinc oxide (ZnO). A co-precipitation method was followed to prepare nano-zinc oxide from aqueous zinc acetate with NaOH. After calcination, it was applied to a jute fabric by the pad-dry-method using potassium methyl siliconate (PMS) as a binder and dispersing medium. Good fire resistance property was achieved when 0.01% ZnO was applied on the jute fabric. Nano-ZnO caused higher char formation, which reduced the production of flammable gases, leading to good fire-retardant properties. The LOI value of the treated jute fabric is 35, with a lower char length of 1 cm. Flame-retardant performance was satisfactory for five cycles of wash. The char production of the treated fabric was more when compared to an untreated fabric where the higher char residue is 5% at 400–500° C. It was observed that 60% of the nano-zinc oxide particles have a size of 30–500 nm. Nano-ZnO was deposited on the surface using a binder, that is PMS. The treated fabric had a lower tenacity and a lower whiteness index than an untreated fabric, whereas stiffness was found to increase with the use of nano-ZnO on the jute fabric [46]. The fire-retardant properties of lightweight wool fabric were enhanced by treating it with nano-kaolinite. The LOI value of the treated wool fabric was 25–33. Nano-kaolinite was applied on a wool fabric by two methods, that is exhaust and pad batch method, and both processes were effective. Applying 2.5% nano-kaolinite to fabrics by the pad batch method showed a decreased char length and burning time, which were found to be 3.3 cm and 8 s, respectively. However, the LOI value was slightly higher by the exhaust method due to a better pick-up by the substrates. The treated wool fabric showed a 5% higher char residue than the untreated sample [47].

The use of nano-magnesium hydroxide (nano-$Mg(OH)_2$) as a flame-retardant material has been studied on silk material. It was found to have good durability. Nano-Mg $(OH)_2$ was prepared by the precipitation method. The average diameter of the particles was 68 nm. This nano-$Mg(OH)_2$ was applied on silk fabric with the help of a cross-linker, that is polycarboxylic acid. The method used for the application was the surface coating technique. The coated silk fabric showed a high flame-retardant efficiency, with an LOI value of more than 28.4%. It also had ability to self-extinguish, and showed the effect even after 15 washes. The coated silk fabric exhibited excellent charring ability as it obtained continuous and dense char residue. This charring effect was the reason for the self-extinguishing ability of the coated silk fabric [48]. The silk fabric was treated with titanium oxide nanoparticles and phytic acid, where BTCA (1,2,3,4-butane tetracarboxylic acid) was used to increase the adhesion of the TiO_2 nanoparticles on the silk fabric. The combination of phytic acid and TiO_2 nanoparticles exhibits significant flame retarding effect. Additionally, phytic acid acts as a catalyst of crosslinking. All three combinations of materials on silk fabric show excellent flame-retardant properties even after 25 washing cycles. Figure 2.6 shows the results of the vertical burning test for all types of treated and untreated silk fabrics. However, there are some disadvantages with this treatment, that is, the whiteness and softness of the silk fabric are decreased slightly [49].

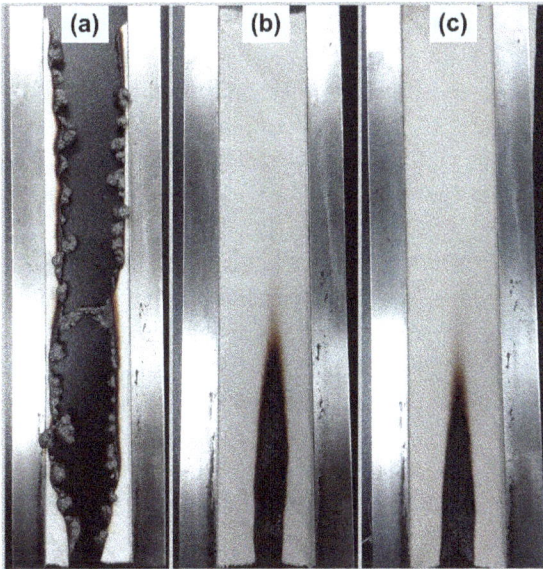

Figure 2.6: Results of the vertical burning test of (a) untreated silk fabric, (b)PA/BTCA-treated silk fabric, and (c) PA/TiO$_2$/BTCA-treated silk fabric (reproduced with kind permission from [49], Copyright 2017, Elsevier).

Application of most nanoparticles on textile enhanced its thermal stability and its flame-retardant properties by the formation of char or non-flammable gases and scavenging free radicals [50]. Although sustainability of the flame retarding properties of nanoparticles even after several washes is of key concern, in some studies, it has been discussed that durability of the flame retarding ability can be maintained even after a few washes.

In addition, the hazardous by-products because of flame retardants on textile must be controlled; else, they could result in health and environmental issues.

2.6 UV-protective textiles

Nowadays, mankind is more aware about the negative impacts of ultraviolet radiations. The rise in global warming results in change in weather, which leads to a high risk of exposure to ultraviolet radiation. This causes modification in the daily lifestyle of human beings causing increased demand of UV protective cosmetics and UV protective clothing. The main source of UV radiation is the Sun, and the UV radiations are classified as UV-A (320–400 nm), UV-B (280–320 nm), and UV-C (100–280 nm). UV-C radiation is totally absorbed by the ozone layer; so human beings are not exposed to it. UV-B radiation is partially absorbed by the ozone layer and the UV-B radiation that reaches the Earth's surface leads to increase in skin diseases, skin cancer, skin burn, and other types of diseases. While, a very small portion of UV-A radiation is absorbed by the ozone layer and major portion of this radiation reaches the Earth's surface. [51]. UV-resistant textiles are very much essential for protection from the harmful UV radiation. Inorganic, organic, and natural UV absorbers are used to modify textiles and make them UV protective. Nowadays, UV shielding nanoparticles are used to achieve UV-resistant property in textiles due to their better performance in comparison to conventional organic UV additives.

In a study reported by Gupta et al., the cotton fabric was dyed with madder along with SiO_2 and TiO_2 nanoparticles. The madder contains several types of phytochemicals that were also responsible for their functional properties. The nanoparticles of TiO_2 and SiO_2 have an average size of less than 200 nm. The colour strength and the fastness properties were enhanced with the addition of nanoparticles. The UPF was higher in the case of nanoparticle-madder-dyed fabric and the control dyed fabric showed higher UPF compared to the bleached fabric. Table 2.1 shows the UPF value of the dyed fabric and the nanoparticle-madder-dyed fabric. When the performance of TiO_2 and SiO_2 nanoparticles were compared, it was observed that TiO_2 nanoparticle-madder-dyed fabric showed better performances such as higher UV resistance, good fastness, and high colour yield, in comparison to the SiO_2 nanoparticle-madder-dyed fabric [52]. In another study, nanocrystalline titanium dioxide (TiO_2) was sono-synthesized and fabricated by the low temperature sol–gel method. Tetrabutyl titanate was used as the precursor for the synthesis of nano-TiO_2. Nano-TiO_2 was synthesized at a low temperature through ultrasonic cavitation, which was simultaneously deposited on the surface of the cotton fibres. The synthesized material improved the ultraviolet resistance property of the cotton fibres significantly. The entrapment of the nanomaterial inside the lumen of the cotton and the mesopores present on the surface offered good protection from UV radiation [53].

Table 2.1: UPF results of the different
nanoparticle treated and dyed fabrics [52].

Types of samples	UPF results
0.1% TiO$_2$ only	35
1.0% SiO$_2$ only	38
Madder dyed	50+
Dye + 0.1% TiO$_2$	50+
Dye + 0.25% TiO$_2$	50+
Dye + 0.5% TiO$_2$	50+
Dye + 1.0% TiO$_2$	50+
Dye + 0.1% SiO$_2$	50+
Dye + 0.25% SiO$_2$	50+
Dye + 0.5% SiO$_2$	50+
Dye + 1.0% SiO$_2$	50+

The surface of the cotton fabric was modified by adhering synthesized ZnO nanoparticles on it. The ZnO nanoparticles were synthesized with two types of reaction media, that is, 1,2-ethanediol and water. The result showed that the obtained UPF value for this modified fabric was very high. The ability to block UV radiation was excellent as indicated by the high UPF value of 320. In addition, it also showed antibacterial property against *E. coli* and *S. aureus* bacteria [54]. In another study by Teli et al., the nylon-knitted fabric was modified by applying synthesized silica nanoparticles via the pad–dry–cure method. Further, zinc oxide was in situ deposited on the silica nanoparticle-coated fabric. This treated fabric could show an excellent ultraviolet resistance property, with a UPF value of 280. This silica nanoparticle and the ZnO-coated fabric also showed hydrophobic property due to the addition of sodium stearate in the process. The fabric showed good durability both in UV protection and hydrophobicity for up to 10 washes [55]. The cotton fabric was oxidized firstly by treating with periodate and then with 4-aminobenzoic acid ligand (PABA). The peroxidization of the cotton fabric resulted in better active site formations in the cotton fabric, which was good for the treatment with PABA. The ZnO nanoparticles grew on the respective appropriate sites created by the PABA treatment. Furthermore, this treated cotton fabric showed significant ultraviolet protection property along with antibacterial property. These properties sustained significantly even after 100 abrasion cycles and 20 cycles of washing [56].

Studies have also reported on the use of a combination of inorganic-organic compound for the modification of the fabric. For example, zinc oxide nanoparticles and cellulose nanofibrils were used for the modification of the cotton fabric. The agglomeration of zinc oxide nanoparticles was reduced by utilizing a suspension of cellulose nanofibril, helping the ZnO nanoparticles to bind to the fibre surface. This developed cotton fabric exhibited excellent UV-resistant properties. The UV-resistant property was so significant that even after 30 washes, the UPF value remained at 50 [57].

Biomolecules like lignin, chitosan, quercetin, and rutin are excellent ultraviolet-resistant agents and when fabrics are treated with these biomolecules, they impart significant UV resistance capabilities. Nanolignin finish treated on cotton and linen fabrics show good UV absorbing properties [58]. Nano-chitosan also offers ultraviolet protection property to finished polyester cotton blend fabrics [59].

Some research studies highlight the use of both natural and inorganic UV-resistant agents, in combination. This type of combination improves the properties along with a reduction in toxicity. During pre-treatment, the extract of Moringa oleifera leaf was applied to bleached cotton fabric so that fabric can act as both UV-resistant and antibacterial. This pre-treated fabric was treated with different nanoparticles separately, that is, nano-titanium dioxide and nano-zinc oxide. A non-formaldehyde crosslinking agent, BTCA (1,2,3,4-butane tetracarboxlyic acid), was used in this treatment. The whole process was carried out by the pad–dry–cure method. The results showed that the application of titanium dioxide nanoparticles to fabrics pre-treated with moringa oleifera have the best UV resistance properties [60].

Many organic, inorganic, and natural UV absorbers have been explored to get efficient UV-resistant clothing. Most of the commercially used UV absorbers are organic in nature and some of these UV absorbers are toxic and not stable. So, research works are in progress to develop UV shielding additives as well as UV-resistant fabrics having less toxicity, more stability, and higher durability. The UV-resistant nanofinish results in a higher efficiency in textiles when compared to traditional UV finishing.

2.7 Wrinkle-resistant textiles

In the traditional process, wrinkle resistance is achieved by adding resin to the textiles. However, several disadvantages arise due to the impregnation of resin in the fabric. The problems are reduction in dyeing capability, abrasion resistance, softness/hand feel, tear strength, and tensile strength [61]. In recent times, resins are replaced by nanoparticle finishing to address these disadvantages.

In a study reported by Ugur et al., the linen fabric was treated with BTCA (1,2,3,4-butanetetracarboxylic acid) in a padding process, where the crosslinking agent used was nano-polyurethane and the catalyst used was nano-Al_2O_3. The wrinkle-resistant property was enhanced after treating the fabric. Moreover, the treated fabric also exhibited flame-retardant property. The modified linen fabric retained properties even after five cycles of washing [62]. The traditional chemicals used for wrinkle-free textiles produce a hydrophobic functionality. However, there are studies that suggest hydrophilic wrinkle-free finishing of textiles. Silica nanoparticles are attached to the cotton fabric via plasma treatment. This leads to the addition of superhydrophilic nature to the cotton fabric along with wrinkle-free property, as shown in Figure 2.7. This treated fabric shows

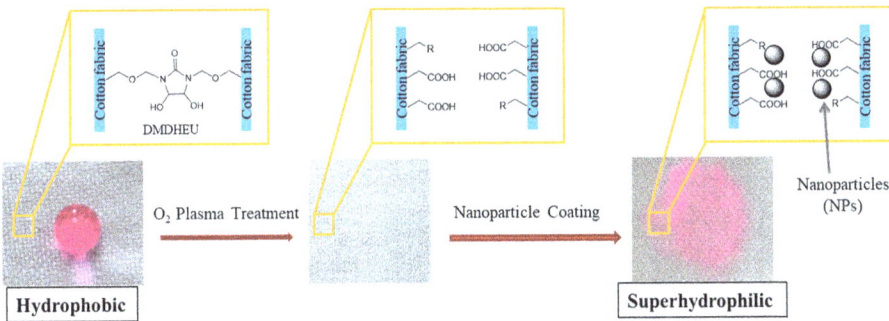

Figure 2.7: Schematic representation of the water contact angle after O_2 plasma treatment and nanoparticle coating (reproduced with kind permission from [63], Copyright 2017, American Chemical Society).

excellent wrinkle recovery property apart from long durability of the exhibiting properties even after 25,000 abrasion cycles and 50 laundering cycles [63].

Gao et al. used nano-silica to coat a silk fabric to increase wrinkle recovery. Additionally, there was an increase in hydrophobicity and ultraviolet resistance because of this treatment [64]. Cotton fabric was modified by treating it with copper nanoparticles where sodium hypophosphite (NaH_2PO_2) was used as a catalyst and maleic acid was used as crosslinking agent. The increase in the concentration of maleic acid results in an increase in the crease recovery property [65].

2.8 Antimicrobial and antiviral textiles

Textile materials, especially textiles made from natural fibres, are a huge source of food for microorganisms. They germinate on the surface of textiles and break down the chains of the fibres by enzyme biocatalysts and produce their food source. The bacterial and fungal growth on the surfaces of textile is the most favourable place, as textile provides food and moisture [66]. Anti-microbial agents in textiles are used to eliminate the microorganisms. Mostly, antibacterial agents penetrate the microorganisms and stop the metabolic pathway, inhibit growth, or damage the cell wall or DNA. These mechanisms of antibacterial agents are shown in Figure 2.8.

The performance of nano-antimicrobial agents on textile materials is excellent. So, maximum research is in progress on antimicrobial textiles with nanofinishes. Antimicrobial and antiviral nanofinishes are obtained by using inorganic, organic, natural, or mixed agents. The different types of nanomaterials that are explored are silver nanoparticle, TiO_2 nanoparticle, Fe_3O_4 nanoparticle, ZnO nanoparticle, cationic NPs of polystyrene sulphate that is covered by a bilayer of dioctadecyldimethylammonium

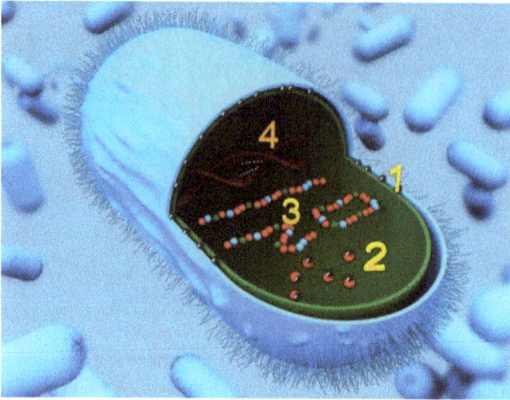

Figure 2.8: Silver as an antibacterial agent: (1) cell wall is perforated, (2) cell respiratory system is damaged, (3) metabolic pathway is hampered, and (4) DNA is damaged [67].

bromide, chitosan nanoparticle, graphene, and so on, whereas antimicrobial properties of many nanoparticles are yet to be explored [68].

Knitted and woven fabrics that are made of pure cotton were used for the application of ZnO nanoparticles to exhibit antibacterial activity. Double jersey fabric showed remarkable antibacterial activity compared to other knitted and woven fabrics. Both spin coater and pad–dry–cure methods were used for the application of ZnO nanoparticles on the fabric where the pad–dry–cure method showed good antibacterial property than the spin coater method. Additionally, the ZnO particles exhibit greater antibacterial property against Gram-positive bacteria, that is, *S. aureus* bacteria, than Gram-negative (*E. coli*) bacteria [69]. Nano-ZnO was synthesized on the cotton fabric where aloe Vera gel was used as a capping agent. Good antimicrobial properties were achieved against both the Gram-positive and Gram-negative bacteria, that is *S. aureus* and *E. coli*, respectively. Additionally, the treated fabric had the ability to block UV radiation [70]. Antibacterial activity and crease resistance were both features that were obtained in a single process. In the recipe, modified glyoxylic resin containing a catalyst was used for crease resistance while nano-aluminium oxide was used to obtain antibacterial property. The addition of nano-aluminium oxide in the recipe showed a significant effect on crease recovery. At a concentration of 10^3 ppm of nano-aluminium oxide, it was only resistant against *S. aureus* bacteria. However, it was resistant to both the *S. aureus* and *E. coli* bacteria at a higher concentration (5×10^4 ppm) of nano-aluminium oxide in the recipe. In addition, the recipe showed a durable effect even after 5 washing cycles [71]. The antimicrobial activity was also excellent for cotton coated with copper oxide nanoparticles, starched cotton coated with copper oxide nanoparticles, CuO-Ag nanocomposites, and cotton coated with Cu(II) curcumin complex. The antimicrobial performance improved by 50% for *E. coli* and by 23% for *S. aureus* [72]. Cotton fabric was modified by treating it with a solution of oxalic acid. The concentration of oxalic acid was responsible for the generation of free carboxyl group mentioned in Table 2.2 and the copper ions from copper (II) sulphate solution binds with the free carboxy group. To form copper-based nanoparticles, sodium borohydride was reduced with copper. This modified cotton textile acted against Gram-positive and Gram-negative bacteria. Antimicrobial property of the co fabric treated with copper nanoparticle is mentioned in Table 2.3 [73].

Table 2.2: Content of free carboxyl group in Co fabrics modified with different concentrations of oxalic acid [73].

Sample	Oxalic acid solutions of different concentrations (w/v %)	Content of free carboxyl groups (μmol/g)
Co + OX4	4	101 ± 7
Co + OX6	6	173 ± 6
Co + OX10	10	218 ± 49

Table 2.3: Microbial-resistant property of the Co fabrics treated with Cu-based nanoparticles [73].

Types of microorganisms	Number of microbial colonies (CFU)					
	Standards	Inoculum	Co + OX10 + Cu	Co + OX6 + Cu	Co + OX4 + Cu	Control Co
E. coli	ATCC 25922	6.7×10^5	<10	<10	<10	6.0×10^5
E. coli	NCTC 13846	1.1×10^8	<10			6.1×10^5
E. coli	ATCC BAA-2469	6.1×10^7	<10			6.0×10^5
S. aureus	ATCC 25923	1.1×10^5	<10	<10	<10	5.4×10^4
S. aureus	ATCC 43300	6.5×10^7	<10			9.8×10^4
P. aeruginosa	ATCC 27853	3.5×10^7	<10			3.2×10^5
C. albicans	ATCC 24433	4.0×10^5	<10			2.0×10^5
K. pneumoniae	ATCC BAA-2146	8.8×10^7	<10			4.8×10^5

Silver nanoparticles have excellent antimicrobial properties. Antimicrobial cotton fabric with good durability was achieved after functionalization with silver nanoparticles and L-cysteine. Antibacterial property was achieved on a cotton fabric coated with silver nanoparticles and the property was retained significantly even after 90 laundering tests [74]. Silver nanoparticles can also be deposited on the surface of cotton-knitted fabrics. Firstly, caustic soda was used to treat the cotton fabrics so that the hydroxyl groups could be activated, and then silver nitrate salt was directly reduced using ascorbic acid to introduce nano-silver particles on the surface of fabrics. Significant antibacterial properties on cotton fabrics were achieved that inhibit the growth of both the Gram-positive and Gram-negative bacteria. However, strength loss was observed in the treated fabric, as NaOH was used in the pre-treatment [75].

The cotton and polyester/cotton blend fabrics were dyed with the extract of onion peel-based nano-emulsion. Antibacterial activity of the nano-emulsion-treated fabric had been tested against Gram-positive and Gram-negative bacteria. The best antibacterial activity obtained in pure cotton with 10% onion peel-based nano-emulsion was 75% against *S. aureus* and 57% against *E. coli* bacteria. The antibacterial property remains significant (38% against *S. aureus* and 31% against *E. coli*) even after home laundry washes of 25 cycles. Additionally, good to very good rating of washing and rubbing fastness was observed for all dyed samples. But, all the dyed samples contained colour stains [76].

The presence of copper nanopowder during the dyeing of the linen fabric by cumin seed extract showed effective antibacterial activity against *E. coli* and *S. aureus* bacteria, while in the absence of copper nanopowder, all the dyed samples exhibited antibacterial activity only against *S. aureus* [77]. In another study, silver nanoparticles had been synthesized by Hybanthusenneaspermus plant extract and tested against *E. coli*, *S. aureus*, *Klebsiella pneumoniae*, *Serratia marcescens*, and *P. aeruginosa*. The silver nanoparticle-treated cotton exhibited significant antibacterial activity against *E. coli* and *S. aureus* even after 40 laundering wash cycles [78].

Viral diseases are infectious and dangerous to human health. So, the development of antiviral textiles is as important as antibacterial fabrics. Sodium pentraborate pentahydrate and triclosan are used to impart antiviral and antibacterial properties to cotton fabrics. Good antiviral results have been obtained from the treated fabric, which acts against the poliovirus type 1 and adenovirus type 5. The antiviral efficacy was manipulated on the basis of viral cell deaths. An optimized combination reduced the viral counts by 60% [79].

The types and the number of microorganisms are growing rapidly, which trigger various kinds of harmful contagious diseases. So, antimicrobial textiles are becoming popular and are commercialized in the market even for daily wear. Several antimicrobial agents lead to toxicity that is hazardous for human health and the environment. So, many studies are in progress to achieve significant antimicrobial properties with less toxicity. The focus is more on the natural antimicrobial agents in the recent years and studies are ongoing to achieve excellent antimicrobial properties with long-term durability.

2.9 Anti-odour and fragrance-finished textiles

Fragrance and aroma finishing of textiles enhance the aesthetic value of the product. They add a feeling of freshness and provide psychological comfort to the wearer. Most of the ingredients used to impart these finishes are volatile. It is therefore a challenging task to impart aroma finish with long standing properties. Techniques like encapsulation have been thought as a good pathway for imparting such finishes. HU et al. incorporated rose fragrance nanocapsules in cotton fabrics. A successful incorporation of theses nanocapsules have been confirmed by various structural analysis techniques. These nanocapsules work on the sustained release mechanism, providing a long-lasting aroma finish to the textiles. The smaller-sized nanocapsules render better sustained release property and also wash durability [80]. In another study, in order to curb the volatilization of the fragrant component, rose fragrance has been encapsulated in biocompatible chitosan through ionic gelification. These have been applied on cotton substrates without the use of any adhesive agents. The fabrics thus showed a sustained release property and wash durability. The use of biodegradable chitosan adds an eco-friendly approach to the technique [81]. Citronella oil is also extensively used in perfumery products. It also has an insect-repellent property. Microencapsulation of citronella oil had also been carried out. It has been found that these microcapsules follow the Fickian diffusion and Avrami equation kinetics [82]. Tourmaline nanomaterial is one of the nanofinishing agents that can be used in textiles as an antiodour finishing agent. The properties of tourmaline are that it has the ability to reduce 90% odour, 99.9% bacteria, and 75% sticky moisture [79, 80].

2.10 Antistatic textiles

Textiles are dielectric materials. They lack the ability to conduct electrons and hence they accumulate these charges, leading to a static charge build-up. When two dissimilar materials come in contact with each other, the flow of electrons commences [85]. On separation, the conducting surfaces equalize these charges by transfer of electrons. The electrically insulated materials, on the other hand, accumulate these charges and cause a static charge build up. Textiles are thus prone to accumulate charges at different stages of processing [86]. This charge accumulation often leads to problems in the subsequent stages of processing as well as during their use. Ballooning and stretching of fibres, attraction of dirt and lint to fabric surface, problems during folding of dyed fabrics, and so on are some of the major issues due to static charge accumulation [87]. The main concept behind antistatic finish is to reduce the charge build-up by easing conduction of charges or by reduction in friction.

One such approach utilizes the sol–gel-based coating consisting of both hydrophobic and hydrophilic components. The hydrophobic component comprises modified alkoxysilanes, while the hydrophilic part comprises the amino-functionalized alkoxysilanes (Figure 2.9). Such a coating provides a hydrophobic solid/air interface while the bulk remains hydrophilic. This results in moisture absorption in the bulk, leading to dissipation of accumulated charges, if any, and imparting an antistatic effect. Textiles with such a coating have an antistatic property along with water repellence capability [88].

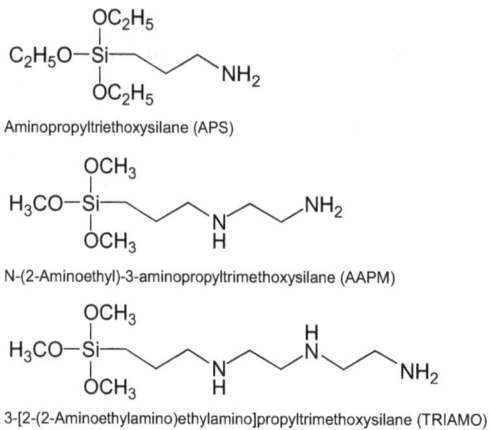

Aminopropyltriethoxysilane (APS)

N-(2-Aminoethyl)-3-aminopropyltrimethoxysilane (AAPM)

3-[2-(2-Aminoethylamino)ethylamino]propyltrimethoxysilane (TRIAMO)

Figure 2.9: Chemical structures of the additives added for antistatic finish (Modified according to [88], Copyright 2010, Elsevier).

In order to further investigate the coexistence of water repellence and antistatic properties, a polyester fabric is finished with nano-silver as an antistatic finish and fluorine-based water repellence finish. These finishes have been applied individually and also in combination to study their compatibility. The application of an antistatic finish followed by a water-repellent finish provided the best performance of both antistatic property and hydrophobicity [89]. Inorganic semiconductor materials like zinc oxide have also been explored as an effective conductive species in polymer composites. A tetrapod-shaped ZnO whisker incorporated into polymer composites help in the dissipation of the accumulated charges. A critical volume fraction of this whisker gives the best antistatic performance. The proposed formula for calculation of the critical volume fraction is affected by parameters such as aspect ratio and the extent of overlap. This critical volume is governed by factors such as aspect ratio, extent of overlap, and the concentration of the charges at the tip of the whiskers, along with the tunnel effect that aid in reducing the electrical resistivity of the polymer system [90]. In another study, polyacrylonitrile fibres were modified with nano-antimony-doped tin oxide particles to impart antistatic properties. These particles were dispersed in water using polyethyleneimine as a dispersant. On passing the as-spun fibres through this suspension, the nanoparticles diffused into the fibres. These particles then formed a conduction path during the drawing and drying process, imparting antistatic property to the polyacrylonitrile fibres. The modified fibres also exhibited good mechanical properties in addition to the imparted functionality [91].

2.11 Superabsorbent finished textiles

Superabsorbent finish for textiles is crucial for end products such as hygiene towels, wipes, and textile reservoirs. These finishes comprise hydrogel-like materials that have the capacity to absorb water and swell up to 100 times their original volume. Generally, superabsorbent finishes show increased affinity toward water absorption. However, materials having absorption affinity towards other organic fluids and gases have also been developed. Superabsorbent materials used for textile finishes can be classified as hydrogels and inorganic materials. Hydrogels form a network-like structure that can absorb huge amount of water without dissolving in it. Their absorption or desorption may be a stimulus-driven response. These stimuli can be pH, temperature, light, and so on [88, 89]. There are also composites of various polymers that have superabsorbent properties. These are increasingly being preferred nowadays because of the added advantages of mechanical properties they offer. They are also biodegradable, based on their constituents, and also cost effective. Inorganic microparticles dispersed in a matrix or grafted on fibres also impart superabsorbent character. The hydrophilic nature of these inorganic molecules helps to improve the wettability of the surface a hundred times over. Such an approach is adopted for polyester fabrics to improve their water absorbency [94]. Based on the chemical composition, the first superabsorbent materials were acrylate-based. However, the search for alternative eco-friendly polymers for superabsorbency brought polysaccharides and other bio-based polymers into the picture [95].

The superabsorbent finish is basically employed to achieve a high fluid absorption, retention, and a controlled rate of absorption. These are highly influenced by the swelling capacity and the gel structure of the superabsorbent. The gel structure depends on the type and extent of crosslinking. These crosslinks can, again, be physical or chemical in nature. The physical crosslinks are established by chain entanglements or electrostatic interactions while the chemical crosslinks are governed by covalent bonds. Physical crosslinks are observed in polyelectrolyte systems while the chemical crosslinking is evident in co-polymerized hydrogels. The mechanism of super absorbency is defined by the phenomenon of hydration, capillarity, diffusion, and swelling. On hydration, the absorbed water molecules are channelled through intra-yarn capillaries and they are then spread over the textile substrate. These water molecules interact with the functional groups of the superabsorbent material and form hydrogen bonds, resulting in the retention of fluid. The difference in the osmotic pressure existing inside and outside the finish matrix results in the diffusion and the subsequent swelling of the gel-structured absorbent. These series of changes may also be stimuli driven. Physical changes like topography of the textile substrate can also enhance the wettability of the substrate, making it superabsorbent. The absorption of fluid by the layer in immediate contact with the fluid leads to the swelling of the layer, blocking the liquid from penetrating to the layers underneath. This creates a pressure on the otherwise dry layers lying underneath and results in the breaking of the gel structure. In order to take care of this issue, superabsorbent matrices are made with varying degrees of crosslinking, from the surface to the core. This helps to ensure proper permeability and uniform swelling throughout the gel structure [95]. Polyester textile surfaces have been plasma modified with films containing nitrogen that changes the textile's characteristics from hydrophobic to hydrophilic. The ratio of the gases used in plasma and the power of plasma influences the nanoporosity of the films deposited on the surface. The use of ammonia influences the accessibility of the amine groups in the film. The coating has excellent durability, is mechanically robust and flexible, and does not affect the comfort properties of the fabric [96]. Initiated chemical vapour deposition (iCVD) technique has also been used to create polymer nanocoating for superhydrophilicity. Such a coating was developed from (2-hydroxyethyl)methacrylate and 1-vinyl-2-pyrrolidone (VP) using ethylene glycol diacrylate as cross linker. In this case, the crosslinking density was controlled by varying the partial pressure of the cross linker [97].

2.12 Multifunctional finished textiles

Nowadays, multifunctional textiles have high demand in the market. Traditionally, multistep treatments are followed and multi-chemicals are used to obtain a finish on textile. However, with the advancement of nanotechnology, multifunctional finishing in textile can be obtained with some special treatments and chemicals. The treatments and chemicals make the fabric more functional and economically valuable. Various natural and inorganic nanoparticles that have sustainable properties are used for achieving a multifunctional finish. Generally, multifunctional textiles include properties such as antimicrobial, UV properties, antioxidant, flame retarding ability, crease recovery, and antistatic.

Single finishing process is followed to achieve water-repellent and antibacterial cotton fabric. Cotton fabric is treated with a solution containing both fluorocarbon and nano-titanium silicon oxide. The treated cotton fabric showed good water repellence along with antibacterial activity against Gram-positive bacteria, that is *S. aureus*. However, there was a limitation in antibacterial activity against *E. coli* bacteria [98]. A combination of TiO_2 and ZnO on the cotton fabric results in a self-cleaning property, with increase in UPF and an increase in anti-bacterial properties. For synthesizing nanoparticles, a sol–gel method was used, and such particles were applied to fabric using a pad–dry–cure method.

The sequence of application of nanoparticle on the cotton fabric was ZnO and then TiO_2, and other sequence was TiO_2, followed by ZnO. When ZnO was applied first, followed by TiO_2, the antimicrobial properties and the durability are affected slightly. As a result, moderate antimicrobial activity and poor wash durability were obtained. However, antimicrobial properties are enhanced and even sustained even after 10 wash cycles when TiO_2 was initially applied, followed by the treatment with ZnO nanoparticles. The drop in air permeability was around 16–27% for the combined

ZnO- and TiO$_2$-coated fabrics [99]. Jhatial et al. prepared a durable bamboo fabric having multifunctional properties using the sol–gel coating method. The sol–gel method was followed to adhere silica nanoparticles and titania nanoparticles to the bamboo cellulose fabrics where citric acid was used as a crosslinking agent. The functionalization of bamboo-knitted fabric was carried out by the dip-pad-dry-pad-cure method. The treated fabric showed soil release, UV protection, washing fastness, and hydrophobicity. The properties on the coated bamboo fabric showed durability for up to five industrial washes [100].

ZnO @SiO$_2$ nanoparticles can be synthesized in situ on polyester fabric. The ZnO @SiO$_2$ nanoparticles enhanced many functional properties and fulfilled the limitations of ZnO nanoparticles. Antibacterial properties, photocatalytic stability, and thermo-catalytic stability were improved in the PET fabric where ZnO @SiO$_2$ nanoparticles were adhered. Additionally, tensile strength was enhanced in the case of in situ-synthesized ZnO @SiO$_2$ nanoparticle fabric, compared to ZnO nanoparticle fabric [101]. The cotton fibres were coated with silver nanoparticles by the pad–dry–cure method in a silver alkylcarbamate solution, followed by a reduction process under microwave irradiation. This treated cotton fibres exhibited UV protection, electrical conduction, antimicrobial, and photocatalytic self-cleaning properties. The silver nanoparticle-treated fabric showed a brownish shade. The AgPU (plasma untreated applied with silver alkylcarbamate) sample showed brown-yellow shade while AgPT (plasma treated applied with silver alkylcarbamate) substrate showed a brown shade. Additionally, improved colour fastness was observed in the treated fabrics. Hydrophobic character can also be observed with the introduction of hexadecyltrimethoxysilane [102].

According to the research, finished polyester/cotton blend fabric was prepared by applying nanochitosan-polyurethane dispersions (NCS-PUs), which was prepared by polymerization. NCS-PUs was applied on both printed and pre-dyed polyester/cotton blend fabric by the pad–dry–cure method. Tensile and tear strength properties, UV protection, and antibacterial properties were measured, and all these properties were improved for both types of samples [59]. In recent studies, nanolignin was prepared by the microbial hydrolysis process, and this prepared lignin nanoparticles were used to treat fabrics to get better performance or more functionality. Multifunctional properties were achieved by applying nanolignin on the surface of cotton and linen fabrics. In this study, nanolignin was synthesized from the bulk of lignin, which was extracted from stalks of cotton. Nanolignin was produced by hydrolysing the extracted bulk lignin in a controlled manner with the help of lignin-degrading fungal isolate, *Aspergillus oryzae*. The size of the synthesized nanolignin particle was less than 50 nm. The characteristics of the prepared nanolignin were compared with the nanolignin produced by two other processes. Ultrasonication processes and high shear homogenization were the other two methods. These two processes provided 62.60% and 79.50% nanolignin, respectively, and on the other side, 45.3% nanolignin was obtained from the microbial process. Pad–dry–cure method was used to apply nanolignin on the surface of the cotton and linen fabric using an acrylic binder. Multifunctional properties were studied after 5 wash and 10 wash cycles, and the properties of nanolignin finish was found to withstand even after 10 washing cycles without any significant loss in multifunctionality. Antibacterial, antioxidant, and UV-absorbing properties were the multifunctional properties obtained from this microbial nanolignin finish on cotton and linen fabric [58]. Bulk lignin was extracted from coconut fibres and nanolignin was produced by microbial hydrolysis process using *Aspergillus nidulans*. The developed nanolignins were applied on cotton and linen fabric by the pad–dry–cure method, using an acrylic binder. The size of the prepared nanoparticle was 27.5 nm ± 2.7 nm, having a yield of 58.4%. The treated fabrics showed excellent UV-resistant properties, antioxidant properties, and enhanced antibacterial properties against *S. aureus* and *K. pneumoniae* bacteria. It was observed that *S. aureus* was more susceptible to nanolignin, compared to *K. pneumoniae*. This might be due to the difference in their cell wall composition. The phenolic groups of lignin helped in UV absorption, which was enhanced in its nanoform due to the increase in surface area [103].

Multifunctional finish was obtained on linen fabric with chitosan and in-situ developed ZnO nanoparticles via the flower extract of *Bombax ceiba*, at three different pH levels of 7, 9, and 12. The antibacterial activity, antioxidant properties, and ultraviolet resistance properties of the treated fabric were evaluated. The antibacterial activity and the antioxidant properties of the treated fabric were satisfactory even after 20 washes. However, the UPF value decreased slightly in the case of 20-washed fabric compared to 0-washed fabric for pH 12. Among all samples, significant properties of the treated fabrics were achieved at a high pH, that is pH 12 [104]. Various functionalities were imparted on linen fabric by treating it with chitosan, citric acid, sodium hypophosphite, and CeO_2 nanoparticles. The properties that were obtained after the treatment are wrinkle resistance, UV resistance, flame retarding ability, and antibacterial properties. The film-forming ability of chitosan helped to improve the crease recovery of the treated fabrics. Chitosan and ceria nanoparticles synergistically contribute in curbing the growth of bacteria, providing antibacterial properties. The nitrogen-rich groups present in chitosan and the thermally stable ceria nanoparticles help to impart flame-retardant properties to the fabric. The extensive crosslinking resulted in the creation of unsaturated ester groups that help in the absorption of UV, to some extent. Excellent flame retarding ability, UV resistance, and antibacterial property was obtained even after 5 washes. However, after 5 washes, the crease recovery angle became 170°, which indicates moderate level of wrinkle resistance. Significant antibacterial properties were achieved against *S. aureus* and *E. coli* bacteria, but after the treatment, tensile strength of the linen fabric reduced [105]. In one research work, chitosan was coated on a cotton fabric and the coated fabric was treated with in-situ-synthesized silver nanoparticles using a natural reducing agent and a stabilizing agent, that is, extract of peanut waste shell. The final treated fabric exhibited antioxidant and antimicrobial properties along with attractive colour [106]. In another study, four types of cotton fabrics were impregnated in a solution of sodium hydroxide with or without a natural stabilizer, that is, grinded date seed waste extract. The pad–dry–cure technique was followed for the finishing of fabrics with zinc hydroxide solution, with or without a stabilizer, where ZnO nanoparticles were formed. The four types of fabrics were greige cotton fabric, bleached cotton fabric, mercerized cotton fabric, and mercerized bleached cotton fabric. In this study, it was found that the extracted date seed waste was an effective capping material and it was used as a stabilizing agent of ZnO nanoparticles. The mercerized bleached cotton fabrics showed better antibacterial properties and UV-resistant properties, compared to mercerized cotton fabric [107]. Another study reports preparation of multifunctional cotton fabric by incorporating silver nanoparticles. AgNP was in-situ-generated using biomolecule extraction of pomegranate peel. Extraction of pomegranate peel acts as a green reducing and capping agent. Antibacterial properties, antioxidant properties, good wash fastness, acceptable light fastness, and colouration are achieved by optimizing factors such as the concentration of $AgNO_3$ and the plant extract. With the increase in $AgNO_3$ concentration, there was increased reduction from Ag^+ to Ag. The increased amount of Ag on the fabric results in better antibacterial efficacy. The antioxidant activity observed on the treated samples was primarily due to the silver nanoparticles and the polyphenolic groups of pomegranate peels that helped to stabilize free radicals. The colony counting method was followed for the evaluation of the antimicrobial activity of silver nanoparticle-treated cotton fabric against *S. aureus* and *E. coli* bacteria. The antibacterial activity against *S. aureus* was around 98% while that against *E. coli* was 99% [108].

Various research studies regarding multifunctional textile have been explored in the recent years and many studies are in progress, as it falls under the category of one of the most highlighted areas. However, some other inherent properties get affected during the treatment, and processes are run to achieve certain multifunctional properties in textiles. So, the treatments and processes must be carried out in a controlled manner. The production of multifunctional textiles is very efficient as it reduces multi-steps into very few steps and multitreatments into few treatments, which leads to less usage of water and energy.

2.13 Future thrust and challenges

Nanotechnology has added a new frontier to textile finishing. These finishes help to overcome the drawbacks experienced in the conventional finishing techniques. They impart the desired functionality to textiles without affecting the aesthetic or comfort properties of the same. The nanofinishing techniques are in their infancy at present. They lack the potential to be upgraded to industrial level. The primary challenge behind its upscaling is the cost that is associated with these techniques. The equipment needed for the setup of such a technological approach is often expensive. The synthesis of the nanomaterials is a lengthy process that makes the overall nanofinishing technique time consuming. There are also concerns about the toxicity index of the nanomaterials. There are reports that suggest the toxicity of the nanomaterials and their suspected negative influence on the lives of humans and animals [109–111]. This is a hindrance in the path of wide acceptance of the nanomaterials for finishing and their subsequent upscaling to commercial and industrial level. Alternative safer approaches are being researched that would solve the problem of toxicity.

However, there are certain positive points of using these nanomaterials in finishing applications. Conventional approaches, on the other hand, require a reasonable time limit for application. The add-on needed in the conventional procedures of finishing is often on the higher side. This affects the handling and feel of the fabric. Nanofinishing takes care of this issue as the add-on required for such approaches is often low. The desired functionality can be imparted to textiles at a much lower add-on value compared to conventional techniques, without altering the porosity, breathability, rigidity, and the aesthetic appeal of the fabric. This low amount of material consumption is expected to take care of the cost effectiveness of the various processes involving such materials.

References

[1] Sawhney APS, Condon B, Singh K V., Pang SS, li G, Hui D. Modern Applications of Nanotechnology in Textiles. Text Res J 2008;78:731–9. https://doi.org/10.1177/0040517508091066.

[2] Kumar Vikram Singh, Paul S. Sawhney and Nozar D. Sachinvala, Guoqiang Li, and Su-Seng Pang, Brian Condon RP. Application and future of nanotechnology in textiles. Beltwide Cott Conf San Antonio, Texas 2006:2497–503.

[3] Ayatullah AKM, Asif H, Hasan MZ. International Journal of Current Engineering and Technology Application of Nanotechnology in Modern Textiles: A Review. 227| Int J Curr Eng Technol 2018;8:227–31.

[4] Hassan B, Islam G, Haque A. Advance Research in Textile Engineering Applications of Nanotechnology in Textiles: A Review. Adv Res Text Eng 2019;4:1–9.

[5] Valverde Aguilar G. Introductory Chapter: A Brief Semblance of the Sol-Gel Method in Research. Sol-Gel Method – Des Synth New Mater with Interes Phys Chem Biol Prop 2019:3–8. https://doi.org/10.5772/intechopen.82487.

[6] Baraket L, Ghorbel A. Control preparation of aluminium chromium mixed oxides by sol-gel process. vol. 118. Elsevier Masson SAS; 1998. https://doi.org/10.1016/s0167-2991(98)80233-4.

[7] Periyasamy AP, Venkataraman M, Kremenakova D, Militky J, Zhou Y. Progress in sol-gel technology for the coatings of fabrics. Materials (Basel) 2020;13. https://doi.org/10.3390/MA13081838.

[8] Chen J, Cheng G, Liu R, Zheng Y, Huang M, Yi Y, et al. Enhanced physical and biological properties of silk fibroin nanofibers by layer-by-layer deposition of chitosan and rectorite. J Colloid Interface Sci 2018;523:208–16. https://doi.org/10.1016/j.jcis.2018.03.093.

[9] Saini S, Gupta A, Singh N, Sheikh J. Functionalization of linen fabric using layer by layer treatment with chitosan and green tea extract. J Ind Eng Chem 2020;82:138–43. https://doi.org/10.1016/j.jiec.2019.10.005.

[10] Safi K, Kant K, Bramhecha I, Mathur P, Sheikh J. Multifunctional modification of cotton using layer-by-layer finishing with chitosan, sodium lignin sulphonate and boric acid. Int J Biol Macromol 2020;158:903–10. https://doi.org/10.1016/j.ijbiomac.2020.04.066.

[11] Kudaibergenov S, Tatykhanova G, Bakranov N, Tursunova R. Layer-by-Layer Thin Films and Coatings Containing Metal Nanoparticles in Catalysis. Thin Film Process – Artifacts Surf Phenom Technol Facet 2017. https://doi.org/10.5772/67215.

[12] Wilson JIB. Textile surface functionalisation by chemical vapour deposition (CVD). Woodhead Publishing Limited; 2009. https://doi.org/10.1533/9781845696689.126.

[13] Campbell SA, Smith RC. Chemical vapour deposition. High-K Gate Dielectr 2003;0123456789:65–88. https://doi.org/10.1016/b978-012524975-1/50009-4.

[14] Li Z, Aik Khor K. Preparation and properties of coatings and thin films on metal implants. vol. 1–3. Elsevier; 2019. https://doi.org/10.1016/B978-0-12-801238-3.11025-6.

[15] Makhlouf ASH. Current and advanced coating technologies for industrial applications. Woodhead Publishing Limited; 2011. https://doi.org/10.1533/9780857094902.1.3.

[16] Mohamed H, EL-HALWAGY A. Plasma-based Nanotechnology for Textile Coating. J Text Color Polym Sci 2021;0:0–0. https://doi.org/10.21608/jtcps.2021.67022.1049.

[17] Borah MP, Jose S, Kalita BB, Shakyawar DB, Pandit P. Water repellent finishing on eri silk fabric using nano silica. J Text Inst 2020;111:701–8. https://doi.org/10.1080/00405000.2019.1659470.

[18] Kwon J, Jung H, Jung H, Lee J. Micro/nanostructured coating for cotton textiles that repel oil, water, and chemical warfare agents. Polymers (Basel) 2020;12. https://doi.org/10.3390/polym12081826.

[19] Ammayappan L, Chakraborty S, Pan NC. Silica nanocomposite based hydrophobic functionality on jute textiles. J Text Inst 2021;112:470–81. https://doi.org/10.1080/00405000.2020.1764779.

[20] Abd El-Hady MM, Sharaf S, Farouk A. Highly hydrophobic and UV protective properties of cotton fabric using layer by layer self-assembly technique. Cellulose 2020;27:1099–110. https://doi.org/10.1007/s10570-019-02815-0.

[21] Bentis A, Boukhriss A, Gmouh S. Flame-retardant and water-repellent coating on cotton fabric by titania–boron sol–gel method. J Sol-Gel Sci Technol 2020;94:719–30. https://doi.org/10.1007/s10971-020-05224-z.

[22] Ye Z, Li S, Zhao S, Deng L, Zhang J, Dong A. Textile coatings configured by double-nanoparticles to optimally couple superhydrophobic and antibacterial properties. Chem Eng J 2021;420:127680. https://doi.org/10.1016/j.cej.2020.127680.

[23] Anjum AS, Ali M, Sun KC, Riaz R, Jeong SH. Self-assembled nanomanipulation of silica nanoparticles enable mechanochemically robust super hydrophobic and oleophilic textile. J Colloid Interface Sci 2020;563:62–73. https://doi.org/10.1016/j.jcis.2019.12.056.

[24] Chirila L, Cinteza LO, Tanase M, Radulescu DE, Radulescu DM, Stanculescu IR. Hybrid materials based on ZnO and SiO2 nanoparticles as hydrophobic coatings for textiles. Ind Textila 2020;71:297–301. https://doi.org/10.35530/IT.071.04.1814.

[25] Aslanidou D, Karapanagiotis I. Superhydrophobic, superoleophobic and antimicrobial coatings for the protection of silk textiles. Coatings 2018;8. https://doi.org/10.3390/coatings8030101.

[26] Cheng YT, Rodak DE, Wong CA, Hayden CA. Effects of micro- and nano-structures on the self-cleaning behaviour of lotus leaves. Nanotechnology 2006;17:1359–62. https://doi.org/10.1088/0957-4484/17/5/032.

[27] Zhang M, Feng S, Wang L, Zheng Y. Lotus effect in wetting and self-cleaning. Biotribology 2016;5:31–43. https://doi.org/10.1016/j.biotri.2015.08.002.

[28] Liu Y, Tang J, Wang R, Lu H, Li L, Kong Y, et al. Artificial lotus leaf structures from assembling carbon nanotubes and their applications in hydrophobic textiles. Journals Mater Chem 2007;17:1071–8. https://doi.org/10.1039/b613914k.

[29] El-Khatib EM. Antimicrobial and Self-cleaning Textiles using Nanotechnology. Res J Text Appar 2012;16:156–74. https://doi.org/10.1108/RJTA-16-03-2012-B016.

[30] Xue CH, Li YR, Zhang P, Ma JZ, Jia ST. Washable and wear-resistant superhydrophobic surfaces with self-cleaning property by chemical etching of fibers and hydrophobization. ACS Appl Mater Interfaces 2014;6:10153–61. https://doi.org/10.1021/am501371b.

[31] Das I, De G. Zirconia based superhydrophobic coatings on cotton fabrics exhibiting excellent durability for versatile use. Nat Publ Gr 2015:1–11. https://doi.org/10.1038/srep18503.

[32] Vasiljević J, Gorjanc M, Tomšič B, Orel B, Jerman I, Mozetič M, Vesel A, Simončič B. The surface modification of cellulose fibres to create superhydrophobic, oleophobic and self-cleaning properties. Cellulose 2013;20:277–89.

[33] Zhang L, Dillert R, Bahnemann D, Vormoor M. Photo-induced hydrophilicity and self-cleaning: Models and reality. Energy Environ Sci 2012;5:7491–507. https://doi.org/10.1039/c2ee03390a.

[34] Giwa A, Nkeonye PO, Bello KA, Kolawole KA. Photocatalytic Decolourization and Degradation of C. I. Basic Blue 41 Using TiO2 Nanoparticles. J Environ Prot (Irvine, Calif) 2012;03:1063–9. https://doi.org/10.4236/jep.2012.39124.

[35] Segundo IR, Freitas E, Landi S, Costa MFM, Carneiro JO. Smart, photocatalytic and self-cleaning asphalt mixtures: A literature review. Coatings 2019;9. https://doi.org/10.3390/coatings9110696.

[36] Gupta KK, Jassal M, Agrawal AK. Functional Finishing of Cotton Using Titanium Dioxide and Zinc Oxide Nanoparticles. Res J Text Appar 2007;11:1–10. https://doi.org/10.1108/RJTA-11-03-2007-B001.

[37] Veronovski N, Rudolf A, Smole MS, Kreže T, Geršak J. Self-cleaning and handle properties of TiO2-modified textiles. Fibers Polym 2009;10:551–6. https://doi.org/10.1007/s12221-009-0551-5.

[38] Afzal S, Daoud WA, Langford SJ. Superhydrophobic and photocatalytic self-cleaning cotton. J Mater Chem A 2014;2:18005–11. https://doi.org/10.1039/c4ta02764g.

[39] Rehan M, Hartwig A, Ott M, Gätjen L, Wilken R. Enhancement of photocatalytic self-cleaning activity and antimicrobial properties of poly(ethylene terephthalate) fabrics. Surf Coatings Technol 2013;219:50–8. https://doi.org/10.1016/j.surfcoat.2013.01.003.

[40] Jiang C, Liu W, Yang M, Liu C, He S, Xie Y, et al. Facile fabrication of robust fluorine-free self-cleaning cotton textiles with superhydrophobicity, photocatalytic activity, and UV durability. Colloids Surfaces A Physicochem Eng Asp 2018;559:235–42. https://doi.org/10.1016/j.colsurfa.2018.09.048.

[41] Kumbhakar P, Pramanik A, Biswas S, Kole AK, Sarkar R, Kumbhakar P. In-situ synthesis of rGO-ZnO nanocomposite for demonstration of sunlight driven enhanced photocatalytic and self-cleaning of organic dyes and tea stains of cotton fabrics. J Hazard Mater 2018;360:193–203. https://doi.org/10.1016/j.jhazmat.2018.07.103.

[42] Afzal S, Daoud WA, Langford SJ. Photostable Self-Cleaning Cotton by a Copper (II) Porphyrin / TiO 2 Visible-Light Photocatalytic System. Appl Mater Interfaces 2013;5:4753–9.

[43] Panwar K, Jassal M. TiO 2 – SiO 2 Janus particles for photocatalytic self-cleaning of cotton fabric. Cellulose 2018;25:2711–20. https://doi.org/10.1007/s10570-018-1698-2.

[44] Cheng XW, Guan JP, Yang XH, Tang RC. Improvement of flame retardancy of silk fabric by bio-based phytic acid, nano-TiO2, and polycarboxylic acid. Prog Org Coatings 2017;112:18–26. https://doi.org/10.1016/j.porgcoat.2017.06.025.

[45] Wang B, Xu YJ, Li P, Zhang FQ, Liu Y, Zhu P. Flame-retardant polyester/cotton blend with phosphorus/nitrogen/silicon-containing nano-coating by layer-by-layer assembly. Appl Surf Sci 2020;509:145323. https://doi.org/10.1016/j.apsusc.2020.145323.

[46] Samanta AK, Bhattacharyya R, Jose S, Basu G, Chowdhury R. Fire retardant finish of jute fabric with nano zinc oxide. Cellulose 2017;24:1143–57. https://doi.org/10.1007/s10570-016-1171-z.

[47] Jose S, Shanmugam N, Das S, Kumar A, Pandit P. Coating of lightweight wool fabric with nano clay for fire retardancy. J Text Inst 2019;110:764–70. https://doi.org/10.1080/00405000.2018.1516529.

[48] Zhang C, Cheng XW, Guan JP, Chen G. Preparation of nano-Mg(OH)2 for surface coating of silk fabric with improved flame retardancy and smoke suppression. Colloids Surfaces A Physicochem Eng Asp 2021;625:126868. https://doi.org/10.1016/j.colsurfa.2021.126868.

[49] Cheng XW, Guan JP, Yang XH, Tang RC. Improvement of flame retardancy of silk fabric by bio-based phytic acid, nano-TiO2, and polycarboxylic acid. Prog Org Coatings 2017;112:18–26. https://doi.org/10.1016/j.porgcoat.2017.06.025.

[50] Norouzi M, Zare Y, Kiany P. Nanoparticles as effective flame retardants for natural and synthetic textile polymers: Application, mechanism, and optimization. Polym Rev 2015;55:531–60. https://doi. org/10.1080/15583724.2014.980427.

[51] Teli MD, Annaldewar BN. Superhydrophobic and ultraviolet protective nylon fabrics by modified nano silica coating. J Text Inst 2017;108:460–6. https://doi.org/10.1080/00405000.2016.1171028.

[52] Gupta V, Jose S, Kadam V, Shakyawar DB. Sol gel synthesis and application of silica and titania nano particles for the dyeing and UV protection of cotton fabric with madder. J Nat Fibers 2021;00:1–11. https://doi.org/10.1080/15440478.2021.1881688.

[53] Morshed MN, Shen X, Deb H, Azad S Al, Zhang X, Li R. Sonochemical fabrication of nanocryatalline titanium dioxide (TiO2) in cotton fiber for durable ultraviolet resistance. J Nat Fibers 2020;17:41–54. https://doi.org/10.1080/15440478.2018.1465506.

[54] Belay A, Mekuria M, Adam G. Incorporation of zinc oxide nanoparticles in cotton textiles for ultraviolet light protection and antibacterial activities. Nanomater Nanotechnol 2020;10:1–8. https://doi.org/10.1177/1847980420970052.

[55] Teli MD, Annaldewar BN. Superhydrophobic and ultraviolet protective nylon fabrics by modified nano silica coating. J Text Inst 2017;108:460–6. https://doi.org/10.1080/00405000.2016.1171028.

[56] Noorian SA, Hemmatinejad N, Navarro JAR. Ligand modified cellulose fabrics as support of zinc oxide nanoparticles for UV protection and antimicrobial activities. Int J Biol Macromol 2020;154:1215–26. https://doi.org/10.1016/j.ijbiomac.2019.10.276.

[57] Li M, Farooq A, Jiang S, Zhang M, Mussana H, Liu L. Functionalization of cotton fabric with ZnO nanoparticles and cellulose nanofibrils for ultraviolet protection. Text Res J 2021;91:2303–14. https://doi.org/10.1177/00405175211001807.

[58] Juikar SJ, Nadanathangam V. Microbial Production of Nanolignin from Cotton Stalks and Its Application onto Cotton and Linen Fabrics for Multifunctional Properties. Waste and Biomass Valorization 2020;11:6073–83. https://doi.org/10.1007/s12649-019-00867-8.

[59] Muzaffar S, Abbas M, Siddiqua UH, Arshad M, Tufail A, Ahsan M, et al. Enhanced mechanical, UV protection and antimicrobial properties of cotton fabric employing nanochitosan and polyurethane based finishing. J Mater Res Technol 2021;11:946–56. https://doi.org/10.1016/j.jmrt.2021.01.018.

[60] El-Bisi MK, Othman R, Yassin FA. Improving antibacterial and ultraviolet properties of cotton fabrics via dual effect of nano-metal oxide and Moringa oleifera extract. Egypt J Chem 2020;63:3441–51. https://doi.org/10.21608/ejchem.2020.39534.2805.

[61] Yetisen AK, Qu H, Manbachi A, Butt H, Dokmeci MR, Hinestroza JP, et al. Nanotechnology in Textiles. ACS Nano 2016;10:3042–68. https://doi.org/10.1021/acsnano.5b08176.

[62] Uğur ŞS, Bilgiç M. A novel approach for improving wrinkle resistance and flame retardancy properties of Linen Fabrics. Bilge Int J Sci Technol Res 2017;1:79–86.

[63] Lao L, Fu L, Qi G, Giannelis EP, Fan J. Superhydrophilic Wrinkle-Free Cotton Fabrics via Plasma and Nanofluid Treatment. ACS Appl Mater Interfaces 2017;9:38109–16. https://doi.org/10.1021/acsami.7b09545.

[64] Gao LZ, Bao Y, Cai HH, Zhang AP, Ma Y, Tong XL, et al. Multifunctional silk fabric via surface modification of nano-SiO2. Text Res J 2020;90:1616–27. https://doi.org/10.1177/0040517519897112.

[65] Nourbakhsh S, Habibi S, Rahimzadeh M. Copper nano-particles for antibacterial properties of wrinkle resistant cotton fabric. Mater Today Proc 2017;4:7032–7. https://doi.org/10.1016/j.matpr.2017.07.034.

[66] Ueda M. Textile Finishing. Sen'i Gakkaishi 2003;59. https://doi.org/10.2115/fiber.59.P_421.

[67] Sim W, Barnard RT, Blaskovich MAT, Ziora ZM. Antimicrobial silver in medicinal and consumer applications: A patent review of the past decade (2007–2017). Antibiotics 2018;7:1–15. https://doi.org/10.3390/antibiotics7040093.

[68] Díez-Pascual AM. Antibacterial activity of nanomaterials. Nanomaterials 2018;8:6–11. https://doi.org/10.3390/nano8060359.

[69] Momotaz F, Siddika A, Shaihan MT, Islam MA. The Effect of Zno Nano Particle Coating and their Finishing Process on the Antibacterial Property of Cotton Fabrics. J Eng Sci 2020;11:61–5. https://doi.org/10.3329/jes.v11i1.49547.

[70] Babiker O, Gibril M. Preparation of an Anti- microbial Cotton Fabric and enhance physical properties Using Synthesize Zinc Nano particles stabilizing by Citric Acid. Int J Eng Inf Syst 2019;3:21–33.

[71] Yılmaz F, Bahtiyari Mİ. Antibacterial Finishing of Linen Fabrics by Combination with Nano-Aluminum Oxide and Crosslinking Agent. Fibers Polym 2021;22:1830–6. https://doi.org/10.1007/s12221-021-0865-5.

[72] El-nahhal IM, Salem J, Kodeh FS, Anbar R, Elmanama A. " Preparation of CuO-NPs Coated Cotton, Starched Cotton and its CuO- Ag Nanocomposite, Cu (II) Curcumin Complex Coated Cotton and their Antimicrobial Activities ". J Nanomed Nanotechnol 2021;12:1–6.

[73] Marković D, Ašanin J, Nunney T, Radovanović Ž, Radoičić M, Mitrić M, et al. Broad Spectrum of Antimicrobial Activity of Cotton Fabric Modified with Oxalic Acid and CuO/Cu2O Nanoparticles. Fibers Polym 2019;20:2317–25. https://doi.org/10.1007/s12221-019-9131-5.

[74] Xu QB, Ke XT, Cai DR, Zhang YY, Fu FY, Endo T, et al. Silver-based, single-sided antibacterial cotton fabrics with improved durability via an l-cysteine binding effect. Cellulose 2018;25:2129–41. https://doi.org/10.1007/s10570-018-1689-3.

[75] Tania IS, Ali M, Azam MS. In-situ synthesis and characterization of silver nanoparticle decorated cotton knitted fabric for antibacterial activity and improved dyeing performance. SN Appl Sci 2019;1:1–9. https://doi.org/10.1007/s42452-018-0068-x.

[76] Joshi S, Kambo N, Dubey S, Shukla P, Pandey R. Effect of Onion (Allium cepa L.) Peel Extract-based Nanoemulsion on Anti-microbial and UPF Properties of Cotton and Cotton Blended Fabrics. J Nat Fibers 2021;00:1–10. https://doi.org/10.1080/15440478.2021.1964127.

[77] Yılmaz F, Bahtiyari Mİ. An Approach for Linen Fabrics Coloring and Antibacterial Activity by Cumin in Combination with Nano Copper and Iron. J Nat Fibers 2021:1–8. https://doi.org/10.1080/15440478.2021.1946884.

[78] M S. Green Synthesis of Silver Nanoparticles using Hybanthusenneaspermus Plant Extract against Nosocomial Pathogens with Nanofinished Antimicrobial Cotton Fabric. Glob J Nanomedicine 2017;1. https://doi.org/10.19080/gjn.2017.01.555554.

[79] Iyigundogdu ZU, Demir O, Asutay AB, Sahin F. Developing Novel Antimicrobial and Antiviral Textile Products. Appl Biochem Biotechnol 2017;181:1155–66. https://doi.org/10.1007/s12010-016-2275-5.

[80] Nanocapsule RF. Properties of Aroma Sustained-release Cotton Fabric with. Chinese J Chem Eng 2011;19:523–8. https://doi.org/10.1016/S1004-9541(11)60016-5.

[81] Hu J, Xiao ZB, Zhou RJ, Ma SS, Li Z, Wang MX. Comparison of compounded fragrance and chitosan nanoparticles loaded with fragrance applied in cotton fabrics. Text Res J 2011;81:2056–64. https://doi.org/10.1177/0040517511416274.

[82] Khounvilay K, Estevinho BN, Sittikijyothin W. Citronella Oil Microencapsulated in Carboxymethylated Tamarind Gum and its Controlled Release. Eng J 2019;23:217–27. https://doi.org/10.4186/ej.2019.23.5.217.

[83] Sarvalkar PD, Barawkar SD, Karvekar OS, Patil PD, Prasad SR, Sharma KK, et al. A review on multifunctional nanotechnological aspects in modern textile. J Text Inst 2022;0:1–18. https://doi.org/10.1080/00405000.2022.2046304.

[84] Saleem H, Zaidi SJ. Sustainable use of nanomaterials in textiles and their environmental impact. Materials (Basel) 2020;13:1–28. https://doi.org/10.3390/ma13225134.

[85] Wang ZL, Wang AC. On the origin of contact-electrification. Mater Today 2019;30:34–51. https://doi.org/10.1016/j.mattod.2019.05.016.

[86] Zhang X. Antistatic and conductive textiles. Woodhead Publishing Limited; 2011. https://doi.org/10.1533/9780857092878.27.

[87] Roy Choudhury AK. Antistatic and soil-release finishes. 2017. https://doi.org/10.1016/b978-0-08-100646-7.00010-2.

[88] Textor T, Mahltig B. A sol-gel based surface treatment for preparation of water repellent antistatic textiles. Appl Surf Sci 2010;256:1668–74. https://doi.org/10.1016/j.apsusc.2009.09.091.

[89] Shyr TW, Lien CH, Lin AJ. Coexisting antistatic and water-repellent properties of polyester fabric. Text Res J 2011;81:254–63. https://doi.org/10.1177/0040517510380775.

[90] Zhou Z, Chu L, Tang W, Gu L. Studies on the antistatic mechanism of tetrapod-shaped zinc oxide whisker. J Electrostat 2003;57:347–54. https://doi.org/10.1016/S0304-3886(02)00171-7.

[91] Wang D, Lin Y, Zhao Y, Gu L. Polyacrylonitrile Fibers Modified by Nano-Antimony-Doped Tin Oxide Particles. Text Res J 2004;74:1060–5. https://doi.org/10.1177/004051750407401206.

[92] Tokarev I, Minko S. Stimuli-responsive hydrogel thin films. Soft Matter 2009;5:511–24. https://doi.org/10.1039/b813827c.

[93] Glampedaki P, Petzold G, Dutschk V, Miller R, Warmoeskerken MMCG. Physicochemical properties of biopolymer-based polyelectrolyte complexes with controlled pH/thermo-responsiveness. React Funct Polym 2012;72:458–68. https://doi.org/10.1016/j.reactfunctpolym.2012.04.009.

[94] Calvimontes A, Dutschk V, Breitzke B, Offermann P, Voit B. Soiling degree and cleanability of differently treated polyester textile materials. Tenside, Surfactants, Deterg 2005;42:17–22. https://doi.org/10.3139/113.100246.

[95] Wenbo Wang,Naihua Zhai AW. Preparation and Swelling Characteristics of a Superabsorbent Nanocomposite Based on Natural Guar Gum and Cation-Modified Vermiculite. J Appl Polym Sci 2010;119:3675–86. https://doi.org/10.1002/app.

[96] Hossain MM, Hegemann D, Fortunato G, Herrmann AS, Heuberger M. Plasma deposition of permanent superhydrophilic a-C:H:N films on textiles. Plasma Process Polym 2007;4:471–81. https://doi.org/10.1002/ppap.200600214.

[97] Martin TP, Lau KKS, Chan K, Mao Y, Gupta M, Shannan O'Shaughnessy W, et al. Initiated chemical vapor deposition (iCVD) of polymeric nanocoatings. Surf Coatings Technol 2007;201:9400–5. https://doi.org/10.1016/j.surfcoat.2007.05.003.

[98] Yılmaz F, Bahtiyari Mİ. An Approach for Multifunctional Finishing of Cotton with the Help of Nano Titanium Silicon Oxide. J Nat Fibers 2020. https://doi.org/10.1080/15440478.2020.1848704.

[99] Butola BS, Garg A, Garg A, Chauhan I. Development of Multi-functional Properties on Cotton Fabric by In Situ Application of TiO2 and ZnO Nanoparticles. J Inst Eng Ser E 2018;99:93–100. https://doi.org/10.1007/s40034-018-0118-3.

[100] Jhatial AK, Khatri A, Ali S, Babar AA. Sol–gel finishing of bamboo fabric with nanoparticles for water repellency, soil release and UV resistant characteristics. Cellulose 2019;26:6365–78. https://doi.org/10.1007/s10570-019-02537-3.

[101] Nozari B, Montazer M, Mahmoudi Rad M. Stable ZnO/SiO2 nano coating on polyester for anti-bacterial, self-cleaning and flame retardant applications. Mater Chem Phys 2021;267:124674. https://doi.org/10.1016/j.matchemphys.2021.124674.

[102] Atta AM, Abomelka HM. Multifunctional finishing of cotton fibers using silver nanoparticles via microwave-assisted reduction of silver alkylcarbamate. Mater Chem Phys 2021;260:124137. https://doi.org/10.1016/j.matchemphys.2020.124137.

[103] Juikar SJ, Vigneshwaran N. Microbial production of coconut fiber nanolignin for application onto cotton and linen fabrics to impart multifunctional properties. Surfaces and Interfaces 2017;9:147–53. https://doi.org/10.1016/j.surfin.2017.09.006.

[104] Gupta M, Sheikh J, Annu, Singh A. An eco-friendly route to develop cellulose-based multifunctional finished linen fabric using ZnO NPs and CS network. J Ind Eng Chem 2021;97:383–9. https://doi.org/10.1016/j.jiec.2021.02.023.

[105] Tripathi R, Narayan A, Bramhecha I, Sheikh J. Development of multifunctional linen fabric using chitosan film as a template for immobilization of in-situ generated CeO2 nanoparticles. Int J Biol Macromol 2019;121:1154–9. https://doi.org/10.1016/j.ijbiomac.2018.10.067.

[106] Shahid-ul-Islam, Butola BS, Kumar A. Green chemistry based in-situ synthesis of silver nanoparticles for multifunctional finishing of chitosan polysaccharide modified cellulosic textile substrate. Int J Biol Macromol 2020;152:1135–45. https://doi.org/10.1016/j.ijbiomac.2019.10.202.

[107] El-Naggar ME, Shaarawy S, Hebeish AA. Multifunctional properties of cotton fabrics coated with in situ synthesis of zinc oxide nanoparticles capped with date seed extract. Carbohydr Polym 2018;181:307–16. https://doi.org/10.1016/j.carbpol.2017.10.074.

[108] Shahid-ul-Islam, Butola BS, Gupta A, Roy A. Multifunctional finishing of cellulosic fabric via facile, rapid in-situ green synthesis of AgNPs using pomegranate peel extract biomolecules. Sustain Chem Pharm 2019;12:100135. https://doi.org/10.1016/j.scp.2019.100135.

[109] Sligo T, Lane A, Yw SF. Toxicity of Nanomaterials: Exposure, Pathways, Assessment, and Recent Advances. ACS Biomater Sci Eng 2018;4:2237–75. https://doi.org/10.1021/acsbiomaterials.8b00068.

[110] Zhu Y, Liu X, Hu Y, Wang R, Chen M, Wu J, et al. Behavior, remediation effect and toxicity of nanomaterials in water environments. Environ Res 2019;174:54–60. https://doi.org/10.1016/j.envres.2019.04.014.

[111] Dugershaw BB, Aengenheister L, Schmidt S, Hansen K, Hougaard KS, Buerki-thurnherr T. Recent insights on indirect mechanisms in developmental toxicity of nanomaterials. Part Fibre Toxic 2020;17:1–22.

Lelona Pradhan, Saptarshi Maiti, Aranya Mallick, Mohammad Shahid,
Sandeep P. More, Ravindra V. Adivarekar

3 Coating- and lamination-based smart textiles: techniques, features, and challenges

Abstract: The application of textile materials is multidimensional. Among the applications, textile materials that have smart properties, enabling them to sense and respond to environmental stimuli, such as chemical, biological, thermal, magnetic, and mechanical stimuli, have attracted huge attention. These smart materials are created by specially functionalizing textile materials. Coating and lamination have always been good tools for imparting functionalization on textile substrates. Application of different polymers and chemicals using various techniques of coating and lamination, chemical vapour deposition, layer-by-layer assembly, surface functionalization, and so on are some of the well-known processes for preparing these smart textile materials. This chapter discusses the various aspects of coating- and lamination-based smart textiles, their preparation techniques, properties, and their advantages and limitations.

Keywords: Coating, lamination, smart textiles, chemical vapour deposition, layer-by-layer assembly, surface functionalization

3.1 Introduction

Textile industry is moving from the conventional to a more performance-intensive sector. This industry has risen far above from just fulfilling the basic human needs of clothing and are expected to deliver diverse functionalities without losing their inherent characteristics. Increased demand for added functionalities has awakened the innovative minds in search of ways that could render traditional textiles with all those desired outputs. In such a scenario, functionalization or modification of the textile substrate comes into play, which can be considered as a worthy bidder to achieve such end results.

The functional textiles market has grown at an incredible growth rate of 33.58% between 2015 and 2020. The global functional textile market reached US $4.72 billion by 2020. India is a prime manufacturer in apparel and textile manufacturing and the fourth largest exporter in the international market. The functional textile sector has seen a compound annual growth rate (CAGR) of 30% from 2015 to 2020 due to strong automotive, fitness, fashion, healthcare, military, and sports textiles [1].

There are several surface functionalization techniques that have been successfully utilized in the field of textiles and many studies are still going on to bring in

https://doi.org/10.1515/9783110759747-003

more advancements. Plasma treatment, sputtering, ozonization, sol–gel method, grafting, electroless depositions are some of them. However, coating and lamination are the age-old techniques of textile surface modification that are still popular due to their easy processability, and little knowhow it requires. Cheaper fabrics can be coated with functional finishes to improve their usability and thus find their suitability for various application areas.

Functionalization of textile substrates can induce properties such as improved hydrophilicity, flame retarding ability, antimicrobial activity, aesthetic look, improved surface properties such as softness and reduced pilling. Apart from these conventional demands, the focus is shifting more toward smart textiles, where textiles can sense and tune to various body environments as well as outside stimuli to perform various activities. These stimuli can range from thermal to mechanical to pressure to magnetic field to many more. The uses of smart textiles are not restricted to only clothing but have found profound applications in the field of automobiles, space crafts, robotics, medicine, and so on. It will not be misleading to call smart textiles, multifunctional textiles. Coating and lamination techniques have been widely employed to produce smart textiles in the form of conductive textiles, in which coating processes have been followed to coat conductive inks onto the surface of traditional textile fabrics to make then conductive. However, when discussing about smart textiles, coating and lamination processes are far different than those used in traditional textiles for conventional uses, as smart textiles require fine adjustments and precision in all parameters to deliver the desired results. Thus, coating and lamination techniques to produce smart textiles form a very broad area that needs to be completely understood to gain the full advantage of these processes for such smart applications.

The goal of this chapter is to introduce readers to the basics of surface modification and functionalization techniques, followed by a detailed discussion on coating and lamination techniques and the materials used for the same. The chapter then proceeds towards nanotechnology-based smart coatings and active coatings where the readers will get idea about the applicability of such techniques in the field of smart textiles. Additionally, the chapter also includes the challenges associated with such approaches and the future scope.

3.2 Surface modification/surface functionalization

The uses of textiles for conventional purposes are well known to everyone. However, with the progress of time, textiles are gradually being challenged to evolve from the conventional means to more advanced ones, delivering multifunctional properties, thus suiting the desired purposes. In such a scenario, the surface of a textile fabric plays an important role in delivering such outputs, and surface modification or functionalization has a pivotal role to play in this step. The surface of a textile can be modified by various means, which can include all types of physical and chemical methods and ways. Some of them are briefly discussed further.

3.2.1 Plasma treatment

This can be described in Figure 3.1 as a treatment where an ionized gas called plasma containing equal densities of positive and negative charges is introduced to the textile substrate, resulting in the modification of the surface at the nanoscale [2]. The ionized gas also contains free radicals, heavy particles, and electrons; these characteristics are highly dependent on the type of gas used. Once in the gas phase, if additional energy is forced into the system, the gas is ionized and it reaches the plasma state. When the plasma encounters the material surface, transfer of the additional energy takes place from the plasma for the subsequent reactions on the material surface, which then leads to the modification of the target surface [3]. Plasma treatment may impose several modifications on the surface of fibres, including cleaning, activation, grafting, etching, and polymerization [4].

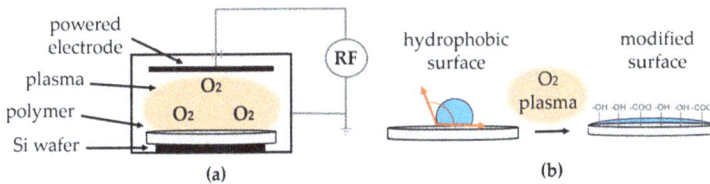

Figure 3.1: Example of plasma surface modification [5].

3.2.1.1 Sputtering

Sputtering can be defined in Figure 3.2 as a phenomenon in which microscopic particles are ejected from a target's surface by the bombardment of very high-energy particles. An exchange of momentum takes place when the energetic ions come in contact with the atoms of the target material. These ions, known as "incident ions", set off collision cascades in the target. If a collision cascade reaches the surface of the target, and its remaining energy is greater than the target's surface binding energy, an atom will be ejected. This process is known as "sputtering" [6]. Magnetron sputtering technology has been successfully utilized for modifying textile surfaces by the deposition of various metals and metal oxides to impart properties such as electromagnetic shielding, UV protection, anti-static, antibacterial, conductive, or waterproof properties. In addition, this technique can also help obtain structural colours through interference and diffraction characteristics of the nanofilms [7]. Various studies have also been undertaken to utilize this technology in textile heat management and in thermal regulation [8].

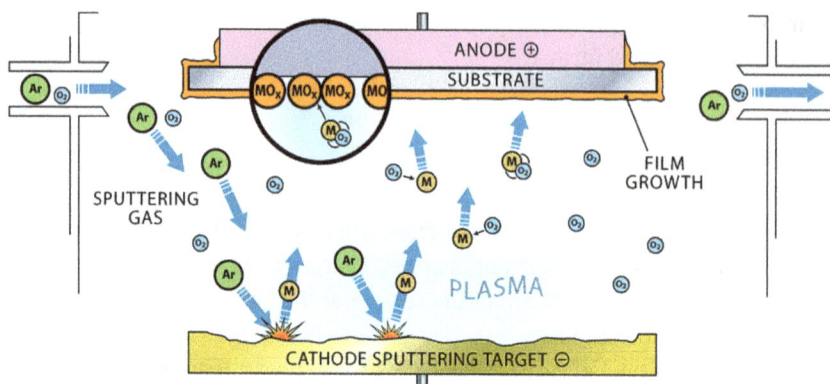

Figure 3.2: Sputtering technique for surface modification [9].

3.2.2 Grafting

A graft copolymer can be defined as a branched copolymer composed of a main chain of a polymer backbone onto which side chain grafts (branches) are covalently attached. The polymer backbone may be a homopolymer or copolymer and differs in chemical structure and composition from the graft material [10].

Surface grafting techniques are especially gaining considerable research attention due to their flexibility in tailoring desirable surface properties using different monomers as well as their precision in conferring the grafts at desired location [11]. Among the various grafting techniques, some of the advanced methods are mentioned further.

3.2.2.1 Photo-induced grafting

This method does not require any additional physical or chemical pre-treatment and can also produce desired and localized active species at the interface between the medium and the target surface. The textile surface can be dipped into a monomer containing a solution onto which UV light can be imparted, which will lead the reaction of monomers with free radicals, forming covalent bonds and hence resulting in surface modification. Parameters that are of great importance are:
- Monomer concentration
- Wavelength of UV light
- Duration of irradiation time

UV-induced grafting has been successfully utilized in the grafting of cellulose nanofibers in the absence of organic solvents to improve its hydrophobicity [12].

3.2.2.2 Plasma-induced grafting

Plasma is essentially a state of matter comprising a mixture of ions, electrons, excited species, and free radicals. Plasma can be generated by direct current, photons, radio frequency waves, microwaves, and so on. The generated plasma is then utilized to graft the targeted surface and induce the required modifications or even result in the functionalization of the substrate. Plasma-induced grafting can be regarded as an environmental-friendly technique as no direct usage of wet and hazardous chemical is involved in the process.

3.2.2.3 Radiation-induced grafting

Radiation-induced grafting is becoming widely popular due to the homogeneous process environment it provides along with the elimination of any external initiator. Numerous sources of radiations can be used, such as electro-magnetic photons (gamma rays and X-ray) and charged particles (electron beam and swift heavy ions). Radiation energy from these sources are utilized to initiate the grafting to change the surface properties and induce modification.

3.2.2.4 Thermal-induced grafting

This technique of grafting mainly utilizes chemical initiators apart from cleavage agents, which serve as the cross-linking modifier. The process of thermal-induced grafting involves three main steps as illustrated in Figure 3.3.

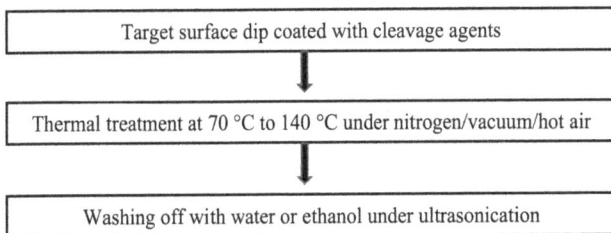

Target surface dip coated with cleavage agents

↓

Thermal treatment at 70 °C to 140 °C under nitrogen/vacuum/hot air

↓

Washing off with water or ethanol under ultrasonication

Figure 3.3: Steps for thermal-induced grafting.

3.2.2.5 Ozone-induced grafting

Dried oxygen is utilized in an ozonizer to produce the required ozone. The substrate along with the chemicals are reacted in the presence of ozone to induce grafting.

3.2.3 Laser ablation

Laser ablation is the thermal or non-thermal process of removing atoms from a solid by irradiating it with an intense continuous wave (CW) or pulsed laser beam. When a solid surface is irradiated by a CW laser beam or a long-pulsed (e.g. nanoseconds pulsed) laser beam, the material is heated by the absorbed laser energy. The thermal motion of some particles is accelerated. Once the absorbed energy exceeds the sublimation energy, these particles evaporate or sublimate and become vaporized particles; that is, part of the target is ablated and this induces surface modifications [13].

3.2.4 Sol–gel

The application of sol–gel technique has been widely known to produce nanoparticles apart from the application of inorganic–organic- or inorganic hybrid-based thin layers on the textile surface. It is the suitable method to tailor various properties in a single processing step. Basically, the process involves three steps, namely, hydrolyzation, application, and curing. Hydrolyzation is carried out in the presence of a solvent in which the precursor is dispersed. After hydrolyzation at specific reaction conditions, the process leads to condensation, and gradually to the formation of nanosols. The conditions that affect the condensation process are as follows:
- Concentration of the precursor
- Type of solvent
- Temperature
- pH
- Presence of external additives or salt

The nanosols then accelerate toward gelation, forming the so-called lyogels (three-dimensional gel networks, where the pores are filled with the solvent). In the third step, curing or drying takes place, where the removal of solvent is carried out, forming the structures called, xerogels. For the application onto textile surfaces, a readily prepared nanosol is mixed with various polymers, pigments, drugs, fragrances, dyestuffs, or nano-powders to impart various properties like self-cleaning, UV protection, X-ray shielding, hydrophobic, flame resistance, improved dyeing, and controlled release [14].

3.2.5 Layer-by-layer

Layer-by-layer or L-b-L is an easy and versatile method to produce systematic, multiple layered, inorganic–organic films on the textile substrates because of surface modification to achieve the various desired properties. It usually works on the principle of electrostatic interactions where the assembly starts with the adsorption of a charged material on the target surface having the opposite charge. Sequential alteration takes places for depositing further layers, resulting in charge neutralization and charge reversal. This process is carried out until the required number of layers is deposited to achieve the desired properties, as depicted in Figure 3.4.

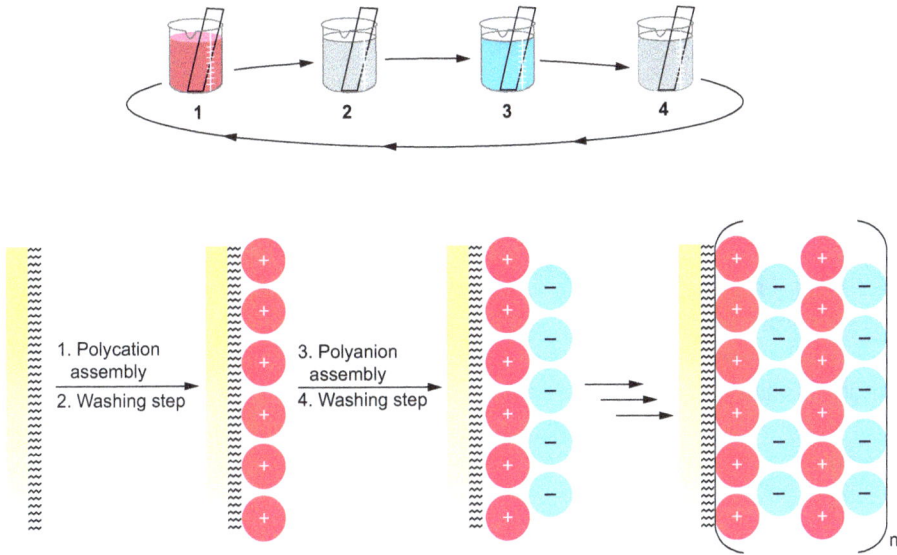

Figure 3.4: Layer-by-layer deposition of cations and anions assembly on the substrate (Modified according to [15], Copyright 2013, Springer).

This method can also deliver surfaces with various morphologies and structures [16]. The conditions that greatly affect this technique are as follows:
-   Polyelectrolyte used
-   Type of substrate
-   pH
-   Ionic strength
-   Drying time after each rinse, and so on

3.2.6 Electroless deposition

Electroless plating dates to the 1940s. Brenner and Riddell developed and patented this process in 1950. Thibodeaux and Baril first reported this process in 1973 as a means of metallizing textiles to make them electrically and thermally conductive. This process is also referred to as metallization and since many of these metals are autocatalytic, it is also called autocatalytic plating [17]. In this process, reduction of metal ions to pure metals takes place on the target surface in the presence of a catalyst before the reaction proceeds. The plating continues as long as the substrate is immersed in the plating bath, as the process is autocatalytic in nature. This helps in delivering an unlimited thickness of metal film. This method primarily works on the principle of localization of metals on the substrate, with the means of a catalyst, a process known as sensitization. Speaking about the type of textile substrate that can be utilized, it can include fibres to yarns to fabrics, without altering any inherent properties of the same, and helping in achieving additional functionalities [18].

3.2.7 Chemical oxidation

Chemical oxidation can be basically defined as a process involving the transfer of electrons from an oxidizing reagent to the chemical species being oxidized [19]. These oxidizing agents can be hydrogen peroxide, chlorine, nitric acid, and so on. Oxidation, as a chemical treatment to induce changes in the surface of textiles and functionalize them, has been widely studied. In this process, various oxidants are used to treat the surface of textiles at appropriate conditions. Oxidation results in the addition of various functional moieties to the substrate, which modifies and enhances the properties of textiles.

3.2.8 Enzyme treatment

The advances of biotechnology, more specifically enzyme technology, are widely spreading in different fields of application, and textiles are not excluded from such advancement. Enzymes are being utilized as superior alternatives to the conventionally used technologies for surface functionalization of textiles due to the high specificity of activity it provides, along with eco-friendliness (Figure 3.5).

Apart from being used for scouring, stone washing, and so on, enzymes are also studied to functionalize the textile surface and bring in desired modifications. Some of the studied functionalizations are listed in Table 3.1.

3.2.9 Coating

Coating can be described as a process of application of a suitable chemical or material onto the top layer of the textile to achieve the desired properties, without altering the inherent properties of the substrate material. A coated substrate can also be called a composite. It basically has two components, the coat and the base fabric. The coating materials can range from chemicals to polymers to nanomaterials to many more. Various techniques have been developed to coat textile substrates, where the resultant fabric could deliver multifunctional properties, such as soil repellency, water repellency,

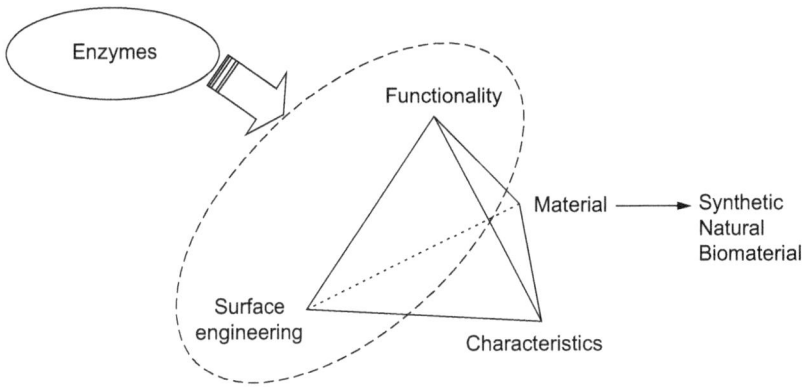

Figure 3.5: Enzyme surface functionalization of textiles (Modified according to [20], Copyright 2009, Elsevier).

Table 3.1: Surface functionalization of textiles with enzymes [20, 21].

Enzymes	Textile substrate	Functionality
Cellulase	– Cotton – Linen – Viscose	– Reduced pilling – Wash proof and non-greasy softening effects – Improved drapability
Pectinase	– Cotton	– Scouring
Xyloglucan endotransglycosylase	– Cellulosic materials	– Abraded look – Microfibrillated cellulose
Tyrosinase	– Chitosan – Protein fibres	– Antioxidant activities – Improved solubility
Cutinase	– Poly(ethylene terephthalate) (PET)	– Improved hydrophilicity

fire retardancy, and conductive fabrics. This chapter will further discuss various coating technologies in detail [22].

3.2.10 Lamination

Lamination is the bonding of two or more substrates, containing at least one textile substrate. These layers can be bonded with the use of an adhesive or using the inherent adhesive properties of the layers. Polymer materials that may not be easily formulated into a resin or a paste for coating can be combined with a fabric by first preparing a film of the polymer and then laminating it to the fabric in a separate process [22].

3.3 Coating techniques

Coating can be simply defined as a process of applying a layer or multiple layers of material onto a substrate. This technique is not only restricted to textiles and has various other application areas. There are several coating techniques and different types of coating formulations used, which depend on the following factors:
- Type of substrate
- End use
- Coating materials
- Concentration and viscosity of coating formulation
- End use and properties required
- Required precision
- Economy and cost

Among the various coating techniques, some of them are discussed further.

3.3.1 Direct coating/knife coating

This is the simplest and oldest technique of coating substrates and is also known as "spread coating" or "floating knife" technique. This technique specifically uses a knife or a doctor blade, which is fixed above the substrate at a distance that typically ranges from 0 to 500 mm, depending on the coating material, required add-on, and the end use of the product. The substrate is laid flat and uniformly spread. The coating solution is placed in front of the knife and as the fabric moves forward the coating material, it is scraped by the knife to layer the substrate with the coating material. Modern equipments are provided with several coating knives and controls for easy facilitation of the process. The parameters that affect the coating applied are [22–24]:
- Substrate type (woven, knitted, etc.)
- Substrate tension
- Knife depression
- Distance between the knife and the substrate
- Sharpness of the blade/knife
- Angle of alignment of the knife to the substrate
- Solution parameters of the coating material (solid content, concentration, viscosity, surface tension, and other rheological aspects)
- Knife blade geometry (sharp, round, shoe, etc.)
- Speed of the machine

Direct coating or knife coating can be classified into two categories – knife-over-air (KOA) and knife-over-roll (KOR), as schematically represented in Figure 3.6. In the *"knife-over-air"* coating, the knife is suspended over the fabric (or air) and it is used to

(A)

(B)

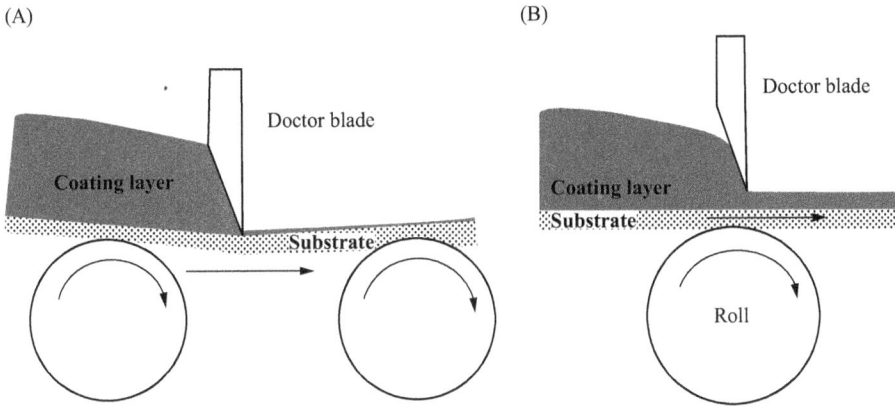

Figure 3.6: Direct coating techniques: (A) floating knife coating and (B) knife over roller (Modified according to [24], Copyright 2019, Elsevier).

spread the coating material on the fabric substrate. On the other hand, in the *"knife-over-roller"* coating, the basic principle is the same; however, the knife is suspended over a roller over which the fabric is passed. The Knife-over-air coating technique is the most extensively used technique in the textile industry, while the Knife-over-roll technique is especially used where a higher coating add-on is required, such as PVC coating.

3.3.2 Metering rod/Mayer rod coating

In this coating technique, the amount of material that is to be coated on the substrate is controlled by a wire bound rod made up of stainless steel, often known as the metering rod/mayer rod/an equalizer bar/a coating rod/a doctor rod, as shown in Figure 3.7. The coating involves a slow rotation of the rod in a direction opposite to that of the web rotation. The coating material stuck between the wires is removed by its continuous rotation thereby keeping the surface of the wire clean throughout the process. The longevity of the rod also increases due to less wear and tear [22]. The geometry and construction of the rod greatly affects the coating process as the cross-sectional area between the wire coils and the rod as well as the wire diameter affects the wet thickness of the film.

Other factors that affect the coating quality are [24]-
– Substrate tension
– Viscosity of the liquid
– Substrate speed
– Metering rod rotation
– Wrap angle
– Coating penetration and so on

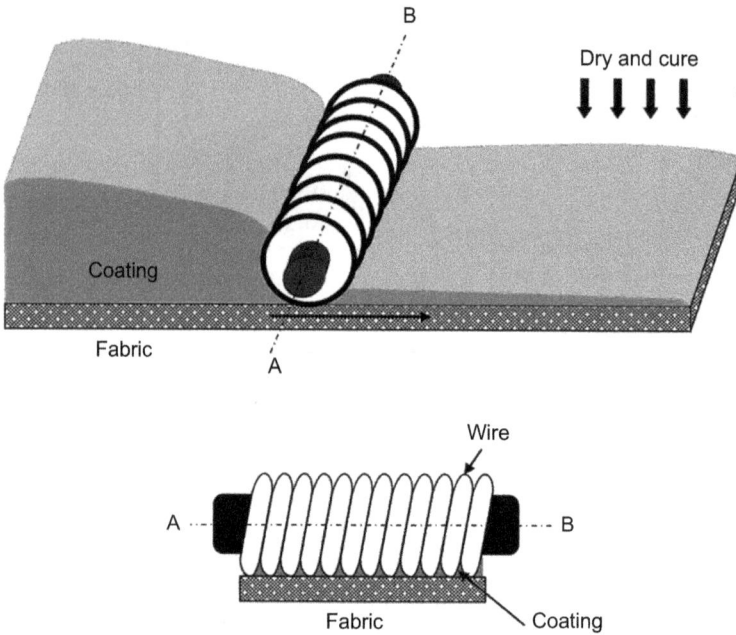

Figure 3.7: Metering rod/Mayer rod coating (Modified according to [25], Copyright 2013, Elsevier).

3.3.3 Spin coating

Spin coating involves the use of centrifugal force to coat the substrate with the coating material (Figure 3.8). The coating material is deposited onto the substrate and the substrate is then subjected to a rotating force at about 1,000–3,000 rpm to generate the centrifugal force, and this causes the coating material to spread across the substrate. In another method, the coating material can be applied to the already rotating substrate to coat it. Due to the volatile nature of the solvent, it evaporates and dilutes the film, delivering the desired thickness (usually in the range of 100–200 nm). Although this technique is able to achieve high production, there is a huge wastage of coating material during the process, which restricts its usage [26].

3.3.4 Meniscus coating

This coating technique has been successfully utilized to coat the substrates to produce low-cost solar panels that can be used in the field of smart textiles. The paint is pumped into the stainless steel applicator roller that has pores. This roller can rotate in the same or in the opposite direction of the textile substrate, depending on the

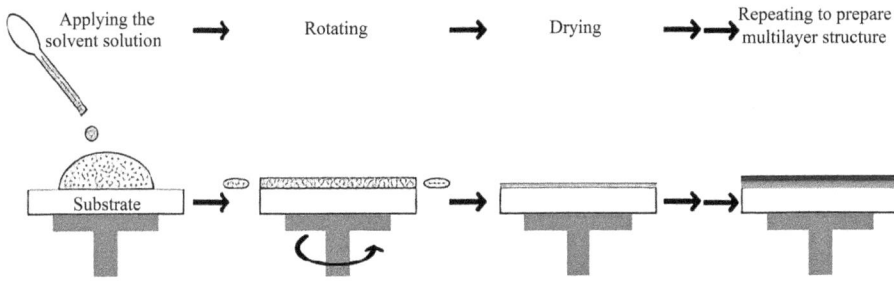

Figure 3.8: Schematic of spin coating technique [27].

requirements. The paint passes through the porous wall of the applicator roll, forming a film on the substrate, which is moved by the coating roller, which is at some distance from the applicator roll. The meniscus formed in the space between the two rollers depends on the distance between the rollers as well as the speed and the direction of the roll [26], as illustrated in Figure 3.9.

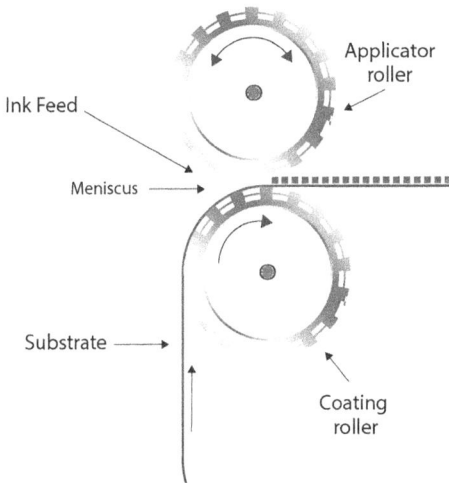

Figure 3.9: Meniscus coating technique (reproduced with permission from [26], Copyright 2020, John Wiley & Sons).

3.3.5 Slot-die coating

Slot-die coating is also referred to as "*hot melt extrusion coating*". Here, the coating material, which is usually polymer granules, is melted and pressed through a sheet of die and is directly coated on the substrate. The textile substrate is supported by a back

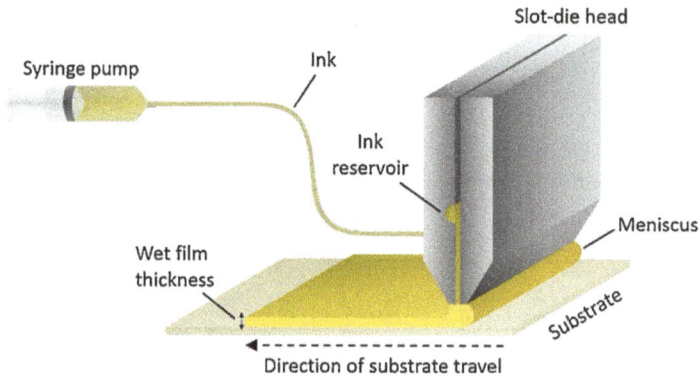

Figure 3.10: Slot-die coating technique for the delivery of ink onto the substrate (reproduced with permission from [28], Copyright 2020, Elsevier).

roller, which imparts pressure to avoid the entrapment of air; thus creating a closed system (Figure 3.10).

The quality of coating depends on following factors [24]:

- Pressure of the die
- Tension on the substrate
- Rate of flow
- Distribution of flow
- Delivery rate of pump
- Line speed of substrate
- Extruder temperature profile
- Extrusion temperature
- Gap or slot thickness of the die
- Required thickness of the film
- Nip pressure of the rolls after coating
- Construction of the substrate
- Melt flow index of the polymer
- Speed of the coating line
- Temperatures of the coating rolls

3.3.6 Spray coating

Spray coating is a completely non-contact coating method where the coating material is atomized to produce a continuous spray through a spray nozzle, as shown in Figure 3.11. Among the various other ways, the spray can be driven pneumatically with the help of a steam of air or pressurized gas. The pressurized gas can be of nitrogen, argon, and so on. This pressurized gas helps in breaking down the droplets of the coating materials and

produce aerosols that are directed toward the target substrate. The breaking down of the droplets to form aerosols is highly dependent on the following parameters:
- Surface tension of the coating material
- Viscosity of coating formulation
- Gas parameters
- Geometry of the nozzle (shape, size, etc.)

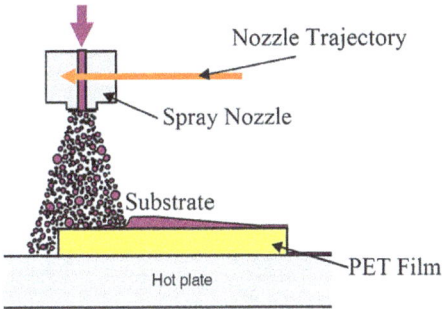

Figure 3.11: Spray coating [29].

Again, the deposition of the coating materials is dependent on
- Pressure of the air
- Distance between the nozzle and target substrate
- Type of solvent used
- Speed of coating
- Solvent used
- Temperature
- Humidity and so on

It is important to note that the distance between the substrate and the nozzle greatly affects the coating layer, as large distance will result in a dry film whereas a small distance will result in moist films. So, this parameter is required to be maintained at an optimum level to obtain the required coating properties [26].

3.3.7 Brush coating

Brush coating can be considered a suitable technique to produce conductive textiles for application areas of smart textiles as high precision in coating can be achieved to obtain defined patterns and many more. The principle involves dipping the brush in the coating material and applying it to the textile substrate (Figure 3.12). The brush used is usually a fibrous one and the material of the brush depends on the solvents used for the production of the coating solution. Chemically stable materials such as nylon are used to prepare the brush. The brush size affects the coating deposition and thus depending on the end use, the brush size should be selected. This method provides a better control over the thickness of the film and can even coat uneven surfaces [26].

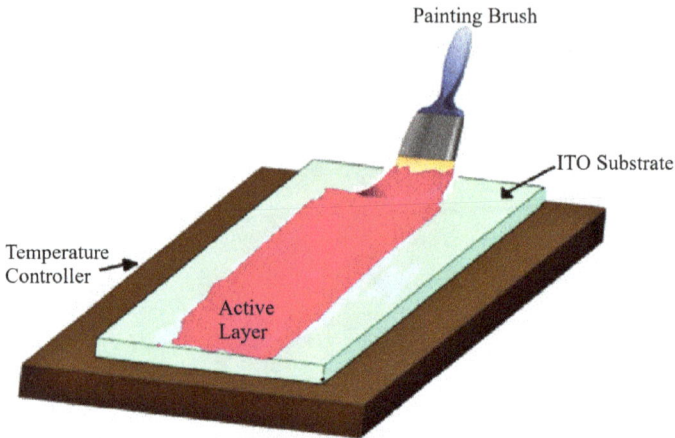

Figure 3.12: Brush coating to deposit the active layer (reproduced with permission from [30], Copyright 2018, Springer Nature).

3.3.8 Dip coating

Dip coating, also known as *"impregnation coating"* or *"saturation coating"*, is a technique where a low viscosity coating liquor is used to coat the fabric. The basic principle involves dipping the textile substrate into the bath containing the coating material and then passing through a nipping roller to squeeze out the excess liquor. Instead of nip rollers, flexible doctor blades can also be used to obtain the required add on. In this coating technique, the pick-up is quite low, and the coating material penetrates onto the interstices of the fabric [22]. The factors that affect the coating properties in this technique are [24]:

–   Geometry of the substrate
–   Viscosity of the coating formulation
–   Surface tension of the liquid
–   Velocity of the coating
–   Geometry of the applicator
–   Nip pressure
–   Number of dip-nip cycles
–   Fibre wettability
–   Duration of the dip
–   Surface roughness of the substrate, and so on

The main forces that get applied during the coating process are gravitational, inertial, viscous, capillary, and intermolecular (disjoining pressure). Figure 3.13 illustrates schematically the function of such forces in the process of dip coating.

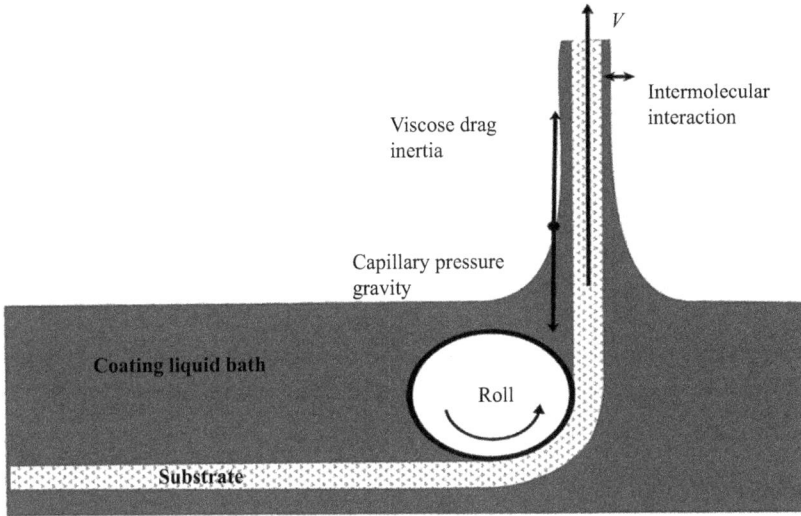

Figure 3.13: Forces involved during dip coating (Modified according to [24], Copyright 2019, Elsevier).

Specifically, inertial forces and hydrodynamic drag pull up the liquid to thicken the film, while the pressure of the capillary maintains the shape of the static meniscus.

3.3.9 Roll coating

Coating the textile substrate with the help of rotating rollers is the basic principle of this coating technique. This technique can be categorized into different types, depending on the number of rollers used, configuration and geometry of the rollers, surface structure of rollers, roller rotating direction, and many more. Some of the most used techniques are discussed further,

3.3.9.1 Direct roll coating

This method uses low viscosity coating materials that are applied to the substrate by the applicator roller, and the deposition is controlled by the doctor knife or a roller placed over the applicator roller, between which the substrate moves as shown in Figure 3.14.

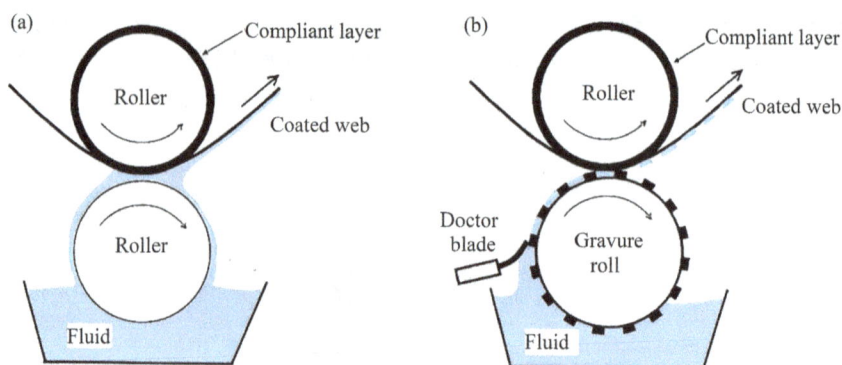

Figure 3.14: Schematic of (a) direct roll coating and (b) gravure coating (reproduced with permission from [31], Copyright 2016, Springer Nature).

The thickness of the coating can be controlled by controlling the nip pressure, coating solution parameters, and absorbency of the substrate [22].

3.3.9.2 Gravure coating

It is also known as **"engraved roll coating"**. Instead of the conventional rollers, rollers with engraved patterns are utilized in this coating process. These engraves can be of different shapes and geometries, and act as the metering device. The engraved roller is partially drowned in the bath of coating, and during its rotation, the engraved pattern gets filled with the coating liquid. An excess of such liquid helps in film formation on the roller's surface. The excess coating is removed through the doctor blade and the roller surface is further pressed against the textile surface to coat it. Various parameters that affect the coating are [24]:

- Geometry of the engraves (shape, depth, area, etc.)
- Coating material properties (viscosity, solid content, etc.)
- Application pressure
- Type of substrate
- Machine speed

3.3.9.3 Kiss coating

In this coating technique, as described in Figure 3.15, two rollers, known as the pick-up roller and the applicator roller, are used. The applicator roller is placed vertically above the pick-up roller, which is kept immersed in the bath containing the coating material. Instead of the substrate moving between these rollers, the substrate passes over the applicator roller. This can be imagined as a two-step procedure, when the pick-up roller first picks up the coating material and transfers it to the applicator roller when they come into contact, which is often referred to as kissing. The applicator roller then transfers the coating material to the substrate passing over it at a constant speed.

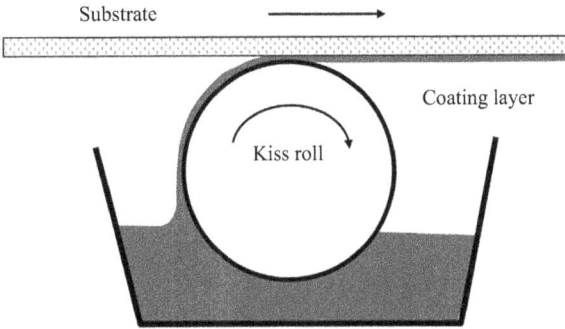

Figure 3.15: Kiss roll coating (Modified according to [24], Copyright 2019, Elsevier).

3.3.9.4 Reverse roll coating

Figure 3.16 describes reverse roll coating, which involves the use of three rollers, namely applicator roll, metering roll, and back-up rubber roll [32]. There is a precise gap, that is, the metering gap, that is needed to be maintained between the applicator roll and metering roll to achieve a uniform coating. The coating material is initially deposited or loaded onto the roll of the applicator, which then transfers the coating material to the substrate. However, before transferring to the substrate, the coating material passes through the metering gap that helps in removing excess material off the applicator. At the application nip, the coating material is transferred to the substrate that is running opposite to the movement of the substrate, resulting in the wiping of the coating.

3.3.10 Transfer coating

In the coating of knitted fabrics, transfer coating plays an important role as knitted fabrics get distorted when stretched over to coat them through the direct coating processes. This distortion could lead to improper and non-uniform coating and even result in coat peel off when the tension is released [22].

The basic principle of transfer coating involves spreading the polymer or coating onto a release paper to form a film and then transferring this film to the target substrate. The steps involved and schematic of the process are illustrated in Figures 3.17 and 3.18, respectively [23]-

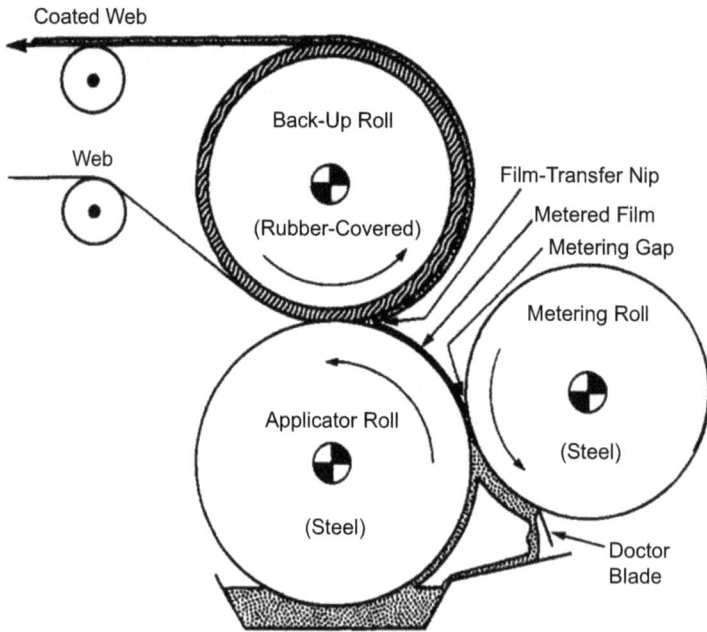

Figure 3.16: Schematic of reverse roll coating [22].

The most common usage of this coating technique is the production of waterproof protective textiles by coating them with polyurethane. These coatings are also in use for achieving the artificial leather look for aesthetic appeal. Other uses include upholstery, luggage, footwear, gloves, waterproof mattress covers, and woven velvet automotive seat fabric [23].

Although this technique is expensive than other conventional techniques, it is still very popular due to the flexibility if offers due to the low penetration of coatings and the ability to overcome the nuisance of stiffness that is usually seen in most coating methods [24].

3.3.11 Rotary screen coating

Rotary screen coating involves the process of coating a textile substrate with the use of a screen, which is seamless, perforated, and nickel-sleeved, as shown in Figure 3.19. The substrate is passed under the meshed roller in which the coating material is present. An array of dots is pushed through the perforated screen by the squeegee bar inside the screen and by centrifugal force onto the fabric. The fabric moves at the same speed as the rotation of the rotary screen and there is, thus, no frictional contact between the two [23]. The coating material is squeezed through the meshed screen and is coated

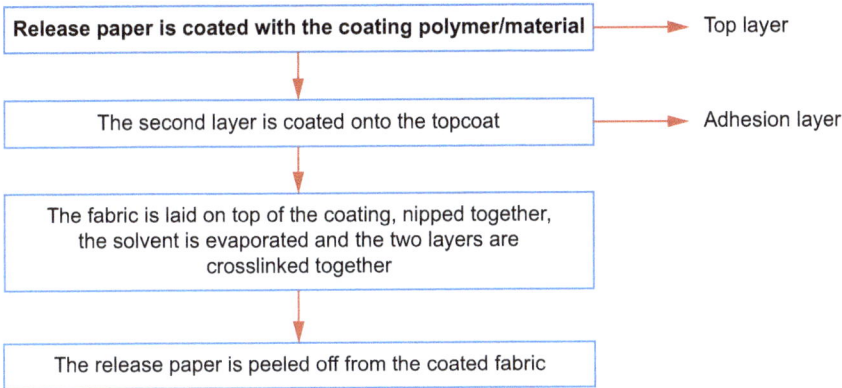

Figure 3.17: Steps for transfer coating.

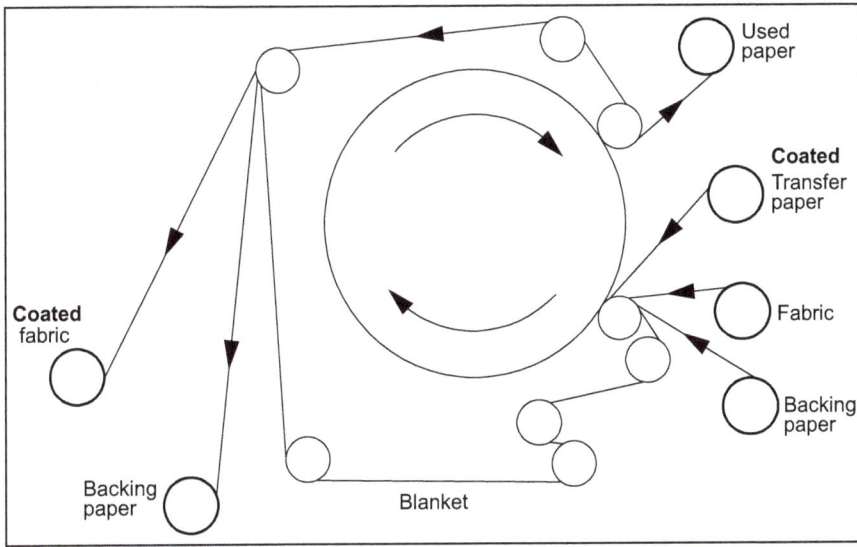

Figure 3.18: Schematic of transfer coating (modified after [33]).

onto the substrate. Little tension or no friction is induced during the coating process; so this technique is suitable to coat soft and delicate fabrics too [22–24].

The amount of coating that is deposited on the substrate depends on following factors-

– Screen mesh number
– Squeeze pressure
– Angle between the squeeze blade and the screen
– Viscosity of the coating fluid
– The squeegee setting with regard to the counter-pressure roller

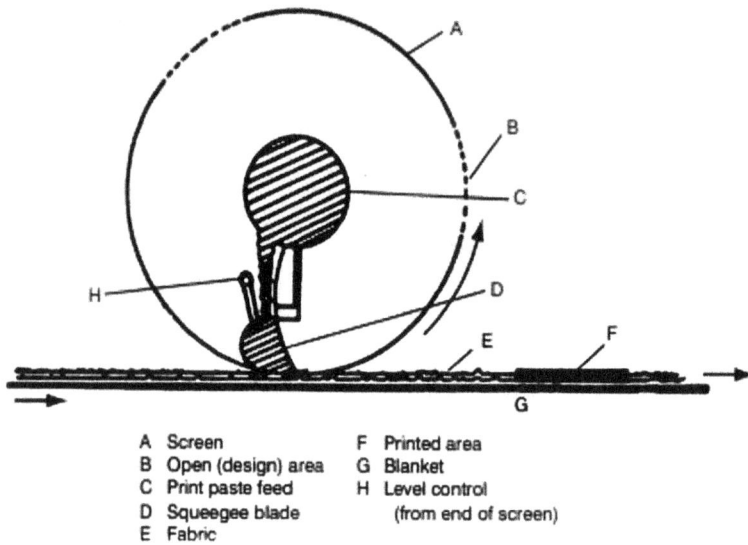

A Screen
B Open (design) area
C Print paste feed
D Squeegee blade
E Fabric

F Printed area
G Blanket
H Level control
    (from end of screen)

Figure 3.19: Schematic of rotary screen coating (modified after [33]).

3.3.12 Foam coating

Foam coating is considered as one of the eco-friendly processes as it involves reduction in solvents as well as chemicals. Foam is prepared from the coating material by dispersing it with water. Instead of dipping the fabric to the coating material, this technique involves direct application of the foam to one side of the textile substrate. Due to the inherent properties of foam, it dries quickly as the water content is usually low; thus coating the fabric with a thin flexible film. Very low add-ons can be achieved (in the range of 2–3%) with this coating technique, leading to cost savings in the required chemicals. This method can be used to apply polymer to the woven fabrics and knitted fabrics and to fabrics produced from spun yarns or fabrics of a general open construction that generally cannot be directly coated [22, 23, 34].

The main advantages of this process are-
- Low chemical requirements
- High production
- Less water usage
- Cost saving
- Suitable for open constructions
- Flexible coatings can be obtained
- High add on is possible

3.3.13 Calender coating

Figure 3.20 depicts the calender coating process, which involves the use of massive rotating rollers, sometimes five or more in different configurations, to crush the dough of the coating material to produce a thin film of uniform thickness. The thickness of the film depends on the distance between the rollers as well as on the properties of materials. This process is also suitable to apply freshly produced film onto the target substrate. More number of rollers provide better homogeneity and more control. Some of these rollers may have in-built heating arrangements to provide heat to the film and substrate, which facilitates better adhesion of the film to the target substrate [22, 23].

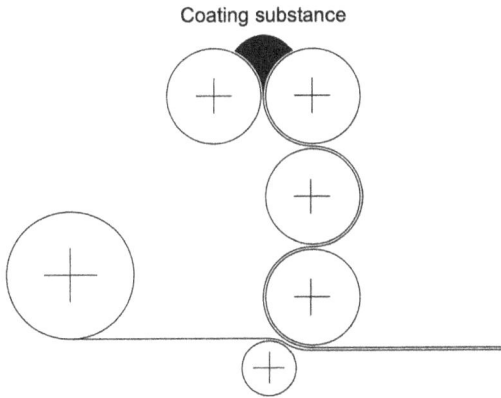

Coating substance

Figure 3.20: Schematic of calender coating (reproduced with permission from [35], Copyright 2012, Springer Nature).

3.3.14 Powder coating

Powder coating is the solid-state coating method, where the coating material does not require to be dissolved in any solvent; thus leading to the generation of low amount of volatile organic compounds and other waste compounds, and thus can be considered as an environmentally friendly process (Figure 3.21). The coating powder is directly applied to the textile substrate. Complete reduction of water or solvent leads to less energy consumption as no drying step is required. Scatter coating is a powder coating method that involves an even spreading of coating powder on the substrate with the help of a scatter roller that is constantly rotating. The amount of powder can be measured by the dosing roller, its rotational speed, as well as by the line speed of the substrate [24].

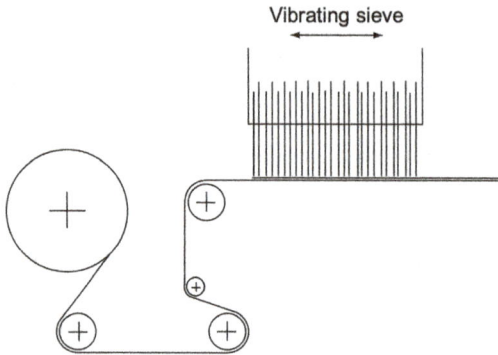

Figure 3.21: Schematic of powder coating using a vibrating sieve (reproduced with permission from [35], Copyright 2012, Springer Nature).

3.3.15 Curtain coating

A freely forming liquid sheet is formed from the coating material and is then made to deposit on the fabric substrate, which is continuously in a moving state. This technique does not utilize any kind of force to impregnate the coating material into the substrate and is a friction-less process (Figure 3.22). The surface roughness does not affect the film thickness and this technique can coat the substrate with thickness as low as few microns. The main issues with this technique are the formation of air bubbles that get entrapped, and the production of a stable curtain of coating material becomes difficult to be achieved. Optimizing these parameters could increase the applicability of this technique to new or advanced products [24].

Principle of application

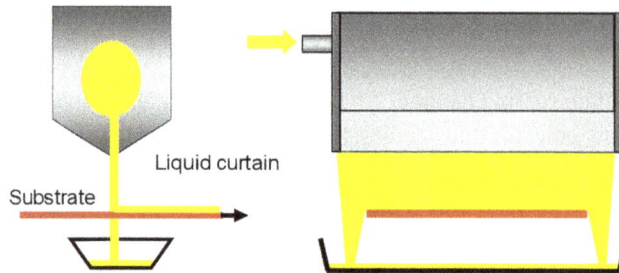

Figure 3.22: Schematic of curtain coating [36].

3.4 Polymers and chemicals used in coating

3.4.1 Coating layers and formulations

Generally, three different types of coating layers are applied on the textile substrate to obtain a coated fabric [37].

3.4.1.1 Primer coat

This is the base coat upon on which all other layers are deposited. This is the first layer that is deposited onto the substrate. Primers are the key to the adhesion of the total coating system. The primer must also provide a proper and compatible base for the top coats.

3.4.1.2 Intermediate coat

Intermediate coat is often used to increase the thickness, in case the end users demand, along with delivering other properties such as strong chemical resistance, resistance to moisture, being waterproof, strong adhesion, and strong bonding between the base and top coat. This is the layer that plays the most important role in delivering the desired properties.

3.4.1.3 Top coat or finish coat

Top coat is the functional layer that is coated onto the textile substrate. Top coat is required mainly for aesthetic property (like appearance, hand feel, and colour). Sometimes, it also contributes to the main properties.

A typical solution coating formulation contains the following chemicals:
i.   Polymer
ii. Solvent and diluent or thinners
iii. Cross-linker or bonding agent
iv. Accelerator
v.   Other chemicals (matting agent, pigment, thickeners, softeners, functional additives [e.g. FR chemicals, antimicrobial chemicals, and UV stabilizer], plasticizers (for PVC coating), heat stabilizer (for PVC coating), fine nano/micropowders [TiO_2, $CaCO_3$], etc.)

Coating formulations contain the following "thinners" or "solvents" [37]:
- **Mineral spirits** for oils or alkyds
- **Aromatics (benzene, xylene, and toluene)** for coal tar epoxies, alkyds, chlorinated rubbers, and polyurethanes
- Dimethylformamide (DMF), methyl ethyl ketone (MEK) for polyurethanes
- **Ketones (MEK, MIBK)** for epoxies and polyurethanes
- **Alcohols (isopropyl alcohol)** for phenolics and inorganic zinc
- **Water** for acrylics, polyurethanes, and some inorganic zinc

3.4.2 Coating materials

Some of the popular coating materials are mainly polymers or its mixtures and various additives. Such polymers include polyacrylic (PA), polyurethane (PU), polyethylene terephthalate (PET), polyvinyl chloride (PVC), silicone rubbers, and other elastomers [38].

The governing factors influencing the quantity of polymers applied are:
- Solution concentration
- Blade profile and its angle
- Fabric speed and tension

The additives for the coating composition can be pigments, flame retardants, fixatives, thickeners, fillers, hydrophobic agents, preserving agents, softeners, and so on [39–41].

3.4.2.1 Polyvinyl chloride (PVC)

Polyvinyl chloride (PVC) is a resin that is white in colour and not soluble in water. Apart from protecting pipes, it is also used for textile coating. It is amorphous in nature with no specific melting point. It shows a transition in properties between 170 and 180 °C. Often, it contains some amount of plasticizers to attain flexibility. However, the main issue with PVC coating is the release of hazardous chlorinated compounds. Therefore, currently, worldwide environmental awareness is restricting the use of PVC coating.

3.4.2.2 Polyacrylic (PA)

Acrylic polymers are mostly used in upholstery due to their numerous properties. However, they are also used in coating formulations owing to their good resistance to water and they do not undergo premature ageing in intense sunlight.

Acrylic polymers are synthesized by the polymerization of methacrylic acid and acrylic esters. They can be prepared by solution, mass, and emulsion polymerization methods. Emulsion polymerization is the preferred method for coating using this polymer [42].

3.4.2.3 Polyurethane (PU)

Polyurethane coating is used most extensively in the textile industry. The merits of polyurethanes in coating are:
– Good abrasion resistance
– Low temperature flexibility
– Good elongation and elastic recovery
– Excellent adhesion property

Polyurethanes are mainly synthesized by the polyaddition reaction between polyols and polyisocyanates. Some of the isocyanates that are used for PU synthesis are:
– Toluene di-isocyanate
– Hexamethylene di-isocyanate
– Diphenyl methane di-isocyanate
– Xylene di-isocyanate
– *p*-Phenylenedi-isocyanate.

In recent years, a blocked isocyanate is being preferred because of its better stability and low reactivity at room temperatures. Polyols show a much higher degree of versatility than di-isocyanates in terms of functionality, chemical structure, and molecular weight. However, there are mainly two types of polyols that are used: polyester polyols and polyether polyols. Tables 3.2 and 3.3 describe the different polymers used for coating along with the properties achieved.

Table 3.2: Various polymers for textile coating [43].

Polymers	Properties/advantages	Application
Polyurethane (PU)	– Solvent and latex form – Good extensibility – Low temperature flexibility – Good abrasion resistance and chemical resistance – Aliphatic PU shows good resistance to weathering – Lamination films available	– Waterproof protective clothing – Life jackets used in aircrafts – Adhesives/binders for various fibres/fabrics and nonwovens
Polyvinylchloride (PVC)	Available in the form of plastisols, powders, and granules, – Compounded to give a wide range of properties – Good inherent improved FR – Good oil, solvent, and abrasion resistance	– Architectural uses – PVC polyester-tent covers – Leather – Tarpaulins

Table 3.2 (continued)

Polymers	Properties/advantages	Application
Polyvinylidene chloride (PVDC)	– Very good fire retardancy (FR) – Very low gas permeability	– Gas barrier-coated/-laminated textiles – Blends with acrylics and PVC to improve FR in coatings
Polyacrylics (PA)	– Large number of variants and co-polymers – Water-based grades available – Blendable with other polymers and chemicals – Good UV resistance and optical clarity, generally cheap	– Tarpaulin lacquers – Carpet backings and upholstery – Wall coverings and exhibition board backing – Blackout coating on curtains
Polyolefin	– High acid, alkali, and chemical resistance – Good flexibility and low water absorption – Low density – Less expensive than other polymers	– PP nonwoven and PE film laminates are used in packaging and garment industry (e.g. for making PPE coverall fabric)

Table 3.3: Functional properties of different coating compositions [44, 45].

Coating chemicals	Functional properties
Perfluorochemical, polyacrylates, silicone-based products, PVC, and vinyl acetate	Stain release Soil release Water repellent Hot oil repellent/resistance Waterproof
Poly-hexamethylene biguanide hydrochloride (PHMB) and cyclodextrin	Deodorant/antibacterial clean room fabric finishes, and chemicals
Activated carbon-based coatings	Chemical odour absorbing
Based on Aramid, Teflon, PTFE, carbon, and neoprene coating	Chemical protective
Hydrophilic, polyacrylamide-based products, and polyurethanes	Breathable coating
PVC, Teflon, carbon coating, and silicon rubber	Thermal resistance and insulating
Polytetrafluoroethylene, perfluoro-octanoic acid, and Teflon	Heat and corrosive-resistant coating

Table 3.3 (continued)

Coating chemicals	Functional properties
LDPE/HDPE/PVC	Stiffness/interlining coating
Polyurethane coating	Solvent resistance, abrasion resistance, ageing resistance, and ozone resistance
Butadiene–polyurethane resin	Waterproofing, electrical encapsulation, and sealants

3.5 Nanotechnology-based smart coatings

Nanotechnology-based coatings are nowadays given importance as a new technique for finishing and surface modification of textiles. Nanoparticles and nanostructured materials play a major role in surface modification as well as in smart functionalization of textiles. The effect of nanoparticles is always size-dependent and surface area-to-volume ratio-dependent. Their importance in the textile arena is quickly growing due to several applications like apparel, industrial, and technical textiles. Such methods involve cross-linking, immobilization, and in situ synthesis of nanostructures and nanoparticles on textile substrates. The primary objectives of scientists are to generate multifunctional moieties such as mothproofing, antimicrobial, superhydrophilicity, self-cleaning properties, flame retardancy, electrical conductivity, and electromagnetic shielding, through nanoparticles.

Because of the huge efforts in textile nanocoatings, several methods have emerged, including transfer printing, spraying, immersion, rolling, padding, and simultaneous exhaust dyeing. Among all of these, pad-dry-cure method is the most commonly used procedure for textile nanocoatings. However, some vital issues concern application of nanoparticles, including their size uniformity, shape of nanomaterials, purity of nanomaterials, dispensability in colloids, health and safety of nanomaterials, as well as their durability in textiles.

A major issue related to the application of nanoparticles on textiles is their ability to form dispersion in colloids before coating. Usually, nanoparticles tend to aggregate into micro clusters owing to Van der Waals and electrostatic double-layer forces. Stable dispersions can be produced by good grinding and surface modification of nanoparticles through covalent bonding and by adding dispersing agents such as surfactants.

Till date, durability is a major problem in nanotechnology and this is being addressed by researchers all over the world. Most nanoparticles lack attraction for textiles because they do not bear any surface functional groups. Thus, in order to build up the bonding power between them and to achieve good durability, surface functionalization through physical/chemical or physico-chemical processes is necessary. All the same, nanoparticles can also be integrated into the polymer matrices to generate nanocomposites for coating on the surfaces of textile materials.

3.5.1 Thin-film deposition technique

Vapour deposition is a method of thin film production on textile substrates. It is classified into two categories: physical vapour deposition (PVD) and chemical vapour deposition (CVD) methods. Some parameters such as uniformity, film thickness, precursor, and the substrate type determine the choice of the method [46].

3.5.1.1 Physical vapour deposition (PVD)

PVD is a process of deposition of a thin layer of coating on the substrate through a vaporization process. A solid material is usually used as the source of vapour, which is then transferred to the substrate surface, forming a uniform coating layer [47].

The vapour atoms or molecules that are produced from a solid material are placed alongside the substrate. PVD involves three main streams: vacuum evaporation, sputter coating, and ion implantation. The coating materials used in vacuum evaporation, sputter coating, and ion implantation techniques have low (a few tenths of eV), moderate (tens to hundreds of eV) and high energy, respectively [48]. The energy applied governs the forces among coatings materials, substrates, and the film.

Vacuum evaporation

In this method, only vacuum is required for the deposition and evaporation of the film on the substrate. Vapour molecules are allowed to directly condense on the substrate surface with the formation of a solid film. This method imparts electromagnetic interference (EMI) shielding to protect textiles by metal nanoparticle deposition on textiles [49–51]. Lai et al. [52] investigated the efficiency of EMI shielding of metals coated with PET film. Metals such as aluminium, copper, titanium, and silver (Ag) were applied to substrates using this technique. Other influencing factors, such as filaments alignment and aperture ratio, on the ultimate efficiency were also studied [52]. However, the coating achieved by this method is generally not stable, especially in the case of synthetic fibres. Therefore, plasma pre-treatment is recommended prior to metallic coatings by the method of vacuum evaporation [53].

Ion implantation

It is an approach to PVD that is generally used for the modification of textile surfaces [54]. This environmentally friendly technique bombards the solid material with popular metallic ions in order to integrate them on substrates such as textiles, polymers, and ceramics, and also to change the surface properties. Ion implanting system contains substrate, accelerator, and ion source. Öktem et al. examined the effects of the integration of Cu ion on the electrostatic characteristics of PET fabrics. He also studied the integration effects of Cr, Ti, and C on the mechanical properties of membranes. A vapour vacuum arc implanter produced the ions. Surface modification of the PET fabric by Cu ions via MEVVA-generated antistatic characteristics resulted in the reduction of the surface resistance of fabrics. As far as PET membrane is concerned, after surface changes with Cr, Ti, and C ions, properties such as coefficient of friction and wear loss of fabrics were reduced [54]. This method involved the ionization of the coating material and energization in an accelerating mode. This resulted in the implementation of ions into the depths next to the surface area of the substrate, generating a thin film [48].

Sputter coating

This method can be used for the deposition of a thin layer of nanoparticles on a textile surface. A low-pressure plasma is generated from an inert gas like argon under vacuum [47, 55]. Plasma contains highly energetic electrons and ions due to which atoms are ejected from the coating layer and they form a bond owing to their high energy. The substrate and the target material are situated on the anode and cathode of the equipment, respectively. The potential difference between them results in the electrons flow from the cathode to the anode, and a glow discharge. The argon gas inside the chamber is ionized by the electrons and accelerated, hitting the cathode-based coating material [55]. Hitting the cathode can produce secondary electrons, whose rate of generation depends on the kinetic energy of the ionized argon gas. The striking ionic energy evaluates the atomic rate of ejection from the target material. In magnetron sputter coating, an attempt was made to keep the secondary electrons near the material via a magnetic field. Though expensive, such a technique can lead to some advantages such as producing more environmentally friendly, durable, compact and uniform layers of coating [56].

3.5.1.2 Chemical vapour deposition (CVD)

Chemical vapour deposition is a technique that can be widely used for the deposition of a thin film on the substrate. It utilizes chemical reactions on the surface of the substrate or in its vicinity to produce the film for deposition. CVD can be divided into various types depending on the processing parameters it utilizes and these can include [57]:

– APCVD (atmospheric CVD)
It utilizes atmospheric pressure or 1 atm in the chamber where the reaction takes place.

– LPCVD (low-pressure CVD)
It utilizes very low pressure and also ensures uniformity in deposition.

– PECVD (plasma-enhanced CVD)
It is used to deposit films on the substrate, from a gas phase to a solid phase.

– LECVD (laser-enhanced CVD)
It utilizes laser beam to fasten the chemical vapour deposition on the substrate.

3.5.2 Electrospun nanofiber coating

When there is an electro spraying of a polymer solution having sufficient viscosity, the starting jet does not tear into drops and form fibres (Figure 3.23). This type of fibre is obtainable from a melt form as well as by a solvent-based solution. Several parameters influence the conversion of solutions of polymers into nanofibres. These include ambient parameters, primary variables, and solution properties [58–60]. Electrospinning is a flexible technique for homogeneous coating of nanofibers on to different substrates. It involves making available very high voltage to the nozzle and the substrate in order to obtain a web of fibres, for example, a mat. Fibres of varying diameters, between 10 and 500 nm, can be produced.

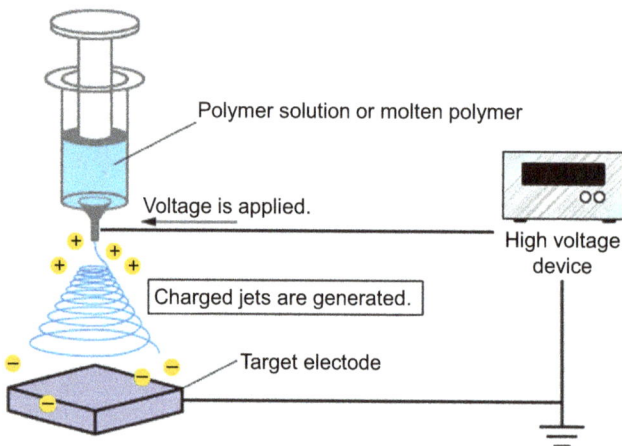

Figure 3.23: Schematic diagram of electrospinning.

3.5.2.1 Solid-phase micro-extraction

Analytical approaches consist of exhaustive preparation of sample-like extractions from solvent and liquid–liquid. Solid-phase micro-extraction (SPME) is a substitutive approach that appeared as a method of preparation of a sample in the early 1990s. This involved a coating of thin film on the textile substrate for the extraction of analytes of interest from a matrix sample. The development of SPME happened as an analytical method of extraction for organic solvents in water, pesticides, flavours, and explosives. It is seen that SPME has always been an efficient extraction method [61, 62]. Electrospun nanofibers are usually thin stainless-steel wire coatings. These are used in the application of SPME [63–65].

SPME coating by the electro-spun method and polyamide is recommended for the extraction of phenolic traces that are aqueously dissolved. It is an inexpensive, basic, and easy method for phenolic extraction, with good reproducibility and sensitivity [64].

3.5.2.2 Electrospun nanofibers in sensors

Coatings of polymers on a sensor surface can improve the sensitivity of sensors. Electrospinning can control the thickness, permeability, and the sensor's porous structure through coating. Nanofibrous coating by electrospinning was pioneered for the promotion of electrodes with a large surface area to volume ratio.

An electro spun nanofibrous membrane was subjected to cove quartz crystal microbalance (QCM) to obtain a gaseous sensor for the NH_3 detection. QCM is an acoustic wave technique for identifying a small quantity. There was an absorption process that was not reversible and there is a strong bonding between the carbonyl groups of polyacrylic acid (PAA) and the NH_3 molecules; so it affected the properties of QCM sensing and caused shifting of frequency. Cross-linkable PAA and PVA were used as a mixed solution for preparing nanofibrous membranes. The rise in rigidity and mean diameter of nanofibers were obtained by raising the conductivity and viscosity of the solution blends and the content of PAA component. The sensing properties of the sensor were remarkably affected by the PAA and NH_3 contents in the blend and relative humidity [66].

3.6 Active coating processes

Recently, invention of active coatings such as "artificial cells" systems have produced materials that can facilitate the healing of damaged substrates. This can be exemplified by a sol–gel system with polymer nano-containers used as reservoirs of active nanoparticles. The mechanical impact causes a steady damage, and the polymeric nano-containers crack and generate a corrosion inhibitor to the affected zone, promoting self-healing.

Research on "smart" coating systems are still in the pioneer stage. However, advancements in this field are primarily increasing the adaptive material fabrication to control their structure and change them, prior to the occurrence of any catastrophic failure [67]. A challenge to material science is to frame synthetic systems that can actively enhance this behaviour, not just by forgetting the localized environmental changes or of a defect, but also by re-developing the continuity and integrity of the affected area or by enhancing the effect of photoelectric phenomenon due to alterations in the irradiation of light. Such "smart" coatings would majorly enhance the utility and durability of the manufactured items. The coatings must be autonomously repaired in the absence of an external hindrance. Furthermore, the coatings should be practically capable of healing damages multiple times for new fissures or cracks that appear. These self-healing effects would result in coatings suitable for automotive, aerospace, and surgical implants applications and allow for structure fabrication.

Of late, an increase in the number of papers and patents has been observed that mainly focus on the specific application or synthetic approach, principally on hybrid inorganic-organic coatings with "smart" behaviour. Such interest comes from the various special features of such materials related to enhancing the pivotal role played by surface molecular chemistry and interfacial forces as the dispersed phase size decreases [68].

The magnetic, catalytic, photochemical, optical, electrical, cohesive, adhesive, and mechanical features of these novel hybrid components are often a synergistic combination. Thus, durable organic polymers that protect from ultraviolet (UV) light radiation and heat, offer hydrophilic or hydrophobic properties [69–71], show elasticity, toughness [72], enhanced hardness [73], and low surface energy [74], and are available at catalytically active host sites or reactive functional groups can be produced by diffusion, inclusion, or dispersion of an inorganic component, where the covalent or coordinative binding provide the required stabilization of the incompatible phases with a large interface area. Entirely new materials can be developed and applications needs can be met when compared to the conventional ones, as in the case of the combination of typical molecular properties, from catalytic, magnetic, photochemical, electrical, biological activity with the adjustable mechanical ones of inorganic and organic polymers, and networks [68].

Sol–gel processes are usually used to fabricate hybrid organic-inorganic materials, including self-assembled films, as shown in Figure 3.24. A sol is a liquid solution containing a colloidal suspension of a material of interest dissolved in an appropriate solvent. Condensation reactions between the dissolved precursor molecules result in structures being formed within the sol. The morphology, growth rate, and size of such structures depend on the kinetics of the reactions within the solvent, which in turn are determined by parameters such as the temperature, pH and agitation of the solvent, amount of water present and solution concentration, and other parameters.

With prolonged time, condensation reactions lead to the aggregation of chains or growing of particles until a gel is formed. It can be visualized as a very large number of cross-linked precursor molecules forming a macroscopic-scale and continuous solid phase, which encloses a continuous liquid phase consisting of the remaining solution. In the end steps of the sol–gel process, the enclosed solvent is removed and the precursor molecules crosslink resulting in the desired solid [70, 71].

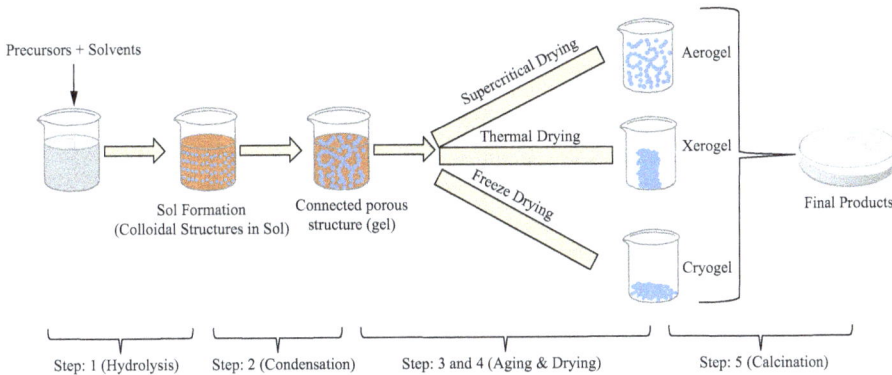

Figure 3.24: Process steps in sol–gel process [75].

Sol–gel process offers various advantages over other synthetic routes that include high magnitude of control over the resulting structure, inexpensive raw materials, and mild processing conditions like mild pH, low temperature and pressure. As regards the final product shape is concerned, there are mandatory merits, due to the easy castability prior to gelling, including micro- or nanoscale particles, thin films, and fibres [70, 71].

3.7 Lamination

Lamination is a process of combining two or more substrates (layers) together, usually by using an adhesive. The main purpose of lamination is to generate a rigid and multiple layered structure having features that are difficult to achieve by a single substrate [76]. The key substrates of textile-based lamination are textile fibre webs, fabrics, nonwovens, polymeric films, foam, and membranes [77, 78].

Lamination, to some extent, differs from coating, but the processes often share many common principles, necessities, and machineries [78]. Strong adhesive force between the layers is necessary in lamination just like in coating processes. In lamination, adhesion is usually obtained by mechanical bonding, thermal, pressure, or by an adhesive [77, 79].

Laminating processes are generally classified in several ways: method used to combine, the number of layers, and substrate type. The next section will cover the different laminating processes classified in terms of bonding methods such as flame lamination, adhesive lamination, and ultrasonic lamination. Adhesive lamination is sub-categorized depending on the adhesive nature: dry heat lamination, hot melt lamination, and wet adhesive lamination.

3.7.1 Flame lamination

It is a process where a very thin, thermoplastic layer of foam such as polyurethane is openly subjected to a flame to produce a thin polymeric molten layer. An adhesive such as molten polymer layers is used to fix the substrates by calendric nip pressure, finally producing a laminated fabric, upon cooling [77, 80]. Usually, a 3-ply laminate is obtained using a two-pass lamination over a dual burner [80], while no drying or curing is required. The processing variables are nip pressure, line speed, foam burn-off, flame height and spread, and type of gas [80]. The flame intensity should be properly adjusted at a constant line speed to ensure proper melting, without burning and keeping the adhesive layer thickness constant [81]. It is an easy and low-cost method, with a high production rate. However, it is not eco-friendly because it produces harmful emissions. This method also produces solid bonding layers with less porosity, resulting in stiffer laminates.

3.7.2 Wet adhesive lamination

Wet adhesives for lamination are either aqueous-based or solvent-based [82]. They are used on one side of the substrate in the form of a liquid by traditional methods of coating, like knife coating, spraying, and roll and gravure roll coating. The web, coated with an adhesive, is further attached to the other side of the substrate at a pressure, followed by drying or curing [78, 83, 84].

Performance of adhesive, such as resistance to heat, durability, and bond strength depends on its chemistry as well as the laminating process. The performance is usually affected by chemical resistance of the fabric, its viscosity, flow characteristics, crystallinity of the adhesive, and machine speed [85]. Its penetration into the structure of the substrate must be under control. A penetration is required for sufficient adhesion. However, high penetration may produce structures with high stiffness, less hand, tear strength, and drapability [84, 85].

A solvent-based adhesive is easy to dry but is not environment-friendly because of the hazardous waste generation and high VOC emission. Owing to the regulations of the government and environmental concerns, traditional adhesives of solvent are only used during coupling with a recovery equipment or for expensive incineration [81, 86]. Though water-based adhesives are eco-friendly, it is difficult to evaporate water, thereby requiring a separate high energy drying process and also occupying a large floor space [81]. However, an infrared (IR) pre-heating system can solve such issues and increase the speed of production. A system based on IR heating can generally heat up the substrate that is coated near to the temperature of evaporation, prior to an entry into the primary dry zone, thereby reducing energy and the time necessary for the evaporation of water or solvents. Such IR-based pre-heating system usually needs less floor space and can be easily integrated with a low capital investment [87, 88].

3.7.3 Hot melt lamination

Hot molten adhesive is used as a binding agent in this process [89]. It is a pure solid and melts to liquid state in a temperature range of 80–200 °C, and solidifies to create linkage upon cooling [90]. These adhesives are mostly polymeric compounds that are specially formulated for lamination purposes [91]. They must have the correct hardness, curing time, melting point, melt viscosity, and sufficient adhesion on the material to generate adequate tensile strength, good peel strength, and shear strength. Based on the applications, washing durability, porosity, lightness, and breathability can be necessary [80]. A good adhesive is chosen to obtain the required flexibility in coating, cost, and quality [89, 91].

Generally, there are two different classes of hot melt adhesives: reactive hot melt adhesive systems and thermoplastic polymer-based systems [90]. Reactive hot molten adhesives are entirely cross-linked through the moisture reaction after getting attached to the surface. After cross-linking, the adhesive hardens upon heating. They show high durability with a good climate and boiling resistance, but are rather expensive. This group mainly includes moisture cross-linking polyurethanes (PUs), called as polyurethane reactive (PUR) adhesives [80, 92].

Thermoplastics melt and become solid, primarily as per the temperate modifications, and involve polyester (PET), polyvinylchloride (PVC), polyethylene (PE), polyamide (PA), and ethylene vinyl acetate (EVA)-based compounds [85, 90]. These are delicate against water and steam, and having limited application due to lower softening points.

Hot molten adhesive is usually melted in a separate unit of melting and produced to a hot molten unit of coating [81]. Several traditional coating processes are followed for applying hot molten adhesives. They include spray coating, knife-over-roll coating, screen coating, slot die coating, and roll and gravure coating. The substrate that is coated is further attached with another substrate to form a multiple layered laminate.

Hot molten adhesives are nowadays a substitute to aqueous- and solvent-based adhesives, owing to their environment friendly processing parameters [90]. They possess merits over dry lamination for coating on substrate that is sensitive to heat because the substrate is not directly exposed to heat during lamination [93]. It eliminates drying, where there is high requirement of energy [89, 90]. Lamination is obtained at lower coating levels; and generates coating with soft flexibility and handling [89]. There is instantaneous formation of bonds with no drying step; so processing speeds are not restricted by the rate of drying, ensuring high productivity [86]. Hot molten adhesive has a longer shelf-life than wet adhesives. However, the hot molten adhesive system has certain limitations. A change in the type of adhesive implies that the total pre-melt and the application system must be neat and clean, and such adhesives are comparatively costly [90].

3.7.4 Dry heat lamination

Dry adhesive is a kind of pure solid adhesive. Unlikely hot molten adhesive, solid form of this method is usually coated on the surface and further enhanced under conditions of heat and pressure to perform lamination. Dry adhesives are powders, films, or webs constructed from PU, PE, EVA, PA, and PET [81, 94]. Large solid content with no toxic emission is always suggested from an eco-friendly viewpoint; however, melting dry adhesives requires a lot of energy as there is a chance of exposure of the substrates to high temperature, which is a limitation [81].

Powder adhesives are finely ground under cryogenic conditions to obtain sizes in the range of 1 to 500 μm. They are subjected to paste dot coating, powder dot coating, or scatter coating. Upon melting, they create a non-continuous bond between the layers – bonding at high permeability, drapability, and softness. Although, there are no harmful emissions, due to fine particles, they may produce airborne dust [81, 85].

The other dry adhesives are mostly found in web and film forms. The adhesive rolls are basically integrated over or between the surfaces for lamination [80]. These are used in the lamination of textiles of open structures where powder adhesives are used to pour the voids rather than sticking on to the surface [81, 84, 95]. However, they are costlier than powder adhesives. Moreover, to control the add-on and the application-width, adhesive films/webs of different thicknesses and widths are required. Dry adhesives are placed between two substrates and conveyed through the heating tunnel by a lower and upper conveyer where the molten adhesive forms a bi-layered bond. Pressure is adjusted by nip rollers after the heating section, upon cooling [96]. It is a continuous process with higher production. It also makes use of the dry adhesive at a comparatively lower temperature so as to eliminate the risk of the substrate being exposed to heat [97]. However, sophisticated substrates such as nanofibrous webs may face changes in the structure during the process. The processing pressure and temperature are required to be adjusted to achieve the desired lamination results [98, 99].

3.7.5 Ultrasonic lamination

Ultrasonic lamination requires sound energy for lamination instead of adhesives or heat [100]. Ultrasonic is basically a sound wave having high frequency, in the range of 20 kHz to 1 GHz. It is a range beyond what humans can hear, but ultrasonic waves possess energy that can produce bonding in various materials. Ultrasonic bonding or welding is adopted for slicing, cutting, slitting, perforating, embossing as well as for lamination [101].

In the welding process of ultrasonic lamination, electric signals with large frequency are transformed into mechanical oscillations in a sonotrode or a weld horn. This results in reiterated friction and compression of the fibres in the material during melting and bonding [102, 103]. The most frequently used frequencies are 20, 30, 35, and 40 kHz. The welding force, the amplitude of the vibrations and, the forces applied by the sonotrode to the substrate, influence the bond strength. The amplitude is dependent on the material to be bonded along with the welding force, and it is required to be attuned as per the required area to be covered by the bonding pattern, substrate feeding speed, welding time, materials, and the bond strength [102]. The speed of the substrate, the gap between the anvil roller and the ultrasonic head, and the pressing area percentage of the engraved roll are the major processing variables [102].

This method particularly does not require temperature as well as adhesive, which make this a safe and environment friendly process. It is a distinct technique as multiple layers (in the range of 6 to 12) can be laminated at a time and various patterns can be used for enabling of tailored properties and appearance. Ultrasonic bonds generate textile-like hand feeling, higher absorption, breathability, pliability, high loft, and open characteristics along with high bending modulus, since the material is only subjected to melt at the pattern point [100, 102, 103]. This method is useful for medical products, sorbents, and multilayer wipes [101, 104, 105].

3.8 Functional polymeric films and adhesives for lamination

3.8.1 Polymeric film

The selection of the adhesive during lamination depends on the physico-chemical properties of the substrates. Polymeric films are becoming the major substrates/materials to be used in lamination for improving the performance of textile /nonwoven fabrics. Common polymeric films used for lamination are as follows:

- Thermoplastic polyurethane (TPU)
- Low-density polyethylene (LDPE)
- High-density polyethylene (HDPE)
- Linear low-density polyethylene (LLDPE)
- Oriented and cast polypropylene
- Oriented and cast nylon
- EVA
- PVC
- Plain and coated cellulose

- PET and Biaxially oriented PET (BOPET)
- Metal-coated film
- Polymer coated film (polyvinylidene chloride or acrylic)

3.8.2 Adhesive

Selection of the right adhesive and proper preparation of the surfaces can ensure a bond that is durable and restricts delamination tendency. Because the adhesive is to be effectively wet bonded to the surface, the surface tension of the adhesive should be practically lower than that of the substrate to be laminated. This may need a pre-bond surface preparation to laminate films that have a lower surface energy, such as fluorocarbons, polyethylene, and polypropylene.

High-density atmospheric pressure plasma is a novel technology of surface modification that is practically suited for continuous lamination of the polymer film. Primer IT (Ciba) is one of the new and multipurpose technologies for enhancing the adhesion of UV-curable adhesives and coatings to polymeric substrates because of the covalent bond formation between the coating and the surface. Such an active layer is obtained by the application of a UV-curable primer to the surface, prior to the application of the laminating adhesive.

Selection of a particular adhesive system is dependent on a number of factors, as illustrated in Table 3.4.

Table 3.4: Factors affecting the selection of laminating adhesives.

Category	Specific factors
Chemical	Mixing ratio of components Shelf-life of resins Pot-life after mixing Functional groups present in the adhesive/substrate Curing time and energy required
Physical	Molecular weight Solid content Solution viscosity and melt viscosity Wetting behaviour and coating ability Drying speed
Performance	Initial bond strength Ultimate bond strength Resistance to service environments
Process	Lamination technique Adaptability to laminating processes Laminating conditions (nip pressure, temperature, speed, etc.)

It must be noted that the adhesive in a laminate is often chosen for more than just its bonding ability. Apart from holding the substrates together for the life of the laminate, the adhesive might have to perform some other necessary functions for developing a good product. These additional functions mostly include chemical and heat resistance, electrical insulation or conductivity, optical clarity, thermoforming capability, flame resistance, and increased or decreased gas permeability.

The adhesive used in lamination should be able to resist a phenomenon called "**tunnelling**". As depicted in Figure 3.25, tunnelling is known as the localized separation or substrate delamination that results due to the stretching or relaxation of two substrates of different extensibility at various rates. The localized stresses generated can show a negative impact on the performance and appearance of the laminates.

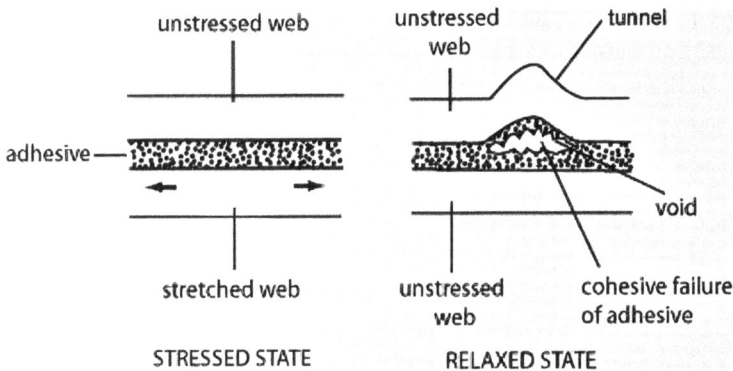

Figure 3.25: Tunnelling in flexible laminates [104].

The factors that are generally considered in the selection of a laminating adhesive are adhesion, heat and chemical resistance, formulating flexibility, and mechanical bond strength. It is important to evaluate the adhesive's ability to uniformly coat on the film's surface and form a smooth continuous coating. It must be experimented so that any water or solvent carrier can be removed either before the nip drying or even later in the case of porous substrates.

3.8.2.1 Waterborne laminating adhesive

These are mainly natural materials, like natural rubber, sodium silicates, and dextrins, as well as organic synthetic polymeric emulsions of acrylic, polyurethane, and vinyl acetate. Natural adhesives are usually used for packaging and labelling. Synthetic adhesives are mostly employed for either dry or wet lamination. The popularity of such a type of adhesive has tremendously increased due to the governmental pressure for lessening the volatile organic contents (VOCs) in the lamination process.

Although these kinds of adhesives usually show lower thermal and moisture resistance than solvent-based, cross-linkers in the formulation have (1) made waterborne adhesive to fill the criteria of the required performance and (2) shortened the performance gap between waterborne and solvent-based adhesives. There are several kinds of such adhesives, with varying performance properties and applications.

Acrylic emulsion is an economical adhesive showing medium performance properties. They are quite versatile because of the wide range of available monomers and resins. Depending on the formulation, acrylic-based adhesives can generate bonds ranging from tough, flexible to hard and rigid. It possesses inherently excellent UV & oxidative stability, and is mostly suitable for outdoor applications.

Waterborne adhesives, based on polyurethane, typically show advanced performance properties. These kinds of adhesives are created for fast line speeds and standard laminating equipment. Generally, they show very good adhesion to various flexible substrates. Upon cross-linking, they can generate higher strength than the substrates attached. The main limitation of polyurethane-based adhesives is that they are not economical.

3.8.2.2 Solvent-borne laminating adhesive

This kind of laminating adhesives can be experimented from various similar polymers used for waterborne systems. For example, acrylic solvent solutions find application when there is a requirement of high degree of non-yellowing and environmental resistance properties. They are mainly found in dry laminating applications but can also be used for wet laminating applications. Owing to the lack of surfactants and emulsifiers, resistance to moisture by this kind of adhesive is comparatively higher than waterborne systems.

In the 1950s polyester was considered a very famous solvent-borne adhesive for lamination purposes. For this, "polyester resin" was utilized that has a very low solid content, ranging from 20–30%. Such materials facilitate very good adhesion to polyester and other polymeric films. The hydroxyl group of the polyester reacts with a polyisocyanate to produce a crosslinked network of adhesive, with excellent chemical and thermal resistance. The laminating adhesives that are based on polyester are fairly fast curing, having high strength and a fast production process.

3.8.2.3 Solvent-less laminating adhesive

The first synthesized solvent-less laminating adhesives were mainly moisture curable polyurethanes. The adhesive gets crosslinked by the reaction of the atmospheric moisture with excess isocyanate groups. The most vital solvent-less adhesives used in laminating flexible packaging are those belonging to the polyurethane family (Table 3.5).

Table 3.5: Polyurethane-based laminating adhesive [106].

Type	Solvent	Curing action
Single component	Ketone or ester	Reaction with moisture on surface of the substrate or in atmosphere
Bicomponent; high solid content	Ketone or ester	Reaction of polyol with isocyanate-terminated resin
Bicomponent; low or medium solid content	Ketone or ester	Reaction of polyol with isocyanate
100% solids	None	Reaction with moisture on the surface of the substrate or in atmosphere
		Reaction of polyol with isocyanate-terminated resin

Two-part solvent-less polyurethanes were produced to overcome certain limitations such as cloudiness at a consistent rate of curing and bubbling. This kind of adhesive generally requires a blending and metering unit since pot life is short. Low initial bond strengths and high residual monomers somewhat restrict the application of such adhesives [106].

An enhanced polyurethane adhesive has been produced, depending on moderately high viscosity polyurethane polymers, which requires an application temperature of 50–70 °C. The enhanced viscosity lessens the curing time to 12–24 h, prior to slitting. Such an adhesive is created by a process that eliminates the excess amount of isocyanate-based monomer from the pre-polymer.

3.8.2.4 Hot melt laminating adhesive

These kinds of adhesives used in lamination are obtained in pure solid form. Thermoplastic and reactive urethane, polyethylene, EVAs, and polyamides are applied at ambient conditions to a substrate and are activated using heat. A second substrate is introduced for laminating to the first after the material is activated.

Dry hot melt adhesive initially undergoes a transforming process, which changes its raw and physical form in terms of pellets or granules to films or webs. Film and web adhesives are produced by the melting of pellets or granules and are then extruded into adhesive rolls.

Hot molten adhesive can be directly applied to the substrate. The adhesive (bulk filled drums, granules, pillows, or pellets) undergoes melting and coating on the surface of a material. It can induce line speeds and result in significant material savings.

3.8.2.5 UV-curable laminating adhesive

Electron beam (EB) or UV light-based curing laminating adhesives are generating widely popular in recent days. Acrylate and methacrylate monomers as well as oligomers and photo-initiators are necessary parts of UV adhesives. Similar monomers and oligomers are also found in EB adhesives, but no photoinitiators are required. Aliphatic urethane acrylates are mostly used in laminations owing to their non-yellowing ability and good adhesion properties to most films.

Such adhesives can be directly coated on the surface of a film, nipped, and cured. Line speed can be adjusted by changing the intensity of UV lamps. These adhesives were developed primarily as low VOC laminating systems but offer great potential advantages over waterborne, solvent-borne, and solvent-less adhesive systems. Some of the advantages are:

– Sufficient bond strength
– A single-component system having long shelf life
– Viscosity adjustments are not required
– The adhesive remains unchanged until cured
– Curing at room temperature
– A non-tacky film upon curing
– Very good adhesive strength to different substrates

3.9 Few smart applications of coated and laminated textiles

Conducting or smart polymers have various application areas such as sensors and sensor arrays, shielding of electromagnetic interferences (EMI), antistatic coatings, metallization of dielectrics, and batteries. Using textile as the base material, a thin coating layer of conducting polymer is capable to possess flexibility along with conductivity. Since degradation and lifetime of conducting polymers are comparable to textiles, the substrates with conductive coatings or laminations are mostly used for smart applications. Such kinds of textiles can find multitudinous applications. Polypyrrole (PPy) coating on textile surfaces can easily show high conductivity in the $k\Omega$/ $\square$ range and can lower down the static voltages by multiple folds. Such kinds of materials can be especially used where antistatic feature is required. Additionally, they find other potential applications that are likely to be microwave filters and absorbers, medical and sports gears, EMI shielding, heating, sensing, and so on.

3.9.1 Electromagnetic interference shielding (EMI shielding)

Electromagnetic interference is known as the disability of the performance of the electronic device caused by electromagnetic radiations from adjacent electronic gadgets. It has been observed that strong waves of electromagnetic field restrict the day-to-day functioning of navigation systems and response to electronic gadgets. For example, some personal electronic devices should not be powered on when a flight takes off and lands. EMI can also result in health issues for humans. Therefore, EMI shielding is very vital in order to save people and electronic gadgets from negative interference. Additionally, microwave reflection is employed to find out and trace some objects or tools in military intelligence. Materials that can absorb microwaves provide effects of camouflage. For the reduction of EMI, radiated EMI is required for filtration. When an incident wave of electromagnetic field strikes a surface, partial transmission takes place along with some amount of absorption and reflection. The corresponding shielding efficiency of absorption and reflection is a vital response for practical aspects.

Metals or surfaces coated with metals integrate into extreme EMI shielding efficiency, depending upon the reflection on the surface. However, textile substrates coated with electro-conductive polymers not only absorb but also reflect electromagnetic waves, thus exhibiting a drastic merit over materials shielded by metals. EMI shielding, based on absorption, can be increased by varying the dielectric constants or conductivities of electro-conductive polymers [105, 107–109]. Apart from the good absorption to EMI signals, textiles that are coated or laminated with conducting polymers also show good moisture and air permeability. Therefore, conductive textiles that are flexible in nature have find huge applications majorly in military clothing, microwave shielding EMI, and in day-to-day life for restricting electromagnetic waves from electronic gadgets such as microwave ovens, microphones, mobiles, and wireless systems [110].

3.9.2 Conductive fabric circuits

Circuits developed from textiles provide opportunities for arranging sensor-based devices and for creating large electronic and electrical systems. An extraordinary merit of textile-based circuits over conventional circuit boards is their high pliability. They are very easily integrated into fabrics and can substitute traditional printed boards in electronic sectors [111]. A technique of patterning can be followed to design and integrate well-structured electronic-based circuits for wearable electronic textiles. Such a pattern is conducting in nature having around 2 mm of resolution, which will allow a margin for chemical migrating along the pattern boundaries and reduce the possibility of loose conductive fibres to short the circuit. Therefore, this method is best suitable for a circuit on a large fabric area. Conducting polymer coating and patterning enable integration of soft circuitry into fabrics, thereby eliminating the requirement for hardwiring and hence increasing wearability and flexibility of the electronic textiles. As conductivity is integrated into the fibre itself, metallic loadings are not required.

3.9.3 Temperature regulation

Textiles coated or laminated with a conducting polymer are often used for the genera-tion of heat. Garments with warmth are required in cold conditions to keep our bodies warm. The heat is generated because the conductive fabric resistively heats when cur-rent flows. While comparing with other heating gadgets, conductive fabrics show sup-pleness, low power density on large surfaces, temperature homogeneity, softness, and flexibility [112]. They can be sewn, cut off, or pasted on substrates for a wide range of applications such as in buildings, curtains, gloves, heating winter sports wear, car seats, and for aircraft de-icing.

Rise in thermal conductivity is important for wool fabrics. Wang et al. [113] ob-served that PPy coating on wool fabrics increased thermal conductivity. The parame-ters of PPy coating affected the thermal conductivity. It was found that thermal conductivity increased with increase in electrical conductivity [113].

Due to the generation of heat by batteries, which is unavoidable, cooling is a tough task for thermal regulative garments, than heating. Hu et al. [114] worked on an idea of cooling conducting-polymer-coated fabrics using the "Peltier effect". The effect occurs when the electric current flow through two dissimilar conductors. Depending on the direction of the current flow, heat will either be absorbed or released at the junction of the two conductors [114]. A significant temperature difference was ob-tained by the fabric coated with a conductive PPy.

3.9.4 Electroless metal plating with high conductivity

By the use of electroless metal plating method, metals can be coated on non-conductive substrates. In the entire process of deposition, metal ions are reduced to only a metal in the presence of catalysts. When an aqueous solution of gold ions comes in contact with PPy nanoparticles or passes through textile substrates coated with PPy, it demonstrates a very good ability for the reduction of ionic gold electrolessly into Au(0), followed by deposition of the gold particles on the substrates [115, 116]. As nanofibers have excellent permeability and a large surface area, PPy-coated electro-spun nanofiber membranes can filter out gold from the aqueous solution stream that continuously passes through the membranes. A higher gold recovery yield can be obtained using a thicker nanofiber membrane. Increased electrical conductivity is also obtained by the metallization of tex-tile materials. Gasana et al. [117] showed that the conductivity of a polyaramide fabric coated with PPy was not found to be good under DC current [117]. To achieve better conduction, they coated the PPy coated structures with a layer of metallic copper through electroless deposition. The DC electro-conductive polyaramide structures that were achieved after depositing copper on top of the prior deposited PPy layer showed excellent electro-conductive properties, comparable with metallic conductivity.

3.10 Challenges and future outlook

Optimization of the coating and lamination processes from laboratory level experiments for the supply of platforms of chain is a vital problem for textile manufacturers. Some of the processes such as electro-spinning coating and thin-film deposition techniques are difficult for large scale production. Thus, extensive research is very much essential for a broad range of commercial applications. Based on this, Nanotex, a subsidiary of the US-based Burlington Industries founded in 1998, is considered to be the leading nanotechnology-based textile coating company for commercial products for household and clothing applications. Control of nano-coating procedures may not be as simple as that of macro-scale coating products. Implementation of advanced pre-treatment machineries for textiles and nanomaterials, such as continuous plasma chambers, equipment for sol–gel, and ball mills, must be developed in industrial areas. There should be safe labour workplace as well as provision of wastewater treatment system at the production line.

Another important issue for the application of nanostructures and nanoparticles on textiles is the effects on human health and environmental that could result from their unintended release from textile products. Use of Ag, alumina (Al_2O_3), silicon dioxide (SiO_2), titanium dioxide (TiO_2), zinc oxide (ZnO), and carbon nanotubes in textile nanocoatings has been commercialized in many developed countries [118, 119]. The effect of biological nanoparticles is extensively scrutinized by scientists based on damage to the central nervous system, crossing and damaging tissue barriers, impairment of DNA, chronic toxicity, acute toxicity, skin effects, the respiratory tract, and gastro-intestinal tract. Exposure of healthy skin to various nanoparticles has always shown long-term effects, where free nanoparticles may enter the lungs, which is one of the most important parts of the body. Electro-spinning method that produces nanomats and nanowebs will be more popular in the near future, based on the coating on technical textiles for smart and sensory applications.

3.11 Conclusion

Coating and lamination are the fundamental technologies for widening the opportunities in textiles, because they can be employed to impart valuable properties and functionalities, and also to develop advanced engineering composite systems. Such technologies have been extensively used in conventional textile applications, such as upholstery and apparel; providing resistance to soil, liquids, and gases; imparting leather-like appearance and fire retardation properties. Coating and lamination are also major tools for technical textiles, such as smart textiles, medical textiles, geotextiles, and automotive and filtration applications.

Coating and lamination technology has been gaining interest in view of environmental concerns and sustainability. Such processes are mainly known for the use of renewable resources or highly recyclable materials. They are less energy intensive and eco-friendly. Increased consumption of high solid content coating, such as hot melt adhesive and powder coating is an example of the trend towards eco-friendly coating and laminating processes.

Other demands originate due to advancements in functional and technical textiles. There happens to be a classic transformation from commodity to highly tailored products with engineering properties, and coating and lamination are prime technologies in this transformation. Precisely, diverse, pliable coating and laminating processes with the handling ability of coating various materials such as substrates, polymers, and films are required. A variety of coating machines to carry out multiple coating processes have been launched commercially, signifying the value of a rapid response to process flexibility and demand.

With advancements in fluid dynamics and engineering, more precise regulation of the coating process is presently possible. Innovations in nanotechnology, materials science, and biotechnology are introducing novel materials to coating and laminating processes, thereby widening the possibilities of advanced functional textiles. This drift is highlighted in the growth of smart and intelligent textiles. There are good opportunities in developing coating and laminating technology that can mix textile structures with materials having high levels of functionality – conductivity, shape memory, self-healing, and in producing integrated systems that can respond to environmental stimuli with high level of engineering.

References

[1] Singh MK. Textiles Functionalization-A Review of Materials, Processes, and Assessment. In: Kumar B, editor. Textiles for Functional Applications: IntechOpen; 2021. http://dx.doi.org/10.5772/intechopen.96936.
[2] Plasma Treatment Technology for Textile Industry. https://www.fibre2fashion.com/industry-article/3884/plasma-treatment-technology-for-textile-industry.
[3] Choudhary U, Dey E, Bhattacharyya R, Ghosh SK. A brief review on plasma treatment of textile materials. Adv Res Text Eng 2018; 3:1–4. http://dx.doi.org/10.26420/ADVRESTEXTENG.2018.1019.
[4] Haji A, Kan C-W. Plasma treatment for sustainable functionalization of textiles. In Ibrahim N, Hussain CM, editors. Green Chemistry for Sustainable Textiles. Elsevier; 2021. p. 265–77https://doi.org/10.1016/B978-0-323-85204-3.00034-8.
[5] Esmail A, Pereira JR, Zoio P, Silvestre S, Menda UD, Sevrin C, et al. Oxygen plasma treated-electrospun polyhydroxyalkanoate scaffolds for hydrophilicity improvement and cell adhesion. Polymers 2021;13:1056. https://doi.org/10.3390/polym13071056.

[6]   Sigmund P. Sputtering by ion bombardment theoretical concepts. In: Behrisch R, editor. Sputtering by Particle Bombardment I: Springer; 1981, p. 9–71 https://doi.org/10.1007/3540105212_7

[7]   Tan X, Liu J, Niu J, Liu J, Tian J. Recent progress in magnetron sputtering technology used on fabrics. Materials 2018;11:1953. https://doi.org/10.3390/ma11101953.

[8]   Hu Q, Wang J, Lu Y, Tan R, Li J, Song W. Sputtering-Deposited Thin Films on Textiles for Solar and Heat Managements: A Mini-Review. Phys Status Solidi A 2022;219:2100572. https://doi.org/10.1002/pssa.202100572.

[9]   Sputter deposition. https://lnf-wiki.eecs.umich.edu/wiki/Sputter_deposition

[10]   Walo M. Radiation-induced grafting. In: Sun Y, Chmielewski AG, editors. Applications of ionizing radiation in materials processing vol 1. Institute of Nuclear Chemistry and Technology;2017, p. 193–210.

[11]   Leea X, Show P, Katsudab T, Chenc W, Chang J. Surface grafting techniques on the improvement of membrane bioreactor: State-of-the-art advances. Biores Technol 2018; 269:489–502. https://doi.org/10.1016/j.biortech.2018.08.090.

[12]   Yang X, Ku T, Biswas S, Yanoa H, Abe K. UV grafting: surface modification of cellulose nanofibers without the use of organic solvents. Green Chem 2019;17:4619–24. https://doi.org/10.1039/C9GC02035G.

[13]   Zhang D, Guan L. Laser Ablation. In: Hashmi S, Batalha GF, Tyne CJV, Yilbas B, editors. Comprehensive Materials Processing, vol 4. Elsevier; 2014, p. 125–169.https://doi.org/10.1016/B978-0-08-096532-1.00406-4

[14]   Textor T. Modification of textile surfaces using the sol-gel technique. In: Wei Q, editor. Surface modification of textiles. Woodhead Publishing; 2009, p. 185–213.https://doi.org/10.1533/9781845696689.185.

[15]   Feifel SC, Kapp A, Lisdat F. Protein Multilayer Architectures on Electrodes for Analyte Detection. In: Gu, M., Kim HS, editors. Biosensors Based on Aptamers and Enzymes. Advances in Biochemical Engineering/ Biotechnology, vol 140. Springer; 2014, p. 253–298. https://doi.org/10.1007/10_2013_236.

[16]   Lu P, Ding B. Nano-modification of textile surfaces using layer-by-layer deposition methods. In: Wei Q, editor. Surface modification of textiles. Woodhead Publishing; 2009, p. 214–37.https://doi.org/10.1533/9781845696689.214.

[17]   Ojstršek A, Plohl O, Gorgieva S, Kurečič M, Jančič U, Hribernik S, Fakin D. Metallisation of textiles and protection of conductive layers: An overview of application techniques. Sensors, **2021**;*21*: 3508. https://doi.org/10.3390/s21103508

[18]   Jiang S, Guo R. Modification of textile surfaces using electroless deposition. In: Wei Q, editor. Surface modification of textiles. Woodhead Publishing; 2009, p. 108–25. https://doi.org/10.1533/9781845696689.108.

[19]   Shammas N, Yang J, Yuan P, Hung Y. Chemical Oxidation. In: Wang LK, Hung YT, Shammas NK, editors. Handbook of Environmental Engineering, Physicochemical Treatment Processes, vol 3. 2005, p. 229–70.https://doi.org/10.1385/1-59259-820-x:229.

[20]   Nierstrasz V. Enzyme surface modification of textiles. In: Wei Q, editor. Surface modification of textiles. Woodhead Publishing; 2009, p. 139–63.https://doi.org/10.1533/9781845696689.139.

[21]   Zhou Q, Rutland M, Teer T, Brumer H. Xyloglucan in cellulose modification. Cellulose 2007;14:625–41. https://doi.org/10.1007/s10570-007-9109-0.

[22]   Joshi M, Butola B. Application technologies for coating, lamination and finishing of technical textiles. In: Gulrajani ML, editor. Advances in the Dyeing and Finishing of Technical Textiles, 2013, p. 355–411.https://doi.org/10.1533/9780857097613.2.355.

[23]   Singha K. A Review on Coating & Lamination in Textiles: Processes and Applications. Am J Polym Sci 2012,2:39–49. https://doi.org/10.5923/j.ajps.20120203.04.

[24] Shim E. Coating and laminating processes and techniques for textiles. In: Smith WC, editor. Smart Textile Coatings and Laminates 2nd ed., 2018, p. 11–45. https://doi.org/10.1016/B978-0-08-102428-7.00002-X

[25] Shim E. Bonding requirements in coating and laminating of textiles. In:Jones I, Stylios GK, editors. Joining Textiles, 2013, 309–51.https://doi.org/10.1533/9780857093967.2.309.

[26] Sampaio P, Gonzalez M, de Oliveira Ferreira P, et al. Overview of printing and coating techniques in the production of organic photovoltaic cells. Int J Energy Res 2020;44:9912–31. https://doi.org/10.1002/er.5664.

[27] Weszka J, Szindler M.M, Szczęsna M, Szindler M. Influence of solvent on the surface morphology and optoelectronic properties of a spin coated polymer thin films, J Achiev Mater ManufEng 2013;61:302–7.

[28] Patidar R, Burkitt D, Hooper K, Richards D, Watson T. Slot-die coating of perovskite solar cells: An overview, Mater Today Commun 2020;22:100808.https://doi.org/10.1016/j.mtcomm.2019.100808.

[29] Nazrin K, Mohamed K, Lee P, Ooi G. A review of roll-to-roll nanoimprint lithography, Nanoscale Res Lett 2014;9:320. https://doi.org/10.1186/1556-276X-9-320.

[30] Kajal P, Ghosh K, Powar S. Manufacturing Techniques of Perovskite Solar Cells. In: Tyagi H, Agarwal AK, Chakraborty PR, Powar S, editors. Applications of Solar Energy, 2018, p 341–36.https://doi.org/10.1007/978-981-10-7206-2_16.

[31] Nickolas P, Thanasis P. Fluid Penetration in a Deformable Permeable Web Moving Past a Stationary Rigid Solid Cylinder. Transp Porous Media 2017;117:393–411.https://doi.org/10.1007/s11242-016-0780-1.

[32] Coyle D, Macosko C, Scriven L. A Simple Model of Reverse Roll Coating. Ind Eng Chem Res 1990;29:1416–19. https://doi.org/10.1021/ie00103a046.

[33] Ingamells W. Colour for Textiles, A User's Handbook, Society of Dyers and Colourists; 1993.

[34] Nakamura M, Yang C, Tajima K, Hashimoto K. High-performance polymer photovoltaic devices with inverted structure prepared by thermal lamination. Sol Energy Mater Sol Cells 2009; 93:1681–84. https://doi.org/10.1016/j.solmat.2009.05.017.

[35] Andreas G. Basic elements of Coating Systems, In:Giessmann A, editor. Coating Substrates and Textiles: A Practical Guide to Coating and Laminating Technologies, 2012, p. 23–76.https://doi.org/10.1007/978-3-642-29160-9_3.

[36] Mendez D. Modelling of the curtain coating process as a basis for the development of coating equipment and for the optimisation of coating technology. Materials Science 2008.

[37] Chapter II Coating characteristics and types of coatings. https://www.rnlkwc.ac.in/pdf/study-material/chemistry/DSE_3_Metal_coating__1__GM.pdf

[38] Akovali, G., 1 Thermoplastic Polymers Used in Textile Coatings, In: Akovali, G, editor. Advances in Polymer Coated Textiles, Smithers RAPRA: 2012 p. 1–24.

[39] Das S, Kumar S, Samal S. K, Mohanty S, Nayak S. K. A review on superhydrophobic polymer nanocoatings: recent development and applications. Ind Eng Chem Res 2018;57:2727–45. https://doi.org/10.1021/acs.iecr.7b04887.

[40] Das A, Deka J, Raidongia K, Manna U. Robust and self-healable bulk-superhydrophobic polymeric coating. Chem Mater 2017;29:8720–28. https://doi.org/10.1021/acs.chemmater.7b02880.

[41] Horrocks AR, Kandola BK, Davies PJ, Zhang S, Padbury SA. Developments in flame retardant textiles–a review. PolymDegrad Stab 2005;88:3–12. https://doi.org/10.1016/j.polymdegradstab.2003.10.024.

[42] Sastri VR. Engineering thermoplastics: acrylics, polycarbonates, polyurethanes, polyacetals, polyesters, and polyamides. In: Sastri VR. Plastics in Medical Devices 2nd ed. Elsevier; 2010, p. 121–72.https://doi.org/10.1016/B978-1-4557-3201-2.00007-0.

[43] Bouasria A, Nadi A, Boukhriss A, Hannache H, Cherkaoui O, Gmouh S. Advances in polymer coating for functional finishing of textiles. In: Shabbir M, Ahmed S, Sheikh JN, editors. Frontiers of Textile Materials: Polymers, Nanomaterials, Enzymes, and Advanced Modification Techniques. Wiley; 2020. https://doi.org/10.1002/9781119620396.ch3.

[44] Krebs F. C. Fabrication and processing of polymer solar cells: A review of printing and coating techniques. Sol Energy Mater Sol Cells 2009;93:394–412. https://doi.org/10.1016/j.solmat.2008.10.004.

[45] Mondal S. Phase change materials for smart textiles–An overview. Appl Therm Eng 2008; 28:1536-1550. https://doi.org/10.1016/j.applthermaleng.2007.08.009

[46] Esen M, Ilhan I, Karaaslan M, Unal E, Dincer F, Sabah C. Electromagnetic absorbance properties of a textile material coated using filtered arc-physical vapor deposition method. J Ind Text 2015;45:298–309. https://doi.org/10.1177/1528083714534710.

[47] Wei Q. Emerging approaches to the surface modification of textiles. In: Wei Q, editor. Surface modification of textiles. Woodhead Publishing; 2009, p. 318–323. https://doi.org/10.1533/9781845696689.318

[48] Bunshah RF. Deposition technologies for films and coatings: developments and applications. Noyes Publications; 1982.

[49] Hong YK, Lee CY, Jeong CK, Sim JH, Kim K, Joo J, Kim MS, Lee JY, Jeong SH, Byun SW. Electromagnetic interference shielding characteristics of fabric complexes coated with conductive polypyrrole and thermally evaporated Ag. CurrAppl Phys 2001;1:439–42. https://doi.org/10.1016/S1567-1739(01)00054-2

[50] Lee CY, Lee DE, Jeong CK, Hong YK, Shim JH, Joo J, Kim MS, Lee JY, Jeong SH, Byun SW, Zang DS. Electromagnetic interference shielding by using conductive polypyrrole and metal compound coated on fabrics. Polym Adv Technol 2002; 13:577–83., https://doi.org/10.1002/pat.227

[51] Lacerda Silva N, Gonçalves L. M, Carvalho H. Deposition of conductive materials on textile and polymeric flexible substrates. J Mater Sci Mater Electron 2013;24:635–43. https://doi.org/10.1007/s10854-012-0781-y.

[52] Lai K, Sun R. J, Chen M. Y, Wu H, Zha A. X. Electromagnetic shielding effectiveness of fabrics with metallized polyester filaments. Text Res J 2007;77:242–6. https://doi.org/10.1177/0040517507074033.

[53] Bula K, Koprowska J, Janukiewicz J. Application of cathode sputtering for obtaining ultra-thin metallic coatings on textile products. Fibre Text East Eur2006;59:75–9.

[54] Öktem T, Özdogan E, Namligöz S, Öztarhan A, Tek Z, Tarakçioglu I, Karaaslan A. Investigating the applicability of metal ion implantation technique (MEVVA) to textile surfaces. Text Res J 2006;76:32–40. https://doi.org/10.1177/0040517506059708.

[55] Sen AK. Coated textiles: principles and applications, 2nd ed. CRC Press; 2007. https://doi.org/10.1201/9781420053463

[56] Yip J, Jiang S, Wong C. Characterization of metallic textiles deposited by magnetron sputtering and traditional metallic treatments. Surf Coat Technol 2009; 204:380–385. https://doi.org/10.1016/j.surfcoat.2009.07.040.

[57] Xia L. Importance of nanostructured surfaces. In: Osaka A, Narayan R, editors. Bioceramics-From Macro to Nanoscale, Elsevier; 2021, p. 5–24. https://doi.org/10.1016/B978-0-08-102999-2.00002-8.

[58] Deitzel JM, Kleinmeyer J, Harris DEA, Tan NB. The effect of processing variables on the morphology of electrospun nanofibers and textiles. Polymer, 2001;42:261-72. https://doi.org/10.1016/S0032-3861(00)00250-0.

[59] Bhardwaj N, Kundu S. C. Electrospinning: a fascinating fiber fabrication technique. Biotechnol Adv 2010;28:325-47. https://doi.org/10.1016/j.biotechadv.2010.01.004.

[60] Huang Z. M, Zhang Y. Z, Kotaki M, Ramakrishna S. A review on polymer nanofibers by electrospinning and their applications in nanocomposites. Compos Sci Technol 2003;63:2223-53. https://doi.org/10.1016/S0266-3538(03)00178-7.

[61] Pawliszyn J. Applications of solid phase microextraction, vol 5. Royal Society of Chemistry; 1999. https://pubs.rsc.org/en/content/ebook/978-0-85404-525-9

[62] Spietelun A, Pilarczyk M, Kloskowski A, Namieśnik J. Current trends in solid-phase microextraction (SPME) fibre coatings. Chem Soc Rev 2010;39:4524–37. https://doi.org/10.1039/C003335A.

[63] Bagheri H, Roostaie A. Electrospun modified silica-polyamide nanocomposite as a novel fiber coating. J Chromatogr A 2014;1324:11–20. https://doi.org/10.1016/j.chroma.2013.11.024.

[64] Bagheri H, Aghakhani A, Baghernejad M, Akbarinejad A. Novel polyamide-based nanofibers prepared by electrospinning technique for headspace solid-phase microextraction of phenol and chlorophenols from environmental samples. Anal Chim Acta 2012;716:34–9. https://doi.org/10.1016/j.aca.2011.03.016.

[65] Bagheri H, Aghakhani A. Novel nanofiber coatings prepared by electrospinning technique for headspace solid-phase microextraction of chlorobenzenes from environmental samples. Anal Method 2011;3:1284–89. https://doi.org/10.1039/C0AY00766H.

[66] Ding B, Kim J, Miyazaki Y, Shiratori S. Electrospun nanofibrous membranes coated quartz crystal microbalance as gas sensor for NH$_3$ detection. Sensors and Actuators B: Chemical, 2004;101:373–380. https://doi.org/10.1016/j.snb.2004.04.008

[67] Balazs AC. Modeling self-healing materials. Mat Today 2007;9(10):18–23.https://doi.org/10.1016/S1369-7021(07)70205-5

[68] Galio A, Muller I. Active Coatings: Examples and Applications. Recent Patents on Mechanical Engineering, 2008;1:68–71.https://doi.org/10.2174/2212797610801010068

[69] Thies J, Mejers G, Pitkin J, Currie E, Tronche C, Southwell J. US2006286305; 2006.

[70] Pantelidis D, Bravman J, Rothbard J, Klein R. WO07092043; 2007.

[71] Pantelidis D, Bravman J, Rothbard J, Klein R:.US2007071789; 2007.

[72] Schneider J, Ragan D, Rechenberg K, Chasser A, Barkac K. WO02081579A2; 2002.

[73] Safta E, Chen F, Forrest R, Muselmann G. US20036641429B2; 2003.

[74] Kron J, Delschmann K, Schottner G. US20020081385A1; 2002.

[75] Dmitry B, Turki J.A, Supat C, Wanich S, Mohammad A, Iman S, Gabdrakhman V. Nanomaterial by Sol-Gel Method: Synthesis and Application, Adv Mater Sci Eng 2021;102014. https://doi.org/10.1155/2021/5102014

[76] Woodruff F. Guide to methods and applications. Int Dyer 2002;187:13.

[77] Scott R. Coated and laminated fabrics. In: Carr CM, editor. Chemistry of the textiles industry. London: Blackie Academic, 1995, p. 210–48.

[78] Swedberg J. Sticking to it: Learning about laminates. Industrial Fabric Products Review 1998;75:30.

[79] Nair G, Pandian S. Spotlight on laminating machines. Colourage 2006;53:72–80.

[80] Mansfield R. Combining nonwovens by lamination and other methods. Textile World 2003;153:22–25.

[81] Gillessen G. Flame, Dry or hot-melt. Int Dryer 2000;185:34.

[82] Walker R. Thermoplastic powder adhesive for lamination. Tech Text Int 1994;3:11–24.

[83] Grant R. Coating and Laminating Applied To New Product Development. J Coated Fabrics 1981;10:232–53.https://doi.org/10.1177/152808378101000307

[84] Stukenbrock K. H. Chemical finishing. In Albrecht W, Fuchs H, Kittelmann W, Lunenschloss J, editors. Nonwoven fabrics. Wiley-VCH; 2003, p. 421–459.https://doi.org/10.1002/3527603344.ch8

[85] Crabtree A. Hot-melt adhesives for textile laminates. Technical Text Int 1999;8:11–24.

[86] Halbmaier J. Overview of Hot Melt Adhesives Application Equipment for Coating and Laminating Full-Width Fabrics. J Coat Fabr 1992; 21:301–310. https://doi.org/10.1177/152808379202100407

[87] O'Boyle C. Infrared delivers for automotive interiors. Process Heating 2013;20.

[88] Walker, J. Infrared technology: Improves finishing processes. Process Heating 2017;24.

[89] Anon. Laminating and coating: Flexible future, Textile Month, March 24 2002.

[90] Glawe A, Reuscher R, Koppe R, Kolbusch T. Hot-Melt application for functional compounds on technical textiles. J Ind Text 2003;33:85–92. https://doi.org/10.1177/152808303038587

[91] Nussli R. The hot-melt coating and laminating of industrial textiles. Nonwovens, Industrial Textiles, 2001;47:14.

[92] Woodruff F. Exciting prospect for PUR compounds. International Dyer, 2002;187:34.

[93] Woodruff F. Laminating and coating: Advanced composites, Textile Month, March 23, 2002.

[94] Anon. Lamination and coating, Textile Month, March 22 2002.

[95] Giessman A. A practical guide to coating and laminating technologies. Springer- Verlag Berlin Heidelberg; 2012. https://doi.org/10.1007/978-3-642-29160-9

[96] Field I. Dry heat lamination technology. International Dyer 2001;186:39–43.

[97] Field I. Why is the industry turning to dry heat lamination technology? Technical Textiles International 2000;9:21.

[98] Kanafchian M, Valizadeh M, Haghi A. Electrospun nanofibers with application in nanocomposites. Korean J Chem Eng 2011; 28:428–439. https://doi.org/10.1007/S11814-010-0376-3

[99] Kanafchian M, Valizadeh M, Haghi A. A study on the effects of laminating temperature on the polymeric nanofiber web. Korean J Chem Eng 2011; 28:445–48. https://doi.org/10.1007/s11814-010-0400-7

[100] McIntyre K. Coatings and laminates move nonwovens forward. Nonwovens Industry, 2006;37:54–61.

[101] Brieger T. Consideration factors for ultrasonic vs. adhesive use in nonwoven products. Nonwovens World, 2006;15:60–66.

[102] Knorre K. Laminating by ultrasonics. Nonwovens, Industrial Textiles, 2001;47:51.

[103] Gil G. Ultrasonic bonding – new possibilities and opportunities. Nonwovens Industry, 1999;30:46.

[104] Brewis, D.M., and Briggs, D., Industrial Adhesion Problems, John Wiley & Sons;1985.

[105] Kaynak A. Electromagnetic shielding effectiveness of galvanostatically synthesized conducting polypyrrole films in the 300–2000 MHz frequency range. Mater Res Bull 1996;31:845–860. https://doi.org/10.1016/0025-5408(96)00038-4

[106] Petrie EM. Laminating Adhesives for Flexible Packaging. https://citeseerx.ist.psu.edu/viewdoc/download?doi=10.1.1.619.9118&rep=rep1&type=pdf#:~:text=Laminating%20adhesives%20for%20flexible%20packaging%20are%20available%20in%20a%20variety,)%20liquid%2C%20and%20hot%20melt.

[107] Kim M, Kim H, Byun S, Jeong S, Hong Y, Joo J, Song K, Kim J, Lee C, Lee J, PET fabric/polypyrrole composite with high electrical conductivity for EMI shielding. Synth Met 2002;126:233–39. https://doi.org/10.1016/S0379-6779(01)00562-8

[108] Håkansson E, Amiet A, Kaynak A. Electromagnetic shielding properties of polypyrrole/polyester composites in the 1–18 GHz frequency range. Synth Met 2006;156:917–25. https://doi.org/10.1016/j.synthmet.2006.05.010

[109] Håkansson E, Amiet A, Nahavandi S, Kaynak A. Electromagnetic interference shielding and radiation absorption in thin polypyrrole films. EurPolym J 2007;43:205–13.https://doi.org/10.1016/j.eurpolymj.2006.10.001

[110] Kuhn HH. Adsorption at the Liquid/Solid Interface: Conductive Textiles Based on Polypyrrole. Text Chem Color 1997;29:17–20.

[111] Locher I, Tröster G. Enabling technologies for electrical circuits on a woven monofilament hybrid fabric. Text Res J 2008;78:583–94.https://doi.org/10.1177/0040517507081314

[112] Boutrois J, Jolly R, Petrescu C. Process of polypyrrole deposit on textile. Product characteristics and applications. Synth Met 1997;85:1405–6. https://doi.org/10.1016/S0379-6779(97)80294-9

[113] Wang J, Kaynak A, Wang L, Liu X. Thermal conductivity studies on wool fabrics with conductive coatings. J Text Inst 2006;97:265–70.https://doi.org/10.1533/joti.2005.0298

[114] Hu E, Kaynak A, Li Y. Development of a cooling fabric from conducting polymer coated fibres: Proof of concept. Synth Met 2005;150:139–43. https://doi.org/10.1016/j.synthmet.2005.01.018

[115] Wang H, Ding J, Lee B, Wang X, Lin T. Polypyrrole-coated electrospun nanofibre membranes for recovery of Au (III) from aqueous solution. J Membrane Sc, 2007;303:119–25.https://doi.org/10.1016/j.memsci.2007.07.012.

[116] Ding J, Wang H, Lin T, Lee B. Electroless synthesis of nano-structured gold particles using conducting polymer nanoparticles. Synth Met 2008;158:585–89.https://doi.org/10.1016/j.synthmet.2008.04.019

[117] Gasana E, Westbroek P, Hakuzimana J, De Clerck K, Priniotakis G, Kiekens P, Tseles D. Electroconductive textile structures through electroless deposition of polypyrrole and copper at polyaramide surfaces. Surf Coat Technol 2006; 201:3547–51. https://doi.org/10.1016/j.surfcoat.2006.08.128

[118] Gashti M, Alimohammadi F, Song G, Kiumarsi A. Characterization of nanocomposite coatings on textiles: a brief review on microscopic technology. CurrMicroscContrib Adv Sci Technol 2012:2:1424–37.

[119] Gashti P. Nanocomposite coatings: state of the art approach in textile finishing. J Text Sci Eng 2014;4:e120.

Subhankar Maity, Kunal Singha, Pintu Pandit

4 Speciality coatings and laminations on textiles

Abstract: Coating and lamination techniques offer great potential in improving the value of technical textiles. These techniques enhance and broaden the spectrum of functional performance of textiles, making them suitable for various smart/technical applications. In this chapter, there is discussion on various polymers and chemicals used in speciality coating and lamination of textiles, preparation of base fabric, coating and laminations methods, their scopes, and end use. This chapter covers fluoropolymer coatings, solvent-based adhesive coating, reactive coating, urea-based adhesive coatings, biopolymer coatings, peelable coating, chemical vapour deposition coatings, flame-retardant polymer coating, electrospun polymeric nanofiber coatings, waterproof coating, electrically conductive coating, active and smart coatings (memory polymer coating), self-healing coating, breathable coatings, biomimetic coatings, seed coatings, and so on. A few speciality lamination techniques such as sheet lamination, thermal lamination of polyurethane films, ethylene-vinyl acetate-based lamination, breathable lamination, transparent film lamination, electrospun nanofibre-based lamination, and decorative laminations are also covered in this chapter.

Keywords: Speciality coating, lamination, polymers, functional coating, functional textiles, surface treatment

4.1 Introduction

Coated textiles comprise a textile fabric as the base and a polymer or adhesive as functional material applied on the textile surface. The coating adds increased functionality to the textile fabric by virtue of the functional polymer or chemicals applied on the textile surface and restricts dust, liquids, and gases from entering the structure. The coating also improves the physical and chemical qualities of the textiles, such as abrasion resistance, and chemical resistance [1]. There is the recent trend of combining various coating processes with various coating materials that regulate a variety of characteristics of coated textiles. Based on a thorough assessment of the necessary features in the final product, the base fabric and polymer must be carefully chosen. The fabric regulates dimensional stability, tensile strength, and elongation, while the polymer enhances abrasion resistance, chemical resistance, liquid or gas permeability, and so on [2]. The type of base fabrics, coating polymer and coating add-on have a significant impact on the mechanical and functional properties of the coated fabric.

https://doi.org/10.1515/9783110759747-004

Laminated textiles are materials made up of two or more layers, and out of these two layers, at least one layer is made of textile fabric. An applied adhesive closely binds the layers together. On the other hand, the coating can be applied on one or both surfaces of a textile fabric. Coating formulations made of suitable grades of polymers such as polyvinyl chloride (PVC), polyurethane (PU), acrylic, and polytetrafluoroethylene (PTFE) are widely used to create advanced textile products such as waterproof clothing, tarpaulin, electrical insulating cloth, and protective clothing [3].

Coating is a physical and chemical technique that involves depositing one or more layers of substance on a textile substrate. Even though solids and gases can be deposited, liquids are the most commonly employed in textile coating. The quality and performance of coated textiles are determined by the three phases of measuring, transferring, and fixing coating materials. The amount of coating material to be deposited is monitored by a metering device. There are three types of metering, namely free meniscus, pre-metering, and post-metering. The most basic meniscus is the one in which the substrate is separated from the coating liquid or vice versa. A displacement pump is used to coat a certain amount of coating liquid using pre-metering. In post-metering, the coating material is saturated on the substrate and the excess materials are removed. In terms of material savings and coating liquid waste, pre-metering is the most efficient.

Suitable combinations of fabric and polymer can provide aesthetic effects on the coated textiles. Coating techniques are quite simple and can give excellent effects on artificial leathers used for seat coverings and clothing [4]. A waterproof garment (worn for rain protection) is one of the most common examples of coated fabric. The coated fabric is frequently used to make a protective element for police, firefighters, postal workers, military, and other emergency personnel. The household items that are produced from coated textiles include mattress ticking, wipe-clean tablecloths, shower curtains, flame-retardant (FR) upholstery, and curtain linings.

Usually, coated textiles used in the industrial sector are flexible containers, awnings, aircraft fuel tanks, protective covers, life rafts, tarpaulins, life jackets, hovercraft skirts, and aircraft safety chutes. The other example of coated textiles is seat upholstery in cars/buses/trains with high abrasion resistance, fire-retardant characteristics, and dust particle penetration. Examples of lamination technique-based products are inflatable products, blackout curtains, and blinds.

The use of proper or apt technology and modern machinery are very crucial to achieve success in textile coating and lamination. In most coating processes, limited machine productivity is a critical issue but is flexible related to production speed, the diversity of coating, and lamination techniques with a high degree of process monitoring, process control, and process automation to meet rigorous technical standards [5].

4.2 Conventional coating methods

There are various methods available to coat textile substrates with suitable agents. Based on the state of applications, the coating processes can be divided into the following four methods: (a) molten or semi-molten state processes, (b) gaseous-state processes, (c) solution-state processes, and (d) solid-state processes. The important features in any coating procedure are the thickness of the coatings, application temperature, application and curing time, and curing temperature [6].

Furthermore, the coating techniques can also differ in the subsequent types, according to their processability, such as
a.   Dip coating
b.   Knife coating
c.   Gravure coating
d.   Rotary screen coating
e.   Hot melt coating
f.   Transfer coating
g.   Flow coating

A typical polymeric coating process of textiles involves preparation of coating formation and the application of polymer-based coating formulation over the surface of textiles, followed by drying and curing process. Such coating process is generally suitable for linear polymers, which can be coated as a polymer melt or polymer solution or as a form of a solid film over the substrate. There are some types of coatings that can be applied in the liquid form and then chemically cross-linked to form a solid film. Table 4.1 shows the few common polymers used for the coating of textiles along with their advantages, limitations, and related products.

4.3 Speciality coating for textiles

Various speciality coatings of textiles are developed in order to achieve special properties. Some of the speciality coating methods and their end-use applications are described in the following sections.

4.3.1 Fluoropolymer coating

Fluoropolymers are special chemicals particularly known for their extremely high thermal, electrical, and chemical resistance. They also offer a low refractive index along with low surface energy attributing to anti-stick and chemical/water-resistant applications [7]. Fluoropolymers are regularly deployed for widely variant industrial uses,

Table 4.1: Polymers used in textile coating, their advantage/disadvantage, and potential applications.

Coating and laminating materials	Advantages	Limitations	Products
Polyvinyl chloride (PVC)	– Used in diverse uses like good inherent FR, excellent oil, solvent resistance-based applications – Heat resistant, radiofrequency wave resistant, lightweight and weldable	– Showing cracks while exposed to moderate heat and cold – Migration issue while applying plasticizer	– Protective clothing, banner, and bunting, tents and architecture use – Tarpaulins, covering, seat upholstery and leather cloths, leisure products, etc.
Polyvinylidene chloride (PVDC)	– Highly efficient as fire-retardant – Low gas permeability – Excellent gloss – Heat weldable – Clear in appearance	– Brittle – Hard	– Blends with acrylics to increase fire-retardant (FR) coating
Polytetrafluroethylene (PTFE) or Teflon®	– Outstanding resistance to acid, alkalis, and chemicals – Excellent resistance to weathering, solvent, oil and oxidation – Non stick property-based usages work in high temperatures even up to 260°C – excellent electrical insulator	– Highly expensive – Unweldable and poor adhesion property	– Calendar belts – Architecture applications – Food and medical uses – Gaskets and seals
Styrene-butadiene rubber	– Better resistance to abrasion and flexing than normal rubber	– Highly viscous and sticky in nature	– Carpet backing
Butyl rubber	– Low gas permeability – Higher resistance to heat and chemicals	– Poor solvent resistance – Poor fire resistance – Difficulty in seaming	– Air cushion, pneumatic spring, protective clothing for chemicals and acid – Lightweight jacket and life rafts
Nitrile rubber	– Excellent oil resistance (that can also be increased by increasing the acrylonitrile content) – High sunlight and heat resistance	– Poor fire resistance – Poor ozone, sunlight, and weather resistance – Limited high temperature resistance	– Air cushion – Protective clothing for chemicals and acid – Lightweight jacket – Life rafts – Pneumatic spring

where these features are essential. This type of fluoropolymer coating is called "smart coating," because of the smart circumstances in which the textiles coated or laminated with fluoropolymers can breathe by allowing passage of water vapour from the body, while blocking comparatively larger water molecules [8]. The common polymers in this class are poly(tetrafluoroethylene) (PTFE), polyvinylidene fluoride (PVDF), tetrafluoro-ethylene-hexafluoropropylene copolymer (FEP), ethylene-tetrafluoroethylene copolymer (ETFE), and so on. Among them, PTFE, FEP, and ETFE are mainly used in anticorrosive and anti-sticking applications. Polyvinylidene fluoride (PVDF) polymers are mainly used in weather-resistant paints for outdoor applications [9]. These processes have certain specific challenges in terms of producing long-lasting or durable products, and an acrylic modifier resin is frequently added to the PVDF resin to overcome these challenges. To optimize coating adherence and pigment dispersion, the acrylic modifier is usually phys-ically blended with the PVDF resin [10].

4.3.2 Solvent-based acrylic/PVA coating

This process is carried out by dipping textile materials in the solvent-based formulation of bonding substances such as acrylic adhesive or polyvinyl alcohol (PVA). Various hy-drophilic or hydrophobic coatings using these bonding substances are available in the market. The major purpose of using a solvent is to dissolve the polymer and disperse suitable additives/other ingredients used for making the coating formulation. The sol-vent also helps provide stability to the solution during storage before the coating pro-cess. In addition, the solvent system should possess the following aspects:
a) Safety considerations
b) Health considerations
c) Environmental considerations
d) Economic considerations

The solvents and their vapours must be non-toxic during the entire coating process. They should be readily available, safe, and should be inexpensive, without any dis-posal problems.Organic solvents have better dissolving power and easy drying and recovery of solvent than aqueous systems. However, the organic solvents and their vapours can be hazardous to operators' health and environment, if they are not ap-propriately collected after evaporation [10]. Organic solvent is always better than in-organic solvent materials because of the following advantages [11, 12] :
a) Provide the best quality level required
b) Provide desired coverage
c) Provide uniformity
d) Provide smooth line/production speed, for example, when using aromatic com-pounds like benzene and toluene as an organic solvent in the coating processes.

4.3.3 Reactive adhesive coating

Reactive coating is generally done by spray coating method over any textile/material surfaces. In modern times, solvent-based polyurethane reactive (PUR) adhesives have been developed, which allow low add-on levels on fabric/film with high bond strengths. The most common issue with reactive spray application is related to getting the uniformity and precision of application, without occasional nozzle blocking. The other relevant issues are the difficulty in achieving a better penetration of the coating spray into the fabric materials and precise control of the amount of spraying liquid. The continuous drying of the solvent or liquid is another main issue related to reactive spray coating [13, 14]. In order to overcome these issues, moisture-cured polyurethane adhesives, which do not require excessive temperature to initiate the cross-linking can be used in reactive spray coating, but they have more issues of nozzle blocking. In addition, hot melt PU adhesives are also available in jelly form with solids content of nearly 100%, which makes them have better adhesion and spray control action over the substrates [5]. In recent times, the hot melt polyurethane reactive (HM-PUR) adhesive has been extensively used for fabric-film or fabric-fabric lamination, which is generally applied by plain or gravure roll systems that have the potential for varying the add-on from very low (5–6 g/m^2) to very high (30–40 g/m^2) levels, as per the requirement (depending on the substrates or required bond strength).

4.3.4 Polyurethane urea-based adhesive coating

Polyurethane ureas are supramolecular materials that have superior mechanical properties, primarily due to their reversible and noncovalent interactions such as intermolecular hydrogen bonding. Polyurethane urea can be synthesized from polyurethane pre-polymers from poly(tetramethylene ether)glycol with a low molecular weight, and isophorone diisocyanates that were chemically bonded with propylamine to create polyurethane urea (shown in Figure 4.1) with high urea content and better cross-linking, molecular weight (3,000 < Mn < 9,000), cutting or shaping ability, improved thermal, mechanical, and morphological properties [12, 15]. Polyurethanes and polyurethane-ureas (especially their water-based dispersions) have emerged as a very versatile field, based on ecologically benign methods. The evolution of their synthesis processes and the nature of the reactants (or substances participating in the process) have positioned these dispersions as a relevant and vital product for a variety of coating application processes [15].

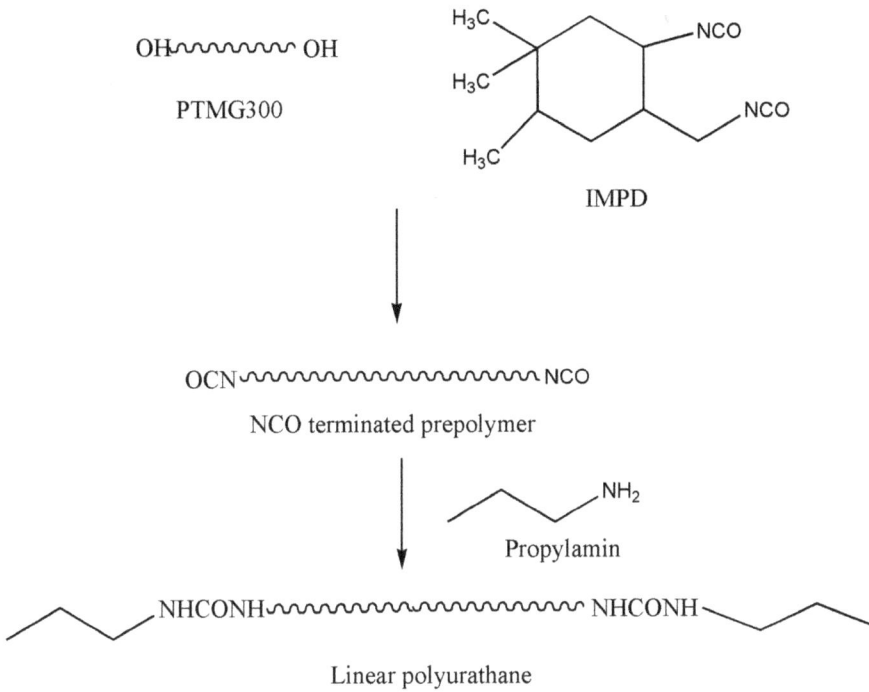

OH〰〰〰 OH

PTMG300

IMPD

OCN〰〰〰〰〰〰 NCO

NCO terminated prepolymer

Propylamin

NHCONH〰〰〰〰〰〰〰 NHCONH

Linear polyurathane

Figure 4.1: Reaction scheme to prepare polyurethane urea.

4.3.5 Biopolymer coating

Biopolymer coatings are generally used in food industry. The combination of higher food quality, longer shelf life, and higher consumer demands have led to increasing interest in alternative packaging materials research. In this field, biopolymers such as starch, chitosan, polylactic acid, alginates, hydrogels, and organic acid-based materials have attracted a lot of interest. These biopolymer coatings have the potential to prevent undesired moisture transfer in food. They provide superior oxygen and oil barriers while also being more biodegradable, making them a better alternative to the present synthetic paper and paperboard coatings used in the food sector [16].

These biopolymer coatings are very popular in various biomedical applications, including drug delivery and tissue engineering. The biopolymer coating can improve the surface properties of the textiles, which is the requirement of many applications, including growth and repair of tissues, the proliferation of cells, delivery of drugs and biomolecules, delivery of antimicrobial agents, active molecules, and growth factors [17].

4.3.6 Peelable coating

Peelable coatings are temporary coatings achieved on the textile surface, which can be peeled off as continuous and sizable sheets from the substrate, at the end of the service period. These coatings are also known as strippable coatings, and they are the most cost-effective and time-saving alternatives for protecting and decontaminating delicate and inaccessible surfaces that require a high precision application procedure (Figure 4.2).

Peelable coating can be applied on textile substrate by roller, brush, spraying, and dipping techniques. Among them, the dipping technique is simple and can be applied to textile substrates of any size or shape.

Release agents such as silicon dioxide or iron oxide are also used in the procedure. Release agents are important in the peelable coating process because they limit the adherence of film to the substrate, while also enhancing the peel ability of film. Peelable coatings are used by coupling reactions with other polymeric release agents like different additives. This can readily generate wide varieties of thickness in the coatings, with a broader spectrum of applications. Peelable coatings can be used in diverse applications such as in electronics protection, food coatings, floor coatings, nuclear reactors, cosmetics, automotive, optics, and labels used for plastics [18].

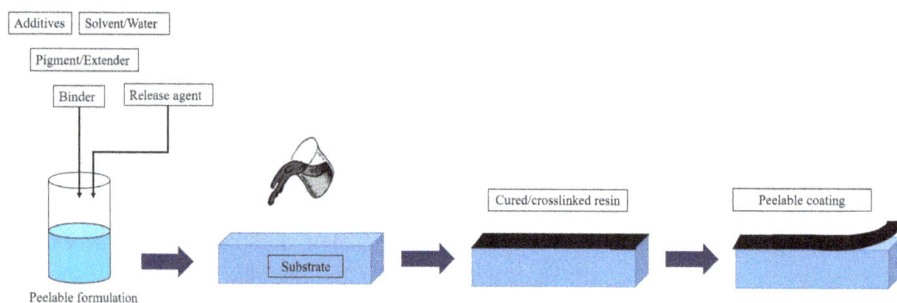

Figure 4.2: Schematic diagram of peelable coating.

4.3.7 Chemical vapour deposition (CVD) coating

The deposition of atoms or molecules by high-temperature-aided molecular reduction or breakdown over any substrate is called chemical vapour deposition (CVD). Coating on a textile surface by means of a vapour-phase chemical reaction with a suitable agent enables a conformal coating to be produced that can bring about a functional material with enhanced mechanical or chemical resistance, electrical conductivity, and/or biochemical activity. A requirement of CVD is the deployment of a chemical reaction to form a coating from a vapour, with the reaction by-products leaving as volatile species, as shown in Figure 4.3. The gas-phase reaction is driven by reactant species in the presence of heat that diffuses to the surface of the textile substrate, migrates over the surface, and finally sticks over the surface. There may be some volatile molecules desorbed from the surface, during the process [19].

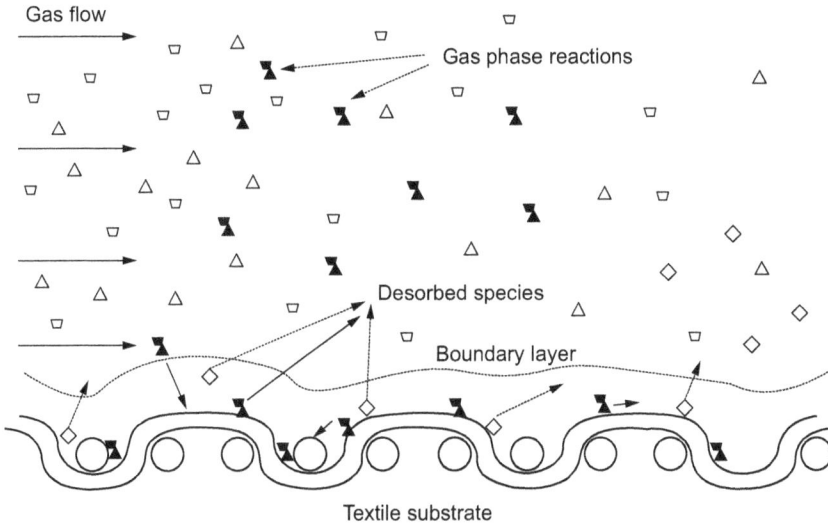

Figure 4.3: Schematic diagram of chemical vapour deposition (CVD) onto a textile substrate (Modified according to [19], Copyright [2009], Elsevier).

Thus, for designing a CVD reactor, the concept of chemical kinetics and chemical thermodynamics is inadequate. The amount of heat flow and mass flow are important considerations, and they depend on temperature gradients within the reactor and coupled effects such as convection.

In situ chemical vapour deposition technique is followed for coating of various conductive polymers (like polypyrrole, polythiophenes, and polyaniline) onto textile substrates. It is a two-step process. In the first step, the textile is impregnated with an oxidant solution (a mixture of suitable oxidant and dopant), and in the second step, the oxidant-enriched sample is exposed to monomer vapour for in situ polymerization, or vice versa [20]. Through this process, various conductive textiles can be prepared by using various natural, manmade and synthetic fibres, such as cotton, wool, viscose, aramid, polyethylene, and polyester as substrates [21, 22]. A schematic diagram of the process is shown in Figure 4.4. The vapour-phase polymer-coated textile fabrics exhibit a uniform coating of the conductive polymer over the textile surface. Therefore, good consistency and uniformity are obtained. The variability in surface resistivity is minimized and fastness to light and washing is improved. However, it is

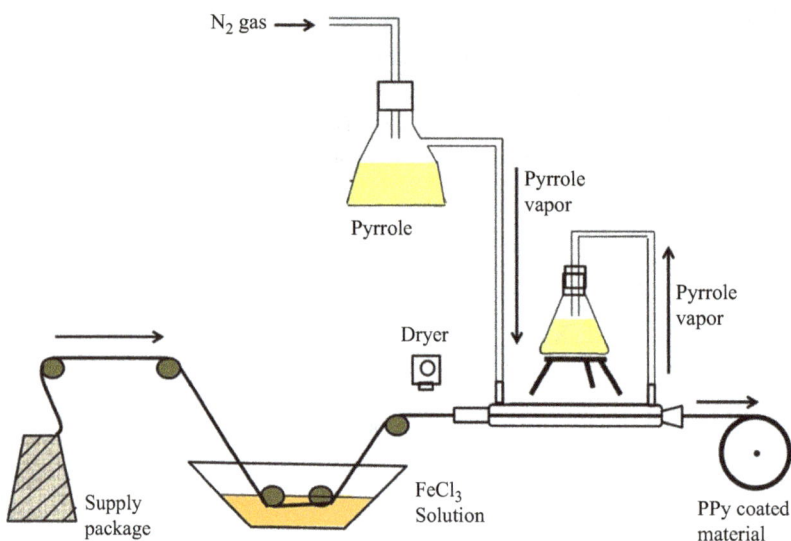

Figure 4.4: In situ chemical vapour deposition technique of polypyrrole onto textile substrate (Modified according to [23], Copyright [2021], Elsevier).

difficult to control the add-on percentage over the textile substrate, and the equipment set-up is complicated [23].

The production of superhydrophobic surfaces for self-cleaning effect can be achieved by this method of CVD coating, without plasma treatment. Polymethylsiloxane coating on cotton textile can be achieved by this technique, where superhydrophobicity (water contact angle greater than 150°) can be achieved. In this method, trichloromethylsilane is exposed at 50 °C in vapour phase, so that trichloromethylsilane can be adsorbed onto fibre surface. The treated fibres are reacted with pyridine solution at room temperature so that hydrolysis of all Si–Cl bonds can be done. The textile samples are then washed and dried at 150 °C for 10 min to achieve nanoscale silicone coating over the textile substrates [24].

4.3.8 Flame-retardant polymer coating

In flame-retardant (FR) coatings, generally, polymeric coating formulation loaded with different fire-retardant additives is applied over the surface of textile substrates. Traditional non-intumescent flame-retardant coatings typically contain compounds of halogens such as chlorine, bromine, or phosphorous, and as well as inorganic metal compounds. The presence of these special chemicals helps to inhibit flame spread via radical quenching and/or the formation of glassy protective layers, instead of voluminous blown char during combustion. An appropriate selection of suitable intumescent components related to physical and chemical qualities is required in order to achieve excellent efficiencies in fire protection for any intumescent FR coatings.

Non-intumescent FR coating solutions often contain less chemicals and are, thus, more compatible with matrices, resulting in improved mechanical and fire performance. Popular halogenated FR systems have come under criticism, in recent years. Most halogen-based compounds function best in the gas phase as free radical scavengers and have been demonstrated to be effective for flame retardation, in most circumstances. Phosphorous (P)-flame retardants are mostly made up of phosphate esters, and they have a lot of promise for application as FR additives in coatings. P-flame retardants can network with coating matrix during heat exposure to increase the surface protection by making more char generation. Antimony oxide is the commonly used synergist for the halogenated flame retardant system. It works by forming antimony trihalide, which, later on, acts as a powerful flame quencher in the gas phase. Moreover, inorganic compounds like zinc borates and aluminium hydroxide (ATH) are used as FR fillers in coating applications to ensure the endothermic and cooling effects during flame propagation.

Flame-retardant coating over textile substrates can also be formulated using silicon dioxide (SiO_2). In order to obtain the flame retardant formulation, tetraethyl orthosilicate (TEOS) is used as precursor, which is hydrolyzed, polycondensated, and then cross-linked over the textile substrate via a sol–gel process [25]. Atmospheric pressure plasma (APP) technique is used for coating of cellulosic materials with this SiO_2 formulation, as shown in Figure 4.5.

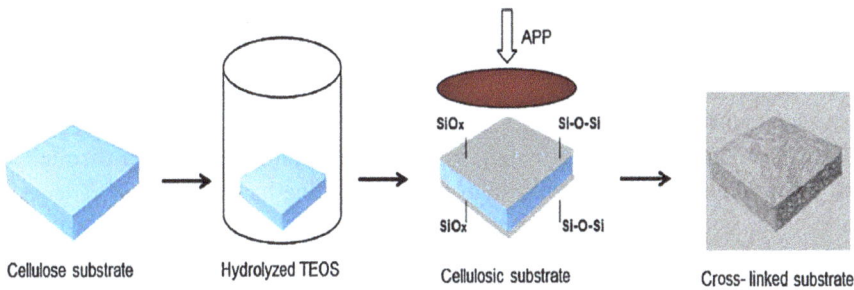

Figure 4.5: Coating of silicon dioxide over cellulose materials by using atmospheric pressure plasma (APP) technique (reproduced/adapted with kind permission [25] Copyright [2022], Elsevier).

The other latest developments in FR coating techniques are inorganic additive incorporated systems, phosphorus-based coating systems, halogen-based formulations, nitrogen-based coating systems, multi-element FR systems, silicon-based coating systems, and nanocomposite-based coating systems [25].

In commercial process, FR-coated fabric is generally prepared by a two-step process, where the first step is pad-dry-cure by passing the fabric through a FR finish solution, and subsequently, the fabric is coated with some FR coating formulation (generally PU- or PVC-based). FR-coated fabrics are used in numerous fields such as in life rafts, life jackets, tarpaulins, awnings, aircraft, hovercraft skirts, safety chutes, protective covers, aircraft fuel tanks, and flexible containers. It is also used in fire barriers, such as ceramic cloth, ceramic coating, intumescent coating, or in the making of other high-temperature foam insulation barriers that can also provide fire resistance and reduce smoke as well as toxicity.

4.3.9 Waterproof coating

Water-resistant (WR) treatment with a waterproof coating is used to increase hydrophobicity and soil release or self-cleaning performance of textile garments. Various microporous membranes, films, and coatings are prepared for the purpose of waterproof breathable applications. These WR-coated textiles have excellent waterproof as well as breathable characteristics, so that they can prevent the penetration of water molecules across the textiles but allow moisture vapour. The size of the micropores generated is in the range of 0.02–1 μm, which is much lower than the size of water drops (100 μm) and larger than the size of vapour molecules (40 × 10 − 6 μm) [26]. PU is the most widely explored coating material for the preparation of waterproof breathable suits because of its ease of availability, ease of processing, flexibility, tailorability, and durability of the film. The first waterproof breathable film was introduced in 1976 by W. Gore and is known as Gore-Tex. This film was prepared using polytetrafluoroethylene (PTFE), which claimed to contain 1.4 billion tiny holes per square centimetre. There are various such membranes prepared with different coating materials on various substrate bases. A few preparation methods that are available in the literature are mechanical fibrillation, coagulation process, foam coating process, solvent extraction method, and so on

Figure 4.6 shows a schematic diagram of water vapour regulation through a coated fabric. The top surface with hydrophobic coating has micropores. The size of micropores is much smaller than that of the smallest raindrops (100 μm), and much larger than the size of a water vapour molecule (0.0004 μm). As a result, water drops cannot penetrate into the coated face exhibiting an excellent showerproof effect. However, water vapour can be comfortably penetrated and transmitted across the pores and capillary system of the coated textile structure, exhibiting an excellent breathable effect. Such coated textiles are popular due to their excellent thermo-regulation behaviour.

Raincoat and luggage fabrics are the ideal examples of waterproof fabrics. In such applications, the fabrics are first treated with water-repellent chemicals (mainly fluorocarbon-based) and then coated with different coating formulations to make waterproof fabric. In luggage fabric, generally, PU or acrylic-based coating is used, while for raincoat application, the fabric is generally coated with PU or PVC to make the fabric highly waterproof. Recently, PU-coated highly waterproof and breathable fabric is attracting huge commercial interest for raincoat applications, because of its better comfort

Water droplets (diameter > 100 µm) too big to
penetrate the coated surface

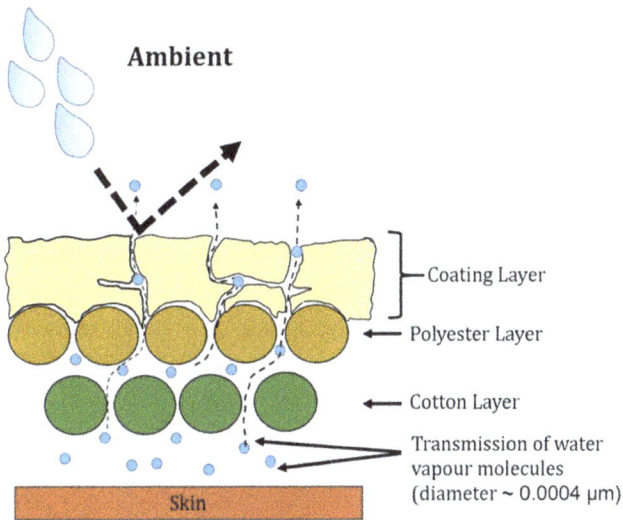

Figure 4.6: Schematic diagram of water vapour transmission through both inter-yarn pores and coating micropores to obtain waterproof breathable effect.

and high waterproofness, protecting the wearer from the heavy rain. In luggage applications, two to four types of fabrics of different counts/constructions may be used with different levels of waterproofness. A few other potential applications of waterproof, coated fabrics are tents, tarpaulins, car covers, and so on.

4.3.10 Anti-biofouling coating

Anti-biofouling coatings are very much used in the marine and ship industry due to the higher fouling and deterioration rate with water. Marine structures such as platforms and ship hulls are subject to the problem of biofouling. Organic and inorganic anti-biofouling routes can be applied for anti-fouling systems in order to provide protection to marine structures. If biofouling accumulates on the hull, it increases propellant consumption and power requirements, which can have a detrimental effect on the hydrodynamics of the hull. Therefore, fouling is a serious issue on surfaces exposed to aquatic environments where marine micro-organisms can bind to the surface and build a conditioning layer, thus providing an easily accessible platform for other aquatic species such as diatoms and algae to adhere and grow. Increased operating and maintenance expenses owing to fouling of water vessels and ship hulls due to biofouling is a major concern in the marine and ship industry.

Wang et al. [27] have claimed that an anti-fouling coating formulation can be obtained by mixing powder of zinc-graphite alloy with epoxy-silicone resin. The coating formulation can be greatly reduced by making $CaCO_3$ fouling deposition. After immersing the composite coating in $CaCl_2$/$NaHCO_3$ solution, just a little amount of fouling was found on the surface.

In another recent study, He et al. [28] have developed a new photochromic, hydrophobic cotton fabric with high sensitivity against ultraviolet (UV) light irradiation, high fatigue resistance, and superior hydrophobic characteristics. In addition, it can swiftly revert to its native hue when exposed to green light, heat, or darkness. Furthermore, to increase the fatigue resistance, the target photochromic spiropyrane molecule was first synthesized and formed into microcapsules as the core and chitosan as the shell. Polydimethylsiloxane (PDMS) was mixed with microcapsules and applied to the surface of cotton fabric in a unique way. The presence of PDMS not only ensures that microcapsules are securely adhered to the fabric surface, but also provides anti-fouling capabilities owing to the hydrophobic effect, long-lasting photochromic features, quick colour-changing properties, and remarkable fatigue resistance. These types of newly developed anti-fouling coatings can be used to make various novel products such as photochromic compounds, photochromic microcapsules, civilian textiles, and particularly, outdoor textiles.

Anti-fouling coatings have gained significant interest due to their unique wettability and self-cleaning properties. However, there are a lot of demerits in this process like low stability, excessive use of fluorinated chemicals during the process, and low transparency. Dong et al. [29] have designed a novel type of smooth anti-fouling coating based on methyltrimethoxysilane. The formulation is created by hydrolytic condensation of methyl trimethoxysilane in isopropanol and then wiping the glass slide with a nonwoven cloth that sucks the stock solution. Any substance coated with a transparent anti-fouling coating (both water- and oil-based) can easily provide high protection from surface deterioration from various external fluids (such as water, diiodomethane, n-hexadecane), regularly encountered liquids (such as milk, red wine, cooking oil, coffee, and soy sauce), artificial fingerprint liquids, mark seals, paints, and so on. Fluids easily slide off 4–30° titled coatings. Additionally, these coatings exhibit strong mechanical (abrasion, bending, scratching, etc.), chemical (acidic, alkali, salt, etc.), and thermal (300 °C) stabilities in terms of smooth sliding behaviour of the probing liquids. Furthermore, due to their remarkable anti-fouling capabilities, high stability, non-fluorinated nature, and simple manufacturing approach, anti-fouling coatings can be applied to a range of substrates, using the same process.

4.3.11 Electrically conductive coating

Electrically conductive coating is done by coating of conductive materials such as metals, conductive polymers, carbon etc., onto textile substrates. This is a well-known process for manufacturing supercapacitors, wearable electronic textiles (e-textiles), sensors, EMI shields, and electrode objects. Fabric-based supercapacitors are developed by coating conductive carbon-based materials like graphene and carbon nanotubes (CNTs) onto non-conductive textile cotton fabrics. Such coating not only improves electrical conductivity of fabric but also their ability to function as active electrode materials [30]. De Falco et al. [31] have confirmed the usage of direct electrospinning to create wearable nickel-coated cotton textiles with multi-walled CNTs (MWCNTs) for wearable electronics applications.

Kim et al. [32] have suggested that woven nylon Lycra® fabric can be coated with nickel/phosphorus and nickel/polypyrrole by electroless method to prepare a flexible and electrically conductive textile fabric. They have also measured the electrochemical behaviour, electrical resistivity, and electro-mechanical response of the coated textiles and have found all of them quite satisfactory. Pre-dyeing of the textile fabrics with poly(2-methoxyaniline-5-sulfonic acid) (PMAS) before electroless metallization with electroless nickel and chemical polymerization of polypyrrole is found to improve the coating durability and stability. The mass gain in nickel coating was observed to rise linearly with deposition time. At longer nickel deposition durations, the surface resistivity of the coated cloth was shown to decrease. These types of materials are widely popular and quite useful as charge storage in electrical and electronic circuits.

The fabrication of a coated textile with a hydrophobic surface that can produce electromagnetic interference shielding capability with low reflection was done using a green process. For this process, firstly, hydroxyl-terminated, polydimethylsiloxane-modified, two-component aqueous polyurethane dispersions are created. Then, in the second step, a high-pressure microfluidizer is used to disperse MWCNT and graphene in water. The dipping approach is used to apply waterborne polyurethane and fillers (80% CNT, 20% graphene) onto fabrics. It is found that the water contact angles of the polyurethane coating and the coated textile are 103.4° and 153.6°, respectively, when the polyurethane coating contained 10% hydroxyl-terminated polydimethylsiloxane. The electromagnetic interference (EMI) shielding effectiveness (EMI SE) of coated textile reached up to 35 dB (decibel) at a thickness of 0.35 mm, with an overall reflectance of around 41.4%. This type of coated textiles offers a lot of potential in EMI shielding advanced applications with good shielding performance and low reflectance. The other most important fact is that it also offers a great way to manufacture the coating in a simple and green process (Dai et al.) [33].

Coating of polypropylene and poly(ethylene terephthalate) on textiles with an organosilicon sol containing scattered graphene particles resulted in electrically conductive composite fabrics. The graphene-modified coating was applied to the fabric by using the sol-gel procedure and the padding approach. The fibre surfaces were stimulated with atmospheric pressure plasma in the form of corona discharge to promote coating bonding. This type of textile coating has a higher electrical conductivity due to the use of after-process, when it is padded with graphene-containing sol. Pre-coating activation was observed in this type of product due to an increase of adhesion between the conducting coating and textiles. This makes such products more suitable for use in electrical appliances and chemical resistance-based products, due to their greater surface stability [34].

4.3.12 Active smart coating (memory polymer coating)

Active smart coatings (memory polymer coating) are performed by using shape memory polymer (SMP). SMP polymers operate based on their molecular mechanism that triggers the two mechanical states as martensites (shape B) and artensites (shape A) (Figure 4.7) [35]. The switching of these two states actually controls the thermally induced shape-memory effect of the active smart coatings. These switching conditions depend on various factors such as the amount of embedded thermoplastic particles into the coated polymeric SMP matrix, melting temperature, and the nature of the semicrystalline SMP thermoplastic particle (Figure 4.8) [36]. The shape recovery of these SMP coating is time-dependent and that might take a total of 10 s at 50 °C. Figure 4.9 shows the time-dependent shape recovery for an SMP-coated tube that was prepared of a poly(ε-caprolactone)dimethacrylate polymer network (M_n of the network was 104 g/mol) [37]. Similarly, Figure 4.10 shows the strain recovery under infrared radiation of curling of the ribbon (made by active smart SMP coating) towards the infrared source within 5 s [38].

A special suit constructed on a membrane using shape memory polymers (SMPs) was invented for Swedish sailors. The proposed suit uses Diaplex smart fabric technology, which is fully windproof and waterproof, but fully breathable. This substance can recall or remember and keep its shape or revert to its former form in a changing environment. A Diaplex garment can detect changes in its surroundings, analyse the situation intelligently, and respond appropriately to guarantee a high degree of comfort. Moreover, the Diaplex membrane undergoes micro-Brownian motion, when the temperature rises over a specific limit, thus resulting in the formation of micropores that enable water vapour and heat to escape. This, finally, protects the human body from extreme hot or cold weather or temperature or any climatic changes [39].

4.3.13 Self-healing coating

Autonomic healing or self-healing materials respond without any outside or external signal/ environmental stimuli in a time-dependent and productive way, and they offer huge prospects for advanced engineering applications. The self-healing coatings can rectify, repair, or rebuild the structure against corrosion, mechanical, and environmental damage. Various mechanisms have been established for achieving self-healing

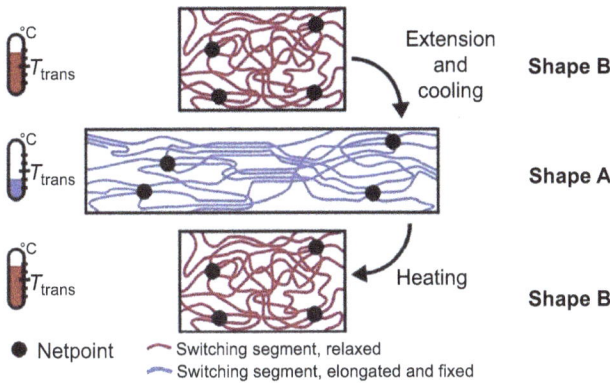

Figure 4.7: Molecular mechanism of SMP polymers involved in thermally induced shape-memory effect and phenomenon involved in T_{trans}-thermal transition temperature-related switching phase (Modified according to [35] Copyright [2007], Elsevier).

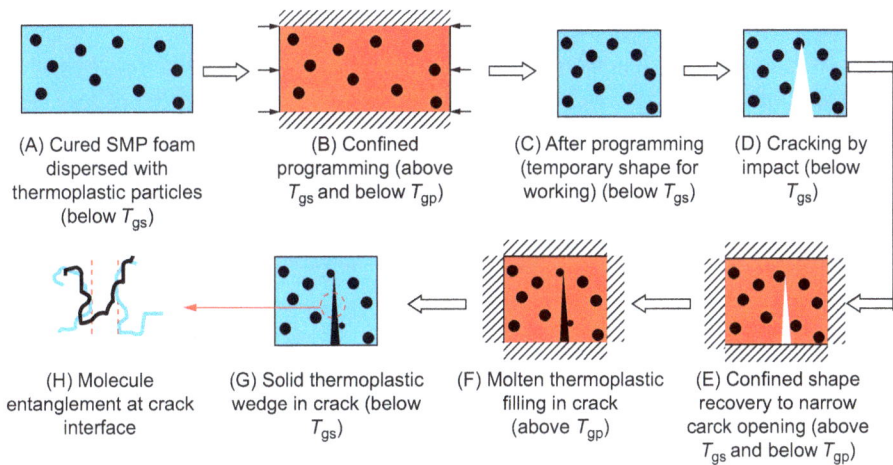

Figure 4.8: Working mechanism of shape memory polymers embedded with thermoplastic particles (T_{gs}: glass transition temperature of SMP; T_{gp} glass transition temperature of thermoplastic particles). (Modified according to [36], Copyright [2010], Elsevier).

ability of the coating materials, including micro-encapsulation, microvascular networks, reversible chemistry, nanophase separation, hollow fibres, polyionomers, and monomer phase separation. They all have been demonstrated as suitable methods of attaining self-healing functionality in textiles. Modern engineered self-healing coatings are highly optimized materials that are unlikely to tolerate significant changes in coating chemistry. Polymeric coatings protect a material or substrate from the outside force or rupture, and when these types of coating fail, the substrate's corrosion is dramatically accelerated. Hence, a self-healing coating needs to be highly stable to stimuli existing in

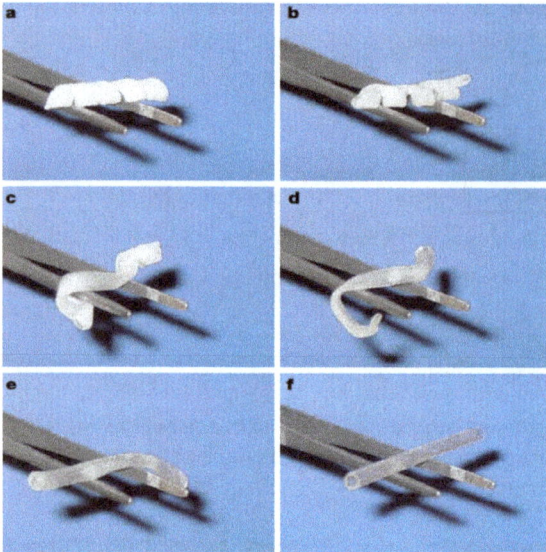

Figure 4.9: Time-dependent shape recovery for an SMP-coated tube prepared of a poly(ε-caprolactone) dimethacrylate polymer network: (a)–(f) Start to finish of the process takes a total of 10 s at 50 °C (reproduced/adapted with kind permission [37], Copyright [2004], Springer Nature).

Figure 4.10: Strain recovery under infrared radiation of curling of the ribbon (made by active smart SMP coating) towards the infrared source within 5 s (exposure from the left) (reproduced/adapted with kind permission [38], Copyright [2004], Springer Nature).

the environment. This healing chemistry is eye-catching, since it is air and water stable and remains active even after exposure to raised temperatures (up to 150 °C) and allows their use in systems for thermal cure [40].

Recently, a new shape-memory-aided self-healing (SMASH) coating has been developed. This material features phase-separated electrospun thermoplastic poly(ε-caprolactone) (PCL) fibres that are arbitrarily distributed inside a shape memory epoxy matrix. Mechanical damage to this SMASH coating can be self-healed by heating. It is also able to control the healing and damage over the coating layers completely, by restoring a higher corrosion resistance [39].

Self-healing coatings can also be applied to produce smart fabrics that require stretchable conductive strands (E-textiles). Shuai et al. [41] have created elastic, conductive, and self-healing hydrogel fibres, using a continuous dry-wet spinning method (Figure 4.11). They have used a process where the physically cross-linked poly(NAGA-co-AAm) (PNA) hydrogel precursor shows a thermally reversible sol-gel transition. It was thoroughly mixed with acrylamide (AAm) and N-acryloylglycinamide (NAGA). The prepared PNA hydrogel fibre showed good tensile strength (2.27 MPa), stretchability (900%), high conductivity (0.69 S/m), and self-healing properties. The PNA/PMA core-sheath fibre has an elastomeric poly (methyl acrylate) (PMA) covering that can provide good resistance to water evaporation and absorption. The strain sensing capabilities of PNA/PMA fibre are proven for tracking human body movements. PNA/PMA fibres can also transform mechanical motion energy into electric power, similar to a triboelectric nanogenerator (TENG) textile woven product. This type of product holds a lot of promise for future multifunctional smart fabrics and wearable electronics.

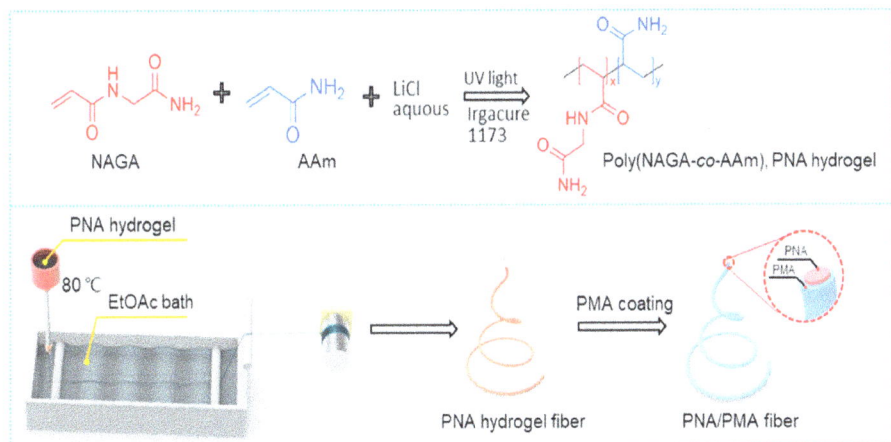

Figure 4.11: Chemical scheme and illustration of method of preparation of hydrogel and fibre from PNA precursor, and PNA/PMA core-sheath fibres (reproduced/adapted with kind permission [41], Copyright [2020], Elsevier).

In another study, Zahid et al. [42] found that the self-healing coating technique can be used to manufacture a textile-based wearable strain sensor with excellent sensitivity and a vast operating range. This product can easily monitor human muscle activity. This kind of self-healing coating is produced by using a gum-like sticky polyvinyl acetate-co-vinyl laurate polymer (commercial name: Vinnapass) combined with 50 wt% carbon nanofibers (CNFs) in an ethyl acetate co-solvent medium to produce electrical conductivity. The Vinnapass-based highly stretchable sensor is used to track human wrist and finger folding and unfolding, bicep muscle contraction/relaxation, and is also designed to track human muscular movements throughout physical activities and exercise.

4.3.14 Biomimetic coatings used in advanced medical applications

Biomimetric coatings are formulated by using biopolymers such as collagen, octacalcium phosphate, and apatite materials for the repairing and restructuring of human dental, bone, joints or other body parts/tissues. Thorough biological integration of dental implants into the surrounding tissues (bone and gingiva) is essential for clinical success. This can be accomplished by developing biomimetic coatings made of collagen type I (for the gingiva region) and hydroxyapatite (HAP) or mineralized collagen (for the bone contact) as acceptable surfaces for the interfaces. In the first step, the biomaterial is dipped into a simulated body fluid at 37 °C and stirred at 150 rpm for 24 h (Figure 4.12a). As a result, an amorphous calcium phosphate (ACaP) is formed over the biomaterial as a dense and fine layer. This layer acts as a substrate for further deposition of more substantial crystalline layer. The crystalline layer is formed in the second step by dipping the ACaP-coated biomaterials in a supersaturated calcium phosphate solution at 37 °C and stirring at 60 rpm for 48 h (Figure 4.12b). The coated biomaterials are then freeze-dried. If certain bioactive agents are required to be added for improving the functionality, they can be added in the calcium phosphate solution with a suitable concentration.

A linear adhesion peptide derived from a laminin sequence is linked to collagen to improve cell attachment in the gingiva area. On the other hand, a cyclic RGD peptide can be biomimetically coated to HAP and mineralized collagen to offer enough anchor systems for bone interface repair. Cell attachment investigations with human keratinocytes and osteoblasts revealed that these biomimetic coatings have the significant biological potential for rapid collagen and bone development and repair [43].

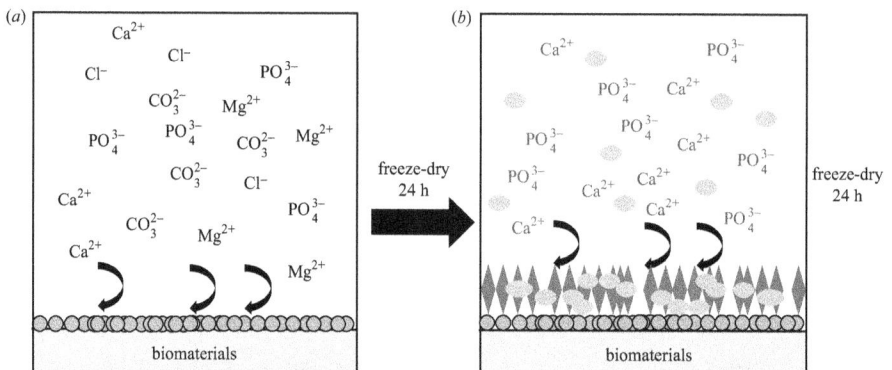

Figure 4.12: The mechanism of the biphasic biomimetic calcium phosphate coating (circles represent amorphous seedling layer; ellipsoids represent BMP-2; diamonds represent crystalline protein-carrying layer). (Modified according to [44], Copyright [2004], Royal Society (Great Britain)).

The repair of critical-sized bone defects is, nevertheless, challenging in the fields of implantology, especially for orthopaedics and maxillofacial surgery. For this purpose, a suitable vehicle was invented by the biomimetic coating, which can deposit on metal implants besides biomaterials. Bone development cannot be triggered by the materials currently used to fill bony defects. This form of biomimetic coating contains bone morphogenetic protein 2 (BMP-2) and can be used to treat major bone defects in the disciplines of dental implantology, maxillofacial surgery, and orthopaedics [44].

A new biomimetic coating method by metal implants allows the fast, homogeneous, and dense development for curing bone calcium phosphate. Titanium alloy (Ti_6Al_4V) disks were coated with a carbonated, thin, amorphous calcium phosphate (ACP) by dipping in a wet solution of calcium, phosphate, magnesium, and carbonate. The ACP-coated disks were managed further by incubation in calcium phosphate mixtures to produce either crystalline octacalcium phosphate (OCP) or carbonated apatite (CA) [45].

4.4 Lamination process

Lamination is the process of joining two or more flexible materials together with a bonding agent, or sometimes, fusing by application of heat. The substrates used for preparing the laminating-webs consist of films, papers, or aluminium foils. An adhesive (bonding agent) is generally applied in between two substrates via pressure in the lamination process to produce a duplex or two-layered final laminate. It is very important to choose appropriate raw materials and web laminating procedure for achieving the targeted properties for intended applications. A variety of technologies are available to cover a wide range of applications in the textile, food, and non-food packaging industries, as well as solar energy and insulating panel industries. The conventional lamination techniques are classified into the following categories [46]:

i.   Wet or solvent lamination
ii. Solventless lamination
iii. Web lamination
iv. Dry lamination
v.   Wax lamination

Several new and recent lamination techniques have been invented for making functional textiles as mentioned in the following sections.

4.4.1 Sheet lamination

Sheet lamination is a process based on additive manufacturing techniques defined by ISO/ASTM 52900–2015 for building a 3D object by stacking and laminating thin sheets of materials. The lamination method can be performed by heat and pressure (for polymer and textile), active adhesive (for polymer and textile), ultrasonic welding (for metals), bonding or brazing (for metals), and the final shape is ensured either by CNC (computer numerical control) machining or laser cutting. The various types of seed laminations include [47]:

a. Selective lamination composite object manufacturing (SLCOM)
b. Plastic sheet lamination (PSL)
c. Selective deposition lamination (SDL)
d. Laminated object manufacturing (LOM)
e. Computer-aided manufacturing of laminated engineering materials (CAM-LEM)
f. Composite-based additive manufacturing (CBAM)
g. Ultrasonic additive manufacturing (UAM)

4.4.2 Thermal lamination of polyurethane films

Polyurethane-based thermal lamination is used for making flexible electronics, where a highly electroconductive, flexible, and durable material is key for several uses. A recent study explains the property changes in graphene nanosheets (GNSs) doped into a conductive silver (Ag) paste that is applied to form a grid-style pattern on a thermoplastic polyurethane film for electromagnetic wave shielding. The results showed that the conductive lamination shows a higher loading, bending ability, good electromagnetic shielding effectiveness (EMSE of 49.9 dB at 1,800 MHz), and a much better electrical conductivity (4.63×10^4 S/cm) than the pure silver paste (1.38×10^4 S/cm). Woven fabrics laminated with PU/Ag/GNS conductive film and doped with different amounts of fillers were successfully manufactured. This product can be readily used for electronic printing, electromagnetic interference (EMI) shielding materials, electrodes for sensor components, wearable elements, smart home textiles, and conductive materials [46].

Nowadays, thermal lamination of TPU films on textiles is very popular for making inflatable products such as life jackets, lifeboats, blood pressure cuffs, and aeroplane escalators. In this technique, TPU films of particular thickness are laminated on one or both sides of the fabric with a suitable tie coat.

4.4.3 Ethylene-vinyl acetate-based lamination

Ethylene-vinyl acetate-based laminations are used to increase the lifespan of timber wood used in the building structure, photovoltaic module cells, and adhesive tapes [48, 49]. A recent study shows the efficiency of this kind of lamination can be further increased by fabricating large complex zirconia structures using a material extrusion (MEX)-built additive manufacturing route or by fused deposition modelling (FDM), or by fused filament fabrication (FFF) [50].

Recent trends towards mass-timber buildings and tall timber structures demand more strict requirements for the bond durability of adhesives for engineered wood composites. Researchers developed a cost-effective method for producing a novel aqueous polyisocyanate adhesive with excellent hygrothermal resistance for engineered wood composites by co-cross-linking with polyamidoamine-epichlorohydrin (PAE) resin and then hybridizing organic-inorganic with montmorillonite clay (MMC). As a result, this novel aqueous polyisocyanate adhesive demonstrated excellent bond durability of plywood with a wet-aged strength of 1.73 MPa and a strength reduction ratio of only 18.4% (after seven boiling-dry hygrothermal ageing cycles). Thus, these materials are able to meet the requirements for manufacturing engineered wood composites as good construction and building materials [51]. The ethylene-vinyl acetate lamination is suitable for textile fabrics substrate, which can be used as floor covering and straps. On the one hand, the laminate serves fixing the materials that form the tread, and on the other hand, determines the pedalling, significantly [52].

4.4.4 Breathable laminated structures

Breathable laminations can provide a waterproof breathable membrane, which can vary based on weather that mimics human skin. This type of lamination technique has excellent ability to transmit water vapour and its structure can be optimized by altering fabrication parameters, in order to achieve waterproof target ability and breathability performance. This functional performance is achieved by lamination of a microporous film by either using physical bonding or by chemical cross-linking with the textile substrate. The physical bonding can be achieved by thermal treatments by applying temperature and pressure, allowing some time. The chemical bonding is done by using suitable primer or cross-linking resins.

Polymeric porous membranes are getting enormous importance in the protective textile sector because of their excellent breathability balanced with waterproofness. These membranes can be laminated using the hot melt adhesive lamination method. Membrane thickness, application temperature, pressure, roller gapping, machine speed, chemistry of adhesive, and add-on of adhesive are all important factors controlling the performance of prepared fabric-membrane laminates. The physical performance of these polymeric porous membrane-based laminates, such as waterproofness, water vapour permeability, and resistance to evaporating heat transfer has been found to be excellent [53].

The mechanical properties and porosity of the laminated product depend on the nature of the bonding between the substrates. The lamination process also affects the hand value of textile materials. The Kawabata evaluation system (KES-FB) can be used to evaluate the objective hand value of the laminated textiles. These breathable laminated textile products can be used in outdoor, medical, and sports applications [12].

4.4.5 Transparent or opaque or pigmented film-based laminates

Transparent or slightly opaque or pigmented polymeric films (TPU, PTFE, polyethylene, etc.) are extensively used for laminating on textiles for making jackets, medical suits, and so on. Laminates made of polyethylene and melt-blown/spun-bonded nonwoven are extensively used in making personal protective equipment (PPE) for medical personnel providing chemical/waterproofness, as well as protection from harmful microbes. The use of breathable TPU films or microporous PTFE films can impart additional features to garments and improve breathability and comfort to the wearer. A triple-layered laminated reversible jacket for military personnel consists of white woven fabric on one side (to wear on snow), olive green coloured woven fabric on the other side (to wear in forest area) and one transparent or slightly opaque film, which is laminated in between these two fabrics. The main properties needed for such type of multi-layered laminated fabrics are good waterproofness, breathability, comfort, and specific interlayer adhesion strength.

Transparent laminations are also used in solar cell and photovoltaic cell manufacturing due to higher mechanical stability. Transparent conductive adhesive (TCA) interlayers can be transparently laminated mechanically to provide a highly stable solar cell with a superior electrical connection between sub-cells and stacked (Figure 4.13). The other added features include adding an ethyl-vinyl acetate (EVA) matrix and silver-coated compliant conductive microspheres, which can provide industry-standard EVA encapsulant

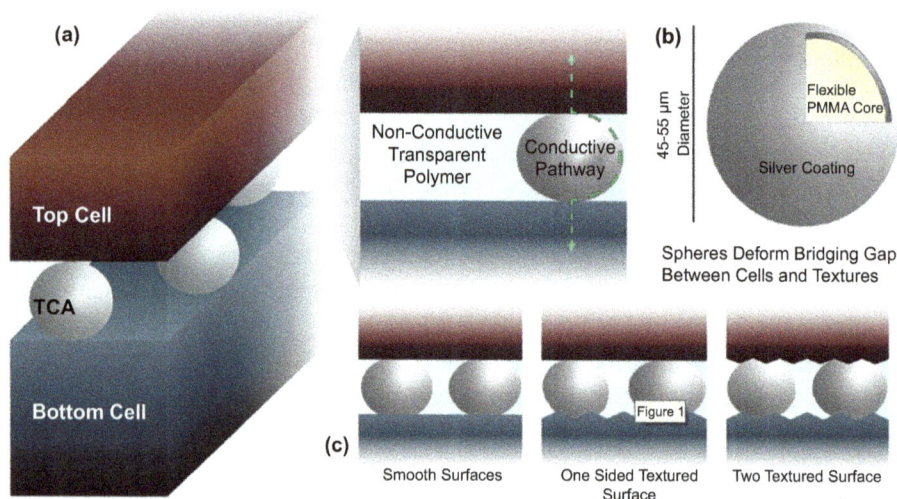

Figure 4.13: (a) Transparent conductive adhesive (TCA) in microspheres size (45–55 μm) is applied in between solar cell stack where top cell is a transparent non-conductive polymer. (b) The TCA microspheres composed of flexible poly(methyl methacrylate) (PMMA) core with a silver coating. (c) Attachment of the TCA microspheres with smooth and textured surface electrically (Modified according to [54], Copyright [1996], IOP Publishing).

sheets with high product durability and lifespan. Transparent laminations can also be applied with blade coating by using 3D printed blades. The testing of vertical conduction through the interlayer showed a good potential for high conductivity, reproducibility, and low-cost fabrication [54].

Transparent laminations can also be applied to produce polymer light-emitting diodes (PLEDs) that are generally fabricated using a wet process. These products are very useful in electronic and radar protective applications [55].

Different coloured/pigmented films are used to laminate on fabric for making attractive products like jackets, snowsuits, and fabrics of extreme cold weathering clothing systems (ECWCS). The main properties needed in these laminated fabrics are good drapability, breathability, comfort, and high waterproofness.

4.4.6 Electrospun polymer nanofibre-based laminates

Electrospinning is a method that uses an electrostatically driven jet of polymer solution or polymer melt to generate polymer nanofibres with diameters less than 100 nm and lengths up to kilometres. In electrospinning, nano-size fibres are formed by electrostatic forces pulling a polymer solution/melt jet. Direct electrospinning on the substrate forms a layer of nanofibres on the surface of the substrate, making a laminate. The jet is first extended along a straight path, and it undergoes a vigorous whirling motion due to bending instability. Since the jet's diameter decreases as it stretches, it solidifies quickly, leading to fibres deposited on the substrate or grounded collector. Fibres with nanometre-scale diameters and with the varying beads, cross-sectional forms, branches, zigzags, and buckling coils are produced by controlling the electrospinning process. The utility of this lamination technique by applying nanofibres is increased by post-treatments such as conglutination, chemical treatment of surfaces, vapour coating, and heat processing [56].

Recently, electrospun nanofibres containing aligned functional nanostructures like nanotubes and nanowires are gaining huge importance in function applications. Over the last several years, significant development has been achieved in this area, and now, this technology has been used in a wide range of applications. The majority of recent electrospinning research has centred on either trying to better apprehend the fundamental characteristics of the process to gain control over nanofiber structure, morphology, surface functionalities, fibre assembly strategies, and determining appropriate electrospinning environments for several polymers and biopolymers [57]. Among different applications of electrospun-nanofibre-based laminates, a few important applications are: (i) electrospun nanofibres and nonwoven-based laminates for filtration applications, (ii) conductive electrospun nanoweb-based textiles for electronic applications such as sensor and supercapacitor [58], (iii) biopolymer-based electrospun nanofibres and nonwoven-based laminates have many potential applications in medical fields such as wound dressing, drug delivery, and tissue engineering, (iv) functional nanofibre-based textiles with multifunctional property, and so on. Figure 4.14 shows the fabrication process of electroconductive carbon fibre nanoweb-laminated cotton fabrics and the process layout. At first, Ni-coated cotton fabric was prepared by a polymer-assisted metal deposition (PAMD) process in liquid medium. The Ni-coated cotton fabric was used as electrically conductive collection surface for an electrospinning instrument, where the Ni-cotton fabric was attached onto the surface of the rotating collector [58]. MWCNT dispersion solution was prepared in N,N'-dimethylformamide (DMF) solvent with a loading 60–80 wt%, and then polyacrylonitrile (PAN) was electrospun directly onto the Ni-cotton surface. As a result, the prepared Ni-cotton fabric coated with carbon fibre web (C-web) was found to be suitable for supercapacitor applications.

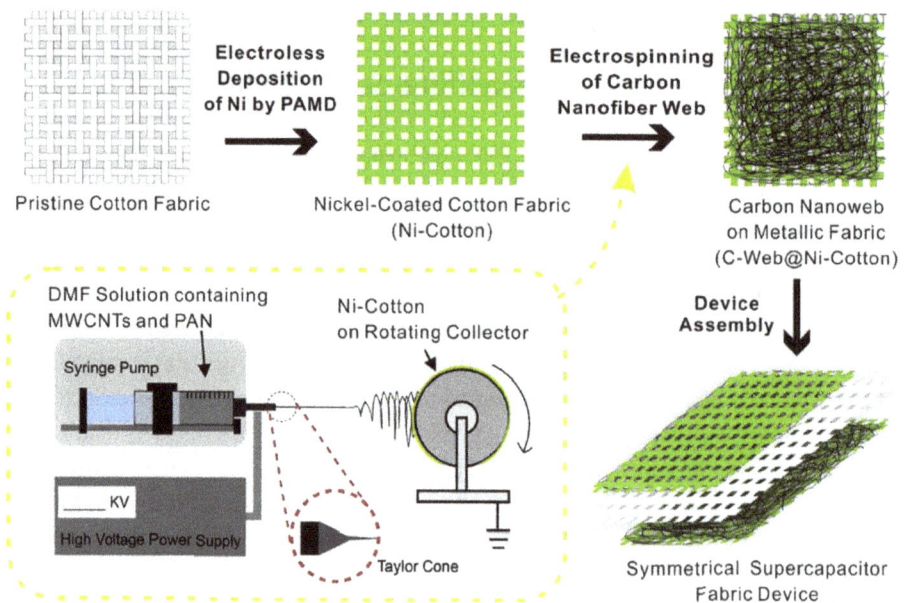

Figure 4.14: Preparation stages of Ni-coated cotton fabrics coated with carbon web, which is suitable for supercapacitor applications (reproduced/adapted with kind permission [58], Copyright [2013], Royal Society of Chemistry).

4.4.7 Decorative laminations

Decorative lamination of textile fabrics can offer manifold creative possibilities for textile and fashion designers to design fashionable sports and leisure garments, costumes, party wears, promotional garments, corporate clothes, and so on. Decorative laminates are composite materials in which a durable and polymer-based decorative material is combined with a textile fabric. During the manufacturing, the decorative polymeric film (as cover or décor sheet) and the textile fabric (as a core sheet) are superimposed and consolidated under heat and pressure to form a single sheet [59]. Based on the composition of the décor sheet and core materials, there are various properties that can be developed for various applications. Some of the properties are mentioned further:

- Imparting colouring effect: plain colours, patterns, glossy, satin, textured, woodgrains, masonry, and so on [60]
- Enhancement of durability, including resistance to staining and abrasion [61]
- Fire-retardant decorative laminates [62]
- Improving cleanability of the laminates [63]

4.4.8 Various testing standards for coated and laminated fabrics

The following tests are performed on the coated/laminated textiles for evaluating different functional properties in the light of applications. Table 4.2 summarizes the various test standards for coated and laminated textiles.

Table 4.2: Standard testing for coated and laminated textiles.

S.no.	Test	Standard methods
1	Water repellence test	AATCC 22–1996, ISO 4920, and IS 390
2	Cone test (for determining water repellence of fabric)	IS 7941
3	Waterproofing	ISO 811
4	Fabric weight (GSM)	ASTM D 3776
5	Thickness	ASTM D 1777
6	Coating mass/area	IS 7016 Part 1 and BS 3424
7	Breaking strength and elongation	ASTM D 5034 and ASTM D 5035
8	Tear strength	ISO 4674–1:2016, BS EN 1875–3
9	Bending length, flexural rigidity, fabric stiffness	ASTM D 1388, ASTM D 4032
10	Breathability (water vapour permeability and resistance)	ISO 11,092, JIS L 1099:2012
11	Peel strength or adhesion strength	ASTM D751, ISO 13,934–1
12	Abrasion resistance	ISO 12,947–2
13	Resistance to bursting strength and puncture	IS 7016 Part 6 and ASTM D751, ASTM 4833
14	Brittleness point test	IS 7016 Part 14, ASTM D2137 and BS 3424

Table 4.2 (continued)

S.no.	Test	Standard methods
15	Flame retardance	IS 11871, ISO 6941:2003, IS:15061:2002
16	Gas barrier	ASTM D 1434
17	Resistance to flexing	ISO 7854
18	Cold cracking	IS 7016–10

4.5 Current challenges and future scope

In the last few decades, tremendous research has been carried out on advanced materials, specifically for the development of high-performance textiles. In the area of technical and functional textiles, coating and lamination technologies as well as their materials are the most attractive topic of research. Beyond the conventional techniques, there are many advanced coating/lamination technologies that are incepted through research and found suitable for the preparation of useful high-performance products for specific applications. The main limitation of the advanced technologies of coating and laminations is that they are not still adopted by industries for the commercial production of coated/laminated textiles. Though the conventional coating technologies have been widely commercialized, there is, still, huge scope for scaling up the advanced technologies in textile industries for mass-scale production of high-end products. Therefore, fabrication, development, and commercialization of coating equipment or machinery suitable for making coated and laminated products with desired functionality are aspects to think about for further research. Material and process control, optimization, and costing are the next stage of the challenge. There is a wide range of materials (such as films, polymers, additives, adhesives, and other chemicals) available for coating and lamination, yet, there is enormous scope for research to develop new and advanced materials targeting multiple applications. Coatings of nanocomposite materials have attracted huge attention for research targeting multifunctional applications, but they are, still, limited mainly to lab scale, and more drive is needed to commercialize them in bulk scale after considering their feasibility in terms of manufacturing process, product performance, and costing. Therefore, the future developments in the advanced materials coating field can be characterized and evaluated in terms of their economic feasibility by looking at different technical support and development strategies – strategies whose efforts are still concentrated on basic research leading up to the discovery of new technologies. These new technological breakthroughs expected in the future will predominantly assist large companies and attract completely new industries. The challenges are as follows:

- Development or selection of advanced materials, processes, and equipment
- Development of control devices and automated robotic handling of materials
- Applications of statistical process control and statistical quality control
- Standardization of testing and characterization of coated/laminated products

Taking into account the recent trend of eco-consciousness, the environmental issues of coating/lamination process and materials is another major challenge for near future. Therefore, researchers are focusing on developing eco-friendly materials and technologies for coating and lamination. Recycling of coated and laminated products is another big challenge that must be taken care of, in the near future. Therefore, researchers are working towards inventing biodegradable materials for coating and lamination.

In summary, the future of the advanced coating and lamination sector appears very bright. There are, however, problems still to be solved. Strict quality control of well-established technology of coating or lamination and close attention to the design and testing of coated/laminated substrate as a single synergistic entity, combined with the development of novel structural and functional coatings/lamination using improved automated equipment and comprehensive databases and expert systems will secure the advanced coating/lamination technologies a substantial market niche in the near future.

4.6 Conclusions

This chapter demonstrates the importance, methodologies, and applications of various speciality coating and lamination techniques as effective surface activation/modification treatments for textiles. These techniques are quite easy and yet sophisticated and impart an extra layer of dissimilar materials over textile substrates in order to obtain significant improvements in terms of resistance to water penetration, fireproofing, antibacterial activity, self-cleaning, self-healing, electromagnetic shielding, antibacterial property, and so on. There are diverse technologies that have been invented by various researchers in recent years for the application of speciality coating and laminations over textile substrates. From the scientific research carried out in recent years, it is evident that continuous development and optimization of these coating techniques could lead to superior quality products, with improved and sustainable performance. Nanotechnologies, including nanoscale coating, nanomaterials, and electrospinning method, are emerging as very promising technologies for achieving high functional surface of the textiles that can achieve breakthroughs in applications. The coating of textiles with nanomaterials, nanofiber, nanowebs, and so on can bring about many multiple properties together, such as breathability, antimicrobial activities, and hydrophobicity in textiles. In the textile industry, recent advancements have also brought the laminating and coating methods to the forefront of technology. Moreover, coated and laminated textiles have become essential value-added products in the textile industry and the global technical textile market.

References

[1]   Venkatraman P. Fabric properties and their characteristics. Mater Technol Sportsw Perform Appar 2015:53–86.

[2]   Jahid MA, Hu J, Zhuo H. Stimuli-responsive polymers in coating and laminating for functional textile. Smart Text. Coatings Laminates, Elsevier; 2019, p. 155–73.

[3]   Kanakannavar S, Pitchaimani J, Thalla A, Rajesh M. Biodegradation properties and thermogravimetric analysis of 3D braided flax PLA textile composites. J Ind Text 2021:15280837211010666.

[4]   Singha K. A review on coating & lamination in textiles: processes and applications. Am J Polym Sci 2012;2:39–49.

[5]   Kadam V, Chattopadhyay SK, Raja ASM, Shakyawar DB. Waste management in coated and laminated textiles. Waste Manag. Fash. Text. Ind., Elsevier; 2021, p. 215–31.

[6]   Holmberg K, Matthews A. Coatings tribology: properties, mechanisms, techniques and applications in surface engineering. Elsevier; 2009.

[7]   Yamabe M. Fluoropolymer coatings. Organofluor. Chem., Springer; 1994, p. 397–401.

[8]   Saleh Alghamdi S, John S, Roy Choudhury N, Dutta NK. Additive manufacturing of polymer materials: Progress, promise and challenges. Polymers (Basel) 2021;13:753.

[9]   Wood KA. Optimizing the exterior durability of new fluoropolymer coatings. Prog Org Coatings 2001;43:207–13.

[10] Iezzi RA, Gaboury S, Wood K. Acrylic-fluoropolymer mixtures and their use in coatings. Prog Org Coatings 2000;40:55–60.

[11] Prest WM, Luca DJ. The alignment of polymers during the solvent-coating process. J Appl Phys 1980;51:5170–4. https://doi.org/10.1063/1.327464.

[12] Gutoff EB, Cohen ED. Water-and solvent-based coating technology. Multilayer Flex. Packag., Elsevier; 2016, p. 205–34.

[13] Fung W. Coated and Laminated Textiles. Vol. 23. Woodhead Publishing; 2002; p. 32–45

[14] Patel J, Patel P, Mahera H, Patel P. Effect of PU and PVC coating on different fabrics for technical textile application. Int J Sci Technol Eng 2015;1:279–84.

[15] Kim YJ, Huh PH, Kim BK. Synthesis of self-healing Polyurethane urea-based supramolecular materials. J Polym Sci Part B Polym Phys 2015;53:468–74.

[16] Khwaldia K, Arab-Tehrany E, Desobry S. Biopolymer coatings on paper packaging materials. Compr Rev Food Sci Food Saf 2010;9:82–91.

[17] Nathanael AJ, Oh TH. Biopolymer coatings for biomedical applications. Polymers (Basel) 2020;12:3061.

[18] Wagle PG, Tamboli SS, More AP. Peelable coatings: A review. Prog Org Coatings 2021;150:106005.

[19] Wilson JIB. Textile surface functionalization by chemical vapour deposition (CVD). Surf. Modif. Text., Elsevier; 2009, p. 126–38.

[20] Maity S, Chatterjee A. Conductive polymer-based electro-conductive textile composites for electromagnetic interference shielding: A review. J Ind Text 2018;47. https://doi.org/10.1177/1528083716670310.

[21] Najar SS, Kaynak A, Foitzik RC. Conductive wool yarns by continuous vapour phase polymerization of pyrrole. Synth Met 2007;157:1–4. https://doi.org/10.1016/j.synthmet.2006.11.003.

[22] Kaynak A, Najar SS, Foitzik RC. Conducting nylon, cotton and wool yarns by continuous vapor
       polymerization of pyrrole. Synth Met 2008;158:1–5.
[23] Maity S, Singha K, Pandit P. Advanced applications of green materials in electromagnetic shielding.
       Appl. Adv. Green Mater., Elsevier; 2021; p. 265–92.
[24] Li S, Xie H, Zhang S, Wang X. Facile transformation of hydrophilic cellulose into superhydrophobic
       cellulose. Chem Commun 2007:4857–9.
[25] Liang S, Neisius NM, Gaan S. Recent developments in flame retardant polymeric coatings. Prog Org
       Coatings 2013;76:1642–65.
[26] Mukhopadhyay A, Midha VK. Waterproof breathable fabrics. Handb. Tech. Text., Elsevier; 2016,
       p. 27–55.
[27] Wang G, Zhu L, Liu H, Li W. Zinc-graphite composite coating for anti-fouling application. Mater Lett
       2011;65:3095–7.
[28] He Z, Bao B, Fan J, Wang W, Yu D. Photochromic cotton fabric based on microcapsule technology
       with anti-fouling properties. Colloids Surfaces A Physicochem Eng Asp 2020;594:124661.
[29] Dong W, Li B, Wei J, Tian N, Liang W, Zhang J. Environmentally friendly, durable and transparent
       anti-fouling coatings applicable onto various substrates. J Colloid Interface Sci 2021;591:429–39.
[30] Chatterjee A, Nivas Kumar M, Maity S. Influence of graphene oxide concentration and dipping cycles
       on electrical conductivity of coated cotton textiles. J Text Inst 2017;108. https://doi.org/10.1080/
       00405000.2017.1300209.
[31] De Falco F, Guarino V, Gentile G, Cocca M, Ambrogi V, Ambrosio L, et al. Design of functional textile
       coatings via non-conventional electrofluidodynamic processes. J Colloid Interface Sci
       2019;541:367–75.
[32] Kim BC, Innis PC, Wallace GG, Low CTJ, Walsh FC, Cho WJ, et al. Electrically conductive coatings of
       nickel and polypyrrole/poly (2-methoxyaniline-5-sulfonic acid) on nylon Lycra® textiles. Prog Org
       Coatings 2013;76:1296–301.
[33] Dai M, Zhai Y, Zhang Y. A green approach to preparing hydrophobic, electrically conductive textiles
       based on waterborne Polyurethane for electromagnetic interference shielding with low reflectivity.
       Chem Eng J 2021;421:127749.
[34] Kowalczyk D, Brzezinski S, Kaminska I, Wrobel S, Mizerska U, Fortuniak W, et al. Electrically
       conductive composite textiles modified with graphene using sol-gel method. J Alloys Compd
       2019;784:22–8.
[35] Behl M, Lendlein A. Shape-memory polymers. Mater Today 2007;10:20–8. https://doi.org/10.1016/
       S1369-7021(07)70047-0.
[36] Li G, Uppu N. Shape memory polymer based self-healing syntactic foam: 3-D confined
       thermomechanical characterization. Compos Sci Technol 2010;70:1419–27.
[37] Langer R, Tirrell DA. Designing materials for biology and medicine. Nature 2004;428:487–92.
[38] Koerner H, Price G, Pearce NA, Alexander M, Vaia RA. Remotely actuated polymer nanocomposites –
       stress-recovery of carbon-nanotube-filled thermoplastic elastomers. Nat Mater 2004;3:115–20.
       https://doi.org/10.1038/nmat1059.
[39] Luo X, Mather PT. Shape memory assisted self-healing coating. ACS Macro Lett 2013;2:152–6.
[40] Cho SH, White SR, Braun P V. Self-healing polymer coatings. Adv Mater 2009;21:645–9.
[41] Shuai L, Guo ZH, Zhang P, Wan J, Pu X, Wang ZL. Stretchable, self-healing, conductive hydrogel
       fibers for strain sensing and triboelectric energy-harvesting smart textiles. Nano Energy
       2020;78:105389.
[42] Zahid M, Zych A, Dussoni S, Spallanzani G, Donno R, Maggiali M, et al. Wearable and self-healable
       textile-based strain sensors to monitor human muscular activities. Compos Part B Eng
       2021;220:108969.
[43] Roessler S, Born R, Scharnweber D, Worch H, Sewing A, Dard M. Biomimetic coatings functionalized
       with adhesion peptides for dental implants. J Mater Sci Mater Med 2001;12:871–7.

[44] Liu Y, Wu G, de Groot K. Biomimetic coatings for bone tissue engineering of critical-sized defects. J R Soc Interface 2010;7:S631–47.

[45] Leeuwenburgh S, Layrolle P, Barrere F, De Bruijn J, Schoonman J, Van Blitterswijk CA, et al. Osteoclastic resorption of biomimetic calcium phosphate coatings in vitro. J Biomed Mater Res an Off J Soc Biomater Japanese Soc Biomater Aust Soc Biomater Korean Soc Biomater 2001;56:208–15.

[46] Cheng H-C, Chen C-R, Cheng K-B, Hsu S. Ag/GNS conductive laminated woven fabrics for EMI shielding applications. Mater Manuf Process 2021;36:1693–700.

[47] Singholi AKS, Sharma A. Review of Materials and Processes Used in 4D Printing. Adv. Eng. Mater., Springer; 2021, p. 677–84.

[48] Gnocchi L, Virtuani A, Vallat-Michel A, Fairbrother A, Li H-Y, Ballif C. Measuring and Modelling the Generation of Acetic Acid in Aged Ethylene-Vinyl Acetate-Based Encapsulants Used in Solar Modules. 36th Eur. Photovolt. Sol. Energy Conf. Exhib, 2019, p. 1069–72.

[49] Hirschl C, Biebl–Rydlo M, DeBiasio M, Mühleisen W, Neumaier L, Scherf W, et al. Determining the degree of crosslinking of ethylene vinyl acetate photovoltaic module encapsulants – A comparative study. Sol Energy Mater Sol Cells 2013;116:203–18.

[50] Hadian A, Koch L, Koberg P, Sarraf F, Liersch A, Sebastian T, et al. Material extrusion based additive manufacturing of large zirconia structures using filaments with ethylene vinyl acetate based binder composition. Addit Manuf 2021;47:102227.

[51] Fan B, Kan H, Kan Y, Bai Y, Han G, Bai L, et al. An aqueous polyisocyanate adhesive with excellent bond durability for engineered wood composites enhanced by polyamidoamine-epichlorohydrin co-crosslinking and montmorillonite hybridization. Int J Adhes Adhes 2022;112:103022.

[52] Müller H, Wormald PS. Vinyl Acetate-Ethylene-Copolymer Dispersions and Textile Web Material Treated herewith 2012.

[53] Anjum AS, Sun KC, Son EJ, Yu JH, Park SH, Park MS, et al. Effect of Process Parameters on Hand Values of PU Laminated Waterproof Breathable Textile. Fibers Polym 2021;22:1853–62.

[54] Klein TR, Young MS, Tamboli AC, Warren EL. Lamination of transparent conductive adhesives for tandem solar cell applications. J Phys D Appl Phys 2021;54:184002.

[55] Morimoto M, Ozawa Y, Naka S, Okada H. Additive color mixing of semitransparent laminated tandem type polymer light-emitting diodes. Mol Cryst Liq Cryst 2021:1–7.

[56] Reneker DH, Yarin AL. Electrospinning jets and polymer nanofibers. Polymer (Guildf) 2008;49:2387–425.

[57] Frenot A, Chronakis IS. Polymer nanofibers assembled by electrospinning. Curr Opin Colloid Interface Sci 2003;8:64–75.

[58] Huang Q, Liu L, Wang D, Liu J, Huang Z, Zheng Z. One-step electrospinning of carbon Nano webs on metallic textiles for high-capacitance supercapacitor fabrics. J Mater Chem A 2016;4:6802–8.

[59] Bemska J. Novel decorative materials for textiles. AUTEX J 2010;10:55–7.

[60] Hunter WM. Decorative Laminates. Plastics, Elsevier; 1971, p. 187–216.

[61] Rhee SH, Dumbleton JH. Abrasion resistance of silica-reinforced decorative laminates Part 1 – examination of the standard testing method. Wear 1976;39:83–100.

[62] Plotkin LG. Fire retardant decorative laminates: technology and properties. Int J Polym Mater 1993;20:59–74.

[63] Badila M, Kohlmayr M, Zikulnig-Rusch EM, Dolezel-Horwath E, Kandelbauer A. Improving the cleanability of melamine-formaldehyde-based decorative laminates. J Appl Polym Sci 2014;131.

Jagadeshvaran P. L., Sampath Parasuram, Suryasarathi Bose

5 Polymer-based nanocomposites for smart and functional textiles

Abstract: This chapter aims to provide a comprehensive outlook and understanding of how nanocomposites can be utilized in the emerging domain of functional and smart textiles. The first section explains the correlations between the structure and properties of nanocomposites. The second section deals with the discussions of the different nanoparticles used in polymer nanocomposites. The third section discusses the different forms in which polymer nanocomposites can be integrated into smart textiles. The final section of the chapter primarily focuses on the applications of such functional and smart textiles in different domains like energy harvesting, improving dyeability of textiles, self-cleaning, antibacterial activity, fire protection, gas barrier, and weather resistance applications, emphasizing the structure-property relationship required for a particular domain. The present challenges that hinder the incorporation of nanomaterials in developing novel functional and smart textile and the future trends in this domain have also been examined.

Keywords: Polymer nanocomposites, smart textiles, nanoparticles, nanocomposite coatings, nanocomposite membranes, melt intercalation

5.1 Introduction

Nanotechnology encompasses the principles of designing and manipulating structures that have at least one dimension in the range 1–100 nm. It deals with nanoscale materials and those that contain nanoscale components like nanoparticles, nanowires, nanoplatelets, and nanofibers. As the dimensions are in the nanorange, nanoparticles usually exhibit significantly different physical and chemical properties than their bulk counterparts. Manufacturing of new structures with improved properties, lower processing costs, additional functionalities, micro-manufacturing, and enhanced durability are now easier with the use of nanotechnology. Nanocomposites are promising for a large domain of applications like transportation, electronics, aerospace, packaging, and biomedical applications [1, 2].

Nanocomposites are a class of composite materials where the dispersed phase has at least one component whose dimensions (at least one) are in the nanoscale, which is called as nano-reinforcement. There must be repeat distances at the nanoscale between the different phases of the composite to be termed as a "nanocomposite". Coincidentally, other than synthesized nanocomposites, there are a few that exist in nature as well. Nanocomposites offer a property profile that can be manipulated according to the

https://doi.org/10.1515/9783110759747-005

end-use requirements, making them suitable for sophisticated and specialty applications [3].

Nanocomposites offer several advantages in terms of design and property combination, which is not possible with conventional "micro" and "macro" composites. The properties of materials change drastically when the dimensions of particles approach a critical level – at nanoscale, the interactions at the interface increase significantly, leading to high surface area to volume ratios. This reason explains the significant improvement in properties in nanocomposites from their conventional counterparts.

The textile industry has been facing an increasing demand for textiles with functionalities, targeted to be used in different applications. With the advent of nanotechnology, this demand can be easily fulfilled using nanocomposites or nanofinishing/nanocoating techniques or by replacing conventional fibers of textiles with nanocomposite fibers. From the advantages that nanocomposites offer, it is clear that they can be used as integral components for functional textiles without hampering comfort and aesthetics [4, 5]. Other than functional textiles, there is an emerging class of textiles called smart or intelligent textiles that can sense and react to environmental conditions or external stimuli like heat, touch, pH, and radiation. They can sense the stimuli and react to it accordingly in a pre-determined way. Smart functionalities require complicated structures that are mostly fulfilled by bulky systems based on conventional materials, but it is challenging. With the use of nanocomposites, it is possible to reduce their size and integrate them easily with textiles – attributed to their lightweight and flexible nature. Deploying nanocomposites in smart textiles can help increase the efficiency and functionalities of textiles [6, 7].

This chapter is a brief study about the use of polymer-based nanocomposites in textiles having smart functionalities. The following section explains the correlations between the structure and properties in polymer nanocomposites. The second section discusses the different nanofillers used in polymer nanocomposites. The third section discusses the different forms in which nanocomposites can be used as (or) integrated into functional and smart textiles. The fourth section explains few selected applications of nanocomposites in smart and functional textiles – namely energy harvesting, improving dyeability of textiles, self-cleaning, antibacterial, fire protection, gas barrier, and weather resistance applications. The last section concludes the chapter, highlighting the potential of nanocomposites in smart textiles for the future. Given the innumerable functionalities that nanocomposites have to offer, it is impossible to cover all of them in a single chapter. Hence, here we have attempted to discuss the recent advancements in the domain, highlighting a few applications.

5.2 Structure–property correlations in nanocomposites

Similar to conventional composites, nanocomposites comprise heterogeneous structures – only that they exist in different length scales. Hence, their properties are governed by the properties of the individual components, composition, structure, and interactions that happen at the interface. Nevertheless, they present more complicated structures than conventional composites due to the several factors involved [8].

The characteristics of the interface have a tremendous influence on the performance of composites. The fluctuation in properties is the highest at the interface, compared to the other component phases. With respect to the polymer, properties like chain mobility, extent of curing, and crystallinity are different at the interface than the other parts. To harness the complete potential of a nanocomposite, the interfacial area must be maximized. With the use of nanoparticles, the surface area to volume ratio is largely increased, leading to higher interfacial area at lower concentrations – giving targeted performance. Thus, the role of interfacial area is very crucial in nanocomposites, compared to conventional composites, as nanoparticles offer a large interfacial area compared to conventional ones [9].

The properties of a nanocomposite are dictated by the dispersion state of the nanoparticles in the polymer matrix [10]. For instance, the mechanical and thermal properties of a composite are strongly correlated with the morphologies obtained (Figure 5.1). Dispersion can be visualized as the distance of separation between the nanoparticles in a polymer matrix. Based on such distance, composites can be classified as conventional composites (microcomposites), intercalated nanocomposites and exfoliated nanocomposites. When the polymer is unable to intercalate between the nanoparticles, the composite thus formed contains separate phases of the nanoparticle and the matrix, resulting in a micro composite. When the polymer chains are able to penetrate the nanoparticle layers to yield a well-ordered multilayer morphology with intercalated layers of polymer and nanoparticle, it is called an intercalated nanocomposite. When the nanoparticles are completely and uniformly dispersed in a continuous polymer matrix, an exfoliated nanocomposite is obtained.

The nanoparticles are subjected to different surface modification techniques to prevent agglomeration and enhance affinity to the matrix. In nanoscale, agglomeration turns out to be a bigger problem compared to their micro-counterparts due to the larger surface areas. Thus, prevention of agglomeration is critical in nanocomposites to ensure a uniform composite structure. A few techniques reported for avoiding agglomeration of nanomaterials in polymer nanocomposites include incorporation of surfactants, chemical modification of nanomaterial or polymer, use of silane coupling agents, ultrasonication, high speed mechanical shearing, high-energy ball milling, master-batch mixing, solution blending, and in situ polymerization over the particles [11]. These techniques enhance the dispersion of nanoparticles by producing repulsion

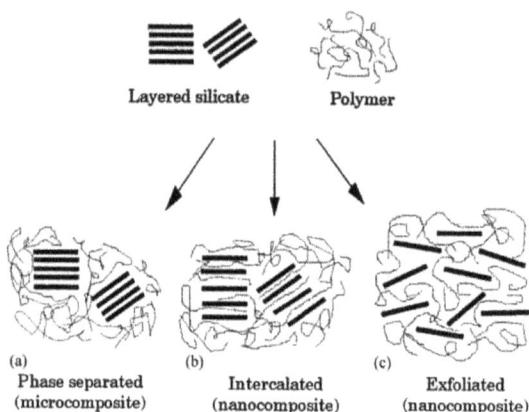

Figure 5.1: Possible dispersion states of nanocomposites using nanoclay as a particle (reproduced from ref. [10] under Creative Commons License).

between particles, thereby promoting interaction with the polymer matrix. Uniform dispersion of particles in the polymer matrix also affects the transparency of the final product, affecting aesthetics [12].

Nevertheless, polymer nanocomposites come with their own set of challenges that need to be addressed. There are difficulties in controlling the elemental composition and stoichiometry of the nanophases as they strongly affect the phase behaviour at the nanoscale. Aggregation- and orientation-related issues are also present in nanocomposites that need attention. Though the interphase plays an important role in determining the properties of the composite, their formation and the related mechanisms are not clearly understood.

5.3 Nanoparticles used in textiles

The advent of nanotechnology has resulted in the development of textiles with a wide range of functionalities. The high surface area to volume ratio in the case of materials at the nanoscale results in textiles with exceptional properties. The properties imparted by nanomaterials are significantly influenced by their shape, form, and size. They are categorized based on the number of dimensions outside the nanoscale (size range between 1 and 100 nm) as shown in Figure 5.2 [13–15].

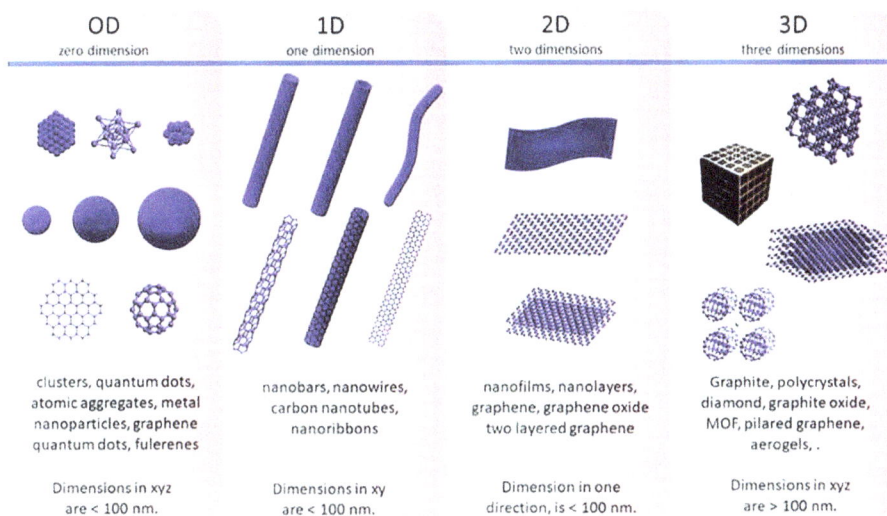

OD zero dimension	1D one dimension	2D two dimensions	3D three dimensions
clusters, quantum dots, atomic aggregates, metal nanoparticles, graphene quantum dots, fulerenes	nanobars, nanowires, carbon nanotubes, nanoribbons	nanofilms, nanolayers, graphene, graphene oxide two layered graphene	Graphite, polycrystals, diamond, graphite oxide, MOF, pilared graphene, aerogels, .
Dimensions in xyz are < 100 nm.	Dimensions in xy are < 100 nm.	Dimension in one direction, is < 100 nm.	Dimensions in xyz are > 100 nm.

Figure 5.2: Classification of nanostructures: zero dimension (0D), one dimension (1D), two dimension (2D), and three dimension (3D) (reproduced with permission from ref. [16] under Creative Commons License).

Different nanomaterials used to develop smart and functional textiles are:

- Carbon-based nanomaterials: Carbon nanotubes (CNTs), carbon nanofibers (CNFs), graphene, graphene oxide (GO), reduced graphene oxide (RGO), and graphite.

  Carbon nanotubes (CNTs) are characterized by their hollow tubular structure, consisting of single or multiple rolled-up graphene sheets. The first case is referred to as single-walled carbon nanotubes (SWCNTs) with a diameter of around 1 nm, and the second is referred to as multi-walled carbon nanotubes (MWCNTs), with diameters reaching 100 nm and lengths in the range of micrometres. CNTs possess exceptional thermal, mechanical, and electrical properties, due to which they are extensively used in textiles to impart functionalities such as energy harvesting, flame retardancy, EMI shielding, and self-cleaning [17–21].

  Graphene is a 2D planar allotrope of carbon that is one atom thick, and the carbon atoms are tightly held in a honeycomb lattice. Graphene has exceptional mechanical properties (Ultimate tensile strength of 130 GPa), electrical conductivity (6000 S/cm), thermal conductivity (5000 W/mK), high specific area (2600 m^2/g), exceptional gas impermeability, and high UV absorption property, making them an ideal choice of nanomaterial for application to textiles to impart additional functionalities. Graphene can be considered the basic building block of other graphitic materials like CNTs. Different forms of graphene, such as Reduced Graphene oxide (RGO), and Graphite, are also widely used in textile applications [22–25].

-   Inorganic nanoparticles
    Nanoclays: Layered silicates or clays are widely used to enhance textiles' flame
    resistance, mechanical properties, gas barrier, dyeability, and corrosion resis-
    tance properties. Phyllosilicates or 2:1 layered silicates such as montmorillonite
    (MMT), Bentonite, and Saponite are the most used nanoclays owing to their high
    aspect ratio and high cation exchange capacity [26–29].

In addition to the nanomaterials briefly discussed here, other nanoparticles such as Metals
(silver (Ag), gold (Au), copper (Cu)), [30, 31] Metal oxide(silicon dioxide (SiO_2), titanium dioxide (TiO_2), zinc oxide (ZnO), cerium oxide (CeO_2), aluminium oxide ($Al2O_3$), [32–34] nano-
hybrid materials (Polyhedral Oligomeric Silsesquioxane (POSS), nanocellulose, h-boron nitride (hBN), [35–37], and molybdenum disulfide (MoS_2) [38, 39] have also been used in textile applications. The effect of these nanomaterials in delivering additional functionalities
to textiles will be discussed in the section "Applications of Nanocomposite-based textiles."

5.4 Synthesis routes to polymer nanocomposites

The preparation of nanocomposites is critical and dictating when it comes to determining the final properties of the nanocomposites, as the dispersion of particles is dependent on the synthesis route adopted. The interfacial interaction between the
particle and the polymer is also controlled by the synthesis. This explains the flexibility in designing composite materials for different applications. It is thus possible to
prepare composites with contrasting properties (given the polymer and particle) just
by changing the synthesis route as they control the dispersion and, consequently, the
interfacial interactions.

5.4.1 Melt intercalation

This is identified as one of the safest (solvent-free) and commercially viable techniques for the preparation of polymer nanocomposites. The particles are dispersed due
to the shearing action of the extruder and the subsequent diffusion in the polymer
matrix (Figure 5.3). As there are no solvents in the process, it is considered to be an
environmentally safe process. However, there are certain drawbacks in this process,
namely, (i) poor dispersion due to the high viscosity of the polymer melts, (ii) possible
thermo-oxidative degradation of the polymer and the additives due to the high processing temperatures involved.

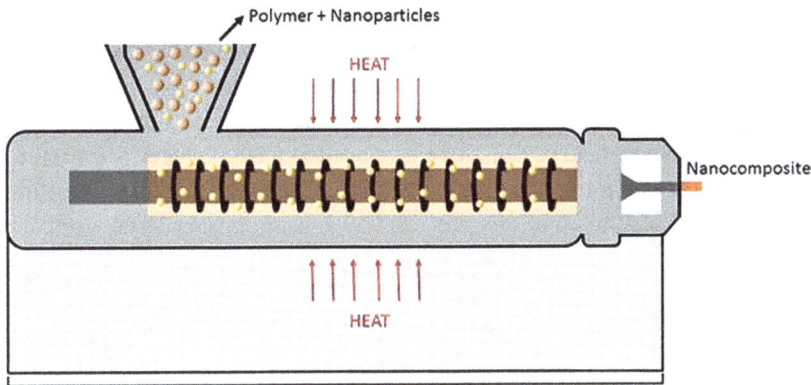

Figure 5.3: Scheme of obtaining polymer nanocomposites by melt intercalation (reproduced with permission from ref. [40] under Creative Commons License).

5.4.2 In situ polymerization

In situ polymerization deploys the particles along with the monomers (starting compounds for polymers) or prepolymers to yield composites. In the course of polymerization, the polymer chains diffuse between the particles, leading to an increase in the interlayer spacing between the particles, causing exfoliation of the particles in the polymer matrix (Figure 5.4). The extent of intercalation or exfoliation happening is dictated by the compatibility between the precursors and the particles. However, the drawbacks of this technique are (i) the use of toxicity solvents, making it environmentally unsafe, (ii) involves a high cost component, making it the last option when no other economically viable techniques are available.

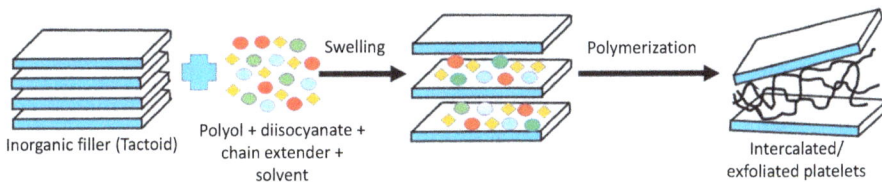

Figure 5.4: Schematic representation of in situ polymerization (Modified according to ref. [27], Copyright 2018 Elsevier).

5.4.3 Solution mixing

In this method, first the polymer pellets are dissolved in a suitable solvent and the particles are subsequently dispersed in the polymer solution by mechanical/shear mixing. However, to obtain a better dispersion, the particles are sonicated in the solvent before mixing in the polymer solution (Figure 5.5). Solution mixing is usually preferred to disperse higher loadings of particles in a better way as it is easy to control the viscosity of the polymer solution. Consequently, exfoliated dispersion of particles is obtained in the polymer matrix. Complete evaporation of the solvent is crucial to obtain optimal properties in the composite. The use of large amounts of solvent impedes the commercial scaling up of this process, which is often considered a setback of this technique.

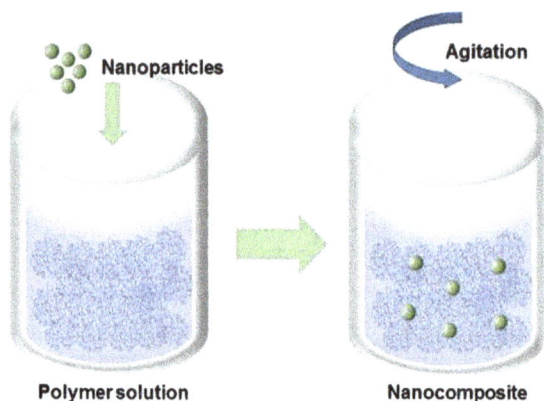

Figure 5.5: Schematic representation of solution mixing
(reproduced with permission from ref. [40] under Creative Commons License).

5.5 Different forms of nanocomposites

Several approaches have been reported in literature to produce smart textiles. A wide spectrum of preparation techniques –from making yarns out of metals or metal alloys to coating metals/polymers on conventional fibers or yarns, and incorporating particles to fibers. From these techniques, it becomes clear that the use of nanocomposites in textiles is not just restricted to coatings. On the contrary, the fibers themselves can get a nanocomposite structure that encompasses smart functionality without the need for a smart coating [4]. Accordingly, nanocomposites for smart textile applications can be in different forms, including laminated nanocomposites, nanocomposite fibers, nanocomposite membranes, and nanocomposite coatings.

5.5.1 Nanocomposite laminates

Layered composites, where one/few of the layers contain nanostructured components, come under this category. A common example of this type is the composite filtration system that usually comprises a nanofibrous web layer, usually produced by electrospinning. Their high porosity ratios and small pore sizes offer advantages in filtration efficiency. Laminated nanocomposites find application in moisture management, especially in sportswear. These fabrics usually consist of two layers. The first layer that is in contact with the body is hydrophobic and the second hydrophilic layer is located outside. Consequently, push-pull effect comes into play where the inner layer pushes moisture to the outside, by capillary activity, causing wicking action [41]. In a similar study, the inner layer was a hydrophobic polydopamine-treated nonwoven layer and the outer layer was a hydrophilic electrospun nanofiber membrane of polyacrylonitrile – silica [42]. Janus fabrics also come under this category as they have superhydrophilic layer on one side and a superhydrophobic layer on the other (Figure 5.6). Other applications of such laminated nanocomposites are now reported in e-textiles. Dry electrodes based on carbon nanotube – thermoplastic polyurethane thin films – prepared by heat lamination are suggested as an alternative to wet electrodes in e-textiles. They employed a "cut" (accomplished by intuitive laser cut) and 'paste' (using heat lamination) method to integrate the CNT film on the fabric [43].

Figure 5.6: SEM micrographs with the corresponding contact angles of the nanofibrous mats (a,c) before and (b, d) after heat treatment (reproduced with permission from ref. [44]. Copyright 2010, American Chemical Society).

5.5.2 Nanocomposite fibers

Another form that is widely reported in literature is fibers, where the nanocomposite itself exists in the form of fibers that is then made into a smart fabric by a suitable process. There are several examples of nanocomposites in their fibrous form reported in literature. In this regard, conducting fibers have gained a lot of attention, especially in e-textiles. A wet spinning technique was used to prepare fibres using polyvinyl alcohol as a matrix filled with silver nanobelts/carbon nanotube hybrid nanoparticles. The prepared fibers show high electrical conductivity, high thermal and chemical stability, and are used in water leakage detection [45]. A core sheath nanocomposite yarn was prepared by row spinning of CNT/cotton roving – prepared by immersing a cotton roving in CNT suspension (Figure 5.7). The prepared yarns were then subjected to electrospinning using a dope solution of polyurethane and thermochromic inks dispersed in DMF. The core sheath yarns showed excellent mechanical properties, electrified heating performance, and thermochromic performance (colour change achieved by change in applied electric current) [46]. An architecture similar to the above was used to prepare a smart textile for biomechanical energy harvesting and personal healthcare monitoring. Polycation-modified carbon dots were incorporated into the PVA matrix and the resulting mixture was subsequently coated on conductive silver-plated nylon yarns. The coated yarns were then encapsulated with a PDMS sheath to give the triboelectric nanogenerators the ability to for efficiently harvest biomechanical energy and demonstrate versatile full range healthcare monitoring [47].

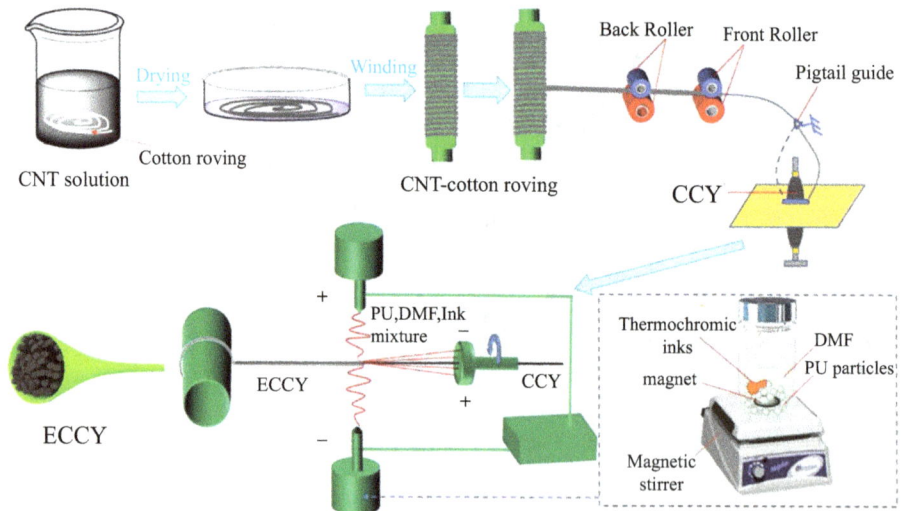

Figure 5.7: Schematic of the fabrication process of the electro-thermochromic yarn (CCY, CNT/cotton composite yarn; ECCY, electrothermochromic composite yarn) (Modified according to ref. [46]. Copyright 2020, Elsevier).

5.5.3 Nanocomposite membranes

Most nanocomposite membranes that are used in smart textiles and several other applications are prepared by electrospinning. They are present as lightweight breathable layers in protective clothing, flexible membranes for high filtration efficiency and low pressure drop, and also in lithium ion batteries [48]. Though electrospinning is the most popular process used to prepare membranes, several other processes like composite spinning, chemical vapour deposition, melt blowing, and template synthesis are also being reported to produce nanofibers. Yet, electrospinning continues to be used due to its versatility and ease of setup. The electrospun mats have low mechanical strength and durability; so they are usually used as one of the layers of a laminated composite. Chalco-Sandoval et al. prepared electrospun webs by encapsulating a phase change material into poly (vinyl alcohol) through emulsion technique for thermoregulation applications. The prepared webs were subsequently subjected to co-axial electrospinning using polycaprolactone to protect the underlying layer from humidity [49]. In a study to develop actuators, nanofibers of EVOH (copolymer of PVA and PE) and cellulose nanocrystals were dispersed and cast onto a PET substrate. This resulted in the build up of hierarchical pores and interlaced nanochannels that improved the sensitivity and the deformation degree of the actuator [50] (Figure 5.8). In another study, carbon-based membranes were prepared by casting a dope solution of cellulose acetate with carbon nanotubes using a film applicator, followed by a non-solvent induced phase separation. The prepared membranes exhibit superior antibacterial activity that can potentially be used for wastewater filtration and biofilm removal

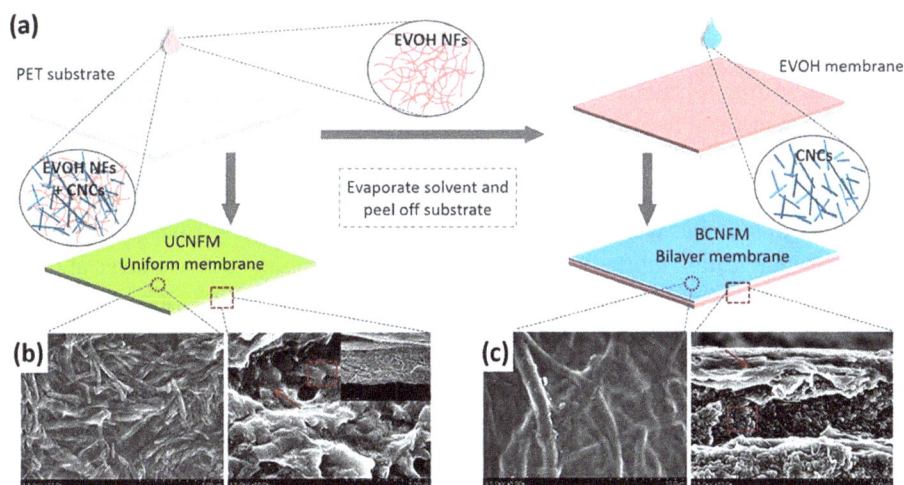

Figure 5.8: (a) Schematic showing the preparation of nanocomposite membranes, and (b and c) their corresponding micrographs (reproduced with permission from ref. [50]. Copyright 2018, American Chemical Society).

[51]. Smart asymmetric Janus membranes were prepared by coating graphene nanoplatelets to one side of a PET fabric (by high temperature and high-pressure method) and phosphoric acid on the other side. The prepared membranes showed high unidirectional water flux (from hydrophobic to hydrophilic) and can also be used in oil-water separation [52].

5.5.4 Nanocomposite coatings

Embedding nanoparticles on textiles by a coating process (without any binder) does not give durable coatings as the particles tend to wear off easily from the textile surface. An effective way to circumvent this issue is by incorporating nanoparticles in a polymer matrix and then coating them onto textiles. This approach enhances the coating quality and durability besides imparting a wide range of properties like UV blocking, fire protection, conductivity, and self-cleaning by an appropriate selection of resin and nanoparticles. The final properties of the nanocomposites are governed by the dispersion of nanoparticles in the polymer system. The nanocomposite approach to coating enhances the binding of particles with the textile, as it is possible to introduce chemical groups having affinity to the textile, besides the already existing functional groups. Consequently, the life of the coating increases in terms of washing cycles and wear and tear. The particles deployed can be subjected to any physical or chemical modification to improve their bonding ability to the substrate. To further enhance the adhesion, a resin having good affinity to the textile can be utilized. The prepared coating formulations can be coated onto fabrics using a variety of processes, depending upon the resin properties and end-use applications [11, 53]. Nanocomposite coatings on textiles mainly contain metal nanoparticles, metal oxides, carbonaceous materials, and phase change materials. In an application of nanocomposite coatings, silver nanowires and fluorosilane were coated onto a cotton fabric using a simple 'dip and dry' process. While the silver nanowires contributed to conductivity and antibacterial activity, fluorosilane was responsible for the superhydrophobic nature of the fabric [54]. In a similar work to develop conductive textiles for electromagnetic shielding application, cotton fabrics were pre-treated with dopamine to give a layer of polydopamine, before subjecting them to carbon nanotube suspension. Subsequently, a fluoropolymer layer was coated to encapsulate the carbon nanotube and give a hydrophobic surface finish. The polydopamine layer assisted the carbon nanotubes to form a conductive network that helped in superior electromagnetic shielding and electrical heating [55] (Figure 5.9). In an attempt to develop sensors, electrophoretic deposition was deployed to deposit polyethyleneimine functionalized carbon nanotubes on the fabric. The resulting thin film of nanocomposite comprises an electrically conductive piezo resistive sensing network that exists as a uniform coating on all the yarns, chemically bonded to the fiber surfaces. Upon integrating this sensing fabric with garments, they exhibit remarkable sensitivity to elbow/knee motion [56].

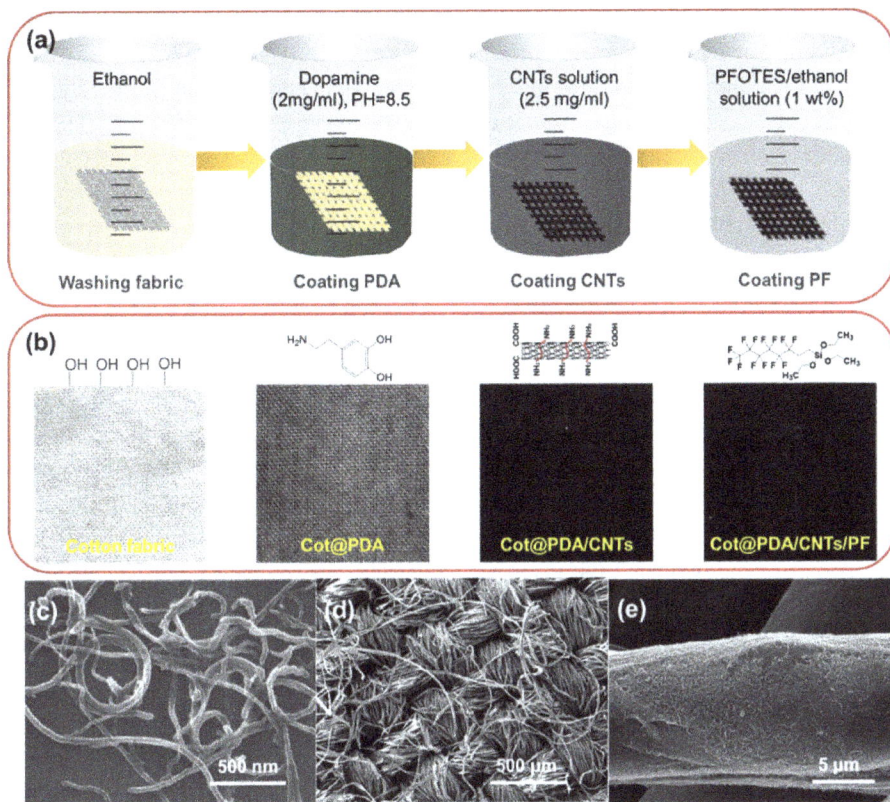

Figure 5.9: (a) Preparation procedures, (b) digital photographs, and SEM micrographs of (c) CNT, and (d and e) the coated fabrics (reproduced with permission from ref. [55]. Copyright 2021, Elsevier).

5.6 Applications of nanocomposite-based textiles

5.6.1 Energy harvesting

Ranging from sensing and actuating to response to an external stimulus, smart textiles have revolutionized the domain of wearable electronics. However, power supply to such intelligent textiles is a significant challenge, given the available materials. A traditional rechargeable battery is used to power the smart textiles, which is heavy and bulky, adding to the product's weight. Further, these batteries are difficult to integrate with the textile. Besides rigidity, limited lifespan and ecological consequences are the other demerits that do not lead to further developments of using batteries in wearable electronics [57]. In the quest for self-powered wearable textiles, researchers discovered that the human body is itself a renewable power source, given its diverse biomechanical movements, body fluids, and environment (Figure 5.10).

Abundant literature available in the energy harvesting domain can be broadly classified as piezoelectric nanogenerators (PENGs), triboelectric nanogenerators (TENGs), pyroelectric generators (PEGs), thermoelectric generators (TEGs), solar cells (SCs), biofuel cells (BFCs), hydro volcanic energy generators (HEGs), and hybrid generators. Despite the availability of different power generators, integrating them into a wearable platform was difficult because of their rigidity and the severe deformation caused by twisting. Thanks to the flexibility and conformability of textiles, different textile-based energy harvesting devices are now available – one of the outstanding accomplishments in personal electronics [58].

Body (biomechanical) movements like walking, breathing, blood flow, and finger motions can release energy that can be harvested. For instance, major body movements occur at the knee, hip, elbow, and shoulders during walking, which are potential energy sources. This biomechanical energy present in different body parts can be harvested using e-textiles, providing a renewable energy source. Piezoelectricity and triboelectricity are the two common principles that help convert biomedical energy from the human body into electricity. Piezoelectricity generates charge by polarization caused by the deformation of crystalline structures. Many PENGs that are made using different piezoelectric materials have been integrated into textiles [59, 60]. However, these PENGs require larger compressive forces to produce considerable electricity, which limits their usage in textiles for energy harvesting. Whereas, TENGs can efficiently harness diverse mechanical energies based on the coupled effect of contact electrification and electrostatic induction. Hence, a very low level of force would suffice for TENGs to extract energy, thereby having an increased efficiency than PENGs [61, 62].

A typical TENG comprises several positive and negative triboelectric materials. Conductive materials like metals, conducting polymers, and carbonaceous particles are used as positive triboelectric electrodes, while, polydimethylsiloxane (PDMS), polytetrafluoroethylene (PTFE), polyvinylidene fluoride (PVDF), silicone rubber and other polymers are used on textiles' surface as negative triboelectric electrode materials [63]. These electrode materials are coated onto a textile that is later integrated into wearable materials like hand gloves to experience mechanical forces, which is converted to energy.

Body heat is another energy source in the human body from which electricity can be constantly generated. Two principles can be used to harness this heat in the form of electricity, namely, thermoelectric effect and pyroelectric effect – respectively corresponding to thermoelectric generators (TEG) and pyroelectric generators (PEG). Textiles that have high electrical and low thermal conductivities can be deployed for this purpose. TEGs use the spatial temperature difference to generate energy, whereas PEGs use the temporal temperature difference [64].

In TEGs, the radiant body heat can be transformed into electricity using thermoelectric materials by using the temperature gradient between the body and the ambient air. The different thermoelectric materials that are most commonly used include

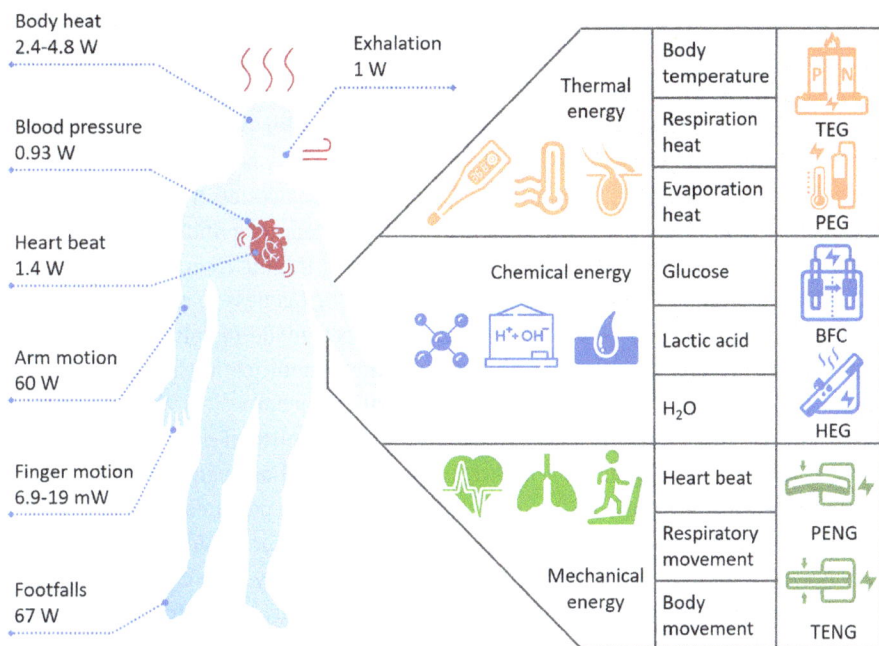

Figure 5.10: Source and distribution of energy in the human body and the respective techniques to harvest them (reproduced from ref. [64] under Creative Commons license).

carbonaceous materials like MWCNT, graphene; PEDOT: PSS, copper and silver (manifesting in different forms), and tellurides. PEGs utilize the changes in the spontaneous polarization due to temperature changes to produce electricity [65]. First-order and second-order pyroelectric materials are deployed for this purpose. First-order pyroelectric effect is usually seen in ferroelectric materials like lead zirconium titanate and barium titanate where charges are generated in the absence of strain. Second-order pyroelectric effect is seen in ZnO, CdS, and other wurtzite materials with piezoelectric effect, where thermal expansion of the material (strain) produces charge [66].

Biochemical energy present in bodily fluids like blood, perspiration, saliva, and tears, is also a source for on-body power generation. Two technologies have been reported to harness chemical energy, namely, biofuel cells (BFC), and hydro voltaic effect generator (HEG). The former technology uses body fluids like glucose and lactate as energy sources, while the latter utilizes the interaction between nanomaterials and water molecules to convert electrical energy through evaporation energy and humidity changes. BFC uses energy changes in redox reactions to generate electricity. They are classified into microbial fuel cells and enzymatic fuel cells, named after the kind of catalyst used [67]. HEG is an emerging concept based on the hydro voltaic effect – conversion of water energy to electricity by the interaction between water molecules and nanoparticles [68]. Readers may be interested in reading Chapter 15 for getting more information about energy harvesting textiles.

5.6.2 Polymer nanocomposite fibers with improved dyeability

Present day textile and apparel industry use several synthetic fibers, namely, polyester or poly (ethylene terephthalate) (PET); polypropylene (PP); polyamide (nylon 6 and nylon 6,6); and thermoplastic polyurethane (TPU). They are either used neat or blended with cellulosic fibers like cotton or viscose, as dictated by the end-use requirements. These fibers are dyed to enhance their appearance and aesthetic appeal in clothes and garments. However, most of the neat synthetic fibers are difficult to dye. Crystallinity in PP, hydrophobicity in PET, and poor fastness to colour in TPU are some of the main reasons. Several attempts have been made to solve this problem, which can be broadly classified as physical/chemical modification of the fiber and changes in the dyeing process. With the advent of nanocomposites, the problem of dyeing synthetic fibers can be easily resolved. The mechanism of dyeing and dyeability of a fiber/filament depends on both its physical as well as chemical structure. In addition to this, incorporation of various nanomaterials in a polymer also has a significant effect on altering the dyeability property of synthetic fibers [69].

Polymer nanocomposites fibers show enhanced dyeability than conventional synthetic fibers. The major reasons for such enhanced dyeability are as follows: (i) abundance of chemical groups in the nanoparticles that can be attached with the dyes, (ii) reduction of crystallinity or crystal size due to particle incorporation, which may enhance the dye intake, and (iii) Tg of nanocomposite fibers are lesser than their neat counterparts, causing easier movement of polymer chains and, consequently, giving a better dye diffusion [70].

There are several reports on nanoclay and polyoligomeric silsequioxane (POSS) being used to enhance the dyeability of textile fibers. Other particles like silica, silver, TiO_2, and ZnO nanoparticles are also reported in few articles covering the same application.

5.6.3 Nanostructured coating for making superhydrophobic textiles

Hydrophobic and liquid-repellent textiles have attracted considerable attention in the last decade to deploy them in applications like self-cleaning, anti-icing, anti-smudge, and oil-water separation. The different applications mentioned above are all based on the principle of superhydrophobicity. Superhydrophobic (SH) surfaces have a water contact angle above 150° and a sliding angle of less than 10°. The lotus-leaf effect has inspired and has been the principal motivation for developing SH surfaces. The architectures of most coatings reported in literature tend to mimic the lotus-leaf structure – predominantly used in self-cleaning applications. The hierarchical micro- and nano-scales of roughness and the low surface energy are usually responsible for the SH properties. The (nano-scale) hydrophobic crystals of wax (low surface energy) embedded in the (microscale) bumps of lotus-leaf are responsible for its SH surface [71]. Dirt and loosely-bound soil attached to such surfaces can be easily removed with a rolling water drop, yielding self-cleaning effects [72].

Several techniques have been reported to engineer SH surfaces over different substrates. Typical ways to produce SH surfaces are: (i) create patterns on a hydrophilic surface, subsequently coat them with a hydrophobic material, or (ii) pattern an inherently hydrophobic material (as shown in Figure 5.11). Further, low surface energy materials like fluoropolymers and siloxanes are often used while fabricating SH surfaces. Nevertheless, low surface energy is not a necessary condition for superhydrophobicity, as inherently SH surfaces have reasonably low energy chemistry on their surface [72–74].

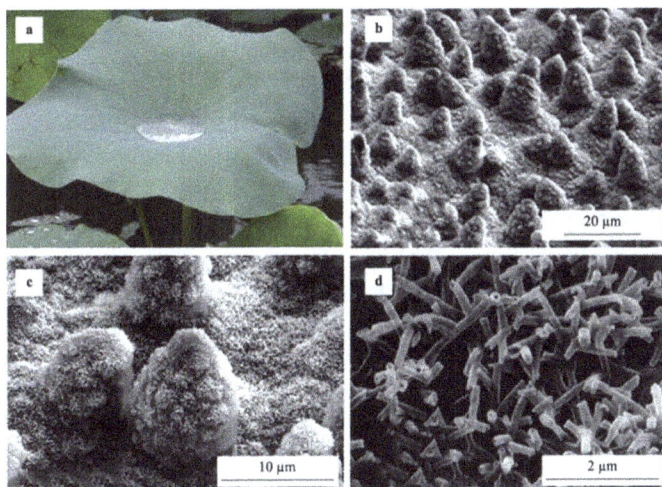

Figure 5.11: (a) Lotus leaves showing superhydrophobic behaviour, (b) the related microstructures as observed by SEM, (c) protrusions and (d) wax tubules coated on their surface (reproduced with permission from ref. [75], under Creative Commons License).

Any solid material, ranging from hard metals and metal oxides to soft and flexible polymers and fiber-based materials like fabrics and paper, can be chosen as a substrate to produce an SH surface. Surface roughening techniques like plasma, laser, and chemical etching yield SH properties by making the surface rough. Photolithography is also a commonly used technique to create SH surfaces by creating a well-defined array of silicon micropillars. Other commonly used techniques include layer-by-layer self-assembly (L-b-L), electrodeposition, electrospinning, CVD, solution immersion, spray coating, and sol–gel method. While techniques like CVD, plasma-etching, electrospinning, and electrodeposition are expensive, others are relatively inexpensive [76, 77].

Of late, nanomaterials have been deployed in different forms to produce SH surfaces due to their unique properties and various processing techniques at disposition.

It is possible to create surface roughness and reduce the surface energy with nanoparticles using appropriate chemistry and preparation methods. The different nanoparticles used to create SH surfaces can be classified into inorganic, organic, and hybrid, based on their nature.

Inorganic materials used here are mainly based on silica, carbon (CNT, CNF, graphene, and fullerene), metals, and metal oxides. Silica, being a dielectric material besides its significant bactericidal properties, is widely used to create SH surfaces,. By an appropriate selection of precursors and processes, it is possible to prepare silica particles in different morphologies that can create surface roughness [78–82]. Carbon-based particles have been used in SH surfaces due to their ability to create rough surfaces and also due to their inherent hydrophobic nature. Further, most carbon-based systems are multifunctional, mainly attributed to their superior mechanical and electrical properties [83–87]. Metals are deployed for this application mostly by electrochemical reduction processes like electroless deposition. Still, the number of instances of such coatings is very minimal as it is hard to use metal as a coating [88–90]. Many metal oxides like TiO_2, ZnO, Al_2O_3, and Fe_3O_4 have been reported to be used in SH coating formulations [91–95].

Several organic materials also confer SH properties to surfaces. The primary aim of adding organic components is to lower the surface energy so as to yield a hydrophobic surface. The organic components mainly used are fluoro/fluorinated polymers [96], resins like PDMS, PU, PP [97], PS [98], chitosan [99], and copolymers of the same to enable enhanced processability. Fluorinated polymers are the widely preferred choice for engineering SH surfaces due to their inherent ability to lower surface energy and cause high C–F bond energy [100], but their usage is limited due to the regulations on fluoropolymers and the use of hazardous organic solvents during their preparation [101]. Several other aliphatic compounds have also been used for the same, but they have a very poor durability to washing/laundering cycles [102]. PDMS and PU have emerged as promising alternatives to fluoropolymers as they can be prepared by several techniques with green solvents [103, 104]. Chitosan, being a natural polymer, has also attracted considerable attention in this area with an added advantage being its antibacterial properties [99].

Hybrid systems – a combination of both organic and inorganic systems – mentioned above are widely being used in the recent days to harness the advantages of both. Two approaches have been adopted in literature to prepare hybrid SH coatings: (i) prepare nanocomposites by dispersing the inorganic nanoparticles into an organic matrix [105–107], or (ii) provide an organic coating over the substrate that contains inorganic coating [108–111].

SH coatings are deployed in several applications, namely, self-cleaning, anti-icing, drag reduction, anti-reflection, and corrosion prevention. Though there are several reports of materials that use SH coatings demonstrating their properties, very little information is available about their durability and long-term properties. It is envisaged that research in this domain will move towards preparation of coatings adopting facile and robust strategies, as there are a lot of materials available in literature.

5.6.4 Antibacterial activity

Textiles have a large surface area, which can serve as a substrate to several processes. Textiles that can hold moisture for a long time, primarily based on cellulose, like cotton, linen, rayon, and viscose, are susceptible to attack by microorganisms like bacteria, fungi, and viruses during their usage. The effect of such organisms on textiles is detrimental – they reduce the mechanical strength, cause staining, lead to unpleasant odours, and leave a possibility for infection. Besides, several applications (primarily healthcare) demand that the fabric be microbe-free and do not facilitate microorganisms' growth.

Antibacterial action is achieved by two strategies: biocidal, which involves killing bacteria, and biostatic, which involves preventing bacterial growth. Most antibacterial agents reported in literature are biocidal –quaternary ammonium compounds, triclosan, chitosan, plant-based compounds, metal & metal oxide nanoparticles, and N-halamines – each having their mechanism of killing. In contrast, the biostatic method involves anti-biofouling treatment of the surface to prevent bacterial attachment – mainly obtained by superhydrophobic coatings. We see that most of the antibacterial finishes reported in literature, of late, encompass both the strategies to synergistically enhance antibacterial activity, besides giving the finish a multifunctional utility.

Among the different biocidal compounds mentioned above, two classes of compounds that are widely reported are metal (silver, gold, copper, cobalt, nickel, zinc, molybdenum, zirconium, etc.) [112] and metal oxides (Ag_2O, CaO, MgO, ZnO, NiO, CoO, CuO, Cu_2O, TiO_2, SiO_2, and Fe_xO_y) [113], especially in their nano-forms, to harness the advantages of nanoparticles, namely, high surface to volume ratio and being effective at lower concentrations – characteristics that do not modify the intrinsic properties of textiles.

Shaheen et al. [114] biosynthesized cupric oxide nanoparticles using the active enzymes/proteins secreted by fungi to identify the safe dosage of particles on cotton fabrics. The use of proteins/enzymes plays an essential role in controlling the size and size distribution of the particles. Using cytotoxicity assessments, they estimated that the safe dosage of CuO for the treatment of cotton fabrics is 100 µg/ml. This concentration was sufficient to show antibacterial activity towards *B. subtilis* and *P. aeruginosa*. (~90% reduction in bacterial growth).

Fu et al. [115] developed a self-cleaning antibacterial nanocomposite coating based on quaternary ammonium salt (QAS), functionalized by fluorinated copolymer and poly (urea-formaldehyde) nanoparticles. The antibacterial action comes from the biocidal activity of QAS through a contact killing mechanism that includes electrostatic and lipophilic interactions with the cell wall of microorganisms. This work relates to the need for water-repellent surfaces in antibacterial applications – demonstrating their use by not allowing the bacteria to grow on surfaces with hierarchical scales of roughness. Though the work does not include any application on fabrics, we see the coating as a potential one for fabrics, given its multifunctional properties.

Hong et al. [116] used polydopamine-assisted deposition to coat silver and copper nanoparticles over polyester fabric, followed by a thiol treatment, to reduce the surface energy. They have used two strategies to harness biocidal and biostatic action for antibacterial activity. The nanoparticles exhibited biocidal action by disrupting the bacterial cell wall with the help of reactive oxygen species (ROS). At the same time, the superhydrophobic surface created as a result of the particles and the thiol coating provided the biostatic action that prevented the attachment of bacteria on the surface. Besides antibacterial and superhydrophobicity, the fabric was also conductive, thanks to the metal particles. In another work by Guo et al. [117], tannic acid was used instead of dopamine to deposit the silver particles, followed by a PDMS (polydimethylsiloxane) coating, for water-repellent properties. The inhibition zone method was used to evaluate the antibacterial property.

Figure 5.12 shows the different mechanisms responsible for antibacterial activity in metal/metal oxide nanoparticle-based systems. The particles are usually dispersed in a suitable resin at specified concentrations and coated on the fabric using a particular process – mostly exhaust or pad-dry-cure method for particle-dispersed systems. Nanoparticles are the most preferred choice for antibacterial finishes because of their different advantages – silver-based systems being the most preferred. Having said this, we notice that the present literature with nanoparticles is moving towards the direction of multifunctional materials. Most of the finishes that reported showing antibacterial activity are superhydrophobic; hence the multifunctional aspect. This way, both the biocidal and biostatic approaches to antibacterial action complement each other. Since most superhydrophobic coatings are self-cleaning, they help prevent biofilm formation due to the bacteria on the surface, besides enhancing the lifetime of the antibacterial finish that contains the biocidal component [118].

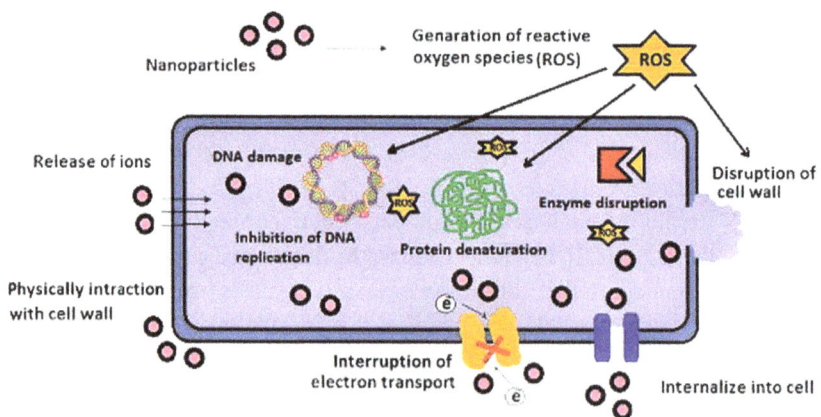

Figure 5.12: Various mechaniplaysms of antibacterial activity exhibited by metal/metal oxide nanoparticles (reprinted with permission from Ref. [118], Copyright 2014 Elsevier).

5.6.5 Fire protection

Textiles are primarily made of organic polymers that are flammable in nature. Given that textiles play a noteworthy role in the daily life of people, their flammability poses a risk of accidents, besides limiting their service temperature. This calls for incorporating flame retardants that prevent or delay ignition, thereby reducing the flame-spreading rate in the textile [119].

Most flame-retardant textiles work by the formation of a char layer on the top surface – characterized by low thermal conductivity that prevents the supply of oxygen and heat. Carbon dioxide and water vapour produced as by-products of combustion that are non-flammable help in diluting the content of flammable gases in the vicinity and reduce heat absorption. These non-flammable gases work as a condensed phase and a gaseous phase, simultaneously, to reduce flames [120–122]. A schematic representing the described mechanism is shown in Figure 5.13.

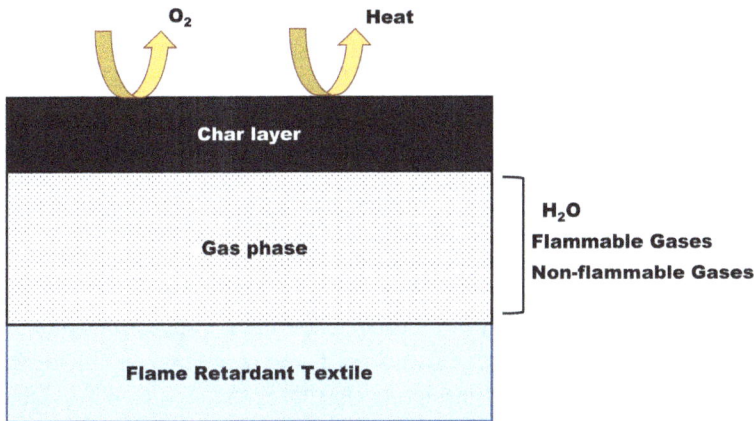

Figure 5.13: Schematic showing mechanism of fire protection in flame-retardant fabrics.

The main test methods used to characterize flammability are limiting oxygen index (LOI) measurements, UL-94 test method, cone calorimetry, and thermogravimetric analysis. For textiles, flammability is also governed by other factors like ignitability, flame-spread rate, and heat release. Hence, for a textile to qualify as flame retardant with an acceptable level of safety, it needs to be characterized by more than one method, depending on the application [119, 123, 124].

The conventional flame-retardant materials that are used in textile finishes are as follows:
(i) Halogen-based compounds (bromine and chlorine-based)
(ii) Nitrogen-based compounds (melamine)

(iii) Phosphorous-based compounds (red phosphorus, ammonium polyphosphate, etc.)
(iv) Inorganic compounds (antimony trioxide, aluminium trihydroxide, etc.)
(v) Boron compounds (borax, boric acid, zinc borate, etc.)

With the advent of nanotechnology, several nanoparticles dispersed in polymers have been used as finishes for textiles. The nanoparticles mainly used for flame-retardant applications include nanoclay, carbon nanotubes, layered double hydroxides, and polyoligomeric silsesquioxanes (POSS), silica, and metal-based nanoparticles.

Nanoclays are layered mineral silicates of several classes primarily used to enhance the barrier properties, mechanical, and thermal properties. The properties of clay-based nanocomposites are usually dictated by their state of dispersion – usually tailored using organic modifiers to give intercalated or exfoliated structures [125]. Exfoliated structures are mostly preferred for flame-retardant applications as they have an extended ignition time, compared to intercalated structures due to their lower decomposition temperatures. The different methodologies adopted with nanoclay to impart flame retardant properties are [126–128]: (i) modification of clay with thermally stable surfactants, based on aromatic compounds that decompose at much greater temperatures than typical alkyl ammonium surfactants; (ii) incorporation of other additives such as POSS that act as an additional protective layer against fire and filling apertures in the silicate-rich barrier; (iii) coupling improvement of silicate layers in char to obtain a continuous, stable, and integrated char by the addition of a slight amount of inorganic additives such as low melting glass and zinc borate; and (vi) promotion of MMT layers migration to the burning surface.

Carbon nanotubes (CNT) are tubular structures composed of a hexagonal network of carbon atoms. They have been used in a broad spectrum of applications due to their remarkable properties. The main mechanism of flame retardancy in polymeric systems is the formation of a thermally insulating char layer that acts as a heat barrier, which re-emits the radiation back to the gas phase, delaying degradation. They also increase the thermal conductivity of the polymeric system, besides acting as a radical scavenger [129–131]. The CNT networks that are formed above a certain concentration (>0.5 wt%) act as a heat-conducting network, dissipating the heat from the sample and delaying the time (and temperature) required to reach ignition point. Further, since thermal degradation of polymers commences with chain scission and radical generation, the radical scavenging property of CNT could delay the thermal degradation and enhance the thermal stability [132, 133]. The degree of improvement in flame retardancy is strongly dependent on the concentration and dispersion of CNT, as these two parameters strongly affect their network formation. In comparison to other carbonaceous structures, multi-layer graphene and reduced graphene oxide performed better than CNT with respect to flame retardant properties. This is attributed to their micron size and better exfoliation, compared to tubular CNT [134–136].

Layered double hydroxides (LDH) are an extensive class of anionic clays with general the formulation, $[M^{2+}_{1-x}M^{3+}_{x}(OH)_2]^{x+}A_x - zH_2O$, where M^{2+} and M^{3+} represent divalent and trivalent metallic cations and A stands for the interlayer anion [137]. They are known as brucite-like compounds. Since the range of metals and anions that can be used in the LDH structure is quite extensive, the LDH properties can be tailored. Two reasons that are attributed to the enhanced flame retardant properties are: (i) the high activation energy of LDH compared to polymers, and (ii) the thermal insulation and the barrier effect caused by LDH [138]. Similar to clay, LDH also require an organic modification to be dispersed well in a polymer matrix [139]. LDH, supplemented with transition metal cations, tend to show synergistic improvements in properties, when used along with intumescent flame retardants [140].

POSS is an inorganic-organic hybrid compound that comprises a typical silicon-oxygen cage structure surrounded by organic groups (R) with a generic formula $(RSiO_{1.5})_n$. POSS can be functionalized by altering the organic groups on the surface [141]. The flame retarding mechanism of POSS is described by the degradation and migration of POSS to the surface during combustion to form a ceramic silica layer that offers protection to the underlying polymer [142]. The presence of metal atoms in the POSS structure aids flame retardation by delaying the polymer degradation [141]. The stage in which POSS is incorporated and the kind of treatment given to POSS also affects the flame retarding behaviour. The nature of the chemical groups attached to POSS further dictates the effect on the final properties, including fire resistance [143].

Silica nanoparticles have been used extensively in flame-retardant applications with an added advantage of hydrophobic surface properties. Several studies have been reported in literature, demonstrating the flame retardant properties of silica, with special reference to textiles [144–146]. Their use in this application is attributed to their high heat resistance, high insulation effect, and mass transport barrier of the silica nanoparticles embedded in the coating. Fumed silica was also reported to be used in this application. However, deterioration in the flame-retardant properties was observed at higher loadings of fumed silica, as the formation of char layers was restricted, consequently reducing their swelling [147, 148].

Metal-based nanoparticles have been used on textiles to improve their flame-retardant properties. Several mechanisms have been hypothesized to explain their action. Metal-based nanoparticles can act as a barrier to heat and mass transfer, alter the degradation pathway of the polymer, restrict the mobility of the polymer chains, and absorb active species such as free radicals. Furthermore, they may increase the heat transfer inside the material, slowing down the migration of bubbles and reducing the heat release and the concentration of the local oxygen due to the oxidation-reduction mechanisms of the oxides [149, 150]. Furthermore, with the aim of enhancing the flame retardant attributes of the nanocomposites and textiles, most researchers have supplemented the nanoparticles with phosphorus compounds [151–153].

It is therefore clear that the use of nanoparticles enhances the flame-retardant properties of the textile fabric. The primary mechanisms of action include the formation of a barrier layer that promotes char formation and free radical trapping. However, the extent of enhancement of properties depends on the type of particle used, its morphology, compatibility with the polymer, dispersion in the matrix, and migration speed to the surface. The present-day research follows a holistic approach to design flame retardant textiles with improved laundering ability, physical properties, and low cost. Further, care is taken to incorporate 'green materials' that do not harm the environment and user health, as most of the flame retardants are toxic compounds. Considerable amount of work is going on in the use of biomacromolecules like whey proteins, caesins, hydrophobins and deoxyribonucleic acids (DNA) [154], extracts from plants [155], and green compounds [156] as flame-retardant finishes for textiles.

5.6.6 Gas barrier applications

Textile fabrics with good gas barrier performance are widely used in inflatable structures such as flexible food packaging, aerostats, stratospheric airships, inflatable waterborne vessels, aircraft evacuation slides, life rafts, and recreational structures [157, 158]. Textile fabric is the preferred choice of material for these inflatable structures because of their advantageous properties such as lightweight, high specific strength, flexibility, and ease of assembly using multiple smaller joints of fabrics [159, 160].

These fabrics generally comprise a strength or load-bearing layer made of textile fabric and a polymeric coating or film lamination component to impart the gas retention capability. The choice of the strength layer can vary from commercial polyester and nylon to high-performance fabric such as Zylon, Vectran, Kevlar, Spectra, Dyneema, and M5 depending on the application requirements of the inflatable structure [161, 162]. The coating or film lamination is generally made of polymers, such as ethylene vinyl alcohol copolymer (EVOH), polyester (BoPET, Mylar), polyvinyl chloride (PVC) and polyvinylidene Chloride (PVDC, Saran), which inherently has good gas barrier properties. In addition to good barrier properties, the polymers used in specific inflatable structures such as aerostats and airships need additional functionalities such as good flex-fatigue, low-temperature flexibility, adhesion and heat sealability, which are met by materials such as Polyurethanes (PU) [163, 164].

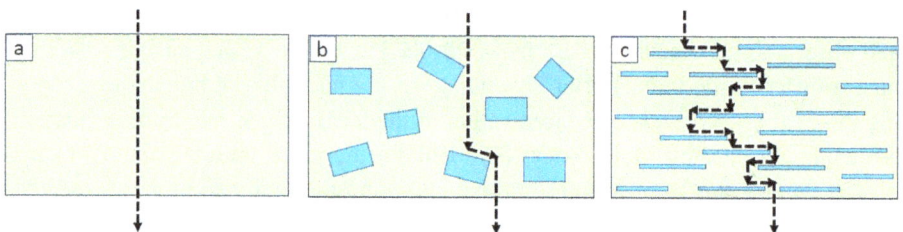

Figure 5.14: Diffusion of gas molecules through (a) polymer, (b) polymer with microfillers, and (c) polymers with nanofillers (reproduced with permission from ref. [163], Copyright 2018, Elsevier).

The gas retention of polymeric coatings and films significantly improves with the inclusion of impermeable 2D platelet nanoparticles such as layered silicates (clay) [26–29], graphene, graphene oxide (GO) [22–25], h-boron nitride (hBN) [35–37], and molybdenum disulfide (MoS_2) [38, 39]. These 2D platelets work as physical barriers, increasing the tortuous path (distance traversed by gas molecules) and thereby decreasing the gas permeability. The very high aspect ratio of these platelets enables them to improve the gas barrier properties at low filler loading rates. In addition to the 2D platelets mentioned above, other nanoparticles such as (silicon dioxide (SiO_2) nanoparticles, titanium dioxide (TiO_2) nanoparticles, aluminium oxide ($Al2O_3$), carbon nanotube (CNT), nano-hybrid materials (polyhedral oligomeric silsesquioxane (POSS) and nanocellulose) have also shown improvement in gas retention properties [163] (Figure 5.14).

Upasana et al. [26] explored the use of an industrially viable technique, ball milling, to exfoliate organo-montmorillonite (OMt) nanoclay effectively. Ball milling of OMt nanoclay was carried out at different combinations of rpm and time. Thermoplastic Polyurethane (TPU)-OMt nanoclay nanocomposites coatings were obtained by continuous solution casting using knife over roller method on PET fabric. A 60% reduction in helium gas permeability was noticed for nanocomposite coating with a 400 rpm/2 h combination, indicating good exfoliation of the stacked layers of OMt nanoclay using the ball milling technique. Bapan et al. [27] developed thermoplastic polyurethane/organo-modified montmorillonite clay nanocomposite films for aerostatic application. They compared two different melt mixing routes (direct mixing versus master batch mixing), and observed that the master batch route involving sonication and high rate mechanical stirring, followed by high-speed shearing in the extruder, was more effective in obtaining an intercalated/partially exfoliated morphology, resulting in a 39% reduction in helium gas permeability for TPU/clay nanocomposite with 3 wt% loading of organoclay (Cloisite 30B).

Hyunwoo Kim et al. [165] studied the influence of two derivatives of graphene oxide – chemically modified (isocyanate-treated GO) and thermally exfoliated graphene oxide (TRGO) – on the nitrogen gas permeability of TPU nanocomposites (Figure 5.15). Three methods of processing, in situ polymerization, melt mixing, and solution mixing, were compared here. About 3 wt% of TRGO reduced nitrogen gas permeability by 90% and also improved the mechanical and electrical properties. Fourier transform infrared (FTIR), wide-angle X-ray diffraction (WAXD), and transmission electron microscopy (TEM) results showed that the solution mixing technique was more efficient in distributing exfoliated TRGO, compared to melt mixing. In situ polymerization fared poorly compared to solution mixing, as the functionalized GO-hindered hydrogen bonding among urethane groups in the TPU matrix.

In addition to nanoparticles like graphene, GO, and nanoclay, recently, hexagonal boron nitride (hBN), an atomically thin 2D graphene analogue, has drawn considerable interest for its ability to improve the gas barrier properties of polymers [35–37]. Muhammad Azeem et al. [37] used liquid phase exfoliation to obtain size-controlled hBN

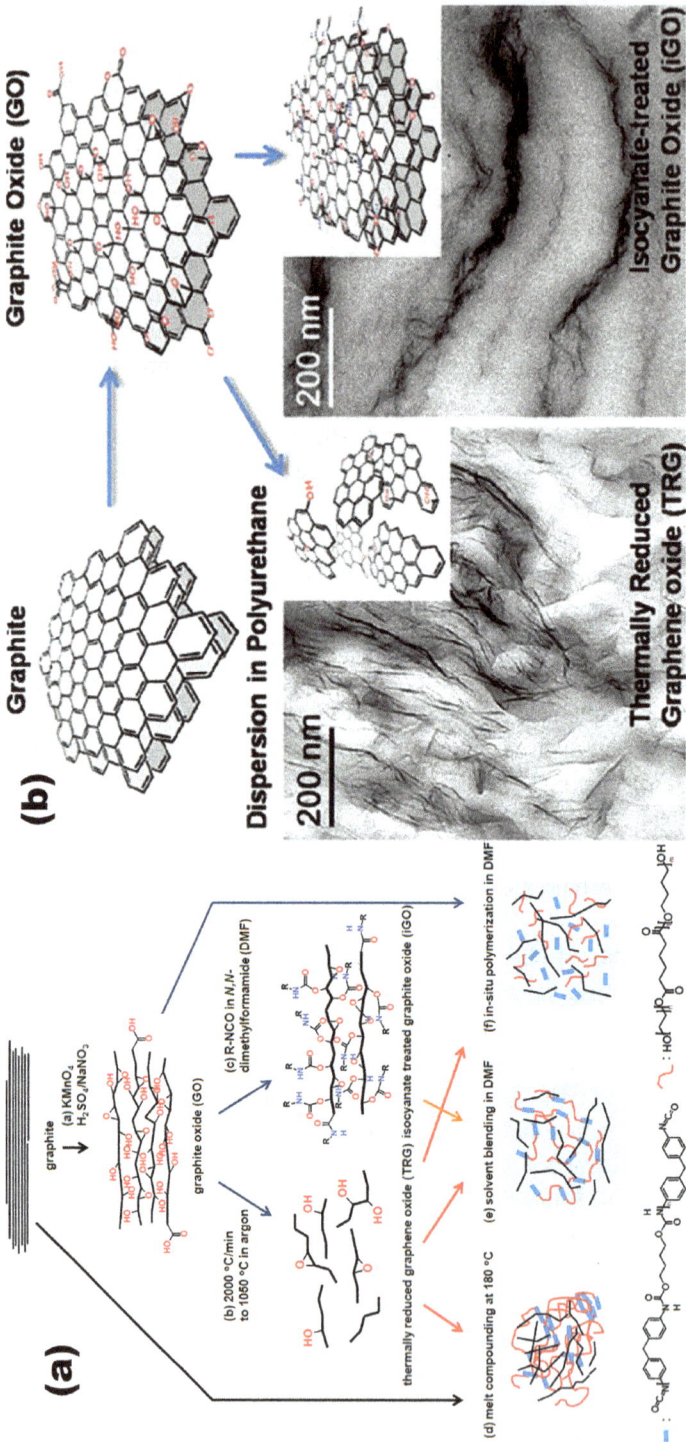

Figure 5.15: (a) Illustration of TPU nanocomposites with two derivatives of graphene; chemically modified (isocyanate-treated GO) and thermally exfoliated graphene oxide (TRGO) with different processing routes and (b) TEM images of TPU/TRG and TPU/iGO nanocomposites [165] (reproduced with permission from ref. [165], Copyright 2010 American Chemical Society).

nanosheets, and synthesized TPU/hBN nanocomposites by solution mixing. They noticed a significant reduction in CO_2 gas permeability at low filler loading of 0.011 vol% hBN and 0.054 vol% hBN, showing 55% and 82%, respectively.

It is evident that the use of nanoparticles enhances the gas retention properties of polymeric coatings and films in gas barrier textiles. These nanoparticles work as physical barriers, increasing the tortuous path (distance traversed by gas molecules). The level of reduction in gas permeability in polymer nanocomposite depends on many factors such as the morphology of nanoparticles, functional compatibility between particles and the polymer matrix, and the dispersion and orientation of nanoparticles. The processing technique considerably impacts the interaction between the nanoparticles and the matrix, and thus the final overall gas permeability.

5.6.7 Weather resistant applications

Use of polymers in specialty applications requires them to be able to perform in a harsh environment containing exposure to ultraviolet (UV) radiation, ozone, oxygen, and water/moisture. On the other hand, such conditions would lead to an accelerated degradation of the polymer, forming radical species by the cleavage of unsaturated bonds. This can be prevented by chemically modifying the polymer or incorporation of organic additives like UV blockers, antioxidants, and light stabilizers. However, the effectiveness of such additives decreases with time, and they affect other properties of the polymers like opacity and toxicity. With the advent of nanotechnology and polymer nanocomposites, there are nano-sized UV shielding particles that can be used to overcome the above problems. Polymer nanocomposite-coated/laminated textiles have huge potential in improving their weather resistance property, which is important for many outdoor applications like aerostat/airship hull, paragliding, and marine applications.

A most commonly used UV stabilizer in polymeric systems is rutile grade TiO_2 – dispersed in the polymer in micro- or nanosizes. Despite being compatible with most polymer matrices, a reduction in weather resistance was observed due to the presence of anatase phase in TiO_2 that causes photocatalytic degradation. However, there are several contradicting opinions from the scientific community regarding the potential of TiO_2 in weather resistance, as its photocatalytic activity might cause degradation of organic polymers to radicals. Zinc oxide is another material that has been used in weather resistance applications due to its excellent UV absorption property. Both zinc oxide and TiO_2 were considered ideal choices for weather resistance application because of their ability to absorb UV and sunlight; further, their effectiveness is dependent on their sizes [166].

Ceria and Zirconia nanoparticles are relatively new entrants in weather resistance applications. The coexistence of Ce^{3+} and Ce^{4+} in CeO_2 not only scavenges the radicals formed upon UV irradiation, but also absorbs the incident UV light. On the contrary, the extra charge in zirconia is trapped in the vacancy site rather than reducing the nearby Zr ions, generating defect states near the centre of the electronic band gap. Zirconia nanoparticles have excellent UV absorption properties and can act as antioxidant that may be useful in enhancing the weather resistance properties of polymers. Graphene has also been used in this application for its excellent UV absorbing ability [22, 167]

5.7 Current challenges and future outlook

Developments in technology have given rise to increased requirements that calls for novel materials that have enhanced properties compared to conventional materials. In this regard, nanocomposites present a tailorable property profile that can outperform monolithic materials as well as conventional composites. From the above discussions, it is becoming clear that the domain of smart textiles is slowly moving to multifunctional textiles to increase the utility of the fabric. Nanocomposites offer advanced multifunctions, without interfering the comfort and aesthetics of textiles. As a part of smart textiles, nanocomposites are used in a variety of applications, of which a few include self-cleaning, fire protection, energy harvesting, and several other niche applications. Nevertheless, the existing solutions for smart textiles still need some more understanding and research towards the durability aspect (like washing stability and abrasion resistance). Recently, many nano-enabled innovative textile products have been commercialized to compete with existing commercial products. However, the high cost and sophistication involved with the production of nanomaterials and the environmental risks involved are a few factors that restrict their widespread usage, despite their advantages. Yet, the rising demand for smart and functional textiles with multiple functionalities targeted towards specialty applications drives the future towards nano-enabled textiles.

References

[1] Bratovčić A, Odobašić A, Ćatić S, Šestan I. Application of polymer nanocomposite materials in food packaging. Croat J Food Sci Technol 2015;7:86–94. https://doi.org/10.17508/CJFST.2015.7.2.06.

[2] Camargo P, Satyanarayana K, Research FW. Nanocomposites: synthesis, structure, properties and new application opportunities. Mater Res 2009;12:1–39.

[3] Majumder DD, Majumder DD, Karan S. Magnetic properties of ceramic nanocomposites. In: Banerjee R, Manna I, editors. Ceram. Nanocomposites, Woodhead Publishing; 2013, p. 51–91. https://doi.org/10.1533/9780857093493.1.51.

[4] Syduzzaman M, Patwary SU, Farhana K, Ahmed S. Smart Textiles and Nano-Technology: A General Overview. J Text Sci Eng 2015;5. https://doi.org/10.4172/2165-8064.1000181.

[5] Cho S, Chang T, Yu T, Lee CH. Smart Electronic Textiles for Wearable Sensing and Display. Biosens 2022;12:222. https://doi.org/10.3390/BIOS12040222.

[6] Liu X, Miao J, Fan Q, Zhang W, Zuo · Xingwei, Tian M, et al. Recent Progress on Smart Fiber and Textile Based Wearable Strain Sensors: Materials, Fabrications and Applications. Adv Fiber Mater 2022;1:1–29. https://doi.org/10.1007/S42765-021-00126-3.

[7] Libanori A, Chen G, Zhao X, Zhou Y, Chen J. Smart textiles for personalized healthcare. Nat Electron 2022 53 2022;5:142–56. https://doi.org/10.1038/s41928-022-00723-z.

[8] Hári J, Pukánszky B. Nanocomposites: Preparation, structure, and properties. In: Kutz M, editor. Appl. Plast. Eng. Handb., 2011, p. 109–42.

[9] Kurahatti R V, Surendranathan AO, Kori SA, Singh N, Kumar A V, Srivastava S. Defence Applications of Polymer Nanocomposites. Def Sci J 2010;60. https://doi.org/10.14429/dsj.60.578.

[10] de Oliveira AD, Beatrice CAG. Polymer Nanocomposites with Different Types of Nanofiller. Nanocomposites – Recent Evol 2018. https://doi.org/10.5772/INTECHOPEN.81329.

[11] Pakdel E, Fang J, Sun L, Wang X. Nanocoatings for smart textiles. In: N. D. Yılmaz, editor. Smart Text. Wearable Nanotechnol., Wiley- Scrivener; 2018, p. 247–300.

[12] Nguyen-Tri P, Nguyen TA, Carriere P, Ngo Xuan C. Nanocomposite Coatings: Preparation, Characterization, Properties, and Applications. Int J Corros 2018;2018:4749501. https://doi.org/10.1155/2018/4749501.

[13] Krifa M, Prichard C. Nanotechnology in textile and apparel research–an overview of technologies and processes. J Text Inst 2020;111:1778–93. https://doi.org/10.1080/00405000.2020.1721696.

[14] Joshi M, Adak B. Advances in nanotechnology based functional, smart and intelligent textiles: A review. vol. 1–5. Elsevier Ltd.; 2019. https://doi.org/10.1016/B978-0-12-803581-8.10471-0.

[15] Gowri S, Almeida L, Amorim T, Carneiro N, Pedro Souto A, Fátima Esteves M. Polymer Nanocomposites for Multifunctional Finishing of Textiles – a Review. Text Res J 2010;80:1290–306. https://doi.org/10.1177/0040517509357652.

[16] María Luisa García-Betancourt, SIR Jiménez, Apsahara González-Hodges, ZEN Salazar, Escalante-García IL, Aparicio JR. Low Dimensional Nanostructures: Measurement and Remediation Technologies Applied to Trace Heavy Metals in Water. Trace Met. Environ., 2021. https://doi.org/10.5772/intechopen.93263.

[17] Sun YP, Fu K, Lin Y, Huang W. Functionalized carbon nanotubes: Properties and applications. Acc Chem Res 2002;35:1096–104. https://doi.org/10.1021/ar010160v.

[18] Rao R, Pint CL, Islam AE, Weatherup RS, Hofmann S, Meshot ER, et al. Carbon Nanotubes and Related Nanomaterials: Critical Advances and Challenges for Synthesis toward Mainstream Commercial Applications. ACS Nano 2018;12:11756–84. https://doi.org/10.1021/acsnano.8b06511.

[19] Ali E, Hadis D, Hamzeh K, Mohammad K, Nosratollah Z, Abolfazl A, et al. Carbon nanotubes: properties, synthesis, purification, and medical applications. Nanoscale Res Lett 2014;9:393.

[20] Popov VN. Carbon nanotubes: Properties and application. Mater Sci Eng R Reports 2004;43:61–102. https://doi.org/10.1016/j.mser.2003.10.001.

[21] Zhang C, Wu L, de Perrot M, Zhao X. Carbon Nanotubes: A Summary of Beneficial and Dangerous Aspects of an Increasingly Popular Group of Nanomaterials. Front Oncol 2021;11:1–12. https://doi.org/10.3389/fonc.2021.693814.

[22] Adak B, Joshi M, Butola BS. Polyurethane/functionalized-graphene nanocomposite films with enhanced weather resistance and gas barrier properties. Compos Part B 2019;176:107303. https://doi.org/10.1016/j.compositesb.2019.107303.

[23] Kim H, Miura Y, MacOsko CW. Graphene/polyurethane nanocomposites for improved gas barrier and electrical conductivity. Chem Mater 2010;22:3441–50. https://doi.org/10.1021/cm100477v.

[24] Checchetto R, Miotello A, Nicolais L, Carotenuto G. Gas transport through nanocomposite membrane composed by polyethylene with dispersed graphite nanoplatelets. J Memb Sci 2014;463:196–204. https://doi.org/10.1016/j.memsci.2014.03.065.

[25] Cui Y. Review polymer-graphene-gas barrier.pdf. Carbon N Y 2016;98. https://doi.org/https://doi.org/10.1016/j.carbon.2015.11.018.

[26] Chatterjee U, Butola BS, Joshi M. High energy ball milling for the processing of organo-montmorillonite in bulk. Appl Clay Sci 2017;140:10–6. https://doi.org/10.1016/j.clay.2017.01.019.

[27] Adak B, Joshi M, Butola BS. Polyurethane/clay nanocomposites with improved helium gas barrier and mechanical properties: Direct versus master-batch melt mixing route. J Appl Polym Sci 2018;135:1–12. https://doi.org/10.1002/app.46422.

[28] Adak B, Butola BS, Joshi M. Effect of organoclay-type and clay-polyurethane interaction chemistry for tuning the morphology, gas barrier and mechanical properties of clay/polyurethane nanocomposites. Appl Clay Sci 2018;161:343–53. https://doi.org/10.1016/j.clay.2018.04.030.

[29] Cui Y, Kumar S, Rao Kona B, Van Houcke D. Gas barrier properties of polymer/clay nanocomposites. RSC Adv 2015;5:63669–90. https://doi.org/10.1039/c5ra10333a.

[30] Ki HY, Kim JH, Kwon SC, Jeong SH. A study on multifunctional wool textiles treated with nano-sized silver. J Mater Sci 2007;42:8020–4. https://doi.org/10.1007/s10853-007-1572-3.

[31] Teli MD, Sheikh J. Modified bamboo rayon-copper nanoparticle composites as antibacterial textiles. Int J Biol Macromol 2013;61:302–7. https://doi.org/10.1016/j.ijbiomac.2013.07.015.

[32] Duan W, Xie A, Shen Y, Wang X, Wang F, Zhang Y, et al. Fabrication of superhydrophobic cotton fabrics with UV protection based on CeO2 particles. Ind Eng Chem Res 2011;50:4441–5. https://doi.org/10.1021/ie101924v.

[33] Yang H, Zhu S, Pan N. Studying the mechanisms of titanium dioxide as ultraviolet-blocking additive for films and fabrics by an improved scheme. J Appl Polym Sci 2004;92:3201–10. https://doi.org/10.1002/app.20327.

[34] Nautiyal A, Shukla SR, Prasad V. ZnO-TiO2 hybrid nanocrystal-loaded, wash durable, multifunction cotton textiles. Cellulose 2022;29:5923–41. https://doi.org/10.1007/s10570-022-04595-6.

[35] Cai W, Zhang D, Wang B, Shi Y, Pan Y, Wang J, et al. Scalable one-step synthesis of hydroxylated boron nitride nanosheets for obtaining multifunctional polyvinyl alcohol nanocomposite fi lms: Multi- azimuth properties improvement. Compos Sci Technol 2018;168:74–80. https://doi.org/10.1016/j.compscitech.2018.09.004.

[36] Xie S, Istrate OM, May P, Barwich S, Bell AP, Khan U, et al. Nanoscale Boron nitride nanosheets as barrier enhancing fillers in melt processed composites † 2015:4443–50. https://doi.org/10.1039/c4nr07228f.

[37] Azeem M, Jan R, Farrukh S, Hussain A. Improving gas barrier properties with boron nitride nanosheets in polymer-composites. Results Phys 2019;12:1535–41. https://doi.org/10.1016/j.rinp.2019.01.057.

[38] Tsai CY, Lin SY, Tsai HC. Butyl rubber nanocomposites with monolayer MoS2 additives: Structural characteristics, enhanced mechanical, and gas barrier properties. Polymers (Basel) 2018;10. https://doi.org/10.3390/polym10030238.

[39]   Wang X, Xing W, Feng X, Yu B, Song L, Yeoh GH, et al. Enhanced mechanical and barrier properties of polyurethane nanocomposite films with randomly distributed molybdenum disulfide nanosheets. Compos Sci Technol 2016;127:142–8. https://doi.org/10.1016/j.compscitech.2016.02.029.

[40]   Tavares MIB, Silva da EO, Silva da PRC, Menezes de LR. Polymer Nanocomposites. In: Seehra MS, editor. Nanostructured Mater. – Fabr. to Appl., IntechOpen; 2017. https://doi.org/10.5772/INTECHOPEN.68142.

[41]   Yoon B, Lee S. Designing waterproof breathable materials based on electrospun nanofibers and assessing the performance characteristics. Fibers Polym 2011;12:57–64. https://doi.org/10.1007/S12221-011-0057-9.

[42]   Ahmed Babar A, Wang X, Iqbal N, Yu J, Ding B, Babar AA, et al. Tailoring Differential Moisture Transfer Performance of Nonwoven/Polyacrylonitrile-SiO2 Nanofiber Composite Membranes. Wiley Online Libr 2017;4. https://doi.org/10.1002/admi.201700062.

[43]   Li BM, Yildiz O, Mills AC, Flewwellin TJ, Bradford PD, Jur JS. Iron-on carbon nanotube (CNT) thin films for biosensing E-Textile applications. Carbon N Y 2020;168:673–83. https://doi.org/10.1016/J. CARBON.2020.06.057.

[44]   Lim HS, Park SH, Koo SH, Kwark YJ, Thomas EL, Jeong Y, et al. Superamphiphilic janus fabric. Langmuir 2010;26:19159–62. https://doi.org/10.1021/LA103829C.

[45]   Shin YE, Cho JY, Yeom J, Ko H, Han JT. Electronic Textiles Based on Highly Conducting Poly(vinyl alcohol)/Carbon Nanotube/Silver Nanobelt Hybrid Fibers. ACS Appl Mater Interfaces 2021;13:31051–8. https://doi.org/10.1021/ACSAMI.1C08175/ASSET/IMAGES/LARGE/AM1C08175_0007. JPEG.

[46]   Pan J, Hao B, Xu P, Li D, Luo L, Li J, et al. Highly robust and durable core-sheath nanocomposite yarns for electro-thermochromic performance application. Chem Eng J 2020;384:123376. https://doi. org/10.1016/J.CEJ.2019.123376.

[47]   Li Z, Xu B, Han J, Huang J, Fu H, Li Z, et al. A Polycation-Modified Nanofillers Tailored Polymer Electrolytes Fiber for Versatile Biomechanical Energy Harvesting and Full-Range Personal Healthcare Sensing. Adv Funct Mater 2022;32:2106731. https://doi.org/10.1002/ADFM.202106731.

[48]   Parvinzadeh Gashti M, Pakdel E, Alimohammadi F. Nanotechnology-based coating techniques for smart textiles. Act Coatings Smart Text 2016:243–68. https://doi.org/10.1016/B978-0-08-100263-6.00011-3.

[49]   Chalco-Sandoval W, Fabra MJ, López-Rubio A, Lagaron JM. Development of an encapsulated phase change material via emulsion and coaxial electrospinning. J Appl Polym Sci 2016;133:43903. https:// doi.org/10.1002/APP.43903.

[50]   Zhu Q, Jin Y, Wang W, Sun G, Wang D. Bioinspired Smart Moisture Actuators Based on Nanoscale Cellulose Materials and Porous, Hydrophilic EVOH Nanofibrous Membranes. ACS Appl Mater Interfaces 2019;11:1440–8. https://doi.org/10.1021/ACSAMI.8B17538/SUPPL_FILE/AM8B17538_SI_012.AVI.

[51]   Silva MA, Felgueiras HP, de Amorim MTP. Carbon based membranes with modified properties: thermal, morphological, mechanical and antimicrobial. Cellulose 2020;27:1497–516. https://doi.org/ 10.1007/S10570-019-02861-8/FIGURES/10.

[52]   Zhang C, He S, Wang D, Xu F, Zhang F, Zhang G. Facile fabricate a bioinspired Janus membrane with heterogeneous wettability for unidirectional water transfer and controllable oil–water separation. J Mater Sci 2018;53:14398–411. https://doi.org/10.1007/S10853-018-2659-8/FIGURES/8.

[53]   Gashti MP, Song G, Kiumarsi A, Alimohammadi F. Characterization of nanocomposite coatings on textiles: A brief review on Microscopic technology Plasticizing polyhydroxyalkanoates View project Glove study: systematic tools for glove engineering View project Characterization of nanocomposite coatings on textiles: a brief review on Microscopic technology. In: Mendez-Vilas A, editor. Curr. Microsc. Contrib. to Adv. Sci. Technol., 2012, p. 1424–37.

[54]   Nateghi MR, Shateri-Khalilabad M. Silver nanowire-functionalized cotton fabric. Carbohydr Polym 2015;117:160–8. https://doi.org/10.1016/J.CARBPOL.2014.09.057.

[55]   Ma J, Zhao Q, Zhou Y, He P, Pu H, Song B, et al. Hydrophobic wrapped carbon nanotubes coated cotton fabric for electrical heating and electromagnetic interference shielding. Polym Test 2021;100:107240. https://doi.org/10.1016/J.POLYMERTESTING.2021.107240.

[56]   Doshi SM, Murray C, Chaudhari A, Sung DH, Thostenson ET. Ultrahigh sensitivity wearable sensors enabled by electrophoretic deposition of carbon nanostructured composites onto everyday fabrics. J Mater Chem C 2022;10:1617–24. https://doi.org/10.1039/D1TC05132F.

[57]   Dolez PI. Energy Harvesting Materials and Structures for Smart Textile Applications: Recent Progress and Path Forward. Sensors 2021;21:6297. https://doi.org/10.3390/S21186297.

[58]   Liman MLR, Islam MT, Hossain MM. Mapping the Progress in Flexible Electrodes for Wearable Electronic Textiles: Materials, Durability, and Applications. Adv Electron Mater 2022;8. https://doi.org/10.1002/AELM.202100578.

[59]   Zaarour B, Zhu L, Huang C, Jin X, Alghafari H, Fang J, et al. A review on piezoelectric fibers and nanowires for energy harvesting. J Ind Text 2019;51:297–340. https://doi.org/10.1177/1528083719870197.

[60]   Gao H, Minh PT, Wang H, Minko S, Locklin J, Nguyen T, et al. High-performance flexible yarn for wearable piezoelectric nanogenerators. Smart Mater Struct 2018;27:95018. https://doi.org/10.1088/1361-665x/aad718.

[61]   Dong K, Peng X, Lin Wang Z, Dong K, Peng X, Wang ZL. Fiber/Fabric-Based Piezoelectric and Triboelectric Nanogenerators for Flexible/Stretchable and Wearable Electronics and Artificial Intelligence. Adv Mater 2020;32:1902549. https://doi.org/10.1002/ADMA.201902549.

[62]   Wang ZL. Nanogenerators for self-powered devices and systems. In: Wang ZL, editor., Georgia Institute of Technology; 2011, p. 131.

[63]   Haque RI, Farine PA, Briand D. Soft triboelectric generators by use of cost-effective elastomers and simple casting process. Sensors Actuators A Phys 2018;271:88–95. https://doi.org/10.1016/J.SNA.2017.12.018.

[64]   Zou Y, Bo L, Li Z. Recent progress in human body energy harvesting for smart bioelectronic system. Fundam Res 2021;1:364–82. https://doi.org/10.1016/J.FMRE.2021.05.002.

[65]   Zoui MA, Bentouba S, Stocholm JG, Bourouis M. A Review on Thermoelectric Generators: Progress and Applications n.d. https://doi.org/10.3390/en13143606.

[66]   Ryu H, Kim SW. Emerging Pyroelectric Nanogenerators to Convert Thermal Energy into Electrical Energy. Small 2021;17. https://doi.org/10.1002/SMLL.201903469.

[67]   Schröder U. From in vitro to in vivo-biofuel cells are maturing. Angew Chemie – Int Ed 2012;51:7370–2. https://doi.org/10.1002/ANIE.201203259.

[68]   Zhang Z, Li X, Yin J, Xu Y, Fei W, Xue M, et al. Emerging hydrovoltaic technology. Nat Nanotechnol 2018 1312 2018;13:1109–19. https://doi.org/10.1038/s41565-018-0228-6.

[69]   Reddy GVR, Joshi M, Adak B, Deopura BL. Studies on the dyeability and dyeing mechanism of polyurethane/clay nanocomposite filaments with acid, basic and reactive dyes. Color Technol 2018;134:117–25. https://doi.org/10.1111/COTE.12332.

[70]   Adak B, Joshi M, Ali SW. Dyeability of Polymer Nanocomposite Fibers. In: Joshi M, editor. Nanotechnol. Text., Jenny Stanford Publishing; 2020, p. 145–78. https://doi.org/10.1201/9781003055815-6.

[71]   Barthlott W, Neinhuis C. Purity of the sacred lotus, or escape from contamination in biological surfaces. Planta 1997;202:1–8.

[72]   Park S, Kim J, Park CH. Superhydrophobic textiles: review of theoretical definitions, fabrication and functional evaluation. J Eng Fiber Fabr 2015;10:155892501501000420. https://doi.org/https://doi.org/10.1177%2F155892501501000401.

[73]   Teisala H, Tuominen M, Kuusipalo J. Superhydrophobic Coatings on Cellulose-Based Materials: Fabrication, Properties, and Applications. Adv Mater Interfaces 2014;1:1300026. https://doi.org/10.1002/ADMI.201300026.

[74] Liu H, Gao SW, Cai JS, He CL, Mao JJ, Zhu TX, et al. Recent Progress in Fabrication and Applications of Superhydrophobic Coating on Cellulose-Based Substrates. Mater 2016;9:124. https://doi.org/10.3390/MA9030124.

[75] Wei DW, Wei H, Gauthier AC, Song J, Jin Y, Xiao H. Superhydrophobic modification of cellulose and cotton textiles: Methodologies and applications. J Bioresour Bioprod 2020;5:1–15. https://doi.org/10.1016/j.jobab.2020.03.001.

[76] Roach P, Shirtcliffe NJ, Newton MI. Progress in superhydrophobic surface development. Soft Matter 2008;4:224–40. https://doi.org/10.1039/B712575P.

[77] Si Y, Guo Z. Superhydrophobic nanocoatings: from materials to fabrications and to applications. Nanoscale 2015;7:5922–46. https://doi.org/10.1039/C4NR07554D.

[78] Yazdanshenas ME, Shateri-Khalilabad M. One-step synthesis of superhydrophobic coating on cotton fabric by ultrasound irradiation. Ind Eng Chem Res 2013;52:12846–54. https://doi.org/10.1021/IE401133Q/SUPPL_FILE/IE401133Q_SI_001.PDF.

[79] Li K, Zeng X, Li H, Lai X, Xie H. Effects of calcination temperature on the microstructure and wetting behavior of superhydrophobic polydimethylsiloxane/silica coating. Colloids Surfaces A Physicochem Eng Asp 2014;445:111–8. https://doi.org/10.1016/J.COLSURFA.2014.01.024.

[80] Seyedmehdi SA, Zhang H, Zhu J. Superhydrophobic RTV silicone rubber insulator coatings. Appl Surf Sci 2012;258:2972–6. https://doi.org/10.1016/J.APSUSC.2011.11.020.

[81] Lee SG, Ham DS, Lee DY, Bong H, Cho K. Transparent superhydrophobic/translucent superamphiphobic coatings based on silica-fluoropolymer hybrid nanoparticles. Langmuir 2013;29:15051–7. https://doi.org/10.1021/LA404005B/SUPPL_FILE/LA404005B_SI_001.PDF.

[82] Lee SE, Lee D, Lee P, Ko SH, Lee SS, Hong SU. Flexible Superhydrophobic Polymeric Surfaces with Micro-/Nanohybrid Structures Using Black Silicon. Macromol Mater Eng 2013;298:311–7. https://doi.org/10.1002/MAME.201200098.

[83] Nagappan S, Park SS, Ha CS. Recent advances in superhydrophobic nanomaterials and nanoscale systems. J Nanosci Nanotechnol 2014;14:1441–62. https://doi.org/10.1166/JNN.2014.9194.

[84] Sohn Y, Kim D, Lee S, Yin M, Song JY, Hwang W, et al. Anti-frost coatings containing carbon nanotube composite with reliable thermal cyclic property. J Mater Chem A 2014;2:11465–71.

[85] Lee J-S, Yoon J-C, Jang J-H. A route towards superhydrophobic graphene surfaces: surface-treated reduced graphene oxide spheres. J Mater Chem A 2013;1:7312–5.

[86] Choi BG, Park HS. Superhydrophobic graphene/nafion nanohybrid films with hierarchical roughness. J Phys Chem C 2012;116:3207–11.

[87] Zhu X, Zhang Z, Ge B, Men X, Zhou X. Fabrication of a superhydrophobic carbon nanotube coating with good reusability and easy repairability. Colloids Surfaces A Physicochem Eng Asp 2014;444:252–6. https://doi.org/10.1016/J.COLSURFA.2013.12.066.

[88] Liang J, Li D, Wang D, Liu K, Chen L. Preparation of stable superhydrophobic film on stainless steel substrate by a combined approach using electrodeposition and fluorinated modification. Appl Surf Sci 2014;293:265–70. https://doi.org/10.1016/J.APSUSC.2013.12.147.

[89] Chapman J, Regan F. Nanofunctionalized Superhydrophobic Antifouling Coatings for Environmental Sensor Applications – Advancing Deployment with Answers from Nature. Adv Eng Mater 2012;14:B175–84. https://doi.org/10.1002/ADEM.201180037.

[90] Lu Y, Xu W, Song J, Liu X, Xing Y, Sun J. Preparation of superhydrophobic titanium surfaces via electrochemical etching and fluorosilane modification. Appl Surf Sci 2012;263:297–301. https://doi.org/10.1016/J.APSUSC.2012.09.047.

[91] Peng S, Tian D, Yang X, Deng W. Highly efficient and large-scale fabrication of superhydrophobic alumina surface with strong stability based on self-congregated alumina nanowires. ACS Appl Mater Interfaces 2014;6:4831–41. https://doi.org/10.1021/AM4057858/SUPPL_FILE/AM4057858_SI_001.ZIP.

[92] Gao L, Zhuang J, Nie L, Zhang J, Zhang Y, Gu N, et al. Intrinsic peroxidase-like activity of ferromagnetic nanoparticles. Nat Nanotechnol 2007 29 2007;2:577–83. https://doi.org/10.1038/nnano.2007.260.

[93] Bao XM, Cui JF, Sun HX, Liang WD, Zhu ZQ, An J, et al. Facile preparation of superhydrophobic surfaces based on metal oxide nanoparticles. Appl Surf Sci 2014;303:473–80. https://doi.org/10.1016/J.APSUSC.2014.03.029.

[94] Roy P, Kisslinger R, Farsinezhad S, Mahdi N, Bhatnagar A, Hosseini A, et al. All-solution processed, scalable superhydrophobic coatings on stainless steel surfaces based on functionalized discrete titania nanotubes. Chem Eng J 2018;351:482–9.

[95] Yap SW, Johari N, Mazlan SA, Hassan NA. Mechanochemical durability and self-cleaning performance of zinc oxide-epoxy superhydrophobic coating prepared via a facile one-step approach. Ceram Int 2021;47:15825–33.

[96] Li H, Zhao Y, Yuan X. Facile preparation of superhydrophobic coating by spraying a fluorinated acrylic random copolymer micelle solution. Soft Matter 2012;9:1005–9. https://doi.org/10.1039/C2SM26689J.

[97] Huovinen E, Takkunen L, Korpela T, Suvanto M, Pakkanen TT, Pakkanen TA. Mechanically robust superhydrophobic polymer surfaces based on protective micropillars. Langmuir 2014;30:1435–43. https://doi.org/10.1021/LA404248D/ASSET/IMAGES/LA404248D.SOCIAL.JPEG_V03.

[98] Hong D, Ryu I, Kwon H, Lee JJ, Yim S. Preparation of superhydrophobic, long-neck vase-like polymer surfaces. Phys Chem Chem Phys 2013;15:11862–7. https://doi.org/10.1039/C3CP51833G.

[99] Ivanova NA, Philipchenko AB. Superhydrophobic chitosan-based coatings for textile processing. Appl Surf Sci 2012;263:783–7. https://doi.org/10.1016/J.APSUSC.2012.09.173.

[100] Zeng X, Ma Y, Wang Y. Enhancing the low surface energy properties of polymer films with a dangling shell of fluorinated block-copolymer. Appl Surf Sci 2015;338:190–6. https://doi.org/10.1016/J.APSUSC.2015.02.134.

[101] Barmentlo SH, Stel JM, Van Doorn M, Eschauzier C, De Voogt P, Kraak MHS. Acute and chronic toxicity of short chained perfluoroalkyl substances to Daphnia magna. Environ Pollut 2015;198:47–53. https://doi.org/10.1016/J.ENVPOL.2014.12.025.

[102] Zhong Y, Netravali AN. 'Green' surface treatment for water-repellent cotton fabrics. http://DxDoiOrg/101680/Jsuin1500022 2016;4:3–13. https://doi.org/10.1680/JSUIN.15.00022.

[103] Zhong X, Zhou M, Wang S, Fu H. Preparation of water-borne non-fluorinated anti-smudge surfaces and their applications. Prog Org Coatings 2020;142:105581. https://doi.org/10.1016/j.porgcoat.2020.105581.

[104] Zhang W, Zou X, Liu X, Liang Z, Ge Z, Luo Y. Preparation and properties of waterborne polyurethane modified by aminoethylaminopropyl polydimethylsiloxane for fluorine-free water repellents. Prog Org Coatings 2020;139:105407. https://doi.org/10.1016/j.porgcoat.2019.105407.

[105] Qing Y, Zheng Y, Hu C, Wang Y, He Y, Gong Y, et al. Facile approach in fabricating superhydrophobic ZnO/polystyrene nanocomposite coating. Appl Surf Sci 2013;285:583–7. https://doi.org/10.1016/J.APSUSC.2013.08.097.

[106] Zhang Y, Li J, Huang F, Li S, Shen Y, Xie A, et al. Controlled fabrication of transparent and superhydrophobic coating on a glass matrix via a Green method. Appl Phys A Mater Sci Process 2013;110:397–401. https://doi.org/10.1007/S00339-012-7176-Z/FIGURES/5.

[107] Chakradhar RPS, Prasad G, Bera P, Anandan C. Stable superhydrophobic coatings using PVDF–MWCNT nanocomposite. Appl Surf Sci 2014;301:208–15. https://doi.org/10.1016/J.APSUSC.2014.02.044.

[108] Rutkevičius M, Pirzada T, Geiger M, Khan SA. Creating superhydrophobic, abrasion-resistant and breathable coatings from water-borne polydimethylsiloxane-polyurethane Co-polymer and fumed silica. J Colloid Interface Sci 2021;596:479–92. https://doi.org/10.1016/j.jcis.2021.02.072.

[109] Liang Y, Zhang D, Zhou M, Xia Y, Chen X, Oliver S, et al. Bio-based omniphobic polyurethane coating providing anti-smudge and anti-corrosion protection. Prog Org Coatings 2020;148:105844. https://doi.org/10.1016/j.porgcoat.2020.105844.

[110] Zhang K, Huang S, Wang J, Liu G. Transparent organic/silica nanocomposite coating that is flexible, omniphobic, and harder than a 9H pencil. Chem Eng J 2020;396:125211. https://doi.org/10.1016/j.cej.2020.125211.

[111] Gee E, Liu G, Hu H, Wang J. Effect of Varying Chain Length and Content of Poly (dimethylsiloxane) on Dynamic Dewetting Performance of NP- GLIDE Polyurethane Coatings 2018. https://doi.org/10.1021/acs.langmuir.8b01965.

[112] Yasuyuki M, Kunihiro K, Kurissery S, Kanavillil N, Sato Y, Kikuchi Y. Antibacterial properties of nine pure metals: a laboratory study using Staphylococcus aureus and Escherichia coli. Biofouling 2010;26:851–8. https://doi.org/10.1080/08927014.2010.527000.

[113] Stanić V, Tanasković SB. Antibacterial activity of metal oxide nanoparticles. Nanotoxicity, Elsevier; 2020, p. 241–74.

[114] Shaheen TI, Fouda A, Salem SS. Integration of Cotton Fabrics with Biosynthesized CuO Nanoparticles for Bactericidal Activity in the Terms of Their Cytotoxicity Assessment. Ind Eng Chem Res 2021;60:1553–63. https://doi.org/10.1021/ACS.IECR.0C04880.

[115] Fu Y, Jiang J, Zhang Q, Zhan X, Chen F. Robust liquid-repellent coatings based on polymer nanoparticles with excellent self-cleaning and antibacterial performances. J Mater Chem A 2016;5:275–84. https://doi.org/10.1039/C6TA06481G.

[116] Hong HR, Kim J, Park CH. Facile fabrication of multifunctional fabrics: use of copper and silver nanoparticles for antibacterial, superhydrophobic, conductive fabrics. RSC Adv 2018;8:41782–94. https://doi.org/10.1039/C8RA08310J.

[117] Guo Z, Wang Y, Huang J, Zhang S, Zhang R, Ye D, et al. Multi-functional and water-resistant conductive silver nanoparticle-decorated cotton textiles with excellent joule heating performances and human motion monitoring. Cellulose 2021;28:7483–95. https://doi.org/10.1007/S10570-021-03955-Y/FIGURES/6.

[118] Dizaj SM, Lotfipour F, Barzegar-Jalali M, Zarrintan MH, Adibkia K. Antimicrobial activity of the metals and metal oxide nanoparticles. Mater Sci Eng C 2014;44:278–84. https://doi.org/10.1016/J.MSEC.2014.08.031.

[119] Islam MS, van de Ven TGM. Cotton-based flame-retardant textiles: A review. BioResources 2021;16:4354–81.

[120] Dai K, Deng Z, Liu G, Wu Y, Xu W, Hu Y. Effects of a Reactive Phosphorus–Sulfur Containing Flame-Retardant Monomer on the Flame Retardancy and Thermal and Mechanical Properties of Unsaturated Polyester Resin. Polym 2020;12:1441. https://doi.org/10.3390/POLYM12071441.

[121] Salmeia KA, Gaan S, Malucelli G. Recent Advances for Flame Retardancy of Textiles Based on Phosphorus Chemistry. Polym 2016;8:319. https://doi.org/10.3390/POLYM8090319.

[122] Yusuf M. A Review on Flame Retardant Textile Finishing: Current and Future Trends. Curr Smart Mater 2018;3:99–108. https://doi.org/10.2174/2405465803666180703110858.

[123] Gou J, Tang Y. Flame retardant polymer nanocomposites. Multifunct Polym Nanocomposites 2011:309–36.

[124] He S, Hu Y, Song L, Tang Y. Fire Safety Assessment of Halogen-free Flame Retardant Polypropylene Based on Cone Calorimeter: http://DxDoiOrg/101177/0734904107067109 2016;25:109–18. https://doi.org/10.1177/0734904107067109.

[125] Zare Y, Garmabi H. Nonisothermal crystallization and melting behavior of PP/nanoclay/CaCO3 ternary nanocomposite. J Appl Polym Sci 2012;124:1225–33. https://doi.org/10.1002/APP.35134.

[126] Onder E, Sarier N, Ersoy MS. The manufacturing of polyamide– and polypropylene–organoclay nanocomposite filaments and their suitability for textile applications. Thermochim Acta 2012;543:37–58. https://doi.org/10.1016/J.TCA.2012.05.002.

[127] Dasari A, Yu ZZ, Mai YW, Liu S. Flame retardancy of highly filled polyamide 6/clay nanocomposites. Nanotechnology 2007;18:445602. https://doi.org/10.1088/0957-4484/18/44/445602.

[128] Isitman NA, Kaynak C. Nanostructure of montmorillonite barrier layers: A new insight into the mechanism of flammability reduction in polymer nanocomposites. Polym Degrad Stab 2011;96:2284–9. https://doi.org/10.1016/J.POLYMDEGRADSTAB.2011.09.021.

[129] Zare Y, Garmabi H. Attempts to Simulate the Modulus of Polymer/Carbon Nanotube Nanocomposites and Future Trends. https://DoiOrg/101080/155837242013870574 2014;54:377–400. https://doi.org/10.1080/15583724.2013.870574.

[130] Parvinzadeh Gashti M, Almasian A. UV radiation induced flame retardant cellulose fiber by using polyvinylphosphonic acid/carbon nanotube composite coating. Compos Part B Eng 2013;45:282–9. https://doi.org/10.1016/J.COMPOSITESB.2012.07.052.

[131] Schartel B, Pötschke P, Knoll U, Abdel-Goad M. Fire behaviour of polyamide 6/multiwall carbon nanotube nanocomposites. Eur Polym J 2005;41:1061–70. https://doi.org/10.1016/J.EURPOLYMJ.2004.11.023.

[132] Shen Z, Bateman S, Wu DY, McMahon P, Dell'Olio M, Gotama J. The effects of carbon nanotubes on mechanical and thermal properties of woven glass fibre reinforced polyamide-6 nanocomposites. Compos Sci Technol 2009;69:239–44. https://doi.org/10.1016/J.COMPSCITECH.2008.10.017.

[133] Wu Z, Xue M, Wang H, Tian X, Ding X, Zheng K, et al. Electrical and flame-retardant properties of carbon nanotube/poly(ethylene terephthalate) composites containing bisphenol A bis(diphenyl phosphate). Polymer (Guildf) 2013;54:3334–40. https://doi.org/10.1016/J.POLYMER.2013.04.051.

[134] Huang G, Wang S, Song P, Wu C, Chen S, Wang X. Combination effect of carbon nanotubes with graphene on intumescent flame-retardant polypropylene nanocomposites. Compos Part A Appl Sci Manuf 2014;59:18–25. https://doi.org/10.1016/J.COMPOSITESA.2013.12.010.

[135] Kashiwagi T, Grulke E, Hilding J, Groth K, Harris R, Butler K, et al. Thermal and flammability properties of polypropylene/carbon nanotube nanocomposites. Polymer (Guildf) 2004;45:4227–39. https://doi.org/10.1016/J.POLYMER.2004.03.088.

[136] Zhang T, Yan H, Peng M, Wang L, Ding H, Fang Z. Construction of flame retardant nanocoating on ramie fabric via layer-by-layer assembly of carbon nanotube and ammonium polyphosphate. Nanoscale 2013;5:3013–21. https://doi.org/10.1039/C3NR34020A.

[137] Shabanian M, Basaki N, Khonakdar HA, Jafari SH, Hedayati K, Wagenknecht U. Novel nanocomposites consisting of a semi-crystalline polyamide and Mg–Al LDH: Morphology, thermal properties and flame retardancy. Appl Clay Sci 2014;90:101–8. https://doi.org/10.1016/J.CLAY.2013.12.033.

[138] Wang L, He X, Lu H, Feng J, Xie X, Su S, et al. Flame retardancy of polypropylene (nano)composites containing LDH and zinc borate. Polym Adv Technol 2011;22:1131–8. https://doi.org/10.1002/PAT.1927.

[139] Wang DY, Leuteritz A, Kutlu B, Landwehr Der MA, Jehnichen D, Wagenknecht U, et al. Preparation and investigation of the combustion behavior of polypropylene/organomodified MgAl-LDH micro-nanocomposite. J Alloys Compd 2011;509:3497–501. https://doi.org/10.1016/J.JALLCOM.2010.12.138.

[140] Zhang M, Ding P, Qu B. Flammable, thermal, and mechanical properties of intumescent flame retardant PP/LDH nanocomposites with different divalent cations. Polym Compos 2009;30:1000–6. https://doi.org/10.1002/PC.20648.

[141] Fina A, Abbenhuis HCL, Tabuani D, Camino G. Metal functionalized POSS as fire retardants in polypropylene. Polym Degrad Stab 2006;91:2275–81. https://doi.org/10.1016/J.POLYMDEGRADSTAB.2006.04.014.

[142] Fina A, Tabuani D, Camino G. Polypropylene–polysilsesquioxane blends. Eur Polym J 2010;46:14–23. https://doi.org/10.1016/J.EURPOLYMJ.2009.07.019.

[143] Devaux E, Rochery M, Bourbigot S. Polyurethane/clay and polyurethane/POSS nanocomposites as flame retarded coating for polyester and cotton fabrics. Fire Mater 2002;26:149–54. https://doi.org/10.1002/FAM.792.

[144] Gashti MP, Alimohammadi F, Shamei A. Preparation of water-repellent cellulose fibers using a polycarboxylic acid/hydrophobic silica nanocomposite coating. Surf Coatings Technol 2012;206:3208–15. https://doi.org/10.1016/J.SURFCOAT.2012.01.006.

[145] Ji Q, Wang X, Zhang Y, Kong Q, Xia Y. Characterization of Poly (ethylene terephthalate)/SiO2 nanocomposites prepared by Sol–Gel method. Compos Part A Appl Sci Manuf 2009;40:878–82. https://doi.org/10.1016/J.COMPOSITESA.2009.04.010.

[146] Erdem N, Cireli AA, Erdogan UH. Flame retardancy behaviors and structural properties of polypropylene/nano-SiO2 composite textile filaments. J Appl Polym Sci 2009;111:2085–91. https://doi.org/10.1002/APP.29052.

[147] Ye L, Wu Q, Qu B. Synergistic effects of fumed silica on intumescent flame-retardant polypropylene. J Appl Polym Sci 2010;115:3508–15. https://doi.org/10.1002/APP.30585.

[148] Carosio F, Di Blasio A, Cuttica F, Alongi J, Frache A, Malucelli G. Flame Retardancy of Polyester Fabrics Treated by Spray-Assisted Layer-by-Layer Silica Architectures. Ind Eng Chem Res 2013;52:9544–50. https://doi.org/10.1021/IE4011244.

[149] Rault F, Pleyber E, Campagne C, Rochery M, Giraud S, Bourbigot S, et al. Effect of manganese nanoparticles on the mechanical, thermal and fire properties of polypropylene multifilament yarn. Polym Degrad Stab 2009;94:955–64. https://doi.org/10.1016/J.POLYMDEGRADSTAB.2009.03.012.

[150] Cinausero N, Azema N, Lopez-Cuesta JM, Cochez M, Ferriol M. Synergistic effect between hydrophobic oxide nanoparticles and ammonium polyphosphate on fire properties of poly(methyl methacrylate) and polystyrene. Polym Degrad Stab 2011;96:1445–54. https://doi.org/10.1016/J.POLYMDEGRADSTAB.2011.05.008.

[151] Wang DY, Liu XQ, Wang JS, Wang YZ, Stec AA, Hull TR. Preparation and characterisation of a novel fire retardant PET/α-zirconium phosphate nanocomposite. Polym Degrad Stab 2009;94:544–9. https://doi.org/10.1016/J.POLYMDEGRADSTAB.2009.01.018.

[152] Bao CL, Song L, Guo Y, Hu Y. Preparation and characterization of flame-retardant polypropylene/α-titanium phosphate (nano)composites. Polym Adv Technol 2011;22:1156–65. https://doi.org/10.1002/PAT.1976.

[153] Lessan F, Montazer M, Moghadam MB. A novel durable flame-retardant cotton fabric using sodium hypophosphite, nano TiO2 and maleic acid. Thermochim Acta 2011;520:48–54. https://doi.org/10.1016/J.TCA.2011.03.012.

[154] Malucelli G, Bosco F, Alongi J, Carosio F, Di Blasio A, Mollea C, et al. Biomacromolecules as novel green flame retardant systems for textiles: an overview. RSC Adv 2014;4:46024–39. https://doi.org/10.1039/C4RA06771A.

[155] Attia NF, Ahmed HE, El Ebissy AA, El Ashery SEA. Green and novel approach for enhancing flame retardancy, UV protection and mechanical properties of fabrics utilized in historical textile fabrics conservation. Prog Org Coatings 2022;166:106822. https://doi.org/10.1016/J.PORGCOAT.2022.106822.

[156] Mayer-Gall T, Plohl D, Derksen L, Lauer D, Neldner P, Ali W, et al. A Green Water-Soluble Cyclophosphazene as a Flame Retardant Finish for Textiles. Mol 2019;24:3100. https://doi.org/10.3390/MOLECULES24173100.

[157] Liao L, Pasternak I. A review of airship structural research and development. Prog Aerosp Sci 2009;45:83–96. https://doi.org/10.1016/j.paerosci.2009.03.001.

[158] Li A, Vallabh R, Bradford PD, Seyam AFM. Textile laminates for high-altitude airship hull materials-a review. J Text Apparel, Technol Manag 2019;11.

[159] Zhai H, Euler A. Material challenges for lighter-than-air systems in high altitude applications. Collect Tech Pap – AIAA 5th ATIO AIAA 16th Light Syst Technol Conf Balloon Syst Conf 2005;3:1756–67. https://doi.org/10.2514/6.2005-7488.

[160] Stockbridge C, Ceruti A, Marzocca P. Airship research and development in the areas of design, structures, dynamics and energy systems. Int J Aeronaut Sp Sci 2012;13:170–87. https://doi.org/10.5139/IJASS.2012.13.2.170.

[161] Adak B, Joshi M. Coated or laminated textiles for aerostat and stratospheric airship. Adv Text Eng Mater 2018:257–87. https://doi.org/10.1002/9781119488101.ch7.

[162] B.S. Butola, S. Parasuram, N. Mandlekar. Testing and Evaluation of LTA Systems. Coat. Laminated Text. Aerostats Airships, 2022, p. 51. https://doi.org/https://doi.org/10.1201/9780429432996.

[163] Joshi M, Adak B, Butola BS. Polyurethane nanocomposite based gas barrier films, membranes and coatings: A review on synthesis, characterization and potential applications. Prog Mater Sci 2018;97:230–82. https://doi.org/10.1016/j.pmatsci.2018.05.001.

[164] Feldman D. Polymer barrier films. J Polym Environ 2001;9:49–55.

[165] Kim H, Miura Y, Macosko CW. Graphene/Polyurethane Nanocomposites for Improved Gas Barrier and Electrical Conductivity 2010:3441–50. https://doi.org/10.1021/cm100477v.

[166] Adak B, Butola BS, Joshi M. Calcination of UV shielding nanopowder and its effect on weather resistance property of polyurethane nanocomposite films. J Mater Sci 2019;54:12698–712. https://doi.org/10.1007/S10853-019-03739-7/FIGURES/9.

[167] Joshi M, Sandhoo R, Adak B. Nano-ceria and nano-zirconia reinforced polyurethane nanocomposite-based coated textiles with enhanced weather resistance. Prog Org Coatings 2022;165:106744. https://doi.org/10.1016/J.PORGCOAT.2022.106744.

Subhankar Maity, Kunal Singha, Pintu Pandit

6 Emerging trends in electroconductive textiles

Abstract: This chapter provides comprehensive information about the various electroconductive textiles prepared from carbon/graphene, carbon nanotube, MXene, and conductive polymers. Preparation, characterization, and applications of these electroconductive textiles are presented in this chapter. Potential applications of these electroconductive textiles, such as heating and cooling garments, electromagnetic interference shielding, antibacterial efficacy, strain sensor, humidity sensor, and pH sensors are described.

Keywords: Electroconductive textiles, graphene, MXene, carbon, conductive polymer, coated textiles

6.1 Electrical conductivity in materials

Electrical conductivity is the measure of a material's ability to permit the flow of electric charges through its body. Its SI unit is Siemens per meter (S/m), named after Werner von Siemens. Depending upon the electrical conductivity of the materials, they can be broadly classified as conductors, semiconductors, and insulators. Conductors have high electrical conductivity (in the range of 10^4 S/m), whereas insulators have inferior electrical conductivity (in the range of 10^{-12} S/m). The electrical conductivities of conductors (e.g. metals) and insulators (e.g. glass and vacuum) generally remain constant. It does not vary much with change in stimuli such as temperature, ions, humidity, And pH. Semiconductors are a class of materials whose electrical conductivity ranges between insulators and conductors ($\sim 10^{-6}$ to 10^2 S/m), and that can be tailored by changing the stimulus, such as temperature, ions, humidity, And pH. Silicon, germanium, and conductive polymers fall under this class.

6.2 Electroconductive textiles

Textile fibres and polymers are insulators by nature. However, for diversified applications of textiles beyond their garmenting use, they are modified to bring about electrical conductivity in textiles. Insertion of metallic fibres and wires into textile yarns and fabrics was the conventional practice to make electroconductive textiles. There are different types of conventional methods like metal coating, inserting the metal wire and metal fibers in the structure of textiles, blending of metal fiber during spinning or weaving, for the preparation of electroconductive textiles. Other than such

https://doi.org/10.1515/9783110759747-006

metalized textiles, various non-metallic electroconductive textiles are developed with much better functional and mechanical properties. Carbon, carbon nanotubes (CNTs), graphene, conductive polymers, MXene, and so on are such non-metallic conductors used for the development of electroconductive textiles. This chapter discusses these non-conventional materials for the preparation of electroconductive textiles, their preparation methodologies, process control, characteristics of the prepared electroconductive fabrics, and applications [1].

6.3 Conducting materials

Metals are known as electrical conductors owing to their free outermost electrons in their atom, which can easily move across the interatomic spaces. Various semiconductors, ionized gases, electrolytes, and carbon can also conduct electricity, but are generally inferior to metals. Graphene, CNTs, MXene, and conductive polymers are newcomers in the class of conductive materials that have been extensively studied in material science in recent years.

6.3.1 Carbon as a conductor

There are various forms of carbon available and they are prepared for various potential applications. Among them, carbon black, graphite, graphene, and CNTs are electrically conductive in nature and are extensively researched for designing electroconductive textiles.

Carbon black is specially prepared by the process of incomplete combustion of carbon-rich material, with a controlled supply of oxygen. Due to lack of affinity to textiles, carbon blacks are not found to be very efficient for the electrification of textiles.

Carbon, in the form of 2-D single layer of carbon atoms, constructed in a honeycomb lattice can conduct electricity. It is called graphene. It has good electrical as well as thermal conductivity. It also possesses excellent optical and mechanical properties, and is therefore suitable for many potential applications. Coating textile materials with graphene brings about many functional characteristics that suit various smart applications. Graphene has good electrical conductivity owing to its sp^2 hybridization, where one s and two p orbitals are superimposed. The sp^2 hybridization of graphene generates free electrons in its ourtermost orbital for flow when an electric field is applied. The hybridization is trigonal in nature as shown in Figure 6.1.

There are two types of CNTs available, namely, single-walled CNTs (SWCNT) and multi-walled CNTs (MWCNT). The SWCNTs have one layer of a cylindrical lattice of carbon atoms, as shown in Figure 6.2a. The MWCNTs have multiple layers of carbon atoms that are arranged concentrically in a helical configuration, as shown in Figure 6.2b.

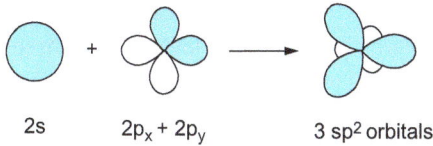

2s   2p$_x$ + 2p$_y$ 3 sp^2 orbitals

Figure 6.1: *sp^2* hybridization.

These CNTs are configured as hollow tubes of rolled graphene sheets, formulated around a suitable substrate, while covering the substrate uniformly. The dimension of the CNTs, particularly their length, depends on the precise control over the process condition and the catalyst used in their fabrication. The diameter of SWCNTs is generally in the range of 0.7–1.0 nm.

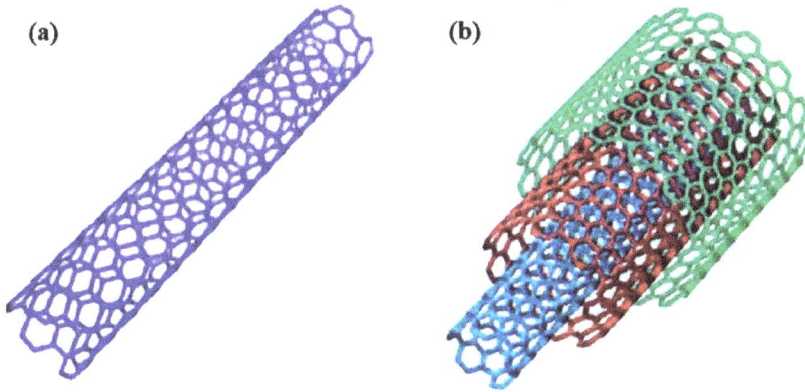

Figure 6.2: Structural geometry of (a) SWCNT (b) and MWCNT.

Owing to the high electrical conductivity of the CNTs, incorporating them into the textile structure imparts electrical conductivity. When sufficient add-on of CNTs are achieved on the textile substrate, a percolated network forms, resulting in high electrical conductivity.

6.3.2 MXene

MXene is a novel 2D material made of transition metal carbides and nitrides. They have drawn remarkable attention from researchers and material scientists since the initial development of Ti$_3$C$_2$. The general formula of MXene is M$_{n+1}$X$_n$T$_x$ (n = 1 – 3), where n + 1 layers of transition metals (M) are interleaved with n layers of carbon or nitrogen (X). The T$_x$ denotes various surface-terminated functional groups, such as −O, −OH, −F, and −Cl. There are varieties of MXenes with different chemical composition, M-elements, and atomic structures that are prepared by various researchers. A top-down approach is followed for the synthesis of MXenes by selective etching of the A-layers from three-dimensional (3D) crystalline-layered carbides and Nitrides. In an MAX phase, Mn + 1Xn layers are bonded with an atomic layer of an A-group element, which is usually a group of 13–16 elements such as Al, Ga, Si, Ge, P, and As. The three types of MAX phase structures are M_2AX, M_3AX_2, and M_4AX_3, shown in Figure 6.3 [2].

Other than these experimentally synthesized MXenes, there are several MXene structures that can be theoretically predicted. The electrical properties of these MXenes can be tailored by changing the nature and proportion of M elements, X elements, and the surface-terminating groups.

Due to their metallic conductivity, suitable chemistry, suitable surface termination groups, electrochemical and optoelectronic properties, and surface hydrophilicity, they are promising materials for many applications such as wearable electronics, EMI shielding, energy storage, conductive electrodes, and biomedicine. They have a large specific surface area, creating active surface functionalities. This results in effective adsorbents for various molecules and ionic species, serving vital applications such as ion sieving, sensory, and catalytic reactions. It is possible to load MXenes on textile surfaces, converting

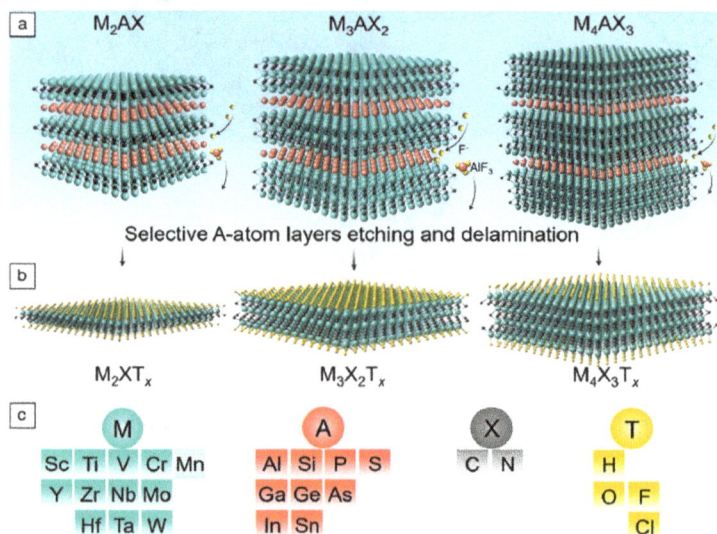

Figure 6.3: Synthesis and structure of MXenes: (a) three types of MAX phases M_2AX, M_3AX_2, M_4AX_3 and the selective etching process of the "A" group layers (red atoms); (b) formation of MXenes by selective etching and formation of surface terminations (yellow atoms) labelled as "T"; (c) possible elements for M, A, X, and T in MAX and MXene phases [2] (reprinted with permission from Springer Nature; Hong et al. [2]. Copyright 2020).

the conventional textiles into high-end smart and functional textiles. It is possible because of the excellent dispensability of MXene in many solvents, such as water, ethanol, dimethylformamide (DMF), dimethyl sulphoxide (DMSO), and N-methyl-2-pyrrolidone, allowing it to be coated over various textile substrates by different techniques such as spraying, nip-n-dip, spinning, coating, and printing [3].

6.3.2.1 Preparation of MXene

Nanoscale MXene-like lateral flakes can be prepared by various methods. Based on the type of precursors and synthesis conditions, the size of the MXene flake largely differs. A minimally intensive layer delamination (MILD) method can be utilized to prepare single-layer MXene flakes with a large distribution of sizes, ranging from 100 nm to ~10 μm, whereas bath or probe sonication methods can be utilized to prepare significantly smaller flake sizes, ranging from 0.1 to 5 μm. The morphology of MXene coating on the textile surface is also investigated. MXene flakes were coated on cotton textiles and it was observed that the smaller-sized flakes penetrate the yarn interstices, while the large MXene flake cover only the surface of the yarn. Therefore, cotton textiles coated with larger flakes give more electrical conductivity than the smaller-sized flakes [4]. The size of the MXene flake is also crucial for the electrochemical performances of the MXene-based electrode systems. While considering the application of MXene onto textile substrates through the aqueous or organic solvent medium, the stability of the MXene system in the medium is essential. The stability of MXene, in the form of aqueous dispersion, can be evaluated at various pH values and NaCl concentrations of the medium. The variation in pH and the salinity of the aqueous solution may cause aggregation of MXene, and as a result, sedimentation occurs, which is not desirable. Organic solvents like DMF and DMSO show better stability of MXene than aqueous dispersion. But the electrical conductivity of the prepared MXene film in these organic solvents is found to be worse than that of the aqueous dispersion [5].

6.3.3 Conductive polymers

Conventionally, polymers are insulators and cannot conduct electricity. However, certain polymers can conduct electricity without using any additives (addition of conductive substances). Such polymers are called intrinsically conductive polymers, or conductive polymers, in simple terms. After the invention of polyacetylene in 1977 by Hideki Shirakawa, Alan J. Heeger, and Alan G. Mac Diarmid, many other conductive polymers have been synthesized that possess conductivity, in a range from 10^{-8} to 10^5 S/cm. These conductive polymers have long π-conjugated chains and hetero atoms in their chemical structure. They are made into good electrical conductors by doping, which can change their π-conjugated chains. This change is possible by adding suitable impurities into the polymer system, which can generate either holes or electrons in sufficient numbers for the conduction of electricity. The holes or electrons can travel across the conjugated chain due to the application of potential difference, and thus conduct electricity. The level of conductivity of such conductive polymers depends on their chemical structure, molecular arrangement, nature of the dopant, level of doping, external stimuli, and so on. Therefore, electrical conductivity of such polymers can be tailored by altering these parameters, and this is the most spectacular advantage of such materials over metallic conductors. Many conductive polymers, including polyacetylene, polyaniline (PANi), polypyrrole (PPy), polythiophenes, polyphenylene, and polyfuran, have been synthesized till date. The chemical structures of some of these conducting polymers are shown in Figure 6.4, and their conductivities are reported in Table 6.1.

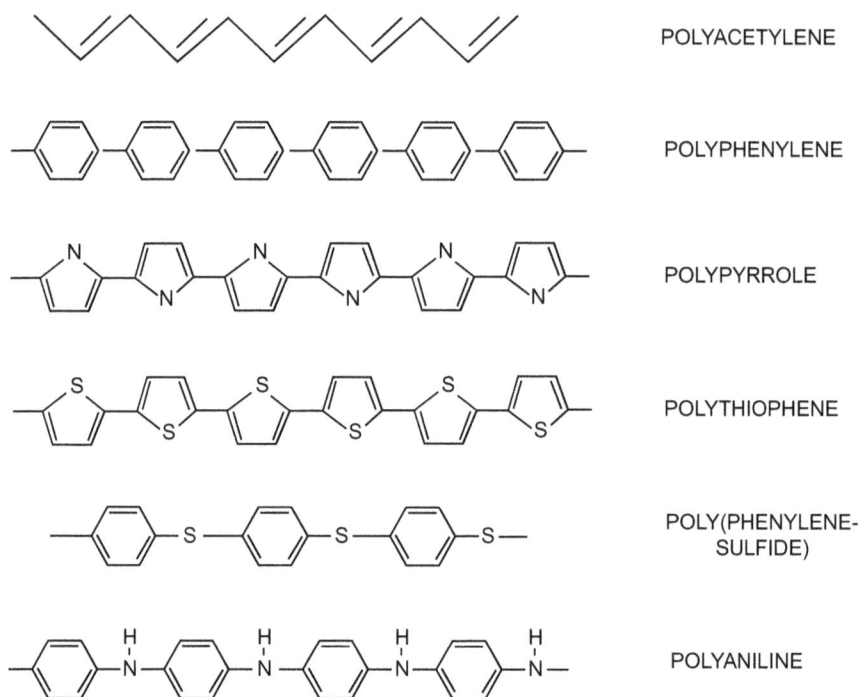

Figure 6.4: Structure of various conducting polymers.

6.3.3.1 Polyacetylene

It is an organic polymer containing the $(CH_2)_n$ repeating unit. In the 1960s, the high electrical conductivity of these polymers sparked interest in using organic compounds in microelectronics.

Natta was the first to conduct substantial research on direct acetylene polymerization. He demonstrated the synthesis of semi-crystalline trans-polyacetylene red powder by bubbling acetylene gas through a Ziegger catalyst solution in a hydrocarbon solvent [6]. Ito et al. [7] made a huge advance in the study of polyacetylene when they observed that exposing the surfaces of Ziegger catalyst solutions to an environment containing acetylene, produced polyacetylene films at the liquid-gas interface [7]. After that, the most extensively used method for making polyacetylene films is based on a catalyst invented by Luttinger [8]. The fundamental advantage of this method is that it eliminates the need for the regular exclusion of moisture that is generally connected with the Ziegler catalyst. Although the Shirakawa and Luttinger approaches appear to be the most popular among researchers studying polyacetylene, there are a variety of alternative catalysts available [9]. It is now widely understood that catalysed acetylene reactions can produce a wide range of products, with many of the principal products being moderately reactive compounds themselves. A catalyst made by reacting phenyl magnesium bromide with ferric chloride in diethyl ether acetylene was introduced as a saturated solution in benzene is an early example. Both the cis and trans variants can be manufactured as silvery, flexible films that can be made free-standing or on various substrates such as glass or metal, and with thicknesses ranging from 10 to 20 cm. The thermodynamically stable form is the trans isomer. At low temperatures, any cis/trans ratio can be maintained. However, following synthesis, the film can be entirely isomerized from cis-$(CH)_x$ to $trans$-$(CH)_x$, by heating it above 15 °C for a few minutes. X-ray studies revealed that the prepared (CH)x films are highly crystalline in nature. $Trans$-$(CH)_x$ is a semiconductor in nature, with a bandgap of around 1.5 eV, owing to bond modification. Polyacetylene is fundamentally distinct from traditional covalent semiconductors because it may be doped after synthesis at ambient temperature, and with a range of dopants. Polyacetylene can be doped by chemical or electrochemical procedures because of its open shape, large specific surface area, and poor interchain binding, which allows dopant ions to diffuse between the polymer chains. Polyacetylene's electrical conductivity may be improved by 12 orders of magnitude by the process of doping. Conductivity, as high as 10 S/cm, has already been reported after suitable doping.

6.3.3.2 Polyphenylenes

Polyphenylenes are generally available in three types: poly(p-phenylene) (PPP), poly (m-phenylene), and poly(p-phenylene sulphide) (PPS).

PPP is particularly an appealing polymer for various reasons. Doping PPP with arsenic fluoride (AsF) can result in high conductivity, of the order of 500 S/cm. Following the formulation of rechargeable batteries based on doped PA, PPP is also explored for the preparation of rechargeable batteries.

Benzene rings connected in the para position make up polyparaphenylene. Carbon-carbon bond lengths inside the rings are around 1.40 Å, while those between the tilted rings are about 1.51 Å, according to crystallographic data on oligomers. Two successive benzene rings are tilted by 23° with respect to each other in the solid state. This tilt angle balances the effects of conjugation and crystal-packing energy, both of which favour a planar structure. PPP has a band gap of 3.4 eV, which is roughly twice that of *trans*-PA. The oxidative coupling reaction of benzene is perhaps the most successful and cost-effective approach for preparing PPP. PPP has also been synthesized using a variety of electrolytic techniques. 0.1% polymer yield was obtained by the electrolysis of phenyl magnesium bromide in ether. Doping of PPP is mainly achieved by employing vacuum exposure of PPP powder or pellets to arsenic pentafluoride (AsF_5) gas at 450 Torr pressure at room temperature. PPP can be doped electrochemically in the same way as polyacetylene is doped electrochemically.

Poly(*m*-phenylene) is a soluble and fusible polymer with little possibility of forming conducting complexes. However, doping with arsenic pentafluoride (AsF_5) results in developing a conducting complex with a conductivity of the order of 10^3 S/cm, owing to the dopant-induced production of carbon-carbon bonds, which results in a polymer with an expanded orbital system.

PPS was the first non-rigid, non fully carbon-backbone polymer and a highly conductive polymer. The crystal structure of PPS has been described by several authors. PPS structure is affected by its preparation methodologies. According to X-ray diffraction data, neighbouring phenyl rings are nearly perpendicular. Despite this structure, the highest occupied band has a large bandwidth of 1.2 eV. Although several procedures have been used to synthesize PPS, Edmonds-Hill process and the Lenz process are found to be most useful [10].

Table 6.1: Conductive polymers with their electrical conductivity at various doping.

Polymer	Dopant	Conductivity (S/cm)
Polyacetylene	I_2, Br_2, Li, Na, AsF_5	10^4
Polypyrrole	BF_4^-, ClO_4^-	500–7.5×10^3
Polythiophene	BF_4^-, $FeCl_4^-$, ClO_4^-, tosylate	10^3
Poly(3-alkylthiophene)	BF_4^-, $FeCl_4^-$, ClO_4^-	10^3–10^4
Polyphenylene sulphide	AsF_5	500
Polyphenylene vinylene	AsF_5	10^4
Polythienylene vinylene	AsF_5	2.7×10^3

6.3.3.3 Polyaniline (PANi)

PANi is a popular conducting polymer among researchers, owing to its tailorable electrical resistivity, sufficient environmental stability, and excellent redox behavior. PANi has been identified as a unique and fascinating member of the class of conducting polymers. It has a nitrogen heteroatom integrated between the constituent phenyl (C_6H_6) rings in the backbone, unlike many other class members, such as polyacetylene, PPy, and polythiophene, whose electrical characteristics are well characterized purely based on their conjugated carbon backbones. The chemical flexibility of the nitrogen heteroatom gives access to several different insulating ground states, each of which is differentiated by its oxidation state. PANi comes in a variety of forms, each with a different level of oxidation. PANi was originally referred to as "aniline black" in 1835 [11]. Any product formed by the oxidation of aniline is referred to as aniline. A few years later, a tentative investigation of the compounds was conducted to synthesize this aromatic amine by chemical oxidation. After that, the ultimate product of anodic oxidation of aniline in aqueous sulphuric acid solution was obtained as a dark brown precipitate at the platinum electrode. Various studies confirmed these findings, and similar observations were found with the oxidation of aniline in aqueous hydrochloric acid solutions. A linear octameric structure of the product was generated by chemical oxidation of aniline, with the quinoneimine type in the para-position. The protonation of nitrogen atoms present in the imine groups in the emeraldine base results in a highly conducting state. The quantity of electrons in the polymer chain structure is unaffected by proton doping, and the pH of the solution determines its conductivity. The environmental stability of PANi is excellent. PANi has fascinating chemistry and physics due to its chemically flexible -NH-group in its backbone. In an aqueous medium, PANi is produced using chemical and electrochemical methods. Quinoid, benzoid, and diimine create an insulator that is doped to a metallic regime (compressed pellet) by dilute aqueous protonic acid, yielding an iminium salt. During the doping procedure, the polymer is not oxidized. In conducting polymers, this is a new form of p-doping phenomenon. In the presence of air or water, both forms of PANi are stable. Treatment with an aqueous alkali reverses the doping process. PANi was synthesized electrochemically in aqueous H_2SO_4. An electrochemical oxidation of aniline in an acidic media was used to make PANi. The resulting pure polymer was doped. To create thin films on plates utilizing certain substrates, PANi was polymerized using an electrochemical cell. By electrochemical and chemical oxidation, conducting PANi was produced from an aqueous solution of oxalic acid (1.0 M) and aniline (0.1 M). On an aluminium surface in H_2SO_4 and electrolytes containing catalytic quantities of H_2IrC_{16} and its salts, production of many types of electrochemical PANi is feasible. Electrochemical polymerization was used to prepare electrically conducting PANi, doped with heteropolyanions (HPA). Reversible redox systems were discovered using cyclic voltammetry in both the polymer and the immobilized HPA. There are various methods of synthesis of PANi in literature and the nature of the final PANi depends on various factors. These are: (1) the medium's nature, (2) the oxidant's concentration, (3) the reaction's duration, and (4) the medium's temperature. In a recent work, it is revealed that the emeraldine base form of PANi is doped by protonic acids to the metallic conducting regime through a process that does not include the polymer being oxidized or reduced, thereby adding a novel idea of doping. In summary, proton-induced conductivity in PANi is a fascinating scientific phenomenon that needs to be explained. The conductivity of PANi that is produced chemically and electrochemically is determined by a variety of factors. Temperature, protonation/pH, humidity, oxidation state, and counterion, for example, all have a significant impact on PANi conductivity. Furthermore, the temperature of the synthesis pressure and the time of compression of PANi particles have a significant impact on PANi conductivity. PANi's capacity to store a significant amount of charge via the redox process has prompted proposals for both non-aqueous and aqueous batteries. PANi films are colourful and highly conductive in their oxidized form. However, they are optically transparent and have minimal conductivity in their reduced state. In fact, doping of the films is linked to coloration and conductivity. PANi has been researched for two applications in photoelectrochemical cells: protection against photo corrosion of inorganic semiconductors and photoresponse of PANi film junction. Catalytic activity has been demonstrated for PANi and modified PANi. The creation of PANi as an indicator is based on repeated colour changes of PANi films on electrodes and chromatic reactions of polyaniline solutions in various pH ranges.

6.3.3.4 Polythiophene (PTh)

Many recent investigations have focused on polythiophenes and, in particular, its derivatives. Chemical and electrochemical processes have typically been employed to manufacture polythiophene and derivatives. Sulphuric acid, iron(III) chloride, and Ziegler catalyst have all been used to polymerize thiophene, furan, and selenphene. Experiments demonstrate that polythiophene is made up of alternating thiophene and tetrahydrothiophene units at an acidic pH. Several studies on the electrochemical synthesis of polythiophene and its derivatives have been reported.

Polythiophene and its derivatives are insoluble and infusible, with densities ranging from 1.4 to 1.6 g/cm^3, as determined by flotation methods. When electron-donating groups are substituted on positions 3 and/or 4, the oxidation potential values of substituted polythiophenes fall. Polythiophenes and their derivatives are widely employed in display technology. Polyhetrocycles are a good option for secondary battery electrodes because of their reversible doping-undoping processes. They have also been employed in photovoltaic applications, such as converting solar energy into electricity using undoped semiconducting polymers as the photoactive material, and grafting thin conducting polythiophene coatings on the surfaces to guard against photo corrosion or photoelectrochemical cells. Polymers are also employed to protect small band gap semiconductors from photodegradation, and physicochemical features of metallic clusters or aggregates have received a lot of attention due to their potential applications in catalysis. A transition metal that is catalysed, dehalogenated, and polymerized is used to make polythiophene [poly(2,5-thienylene)] (PT) from di-halogenated thiophene. However, this technique produces a polymer with a relatively low molecular weight –a chain length of only 6–15 monomer units.

6.3.3.5 Polypyrrole (PPy)

In recent years, PPy has received remarkable attention from researchers due to its high electrical conductivity, ease of preparation, environmental stability, and good mechanical properties. Owing to these spectacular properties, PPy has been found suitable for various novel applications such as electrochromic devices, capacitors, batteries, membrane separators, and sensors. Several aspects contribute to the appeal of PPy systems. The chemical and thermal stability of these polymers, in comparison to $(SN)_x$ and $(CH)_x$, was obviously an essential consideration, but the ease of preparation was also enticing. Its precursor is the pyrrole monomer that can be easily polymerized to synthesize PPy by oxidative process. This chemistry is extremely simple, involving suitable oxidizing agents. For many years, the resulting conducting powder was known as pyrrole black.

Polymerization can occur, both electrochemically and chemically. High-quality PPy films were prepared using a variation of the electrochemical process. PPy is one of the few conducting polymers that can be synthesized in aqueous media. Chemical polymerization is a straightforward and quick procedure that does not necessitate the use of any specific equipment. PPy also can be synthesized in the vapour phase where the pyrrole vapour and the oxidant vapour can be mixed in the reaction chamber to prepare PPy in powder form. Chemical polymerization of iron(III) chloride in liquid media, on the other hand, has been discovered to be the best chemical oxidant, while water is the optimum solvent where both pyrrole and oxidant can be dissolved. The electrochemical method for producing electroactive/conductive films is extremely adaptable, and it allows for easy modification of the film's properties by changing the electrolysis conditions (e.g. electrode potential, current density, solvent, and electrolyte) in a controlled manner. Furthermore, electrosynthesis enables precise control over the film thickness and morphology. Pyrrole is also easily electropolymerized in both aqueous and non-aqueous solvents due to its excellent solubility in a wide range of solvents. The polymerization reaction is extremely complex, and the mechanism of electropolymerization is still a mystery. Due to the polymer's persistent structural disorder, the mechanism of electrical conduction in PPy has yet to be conclusively established. Charge transport via polymer chains as well as carrier hopping are the most widely accepted theories of conductivity in these systems (holes, bipolarons, etc.) The number of carriers (e-or holes) and charge carrier mobility combine to determine its electrical conductivity. More crystalline, better oriented, and defect-free materials will have higher mobilities. Preparation variables such as the nature and concentration of the electrolyte or counterion, doping level, synthesis temperature, and solvent have a significant influence on the electrical conductivity of PPy films. By selecting the appropriate electrochemical polymerization conditions, highly conductive PPy films, with electrical conductivity in the range of 500 S/cm can be prepared [12]. When compared to analogous films synthesized by the constant potential mode, applying a pulsed potential technique to form PPy films increases electrical conductivity, molecular anisotropy, and surface smoothness. Synthesis temperature has a significant influence on the electrical conductivity of the PPy films. The PPy synthesized at a lower temperature has a longer conjugation length, better molecular order, fewer structural defects, and higher conductivity. Doping enhances the mechanical strength, stability, and conductivity of the PPy films. But, undoping of the polymer due to chemical reactions causes loss of electrical conductivity and mechanical strength. In the presence of an alkaline atmosphere such as NaOH solution, undoping occurs and significant loss in conductivity is observed. The surface morphology of the PPy varies after base or acid treatment, according to scanning electron microscopy (SEM) studies [13]. The redox properties and the electroactive nature of PPy make it suitable for various promising applications, including sensors, separation devices, rechargeable batteries, and capacitors. Unlike polythiophenes, which are stable in the air even when dedoped, PPy is less stable in its reduced or undoped form, and in the presence of a suitable oxidant, auto-oxidation occurs very quickly and permanently, resulting in a black highly conductive film. The undoped PPy interacts readily with O_2, resulting in polymer oxidation and a reduction in mechanical strength. The reduced state of PPy films, in contrast to the oxidized state, is unstable to oxygen and water. It is susceptible to moisture because it causes the counterion to seep out, resulting in a drop in conductivity. Using proper hydrophobic or polymeric counterions such as camphor sulphonic acid and poly(styrene-sulphonic acid) as dopant, this can be prevented [14]. The molecular structure of conducting polymers is known to alter when they are exposed to high temperatures. Some structural alterations occur between the charged polymer backbone and the counterion at high temperatures. It has been found that the conductivity of PPy doped with aryl sulphonates is extremely stable in inert environments but is slightly less stable in dry or humid environments.

Advantages of using PPy:
- It is widely available.
- It is easy to synthesize in the laboratory.
- It has a good affinity to cellulosic substances.
- Its adsorption capability is good.
- It has good environmental stability.

6.3.3.6 Poly(phenylene vinylene)

It is a composite polymer of polyphenylene and polyacetylene, doped with AsF_5. The maximum conductivity achieved with poly (phenylene vinylene) is significantly lower (3 S/cm), as compared to polyphenylene or polyacetylene. This may be due to a low degree of polymerization. The calculated band gap of the polymer is around 2.8 eV [15].

6.3.3.7 Polydiacetylenes

Polydiacetylens (PDAs) are usually made by combining single crystal, substituted diac-etylenes – RC'C-C'CR – in a solid-state process. This is a single crystal polymer possess-ing extremely one-dimensional electrical characteristics that us generated under favourable packing circumstances and dominated by R groups. Bloor examined the characteristics of PDAs in 1985. Since then, there has been a lot of interest in PDAs, and a lot of research has been carried out to develop new characteristics and insights. Various researchers have developed different PDA structures, namely, bulk single crystal, monolayer, and multilayer films in both Langmuir and self-assembled forms, nanocomposites, and so on. There has been a great research interest on this polymer regarding its nonlinear optical susceptibility, faster optical response, structural anisot-ropy attributed to high electrical conductivity and electronic properties, chromatic transitions, and so on [16].

6.4 Mechanism of electricity conduction in conductive polymers

Generally, polymers that are used for manufacturing plastics and filaments do not have free electrons in their conduction band. Therefore, their valence band electrons creates stable sigma bonds between the adjacent atoms. The electrons in sigma bonds are not free to move and hence do not conduct electricity. But, the conducting poly-mers have a common significant overlap of delocalized π-electrons along their molec-ular chain. These π-electrons can travel across the polymer chains by doping and application of the potential difference. When the conductivity of the conducting poly-mer is achieved only by applying an external electrical field, then the conducting polymer is called intrinsically conductive polymer(ICP). The ICPs can be converted into a conductor by doping with either an electron donor or acceptor [17].

The doping of ICPs can be done during their synthesis process either by oxidation or by reduction in the presence of a suitable oxidant or reducing agent, respectively. Typical oxidants are ferric chloride, iodine vapour, arsenic pentachloride, hydrogen peroxide, and so on, and a typical reducing agent is sodium naphthalide. The suitabil-ity of the dopant is decided based on the ability to polymerize the precursor and en-hance the stability.

6.5 Electroconductive textiles

6.5.1 Metal-based electroconductive textiles

Electroconductive textiles are prepared from metal fibers by blending them with common textile fibers during spinning. Such electroconductive yarns are used to manufacture woven fabrics that are suitable for electromagnetic interference (EMI) shielding and electrostatic charge dissipation. Owing to flexibility and comfort, these woven electroconductive textiles are increasingly used for these potential applications. K. B. Cheng, with other authors, investigated the electromagnetic shielding effectiveness of the twill copper woven fabrics [18]. They conclude that in order to have a better shield, the fabric porosity should be as low as possible and more conductive material should be available. Demand for these products has increased tremendously. However, the preparation process of these metallic yarns is difficult and even preparation of the fabrics either by weaving or by the knitting process is difficult with the conventional setup. Even the prepared fabrics lack textiles properties like flexibility, drape, and comfort [19, 20]. These limitations of metallic electroconductive textile can be successfully overcome by coating conducting polymers on textile fibres, yarns, and fabrics. Among the conducting polymers, PPy has been mostly used due to its conductivity, high environmental stability, and low toxicity [21, 22].

6.5.2 Graphene-based electroconductive textiles

As an allotrope of carbon, graphene is one of the popular non-metallic conductors explored by many researchers, nowadays. It has a two-dimensional hexagonal lattice structure in its atomic scale. It has flat polycyclic aromatic hydrocarbons, with infinitely large aromatic molecules. Graphene is synthesized from graphite. First, graphite is converted to graphene oxide (GO) by using one of the three available techniques, namely, Hummer's method, modified Hummer's method, or Improved Hummer's method. The GO particles are dispersed in aqueous or other media with the help of an ultrasonicator and then a conventional "Dip and Dry" techniques is followed for coating GO particles onto textile substrates. Multiple "Dip and Dry" processes can be followed for obtaining a higher GO add-on onto textile substrates. After successful add-on of GO onto textile substrates with the required quantity of loading, the GO-loaded textiles are treated with an aqueous solution of $Na_2S_2O_4$ at 95 °C for 30 min for converting the immobilized GO into graphene [23].

Graphene has high electrical and thermal conductivity, good thermal and chemical stability, and good affinity and flexibility over the textile surface [24, 25]. Its major advantages are ease of preparation, low cost, high specific surface area, high functionality, and flexible two-dimensional geometry, in comparison with other conductive materials, including carbon black, CNTs, and conductive polymers [25, 26]. Textile substrates are strong, lightweight, flexible, and stretchable. The successful and durable coating of graphene onto textile substrates by a suitable methodology will produce electroconductive textiles, which possess synergistic properties of both the substrates: textiles and graphene [24, 27]. These electroconductive textiles may have many potential applications such as energy storing devices, drug delivery systems, EMI shielding, flexible heating pad, heating garment, cooling garment, sensors, actuators, anti-static clothing, and antibacterial clothing [28].

There are plenty of studies reported on the preparation of graphene-based electroconductive textiles by using the conventional dip-n-dry process. All these researches address the difficulties in the preparation process, suitable substrates, process parameters, and their optimization in order to obtain the best electroactive properties of the coated textiles.

It has been reported that graphene coating over cotton fabric has significantly improved its mechanical strength and electrical conductivity [27]. The graphene-coated cotton fabric exhibits a linear response due to the application of strain on it [27]. Though other reducing agents are tried for the in situ reduction of GO over textile surface, the best conductivity and mechanical performance are obtained using $Na_2S_2O_4$. The treatment time of 30 min at 95 °C is found sufficient for a complete reduction of the GO with this reducing agent. The electrical conductivity of cotton textiles is enhanced by approximately three times with the increase in the number of nip-n-dip coating cycles to 20 [26]. The concentration of GO in the dispersion solution, the number of dipping cycles, treatment time, and temperature are found to be the important process parameters that influence graphene add-on onto textile substrates. Similarly, for the reduction process, the choice of the reducing agent, the concentration of the reducing solution, temperature, and time of the treatment are important process parameters for converting GO into graphene.

6.5.3 MXene-based electroconductive textiles

For a successful and durable coating of MXene flakes on the textile surface, good adhesion is required. It is observed that fibres coated with larger flakes show better electrical conductivity but poor flexibility and adhesion. So, MXene is functionalized with suitable binders such as poly(3,4-ethylenedioxythiophene) polystyrenesulphonate (PEDOT:PSS) to enhance the affinity with the textile surface. Another procedure is plasma treatment of the textile substrate, creating oxygen functionalized groups for enhancing binding power. The plasma treatment can make the textile surface hydrophilic in nature, thereby attracting the MXene flakes. The hydrophobic polyester surface is functionalized with 3-aminopropyl triethoxysilane (APTES) for better adhesion [2]. The APTES-treated polyester fibres produce sufficient electrostatic attraction for MXene flakes, driving the self-assembly of the flakes on the polyester surface. Again, the vacuum-assisted filtration (VAF) method can be used to assemble MXene over the polymer matrices to prepare MXene-coated polymer composites. Tempo-oxidized cellulose nanofibers (TOCNF) are coated with MXene to prepare TOCNF/MXene composite papers by the VAF process. The active sites present in TOCNF help to form binding sites with MXene particles. These cellulose/MXene composites possess excellent mechanical properties and electromagnetic shielding efficiency [2]. In another study, cellulose/MXene films are prepared by mixing the cellulose material and MXenes in a polar solvent, followed by vigorous stirring or sonication, and further followed by vacuum filtration.

6.5.4 CNT-based electroconductive textiles

CNTs are one of the major components in the field of nanoscience and nanotechnology. The length to diameter ratio of CNTs is in the range of 1,000,000. Owing to their exceptional functional, mechanical, and physical properties, they have been used in several applications recently. In textile science, various textile materials are modified with the loading of CNTs over the textile surface to bring about many functionalities, including electrical conductivity. CNT-based textiles are found to be suitable for many applications, including microwave shielding, protective clothing, automotive textiles, heating equipment, building covering, solar cell, thermoelectric generators, geotextiles, biomedical applications, and antimicrobial textiles [29].

MWCNTs-reinforced regenerated cellulose fibers are prepared by solution spinning method [30]. High electrical conductivity, as high as 2.7 S/cm, was achieved with a MWNT loading of 30 wt%. MWCNT is used as reinforcement during the carbonization of the polyacrylonitrile (PAN) fibre, and a composite fibre structure is prepared that has an excellent electrical conductivity of 35 S/cm with only a 10% MWCNT loading [31]. Cellulosic core-sheath type of fibre structure is produced by electrospinning using MWCNT as the core [32]. Electrically conductive elastic film is prepared by dispersing SWNTs in a vinylidene fluoride-hexafluoropropylene (PVDF-HFP) copolymer matrix, followed by a coating with dimethylsiloxane–based rubber [33]. Film conductivity of 57 S/cm is achieved as. SWCNT and polyvivyle alcohol (PVA) is wet spun to produce composite fibre with 60% loading of SWCNT. The prepared fibre exhibits exceptional electrical and mechanical properties [34]. MWCNT-coated silk fabric is prepared by a surface micro dissolution strategy where silk fabric is micro-dissolved, followed by treatment in a MWCNT suspension solution for the coating. A uniform thin film of MWCNT is created over the silk fabric, resulting in good electrical conductivity, in the range of 468 Ω/cm [35]. Owing to the good environmental, thermal stability, and excellent electrical properties of the coated silk fabrics, they are suitable for applications in flexible electronics and heating materials. MWCNT is deposited on the polyester fabric from its colloidal suspension by a simple tape casting method, followed by air drying [36]. The fabric shows surface resistivity, as low as 15 $\Omega/\square$. This polyester fabric has strong adhesion with MWCNT and demonstrates stable conductivity against bending cycles and durability. In another study, amine-functionalized MWCNT (NH$_2$-MWCNT) is coated over polyester fabric to obtain microwave shielding efficiency [37]. The fabric shows reflection loss of ~ −18.2 dB at a frequency level of 11 GHz. There is a strong interaction between NH$_2$-MWCNT and PET fibres due to the formation of intermolecular hydrogen bonds. A high-performance stretchable strain sensor is developed using nylon fabric as the substrate for coating SWNT [38]. The fabric exhibits a high sensitivity of 72 gauge factor at 100% strain, with a swift response and excellent durability.

6.5.5 Conductive polymer-based electroconductive textiles

Electroconductive textiles are prepared by coating various textile materials like fibres, yarns, and fabrics with various conductive polymers such as PANi, PPy, and polythiophenes. The most common and familiar method of coating is in situ chemical polymerization, where the textile substrate is first soaked with a monomer, followed by treatment in an oxidant atmosphere or vice versa. The polymerization reaction can occur in both liquid and vapour phases. The methodology requires adequate adsorption of the monomer or oxidant by the textile fibres so that sufficient polymer can be loaded in situ on the fibre surface. Solid–liquid and solid–vapor phase adsorption occur in en situ chemical and in situ vapour phase polymerization, respectively. Different textiles fibres, including cotton, wool, silk, polyester, nylon, and aramid are coated with conductive polymers by these methods. Almost all these fibres respond more or less similarly in the coating process. But it is suspected that the morphology of PPy and PANi film of hydrophobic and hydrophilic fibre surfaces are different. It is reported that the conductivity of the film deposited on hydrophobic surface is about 10^4 times better than that of film deposited on the hydrophilic surface, since the films deposited on the hydrophobic surface are more uniform, continuous and granular in nature. Whereas, the polymer molecules on the hydrophilic surface are deposited as particulates, such as spherical, in the case of PPy, and rod-like in the case of PANi. It has been noticed that the deposition of PPy and PANi on a hydrophobic glass surface is more rapid than that of the hydrophobic glass substrate. On a hydrophobic surface, the morphology of PPy and PANi is found to be more extended and continuous in nature, giving rise to better electrical conductivity. It is later revealed that the nature of functional groups that are present on the surface of textile fibres plays a significant role in the add-on of the polymers on the fibre surface. The functional groups like hydroxyl, amide, and carboxyl act as anchoring sites for the adsorption of monomer and polymer molecules on the textile surfaces for the formation of self-assembled layers. Owing to these active anchoring groups and their densities on the fibre surface, various levels of polymer add-on and, thereby, electrical conductivity can be achieved.

6.6 Methods of preparation of MXene-based conductive textiles

Coating is the most widely explored techniques for the preparation of electroconductive textiles using MXene due to ease of operation, experimental setup, process control, and lower cost. Various cotton fibres and yarns, and fabrics such as cotton and nylon polyesters are coated with MXene [2]. In this method, a homogeneous MXene dispersion solution was first prepared with a suitable solvent, and then the textile substrate was soaked, or the solvent was sprayed, or drop-casted onto textile substrate, followed by solvent evaporation. By this approach, cellulose yarns are coated with MXene with a 77% add-on and the coated yarns are flexible enough to prepare knitted fabrics from them with different stitch designs without losing electrical conductivity [39]. In another study, MXene is drop cased over the individual nylon fibre surface to prepare electroconductive nylon for the fabrication of a super-capacitor [40]. In all these cases, it has been observed that the fibres do not lose their textile properties after coating.

6.6.1 Electrospinning methods

Various solvent systems of MXene have been developed with homogeneous solutions to electrospun MXene-based polymer composite fibers. Poly(vinyl alcohol) (PVA), polycaprolactone (PCL), polyacrylonitrile (PAN), and so on are the solvent systems successfully used for electrospinning of MXene composite filaments. Recently, MXene-infiltrated nano yarns from nylon and polyurethane are prepared by the addition of up to 90% MXene content by weight. The high electrical conductivity of MXene is utilised to prepare highly conductive nanofibers by the electrospinning technology, as shown in Figure 6.5 [41].

6.6.2 Wet spinning

In this approach, a polymer solution is injected through a spinneret into a coagulation bath where polymer mass coagulates and solidifies to form filaments. The formed filaments are dried and collected on a winder. As MXene can be dispersed homogeneously in many polar solvents, it can be doped with many polymer solutions for wet spinning to prepare hybrid yarns with high electrical conductivity. Highly conductive reduced graphene oxide (rGO) and MXene are deposited on the fibre surface by a scalable wet-spinning method to prepare electroconductive composite yarns, with electrical conductivity in the range of 743.1 S cm–1 [42].

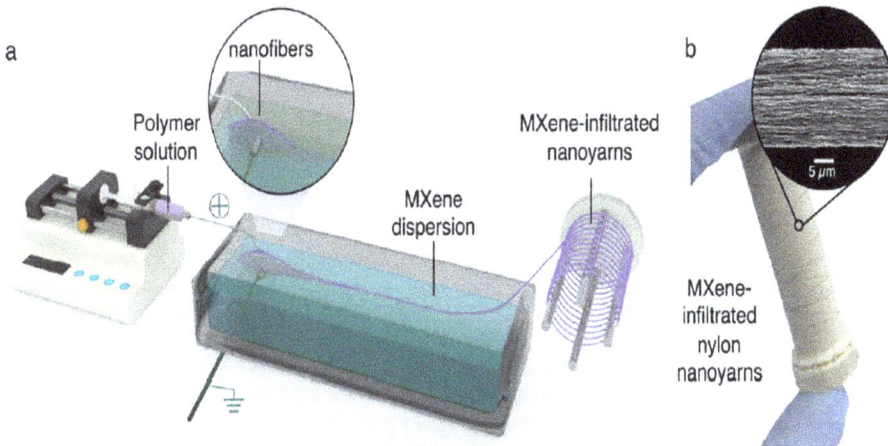

Figure 6.5: Electrospinning method of preparation of MXene-based nanoyarns. (a) Schematic diagram of the electrospinning set-up and (b) microscopic images of MXene/nylon nanoyarns (~50 cm, produced using ~ 220 nm MXene flakes and a 10 mg/mL MXene dispersion) [41] (reprinted with permission from John Wiley and Sons; Levitt et al. [41]. Copyright 2020).

6.7 Applications of MXene-based conductive textiles

6.7.1 Heat generation

MXene and its composites possess high electrical conductivity, which makes them suitable for electrical heating. Unlike conductive polymers and graphene, they do not require any chemical and thermal reduction or oxidation processes. Moreover, they have a good affinity to textile substrates that promote durable applications on textile substrates, leading to the development of intelligent textiles. The presence of sufficient functional groups such as -OH, O, and F groups in MXenes are liable to bond formation with textiles substrates.

The achieved electrical conductivity of MXene is in the range of 1,000–6,500 S/cm, depending on the size and defects of MXene flakes, and the gap between the MXene layers on the textile composites. MXene-loaded nonwoven textiles are found suitable for heat generation applications. Upon heating, the molecular water extracted from these coated nonwovens resulted in an electrical response that acts as a temperature alarm, addressing real-time temperature monitoring for thermotherapy [43]. Thermotherapy effect is found suitable for wound healing. The coated composite fabrics act as smart materials that are ideal for making wearable respiration monitoring systems and humidity sensors, owing to water-induced swelling or contraction. Such smart textiles are the future of the recent smart textiles for healthcare monitoring and biomedical therapy. A flexible textile heater is prepared by coating MXene over PET fabrics. The electrical conductivity of the PET fabrics is found to be directly proportional to the coating time, and the surface temperature of the coated fabric increases at the same time [44]. MXene-based electroconducting thread are inserted in cotton gloves for heat generation and therapeutic use. The gloves show a steady increase in the surface temperature, up to 53.5 °C, due to increase in the voltage supply of 3.3 V/cm. MXene-coated knitted fabrics showed excellent heating performance as well as wearability and comfort, suggesting MXene is an ideal material for thermotherapy. Poly(ethylene terephthalate) textiles are coated with PPy and MXene by in situ polymerization, followed by silicon coating, to achieve high electrical conductivity with sufficient durability for a probable application of EMI shielding and Joule's effect of heat generation, as shown in Figure 6.6 [45]. This textile delivers excellent Joule heating in the presence of moderate supply voltage along with excellent water repellency due to silicon coating. The Joule heating performances of the silicon-coated textile are shown in Figure 6.7. The current–voltage (I–V) characteristic curve is nearly linear and reveals the textile's low resistance, enabling moderate-voltage-driven heating, which is required for safe use. Such textiles can generate temperatures as high as 40 °C due to the application of 2 V supply, and the temperature can be further elevated to 57 and 79 °C at 3 and 4 V, respectively.

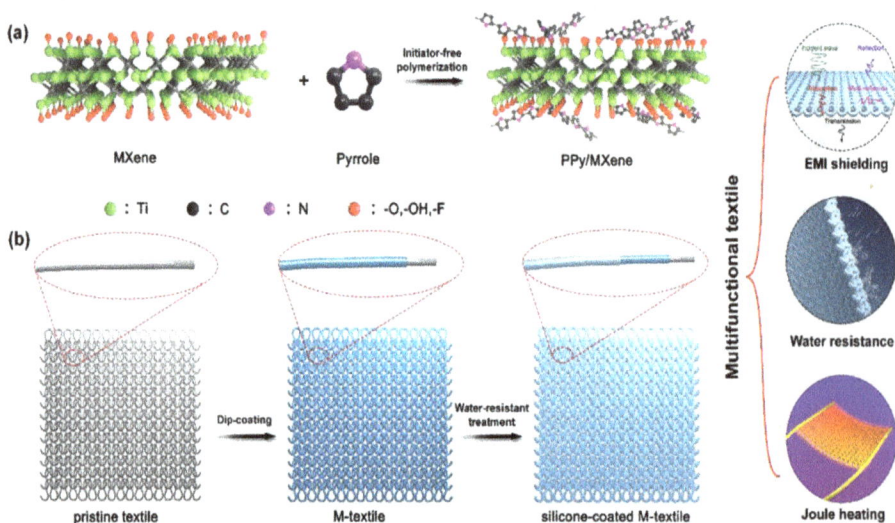

Figure 6.6: Schematic illustration of preparation of PPy/MXene/silicon-coated water-repellent polyester. (a) First step: hybridization of MXene with in situ-polymerized PPy; and (b) second step: preparation of PPy/MXene/silicon-coated polyester textile [45] (reprinted with permission from John Wiley and Sons; Wang et al. [45]. Copyright 2019).

6.7.2 EMI shielding

MXenes could be effective alternative materials for EMI shielding as they have excellent electrical conductivity, are lightweight, and have a large specific surface area.

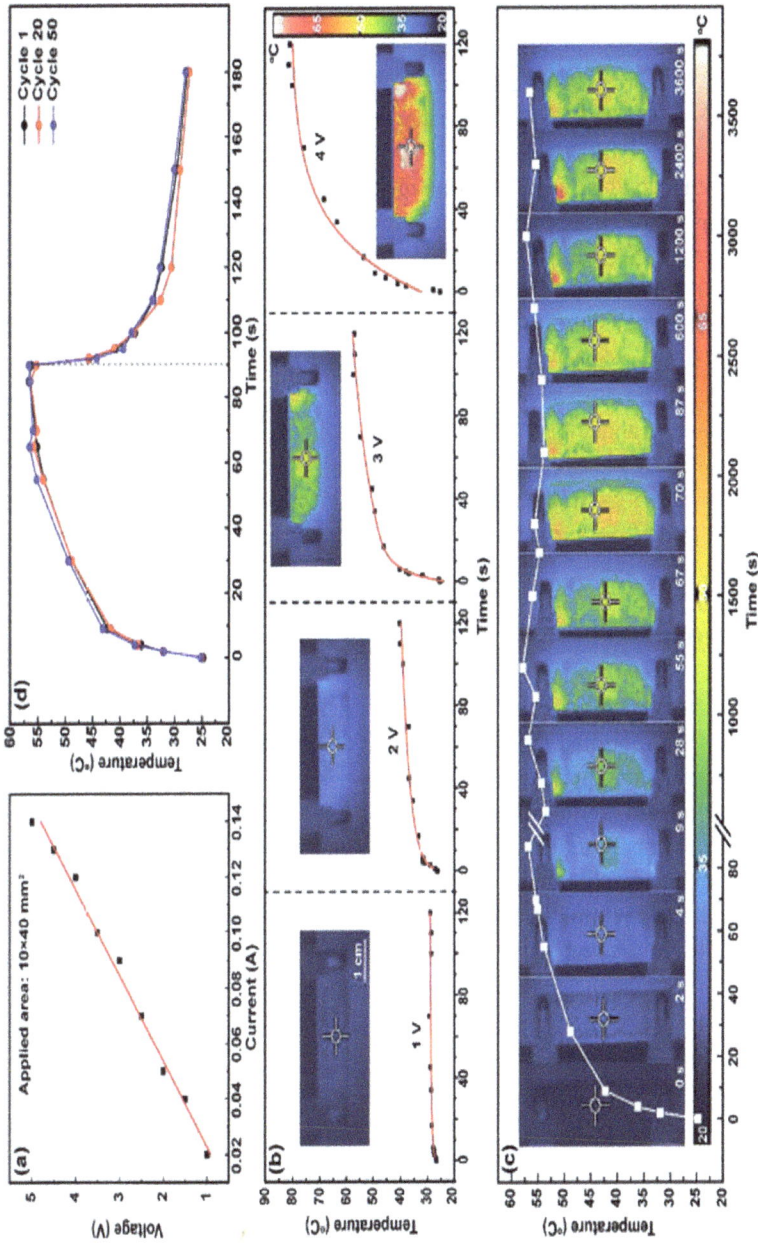

Figure 6.7: Electrical performance characteristics of silicon-coated PPy/MXene textiles: (a) current–voltage (*I*–*V*) characteristic curve; (b) Joule heating performances; (c) time–temperature curves, and (d) temperature-stability characteristics [45] (reprinted with permission from John Wiley and Sons; Wang et al. [45]. Copyright 2019.

Recently, MXene-based 2D films, textile composites, 3D MXene aerogels, and so on proved to be excellent EMI shielding materials. A layer-by-layer structured conductive textile is prepared by spraying MXene and AgNW dispersions alternatively on both sides of the silk fabric (Figure 6.8). As the concentration of AgNW and number of spraying cycles increases, the EMI shielding efficiency of the textile increases linearly. The textile material, with a conductive MXene film of 480 μm thickness shows EMI shielding efficiency as high as ~90 dB at 12.4 GHz frequency [46].

PET fabric is coated with PPy-functionalized MXene by multiple dipping and drying in its dispersion solution. The coated fabric possesses an electrical conductivity of 150 S/m and an EMI SE of 25 dB. With the increase in the number of dipping cycles, a linear increase in SE was observed, reaching up to 66 dB with a conductivity level of 4,000 S/m. When three pieces of such composite textiles (thickness: 0.43 mm, 1,000 S/m) were laminated, the EMI shielding

performance increased to 90 dB [20]. In another study, cotton fabrics are coated with 2D MXene sheets, where a vertically interconnected conductive network of MXene is created on the cotton surface, resulting in a high EMI SE, in the range of 36 dB [47].

6.7.3 Strain sensor

Various wearable strain sensors are prepared for real-time tacking of human activities. These sensors are mounted either on the skin or on some clothing for real-time monitoring. Like conductive polymers and graphene, MXenes are also utilized for the fabrication of such strain sensors for these applications. The electrical resistance of MXene drastically increases under applied strains due to a mechanism of crack propagation, leading to the measurement of applied strain with high sensitivity. Such sensors are beneficial for tracking limb movements, gestures, joint movements, certain physiological signals, and so on. The sensitivity for such strain sensors is defined by the gauge factor. The gauge factor is expressed as follows:

$$((R - R_0)/R_0)/\varepsilon \tag{6.1}$$

where R_0 is initial resistance, R is resistance under strain, and ε is applied strain. A wearable strain sensor is fabricated with ultra-high strain sensitivity, based on the hierarchical "brick-and-mortar" architecture [48]. The strain sensor has high stretchability and is composed of hydrophilic two-dimensional titanium carbide ($Ti_3C_2T_x$) MXene nanosheets and one-dimensional silver nanowires (AgNWs) as the "brick" and poly(dopamine) (PDA)/Ni^{2+} as the "mortar". The structure is bioinspired and prepared through a simple screen-printing method as shown in Figure 6.9 [48]. The sensor has excellent stretchability, ultra-high sensitivity, broad working range, good linearity, durability, and repeatability.

A polyurethane (PU) fiber-based multilayers strain sensor is developed and embedded in a garment for the monitoring of limbs and body postures [49]. The sensory mechanism proposed for this sensor is crack propagation, as mentioned earlier. The gauge

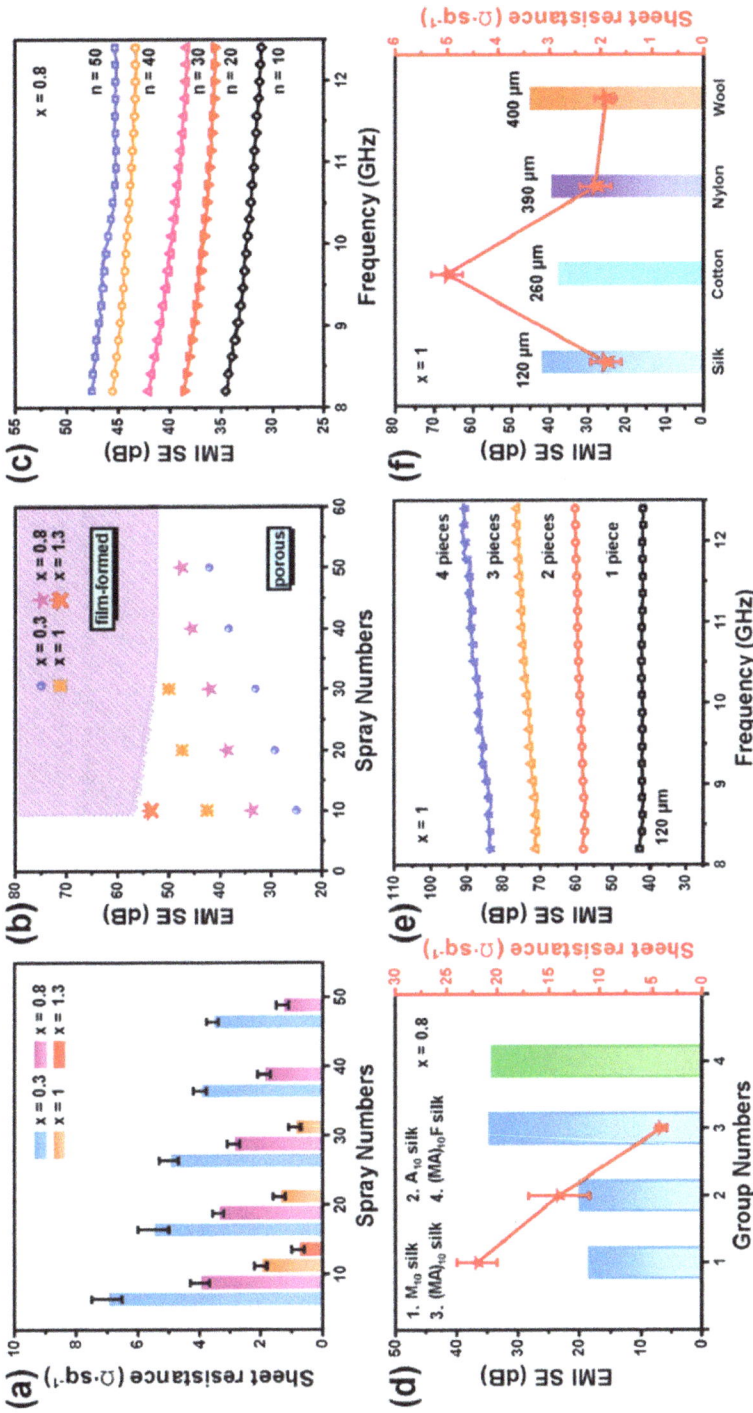

Figure 6.8: Performance of MXene silk textile with different spray cycles and concentrations: (a) electrical resistances, (b) EMI shielding performances, (c) plots of EMI SE versus spray cycles for MXene silk textile, (d) synergistic effect of AgNW and MXene on the EMI SE and the sheet resistances, (e) EMI shielding performances for different coating thicknesses, and (f) EMI shielding performances and sheet resistances for different textile substrates (cotton, silk, wool, and nylon) [46] (reprinted with permission from John Wiley and Sons; Liu et al. [46]. Copyright 2019.

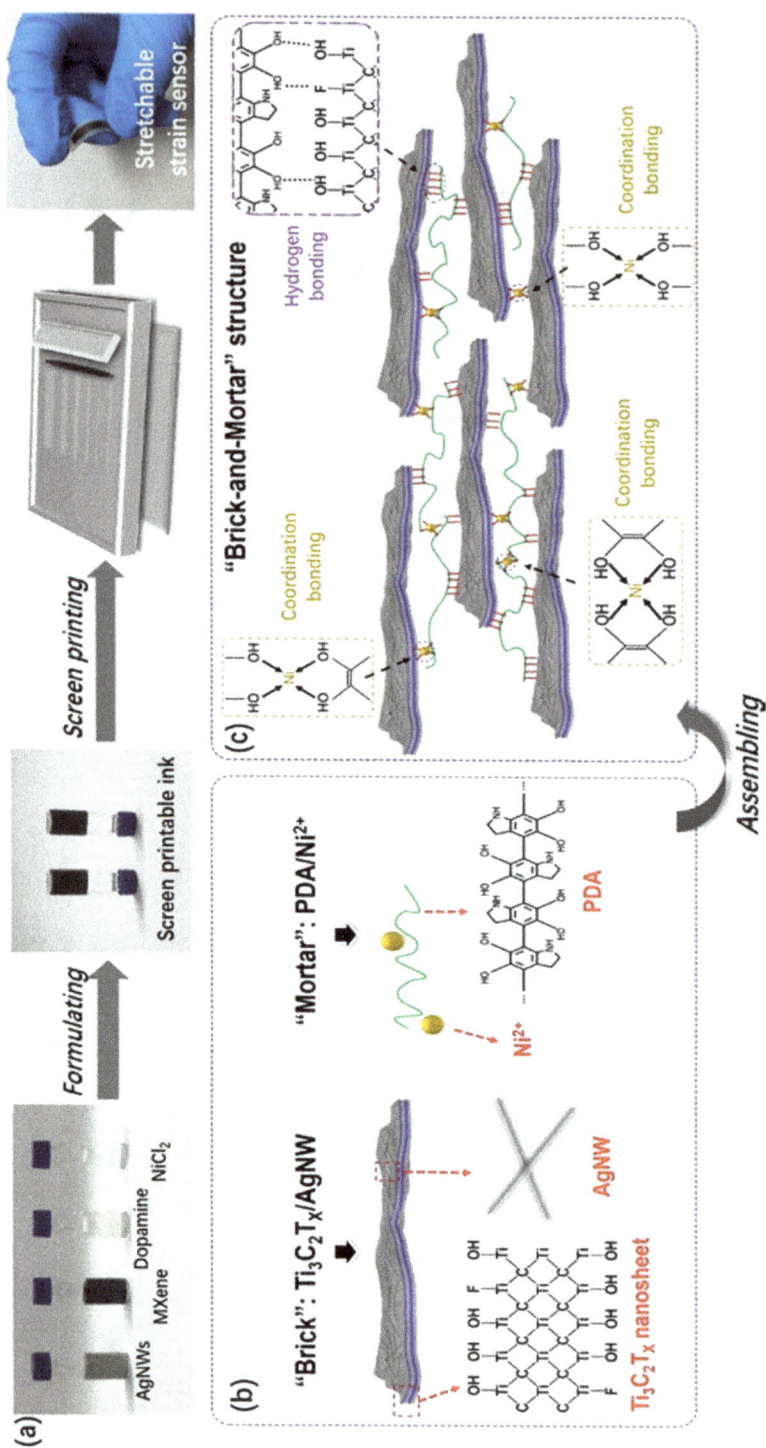

Figure 6.9: (a) The method of preparation of MXene-AgNW-PDA/Ni^{2+} strain sensor by one-step screen printing method. (b) The "brick" and "mortar" material of the sensor; (c) The "brick-and-mortar" architecture and their interfacial interactions, including hydrogen and coordination bonding [48] (reprinted with permission from American Chemical Society; Shi et al. [48]. Copyright 2019).

factor of this sensor is found to be greater than 100 for a large operating strain range, up to 100%. The same sensor can perform with high durability, showing a gauge factor as high as 1,500 at a strain level of 25% and as high as 1.6×10^7 at a strain level of 85–100%.

6.8 Applications of conductive polymer-based electroconductive textiles

The preparation and characterization of electroconductive textiles from conductive polymers have been a research topic by many researchers in recent times due to their several applications. Some of the important applications of these electroconductive textiles will be discussed here.

6.8.1 Heat generation

Electroconductive textiles based on conductive polymers such as PANi, PPy, and PTh are prepared by coating or applying them on the surface of various textile substrates. These electroconductive textiles are polymeric, non-metallic, flexible, durable, strong and light in weight. They possess both the properties of the conductive polymers and textile substrates and many researchers have termed them as electroconductive composites. These materials are found suitable for Joule's effect of heat generation. These heat generators are found superior in terms of temperature homogeneity, low power density, lightweight, portable, flexible, washable, suppleness, and fineness. They can be tailored and therefore can be cut-off, sewed, and pasted with other parts for making large applications in a variety of shapes. PPy is coated on various cellulosic fibre substrates like cotton, viscose, cupro, and lyocell by in situ vapour phase polymerization and tested for heat generation [50]. The PPy-coated cellulosic textiles show good performance in heat generation due to the supply of electricity. Similarly, cotton woven fabrics are coated with PPy by other researchers and they found that the coated cotton fabric is developed with anti-static, anti-microbial properties, including heat generation behavior [51]. The voltage-current (V–I) and temperature-current (T–I) behaviors of the electroconductive cotton fabric are shown in Figure 6.10. It can be observed that the V–I characteristic curve follows a power curve, whereas the T–I characteristic follows an exponential trend.

In a separate study, cotton fabric is coated with PPy and a constant direct voltage of 9 V is then supplied across it for 10 min. As a result, the surface temperature of the fabric rises up to 90 °C and this causes heat generation by the fabric, which is sustained for many heating cycles [26]. Later on, Maity et al. prepared PPy-coated polyester woven and nonwoven fabrics by in situ chemical polymerization in aqueous media and evaluated them for heat generation performance [1]. The time–temperature (t–T) characteristics of the electroconductive needlepunched nonwoven fabrics for various applied

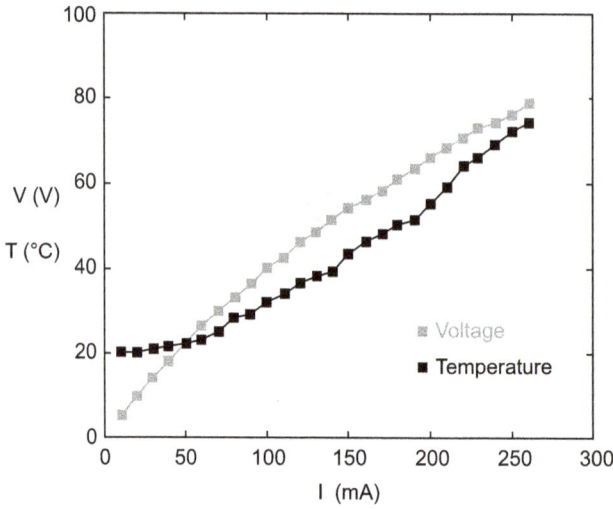

Figure 6.10: The voltage and temperature characteristic of PPy-coated cotton fabric as a function of applied current [51] (adapted from The Society of Fiber Science and Technology, Japan; Seshadri and Bhat [51]. Copyright 2005).

voltage levels are shown in Figure 6.11. It can be seen that the surface temperature of the PPy-coated polyester nonwoven fabric rises very sharply and quickly at the initial level, and then levels off at a constant temperature. The stable temperature achieved depends upon the applied voltage. This particular performance of the fabrics is proposed to be suitable for the fabrication of heating pads for therapeutic use.

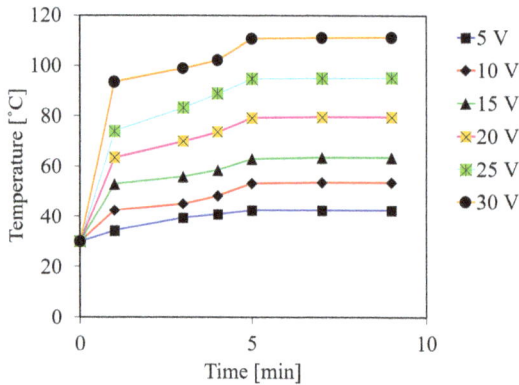

Figure 6.11: Time–temperature characteristics of PPy coated polyester needlepunched nonwoven fabric at various applied voltages [1] (reprinted with permission from Taylor & Francis; Maity et al. [1]. Copyright 2014).

It is reported that the V–T characteristic of the cellulosic fabrics prepared by in situ vapour phase polymerization follows an exponential trend:

$$T = ae^{bV} + C \qquad (6.2)$$

where T is surface temperature, a is coefficient, b is exponent, and c is a constant, whereas V–T characteristics of polyester/PPy fabrics prepared by in situ chemical polymerization follow a different form of exponential trend:

$$T = T_0 + ae^{-bV} \qquad (6.3)$$

where T is the final temperature, T_0 is the initial temperature, and V is the applied potential. The V–I and V–T characteristics of such PPy-coated textiles depend on the FeCl$_3$ concentration used during the in situ polymerization of pyrrole. At a low FeCl$_3$ concentration, the trend of V–I characteristic is almost linear, but at high FeCl$_3$ concentration, the trend would change to an exponential one, as shown in Figure 6.12 [50].

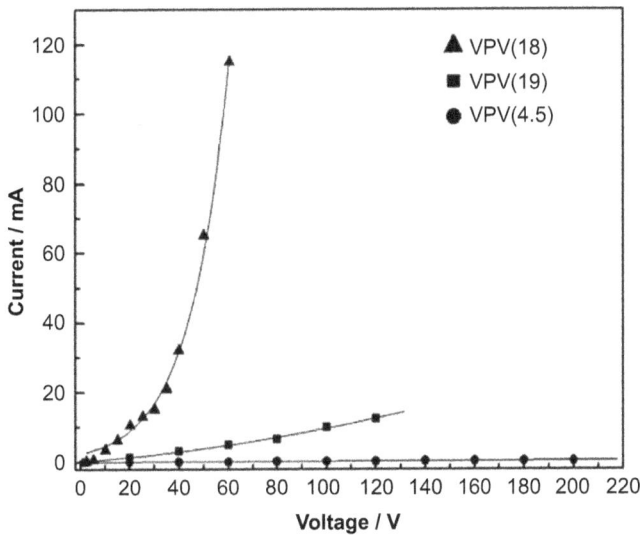

Figure 6.12: Voltage–current behavior of in situ vapour-polymerized PPy-coated cellulosic fabrics with different FeCl$_3$ concentrations [50] (Modified according to Elsevier; Dall'Acqua et al. [50]. Copyright 2006).

It is suggested that PPy-coated pure polyester fabrics are practically useful for the fabrication of a flexible, durable, portable heating pad for medical use [52, 53]. Polyester-/lycra-blended fabrics are coated with PPy and characterized for heat generation. The fabric exhibits temperature generation, as high as 40.55 °C, due to the supply of only 24 V [54]. There is one similarity that has been observed in all such heat generators – there is a quick sharp rise in the temperature in the initial phase, followed by levelling-off to a plateau, similar to PPy-coated polyester fabrics, as mentioned earlier [55]. The electrical resistivity and heat generation efficiency of such electroconductive textiles depends on the types of dopant anions like chloride (Cl⁻) or others [56]. A sequential high temperature and high pressure (HTHP) chemical and electrochemical polymerizations methods are used for imparting electrical conductivity nylon fabrics by PPy coating [57]. The coated nylon fabrics show electrical resistance of 5 $\Omega/\square$ and are liable to heat up quickly, up to 55 °C within 2 min, due to the application of only 3.6 V DC supply. This heating performance is found to be stable even up to 10 repeat cycles of tests. Similarly, E-glass fabrics are coated with PPy and are found effective for stable heat generation behavior. During a constant voltage supply to these fabrics, the surface temperature is found to be increasing, whereas power consumption is found to be decreasing [58]. PPy-coated silk fabrics are also tested for heat generation performance and found competent enough [59]. The PPy-coated polyester woven, spunlaced and needlepunched nonwoven fabrics are prepared by in situ chemical polymerization and tested for heat generation. Exponential rise of surface temperature has been observed for all these three electroconductive textiles on an application of voltage, and the heating behaviour is reported to be a function of the time duration of the applied voltage [1].

6.8.2 Sensory applications

Conductive polymers have drawn considerable attention from various researchers due to their spectacular properties like tailorable electronic properties, low-cost, environmental stability in the doped stage, and ease of their synthesis. The electronic properties of the conductive polymers depend on the surrounding atmosphere (like pH, gas, humidity) and the physical strain, owing to which these polymers can be used as sensors for these stimuli. As a result, a new dimension of applications of these polymers opens up, as discussed in the following.[60].

6.8.2.1 Strain sensor

Conductive polymer-coated textiles are very promising materials as flexible strain sensors and many researchers demonstrate their ability with good sensitivity against applied strain. Wang et al. have demonstrated that the PPy-coated lycra filaments have very good sensitivity against strain, showing a non-linear increase in electrical resistivity due to the increment of applied strain, up to 0.5 [61]. Initially, a slow increase in electrical resistivity is observed, up to a strain of 0.2, but after that, an abrupt increase in the electrical resistance is observed as shown in Figure 6.13 [61]. An elastic electroconductive textile composite is prepared by the sequential chemical and electrochemical polymerization of PPy and PEDOT. The flexible electroconductor shows a gradual increase in the electrical resistance, with increase in the elongation up to 50%. Therefore, this flexible electroconductor is suitable for the detection of large strain. PPy-coated nylon/spandex-blended fabric is prepared for the possible application of monitoring of various movements of the human body [62]. The fabric exhibits good performance for electrotherapy for repetitive cycles. In another study, cotton rotor yarns of various linear densities were coated with PPy by in situ chemical polymerization and explored for their strain sensing behavior. It has been observed that electrical resistivity of the yarns decreases gradually as strain increases, as shown in Figure 6.14 [63]. The finer yarn shows better strain sensitivity than the coarser yarns. These electroconducive yarns also responded well due to the imparted twist in the yarn structure, and the twist sensitivity of the finer yarns is found to be better than coarser yarns. Similarly, in another study involving PPy-coated spandex/nylon fabrics, it was observed that electrical conductivity increases when the fabric is stretched by 50%, as shown in Figure 6.15. The increase in electrical conductivity due to the increase in the applied strain is due to the progressive increase in the fibre-to-fibre and polymer-to-polymer contact at higher strain rates, and their rearrangements and parallelization in the direction of the applied load.

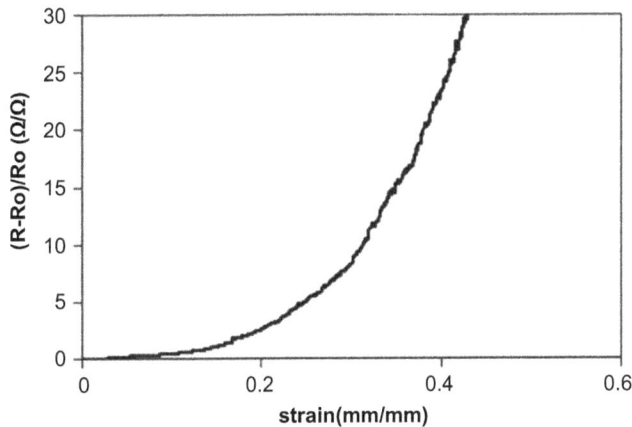

Figure 6.13: The strain response curve of PPy-coated Lycra fibers [61] (Modified according to Elsevier; Wang et al. [61]. Copyright 2011).

PPy-coated nylon fabric is prepared for measuring urine volume and bladder dysfunction in the bladder of patients. The fabric is able to measure urine volume and bladder dysfunction due to its change in electrical resistance when it is mounted around a phantom bladder [64]. Natural rubber is coated with PPy with a very thin layer by vapour phase polymerization. It shows a gauge factor of up to 1.86, and, therefore, found suitable for measurement of large strain deformation [65]. It is suggested by Wu et al. that the conductive polymer-coated lycra fabrics are very suitable for integration with

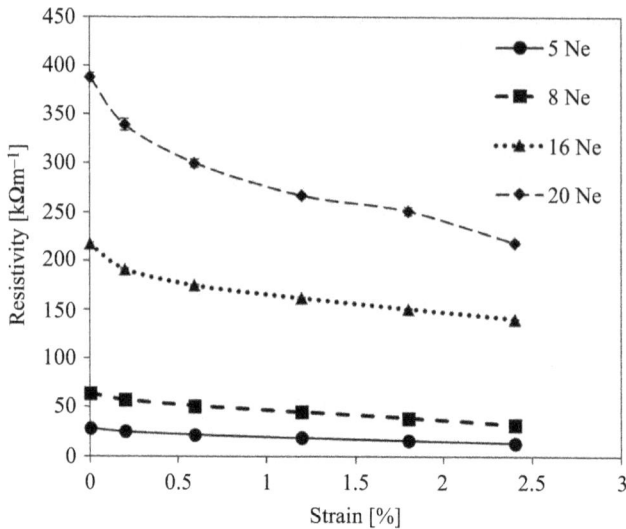

Figure 6.14: Strain response characteristics of PPy-coated cotton yarns [63] (Modified according to Springer Nature; Maity and Chatterjee [63]. Copyright 2013).

wearable clothing and garments for real-time monitoring of human movements in a wide dynamic range [66]. PPy-coated nylon-6 fabrics are prepared and tested for strain sensitivity. It is reported that the electrical resistivity of the fabric initially increases slowly at a lower strain rate but it is rapid at a higher strain rate [67].

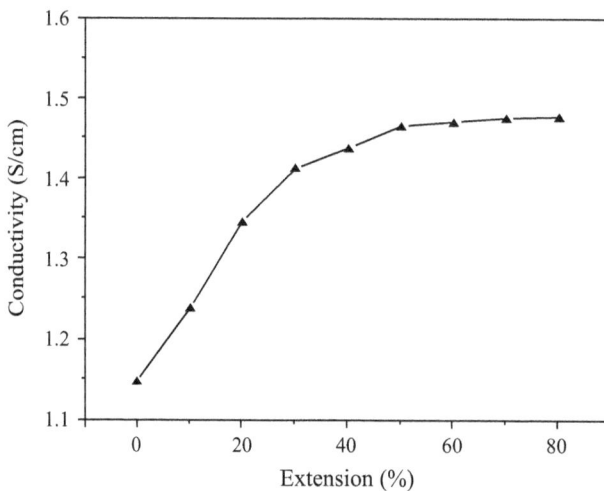

Figure 6.15: Effect of extension on electrical conductivity of PPy-coated spandex/nylon fabric [62] (Modified according to John Wiley and Sons; Oh et al. [62]. Copyright 2003).

6.8.2.2 Gas sensor

Owing to the ion exchange and redox properties of the conductive polymers and their derivatives, they have been largely explored for gas sensing applications since the 1980s. The conductive polymers and their derivatives exhibit rapid change in electrical resistivity upon exposure to certain gas and vapours. Therefore, textile materials coated with these conductive polymers can present a higher specific surface area and are found suitable for gas sensing applications. The conventional gas sensors that are based on metal oxides can operate at high temperatures and, therefore, have limitations in many applications. These conductive polymer-based textile sensors can operate at normal temperatures and have many advantages. The detection level of gas and vapours by these conductive polymers is at parts per million (ppm) level or less. PANi is a pioneering conductive polymer in this sensory application for detecting various acidic chemical agents or gases that originate from various chemical reactions, pyrolysis, degradation, and so on. Protonated emeraldine salt of PANi is suitable for detecting basic vapours like ammonia or organic amines. Upon adsorption, these basic vapours deprotonate PANi and reduce its electrical resistivity, which is the indicator of the presence of the vapour in the surroundings. PANi-coated PMMA substrate is found very suitable for the detection of NH_3 gas at very low concentrations of less than 10 ppm [68]. Acrylic acid-domed PANi is a good detector of NH_3 gas over a wide range of concentrations – 1–600 ppm. PANi films are good detectors of NO_2 gas. Amine-PANi nanofibers are found to be good detectors of H_2S gas [69]. Gold and platinum electrodes are coated with PANi nanofibers and used as resistive sensors for the detection of H_2 gas [70]. The H_2 gas detection sensitivity (lower than 10 ppm) is found to be better in the case of platinum electrode than gold electrode. Other than PANi, PTh is also explored as gas sensors. Boron trifluoride etherate-doped polythiophene is observed to be highly sensitive against ammonia gas, with detection levels of 50 mmol/L sensor [71]. Poly(phenylene ethynylene) is another conductive polymer that is found suitable for real-time detection of 2,4,6-trinitrotoluene (TNT) in a complex environment [72]. The mechanism of TNT detection is established as transfer of electrons from the polymer backbone due to photoinduced excitation of the TNT. PPy, which is the most widely used polymer in recent times, is also explored for gas sensors and found to be an excellent performer due to its high electrical conductivity and environmental stability. PPy films are prepared by in situ chemical polymerization and found suitable for the detection of NH_3 and NO gas at room temperature. Increase in electrical resistivity of PPy film is observed upon adsorption of basic gases like NH_3, and a decrease in electrical resistivity is observed upon adsorption of acidic gases like NO. However, in both cases, a good sensitivity, with a linear response in a range of 4–80 ppm, is observed, as shown in Figure 6.16 [73]. The sensitivity of the PPy film is defined as follows:

$$\text{Sensitivity} = \frac{\text{Resistance}_{gas} - \text{Resistance}_{air}}{\text{Resistance}_{air}} \times 100\% \qquad (6.4)$$

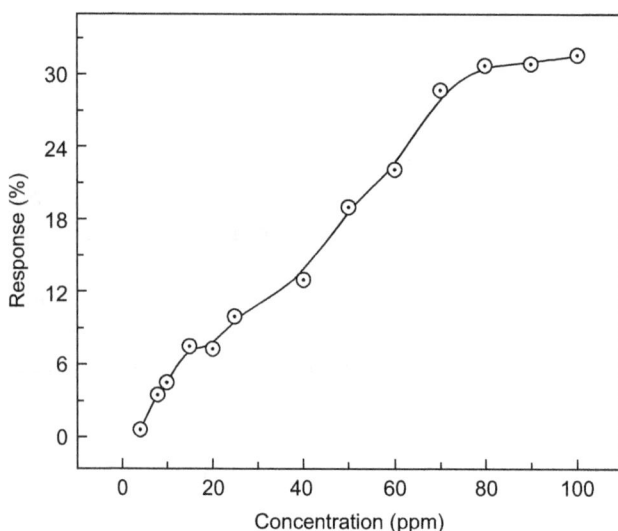

Figure 6.16: Response of PPy film against NH_3 gas for various concentrations [73] (Modified according to Elsevier; Joshi et al. [73]. Copyright 2011).

The PPy film exhibits a higher response toward detection of NH_3 gas, in comparison to NO, and humidity. This may be due to the difference in the interaction between PPy and the gas molecules, resulting in doping or de-doping reactions. PPy is a p-type material, and when it adsorbs alkaline gases like NH_3, the gas molecules act as a reducing agent and reduce the charge carrier densities, thereby reducing the electrical conductivity. Whereas, in case of oxidizing gases like NO, it acts as electron acceptors and the electrical conductivity of PPy film increases upon adsorbing NO molecules. In another study, PPy was doped with $LiClO_4$, p-toluene sulphonic acid (PTSA), and naphthalenesulphonic acid (NSA), and tested for the detection of NH_3 [74]. The NH_3 sensitivities are reported as 0.54, 1.28, 2.08, and 2.86 for pure PPy, PPy/ $LiClO_4$, PPy/PTSA, and PPy/NSA, respectively. PPy-coated pristine and Fe_3O_4 composite films are prepared and found suitable for NH_3 sensing [75]. In another study, emulsion polymerization process is followed to prepare phenylalanine/PPy film for NH_3 gas sensor. PPy is deposited on glass and ITO surface, and used for the detection of various volatile substances. PPy-coated textile sensors are prepared for the detection of both NH_3 and HCl gases. During the sequential exposure of PPy-coated textiles to NH_3 and HCl gases, it has been observed that when HCl is replaced with CO_2, there is a twofold decrease in electrical conductivity. Water-soluble PPy is synthesized especially for the detection of trimethylamine gas. PPy/ZnO nanocomposite, with different levels of ZnO loading (10%, 20%, 30%, 40%, and 50%). It can detect NO_2 gas at room temperature and at low concentrations with very high sensitivity [76]. PPy is even explored for detecting aerosol particles, optically. A reflectance probe made of an optical fibre is prepared and coated with PPy nanolayer at the end of the optical fibre. The refractive index of the PPy nanofilm-coated optical fibres changes due to the interaction with aerosol nanoparticles. Depending upon the kinds of nanoparticles, like sodium chloride, carbon black, and polystyrene latex, distinct variation in the reflective index is observed. This optical sensing approach may promote the use of conductive polymers for the detection of atmospheric aerosols.

6.8.2.3 pH sensor

Protonation and deprotonation ability of the conductive polymers, by adsorbing various acid and alkali species, is very useful for the measurement of pH of an unknown medium. The electrical resistivity of the conductive polymers will change upon protonation or deprotonation, which is the measurement of sensitivity of the conductive polymers toward detection of pH. Protonation reduces electrical resistivity of the polymers and deprotonation enhances the electrical resistivity. Therefore, with the increase in pH of the environment, PPy-coated polyester fabrics show an increase in surface resistivity. After a certain level of basic pH, a drastic increase in resistivity is observed, and at high levels of basic pH, total loss of counter-ion occurs, resulting in the conductor behaving as an insulator. Optical properties of PPy film will also change with different atmospheric pH, and the results are shown in Figure 6.17 [77].

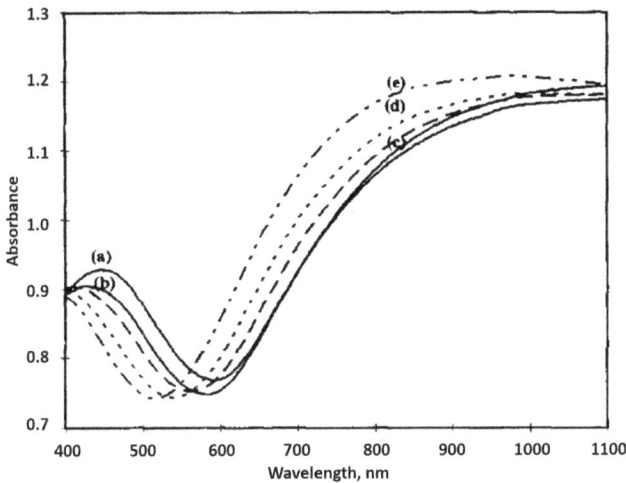

Figure 6.17: Effect of atmospheric pH on the absorption spectra of a PPy film: (a) pH 3.0; (b) pH 6.0; (c) pH 7.0; (d) pH 9.0; and (e) pH 12.0 [77] (reprinted with permission from Elsevier; de Marcos and Wolfbeis [77]. Copyright 1996).

It can be seen from Figure 6.17 that the absorbance between 600 and 900 nm increases with pH increasing from 6 to 12. However, the minima is shifted from 600 nm (at pH 6) to 500 nm (at pH 12). The maxima is observed at around 400 nm, and it is found to be shifted to a shorter wavelength when pH increases from 6 to 12. Various films of conductive polymers are prepared from their precursors, such as pyrrole, aniline, 1,3-diaminopropane, p-phenylenediamine, and diethylenetriamine, by electrochemical polymerization [78]. The amino groups that are anchored with these monomers are responsible for the potentiometric responses toward pH. Their responses are found to be linear in the whole range of pH from 2 to 11. The sensors have good regeneration ability and long life for repetitive use. In another study, PPy-coated cellulosic Palmyra fibres are prepared and found to be suitable for the possible application of pH sensor [79].

6.8.2.4 Humidity sensor

Water molecules are readily absorbed by conductive polymers, and their electrical conductivity alters. Electrical resistivity of PPy-coated polyester and nylon-6 fibres decreases gradually when relative humidity of the atmosphere increases from 40% to 85% at 25 °C temperature. This relationship is found to be linear within this range of relative humidity.

In another study, a detailed work is done to understand the effect of moisture on electrical resistivity of PPy-coated textile fabrics. Cotton woven, cotton knitted, wool woven, polyester spunlace nonwoven, polyester woven and polyester needlepunched nonwoven fabrics are coated with PPy and the change in resistivity with moisture content is evaluated, at a fixed temperature [79]. The results are shown in Figure 6.18. It is observed that as the moisture content of the fabrics increases, surface resistivity decreases with a sigmoid trend for all these kinds of fabrics [79].

In another study, PPy-coated cellulose film is prepared. It is reported that its resistivity decreases almost linearly from 23.3 to 700 Ωcm when relative humidity of the surrounding increases from 30% to 90% [80].

6.8.2.5 EMI shielding

Electroconductive textiles based on conductive polymers are suitable for effective EMI shielding. Various textile substrates made of different fibres like cotton, wool, viscose, lyocell, cupro, nylon, and polyester are coated with conductive polymers by in situ polymerization methods and explored for EMI shielding applications. These coated textiles are found to be superior to that of metals in terms of EM radiation absorption. The metalized textiles or metal composites suffer from galvanic corrosion and loss of conductivity due to rubbing and they mostly reflect EM waves rather than absorb. Whereas, the conductive polymer-coated textiles have the ability to absorb EM

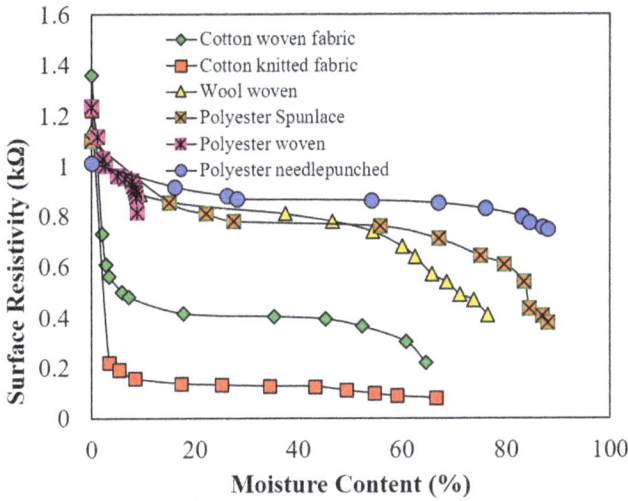

Figure 6.18: Change in surface resistivity of PPy-coated textile fabrics with moisture content [79] (adapted from Hindawi Publishing Corporation; Maity and Chatterjee [79]. Copyright 2015).

radiation and prevent proliferation [80]. It has been reported that the minimum surface resistivity of the PPy-coated textile fabric is required to be 100 Ω/cm^2 for effective EMI shielding [81]. However, PANi-coated polyester fabric with a resistivity of 5 k$\Omega/\square$ exhibits good shielding. Polythiophene and PANI is coated on polyester, glass, and high silica fabric by in situ chemical polymerization of EMI shielding applications [81]. The PANi-coated polyester fabric exhibits EMI SE in the range of 30–40 dB in the frequency range of 100–1,000 MHz, as shown in Figure 6.19. The PANi-coated fabric absorbs about 98% of the electromagnetic rays and reflects only 2%. Whereas PPy- and PTh-coated fabric can absorb about 96% and 82% electromagnetic radiation, respectively, and reflect only 4% and 18%, respectively [82]. An ideal shield should absorb all radiation and reflect nothing back or transmit across. Nylon/lycra-blended fabric is coated with PPy using anthraquinone-2-sulphonic acid (AQSA) as dopant and tested. The EMI SE was 89.9% at 18 GHz. A sequential chemical and electrochemical polymerization of PPy onto woven polyester fabric is conducted to achieve electrical resistivity as low as 0.2 Ωcm [83]. This highly conductive fabric shows EMI SE of about 36 dB for a frequency of 1.5 GHz, as shown in Figure 6.20. In all these, conductive polymer-coated textile fabrics restrict electromagnetic waves largely by absorption rather than by reflection or transmission [84].

PPy-coated epoxy film shows EMI SE of about 30 dB for a frequency range of 30–1,500 MHz, as shown in Figure 6.21 [59]. PPy-coated polyester and silica fabric exhibit EMI SE of 21.48 and 35.51 dB, respectively at a frequency of 101 GHz. PPy-coated glass fabrics with low resistivity of 500 $\Omega/\square$ show EMI SE of 98.67–99.23% in the frequency range of 800–2,400 MHz [82]. Therefore, it can be concluded that the conductive polymer-coated textiles can be successfully used for EMI shielding material for various household applications, shielding of wireless phones, FM/AM radio, computers, laptops,

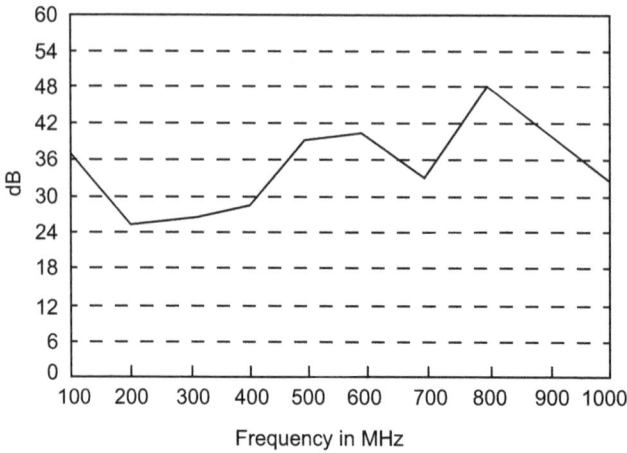

Figure 6.19: Shielding effectiveness of PANi-coated fabrics in the frequency range 100–1,000 MHz [82] (Modified according to Elsevier; Dhawan et al. [82]. Copyright 2002).

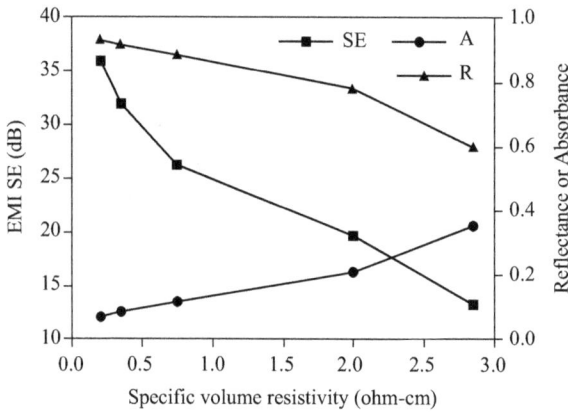

Figure 6.20: Electromagnetic shielding efficiency, absorbance (A), and reflectance (R) of PPy-coated PET fabrics [84] (Modified according to Elsevier; Kim et al. [84]. Copyright 2002).

building, rooms, and various other electronic gadgets and secret places that operate up to 2.4 GHz frequency [84–86].

6.8.2.6 Thermo-electric effect

The conversion of thermal energy to electrical energy or vice versa is called thermo-electric effect. Materials that can convert thermal energy to electrical energy are called thermo-electric materials, and the device for the same is called thermo-electric device.

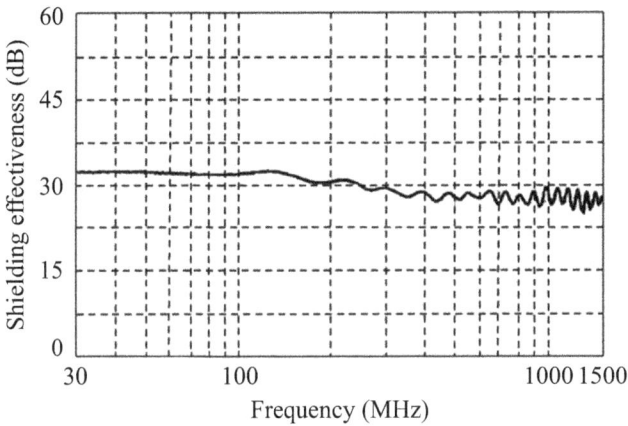

Figure 6.21: Shielding effectiveness of PPy film [85] (Modified according to Springer Nature; Qiao et al. [85]. Copyright 2010).

Thermo-electrical materials are used to design thermocouples in a close circuit and many thermocouples are joined together to make a thermo-electric device that can generate electromotive force (emf) due to the creation of temperature difference at its hot and cold junction. This phenomenon was discovered by Seebeck in 1821 and therefore, it is also called the Seebeck effect. The reverse effect of this is also there, that is, if emf is applied across the electric circuit of such a device, a hot and cold junction will be created. It is called the Peltier effect. Semiconductors and metals are conventionally used for the preparation of thermocouples and thermoelectric devices. Organic conductors were not so attractive materials for the thermo-electric effect due to their low electrical conductivity. Since the discovery of conductive polymers, researchers have been exploring their potential for exhibiting the thermo-electric effect. High electrical conductivity, ability to dope, and good thermal stability are the positives of conductive polymers, making them suitable for this application. PPy-coated polyester fabrics, with very low surface electrical resistivity of 306 $\Omega/\square$, exhibit low thermal diffusivity, low thermal conductivity, and excellent thermo-electric performance. A low thermal conductivity of the conductive textiles is required for obtaining high thermoelectric figure-of-merit ZT [52]. The thermoelectric figure-of-merit ZT is defined by eq. (6.5). Compared to metallic conductors and other inorganic semiconductors, PPy-coated textiles have lower thermal conductivity but good electrical conductivity, ultimately leading to high ZT values:

$$ZT = \left(\frac{S^2\sigma}{k}\right) \tag{6.5}$$

where S is Seebeck coefficient, σ is electric conductivity, k is thermal conductivity, and T is absolute temperature.

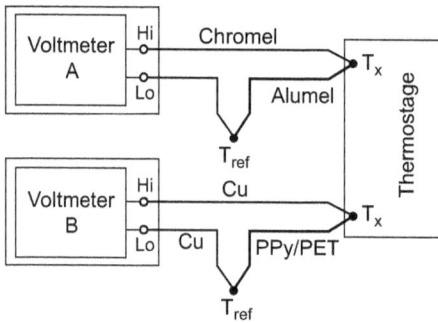

Figure 6.22: The experimental set-up for measuring the Seebeck effect of PPy/PET-Cu thermocouple.

One thermocouple circuit is prepared using PPy-coated PET fabric and copper wire as shown in Figure 6.22 [52]. The two dissimilar conductive materials, that is, PPy/PET fabric and the copper wire are electrically connected by gripping firmly with small silver clips using sufficient pressure. The hot junction is placed in a thermostat with reference to a chromel-alumel thermocouple and the cold junction is at room temperature (26 °C) [52]. The thermo-electric emf generated by this thermocouple is plotted against temperature difference and the result is shown in Figure 6.23.

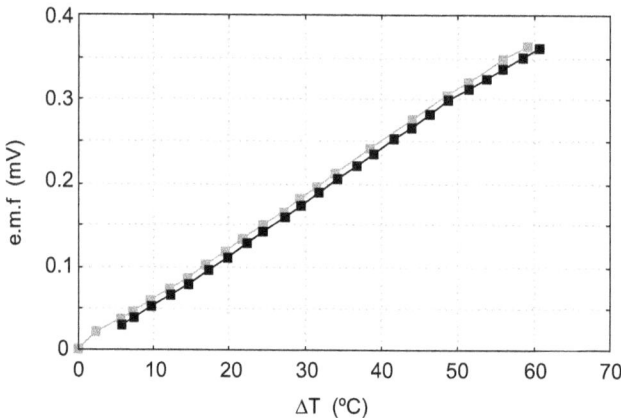

Figure 6.23: The thermo-electric effect of PPy-PET/copper thermocouple [52] (adapted from Scientific Research; Sparavigna et al. [52]. Copyright 2010).

Electrical conductivity and Seebeck coefficient of PANi and PPy at different doping levels are measured and reported as shown in Figure 6.24 [87]. It can be seen that as the electrical conductivity of PPy and PANi decreases, Seebeck coefficient increases almost linearly. According to Kelvin, the measurement of Seebeck coefficient can estimate the cooling power of the thermocouple. Though the Peltier coefficient is directly related to the cooling power, Kelvin establishes a relationship between Seebeck and Peltier coefficients, which can help in estimating the cooling power from the Seebeck coefficient. The conductive-coated fabrics are also useful for making cooling garments.

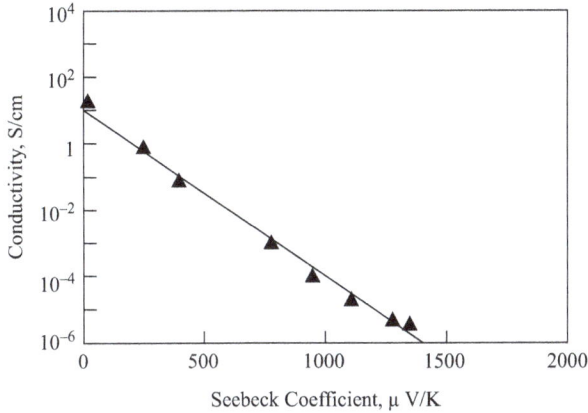

Figure 6.24: Effect of electrical conductivity on Seebeck coefficient of PANi [87] (reprinted with permission from AIP Publishing; Mateeva et al. [87]. Copyright 1998).

Conductive -coated textile thermocouples can achieve a Seebeck coefficient of about 10 V/°C. Cooling performance of the thermocouple can be achieved with this result, but the performance is found to be unsteady [88]. PPy is used as the P-type material and a MWCNT is used as the N-type conductive material, to prepare an active cooling e-textile [89]. The experimental set-up is shown in Figure 6.25.

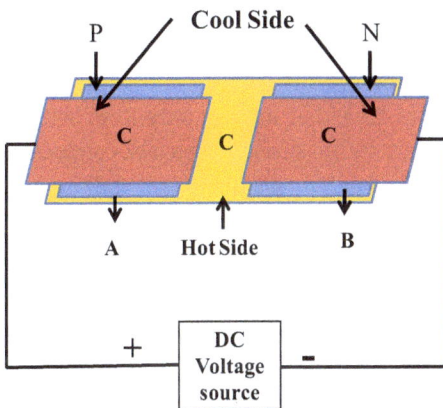

Figure 6.25: An experimental set-up for active cooling e-textile: (A) P-type PPy-coated fabric, (B) N-type MWNT-coated fabric, and (C) copper plates.

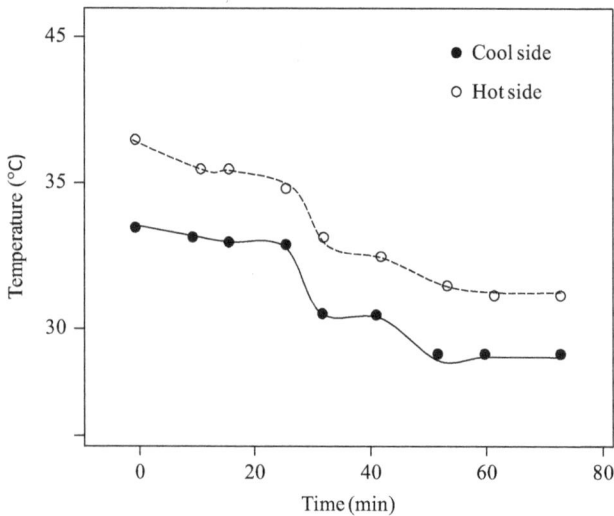

Figure 6.26: Peltier effect experiments with temperature versus time.

By this cooling e-textiles, a temperature difference of 5 °C is achieved between the top and bottom surfaces as shown in Figure 6.26. The temperature difference can be enhanced by improving the insulation between the top and bottom surfaces.

6.8.2.7 Corrosion protection

Metals and alloys of metals suffer from corrosion due to the attack of moisture and oxygen in the atmosphere. The most common example of environmental corrosion is rusting in iron. Even metals used in various chemical environments suffer from this surface degradation. The coating of conductive polymers over metals is found to be a protective layer to prevent corrosion. Conductive polymers are either in their pure form or they are blended with some paints or primers to apply on the surface of metals. Conductive polymers have higher redox potential than metals like iron and aluminium and can prevent corrosion. PANi is found suitable for this application for passivated steel and carbon steel. PPy and poly(3-methylthiophene) are found suitable for protecting stainless steel against 1 N sulphuric acid solutions [90].

6.8.2.8 Wastewater treatment

Owing to good ion exchange capacity, conductive polymers can remove heavy metal traces, colour, anion, turbidity, and so on from wastewater. Even conductive polymer-coated textile and other cellulosic substrates can reduce chemical oxygen demand

(COD) and biological oxygen demand (BOD) of water after treatment. After one cycle of water treatment, the ion exchange capacity of conductive polymers can be regenerated by a simple alkali treatment and reused [91].

6.8.2.9 Antistatic properties

Successful coating of the conductive polymers over textile fibre, yarns, and fabrics impart sufficient electrical conductivity that allows the dissipation of static charge. Therefore, the coating acts as an anti-static coating for the textile materials. PANi and PPy are already tried for the preparation of anti-static industrial textiles. PPy is found to be a better material for anti-static finish than PANi due to its higher electrical conductivity and stability. Other than PPy, graphene is also used as an anti-static material with about 1.5% weight add-on, which is found to be sufficient for static dissipation [92].

6.8.2.10 Antimicrobial properties

PPy has charged nitrogen and chloride ions that can attack the cell walls of microorganisms and dismiss them. Therefore, PPy is found to be an anti-microbial agent. PPy-coated cotton fabrics show 65%, 59%, and 75% reduction of gram-positive bacteria, gram-negative bacteria, and fungi, respectively. The anti-microbial activity is boosted further when $CuCl_2$ salt is added as an additive during the coating of PPy. The results are shown in Table 6.2.

Table 6.2: Anti-microbial properties of PPy-treated cotton fabrics.

Sample	Microbial reduction (%)		
	Staphylococcus aureus	*Escherichia coli*	*Candida albicans*
Cotton + PPy	64.86	59.14	73.07
Cotton + PPy + $CuCl_2$	92.53	97.60	100.00

The advantage of using PPy over conventional antimicrobial agents like quaternary ammonium salt and halamines is that it is insoluble in water. Its durability over the textile surface and its regeneration ability are good [51].

6.9 Conclusion

Inherently insulating textile materials can be made electrically conductive by coating/modifying them with various conductive materials like metals, carbon, graphene, MXene, and conductive polymers. Conventionally, metal fibers are blended with textile fibers in different stages of textile mechanical processing for the preparation of electroconductive textiles. But, processing of those metallic textiles is difficult and they lose their textile properties. These metallic textiles are not satisfactory for their practical use due to their stiffness, poor recovery, and breakage. Also, the metal coating peels off the fiber surface during use and processing due to lack of affinity. The use of graphene, MXene, and conductive polymers is a relatively novel approach for the preparation of electroconductive textiles. These non-metallic electroconductive textiles are flexible, durable, strong, easy to synthesize, and cost less. Most spectacularly, their electrical conductivity can be tailored as per the required application. The redox properties, ion exchange ability, doping-dedoping ability of the conductive polymers make them the most interesting material for today's research. They are considered for many novel applications, such as electromagnetic interference shielding, heat generation, thermo-electric generator, sensors, antibacterial efficacy, wastewater treatment, corrosion prevention, and anti-static textile. Though the durability and mechanical properties of the coated electroconductive textiles are excellent, their environmental stability, thermal stability, and launderability are still a challenge for the present researchers.

References

[1] Maity S, Chatterjee A, Singh B, Pal Singh A. Polypyrrole based electro-conductive textiles for heat generation. J Text Inst 2014;105:887–93.
[2] Hong W, Wyatt BC, Nemani SK, Anasori B. Double transition-metal MXenes: Atomistic design of two-dimensional carbides and nitrides. MRS Bull 2020;45:850–61. https://doi.org/10.1557/mrs.2020.251.
[3] Ahmed A, Hossain MM, Adak B, Mukhopadhyay S. Recent Advances in 2D MXene Integrated Smart-Textile Interfaces for Multifunctional Applications. Chem Mater 2020;32:10296–320.
[4] Uzun S, Seyedin S, Stoltzfus AL, Levitt AS, Alhabeb M, Anayee M, et al. Knittable and Washable Multifunctional MXene-Coated Cellulose Yarns. Adv Funct Mater 2019;29. https://doi.org/10.1002/adfm.201905015.
[5] Levitt A, Zhang J, Dion G, Gogotsi Y, Razal JM. MXene-Based Fibers, Yarns, and Fabrics for Wearable Energy Storage Devices. Adv Funct Mater 2020;30. https://doi.org/10.1002/adfm.202000739.
[6] Shirakawa H. The discovery of polyacetylene film+IBM-the dawning of an era of conducting polymers. Curr Appl Phys n.d.;1:281–6.
[7] Ito T, Shirakawa H. Ikeda S. Thermal cis+IBM-trans isomerization and decomposition of polyacetylene. J Polym Sci Polym Chem Ed n.d.;13:1943–50.
[8] chemistry CJCP. physics, and material science. Elsevier; 2012.
[9] Synthesis SH. and characterization of highly conducting polyacetylene. Synth Met n.d.;69:3–8.
[10] Sulfide) CJWP. In: Culbertson B. In: McGrath JE, editor. M, 1985+ADs- Vol 31. Springer, Boston, MA: Advances in Polymer Synthesis. Polymer Science and Technology; n.d.

[11] Green AG, CCXLIII.+IBQ-Aniline-black WAE. and allied compounds. Part I.+AKA-Journal of the Chemical Society. Transactions,+AKA- n.d.;1910:2388–403.

[12] Ansari R. Thermal studies of conducting electroactive polymers. The University of Ullongong, 1995.

[13] Hwang JH, pH-induced mass PM. and volume changes of perchlorate-doped polypyrrole. Synth Met n.d.;157:155–9.

[14] Ansari R. Polypyrrole conducting electroactive polymers: synthesis and stability studies. E-Journal Chem n.d.;3:186–201.

[15] Antoniadis H, Abkowitz MA, BR. H. Carrier deep+IBA-trapping mobility+IBA-lifetime products in poly (p+IBA-phenylene vinylene). Appl Phys Lett n.d.;65:2030–2.

[16] Carpick RW, Sasaki DY, Marcus MS, Eriksson MA. +ACY- Burns AR. Polydiacetylene Film a Rev Recent Investig into Chromogenic Transitions Nanomechanical Prop n.d.; 16:23.

[17] Reynolds JR, Baker CK, Jolly CA, Poropatic PA, conductive polymers RJPE. In Conductive Polymers and Plastics, Springer, Boston, MA.1989+ADs-. P n.d.:1–40.

[18] Cheng KB, Ramakrishna S, Lee KC. Electromagnetic shielding effectiveness of copper/glass fiber knitted fabric reinforced polypropylene composites. Compos Part A Appl Sci Manuf 2000;31:1039–45.

[19] Power EJ, Dias T. Knitting of electroconductive yarns. 2003 IEE Eurowearable, IET; 2003, p. 55–60.

[20] Alagirusamy R, Eichhoff J, Gries T, Jockenhoevel S. Coating of conductive yarns for electro-textile applications. J Text Inst 2013;104:270–7.

[21] Kuhn HH, Child AD, Kimbrell WC. Toward real applications of conductive polymers. Synth Met 1995;71:2139–42.

[22] Zhang Z, Roy R, Dugré FJ, Tessier D, Dao LH. In vitro biocompatibility study of electrically conductive polypyrrole-coated polyester fabrics. J Biomed Mater Res An Off J Soc Biomater Japanese Soc Biomater Aust Soc Biomater Korean Soc Biomater 2001;57:63–71.

[23] Zhou T, Chen F, Liu K, Deng H, Zhang Q, Feng J, et al. A simple and efficient method to prepare graphene by reduction of graphite oxide with sodium hydrosulfite. Nanotechnology 2010;22:45704.

[24] Yun YJ, Hong WG, Kim W, Jun Y, Kim BH. A novel method for applying reduced graphene oxide directly to electronic textiles from yarns to fabrics. Adv Mater 2013;25:5701–5.

[25] Shateri-Khalilabad M, Yazdanshenas ME. Fabricating electroconductive cotton textiles using graphene. Carbohydr Polym 2013;96:190–5.

[26] Shateri-Khalilabad M, Yazdanshenas ME. Preparation of superhydrophobic electroconductive graphene-coated cotton cellulose. Cellulose 2013;20:963–72.

[27] Gan L, Shang S, Yuen CWM. Conductive and stretchable graphene nanoribbon coated textiles. Int. Conf. Compos. Mater. [ICCM], 2015.

[28] Krishnamoorthy K, Navaneethaiyer U, Mohan R, Lee J, Kim S-J. Graphene oxide nanostructures modified multifunctional cotton fabrics. Appl Nanosci 2012;2:119–26.

[29] Shahidi S, nanotube MBC. and its applications in textile industry+IBM-A review. J Text Inst n.d.;109:1653–66.

[30] Lee TW, Han M, Lee SE, conductive JYGE. and strong cellulose-based composite fibers reinforced with multiwalled carbon nanotube containing multiple hydrogen bonding moiety. Compos Sci n.d.;123:57.

[31] Ra EJ, An KH, Kim KK, Jeong SY, YH. L. Anisotropic electrical conductivity of MWCNT/PAN nanofiber paper. Chem n.d.;413:188.

[32] Miyauchi M, Miao J, Simmons TJ, Lee JW, Doherty T V, Dordick JS, et al. Conductive Cable Fibers with Insulating Surface Prepared by Coaxial Electrospinning of Multiwalled Nanotubes and Cellulose. Biomacromolecules n.d.;11:2440.

[33] Sekitani T, Noguchi Y, Hata K, Fukushima T, Aida T, Someya, T. Rubberlike Stretchable Active Matrix Using Elastic Conductors. Science (80-) 2008;321:1468.

[34] Dalton A, Collins S, Mu+APE-oz E, Razal J, Ebron VH, Ferraris JP, et al. Nat. Cell Biol n.d.; 423:703.

[35] Zhou J, Zhao Z, Hu R, Yang J, Xiao H, Liu Y, et al. Multi-walled carbon nanotubes functionalized silk fabrics for mechanical sensors and heating materials. Mater Des 2020;191:108636. https://doi.org/10.1016/j.matdes.2020.108636.

[36] Arbab AA, Sun KC, Sahito IA, Qadir MB. +ACY- Jeong SH. Multiwalled carbon nanotube coated polyester fabric as textile based flexible counter electrode for dye sensitized solar cell. Phys Chem Chem Phys n.d.;17:12957–69.

[37] Haji A, Rahbar RS, AM. S. Improved microwave shielding behavior of carbon nanotube-coated PET fabric using plasma technology. Appl Surf Sci n.d.;2014:593–601.

[38] Lee Y, Kim J, Hwang H, stretchable JSHH. and sensitive strain sensors based on single-walled carbon nanotube-coated nylon textile. Korean J Chem Eng n.d.;36:800–6.

[39] Uzun S, Seyedin S, Stoltzfus AL, Levitt AS, Alhabeb M, Anayee M, et al. Adv. Funct Mater n.d.;29:45.

[40] Hu M, Li Z, Li G, Hu T, Zhang C, Wang X. All-Solid-State Flexible Fiber-Based MXene Supercapacitors. Adv Mater Technol 2017;2:1700143.

[41] Levitt A, Seyedin S, Zhang J, Wang X, Razal JM, Dion G, et al. Bath Electrospinning of Continuous and Scalable Multifunctional MXene-Infiltrated Nanoyarns. Small 2020;16:2002158.

[42] He N, Patil S, Qu J, Liao J, Zhao F, Gao W. Effects of Electrolyte Mediation and MXene Size in Fiber-Shaped Supercapacitors. ACS Appl Energy Mater 2020;3:2949–58.

[43] Zhao X, Wang LY, Tang CY, Zha XJ, Liu Y, Su BH, et al. Smart Ti3C2TxMXene Fabric with Fast Humidity Response and Joule Heating for Healthcare and Medical Therapy Applications. ACS Nano 2020;14:8793–805. https://doi.org/10.1021/acsnano.0c03391.

[44] Zhang J, Seyedin S, Gu Z, Yang W, Wang X, Razal JM. MXene: a potential candidate for yarn supercapacitors. Nanoscale 2017;9:18604–8.

[45] Wang QW, Zhang H Bin, Liu J, Zhao S, Xie X, Liu L, et al. Multifunctional and Water-Resistant MXene-Decorated Polyester Textiles with Outstanding Electromagnetic Interference Shielding and Joule Heating Performances. Adv Funct Mater 2019;29:1–10. https://doi.org/10.1002/adfm.201806819.

[46] Liu LX, Chen W, Zhang H Bin, Wang QW, Guan F, Yu ZZ. Flexible and Multifunctional Silk Textiles with Biomimetic Leaf-Like MXene/Silver Nanowire Nanostructures for Electromagnetic Interference Shielding, Humidity Monitoring, and Self-Derived Hydrophobicity. Adv Funct Mater 2019;29:1–10. https://doi.org/10.1002/adfm.201905197.

[47] Zhang X, Wang X, Lei Z, Wang L, Tian M, Zhu S, et al. Flexible MXene-decorated fabric with interwoven conductive networks for integrated joule heating, electromagnetic interference shielding, and strain sensing performances. ACS Appl Mater Interfaces 2020;12:14459–67.

[48] Shi X, Wang H, Xie X, Xue Q, Zhang J, Kang S, et al. Bioinspired Ultrasensitive and Stretchable MXene-Based Strain Sensor via Nacre-Mimetic Microscale "brick-and-Mortar" Architecture. ACS Nano 2019;13:649–59. https://doi.org/10.1021/acsnano.8b07805.

[49] J-h P, Zhao X, X-j Z, Bai L, Ke K, R-y B, et al. No Title. Multilayer structured AgNW/WPU-MXene fiber strain sensors with ultrahigh sensitivity and a wide operating range for wearable monitoring and; n.d.

[50] Dall'Acqua L, Tonin C, Varesano A, Canetti M, Porzio W, Catellani M. Vapour phase polymerisation of pyrrole on cellulose-based textile substrates. Synth Met 2006;156:379–86.

[51] Seshadri DT, Bhat N V. Synthesis and properties of cotton fabrics modified with polypyrrole. Sen'i Gakkaishi 2005;61:103–8.

[52] Sparavigna AC, Florio L, Avloni J, Henn A. Polypyrrole-coated PET fabrics for thermal applications. Mater Sci Appl 2010;1:253.

[53] Macasaquit AC, Binag CA. Preparation of conducting polyester textile by in situ polymerization of pyrrole. Philipp J Sci 2010;139:189–96.

[54] Kaynak A, Håkansson E. Generating heat from conducting polypyrrole-coated PET fabrics. Adv Polym Technol 2005;24:194–207. https://doi.org/10.1002/adv.20040.

[55] Hakansson E, Kaynak A, Lin T, Nahavandi S, Jones T, Hu E. Characterization of conducting polymer coated synthetic fabrics for heat generation. Synth Met 2004;144:21–8.

[56] Rodriguez J, Otero TF, Grande H, Moliton JP, Moliton A, Trigaud T. Optimization of the electrical conductivity of polypyrrole films electrogenerated on aluminium electrodes. Synth Met 1996;76:301–3.

[57] Lee JY, Park DW, Lim JO. Polypyrrole-coated woven fabric as a flexible surface-heating element. Macromol Res 2003;11:481–7.

[58] Abbasi A. MR, Militky J, Gregr J. Heat Generation by Polypyrrole Coated Glass Fabric. J Text 2013;2013:1–5. https://doi.org/10.1155/2013/571024.

[59] Malhotra U, Maity S, Chatterjee A. Polypyrrole-silk electro-conductive composite fabric by in situ chemical polymerization. J Appl Polym Sci 2015;132. https://doi.org/10.1002/app.41336.

[60] Li X, Wang Y, Yang X, Chen J, Fu H, Cheng T. Conducting polymers in environmental analysis. TrAC Trends Anal Chem 2012;39:163–79.

[61] Wang JP, Xue P, Tao XM. Strain sensing behavior of electrically conductive fibers under large deformation. Mater Sci Eng A 2011;528:2863–9.

[62] Oh KW, Park HJ, Kim SH. Stretchable conductive fabric for electrotherapy. J Appl Polym Sci 2003;88:1225–9.

[63] Maity S, Chatterjee A. Preparation and characterization of electro-conductive rotor yarn by in situ chemical polymerization of pyrrole. Fibers Polym 2013;14. https://doi.org/10.1007/s12221-013-1407-6.

[64] Rajagopalan S, Sawan M, Ghafar-Zadeh E, Savadogo O, Chodavarapu VP. A polypyrrole-based strain sensor dedicated to measure bladder volume in patients with urinary dysfunction. Sensors 2008;8:5081–95. https://doi.org/10.3390/s8085081.

[65] Tjahyono AP, Aw KC, Travas-Sejdic J. A novel polypyrrole and natural rubber based flexible large strain sensor. Sensors Actuators B Chem 2012;166:426–37.

[66] Wu J, Zhou D, Too CO, Wallace GG. Conducting polymer coated lycra. Synth Met 2005;155:698–701.

[67] Xue P, Tao XM, Kwok KWY, Leung MY, Yu TX. Electromechanical Behavior of Fibers Coated with an Electrically Conductive Polymer. Text Res J 2004;74:929–36. https://doi.org/10.1177/004051750407401013.

[68] Nicho ME, Trejo M, García-Valenzuela A, Saniger JM, Palacios J, Hu H. Polyaniline composite coatings interrogated by a nulling optical-transmittance bridge for sensing low concentrations of ammonia gas. Sensors Actuators B Chem 2001;76:18–24.

[69] Virji S, Huang J, Kaner RB, Weiller BH. Polyaniline nanofiber gas sensors: examination of response mechanisms. Nano Lett 2004;4:491–6.

[70] Fowler JD, Virji S, Kaner RB, Weiller BH. Hydrogen detection by polyaniline nanofibers on gold and platinum electrodes. J Phys Chem C 2009;113:6444–9.

[71] Ma X, Li G, Xu H, Wang M, Chen H. Preparation of polythiophene composite film by in situ polymerization at room temperature and its gas response studies. Thin Solid Films 2006;515:2700–4.

[72] Yang J-S, Swager TM. Fluorescent porous polymer films as TNT chemosensors: electronic and structural effects. J Am Chem Soc 1998;120:11864–73.

[73] Joshi A, Gangal SA, Gupta SK. Ammonia sensing properties of polypyrrole thin films at room temperature. Sensors Actuators B Chem 2011;156:938–42.

[74] Chitte HK, Bhat N V, Gore MA V, Shind GN. Synthesis of polypyrrole using ammonium peroxy disulfate (APS) as oxidant together with some dopants for use in gas sensors. Mater Sci Appl 2011;2:1491.

[75] Wu Y, Xing S, Jing S, Zhou T, Zhao C. Examining the use of Fe3O4 nanoparticles to enhance the NH3 sensitivity of polypyrrole films. Polym Bull 2007;59:227–34.

[76] Chougule MA, Dalavi DS, Mali S, Patil PS, Moholkar A V, Agawane GL, et al. Novel method for fabrication of room temperature polypyrrole–ZnO nanocomposite NO2 sensor. Measurement 2012;45:1989–96.

[77] de Marcos S, Wolfbeis OS. Optical sensing of pH based on polypyrrole films. Anal Chim Acta 1996;334:149–53.

[78] Lakard B, Herlem G, Lakard S, Guyetant R, Fahys B. Potentiometric pH sensors based on electrodeposited polymers. Polymer (Guildf) 2005;46:12233–9.

[79] Maity S, Chatterjee A. Textile/polypyrrole composites for sensory applications. J Compos 2015;2015.

[80] Mahadeva SK, Kim J. Enhanced electrical properties of regenerated cellulose by polypyrrole and ionic liquid nanocoating. Proc Inst Mech Eng Part N J Nanoeng Nanosyst 2011;225:33–9.

[81] Avloni J, Ouyang M, Florio L, Henn AR, Sparavigna A. Shielding effectiveness evaluation of metallized and polypyrrole-coated fabrics. J Thermoplast Compos Mater 2007;20:241–54.

[82] Dhawan SK, Singh N, Venkatachalam S. Shielding behaviour of conducting polymer-coated fabrics in X-band, W-band and radio frequency range. Synth Met 2002;129:261–7.

[83] Krupa I, Miková G, Novák I, Janigová I, Nógellová Z, Lednický F, et al. Electrically conductive composites of polyethylene filled with polyamide particles coated with silver. Eur Polym J 2007;43:2401–13.

[84] Kim MS, Kim HK, Byun SW, Jeong SH, Hong YK, Joo JS, et al. PET fabric/polypyrrole composite with high electrical conductivity for EMI shielding. Synth Met 2002;126:233–9.

[85] Qiao Y, Shen L, Dou T, Hu M. Polymerization and characterization of high conductivity and good adhesion polypyrrole films for electromagnetic interference shielding. Chinese J Polym Sci 2010;28:923–30.

[86] Abbasi AMR, Militky J. EMI shielding effectiveness of polypyrrole coated glass fabric. J Chem Chem Eng 2013;7:256.

[87] Mateeva N, Niculescu H, Schlenoff J, Testardi LR. Correlation of Seebeck coefficient and electric conductivity in polyaniline and polypyrrole. J Appl Phys 1998;83:3111–7.

[88] Hu E, Kaynak A, Li Y. Development of a cooling fabric from conducting polymer coated fibres: Proof of concept. Synth Met 2005;150:139–43. https://doi.org/10.1016/j.synthmet.2005.01.018.

[89] Lee S-A, Kim J. Active cooling e-Textiles for smart clothing. J Donghua Univ (English Ed 2006;23:24–6.

[90] Zarras P, Anderson N, Webber C, Irvin DJ, Irvin JA, Guenthner A, et al. Progress in using conductive polymers as corrosion-inhibiting coatings. Radiat. Phys. Chem., vol. 68, 2003, p. 387–94. https://doi.org/10.1016/S0969-806X(03)00189-0.

[91] Ghorbani M, Esfandian H, Taghipour N, Katal R. Application of polyaniline and polypyrrole composites for paper mill wastewater treatment. Desalination 2010;263:279–84.

[92] Lee Y, Cho J, Park YH, Son Y, Baik DH. Electrostatic Interactions in Conducting Polymer Composite PAN/PPy. Mol Cryst Liq Cryst Sci Technol Sect A Mol Cryst Liq Cryst 1998;316:313–6. https://doi.org/10.1080/10587259808044517.

Arobindo Chatterjee and Vinit Kumar Jain

7 Graphene-based functional textile materials

Abstract: Functional textile materials have gained a lot of interest in recent years for the preparation of next-generation advanced materials. Various approaches are used to introduce functionalities in the textile material as per the intended application. In this context, graphene is, currently, the focus of research because of its unique electrical, mechanical, thermal, and optical properties. Graphene-based functional textiles are advantageous due to their high strength, lightweight, uniform coatability, excellent flexibility, good electrochemical/thermal stability, great optical property, and good process-ability. This chapter provides comprehensive information on the development strategies of graphene-based functional textiles and their diverse applications. Different approaches for preparing functional textiles, fundamental aspects of graphene, graphene synthesis routes, and graphene application techniques on textile substrates, as well as the influence of graphene treatment parameters and substrate types on the functional properties of graphene-based textiles are summarized. Various graphene synthesis routes, such as mechanical exfoliation, chemical exfoliation, chemical vapor deposition, epitaxial growth, and chemical synthesis, as well as different application techniques of graphene on textile substrates namely vacuum filtration, brush coating, printing, wet transfer of monolayer films, and dip coating are critically reviewed. The application perspective of graphene-based textiles having different functionalities, such as being electrically and thermally conductive, and properties such as electrical heating, UV-protecting, hydrophobic, antimicrobial, photocatalytic, EMI shielding, capacitors, energy storage, and piezo-resistive sensors are highlighted.

Keywords: Graphene, conductivity, sensor, UV protection, electrical heating, EMI shielding

7.1 Introduction

Polymers are inherently non-conductive in nature. Being a subcategory of polymers, textile materials are also nonconductive. Researchers are working on the development of conductive polymers as they have numerous advantages over their metal counterparts. Conductive polymers developed so far lack processability; hence conversion of these polymers into fibres and subsequently conductive textile is not possible. The only alternative is to incorporate those conductive polymers into conventional textile substrates by some techniques to impart conductivity to otherwise nonconductive textiles. In the past, metals have been extensively used to prepare E-textiles or smart textiles for

https://doi.org/10.1515/9783110759747-007

a wide range of applications. Metal-based E-textiles are prepared by metal coating on textile substrate or insertion of metal threads during yarn spinning, weaving, and embroidery. Excellent electroconductive properties have been reported, with good stability and durability. However, they require complex and costlier techniques for incorporation in the textile substrate. In addition, the metal-based E-textiles are heavy, rigid, often toxic, non-biodegradable, expensive, and have compatibility issues. These limitations can be overcome by the application of conductive polymers (such as polypyrrole and polyaniline) and carbon-based conductive materials (such as carbon nanotube and graphene) on textile substrates [1, 2]. Conductive polymer-based E-textiles can be prepared by direct solution coating on fabric, or coating on fibres or yarn and subsequently converting them in fabric form. Conductive polymer-based E-textiles are advantageous over metal-based E-textile due to their flexibility, and lightweight. However, the electroconductive properties of conductive polymers-based E-textiles are inferior compared to the metal-based E-textiles. Moreover, the processing of the conductive polymer is difficult and often unstable towards atmospheric ageing because of de-doping [1, 2]. In this scenario, carbon-based conductive materials emerge as a favourable candidate for preparation of E-textiles, because they have excellent electrical conductivity, chemical inertness and, stability. Graphene is the basic structural element of all other carbon-based materials, and it is far superior to any other form of carbon material in terms of electrical, mechanical, thermal, and optical properties. Unlike metals-based E-textiles, graphene-based E-textiles can be twisted, bent, compressed, and stretched into complicated shapes, while maintaining the same levels of performance. Moreover, graphene-based E-textiles are comparatively lighter due to the lower specific gravity of graphene than metals [3–6]. Graphene-based E-textiles have been explored for numerous applications such as electrical conduction [7, 8], water-repellent products [7], UV-blocking [7], flame-retardant products [9], antibacterial activity [10–12], resistive heating [8, 13, 14], tissue scaffolds [15], flexible electrodes [8], biosensors [8], capacitors [16], and wearable biomedical devices [8, 17].

This chapter is designed based on the following perspectives:
– Brief discussion about the mechanism of electrical conductivity in different conductive materials and their relative merits and demerits
– Electroconductive textiles with special emphasis on graphene-based conductive textiles
– Fundamental aspects of graphene, graphene synthesis routes, and graphene application techniques on textile substrates, as well as the influence of graphene treatment parameters and substrate types on the functional properties of graphene-based textiles
– Properties such as thermally conductive, electrical heating, UV-protecting, hydrophobic, antimicrobial, photocatalytic, EMI shielding, capacitors, and energy storage, and piezo-resistive sensors, from the application perspective.

7.2 Electrical conductivity

Electrical conductivity is a fundamental property of electroconductive materials. It is a quantitative measure of a material's capacity to carry electrical current. Based on electrical conductivity, materials can be divided into conductors, semiconductors, and insulators. Conductors have high electrical conductivity, which allows current to flow easily through them. On the other hand, insulators have low electrical conductivity, which restricts the flow of current through them. Materials whose conductivity lies between conductors and insulators are called semiconductors. Electrical conductivity is measured in siemens per metre (S/cm, SI units) and is generally denoted by the symbols σ (sigma), and κ (kappa) (especially in electrical engineering). The electrical conductivity of conductors is more than 10^3 S/cm and that of semiconductors between 10^3 and 10^{-12} S/cm. Materials with conductivity lower than 10^{-12} S/cm are considered insulators [18].

Electrical conductivity depends on the charge carrier mobility (or ease of movement) and concentration. In the case of conductors, charge carriers are electrons only, and in the case of semiconductors, charge carriers are electrons and holes [18, 19].

Electrical resistivity, which is the reciprocal of electrical conductivity, is defined as the ability of a material to resist the flow of electric current. Electrical resistivity is measured in ohm-metre (Ω.m, SI units) and denoted by the symbol ρ (rho):

$$\rho = \frac{R\,A}{\ell} \qquad (7.1)$$

where R denotes electrical resistance in Ω; A denotes the area of the material in m^2; and ℓ is the length of the material in metre [20].

7.3 Band structure theory of solids

Band structure theory is the most accepted theory used to explain the mechanism of electric conduction in solids. According to the Pauli Exclusion Principle, no two electrons in a single atom can have the same quantum numbers or energy levels. In a solid, large numbers of identical atoms are combined to form a portion of it. Since the number of atoms in a solid is high, the number of energy levels is also very high. A set of energy levels is tightly packed together to form an energy band. Generally, there are two types of band forms, namely valence band and conduction band. The valence band represents lower energy states, where a large number of electrons are tightly bound and cannot move freely, whereas the conduction band is at a higher energy state, where a small number of electrons are loosely bound and move freely. Bandgap or Fermi level is an energy level that is not filled by any of the bands. The conduction

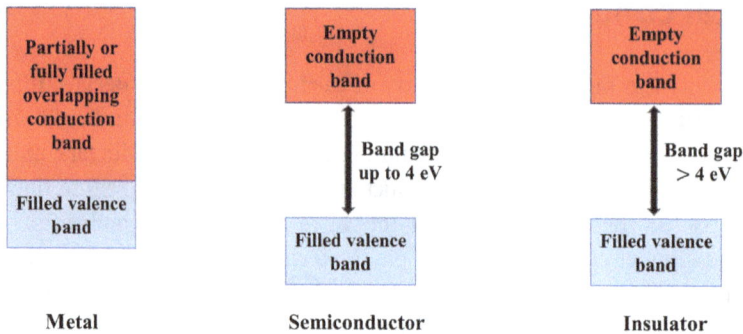

Figure 7.1: Band structure for conductor, semiconductor, and insulator.

band is the closest band above the Fermi level, while the valence band is the closest band below the Fermi level.

The outermost electrons (valence electrons) in the atom's orbit are subjected to the least amount of attraction force. When external heat or electrical energy is applied, they are easily detached and moved to a higher energy level (conduction band). The amount of energy required for electron excitation is termed bandgap energy, and this depends on the width of the bandgap. Band structure of different materials is shown in Figure 7.1. Materials with a bandgap larger than 4.0 eV are classified as insulators, whereas those with a bandgap less than 4.0 eV are classified as semiconductors. Metals do not have a bandgap [19].

7.4 Electrical conductivity in conductors

Metals and alloys are primary examples of conductors and electrons act as the charge carriers. The atoms in metals are so closely packed that an electron from one atom is subjected to a large force from other nearby atoms. As a result, the valence band and conduction band overlap; and the electrons move easily to conduction bands by gaining a small amount of external energy. When an electric source is attached to metal, the free electrons begin to flow towards the higher potential terminal of the source, and as a consequence, current flows through the metal. Metals show very good electrical conductivity because of the availability of a large number of free electrons in the conduction band.

Table 7.1: Electrical conductivity of different materials at temperature of 300 K [1, 18, 19].

Material category	Material	Electrical conductivity (S/cm)
Conductors	Graphene	0.2310^8
	Silver	6.8310^5
	Copper	5.9810^5
	Gold	4.2610^5
	Aluminium	3.9810^5
	Zinc	6.8310^5
	Nickel	1.510^5
	Iron	1.010^5
	Gallium	0.6610^5
	Expanded graphite	1.2510^5 (perpendicular to carbon axis)
	Carbon nanotube	10^4 to 10^7
	Graphite	2 to 310^3 (parallel to carbon axis) and 3.3 (perpendicular to carbon axis)
Semiconductors	Conducting polymers	10
	Germanium	0.02
	Silicon	410^{-6}
	Gallium arsenide	10^{-9}
	Silicon carbide	10^{-10}
Insulators	Polyester	10^{-12}
	Nylon	10^{-14}
	Polyethylene	10^{-15}
	PTFE	10^{-18}
	Polystyrene	10^{-17} to 10^{-19}

7.5 Electrical conductivity in semiconductors

According to band structure theory, a bandgap exists between the valence and conduction bands in semiconductors. At low temperatures, no electron has enough energy to cross the band gap and enter the conduction band. As a result, semiconductors act as insulators at low temperatures, preventing the flow of current. When thermal energy is applied, some of the electrons are excited and enter the conduction band from the valence band, and electric conduction starts. These annihilated electrons create vacancies or holes in the valence band, and these holes also contribute to conductivity by wandering in the valence band. Therefore, the charge carriers in semiconductors are electrons and holes, in which electrons are negative and holes are positive. The conductivity of a pure semiconductor rises exponentially with temperature; it produces heat as it operates, which is undesirable. Consequently, impurities are purposefully introduced into semiconductors, in order to precisely control the conductivity of semiconductors, a process known as doping. On the basis of the predominance of charge carriers, they are termed p-type semiconductor (predominance of the holes) and n-type semiconductor (predominance of the electron). Here, p stands for positive and n stands for negative. In contrast to metals, the conductivity of semiconductors increases, with an increase in temperature.

7.6 Electrical conductivity in conducting polymers

Organic polymers that conduct electricity are referred to as conductive polymers or intrinsically conducting polymers (ICPs). Based on electrical conductivity, they are semiconductors, and in some situations, their conductivity may be similar to metals. Several types of ICPs are reported such as polyaniline (PANi), polypyrrole (PPy), polyacetylene, polythiophene (PTh), and poly-3,4-ethylene dioxythiophene (PEDOT). The valence electrons in insulating polymers such as polyethylene are closely packed in sp^3-hybridized covalent bonds (σ, sigma bond). The mobility of such σ-bond electrons is limited, and they do not allow current to flow through them. On the other hand, conjugated materials or conductive polymers have sp^2-hybridized structure, in which carbon atoms are bonded with localized σ-bonds and one delocalized pie (π) bond. The π-bonds are weaker than the σ-bonds. As the external energy is applied, these π-electrons easily move with high mobility; as a result, electrical conduction is initiated. The type of dopant used during doping determines whether the electrons or the holes will be the charge carriers. Doping can be done by oxidation or reduction of material. In oxidative doping, some of the electrons are removed from the conduction band whereas, in reductive doping, some of the electrons are added to the conduction band. Variations in chemical structure, chemical composition, and doping significantly influence electrical conductivity [1, 18]. Table 7.1 represents the typical electrical conductivity of different materials.

7.7 Graphene: structure and properties

The ideal graphene is a single layer of carbon atoms densely packed in the two-dimensional honeycomb crystal lattice. The single layer graphene shows superior properties compared to the few-layered graphite or multi-layered graphene. It is a fundamental structure of all other dimensionalities of carbon materials such as fullerenes, carbon nanotubes, and graphite (Figure 7.2).

Naturally occurring graphite is a multi-layered structure of carbon atoms held together by van der Waals bonds. Graphite has inferior properties in perpendicular directions compared to the in-plane direction, due to interlayer spacing of 0.335 nm. A large number of experiments were conducted in the past to separate a single layer of graphite. In 1961–1962, Hanns-Peter Boehm hypothesized a single-layer structure of graphite and labelled it "graphene". In the year 2004, physicists, Andre Geim and Konstantin Novoselov, succeeded in isolating single-layer graphene by micromechanical exfoliation technique at the University of Manchester. They were awarded the Nobel Prize in Physics

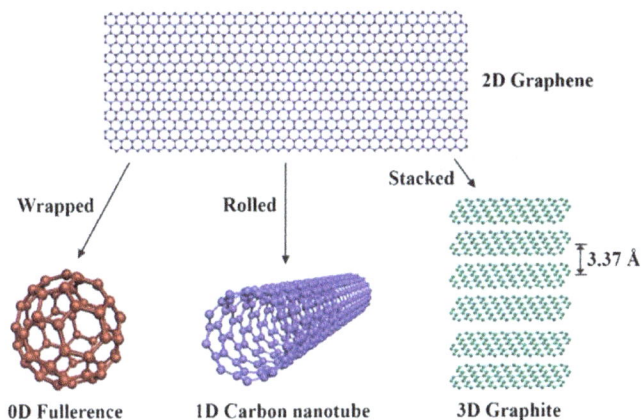

Figure 7.2: From 2D graphene to 0D fullerene, 1D carbon nanotube, and 3D graphite (reprinted from reference [21] with permission, Copyright (2020) Elsevier).

in 2010 for the discovery of the thinnest and strongest known material [22]. Since then, graphene has become one of the most popular scientific topics, receiving tremendous attention and investment from all over the world.

Graphene exhibits electrical conductivity comparable to metals; therefore, it can replace metals in nano-electronic or electrical devices. Graphene has a high electron mobility of about 200,000 cm^2 V/s, which is about 100 times faster than silicon. Its electrical conductivity is like copper, although its density is four times lower than copper (graphene = 2.267 g/cm^3 and copper = 8.95 g/cm^3). Graphene shows fascinating thermal conductivity (3,500–5,300 W m/K) at room temperature, which is five times higher than that of copper. It has a high specific surface area (2,630 m^2/g) and excellent moisture barrier properties. Young's modulus of graphene is around 1,100 GPa. The monolayer graphene is practically transparent (>97.7% transmittance) because of its extreme thinness. Due to its outstanding electronic, optical, thermal, and mechanical properties, it has attracted great interest as an apt candidate for the production of advanced materials with much potential in various applications [23–25].

7.8 Source of conductivity in graphene

Graphene is made of carbon that has six electrons in three different orbitals, and the electron configuration is $1s^2\ 2s^2\ 2p^2$. Out of these two electrons will occupy the 1s orbital, the next two electrons will occupy the 1s orbital, and the remaining two will occupy two different p orbitals, namely, $2p_x$ and $2p_y$. When two carbon atoms combine, in each carbon atom, one electron is excited from the 2s orbital to the empty $2p_y$ orbital, and three new orbitals will form from the hybridization of 2s orbital, $2p_x$, and $2p_y$, called

sp^2-hybridized orbital; p_z orbital will remain unhybridized. These three sp^2-hybridized orbitals form in-plane σ-bonds with the neighbouring three atoms with bond length of 0.142 nm. The remaining unhybridized p_z orbital of each atom is hybridized via interplane π-bond. These interplane π-bonds are relatively weaker than the in-plane σ-bonds.

After the excitation, this π-electron (in valence band) and π*-electron (in conduction band) freely move and govern the electroconductive properties of graphene. The chemical structure of graphene is shown in Figure 7.3.

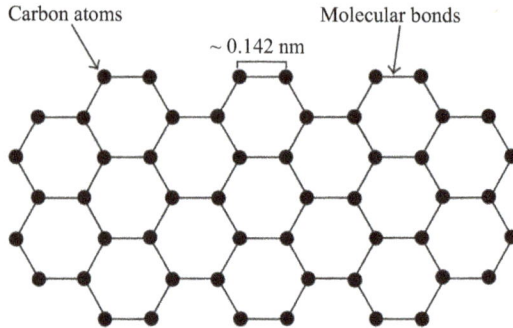

Figure 7.3: Structure of a graphene sheet [26].

7.9 Band structure of graphene

Near the Fermi level, the conduction and valence bands of graphene converge at six points called Dirac points (Fermi energy at $E = 0$), shown in Figure 7.4. At the Dirac points, the charge carriers (electrons and holes) behave like massless Dirac fermions

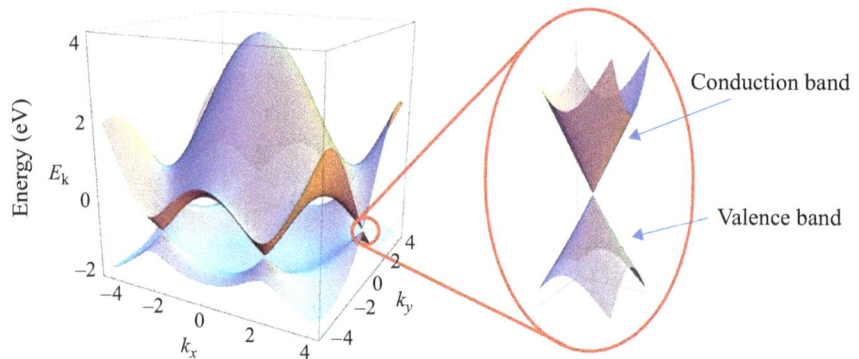

Figure 7.4: Left: Electronic band structure of graphene. Right: Magnified image at meeting point of conduction band and valence band (reprinted with permission from reference [27], Copyright (2009) by the American Physical Society).

and show high carrier mobility. This unique and unusual band structure of graphene is responsible for its outstanding electroconductive properties and makes it a zero-gap semiconductor. The number of layers in the graphene structure significantly influences the electronic properties. Only monolayer and bilayer graphene are considered zero-gap semiconductors [27].

7.10 Synthesis of graphene

Several graphene synthesis methods are reported in the literature including mechanical exfoliation, chemical vapor deposition, chemical exfoliation, epitaxial growth, and chemical synthesis. These methods are detailed in this section with their merits and demerits.

7.10.1 Mechanical exfoliation

Nobel laureates, Andre Geim and Konstantin Novoselov, used the mechanical exfoliation method to peel off single-layer graphene from highly oriented pyrolytic graphite (HOPG) (Figure 7.5). In mechanical exfoliation method, peeling can be done by the use of scotch tape, ultrasonication, electric field, and so on. This method produces the highest quality graphene and does not require any special equipment. However, this method does not have commercial viability for the large-scale production of graphene because of its low yield and reproducibility. Moreover, the monolayer or few layers of graphene prepared by this route are weak and unevenly distributed. Several other

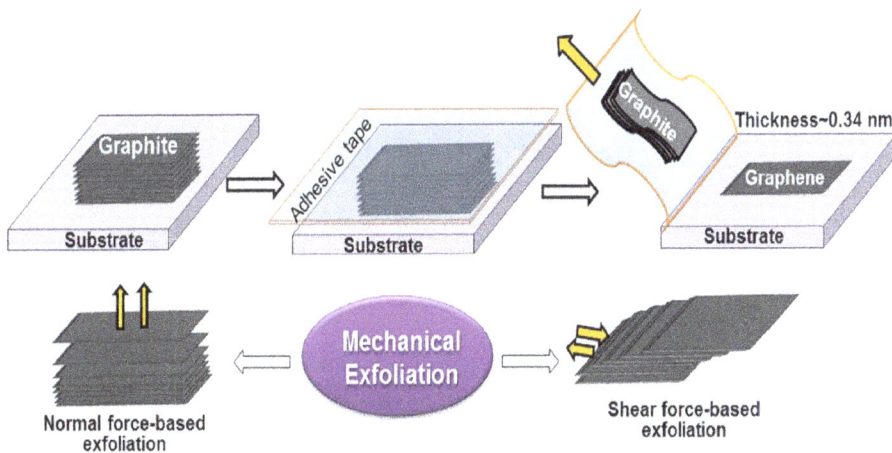

Figure 7.5: Mechanical exfoliation method for graphene synthesis (reprinted from reference [31] with permission, Copyright (2020) Elsevier).

graphene synthesis approaches have been reported in the literature to address this issue. These approaches include chemical vapor deposition (CVD), chemical exfoliation, chemical synthesis, and epitaxial growth on SiC. Some other novel methods are also proposed, such as microwave synthesis and unzipping of carbon nanotubes (CNTs); however, these methods need further investigation [24, 28–30].

7.10.2 Chemical exfoliation

Chemical exfoliation uses alkali metals to intercalate with the graphite structure and separate graphite layers. Alkali metals such as potassium (K), caesium (Cs), lithium (Li), and sodium (Na) were reported in the literature to intercalate with graphite. Alkali metals easily attach in between the graphite layers because of their smaller ionic radii than interlayers spacing. Kaner and co-workers reacted potassium with graphite at 200 °C under an inert helium atmosphere, and an intercalated compound KC_8 was formed. When KC_8 reacted with ethanol (CH_3CH_2OH), potassium ethoxide was formed and hydrogen liberated, which assist in the isolation of few-layer graphene [28]. The process flow of chemical exfoliation method is shown in Figure 7.6.

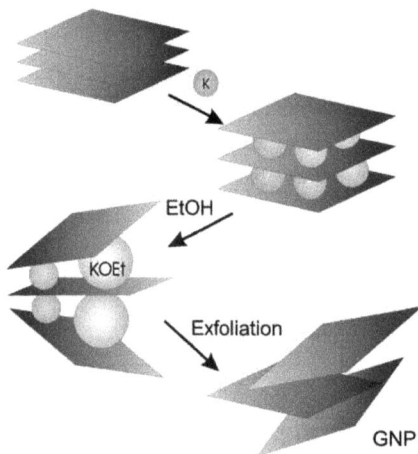

Figure 7.6: Process flow of chemical exfoliation method (reprinted from reference [32] with permission of The Royal Society of Chemistry).

7.10.3 Chemical vapor deposition

Somani and coworkers reported the first successful CVD synthesis of few-layers graphene films using camphor ($C_{10}H_{16}O$) as the precursor on Ni foils [33]. Since then, CVD has become the most popular method of graphene production.

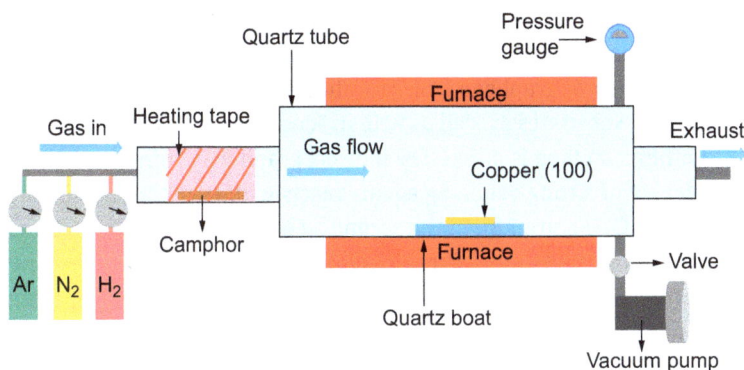

Figure 7.7: Schematic diagram of chemical vapour deposition setup Modified according to [34].

In this method, carbon precursors such as methane, ethylene, and acetylene are thermally dissociated at high temperatures and deposited on transition metal surfaces such as nickel, gold, copper, cobalt, and ruthenium. The transition metal is then etched out and the layers of graphene deposited on top of it are transferred to another substrate. In thermal CVD, the production of graphene is carried out in a resistive heating furnace, and when this process is assisted by plasma, it is termed plasma-enhanced CVD (PECVD) [28, 29, 34]. Graphene preparation by CVD method is schematically represented in Figure 7.7.

7.10.4 Epitaxial growth of graphene

Epitaxial growth is a recognized method to fabricate semiconducting devices. Epitaxy is derived from the Greek word epi, which means "above," and taxis, which means "arrangement". It literally means "to arrange above". Epitaxial growth is defined as the orderly arrangement of single-crystalline film on a single crystalline substrate such as silicon carbide (SiC). As a consequence of this method, silicon atoms sublimate on silicon substrates under ultrahigh vacuum, leaving carbon atoms to arrange into graphene layers. In contrast to CVD, transition metals are not required in this method. The heating time and temperature determine the number of layers of graphene. Epitaxial growth has been extensively reported to prepare monolayer or few layers of graphene on SiC for light-emitting devices and high-frequency electronics [28, 29].

7.10.5 Chemical synthesis

In the context of textile fibre and graphene, the main problem associated with the use of pristine graphene is that it does not have affinity towards textile fibre due to the absence of functional groups. As a result, it requires intermediaries, that is, cross-linking or binding agents for deposition on textile fibres. This problem can be overcome by the functionalization of graphene [35]. The attached functional groups facilitate interaction with other substrates as well as help in separation of graphene layers.

The chemical synthesis method is a two-step indirect graphene synthesis method. In the first step, a mixture of strong oxidizing agents reacts with graphite (precursor) to form graphene oxide (GO) (Figure 7.8). In the second step, the prepared GO reduces to convert it into graphene. Oxygenated functionalities such as epoxides (bridging oxygen atoms), carbonyls (C = O), hydroxyls (–OH) are attached to the graphene layers during the oxidation, making the GO hydrophilic in nature. Graphene oxide is sp^3-hybridized carbon atom network, which is electrically insulating in nature. Different methods such as thermal, electrochemical, ultraviolet, and chemical methods can be employed to restore the sp^2-hybridized carbon atom network from sp^3-hybridized GO structure, and this is termed reduced graphene oxide (rGO) [24, 29]. If the layers are not properly separated from each other, they are likely to agglomerate or even restack to form a few layers of graphite, via π-π and van der Waals interactions. The properties of stacked graphene are inferior, compared to the monolayer graphene.

Figure 7.8: (a) Chemical synthesis of graphene and (b) images of graphite powder and GO solution (reprinted from reference [36] with permission, Copyright (2019) Springer Nature).

7.10.6 Some other novel routes for graphene synthesis

A recently developed graphene production technique employed chemical and plasma-etched processes to unzip carbon nanotubes. Initially, a sonicated dispersion of multi-walled carbon nanotubes (MWCNTs) was deposited on Si substrate, and then, a thick film of Poly (methyl methacrylate) (PMMA) was spin-coated on the top of MWCNTs (Figure 7.9b). The sidewalls of MWCNTs were etched by air plasma, which produces a thin elongated strip of graphene with straight edges, termed "graphene nanoribbon" (GNR) (Figure 7.9c). Subsequently, the attached PMMA film was removed by acetone vapor. The number of layers in GNR depends on the width and layers of the precursor material [29, 37].

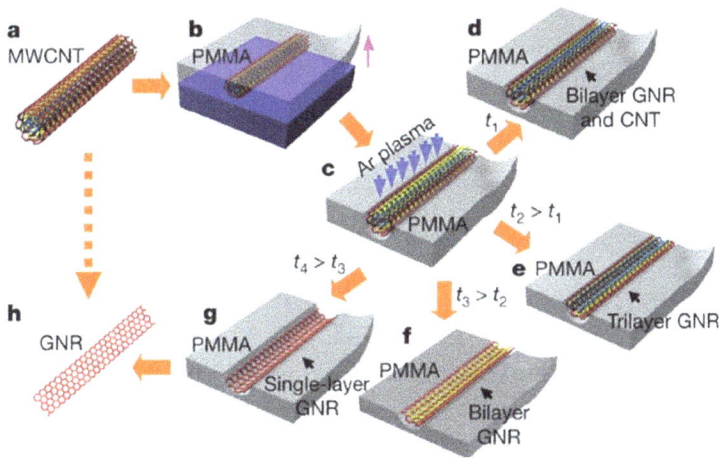

Figure 7.9: Synthesis of GNRs from CNTs (reprinted from reference [37] with permission, Copyright (2009) Springer Nature).

Sridhar, Jeon, and Oh reported a green synthesis route for graphene production assisted by microwaves. Graphite powder was intercalated and expanded by a mixture of hydrogen peroxide and ammonium peroxydisulphate, under microwave radiation. As a result, a dispersion of a few layers of graphene was formed, which subsequently got converted to monolayer graphene by sonication [38]. Al-Hazmi and co-workers used a combination of glutaric acid and methanol to intercalate and expand graphite powder in a microwave oven [39].

Each method has certain advantages and disadvantages depending on the applications, which are briefly summarized here. The mechanical exfoliation method produces superior quality graphene and does not require special equipment. However, this method is not commercially viable for the large-scale production of graphene because of the low yield and lack of reproducibility. CVD and epitaxial growth on SiC methods also produce few layers of high-quality graphene, but these methods are not scalable due to low yield and lack of reproducibility; they demand expensive and complex machinery, as well. In the epitaxial growth on SiC method, it is challenging to control the thickness and rotational stacking of graphene layers in the routine manufacturing of graphene. In comparison to the mechanical exfoliation method, few layers GNRs produced by the unzipping process have lower electronic properties and require expensive starting material (CNTs). The chemical exfoliation method can be used for the high-volume production of graphene. However, the production of mono-layer graphene is still not feasible by this route, and quality of the produced graphene is also inferior due to large numbers of defects. The chemically synthesized graphene displays properties below the theoretical potential of pristine graphene due to the in-complete reduction of GO. In addition, the chemical synthesis route comprises several time-consuming processes and utilizes some potentially dangerous and explosive compounds such as hydrazine. However, the chemical synthesis route has numerous advantages such as bulk production, low-temperature process, and cheaper raw mate-rial. This technique produces functionalized graphene (GO) that can readily disperse in water and be deposited on textile substrates, without the need of any binders or intermediaries. Several eco-friendly reduction techniques were also reported to avoid the use of hydrazine, which are mentioned in Section 7.13.3 [24, 25, 28, 29, 40–42].

7.11 Chemical synthesis of graphene

The chemical synthesis method is the most favoured graphene production method for the preparation of graphene-based electroconductive textile. In 1859, Brodie first syn-thesised GO by oxidizing graphite with a mixture of potassium chlorate and nitric acid. This process was advanced by Staudenmaier in 1898, by adding sulfuric acid in the mix-ture of potassium chlorate and nitric acid. This modification considerably improved the oxidation of graphite [28, 29, 41]. However, this process comprised many stages that were tedious and unsafe. To address these issues, Hummers, and Offeman introduced an effective combination of oxidizing agents, viz., sodium nitrite, sulfuric acid, and po-tassium permanganate, and the method was known as Hummers method [43]. The mix-ing of sulfuric acid and potassium permanganate produces a highly reactive bimetallic heptoxide, viz., diamanganese heptoxide (Mn_2O_7), which is responsible for graphite ex-foliation [44]. Hummers method is classified into 3 categories: Hummers method, modi-fied Hummers method, and improved Hummers method [40].

The Hummers method involves the addition of concentrated H_2SO_4 (69 mL) to a mixture of graphite flakes (3.0 g) and $NaNO_3$ (1.5 g), followed by cooling the mixture to 0 °C. About 9.0 g of $KMnO_4$ is gradually added to the mixture in small portions to keep the reaction temperature below 20 °C. Thereafter, the reaction temperature increases to 35 °C along with agitation for 30 min, and then, water (138 mL) is gradually added to it. Due to the exothermic reaction, the reaction bath temperature rises, and it is maintained at 98 °C for 15 min, by external heating. The reaction bath is cooled down and additional water of 420 mL is added to it, along with 3 mL of 30% H_2O_2. After cool-ing the reaction bath to room temperature, the mixture is purified by filtration, multi-ple washings, centrifugation, and vacuum drying, which yield 1.2 g of GO [40].

The modified Hummers method uses additional $KMnO_4$ to the Hummers method mixture. In this method, concentrated H_2SO_4 (69 mL) is added to a mixture of graphite flakes (3.0 g) and $NaNO_3$ (1.5 g), followed by cooling the mixture to 0 °C. About 9.0 g of $KMnO_4$ is gradually added to the mixture in small portions to keep the reaction temperature below 20 °C. Thereafter, the reaction temperature is raised to 35 °C and agitated for 7 h. Additional $KMnO_4$ of 9.0 g is added to the reaction bath in small portions and heated at 35 °C for 12 h. The reaction bath is cooled down and water (400 mL) and 30% H_2O_2 (3 ml) are added to the mixture. After cooling the reaction bath to room temperature, the mixture is purified by filtration, multiple washings, centrifugation, and vacuum drying which yield 4.2 g of GO [40].

All the methods discussed above generate toxic NO_x gases, and yield is also limited. Marcano and co-workers adopted a novel approach in which $NaNO_3$ was replaced by H_3PO_4, and this method known as the improved Hummers method or Tour method. This approach produces highly hydrophilic GO with a greater yield, while eliminating hazardous NO_x gas emissions.

In the improved Hummers method, 3.0 g graphite flakes are added to a mixture of concentrated H_2SO_4:H_3PO_4 (360:40 mL). $KMnO_4$ (18.0 g) is added gradually in six equal parts to the mixture [45]. The addition of $KMnO_4$ elevates the reaction temperature to 40 °C, after which it is externally heated to 50 °C and agitated for 12 h. Thereafter, the reaction bath is cooled to room temperature and poured onto 400 mL ice, with 3 mL 30% H_2O_2. The mixture is purified by filtering, several washings, centrifugation, and vacuum drying, yielding 5.8 g of GO [40].

The procedures of Hummers (HGO), Hummers modified (HGO+), and improved Hummers (IGO) methods are shown in Figure 7.10 with their unoxidized hydrophobic carbon residuals.

7.12 Application techniques of graphene on textile

Despite the fact that GO has been studied for almost a century, the exact chemical structure of GO is still unknown, which is due to its complex partial amorphous nature. Several structural models of GO are suggested, and among them the Lerf–Klinowski model is widely accepted [46]. According to the Lerf-Klinowski model, GO contains epoxide and hydroxyl groups on their basal planes, as well as carbonyl and carboxyl groups near the edges. The structure of GO as per the Lerf–Klinowski model is presented in Figure 7.11.

The presence of oxygenated functional groups on GO facilitates interaction with textile material via electrostatic interaction, van der Waals, hydrogen bonding, π-π interactions, hydrophobic interactions, and so on [3]. There are several ways to deposit graphene or graphene oxide on textile material, such as vacuum filtration, brush coating, printing, wet transfer of graphene layers, and dip coating.

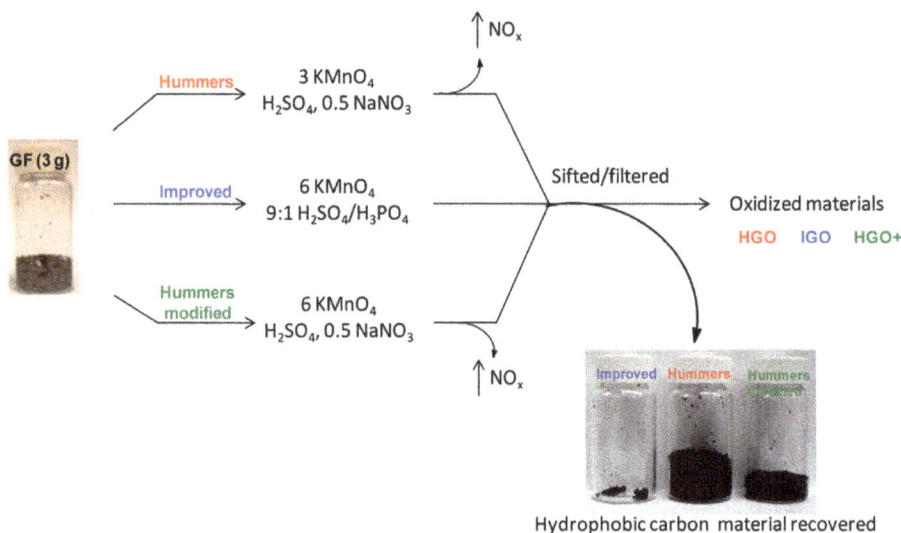

Figure 7.10: Representation of the different Hummers methods with their residual unoxidized hydrophobic carbon material (reprinted from reference [40] with permission, Copyright (2010) American Chemical Society).

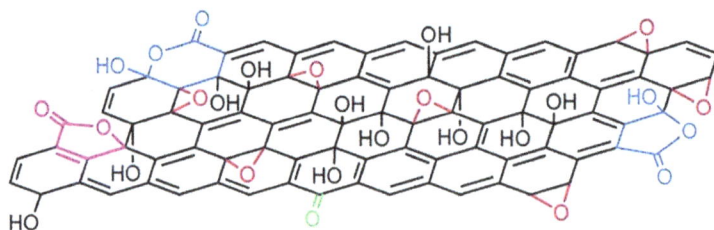

Figure 7.11: Chemical structure of graphene oxide (reprinted from reference [47] with permission, Copyright (2009) Springer Nature).

7.12.1 Vacuum filtration deposition

In vacuum filtration deposition, graphene oxide dispersion is poured on the textile media, which is supported by filter paper underneath, and the whole process is carried out in filtration assembly. Graphene oxide dispersion is passed through the textile media by the applied vacuum. Liang et al. prepared GO-silk composite by passing an ultrasonicated dispersion of 50 mg silk fibres in 20 mL GO dispersion (0.5 g/L), through a 0.22 μm pore size filter [8]. Graphene oxide-enriched cotton was prepared by filtering 5 mg/mL aqueous dispersion of GO through the cotton woven fabric, as shown in Figure 7.12a. This technique is limited to miniature sample preparation [48, 49].

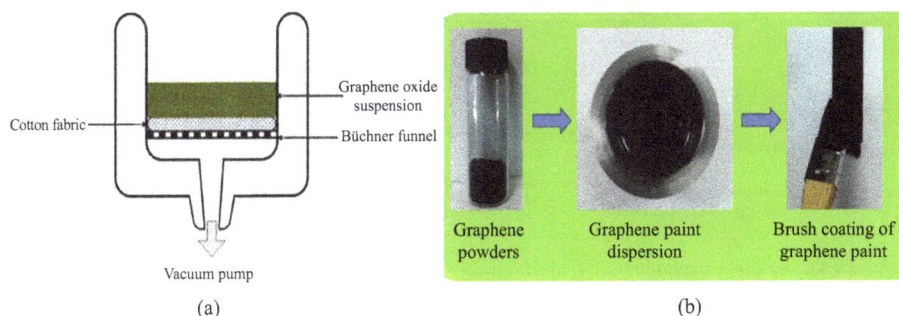

Figure 7.12: (a) Vacuum filtration deposition and (b) brush coating technique [(a) Modified according to reference [48] with permission, Copyright (2015) Elsevier; and (b) Modified according to reference [50] with permission, copyright (2018) John Wiley and Sons Inc].

7.12.2 Brush coating technique

In brush coating techniques, sonicated GO or graphene dispersion is painted onto the fabric surface by brush and then dried. Afterward, the GO-enriched fabric can be reduced to restore conductive properties, if needed. The amount of GO mass loading can be tailored by varying the number of brush coatings. Liu et al. painted 2 mg/mL GO dispersion on cotton woven fabric, using brush coating technique [51]. Later, Javed et al. painted 2 mg/mL GO dispersion on cotton and wool fabrics [52]. Krishnamoorthy et al. prepared graphene nano-paint by ball milling the chemically synthesized graphene with alkyd resin (binder) and other additives. Thereafter, the prepared graphene nano-paint was painted on glass substrate by brush coating technique (Figure 7.12b) [50].

Fabrication of graphene on textile is relatively easy; however, the penetration of graphene inside the porous structure of fabric is limited. Therefore, the amount of graphene mass-loading on substrate is poor [51, 52].

7.12.3 Printing technique

Printing technique is inspired by the traditional printing approach. In the context of nanoparticle fabrication on textile media, it is an approach that is trending. Printing technique is widely adopted to print subtle conductive circuits on fabric. In this technique, GO or graphene ink is sonicated with a suitable solvent to form a printing paste and then transferred to fabric, followed by curing. Graphene-based electroconductive textiles were developed via inkjet printing, screen printing, or transfer printing techniques [16, 53–56]. Cao and co-workers used hot press transfer printing to paste GO (20 mg/mL) on silk fabric at 190–230 °C for 0–240 s, as shown in Figure 7.13 [57].

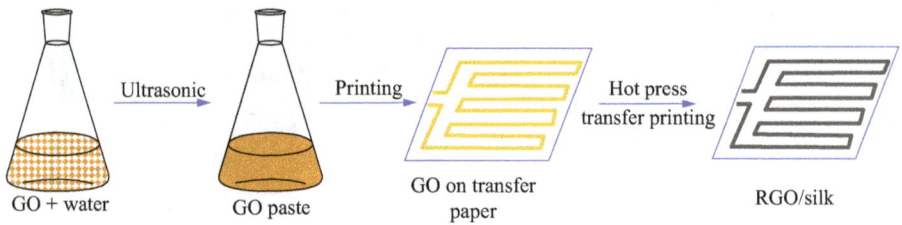

Figure 7.13: Hot press transfer printing of GO on silk fabric (Modified according to reference [57] with permission, Copyright (2018) Elsevier).

Skrzetuska, Michałand, and Krucińska introduced graphene on cotton woven fabric via the screen-printing technique. Printing paste was prepared from addition of graphene pellets (1%wt and 3%wt) and aliphatic urethane acrylate in the aqueous dispersion of carbon nanotubes. Aliphatic urethane acrylate is a cross-linking that was used to enhance the fixation of these conductive particles on fabric [54]. Abdelkader et al. used screen-printing techniques to deposit 5 mg/mL GO on cotton woven fabric for the preparation of flexible supercapacitors [16]. Pepłowski et al. developed a novel method of printing graphene on textile material at room temperature, without using any solvents. They mixed the graphene nano platelets with silicon rubber in varying proportions and printed them on PET foil, and reported a low percolation threshold at 0.147 vol% [55]. Printing techniques are preferred for localized conductive pattern printing on textile surfaces. However, it is not advisable for other applications, where the entire fabric needs to be coated homogeneously with graphene.

7.12.4 Wet transfer technique

In this technique, graphene was synthesized via CVD on copper foil using a carbon source. Thereafter, the prepared graphene/copper composite was transferred on a fabric, where copper was etched out and monolayer film of graphene transferred on a textile substrate. This technique has been used to deposit monolayer graphene sheets on polyethylene terephthalate (PET), polypropylene, polylactic acid, polyethylene, and nylon [58–61].

Verma et al. schematically transferred CVD-grown graphene layers on PET (Figure 7.14). Initially, graphene was grown over Cu foil by the CVD method, and then deposited on top of the PET substrate by hot press lamination. In the next step, Cu foil was etched by dipping in an aqueous solution of iron chloride. However, it was challenging to transfer the fragile one-atom thick graphene layers on a PET substrate, without any supporting film [60]. In 2015, Paquin and co-workers spin-coated PMMA film on CVD-grown graphene layers to provide support to graphene layers during deposition. Thereafter, Cu foil was etched by dipping in an aqueous solution of iron

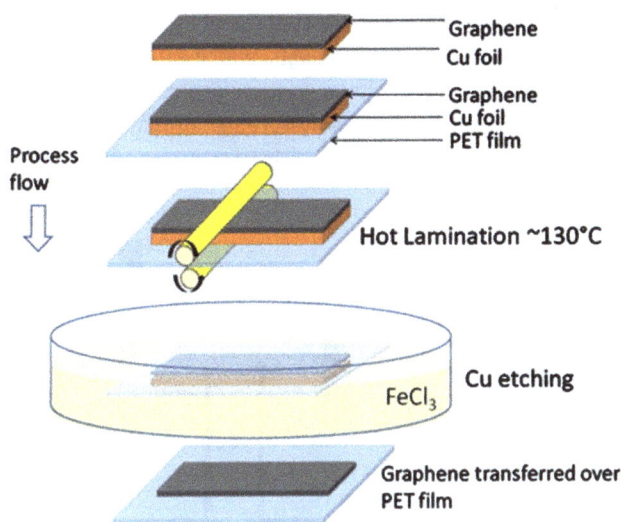

Figure 7.14: Wet transfer of graphene layer on PET substrate (reprinted from reference [60], with the permission of AIP Publishing).

chloride, and the cleaned graphene-PMMA layers were reversely transferred onto a PET substrate [61].

This technique transfers high quality and defect-free graphene monolayer sheets to textiles. However, this technique is not economically scalable to produce graphene-based E-textiles due to the tedious steps and expensive material (copper) and machinery it calls for. In addition, it is challenging to homogeneously transfer clean monolayer or few-layer graphene sheets on uneven surfaces of fibres without wrinkles, folds, tears, and cracks.

7.12.5 Dip coating technique

The dip coating technique (as schematically explained in Figure 7.15) is widely reported to prepare graphene-based electroconductive textile. Fabric is immersed in a GO or graphene dispersion as a function of time, temperature, pH, and concentration of GO. After dipping, the material can be padded to enhance the graphene add-on fabric surface, and then, the fabric is dried. If needed, it is subsequently reduced to restore the conductive structure of graphene. The number of dipping cycles can be varied to achieve the required level of GO mass loading for the respective functional applications [57, 62]. Dip coating was followed for the most of the textile fibres such as cotton [2, 4, 6, 63–65], wool [52], silk [7, 66, 67], nylon [11], polyester [4, 68, 69], jute [70], spandex (lycra®) [71], para-aramid (Kevlar®) [72], and acrylic [73]. Dip coating technique is convenient, scalable, and economical. It does not require any special equipment for graphene deposition. However, there are chances of aggregation of GO or RGO sheets due to strong interaction between these sheets, which results in uneven deposition of graphene onto the textile surface. The aggregation of rGO can be prevented by the addition of dispersing agents such as polystyrene sulfonate (PSS) and sodium dodecylbenzenesulfonate (SDBS) in aqueous dispersion [2, 14].

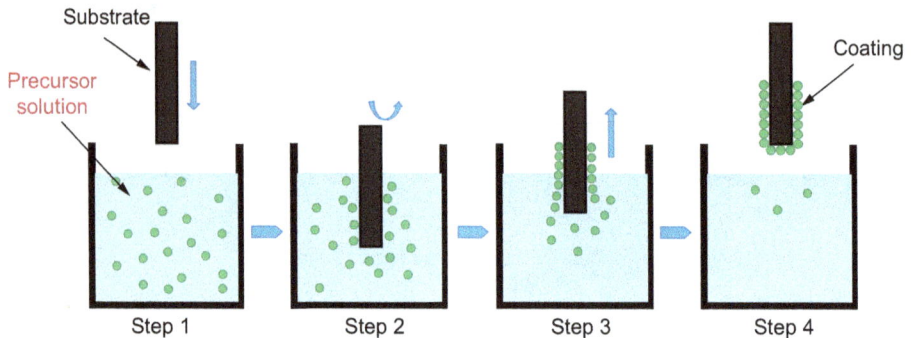

Figure 7.15: Dip coating technique of graphene deposition (Modified according to reference [74] with permission, Copyright (2016) Elsevier).

7.13 Reduction methods of graphene oxide-coated textile

Electric conduction in graphene-based textile is due to graphene only. However, the oxidized form of graphene (GO) is an electrical insulator due to its sp^3-hybridized structure and, thus, cannot be used for electro-conductive applications. Different reduction methods such as thermal, chemical, and ultraviolet (UV) were used to restore the conductive sp^2-hybridized carbon atom network (rGO) from insulating sp^3-hybridized GO structure [24]. All the reported studies related to the reduction of GO suggested that reduction of GO was partial, and the level of reduction can be analysed by carbon to oxygen atomic ratio (C/O ratio) and electrical conductivity. The C/O ratio of GO depends on the synthesis method followed and can be measured by X-ray photoelectron spectrometry (XPS) analysis and Energy dispersive X-ray analysis (EDX). Both C/O ratio and electrical conductivity increase after the reduction.

7.13.1 Thermal reduction

In this method, thermal annealing of GO is usually carried out at high temperatures in an inert atmosphere for deoxygenation. Thermal reduction method is an effective and chemical-free method. Thermal annealing of GO can be done by thermal, microwave, and photo-irradiation. Heating temperature significantly influences the extent of GO reduction. The heating temperature should be above 500 °C for efficient reduction. However, most of the textile materials degrade at these temperature ranges; that's why thermal reduction is not a preferred method for the reduction of GO when it is deposited on a textile material [3, 75]. Cai et al. thermally reduced GO-functionalized cotton at three different temperatures, viz., 160, 200, and 250 °C for 2 h in a nitrogen atmosphere. After thermal reduction, the surface resistivity of GO-functionalized cotton substantially decreases from 10^9 to 10^4 Ω/m^2. However, it is inadequate for high-end electro-conductive applications [76]. Another demerit associated with this method is that it requires expensive equipment and a critical environment.

7.13.2 UV reduction

This method uses UV irradiation to deoxygenate GO. Javed and co-workers treated cotton and wool fabrics with GO dispersion of 2 mg/mL concentration five times, and then dried them. Even after reduction by eight times in the UV curing chamber (365 nm wavelength), the obtained values of surface resistivity for cotton (100.8) and wool (45.0) were very high [52]. The UV-reduction method is a cost-effective and chemical-free process; however, it is time-consuming and offers low electrical conductivity. Therefore, this method is inefficient to reduce GO-coated textiles on a wide scale.

7.13.3 Chemical reduction

Chemical reduction method is the most suitable reduction method for GO-coated textiles, because the reduction is carried out at below 100 °C. In addition, it is an economical method and does not require expensive equipment. These advantages promote the large-scale reduction viability of this method. Chemical reduction can be done by electrochemical process and chemical reagents. Electrochemical reduction of GO is carried out in an electrochemical cell, where reduction is caused by electron exchange between the GO and electrodes.

This method has the advantage of avoiding the use of hazardous reductants (such as hydrazine) as well as the generation of by-products. Abdelkader et al. followed an electrochemical method to reduce GO-coated cotton fabric [16]. Chemical reagent-based reduction is based on chemical reactions between the GO and reductant, resulting in the restoration of the sp^2-hybridized graphene network. Prior to the discovery of graphene, hydrazine (N_2H_4) was usually used to reduce graphite oxide [77], and Stankovich et al. [78] were the first to use hydrazine to reduce GO for the synthesis of graphene. Since then, hydrazine and its derivatives such as hydrazine hydrate and dimethylhydrazine gained popularity for GO reduction [79].

Reduction of Vat dye during dyeing is a well-known process. This inspired researchers to explore similar reductants for GO reduction that are known for vat dye reduction. These chemical reductants include sodium hydrosulfite ($Na_2S_2O_4$), sodium hydroxide (NaOH), sodium borohydride ($NaBH_4$), and lithium aluminium hydride ($LiAlH_4$) [80] Shateri-Khalilabad and Yazdanshenas used five different types of reductants, namely, $NaBH_4$, N_2H_4, $C_6H_8O_6$ (ascorbic acid), NaOH, and $Na_2S_2O_4$ to reduce GO-coated cotton fabric and suggested that $Na_2S_2O_4$ was the best reductant [80].

The majority of reported reductants are hazardous or explosive, as well as difficult to handle on an industrial scale. To address this issue, some novel eco-friendly reductants were also reported, such as potassium hydroxide [81], ascorbic acid [80, 82, 83], sugar [84], baker's yeast [85], melatonin [86], glucose [87], green tea [88], Escherichia coli biomass [89], and Bacillus marisflavi biomass [90]. However, these green reduction routes need further investigation before establishment. In a nutshell, the chemical reagent-based reduction method has huge potential for large-scale economical reduction of GO-coated textile material.

7.14 Effect of graphene coating variables on graphene add-on and electrical resistivity

The electroconductive property of graphene-based textile substrate depends on the level of graphene add-on achieved on the textile substrate; higher the graphene add-on, smoother the electric conduction. On the other hand, the level of graphene add-on is influenced by the graphene dipping or coating variables and type of substrate. In this section, the effect of dipping variables on the electrical resistivity of graphene-coated textiles is reviewed, along with the graphene add-on. Dipping variables in the dip coating techniques are GO concentration, pH of GO dispersion, dipping time, dipping temperature, and reductant parameters [80].

7.14.1 Effect of GO concentration

Concentration of GO has a strong influence on the graphene add-on and resultant properties of the graphene-coated textiles. Higher amounts of graphene nanoparticles deposited at the higher GO concentrations facilitate smoother electric conduction [3, 65]. Cotton woven fabric was made electroconductive by coating with three different GO concentrations, viz., 0.75%, 1.5%, and 2.25%, up to 15 dipping cycles. The authors made an attempt to relate the amount of graphene add-on with the obtained surface resistivity. With the increase in GO concentration from 0.75% to 2.25%, the graphene add-on increased from 2.41% to 3.31%; as a result, the surface resistivity decreased from 13.68 to 0.26 MΩ/square (Figure 7.16) [6].

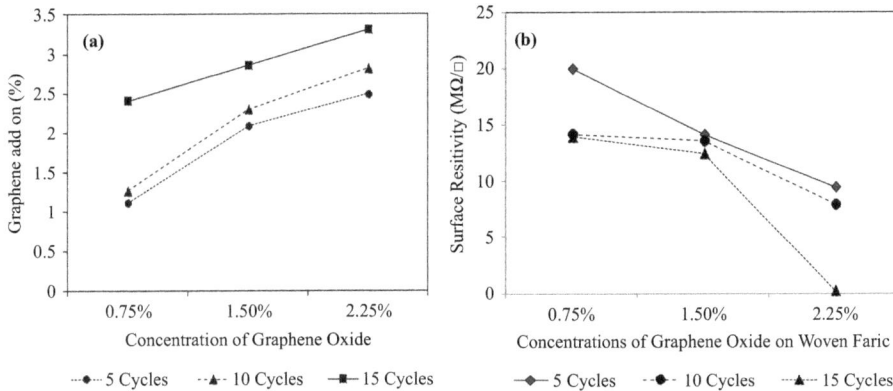

Figure 7.16: (a) Effect of GO concentration on graphene add-on and (b) effect of GO concentration on surface resistivity [Modified according to reference [6] with the permission of Taylor & Francis (https://www.tandfonline.com)].

Karimi et al. treated cotton fabric with various concentrations of GO (0.02, 0.05, 0.1, 0.2, and 0.5% w/v) and subsequently reduced the treated fabric with varying amounts of titanium trichloride aqueous solution. The electrical resistivity decreases from 6.6×10^6 Ω/square to 6.3×10^3 Ω/square, as the GO concentration increases from the 0.02% to 0.5% (w/v) [91]. Liu et al. coated rGO on polyurethane (PU)-treated PET nonwoven fabrics with different concentrations of rGO, ranging from 0.001% to 0.150%. Here, PU was used as a cross-linking layer in between the PET fabric and rGO to improve the rGO loading on PET fabric. Once the concentration of rGO crossed the electrical percolation threshold (0.0012%), the conductivity rapidly increased up to 0.080% rGO concentration, and then stabilized. Electrical conductivity of 2.0×10^{-5} S/sq was reported at 0.080% rGO concentration.

Figure 7.17 (SEM image-III) shows that, at higher concentrations, a large number of wrinkled and overlapped graphene sheets were deposited on the PU-treated PET fabric and formed a continuous conducting network that enhanced the electrical conductivity [92].

The surface resistivity of graphene/WPU (waterborne polyurethane) composite film decreases from $2.5 \times 10^{10} \pm 1.9 \times 10^{10}$ Ω/sq to $4.0 \times 10^3 \pm 1.9 \times 10^3$ Ω/sq, when the graphene content increases from 0 wt% to 16 wt%. However, at the 16 wt% of graphene, the graphene/WPU composite film became brittle and crushed after the annealing treatment, due to the irreversible agglomeration of graphene sheets through van der Waals interaction [93]. The GO add-on on the unit mass of silk was significantly increased from 5.19 to 17.39 mg/g, as the initial concentration of the GO dispersion increased from 100 to 1,000 mg/L [94].

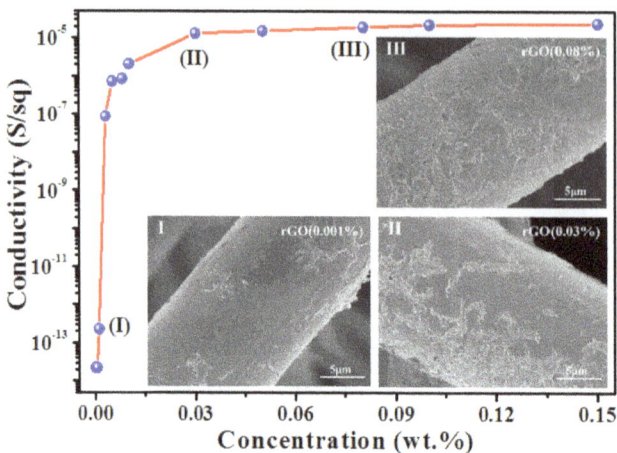

Figure 7.17: Effect of rGO concentration on the surface conductivity of the composite fabrics (reprinted from reference [92] with permission of The Royal Society of Chemistry).

7.14.2 Effect of the pH of GO dispersion

According to the Lerf -Klinowski model, GO carries hydroxyl and epoxy groups on the basal plane and carboxyl groups on the edge of the GO sheets [46]. These attached oxygenated functionalities on the carbon backbone are responsible for the hydrophilic nature of GO. The surface chemistry of GO is strongly influenced by the pH of dispersion. GO displays a negative surface charge throughout the pH range, due to the deprotonation of the attached carboxyl groups ($-COO^-$) on GO edges [63, 95–98]. Konkena and Vasudevan measured the electrokinetic potential or Zeta potential of GO and rGO in aqueous dispersions at different pH (Figure 7.18).

It can be observed from Figure 7.18 that as the pH of GO dispersion increases up to 10, the negative zeta potential increases, and further increase in pH results in slight reduction in negative zeta potential [97].

WU et al. measured the optical transmittance and carbon-to-oxygen (C/O) ratio of an aqueous dispersion of GO as a function of pH. The C/O ratio gradually increases with the rise in pH of the GO dispersion up to 7, due to the aggregation of GO sheets. The C/O ratio of GO dispersion rapidly increases in the alkaline pH range, due to deoxygenation and reduction of oxygen-containing functional groups ($-COO^-$) (Figure 7.19a) [99]. Addition of alkali separates the highly oxidized low molecular weight material called oxygen debris, which were attached to GO; as a result, graphene sheets start to coagulate and quality of GO deteriorates. When pH of GO dispersion rises from 9 to 12, the C/O ratio increases dramatically, due to the elimination of the majority of the oxygen-containing functional groups of GO [99, 100]. The transmittance of the GO dispersion also decreases,

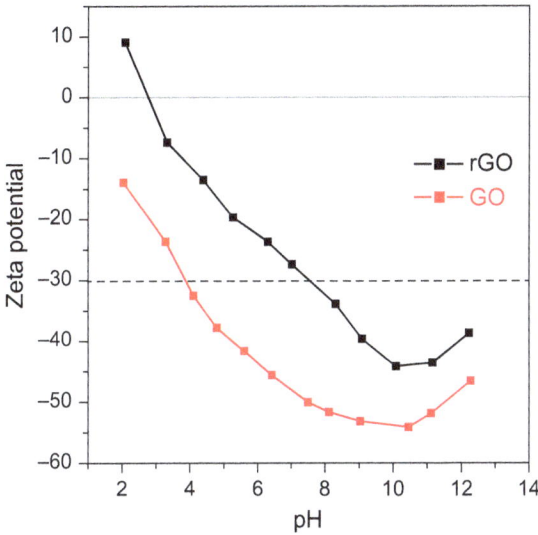

Figure 7.18: Zeta potential of GO hydrosol at various pH (Modified according to reference [97] with permission, Copyright (2012) American Chemical Society).

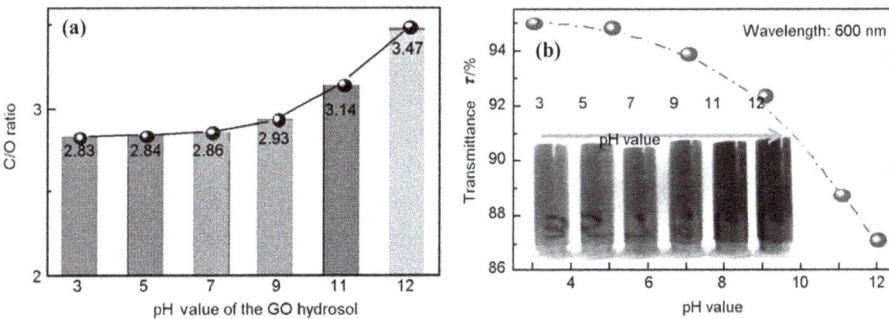

Figure 7.19: (a) Carbon-to-oxygen ratio and (b) optical transmittance of the GO dispersion at different pH (reprinted from reference [99] with permission, Copyright (2013) Elsevier).

as the pH of GO dispersion increases, which was validated by the change in colour from brown to black (Figure 7.19b).

Shih et al. investigated the pH-dependent behaviour of GO aqueous dispersions (Figure 7.20). At pH 1, the edge carboxyl groups of GO nanoparticles protonated strongly, which resulted in an aggregation of GO nanoparticles and colloidal instability. At pH 14, two different particle sizes, viz., 200 nm and 3 µm of GO, were observed with dark brown color dispersion [95]. In a nutshell, the ideal pH range of GO dispersion for coating on textile fibre lies between 2 and 7 [95, 99, 101].

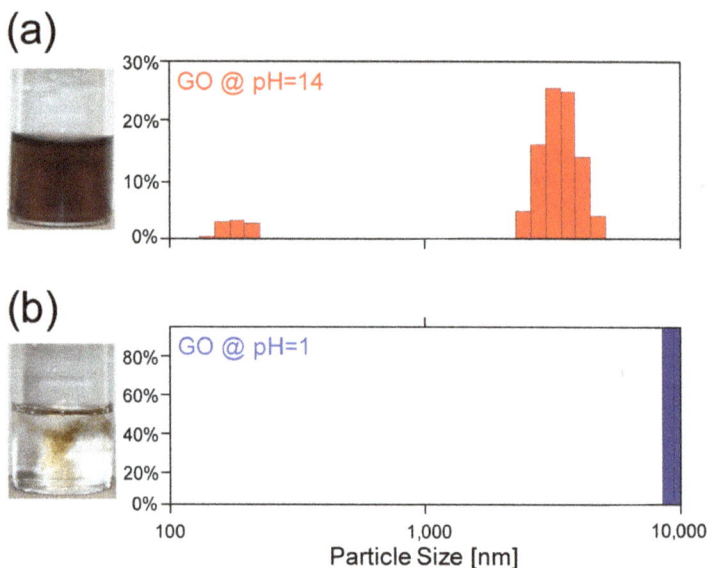

Figure 7.20: Photographs of a 1 mg/mL GO dispersion (left) and its colloidal size distribution (right), measured using the DLS technique at (a) pH 14 and (b) pH 1 (reprinted from reference [95] with permission, Copyright (2012) American Chemical Society).

In neutral aqueous media, most of the textile fibres acquire negative surface charge due to the dissociation of functional groups, which increases further, with the increase in pH (Figure 7.21). The number of positively charged residues equals the number of negatively charged residues, or they are electrically neutral at a specific pH value, which is known as the isoelectric point and denoted by IEP. Above the IEP, the fibre surface has a net negative charge, and below the IEP, the fibre surface has a net positive charge. The IEP of fibre is a crucial parameter in textile wet processing [102, 103]. The IEP of commonly used textile fibres is shown in Table 7.2.

Table 7.2: Isoelectric point of textile fibres [102–105].

S. no.	Fibre	IEP	S. no.	Fibre	IEP
1	Cotton	2.9	5	Nylon 6.6	4–4.7
2	Viscose	2.8	6	Polyester	<2.5
3	Wool	4.7	7	Acrylic	3.0
4	Silk	3.8–3.9	8	Spandex	7.9

When a fibre whose IEP lies below 4.7 is dispersed in a GO dispersion at neutral pH, a strong electrostatic repulsion arises between the fibre and GO, since both the fibre and GO are anionic at neutral pH. As a consequence, the amount of GO deposition on the fibre surface will be lower, and the resultant functional properties of grapheme-

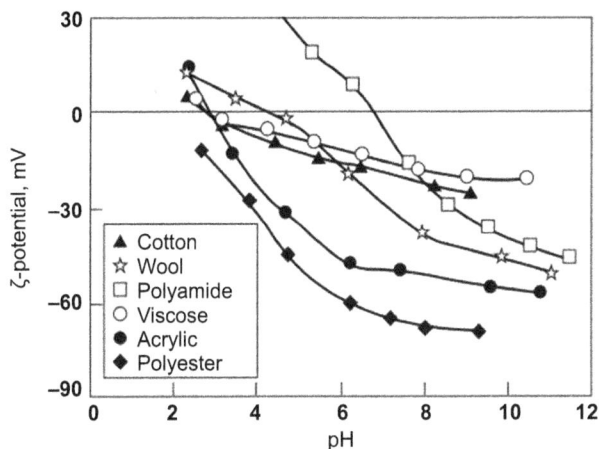

Figure 7.21: Zeta potential of the standard fabrics at various pH (Modified according to reference [103] with permission, copyright (2005) John Wiley and Sons Inc.).

coated fibre will be poor. To overcome this issue, researchers adjusted the pH of the GO dispersion below the IEP of fibre. Chatterjee and Jain obtained a maximum GO adsorption capacity of 11.07 mg/g on silk fibre, which decreased with an increase in pH from 2 to 7, due to the increased repulsion between GO and silk as well an increase in the C/O ratio of GO [94]. Javed et al. painted 2 mg/mL aqueous dispersion of GO on wool fabric at pH 4.5 [52]. Sahito et al. dipped cotton fabric in 0.1% aqueous dispersion of GO at pH 2 and reported 3.1 mg/g GO loading. After the Bovine Serum Albumin (cationic agent) treatment on cotton fabric, the amount of GO loading was enhanced to 5.2 mg/g [63].

7.14.3 Effect of dipping temperature

Dipping temperature or coating temperature is particularly applicable to the dip coating technique. The dipping temperature mainly influences the carbon-to-oxygen ratio of GO. With the rise in dipping temperature, the active sites of GO are degraded or deoxygenated, and as a result, the number of available surface-active sites for interaction with textile substrate decreases [13]. As a result, the graphene add-on and electroconductive properties would be downgraded. According to Sahito et al., the dipping temperature should be less than 80 °C; above this temperature, undesired thermal reduction of GO initiates, which limits the restoration of the sp^2-hybridized graphene structure during the reduction stage [63]. Chatterjee and Jain also made similar observations during GO adsorption on silk as a function of temperature [94].

During the GO coating on textile substrates, arbitrary dipping temperatures were used. Majority of researchers maintained room temperature of GO dispersion during dipping of textile substrates [6, 64, 80, 106, 107]. Sahito et al. maintained 80 °C bath temperature during dipping cotton in GO dispersion [63]. Even during the drying of GO-coated textiles, the recommended drying temperature has been below 80 °C [6, 63, 69, 76, 80, 92, 106, 108].

7.14.4 Effect of dipping time

The dipping time or contact time refers to how long the substrate is immersed in the GO dispersion. The deposition of GO onto textile fibre is an adsorption process in which dipping time has a significant impact. As time proceeds, GO nanoparticles transfer from the GO dispersion to the fibre surface and attain equilibrium. The optimization of dipping time is vital to minimize the loss of mechanical properties of fibre due to prolonged dipping as well as energy consumption. Researchers used arbitrary dipping time ranging from 2 to 60 min, and most of the studies employed 30 min of dipping time to adsorb an adequate amount of GO on the substrate [6, 63, 69, 80, 92, 106, 108–110]. Chatterjee and Jain made an effort to investigate the effect of dipping time on the GO add-on for silk fibre, and they found 3 min of dipping time was sufficient for efficient adsorption of GO on silk [94]. Ji, Chen, and Xing studied the effect of coating time on the K/S value and whiteness index of silk (Figure 7.22).

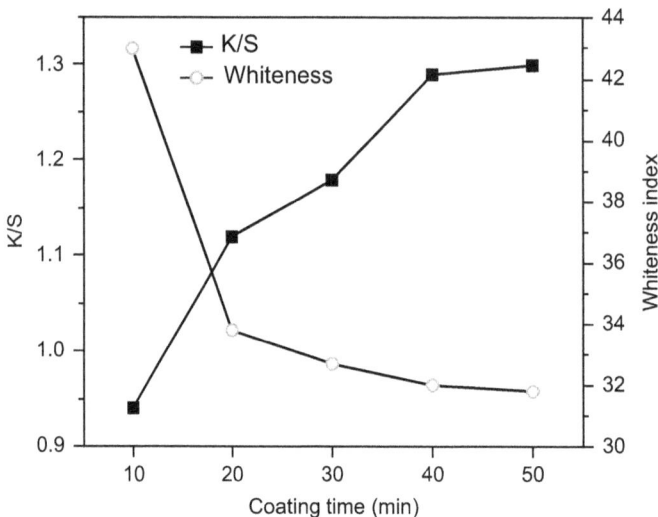

Figure 7.22: Effect of coating time on the K/S and whiteness of GO-coated silk fabric (Modified according to reference [9] with permission, Copyright (2019) Elsevier).

The K/S value of silk fabric rapidly increased up to 40 min of coating time and then stabilized. On the other hand, the whiteness index of silk fabric decreased with increasing dipping time and attained equilibrium after 40 min of coating time [9].

7.14.5 Effect of reductant types and parameters

Reduction of GO-treated textile substrate is an essential step to restore the electrocon-ductive properties of graphene by removing the oxygen-containing functional groups of the attached GO. The majority of reported studies followed chemical reagents-based re-duction methods, when GO was coated on a textile substrate. The widely used chemical reductants were hydrazine (N_2H_4) [63, 80, 92, 111, 112], hydroiodic acid (HI) [72, 113–117], sodium hydrosulfite ($Na_2S_2O_4$) [2, 6, 7, 69, 73, 80, 118–120], sodium hydroxide (NaOH) [80, 121], sodium borohydride ($NaBH_4$) [80, 122], ascorbic acid ($C_6H_8O_6$) [8, 9, 71, 80, 106], tita-nium(III) chloride ($TiCl_3$) [91, 123], and so on.

The level of conductivity achieved on a GO-coated textile material is influenced by the type of reductant and its parameters, viz., reductant concentration, reduction temperature, and reduction time. The reduction time and temperature are deter-mined on the basis of the type of reductant used during the reduction process.

Shateri-Khalilabad and Yazdanshenas earned a great number of citations for a de-tailed study on the effect of different types of reducing agents on the electrical and me-chanical properties of GO-cotton fabrics. The GO-cotton fabric was reduced with $NaBH_4$, N_2H_4, $C_6H_8O_6$, $Na_2S_2O_4$, and NaOH while keeping the reduction parameters constant (25 mM concentration, 95 °C, and 60 min). Least surface resistance of 0.374 kΩ/cm was reported with $Na_2S_2O_4$ reduction and recommended concentration of 50 mM, reduction time of 30 min, and reduction temperature of 95 °C. In addition, the mechanical proper-ties of cotton fabric were unaffected due to the reduction with $Na_2S_2O_4$ [80]. According to Zhou et al., $Na_2S_2O_4$ is an efficient reductant for GO because of its low electrode po-tential (– 1.12 V) in an alkaline solution. In an alkaline solution, it easily donates two protons and reduces the epoxide and hydroxyl groups of the GO, and restores the gra-phene structure [124]. Jain and Chatterjee optimized the $Na_2S_2O_4$ concentration for cot-ton nonwoven fabric using Box–Behnken response surface design, in order to achieve minimum surface resistivity [125].

The reduction parameters maintained during the reduction of GO-coated silk (GO-silk) fabric by L-ascorbic acid were as follows: 0.2 mol/L concentration, 90 °C temperature, and 90 min of reduction time [9]. Sahito et al. reduced GO-coated cotton (GO-cotton) fab-ric in 0.1 M hydriodic acid vapour at 90 °C for 20 min to restore the conductive properties and reported minimum surface resistivity of 114 Ω/sq, after 20 coating cycles [109]. Xu et al. studied the effect of $NaBH_4$ concentration (0.1–0.5 mol/L) and reduction time (2–12 h) on the sheet resistance of graphene/cotton composite fabric. The sheet resistance of fabric rapidly decreases as the concentration of $NaBH_4$ increases from 0.1 to 0.3 mol/L and then gradually decreases up to 0.5 mol/L (Figure 7.23a). Lowest sheet resistance of

560 Ω/sq was reported at 0.5 mol/L concentration of $NaBH_4$. The sheet resistance of fabric rapidly decreases as the reduction time rises from 2 to 6 h and then stabilizes, because most of the oxygenated surface functionalities of GO were already removed in the 6 h of reduction time (Figure 7.23b) [122].

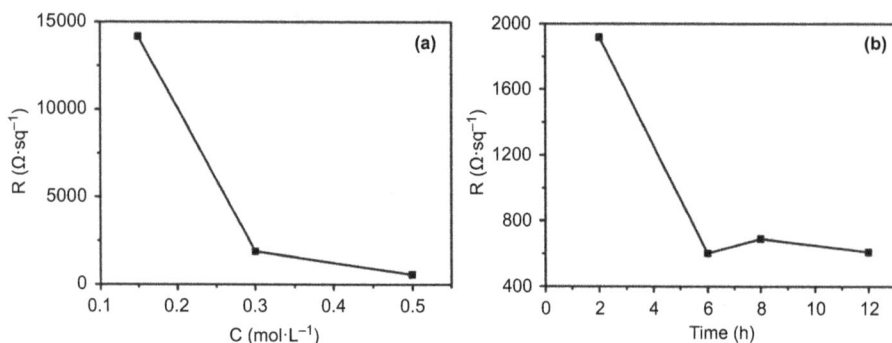

Figure 7.23: The effects of (a) $NaBH_4$ concentration and (b) $NaBH_4$ reduction time on the sheet resistance of graphene/cotton composite fabrics (Modified according to reference [122] with permission of The Royal Society of Chemistry).

7.15 Effect of dipping cycle on graphene add-on and electrical resistivity

The dipping cycle, also known as the coating cycle, refers to the number of times the substrate is treated with graphene. The purpose of repetitive dipping is to improve graphene loading on the substrate. In initial dippings, graphene is mainly adsorbed on a substrate due to the interaction between the graphene and substrate. In subsequent dippings, the graphene adsorption enhances mainly due to the interaction between the graphene practices via hydrophobic force and $\pi-\pi$ interaction [7, 52, 125].

Repetitive dipping can be done in two ways, namely, consecutive dipping and dip–dry–reduce methods. In the consecutive dipping method, the number of dipping cycles can be increased by repeating the dipping and drying processes (processes "b" and "c" in Figure 7.24) and then reduced.

In the dip–dry–reduce method, a substrate is dipped in GO dispersion, dried, and reduced. Here, one dipping is completed. These processes are repeated n times for n number of dippings (Figure 7.25).

In the consecutive dipping method, the already deposited negatively charged GO on the substrate repels the available GO in the dispersion during the next dippings, and thus, the graphene add-on is relatively lower. In the dip–dry–reduce method, the reduction step after each dipping minimizes or even neutralizes the negatively charged GO

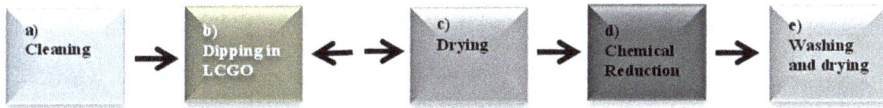

Figure 7.24: Representation of the consecutive dipping method (reprinted from reference [68] with permission of the Royal Society of Chemistry).

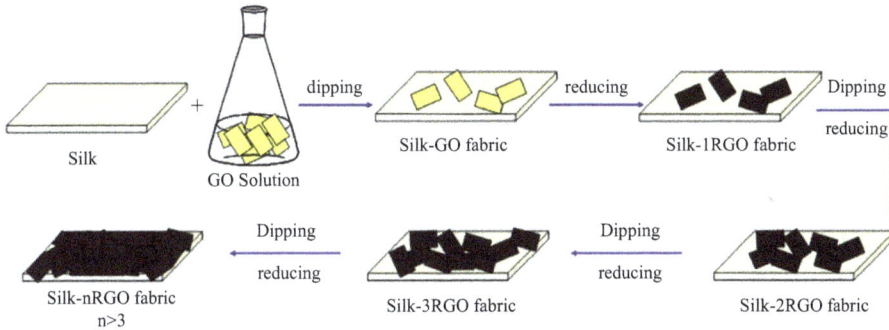

Figure 7.25: Representation of the dip–dry–reduce method (Modified according to reference [7] with permission, Copyright (2017) Elsevier).

deposited on substrate, that is, reduced to rGO. As a result, in subsequent dipping, GO present in the dispersion easily interacts with the rGO deposited on the substrate [7, 125].

Shateri-Khalilabad and Yazdanshenas followed the consecutive dipping method to coat GO on cotton fabric. With the increase in the dipping cycle, the percentage reflectance of GO-coated cotton decreases due to the improvement in GO loading [80]. Kongahge et al. made conductive polyester nonwoven fabric by consecutively dipping in 0.6% liquid-crystalline GO dispersion and subsequent reduction with a mixture of sodium dithionite and sodium bisulfite (Figure 7.26).

The surface resistivity of polyester fabric decreases with the increase in coating cycle, due to the improved GO add-on [68].

Kim et al. dipped para-aramid knit fabric in a composite solution of 8 wt% graphene and 15 wt% WPU. With the increase in the dipping cycle from 1 to 5, the depth of black colour increases (Figure 7.27a), and the porosity of fabric also decreases (Figure 7.27b), due to the abrupt increase in graphene add-on from 38.70% to 125.50%. As a consequence, the surface resistivity decreases to $3.1 \times 10^4 \pm 1.5 \times 10^3$ Ω/square after five dipping cycles (Figure 7.28) [13].

The surface resistance of graphene-coated cotton fabric decreases from 201.1 to 0.374 kΩ/cm when the dipping cycle rises from 1 to 20, due to the improvement in the layered structure of graphene on cotton [80]. With the rise in the dipping cycle from 5 to 15, the surface resistivity of cotton woven fabric drastically decreases from 9.48 to 0.26 MΩ/square due to the improvement in graphene add-on from 2.48% to 3.31% [6].

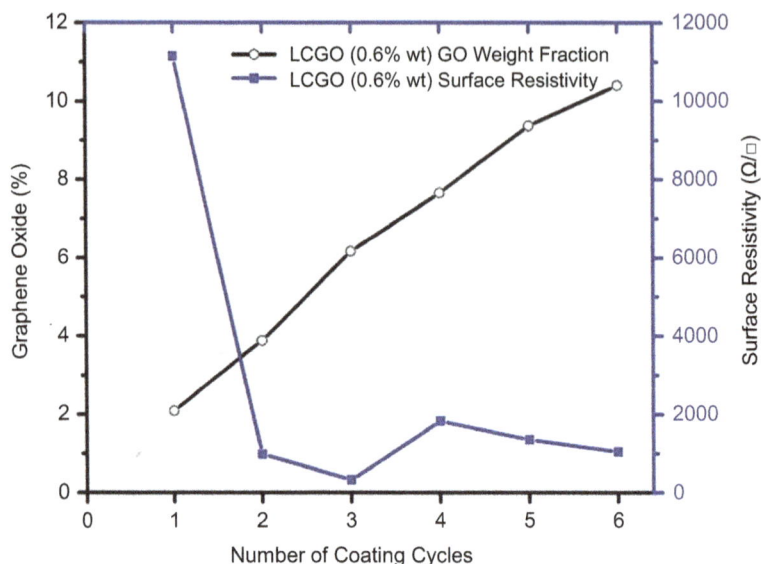

Figure 7.26: Effect of coating cycles on GO add-on and surface resistivity (Modified according to reference [68] with permission of The Royal Society of Chemistry).

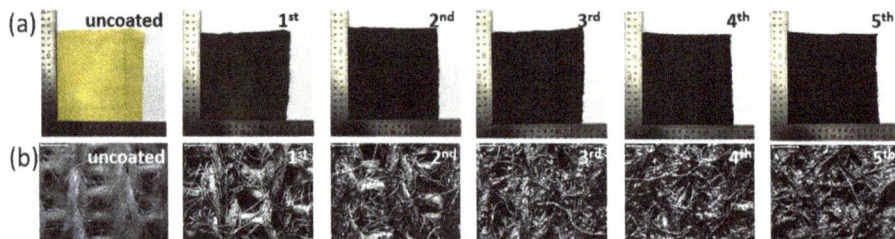

Figure 7.27: (a) Digital images and (b) morphology of graphene/WPU dip-coated para-aramid knits [13].

Similarly, the graphene add-on on knitted fabric improved from 2.75% to 3.96%, after enhancing the dipping cycle from 5 to 15. As a consequence, the surface resistivity of fabric lowered from 8.02 to 0.19 MΩ/square [6]. The electrical conductivity of silk fabric improved after multiple graphene coatings and a lowest surface resistance of 3.24 kΩ/cm was obtained after nine graphene coatings [7]. Along with electrical conductivity, hydrophobicity and anti-ultraviolet properties were also improved. Similarly, the surface resistance of bovine serum albumin-functionalized silk fabric decreased after the multiple graphene coatings, and a lowest surface resistance of 1.5 kΩ/cm was achieved after seven coatings [111].

Xu et al. dipped cotton fabric in 2 mg/mL dispersion of GO and then chemically reduced it with $NaBH_4$ to prepare graphene/cotton flexible electrodes. The surface

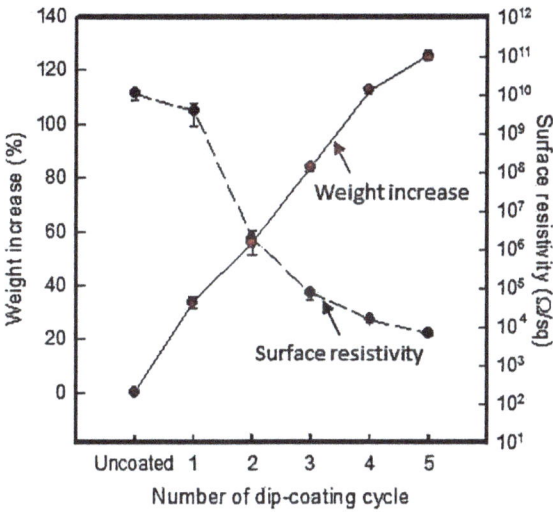

Figure 7.28: Weight increase and surface resistivity of graphene/WPU dip-coated para-aramid knits [13].

resistance of fabric drastically decreases as the number of coating cycles rises, and a lowest surface resistance of 611 Ω/sq was obtained after 20 coating cycles (Figure 7.29a). However, the tensile strength of cotton fabric was decreased after repetitive graphene coating, maybe due to the destruction of the cotton structure (Figure 7.29b) [122].

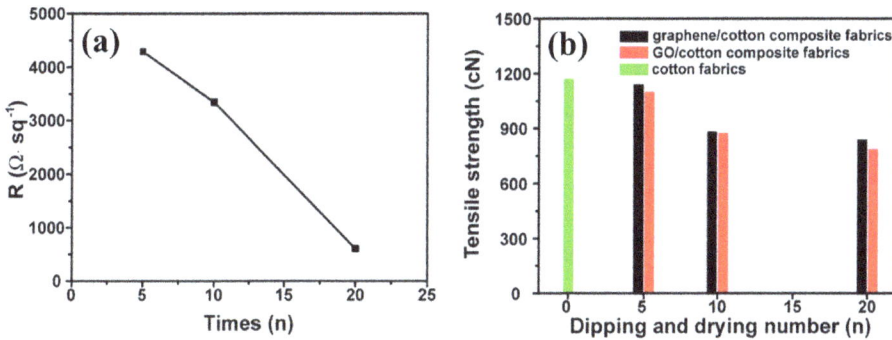

Figure 7.29: The effects of dipping-drying process on (a) sheet resistance and (b) tensile strength (reprinted from reference [122] with permission of The Royal Society of Chemistry).

7.16 Effects of substrate types

The presence of oxygenated surface functional groups on GO facilitates interaction with the textile substrate as well as in the dispersion. However, other forms of graphene such as rGO and pristine graphene lack surface functional groups and need dispersants to coat on textile substrates [97]. As discussed previously, GO displays a strongly negative surface charge throughout the pH range due to the deprotonation of the attached carboxyl groups ($-COO^-$) on GO edges [63, 95–98].

Textile fibres also carry a surface charge that influences its sorption properties. The sorption properties of fibres are influenced by their chemical structure, number and type of accessible functional groups, degree of crystallinity, amorphous regions, size and shape of voids, and so on [103]. It is well known that different textile fibres have unique chemical structures and morphology. Therefore, the type of fibre influences graphene adsorption. Several types of fibres are explored for graphene-based E-textile preparation. Chatterjee and Jain [94] conducted equilibrium isotherms, kinetics, and thermodynamics studies for GO adsorption on silk, in order to understand the adsorption mechanism. The adsorption of GO onto silk is best described by the Freundlich isotherm, and the kinetics of GO adsorption is best described by the pseudo-second-order kinetic model. According to thermodynamic analyses, GO adsorption is spontaneous and exothermic.

Javed and co-authors coated GO on cotton and wool fabrics, while maintaining similar process conditions for both fabrics. The wool fabric showed lower surface resistivity (139.5 kΩ/sq) than cotton (331 kΩ/sq), owing to the comparatively higher graphene add-on on wool [52]. Chatterjee and co-authors reported higher graphene add-on on knitted fabric compared to woven fabric, due to the higher porosity of knitted fabric [6]. Jain and Chatterjee studied the influence of fibre type on rGO add-on and the resultant mass-specific resistance (electrical conductivity).

The authors observed that the type of fibre strongly influenced the graphene add-on. Highest rGO add-on was obtained on cotton fibre followed by viscose, silk, nylon, wool, polyester, and acrylic. In terms of mass-specific resistance, lowest mass-specific resistance was achieved in cotton fibre followed by viscose, silk, nylon, wool, polyester, and acrylic [126].

7.17 Application of graphene-based functional textiles

Graphene was widely employed to impart various functional qualities to textile materials. This section discusses important applications along with their working mechanisms.

7.17.1 Thermotherapy

In thermotherapy, external heat is applied to enhance blood circulation to the affected area. The increased blood flow to the affected tissue provides proteins, nutrients, and oxygen for accelerated healing [127, 128]. Electrically driven heaters prepared by conductive textiles attracted great interest in personal heating and thermotherapy applications.

The electrothermal or resistive heating principle of conductive textile material is based on Joule's first law. Assuming that the heating element acts as a perfect resistor and all of the power (P) is transmitted to heat (Q) in a time interval Δt, the Joule heating equation can be written as [129, 130]

$$P = \frac{Q}{\Delta t} \tag{7.2}$$

or

$$Q = P \, \Delta t \tag{7.3}$$

using Ohm's law ($V = I \, R$) in conjunction with the definition of electrical power in terms of current I and electric potential difference V. The result is:

$$P = V \, I = I^2 \, R \tag{7.4}$$

The Joule heating equation can be rearranged as follows:

$$Q = V \, I \, \Delta t = I^2 \, R \, \Delta t \tag{7.5}$$

This implies that the heat generation capacity of conductive textile material is governed by the current flow through the material, the resistance of the material, and the time duration for the voltage supply applied.

Electrically driven heaters have recently used metals (copper and silver) [131–134], conductive polymers [135–144], carbon nanomaterials (graphene and carbon nanotubes) [13, 92, 145–150], and their hybrid nanocomposites [14, 151–154]. Metal-based heating pads are widely used for heat therapy. Metal-based heating pads are prepared by metal coating on textile materials or insertion of metal thread in textile materials during sewing, weaving, knitting, or embroidery. Metal-based heating pads exhibit poor flexibility, are often toxic, non-biodegradable, and expensive, and do not maintain a constant temperature over an extended time [4, 68, 142, 155]. These issues can be overcome by the application of conductive polymers and carbon-based materials on textile substrates. The processability of conductive polymers is difficult and often unstable towards atmospheric ageing. Graphene is far superior to any other form of carbon material due to its unique 2D structure, which exhibits exceptional electrical, mechanical, optical, and thermal properties. The graphene-based textile materials are highly conductive, flexible, and stable towards washing, rubbing, and atmospheric aging [3–6].

Stretchable heaters were prepared by coating graphene/carbon black on cotton (CC) and wool (CW) fabrics; thereafter, these conductive fabrics were fixed on an Eco-flex, which is a highly stretchable elastomer. The surface temperature of these prepared heaters reached up to 103 °C at 20 V (Figure 7.30a and b). The authors also evaluated the cyclic electrothermal performance of CW and CC heaters at an applied current of 50 mA (Figure 7.30c and d) [14].

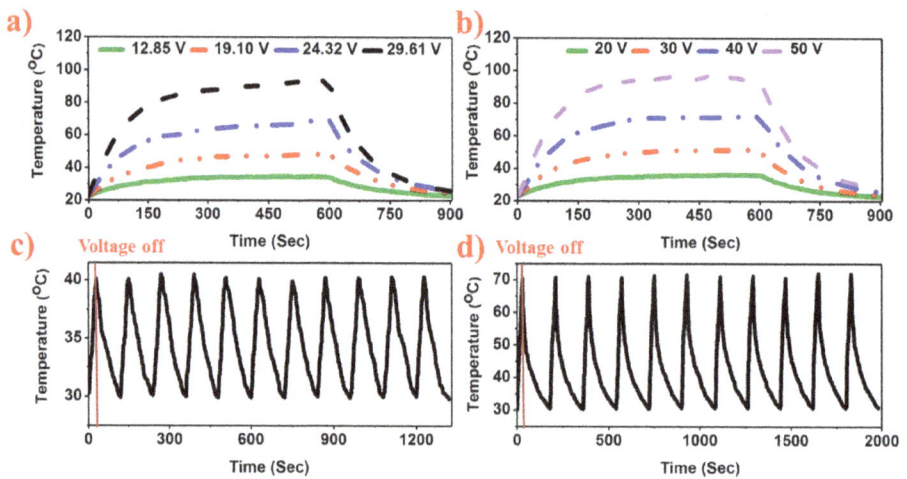

Figure 7.30: Time-dependent temperature changes of (a) CW and (b) CC wearable heaters under various input powers; cyclic electrothermal performance of (c) CW and (d) CC heaters (reprinted from reference [14] with permission, Copyright (2018) American Chemical Society).

X. Liu et al. coated rGO on polyurethane-treated polyester nonwoven fabric (5 cm × 5 cm), and surface temperature of 59 °C was reported at 30 V voltage supply in 30 min of heating time (Figure 7.31) [92]. Kim et al. treated para-aramid knit fabric with 8 wt% graphene/WPU and reported a surface temperature of 54.8 °C at 50 V in 20 min of heating time [13]. Graphene-based flexible heaters were prepared by the brush (2 wt%) and dip coatings (4.5 wt%) of few-layer graphene (FLG) on cotton cloth (CC) of 15 cm × 15 cm. The corresponding surface temperatures of FLG-coated CC (FLG@cotton) fabrics at different voltage supplies are shown in Figure 7.32, with their corresponding thermal images [156].

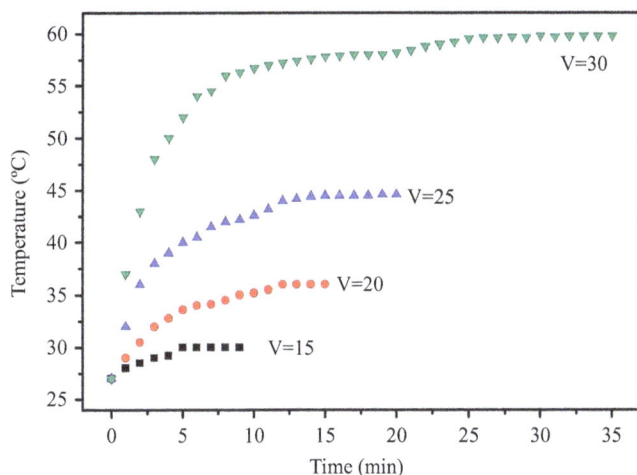

Figure 7.31: Change in surface temperature of composite fabrics as a function of time at various voltage supplies (Modified according to reference [92] with permission of The Royal Society of Chemistry).

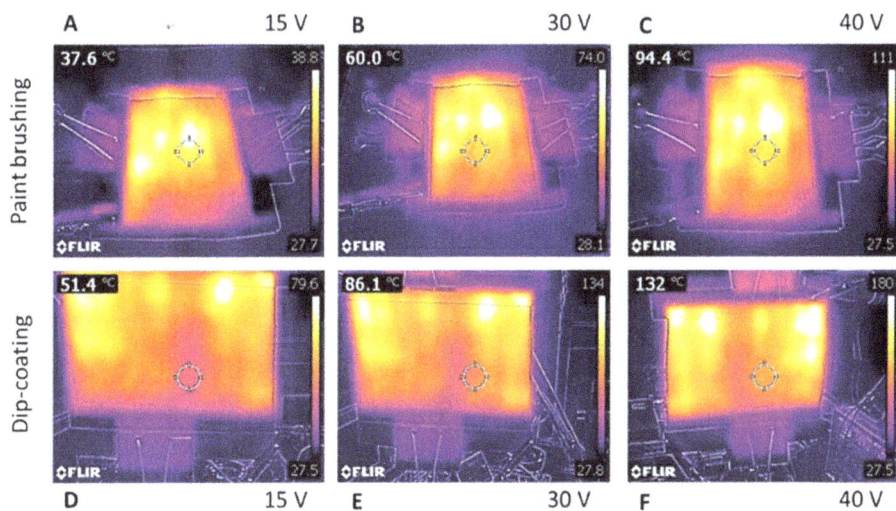

Figure 7.32: (A–C) Surface temperature of painted FLG@cotton fabric and (D–F) dip-coated FLG@cotton fabric under the same applied voltage (reprinted from reference [156] with permission, Copyright (2020) American Chemical Society).

7.17.2 UV Protective textiles

UV protection properties are imparted to textiles to safeguard the textiles and the person from overexposure of UV rays. UV protection properties of textile material can be improved by designing coating with UV blockers and fabric structures. According to their chemical structures, UV blockers are classified into two categories: organic and inorganic UV blockers. Benzophenone [157] and natural dyes [158] are widely known organic UV blockers. Inorganic UV blockers include several semiconductors such as titanium dioxide (TiO_2) [159, 160], silicon dioxide (SiO_2) [161], zinc oxide (ZnO) [162, 163], as well as graphene [64].

UV absorption and reflection are the most acceptable UV blocking mechanisms of graphene [164–166]. Graphene shows strong UV absorption below 280 nm s due to $\pi \rightarrow \pi^*$ transitions [35], which are a result of its unique two-dimensional structure that reflects a part of incident UV radiation at higher wavelengths [167]. UV protection factor (UPF) shows a material's UV protection performance; higher the UPF, better the UV protection. Tian and co-workers deposited layer-by-layer self-assembly of GO and chitosan on cationized cotton fabric and reported UPF values of 88.93 and 452 after one and 10 coatings, respectively (Figure 7.33). In addition, the prepared fabric exhibited excellent UV protection properties with wash stability [64]. A UPF value of 97 was achieved on GO/polymeric N-halamine-coated cotton fabric, which was further enhanced to 187, after the reduction with L-ascorbic acid [12].

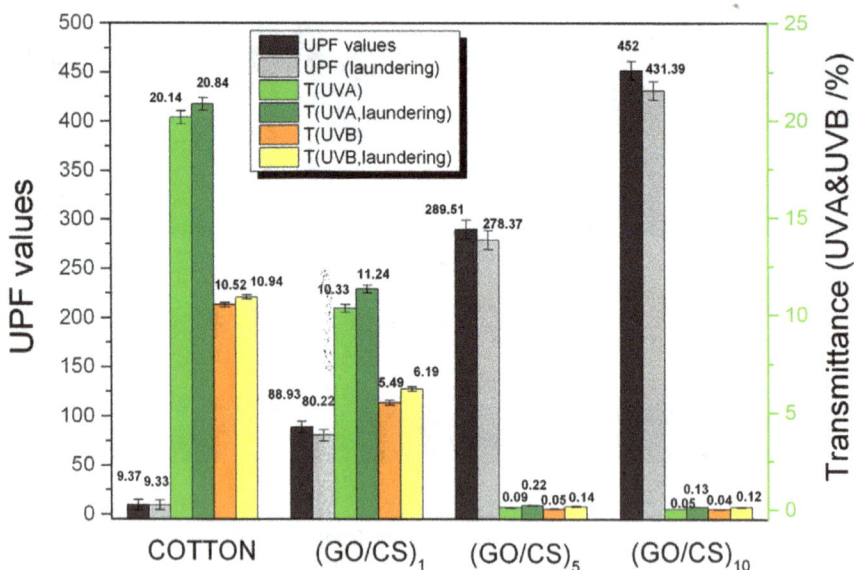

Figure 7.33: UPF values and UVA & UVB transmittance of control and GO/CS-deposited fabrics, before and after 10 times water laundering (reprinted from reference [64] with permission, Copyright (2016) Elsevier).

Ji and co-authors reported a UPF of 55.19 for silk woven fabric after nine rGO coatings [9]. The UV protection factor of cotton and linen woven fabrics considerably improved to 86.5 and 83.3, after the deposition of rGO and Ag nanoparticles [168]. The UV protection factor of untreated cotton woven fabric increases from 13.5 to 54.1, after five coatings of

GO spray [169]. Jain and Chatterjee reported excellent UV blocking properties (UPF = 89.38) on 102.60 g/m² GSM cotton nonwoven fabric with a single rGO coating. The UPF value of cotton nonwoven fabric was dramatically improved to 18,702.56 after the second coating, which is far above the excellent range [125].

7.17.3 Antibacterial textiles

Graphene and its derivatives have been identified as effective antibacterial agents, in recent years [170]. Physical contact with bacterial membranes, efflux of cytoplasmic materials, lowering metabolism, ruinous extraction of lipid, photothermal ablation, oxidative stress-producing reactive oxygen species (ROS), and glutathione loss are all the proposed mechanisms used by graphene and its derivatives to exhibit good antibacterial activity, as represented in Figure 7.34 [171–173].

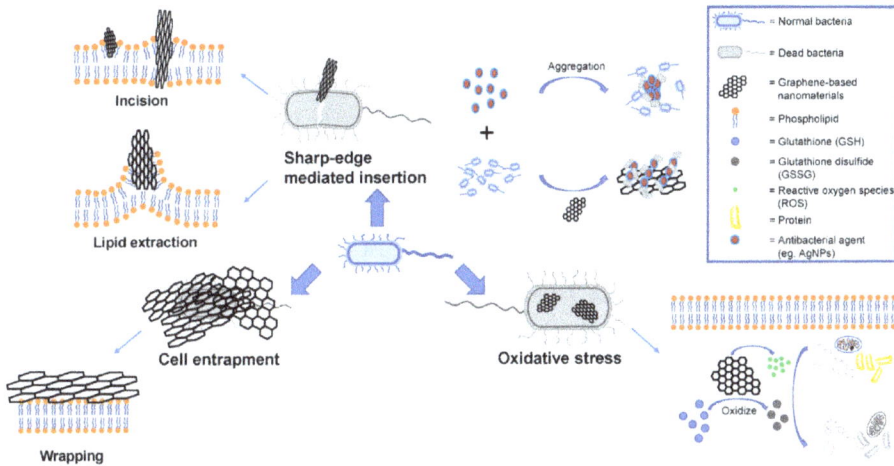

Figure 7.34: Antimicrobial behaviour of graphene materials (reprinted from reference [174] with permission, Copyright (2019) Elsevier).

Staphylococcus aureus bacterial infection is common during thermotherapy due to perspiration. Mahfam Hasani and Montazer reported an excellent antibacterial activity of GO-coated cellulosic/polyamide fabric towards gram-positive (*S. aureus* and *E. faecalis*) and gram-negative bacteria (*E. coli* and *P. aeruginosa*) [175].

rGO/polymeric *N*-halamine-coated cotton fabric exhibited antibacterial activity of 87.4% against *S. aureus* which was further enhanced to 100% after chlorination [12]. Mahfam Hasani and Montazer coated rGO on cotton/nylon fabric and reported antibacterial activity of 77.8% against *S. aureus* bacterium which was enhanced to 100% after cationization of fabric with 3-chloro-2-hydroxy propyl trimethyl ammonium chloride (CHPTAC) [11].

7.17.4 Textile pressure sensors

With the technological advancements in wearable electronics in recent years, the use of flexible wearable sensors has drawn the attention of many researchers. Pressure sensors can detect applied strain through controlled compression. These devices find potential application in various fields such as wearable healthcare monitors, aerospace, automotive, and heavy machinery. Based on the working principle, pressure sensors can be classified as capacitive- and resistive-based pressure sensors [176]. Capacitive pressure sensors are made of two electrodes, separated by a gap that changes with the exerted pressure, and the relative changes in capacitance are monitored. Capacitive pressure sensors work in an elastic region without contact, which offers high sensitivity, stability, and repeatability with fair response speed [176, 177]. However, this type of pressure sensor is thick and shows poor repeatability when used in a wearable healthcare monitor, due to fluctuation in capacitance caused by the human body (acts as a conductor) [177]. Resistive (piezo-resistive) pressure sensors measure the relative changes in electrical resistance when an external force is applied. The piezo-resistive sensor shows high sensitivity, stability, and repeatability with excellent response speed. Gauge factor and pressure sensitivity are the two important parameters for defining the performance of a pressure sensor [176]. Pressure sensitivity is defined as the fractional change in resistance per unit pressure increment, whereas the gauge factor is defined as the fractional change in resistance per unit elastic strain [178]. Silicone-based pressure sensors are widely recognized due to their high gauge factor. However, their reliability is undermined by environmental factors such as temperature and moisture. Carbon nanotube (CNT)-based piezo-resistive sensors have been proposed as an alternative, due to their high sensitivity and gauge factor that ranges from 100 to 300 [179]. However, certain issues associated with CNT such as poor dispersion, variation in chirality [180], and expensive production process [20] restrict their extensive use.

Graphene is currently a focus of research for the next generation of piezo-resistive materials because of its unique electronic, thermal, mechanical, and piezo-electrical properties [20]. In the context of graphene-coated flexible textile sensors, Ren and co-workers reported excellent bending stability of rGO-coated cotton fabric under compressive strain [49]. In another work, a double-layered graphene structure was transferred on PDMS (polydimethylsiloxane) and, then, the whole assembly was spin-coated on a PET substrate. The prepared sensor was examined for tactile sensing and reported a pressure sensitivity of -0.24 kPa^{-1} for 250 Pa and 0.034 kPa^{-1} for 1–8 kPa [49]. A high gauge factor of 6.2 was reported for a pure compression-based magnetite-decorated graphene/polyurethane flexible piezo-resistive pressure sensor. Additionally, piezo-resistive responses of 0.1 and 0.4 were reported for pressure variations of 3 kPa and 100 kPa [178]. Y. Liu et al. arranged rGO-coated silk fabric up to nine layers in order and examined it for pressure sensing application. The pressure sensitivity of the rGO-coated silk fabric increased as the number of conductive fabric layers increased.

Maximum pressure sensitivity of 0.4 kPa^{-1} was reported at nine fabric layers [181]. P. Li et al. prepared a pressure sensor by sandwiching rGO-coated cotton fibres between copper electrodes and fixing it with silver paste (Figures 7.35a and 7.35b).

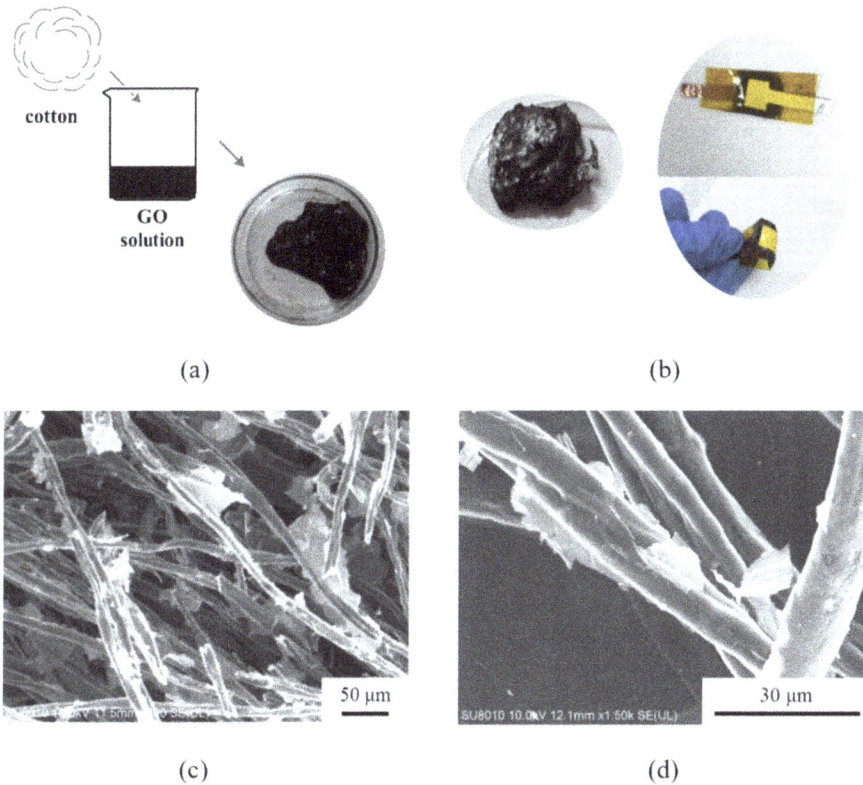

(a)   (b)

(c)   (d)

Figure 7.35: (a) Schematic of degreasing cotton pieces soaked in GO solution; (b) conductive rGO cotton and pressure sensors; (c, d) SEM image showing how rGO was conformably adhered to cotton fibres [182].

The authors measured the pressure sensitivity up to 500 kPa with three different amounts of rGO-coated cotton fibres, namely, 0.08 g (g1), 0.12 g (g2), and 0.14 g (g3) (Figure 7.36). Pressure sensitivity decreases with the increase in the pressure range, and maximum pressure sensitivity of 0.21 kPa^{-1} was reported in the pressure range of 0 to 2 kPa for the g3 sensor [182].

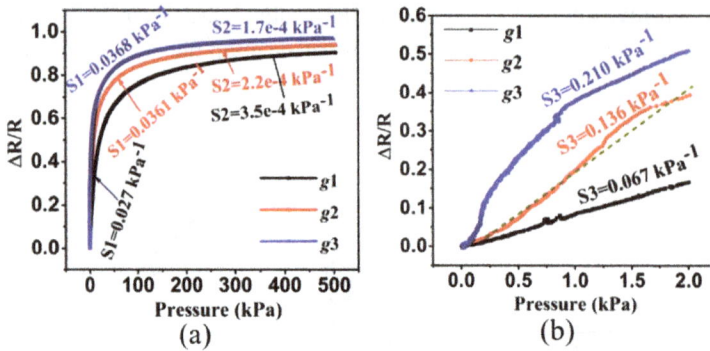

Figure 7.36: Relative resistance change versus pressure for different sensors (g1, g2, and g3) (a) from 0 to 500 kPa and (b) from 0 to 2 kPa [182].

7.17.5 Hydrophobic textiles

Naturally occurring hydrophobic surfaces such as lotus leaves and rose petals inspired researchers to develop similar kinds of artificial hydrophobic surfaces. Hydrophobic surfaces have been applied as anti-sticking, anti-fouling, oil-water separation, microfluidic channels, Janus interface materials, high-performance optics, and super waterproof textiles, in addition to their core focus on self-cleaning. The difference in interfacial energy between the surface and the water droplet determines surface hydrophobicity. This differential interfacial energy phenomenon has been used to develop hydrophobicity on surfaces using two approaches: reducing the surface free energy or changing the surface roughness. Water contact angle measurement is commonly used to define the wettability of a solid surface. A surface is hydrophilic when the water contact angle is less than 90°, hydrophobic when it is higher than 90°, and superhydrophobic when it is higher than 150° [183]. Graphene is also used to impart hydrophobicity in textile materials due to the presence of a hydrophobic carbon backbone in its structure.

Tissera et al. observed an increase in water contact angle of GO-treated cotton fabric with increasing GO concentration due to thicker GO coatings at the higher concentrations [65]. Shateri-Khalilabad and Yazdanshenas reported an average water contact angle of 143.2° on rGO-coated cotton fabric, which was increased to 163.0° after the polymethyl siloxane (PMS) treatment. In this research, PMS was used as a rough low-surface energy layer to prepare superhydrophobic cotton fabric [106]. An rGO-modified polymeric *N*-halamine precursor-coated cotton fabric displayed a water contact angle of 130°, which was raised to 140° after chlorination. According to the authors, the enhanced hydrophobicity of the cotton fabric is attributable to the addition of surface roughness and the removal of oxygenated functional groups after the treatment [12]. The hydrophobicity of cotton fabric enhanced in terms of water contact angle from 85° to 141°, after 10–12 PEDOT:PSS/rGO coatings [184].

7.17.6 Photocatalytic fabrics

Photocatalysis is a green technology that converts absorbed photon energy into chemical energy for various applications including pollutant degradation, photocatalytic hydrogen evolution, and CO_2 reduction. When a photon with energy equal to or greater than the bandgap energy of the semiconductor is absorbed by the semiconductor, electrons are excited from the valence band to the conduction band; as a result, positive holes are created in the valence band. These photogenerated holes in the valence band and photogenerated electrons in the conduction band migrate towards the surface of the semiconductor and cause chemical oxidation and reduction of the reactants[185]. Krishnamoorthy et al. prepared GO-coated cotton fabric and analysed the photoreduction of resazurin under UV light irradiation. When UV light was illuminated on GO-coated cotton fabric with energy of 3.54 eV, greater than the bandgap energy of GO (3.26 eV), photogenerated holes and electrons pairs were created on the fabric surface, which reduced the resazurin (blue) to resorufin (pink) [185].

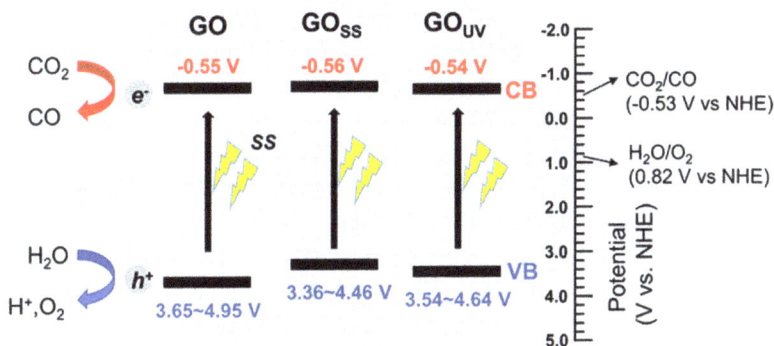

Figure 7.37: Scheme of the photocatalytic CO_2 reduction mechanism on GO and irradiated GO samples (reprinted from reference [186] with permission, Copyright (2020) American Chemical Society).

Sahito et al. reported photocatalytic degradation efficiency of 27% and 45% towards methylene blue (MB) for rGO-coated standard cotton fabric and rGO-coated cationized cotton fabric [63]. Kuang et al. improved the photocatalytic performance of GO for CO_2 reduction by using light irradiation and proposed a scheme for photocatalytic CO_2 reduction mechanism on GO and irradiated GO samples (Figure 7.37) [186].

Numerous metal oxide semiconductors have been reported as photocatalysts, notably TiO_2, ZnO, CdS, Bi_2WO_4, Fe_2O_3, and WO_3 [187]. Inhibiting photogenerated charge carrier recombination and maximizing the whole solar spectrum usage are two major challenges that should be overcome to make this process commercially viable. Graphene is an ideal choice for composite semiconductor fabrication for photocatalytic application, because it can extend light absorption and electron/hole pair life. Furthermore, graphene has a strong adsorption trait for pollutants which enhances the interaction between pollutants and the photocatalytic material. Graphene/TiO_2 nanocomposite-treated cotton fabric displayed higher photocatalytic activity for MB, compared to TiO_2-coated cotton fabric [91]. Another study showed that the addition of rGO improved the photocatalytic performance of ZnO particles [188]. Mandal and co-authors published a detailed review on various graphene-metal oxide semiconductor nanocomposites for photocatalytic application [187].

7.17.7 Thermally conductive textiles

Since graphene has high thermal conductivity (4,840–5,300 W/m K) [189] at room temperature, it is used to improve heat conduction properties in textiles. Abbas et al. coated graphene, MWCNTs, and fine boron nitride particles on cotton fabric and measured thermal conductivity. Graphene-coated cotton fabric showed higher thermal conductivity due to the deposition of highly thermal conductive graphene. Furthermore, graphene coating grammatically reduced the air permeability of the fabric, which also facilitated the improvement of the thermal conductivity of the fabric [190].

Hu et al. observed an increase in thermal conductivity of graphene/WPU-coated cotton fabric with the increase in graphene loading [191]. Similar findings were observed by Manasoglu and co-authors, where thermal conductivity of graphene-coated polyester fabric increases with the increase in graphene concentration and coating thickness [192]. Gong et al. prepared graphene woven fabrics (GWFs)/polyimide (PI) composite films by reinforcing 10 layers of CVD-grown GWFs on the PI matrix. GWFs/PI composite films displayed an in-plane thermal conductivity of 3.73 W/m K, which was 1,418% higher than that of the pure PI (0.25 W/m K) [193]. In order to prepare rGO/CNTs/NWF, Tang et al. used nano-soldering to anchor CNTs on non-woven fabric (NWF) and then coated rGO on it. The obtained rGO/CNTs/NWF showed thermal conductivity of 2.90 W/m K, which was substantially greater than the NWF (0.03 W/m K) (Figure 7.38). The deposition on rGO improved CNTs connection, resulting in enhanced thermal conductivity [194].

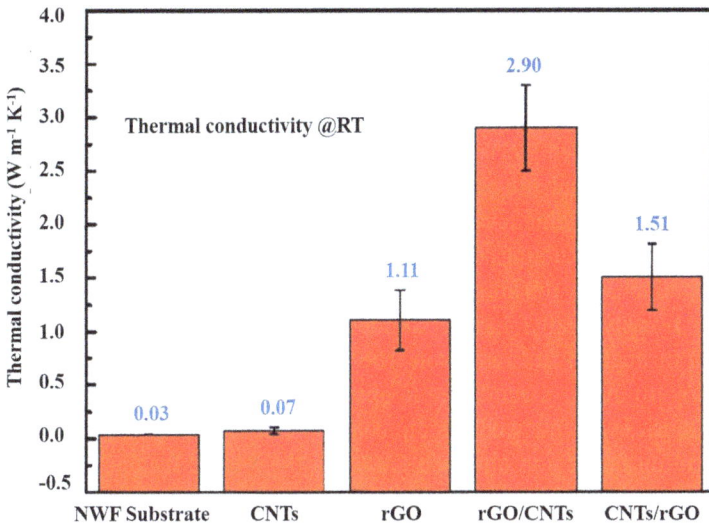

Figure 7.38: Thermal conductivity of untreated and treated non-woven fabrics (reprinted from reference [194] with permission of The Royal Society of Chemistry).

7.17.8 EMI shield textiles

The rapid growth of electronics resulted in alarming levels of electromagnetic pollution. Various conductive textiles with EMI shielding functionalities have been developed to protect ecology and sensitive electronic devices from electromagnetic pollution. Metal-based E-textiles are well-known for EMI shielding applications with a reflection-based shielding principle. However, reflection-based shielding devices are not encouraged, since they may cause damage to other nearby devices. Unlike metal-based E-textile, graphene-based E-textiles are lightweight and flexible, and they work on the absorption-dominated principle [195]. The EMI shielding effectiveness (EMISE) is defined from the ratio of the incident (input) and transmitted (output) energy or fields in decibels (dB), and it can be expressed as eq. (7.6) [196]:

$$SE_{Total} = 10 \log \left(\frac{P_{input}}{P_{output}} \right) = 20 \log \left(\frac{E_{input}}{E_{output}} \right) = 20 \log \left(\frac{H_{input}}{H_{output}} \right) \; (dB) \qquad (7.6)$$

where P, H, and E represent the power, magnetic field, and electric field, respectively.

When electromagnetic waves strike a shield, the reflection, absorption, and multiple reflection phenomena can be observed. As per the Schelkunoff's theory, the total EMISE (SE_{Total}) is evaluated by reflection (SE_R), absorption (SE_A), and multiple reflections (SE_M) [197].

$$SE_{Total} = SE_R + SE_A + SE_M \qquad\qquad (7.7)$$

Tin et al. used the layer-by-layer (LbL) electrostatic self-assembly (ESA) method to enhance graphene add-on on cotton fabric (Figure 7.39). In this method, cotton fabric was soaked in negatively charged PSS solution, and thereafter, dried and dipped in a positively charged chitosan/graphene solution. The EMISE of the prepared fabric increases with the increase in coating layers, and a maximum EMISE value of 30.04 dB was obtained after 10 layers of coatings (PCSG10). The EMI shielding mechanism of the prepared fabric is dominated by the absorption phenomenon [198]. Wang et al. cationized polyester fabric by poly (diallyl dimethylammonium chloride) (PDDA), thereafter treated with Ag/rGO through in situ reduction. The prepared fabric displayed EMISE ranging from 52 to 57 dB in the X-band, due to absorption-dominated EMI shielding phenomena [199]. Ghosh et al. followed the dip coating technique to deposit AgNPs-decorated rGO on cotton fabric and analysed EMISE in the X-band frequency range (8.2–12.4 GHz). EMI shielding effectiveness of developed fabric increases with the increase in coating cycle and maximum effectiveness of ~ 99 % (27.36 dB) was obtained at 30 coating cycles. The improvement in EMISE with coating cycles is attributed to the increase in conductive filler thickness [151]. Cotton fabric with EMISE was made by a uniform coating of ZnO and rGO nanoparticles by sol-gel and spraying techniques, respectively. The ZnO/rGO-coated cotton exhibited excellent EMISE of 57 dB (99.999%) in X-band (8.2–12.4 GHz) due to the ~ 82.216% of absorption and ~ 17.783% of reflection [195]. Chen and co-workers prepared rGO and hematite (Fe_2O_3) hollow microsphere-coated PU sponge and examined EMISE in X-band (8.2–12.4 GHz). The rGO-coated PU sponge showed EMISE of 37.45 dB. The addition of hematite (Fe_2O_3) hollow microspheres further improved the EM wave absorption to 94.5%. The ultra-strong EMISE was owing to conduction loss, low density, multiple reflections, and scattering of electromagnetic waves [200].

7.17.9 Flexible capacitors and energy storage devices

Fossil fuels-based energy sources are associated with energy shortages and environmental concerns. Several clean and renewable energy sources such as solar, hydrogen, and wind are encouraged by the energy sector as alternatives to fossil fuels. The realistic usage of clean and renewable energy relies on the energy storage devices such as batteries that need high power density and long cycle life. Supercapacitors are high-capacity energy storage devices with quick charging and discharging, high energy density, and extended cycle life properties. Wearable and flexible electronic energy storage devices require flexibility and light weight, in addition to the above-mentioned properties. Graphene-based E-textiles are an excellent contender for this use, as evidenced by extensive research [51, 71, 117, 122, 201–208].

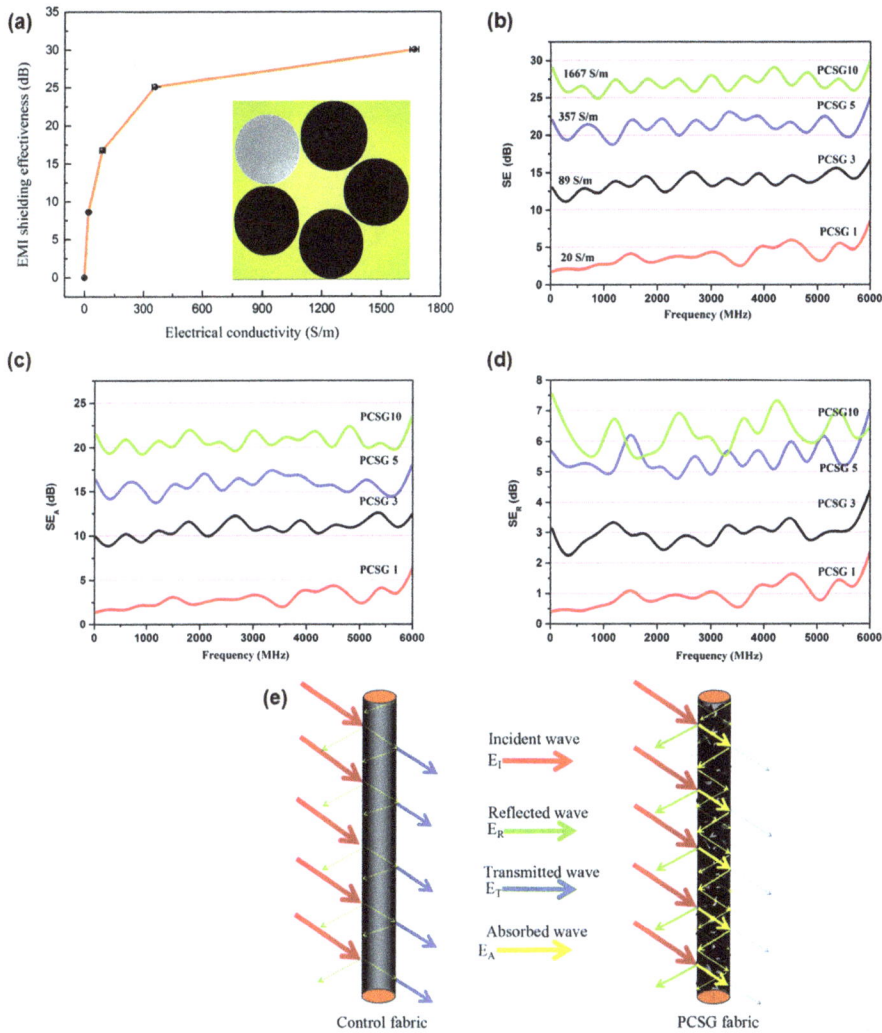

Figure 7.39: (a) EMI shielding effectiveness changes with electrical conductivity, (b) SE_{total} of PCSG fabrics, (c) SE_A of PCSG fabrics, (d) SE_R of PCSG fabrics, and (e) contrast of shielding mechanism before and after LbL self-assembly [198].

Liu and co-workers prepared lightweight and flexible supercapacitor electrode by brush coating and drying (50 coatings) of GO onto cotton fabric (Figure 7.40). For the preparation of the supercapacitor electrode, untreated cotton fabric was sandwiched between the GO-treated cotton fabrics and then, the final composite structure was annealed at 300 °C for 2 h in an argon atmosphere. The resulting electrode achieved a high specific capacitance of 81.7 F/g in an aqueous electrolyte [51].

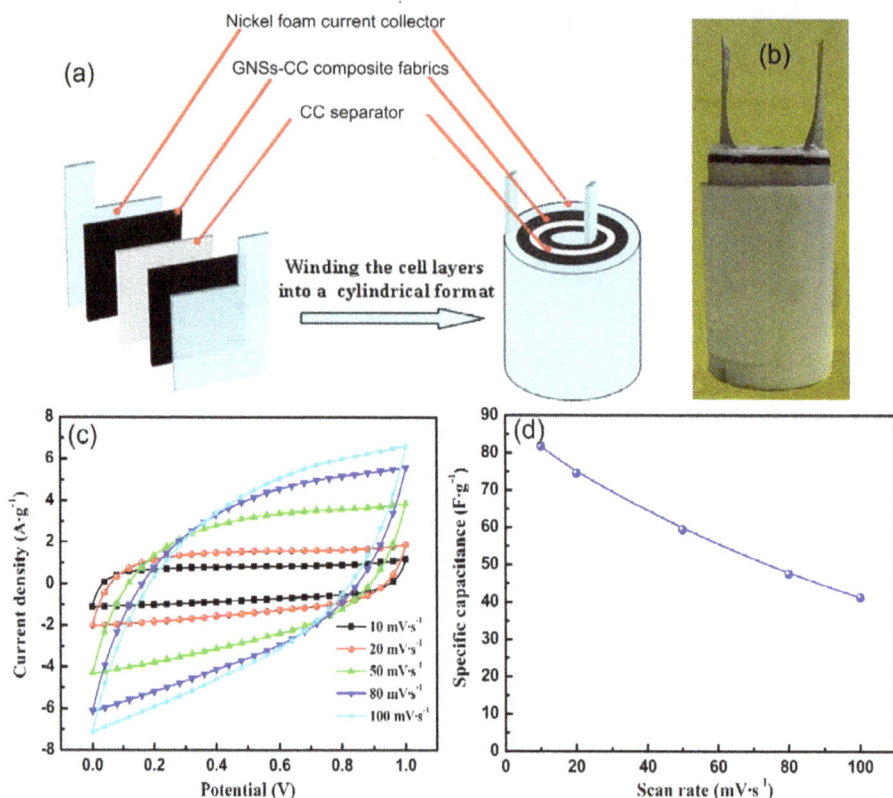

Figure 7.40: (a) Structure and (b) digital photograph of the prepared supercapacitor; (c) CV curves and (d) specific capacitances at various scan rates (reprinted from reference [51] with permission of The Royal Society of Chemistry).

Carbon materials are attractive for capacitor manufacturing, because their large surface area enables them to absorb more electrolyte ions, thereby increasing the double-layer capacitance. Cao et al. observed the addition of electrochemically reduced graphene oxide (ERGO) on carbon fibre enhanced the electrochemical energy storage performance of carbon fibre, due to the formation of excellent electrochemical double-layers (Figure 7.41) [209]. Multi-layered conducive cotton yarns were prepared by immobilizing rGO nanoparticles on the surface of Ni-coated cotton yarns, which were created via highly scalable electroless Ni deposition. These conductive yarns were examined for energy storage electrodes, where they exhibited an outstanding power density of 1,400 mW/cm^3 and volumetric energy density of 6.1 mWh/cm^3 [112].

Other pseudo-capacitative materials were also combined with graphene to enhance capacitance such as PEDOT [204], PANI [117, 205], PPy [71, 206], CNTs [207, 208], vanadium oxide (V_2O_5) [208], and MnO_2 [201, 202]. Zhao et al. prepared a stretchable supercapacitor by coating rGO on nylon lycra fabric and obtained a specific capacitance of

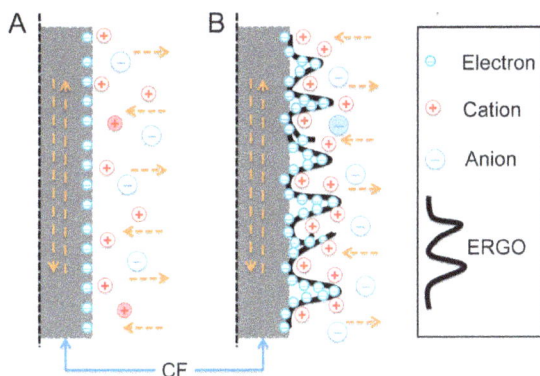

Figure 7.41: Mechanism of (A) pure CF and (B) ERGO@CF electrodes (reprinted from reference [209] with permission of The Royal Society of Chemistry).

12.3 F/g at a scan rate of 5 mV/s in 1.0 M lithium sulfate aqueous solution. The obtained capacitance dramatically enhanced to 114 F/g after the PPy coating, due to the double-layer coating and subsequent increase in electrical conductivity [71]. Yaghoubidoust et al. also reported improvement in capacitive behaviour of GO-coated cotton fabric after the PPy coating [206]. Zang et al. transferred CVD-grown GWF on PET fabric, and then, this composite was treated with PANI. The addition of PANI improved the electro-chemical characteristics of the prepared supercapacitors to as high as 23 mF/cm^2, along with good cycling stability of 2,000 cycles [205]. Yu et al. prepared rGO/PANI/eCFC (nitrogen-doped carbon cloth) composites by treating eCFC with PANI, followed by rGO. The resultant composites exhibited enhanced capacitive behaviour with a maximum power density of 92.2 kW/kg, energy density of 25.4 Wh/kg, and a specific capacitance of 1,145 F/g. In addition, the prepared composites displayed excellent capacitance retention, even after 5,000 cycles [117]. Zang et al. constructed thin-film super-capacitors by transferring CVD-grown GWF on PET and reported area-specific capacitance of as high as 8 mF cm^{-2} [210]. Xu et al. prepared graphene/cotton composite fabrics by 20 coatings and obtained sheet resistance of 611 Ω/sq due to the formation of a 3D conductive network of graphene sheets on cotton fabric. The prepared graphene/cotton composite fabric showed an areal capacitance of 40 F/g within 0.0–1.0 voltage range [122]. Zhou et al. prepared rGO-coated carbonized cotton fabric by treating the cotton fabric with GO via the dip coating technique, followed by thermal reduction at 300 °C for 2 h under argon atmosphere. The prepared flexible supercapacitor demonstrated a high capacitance of 87.53 mF/cm at 2 mV/s, along with excellent electrochemical stability and cycling stability [203]. Shakir et al. inserted layer-by-layer assembly of graphene as a conductive spacer between V$_2$O$_5$-coated MWCNT films to improve their capacitive behaviour. The addition of graphene layer prevented MWCNT agglomeration and increased the specific capacitance by 67% and obtained values as high as 2,590 F/g. Moreover, its cyclic performance was also improved [208].

7.18 Conclusions and perspectives

Textile materials have distinct advantages over sheet materials, such as high surface area, mechanical properties, flexibility, and so on, making them favourable substrates onto which other functional materials can be deposited. In this chapter, we had discussed the source of conductivity in various types of conductive materials that have been used to impart electroconductive properties in textiles. Various methods for graphene synthesis were presented, such as mechanical exfoliation, CVD, chemical exfoliation, epitaxial growth of graphene, and chemical synthesis. CVD and epitaxial growth methods produce a few layers of high-quality graphene, but these methods are not scalable due to low yield and lack of reproducibility. The chemical exfoliation method is easier for the large-scale production of graphene. However, the production of monolayer graphene is still challenging by this route, and the quality of the produced graphene is also inferior due to large numbers of defects. In addition, this method comprises several time-consuming processes and utilizes some potentially dangerous and explosive compounds, which need to be resolved. Meanwhile, we also reported some application techniques of graphene onto textile materials such as vacuum filtration, brush coating, printing, wet transfer of graphene layers, and dip coating. Dip coating technique is convenient, scalable, and economical.

The functional properties and applications of graphene-coated textiles depend on the magnitude of graphene add-on, which, in turn, is influenced by the dipping variables such as GO concentration, pH of GO dispersion, dipping time, dipping temperature, and reductant parameters. In the future, greater work should be devoted to the optimization of dipping variables for different textiles substrates, in order to improve the graphene add-on.

The oxidized form of graphene, namely, GO is an electrical insulator due to its sp^3-hybridized structure and, thus, cannot be used for electro-conductive applications. Different reduction methods such as thermal, chemical, and UV, partially restore the conductive sp^2-hybridized carbon atom network (rGO) from the insulating sp^3-hybridized GO structure. Although the UV reduction approach is cost-effective and chemical-free, it is time-consuming and offers low electrical conductivity. To make this technique realistic, additional effort needs to be put in towards the optimization of UV reduction parameters. More controlled strategies are required to develop cost-effective, large-scale graphene-based textiles, coupled with other conductive materials for various functional applications with enhanced performance and environmental friendliness.

References

[1]   Chatterjee A, Maity S. Electroconductive Textiles. In: Shahid-ul-Islam, Butola BS, editors. Adv. Text. Eng. Mater., vol. 20, Hoboken, NJ, USA: John Wiley & Sons, Inc.; 2018, p. 177–255. https://doi.org/10.1002/9781119488101.ch6.

[2]   Karim N, Afroj S, Tan S, He P, Fernando A, Carr C, et al. Scalable Production of Graphene-Based Wearable E-Textiles. ACS Nano 2017;11:12266–75. https://doi.org/10.1021/acsnano.7b05921.

[3]   Molina J. Graphene-based fabrics and their applications: a review. RSC Adv 2016;6:68261–91. https://doi.org/10.1039/C6RA12365A.

[4]   Afroj S, Tan S, Abdelkader AM, Novoselov KS, Karim N. Highly Conductive, Scalable, and Machine Washable Graphene-Based E-Textiles for Multifunctional Wearable Electronic Applications. Adv Funct Mater 2020;30:2000293. https://doi.org/10.1002/adfm.202000293.

[5]   Cataldi P, Ceseracciu L, Athanassiou A, Bayer IS. Healable Cotton–Graphene Nanocomposite Conductor for Wearable Electronics. ACS Appl Mater Interfaces 2017;9:13825–30. https://doi.org/10.1021/acsami.7b02326.

[6]   Chatterjee A, Nivas Kumar M, Maity S. Influence of graphene oxide concentration and dipping cycles on electrical conductivity of coated cotton textiles. J Text Inst 2017; 108:1910–6. https://doi.org/10.1080/00405000.2017.1300209.

[7]   Cao J, Wang C. Multifunctional surface modification of silk fabric via graphene oxide repeatedly coating and chemical reduction method. Appl Surf Sci 2017;405:380–8. https://doi.org/10.1016/j.apsusc.2017.02.017.

[8]   Liang B, Fang L, Hu Y, Yang G, Zhu Q, Ye X. Fabrication and application of flexible graphene silk composite film electrodes decorated with spiky Pt nanospheres. Nanoscale 2014;6:4264–74. https://doi.org/10.1039/C3NR06057H.

[9]   Ji Y, Chen G, Xing T. Rational design and preparation of flame retardant silk fabrics coated with reduced graphene oxide. Appl Surf Sci 2019;474:203–10. https://doi.org/10.1016/j.apsusc.2018.03.120.

[10]   Wang S, Ma Q, Wang K, Chen H-W. Improving Antibacterial Activity and Biocompatibility of Bioinspired Electrospinning Silk Fibroin Nanofibers Modified by Graphene Oxide. ACS Omega 2018;3:406–13. https://doi.org/10.1021/acsomega.7b012 10.

[11]   Hasani M, Montazer M. Cationization of cellulose/polyamide on UV protection, bio-activity, and electro-conductivity of graphene oxide-treated fabric. J Appl Polym Sci 2017;134:45493. https://doi.org/10.1002/app.45493.

[12]   Pan N, Liu Y, Ren X, Huang TS. Fabrication of cotton fabrics through in-situ reduction of polymeric N-halamine modified graphene oxide with enhanced ultraviolet-blocking, self-cleaning, and highly efficient, and monitorable antibacterial properties. Colloids Surfaces A Physicochem Eng Asp 2018;555:765–71. https://doi.org/10.1016/j.colsurfa. 2018.07.056.

[13]   Kim H, Lee S, Kim H. Electrical Heating Performance of Electro-Conductive Para-aramid Knit Manufactured by Dip-Coating in a Graphene/Waterborne Polyurethane Composite. Sci Rep 2019;9:1511. https://doi.org/10.1038/s41598-018-37455-0.

[14]   Souri H, Bhattacharyya D. Highly Stretchable Multifunctional Wearable Devices Based on Conductive Cotton and Wool Fabrics. ACS Appl Mater Interfaces 2018;10:20845–53. https://doi.org/10.1021/acsami.8b04775.

[15]   Wang L, Lu C, Zhang B, Zhao B, Wu F, Guan S. Fabrication and characterization of flexible silk fibroin films reinforced with graphene oxide for biomedical applications. RSC Adv 2014;4:40312–20. https://doi.org/10.1039/C4RA04529G.

[16]   Abdelkader AM, Karim N, Vallés C, Afroj S, Novoselov KS, Yeates SG. Ultraflexible and robust graphene supercapacitors printed on textiles for wearable electronics applications. 2D Mater 2017;4:035016. https://doi.org/10.1088/2053-1583/aa7d71.

[17] Karim N, Afroj S, Malandraki A, Butterworth S, Beach C, Rigout M, et al. All inkjet-printed graphene-based conductive patterns for wearable e-textile applications. J Mater Chem C 2017;5:11640–8. https://doi.org/10.1039/c7tc03669h.

[18] Taherian R. The Theory of Electrical Conductivity. Electr. Conduct. Polym. Compos. Exp. Model. Appl., Elsevier; 2019, p. 1–18. https://doi.org/10.1016/B978-0-12-812541-0.00001-X.

[19] Askeland DR, Fulay PP, Wright WJ. Electronic Materials. Sci. Eng. Mater. 6th ed., Stamford: Cengage Learning; 2011, p. 719–65.

[20] Tung TT, Robert C, Castro M, Feller JF, Kim TY, Suh KS. Enhancing the sensitivity of graphene/polyurethane nanocomposite flexible piezo-resistive pressure sensors with magnetite nano-spacers. Carbon N Y 2016;108:450–60. https://doi.org/10.1016/j.carbon.2016.07.018.

[21] Zhao S, Zhao Z, Yang Z, Ke L, Kitipornchai S, Yang J. Functionally graded graphene reinforced composite structures: A review. Eng Struct 2020;210:110339. https://doi.org/10.1016/j.engstruct.2020.110339.

[22] Geim AK, Novoselov KS. The rise of graphene. Nat Mater 2007;6:183–91. https://doi.org/10.1038/nmat1849.

[23] Allen MJ, Tung VC, Kaner RB. Honeycomb Carbon: A Review of Graphene. Chem Rev 2010;110:132–45. https://doi.org/10.1021/cr900070d.

[24] Kuila T, Bose S, Mishra AK, Khanra P, Kim NH, Lee JH. Chemical functionalization of graphene and its applications. Prog Mater Sci 2012;57:1061–105. https://doi.org/10.1016/j.pmatsci.2012.03.002.

[25] Novoselov KS, Fal'Ko VI, Colombo L, Gellert PR, Schwab MG, Kim K. A roadmap for graphene. Nature 2012;490:192–200. https://doi.org/10.1038/nature11458.

[26] Clemons CB, Roberts MW, Wilber JP, Young GW, Buldum A, Quinn DD. Continuum plate theory and atomistic modeling to find the flexural rigidity of a graphene sheet interacting with a substrate. J Nanotechnol 2010. https://doi.org/10.1155/2010/868492.

[27] Castro Neto AH, Guinea F, Peres NMR, Novoselov KS, Geim AK. The electronic properties of graphene. Rev Mod Phys 2009;81:109–62. https://doi.org/10.1103/RevModPhys.81.109.

[28] Wang M, Webster TJ. Nano-Biomaterials and their Applications. vol. 1–3. Elsevier; 2019. https://doi.org/10.1016/B978-0-12-801238-3.99871-4.

[29] Singh V, Joung D, Zhai L, Das S, Khondaker SI, Seal S. Graphene based materials: Past, present and future. Prog Mater Sci 2011;56:1178–271. https://doi.org/10.1016/j.pmatsci.2011.03.003.

[30] Ray SC. Application and Uses of Graphene. Appl. Graphene Graphene-Oxide Based Nanomater., Elsevier; 2015, p. 1–38. https://doi.org/10.1016/B978-0-323-37521-4.00001-7.

[31] Agrawal A, Yi G-C. Sample pretreatment with graphene materials. In: Hussain CM, editor. Compr. Anal. Chem., Elsevier; 2020, p. 21–47. https://doi.org/10.1016/bs.coac. 2020.08.012.

[32] Viculis LM, Mack JJ, Mayer OM, Hahn HT, Kaner RB. Intercalation and exfoliation routes to graphite nanoplatelets. J Mater Chem 2005;15:974–8. https://doi.org/10.1039/b413029d.

[33] Somani PR, Somani SP, Umeno M. Planer nano-graphenes from camphor by CVD. Chem Phys Lett 2006;430:56–9. https://doi.org/10.1016/j.cplett.2006.06.081.

[34] Tsai D-S, Chiang P, Tsai M, Tu W, Chen C, Chen S-L, et al. Camphor-Based CVD Bilayer Graphene/Si Heterostructures for Self-Powered and Broadband Photodetection. Micromachines 2020;11:812. https://doi.org/10.3390/mi11090812.

[35] Qu L, Tian M, Hu X, Wang Y, Zhu S, Guo X, et al. Functionalization of cotton fabric at low graphene nanoplate content for ultrastrong ultraviolet blocking. Carbon N Y 2014;80:565–74. https://doi.org/10.1016/j.carbon.2014.08.097.

[36] Singh A, Sharma N, Arif M, Katiyar RS. Electrically reduced graphene oxide for photovoltaic application. J Mater Res 2019;34:652–60. https://doi.org/10.1557/jmr. 2019.32.

[37] Jiao L, Zhang L, Wang X, Diankov G, Dai H. Narrow graphene nanoribbons from carbon nanotubes. Nature 2009;458:877–80. https://doi.org/10.1038/nature07919.

[38] Sridhar V, Jeon JH, Oh IK. Synthesis of graphene nano-sheets using eco-friendly chemicals and microwave radiation. Carbon N Y 2010;48:2953–7. https://doi.org/10.1016/j.carbon.2010.04.034.
[39] Al-Hazmi FS, Al-Harbi GH, Beall GW, Al-Ghamdi AA, Obaid AY, Mahmoud WE. One pot synthesis of graphene based on microwave assisted solvothermal technique. Synth Met 2015;200:54–7. https://doi.org/10.1016/j.synthmet.2014.12.028.
[40] Marcano DC, Kosynkin D V., Berlin JM, Sinitskii A, Sun Z, Slesarev A, et al. Improved Synthesis of Graphene Oxide. ACS Nano 2010;4:4806–14. https://doi.org/10.1021/nn1006368.
[41] Bhuyan MSA, Uddin MN, Islam MM, Bipasha FA, Hossain SS. Synthesis of graphene. Int Nano Lett 2016;6:65–83. https://doi.org/10.1007/s40089-015-0176-1.
[42] Ghany NAA, Elsherif SA, Handal HT. Revolution of Graphene for different applications: State-of-the-art. Surfaces and Interfaces 2017;9:93–106. https://doi.org/10.1016/j.surfin.2017.08.004.
[43] Hummers WS, Offeman RE. Preparation of Graphitic Oxide. J Am Chem Soc 1958;80:1339–1339. https://doi.org/10.1021/ja01539a017.
[44] Dreyer DR, Park S, Bielawski CW, Ruoff RS. The chemistry of graphene oxide. Chem Soc Rev 2010;39:228–40. https://doi.org/10.1039/B917103G.
[45] Marcano DC, Kosynkin D V., Berlin JM, Sinitskii A, Sun Z, Slesarev AS, et al. Correction to Improved Synthesis of Graphene Oxide. ACS Nano 2018:acsnano. 8b00128. https://doi.org/10.1021/acsnano.8b00128.
[46] Lerf A, He H, Forster M, Klinowski J. Structure of graphite oxide revisited. J Phys Chem B 1998;102:4477–82. https://doi.org/10.1021/jp9731821.
[47] Gao W, Alemany LB, Ci L, Ajayan PM. New insights into the structure and reduction of graphite oxide. Nat Chem 2009;1:403–8. https://doi.org/10.1038/nchem.281.
[48] Tang X, Tian M, Qu L, Zhu S, Guo X, Han G, et al. Functionalization of cotton fabric with graphene oxide nanosheet and polyaniline for conductive and UV blocking properties. Synth Met 2015;202:82–8. https://doi.org/10.1016/j.synthmet.2015.01.017.
[49] Ren J, Wang C, Zhang X, Carey T, Chen K, Yin Y, et al. Environmentally-friendly conductive cotton fabric as flexible strain sensor based on hot press reduced graphene oxide. Carbon N Y 2017;111:622–30. https://doi.org/10.1016/j.carbon.2016.10.045.
[50] Krishnamoorthy K, Pazhamalai MP, Lim JH, Choi KH, Kim S-J.Mechanochemical Reinforcement of Graphene Sheets into Alkyd Resin Matrix for the Development of Electrically Conductive Paints. ChemNanoMat 2018;4:568–74. https://doi.org/10.1002/cnma.201700391.
[51] Liu W, Yan X, Lang J, Peng C, Xue Q. Flexible and conductive nanocomposite electrode based on graphene sheets and cotton cloth for supercapacitor. J Mater Chem 2012; 22:17245. https://doi.org/10.1039/c2jm32659k.
[52] Javed K, Galib CMA, Yang F, Chen CM, Wang C. A new approach to fabricate graphene electro-conductive networks on natural fibers by ultraviolet curing method. Synth Met 2014;193:41–7. https://doi.org/10.1016/j.synthmet.2014.03.028.
[53] Han X, Chen Y, Zhu H, Preston C, Wan J, Fang Z, et al. Scalable, printable, surfactant-free graphene ink directly from graphite. Nanotechnology 2013;24:205304. https://doi.org/10.1088/0957-4484/24/20/205304.
[54] Skrzetuska E, Puchalski M, Krucińska I. Chemically Driven Printed Textile Sensors Based on Graphene and Carbon Nanotubes. Sensors 2014;14:16816–28. https://doi.org/10.3390/s140916816.
[55] Pepłowski A, Walter P, Janczak D, Górecka Ż, Święszkowski W, Jakubowska M. Solventless Conducting Paste Based on Graphene Nanoplatelets for Printing of Flexible, Standalone Routes in Room Temperature. Nanomaterials 2018;8:829. https://doi.org/10.3390/nano8100829.
[56] He H, Akbari M, Sydänheimo L, Ukkonen L, Virkki J. 3D-Printed Graphene Antennas and Interconnections for Textile RFID Tags: Fabrication and Reliability towards Humidity. Int J Antennas Propag 2017;2017:1–5. https://doi.org/10.1155/2017/ 1386017.

[57] Cao J, Huang Z, Wang C. Natural printed silk substrate circuit fabricated via surface modification using one step thermal transfer and reduction graphene oxide. Appl Surf Sci 2018;440:177–85. https://doi.org/10.1016/j.apsusc.2018.01.094.

[58] Neves AIS, Bointon TH, Melo L V, Russo S, de Schrijver I, Craciun MF, et al. Transparent conductive graphene textile fibers. Sci Rep 2015;5:9866. https://doi.org/10.1038/srep09866.

[59] Neves AIS, Rodrigues DP, De Sanctis A, Alonso ET, Pereira MS, Amaral VS, et al. Towards conductive textiles: coating polymeric fibres with graphene. Sci Rep 2017;7:4250. https://doi.org/10.1038/s41598-017-04453-7.

[60] Verma VP, Das S, Lahiri I, Choi W. Large-area graphene on polymer film for flexible and transparent anode in field emission device. Appl Phys Lett 2010;96:203108. https://doi.org/10.1063/1.3431630.

[61] Paquin F, Rivnay J, Salleo A, Stingelin N, Silva C. Multi-phase semicrystalline microstructures drive exciton dissociation in neat plastic semiconductors. J Mater Chem C 2015;3:10715–22. https://doi.org/10.1039/b000000x.

[62] Dhineshbabu NR, Bose S. Smart Textiles Coated with Eco-Friendly UV-Blocking Nanoparticles Derived from Natural Resources. ACS Omega 2018;3:7454–65. https://doi.org/10.1021/acsomega.8b00822.

[63] Sahito IA, Sun KC, Arbab AA, Qadir MB, Jeong SH. Integrating high electrical conductivity and photocatalytic activity in cotton fabric by cationizing for enriched coating of negatively charged graphene oxide. Carbohydr Polym 2015;130:299–306. https://doi.org/10.1016/j.carbpol.2015.05.010.

[64] Tian M, Hu X, Qu L, Du M, Zhu S, Sun Y, et al. Ultraviolet protection cotton fabric achieved via layer-by-layer self-assembly of graphene oxide and chitosan. Appl Surf Sci 2016;377:141–8. https://doi.org/10.1016/j.apsusc.2016.03.183.

[65] Tissera ND, Wijesena RN, Perera JR, Silva KMN De, Amaratunge GAJ. Hydrophobic cotton textile surfaces using an amphiphilic graphene oxide (GO) coating. Appl Surf Sci 2015;324:455–63. https://doi.org/10.1016/j.apsusc.2014.10.148.

[66] Song J, Xu S, Chen T, Yamanaka S, Morikawa H. Preparation of graphene oxide-coated silk fibers through HBPAA [a molecular glue]-induced layer-by-layer self-assembly. J IRAN CHEM SOC 2018;15:101–9. https://doi.org/10.1007/s13738-017-1213-y.

[67] Shen G, Hu X, Guan G, Wang L. Surface modification and characterisation of silk fibroin fabric produced by the layer-by-layer self-assembly of multilayer alginate/regenerated silk fibroin. PLoS One 2015;10:1–19. https://doi.org/10.1371/journal.pone.0124811.

[68] Kongahge D, Foroughi J, Gambhir S, Spinks GM, Wallace GG. Fabrication of a graphene coated nonwoven textile for industrial applications. RSC Adv 2016;6:73203–9. https://doi.org/10.1039/C6RA15190F.

[69] Ouadil B, Cherkaoui O, Safi M, Zahouily M. Surface modification of knit polyester fabric for mechanical, electrical and UV protection properties by coating with graphene oxide, graphene and graphene/silver nanocomposites. Appl Surf Sci 2017;414:292–302. https://doi.org/10.1016/j.apsusc.2017.04.068.

[70] Sarker F, Karim N, Afroj S, Koncherry V, Novoselov KS, Potluri P. High-Performance Graphene-Based Natural Fiber Composites. ACS Appl Mater Interfaces 2018;10:34502–12. https://doi.org/10.1021/acsami.8b13018.

[71] Zhao C, Shu K, Wang C, Gambhir S, Wallace GG. Reduced graphene oxide and polypyrrole/reduced graphene oxide composite coated stretchable fabric electrodes for supercapacitor application. Electrochim Acta 2015;172:12–9. https://doi.org/10.1016/j.electacta.2015.05.019.

[72] Samad YA, Li Y, Alhassan SM, Liao K. Non-destroyable graphene cladding on a range of textile and other fibers and fiber mats. RSC Adv 2014;4:16935–8. https://doi.org/10.1039/C4RA01373E.

[73] Fugetsu B, Sano E, Yu H, Mori K, Tanaka T. Graphene oxide as dyestuffs for the creation of electrically conductive fabrics. Carbon N Y 2010;48:3340–5. https://doi.org/10.1016/j.carbon.2010.05.016.

[74] Neacșu IA, Nicoară AI, Vasile OR, Vasile BȘ. Inorganic micro- and nanostructured implants for tissue engineering. Nanobiomaterials Hard Tissue Eng., Elsevier; 2016, p. 271–95. https://doi.org/10.1016/B978-0-323-42862-0.00009-2.

[75] Pei S, Cheng H-M. The reduction of graphene oxide. Carbon N Y 2012;50:3210–28. https://doi.org/10.1016/j.carbon.2011.11.010.

[76] Cai G, Xu Z, Yang M, Tang B, Wang X. Functionalization of cotton fabrics through thermal reduction of graphene oxide. Appl Surf Sci 2017;393:441–8. https://doi.org/10.1016/j.apsusc.2016.10.046.

[77] Kotov NA, Dékány I, Fendler JH. Ultrathin graphite oxide-polyelectrolyte composites prepared by self-assembly: Transition between conductive and non-conductive states. Adv Mater 1996;8:637–41. https://doi.org/10.1002/adma.19960080806.

[78] Stankovich S, Dikin DA, Piner RD, Kohlhaas KA, Kleinhammes A, Jia Y, et al. Synthesis of graphene-based nanosheets via chemical reduction of exfoliated graphite oxide. Carbon N Y 2007;45:1558–65. https://doi.org/10.1016/j.carbon.2007.02.034.

[79] Thema FT, Moloto MJ, Dikio ED, Nyangiwe NN, Kotsedi L, Maaza M, et al. Synthesis and Characterization of Graphene Thin Films by Chemical Reduction of Exfoliated and Intercalated Graphite Oxide. J Chem 2013;2013:1–6. https://doi.org/10.1155/2013/ 150536.

[80] Shateri-Khalilabad M, Yazdanshenas ME. Fabricating electroconductive cotton textiles using graphene. Carbohydr Polym 2013;96:190–5. https://doi.org/10.1016/j.carbpol. 2013.03.052.

[81] Fan X, Peng W, Li Y, Li X, Wang S, Zhang G, et al. Deoxygenation of Exfoliated Graphite Oxide under Alkaline Conditions: A Green Route to Graphene Preparation. Adv Mater 2008;20:4490–3. https://doi.org/10.1002/adma.200801306.

[82] Zhang J, Yang H, Shen G, Cheng P, Zhang J, Guo S. Reduction of graphene oxide via L-ascorbic acid. Chem Commun 2010;46:1112–4. https://doi.org/10.1039/B917705A.

[83] Emiru TF, Ayele DW. Controlled synthesis, characterization and reduction of graphene oxide: A convenient method for large scale production. Egypt J Basic Appl Sci 2017;4:74–9. https://doi.org/10.1016/j.ejbas.2016.11.002.

[84] Zhu C, Guo S, Fang Y, Dong S. Reducing Sugar: New Functional Molecules for the Green Synthesis of Graphene Nanosheets. ACS Nano 2010;4:2429–37. https://doi.org/10.1021/nn1002387.

[85] Khanra P, Kuila T, Kim NH, Bae SH, Yu D, Lee JH. Simultaneous bio-functionalization and reduction of graphene oxide by baker's yeast. Chem Eng J 2012;183:526–33. https://doi.org/10.1016/j.cej.2011.12.075.

[86] Esfandiar A, Akhavan O, Irajizad A. Melatonin as a powerful bio-antioxidant for reduction of graphene oxide. J Mater Chem 2011;21:10907. https://doi.org/10.1039/c1jm10151j.

[87] Akhavan O, Ghaderi E, Aghayee S, Fereydooni Y, Talebi A. The use of a glucose-reduced graphene oxide suspension for photothermal cancer therapy. J Mater Chem 2012;22:13773. https://doi.org/10.1039/c2jm31396k.

[88] Abdulmageed HA, Judran AK, Noori FTM. Green synthesis of reduce graphene oxide by green tea leaves. J Phys Conf Ser 2021;1795:012070. https://doi.org/10.1088/1742-6596/1795/1/012070.

[89] Gurunathan S, Han JW, Eppakayala V, Kim J. Microbial reduction of graphene oxide by Escherichia coli: A green chemistry approach. Colloids Surfaces B Biointerfaces 2013;102:772–7. https://doi.org/10.1016/j.colsurfb.2012.09.011.

[90] Gurunathan S, Woong Han J, Eppakayala V, Kim J. Green synthesis of graphene and its cytotoxic effects in human breast cancer cells. Int J Nanomedicine 2013;8:1015. https://doi.org/10.2147/IJN.S42047.

[91] Karimi L, Yazdanshenas ME, Khajavi R, Rashidi A, Mirjalili M. Using graphene/TiO_2 nanocomposite as a new route for preparation of electroconductive, self-cleaning, antibacterial and antifungal cotton fabric without toxicity. Cellulose 2014;21:3813–27. https://doi.org/10.1007/s10570-014-0385-1.

[92]   Liu X, Qin Z, Dou Z, Liu N, Chen L, Zhu M. Fabricating conductive poly(ethylene terephthalate) nonwoven fabrics using an aqueous dispersion of reduced graphene oxide as a sheet dyestuff. RSC Adv 2014;4:23869–75. https://doi.org/10.1039/C4RA01645A.

[93]   Kim H, Lee S. Electrical properties of graphene/waterborne polyurethane composite films. Fibers Polym 2017;18:1304–13. https://doi.org/10.1007/s12221-017-7142-7.

[94]   Chatterjee A, Jain VK. Isothermal, Kinetic, and Thermodynamic Studies of Graphene Oxide Adsorption on Silk. AATCC J Res 2021;8:18–29. https://doi.org/10.14504/ajr.8.5.3.

[95]   Shih C, Lin S, Sharma R, Strano MS, Blankschtein D. Understanding the pH-Dependent Behavior of Graphene Oxide Aqueous Solutions: A Comparative Experimental and Molecular Dynamics Simulation Study. Langmuir 2012;28:235–41. https://doi.org/10.1021/la203607w.

[96]   Dimiev AM, Alemany LB, Tour JM. Graphene Oxide. Origin of Acidity, Its Instability in Water, and a New Dynamic Structural Model. ACS Nano 2012;7:576–88. https://doi.org/10.1021/nn3047378.

[97]   Konkena B, Vasudevan S. Understanding Aqueous Dispersibility of Graphene Oxide and Reduced Graphene Oxide through p K a Measurements. J Phys Chem Lett 2012;3:867–72. https://doi.org/10.1021/jz300236w.

[98]   Ersan G, Apul OG, Perreault F, Karanfil T. Adsorption of organic contaminants by graphene nanosheets: A review. Water Res 2017;126:385–98. https://doi.org/10.1016/j.watres.2017.08.010.

[99]   WU H, LU W, SHAO J, ZHANG C, WU M, LI B, et al. pH-dependent size, surface chemistry and electrochemical properties of graphene oxide. New Carbon Mater 2013;28:327–35. https://doi.org/10.1016/S1872-5805(13)60085-2.

[100]   Thomas HR, Day SP, Woodruff WE, Vallés C, Young RJ, Kinloch IA, et al. Deoxygenation of graphene oxide: Reduction or cleaning? Chem Mater 2013;25:3580–8. https://doi.org/10.1021/cm401922e.

[101]   Moleon JA, Ontiveros-Ortega A, Gimenez-Martin E, Plaza I. Effect of N-cetylpyridinium chloride in adsorption of graphene oxide onto polyester. Dye Pigment 2015;122:310–6. https://doi.org/10.1016/j.dyepig.2015.07.004.

[102]   Ripoll L, Bordes C, Marote P, Etheve S, Elaissari A, Fessi H. Electrokinetic properties of bare or nanoparticle-functionalized textile fabrics. Colloids Surfaces A Physicochem Eng Asp 2012;397:24–32. https://doi.org/10.1016/j.colsurfa.2012.01.022.

[103]   Grancaric AM, Tarbuk A, Pusic T. Electrokinetic properties of textile fabrics. Color Technol 2005;121:221–7. https://doi.org/10.1111/j.1478-4408.2005.tb00277.x.

[104]   Ayub ZH, Arai M, Hirabayashi K. Mechanism of the Gelation of Fibroin Solution. Biosci Biotechnol Biochem 1993;57:1910–2. https://doi.org/10.1271/bbb.57.1910.

[105]   Giménez-Martín E, López-Andrade M, Moleón-Baca JA, López MA, Ontiveros-Ortega A. Polyamide Fibers Covered with Chlorhexidine: Thermodynamic Aspects. J Surf Eng Mater Adv Technol 2015;05:190–206. https://doi.org/10.4236/jsemat.2015.54021.

[106]   Shateri-Khalilabad M, Yazdanshenas ME. Preparation of superhydrophobic electroconductive graphene-coated cotton cellulose. Cellulose 2013;20:963–72. https://doi.org/10.1007/s10570-013-9873-y.

[107]   Wang D, Li D, Lv P, Wang Q, Xu Y, Wei Q. Graphene Oxide/Polyester Fabric Composite by Electrostatic Self-Assembly as a New Recyclable Adsorbent for the Removal of Methylene Blue. Fibers Polym 2018;19:1726–34. https://doi.org/10.1007/s12221-018-7876-x.

[108]   Berendjchi A, Khajavi R, Yousefi AA, Yazdanshenas ME. Improved continuity of reduced graphene oxide on polyester fabric by use of polypyrrole to achieve a highly electro-conductive and flexible substrate. Appl Surf Sci 2016;363:264–72. https://doi.org/10.1016/j.apsusc.2015.12.030

[109]   Sahito IA, Sun KC, Arbab AA, Qadir MB, Jeong SH. Graphene coated cotton fabric as textile structured counter electrode for DSSC. Electrochim Acta 2015;173:164–71. https://doi.org/10.1016/j.electacta.2015.05.035.

[110] Sahito IA, Sun KC, Arbab AA, Qadir MB, Choi YS, Jeong SH. Flexible and conductive cotton fabric counter electrode coated with graphene nanosheets for high efficiency dye sensitized solar cell. J Power Sources 2016;319:90–8. https://doi.org/10.1016/j.jpowsour.2016.04.025.

[111] Lu Z, Mao C, Zhang H. Highly conductive graphene-coated silk fabricated via a repeated coating-reduction approach. J Mater Chem C 2015;3:4265–8. https://doi.org/10.1039/c5tc00917k.

[112] Liu L, Yu Y, Yan C, Li K, Zheng Z. Wearable energy-dense and power-dense supercapacitor yarns enabled by scalable graphene–metallic textile composite electrodes. Nat Commun 2015;6:7260. https://doi.org/10.1038/ncomms8260.

[113] Yun YJ, Hong WG, Kim WJ, Jun Y, Kim BH. A novel method for applying reduced graphene oxide directly to electronic textiles from yarns to fabrics. Adv Mater 2013;25:5701–5. https://doi.org/10.1002/adma.201303225.

[114] Hsiao S-T, Ma C-CM, Tien H-W, Liao W-H, Wang Y-S, Li S-M, et al. Preparation and characterization of silver nanoparticle-reduced graphene oxide decorated electrospun polyurethane fiber composites with an improved electrical property. Compos Sci Technol 2015;118:171–7. https://doi.org/10.1016/j.compscitech.2015.05.017.

[115] Yapici MK, Alkhidir T, Samad YA, Liao K. Graphene-clad textile electrodes for electrocardiogram monitoring. Sensors Actuators, B Chem 2015;221. https://doi.org/10.1016/j.snb.2015.07.111.

[116] Yun YJ, Hong WG, Choi NJ, Kim BH, Jun Y, Lee HK. Ultrasensitive and highly selective graphene-based single yarn for use in wearable gas sensor. Sci Rep 2015;5:1–7. https://doi.org/10.1038/srep10904.

[117] Yu P, Li Y, Zhao X, Wu L, Zhang Q. Graphene-wrapped polyaniline nanowire arrays on nitrogen-doped carbon fabric as novel flexible hybrid electrode materials for high-performance supercapacitor. Langmuir 2014;30:5306–13. https://doi.org/10.1021/la40 4765z.

[118] Molina J, Fernández J, Inés JC, Del Río AI, Bonastre J, Cases F. Electrochemical characterization of reduced graphene oxide-coated polyester fabrics. Electrochim Acta 2013;93:44–52. https://doi.org/10.1016/j.electacta.2013.01.071.

[119] Molina J, Fernandes F, Fernández J, Pastor M, Correia A, Souto AP, et al. Photocatalytic fabrics based on reduced graphene oxide and TiO₂ coatings. Mater Sci Eng B 2015;199:62–76. https://doi.org/10.1016/j.mseb.2015.04.013.

[120] Molina J, Fernández J, del Río AI, Bonastre J, Cases F. Chemical and electrochemical study of fabrics coated with reduced graphene oxide. Appl Surf Sci 2013;279:46–54. https://doi.org/10.1016/j.apsusc.2013.04.020.

[121] Huang Y, Hu H, Huang Y, Zhu M, Meng W, Liu C, et al. From industrially weavable and knittable highly conductive yarns to large wearable energy storage textiles. ACS Nano 2015;9:4766–75. https://doi.org/10.1021/acsnano.5b00860.

[122] Xu LL, Guo MX, Liu S, Bian SW. Graphene/cotton composite fabrics as flexible electrode materials for electrochemical capacitors. RSC Adv 2015;5:25244–9. https://doi.org/10.1039/c4ra16063k.

[123] Karimi L, Yazdanshenas ME, Khajavi R, Rashidi A, Mirjalili M. Optimizing the photocatalytic properties and the synergistic effects of graphene and nano titanium dioxide immobilized on cotton fabric. Appl Surf Sci 2015;332:665–73. https://doi.org/10.1016/j.apsusc.2015.01.184.

[124] Zhou T, Chen F, Liu K, Deng H, Zhang Q, Feng J, et al. A simple and efficient method to prepare graphene by reduction of graphite oxide with sodium hydrosulfite. Nanotechnology 2011;22:045704. https://doi.org/10.1088/0957-4484/22/4/045704.

[125] Jain VK, Chatterjee A. Graphene coated cotton nonwoven for electroconductive and UV protection applications. J Ind Text 2021:152808372110592. https://doi.org/10.1177/15280837211059202.

[126] Jain VK, Chatterjee A. Influence of fibre type and dipping cycle on graphene adsorption and electrical conductivity of fibres. Indian J Fibre Text Res 2021;46:149–57. http://nopr.niscair.res.in/handle/123456789/57641.

[127] Lin Y-F, Lin D-H, Jan M-H, Lin C-HJ, Cheng C-K. Orthopedic Physical Therapy. In: Anders Brahme, editor. Compr. Biomed. Phys., vol. 10, Elsevier; 2014, p. 379–400. https://doi.org/10.1016/B978-0-444-53632-7.01024-8.

[128] Zhang M, Wang C, Liang X, Yin Z, Xia K, Wang H, et al. Weft-Knitted Fabric for a Highly Stretchable and Low-Voltage Wearable Heater. Adv Electron Mater 2017;3:1–8. https://doi.org/10.1002/aelm.201700193.

[129] Kim H, Lee S. Characterization of Electrical Heating Textile Coated by Graphene Nanoplatelets/PVDF-HFP Composite with Various High Graphene Nanoplatelet Contents. Polymers (Basel) 2019;11:928. https://doi.org/10.3390/polym11050928.

[130] Faruk MO, Ahmed A, Jalil MA, Islam MT, Shamim AM, Adak B, et al. Functional textiles and composite based wearable thermal devices for Joule heating: progress and perspectives. Appl Mater Today 2021;23:101025. https://doi.org/10.1016/j.apmt. 2021.101025.

[131] Kim HJ, Kim Y, Jeong JH, Choi JH, Lee J, Choi DG. A cupronickel-based micromesh film for use as a high-performance and low-voltage transparent heater. J Mater Chem A 2015;3:16621–6. https://doi.org/10.1039/c5ta03348a.

[132] Hong S, Lee H, Lee J, Kwon J, Han S, Suh YD, et al. Highly Stretchable and Transparent Metal Nanowire Heater for Wearable Electronics Applications. Adv Mater 2015;27:4744–51. https://doi.org/10.1002/adma.201500917.

[133] Seo KW, Kim MY, Chang HS, Kim HK. Self-assembled Ag nanoparticle network passivated by a nano-sized ZnO layer for transparent and flexible film heaters. AIP Adv 2015;5. https://doi.org/10.1063/1.4939139.

[134] An BW, Gwak E-J, Kim K, Kim Y-C, Jang J, Kim J-Y, et al. Stretchable, Transparent Electrodes as Wearable Heaters Using Nanotrough Networks of Metallic Glasses with Superior Mechanical Properties and Thermal Stability. Nano Lett 2016;16:471–8. https://doi.org/10.1021/acs.nanolett.5b04134.

[135] Hakansson E, Kaynak A, Lin T, Nahavandi S, Jones T, Hu E. Characterization of conducting polymer coated synthetic fabrics for heat generation. Synth Met 2004;144:21–8. https://doi.org/10.1016/j.synthmet.2004.01.003.

[136] Kaynak A, Håkansson E. Generating heat from conducting polypyrrole-coated PET fabrics. Adv Polym Technol 2005;24:194–207. https://doi.org/10.1002/adv.20040.

[137] Boschi A, Arosio C, Cucchi I, Bertini F, Catellani M, Freddi G. Properties and Performance of Polypyrrole (PPy)-coated Silk Fibers 2008;9:698–707.

[138] Kaynak A, Håkansson E. Short-term heating tests on doped polypyrrole-coated polyester fabrics. Synth Met 2008;158:350–4. https://doi.org/10.1016/j.synthmet. 2008.02.004.

[139] Sparavigna AC, Florio L, Avloni J, Henn A. Polypyrrole Coated PET Fabrics for Thermal Applications. Mater Sci Appl 2010;01:253–9. https://doi.org/10.4236/msa. 2010.14037.

[140] Macasaquit AC, Binag CA. Preparation of conducting polyester textile by in situ polymerization of pyrrole. Philipp J Sci 2010;139:189–96.

[141] Abbasi AMR, Militky J, Gregr J. Heat Generation by Polypyrrole Coated Glass Fabric. J Text 2013;2013:1–5. https://doi.org/10.1155/2013/571024.

[142] Maity S, Chatterjee A, Singh B, Pal Singh A. Polypyrrole based electro-conductive textiles for heat generation. J Text Inst 2014;105:887–93. https://doi.org/10.1080/00405000.2013.861149.

[143] Dall'Acqua L, Tonin C, Varesano A, Canetti M, Porzio W, Catellani M. Vapour phase polymerisation of pyrrole on cellulose-based textile substrates. Synth Met 2006;156:379–86. https://doi.org/10.1016/j.synthmet.2005.12.021.

[144] Hao D, Xu B, Cai Z. Polypyrrole coated knitted fabric for robust wearable sensor and heater. J Mater Sci Mater Electron 2018;29:9218–26. https://doi.org/10.1007/s10854-018-8950-2.

[145] Sui D, Huang Y, Huang L, Liang J, Ma Y, Chen Y. Flexible and transparent electrothermal film heaters based on graphene materials. Small 2011;7:3186–92. https://doi.org/10.1002/smll.201101305.

[146] Bae JJ, Lim SC, Han GH, Jo YW, Doung DL, Kim ES, et al. Heat dissipation of transparent graphene defoggers. Adv Funct Mater 2012;22:4819–26. https://doi.org/10.1002/adfm.201201155.

[147] Kang J, Kim H, Kim KS, Lee SK, Bae S, Ahn JH, et al. High-performance graphene-based transparent flexible heaters. Nano Lett 2011;11:5154–8. https://doi.org/10.1021/nl202311v.

[148] Liu P, Liu L, Jiang K, Fan S. Carbon-nanotube-film microheater on a polyethylene terephthalate substrate and its application in thermochromic displays. Small 2011;7:732–6. https://doi.org/10.1002/smll.201001662.

[149] Jung D, Kim D, Lee KH, Overzet LJ, Lee GS. Transparent film heaters using multi-walled carbon nanotube sheets. Sensors Actuators, A Phys 2013;199:176–80. https://doi.org/10.1016/j.sna.2013.05.024.

[150] Vertuccio L, De Santis F, Pantani R, Lafdi K, Guadagno L. Effective de-icing skin using graphene-based flexible heater. Compos Part B Eng 2019;162:600–10. https://doi.org/10.1016/j.compositesb.2019.01.045.

[151] Ghosh S, Ganguly S, Das P, Das TK, Bose M, Singha NK, et al. Fabrication of Reduced Graphene Oxide/Silver Nanoparticles Decorated Conductive Cotton Fabric for High Performing Electromagnetic Interference Shielding and Antibacterial Application. Fibers Polym 2019;20:1161–71. https://doi.org/10.1007/s12221-019-1001-7.

[152] Kang J, Jang Y, Kim Y, Cho SH, Suhr J, Hong BH, et al. An Ag-grid/graphene hybrid structure for large-scale, transparent, flexible heaters. Nanoscale 2015;7:6567–73. https://doi.org/10.1039/c4nr06984f.

[153] Zhang W, Yin Z, Chun A, Yoo J, Kim YS, Piao Y. Bridging Oriented Copper Nanowire-Graphene Composites for Solution-Processable, Annealing-Free, and Air-Stable Flexible Electrodes. ACS Appl Mater Interfaces 2016;8:1733–41. https://doi.org/10.1021/acsami.5b09337.

[154] Ahn J, Gu J, Hwang B, Kang H, Hwang S, Jeon S, et al. Printed fabric heater based on Ag nanowire/carbon nanotube composites. Nanotechnology 2019;30:455707. https://doi.org/10.1088/1361-6528/ab35eb.

[155] Malhotra U, Maity S, Chatterjee A. Polypyrrole-silk electro-conductive composite fabric by in situ chemical polymerization. J Appl Polym Sci 2015;132:1–10. https://doi.org/10.1002/app.41336.

[156] Ba H, Truong-Phuoc L, Papaefthimiou V, Sutter C, Pronkin S, Bahouka A, et al. Cotton Fabrics Coated with Few-Layer Graphene as Highly Responsive Surface Heaters and Integrated Lightweight Electronic-Textile Circuits. ACS Appl Nano Mater 2020;3:9771–83. https://doi.org/10.1021/acsanm.0c01861.

[157] El-Tahlawy K, El-Nagar K, Elhendawy AG. Cyclodextrin-4 Hydroxy benzophenone inclusion complex for UV protective cotton fabric. J Text Inst 2007;98:453–62. https://doi.org/10.1080/00405000701556327.

[158] Grifoni D, Bacci L, Di Lonardo S, Pinelli P, Scardigli A, Camilli F, et al. UV protective properties of cotton and flax fabrics dyed with multifunctional plant extracts. Dye Pigment 2014;105:89–96. https://doi.org/10.1016/j.dyepig.2014.01.027.

[159] Hsieh S-H, Zhang F-R, Li H-S. Anti-ultraviolet and physical properties of woollen fabrics cured with citric acid and TiO_2/chitosan. J Appl Polym Sci 2006;100:4311–9. https://doi.org/10.1002/app.23830.

[160] Yang H, Zhu S, Pan N. Studying the mechanisms of titanium dioxide as ultraviolet-blocking additive for films and fabrics by an improved scheme. J Appl Polym Sci 2004;92:3201–10. https://doi.org/10.1002/app.20327.

[161] Gao Y, Gereige I, El Labban A, Cha D, Isimjan TT, Beaujuge PM. Highly Transparent and UV-Resistant Superhydrophobic SiO_2-Coated ZnO Nanorod Arrays. ACS Appl Mater Interfaces 2014;6:2219–23. https://doi.org/10.1021/am405513k.

[162] Ates ES, Unalan HE. Zinc oxide nanowire enhanced multifunctional coatings for cotton fabrics. Thin Solid Films 2012;520:4658–61. https://doi.org/10.1016/j.tsf.2011.10.073.

[163] Ibrahim NA, El-Zairy EMR, Abdalla WA, Khalil HM. Combined UV-protecting and reactive printing of Cellulosic/wool blends. Carbohydr Polym 2013;92:1386–94. https://doi.org/10.1016/j.carbpol.2012.09.063.

[164] Hasani M, Mahdavian M, Yari H, Ramezanzadeh B. Versatile protection of exterior coatings by the aid of graphene oxide nano-sheets; comparison with conventional UV absorbers. Prog Org Coatings 2018;116:90–101. https://doi.org/10.1016/j.porgcoat. 2017.11.020.

[165] Mistretta MC, Botta L, Vinci AD, Ceraulo M, La Mantia FP. Photo-oxidation of polypropylene/graphene nanoplatelets composites. Polym Degrad Stab 2019;160:35–43. https://doi.org/10.1016/j.polymdegradstab.2018.12.003.

[166] Karimi S, Helal E, Gutierrez G, Moghimian N, Madinehei M, David E, et al. A review on graphene's light stabilizing effects for reduced photodegradation of polymers. Crystals 2021;11:1–22. https://doi.org/10.3390/cryst11010003.

[167] de Oliveira YDC, Amurin LG, Valim FCF, Fechine GJM, Andrade RJE. The role of physical structure and morphology on the photodegradation behaviour of polypropylene -graphene oxide nanocomposites. Polymer (Guildf) 2019;176:146–58. https://doi.org/10.1016/j.polymer.2019.05.029.

[168] Farouk A, Saeed SES, Sharaf S, Abd El-Hady MM. Photocatalytic activity and antibacterial properties of linen fabric using reduced graphene oxide/silver nanocomposite. RSC Adv 2020;10:41600–11. https://doi.org/10.1039/d0ra07544b.

[169] He X, Liu Q, Zhou Y, Chen Z, Zhu C, Jin W. Graphene oxide-silver/cotton fiber fabric with antibacterial and anti-UV properties for wearable gas sensors. Front Mater Sci 2021. https://doi.org/10.1007/s11706-021-0564-6.

[170] Hu W, Peng C, Luo W, Lv M, Li X, Li D, et al. Graphene-Based Antibacterial Paper. ACS Nano 2010;4:4317–23. https://doi.org/10.1021/nn101097v.

[171] Lukowiak A, Kedziora A, Strek W. Antimicrobial graphene family materials: Progress, advances, hopes and fears. Adv Colloid Interface Sci 2016;236:101–12. https://doi.org/10.1016/j.cis.2016.08.002.

[172] Yousefi M, Dadashpour M, Hejazi M, Hasanzadeh M, Behnam B, de la Guardia M, et al. Antibacterial activity of graphene oxide as a new weapon nanomaterial to combat multidrug-resistance bacteria. Mater Sci Eng C 2017;74:568–81. https://doi.org/10.1016/j.msec.2016.12.125.

[173] Bhattacharjee S, Joshi R, Chughtai AA, Macintyre CR. Graphene Modified Multifunctional Personal Protective Clothing. Adv Mater Interfaces 2019;6:1–27. https://doi.org/10.1002/admi.201900622.

[174] Xia M, Xie Y, Yu C, Chen G, Li Y, Zhang T, et al. Graphene-based nanomaterials: the promising active agents for antibiotics-independent antibacterial applications. J Control Release 2019;307:16–31. https://doi.org/10.1016/j.jconrel.2019.06.011.

[175] Hasani M, Montazer M. Electro-conductivity, bioactivity and UV protection of graphene oxide-treated cellulosic/polyamide fabric using inorganic and organic reducing agents. J Text Inst 2017;108:1777–86. https://doi.org/10.1080/00405000. 2017.1286700.

[176] Huang Y-C, Liu Y, Ma C, Cheng H-C, He Q, Wu H, et al. Sensitive pressure sensors based on conductive microstructured air-gap gates and two-dimensional semiconductor transistors. Nat Electron 2020;3:59–69. https://doi.org/10.1038/s41928-019-0356-5.

[177] Choudhry NA, Rasheed A, Ahmad S, Arnold L, Wang L. Design, Development and Characterization of Textile Stitch-Based Piezo-resistive Sensors for Wearable Monitoring. IEEE Sens J 2020;20:10485–94. https://doi.org/10.1109/JSEN.2020.2994 264.

[178] Chun S, Kim Y, Oh H-S, Bae G, Park W. A highly sensitive pressure sensor using a double-layered graphene structure for tactile sensing. Nanoscale 2015;7:11652–9. https://doi.org/10.1039/C5NR00076A.

[179] Li J, Orrego S, Pan J, He P, Kang SH. Ultrasensitive, flexible, and low-cost nanoporous piezo-resistive composites for tactile pressure sensing. Nanoscale 2019;11:2779–86. https://doi.org/10.1039/c8nr09959f.

[180] Lee K, Lee SS, Lee JA, Lee K-C, Ji S. Carbon nanotube film piezo resistors embedded in polymer membranes. Appl Phys Lett 2010;96:013511. https://doi.org/10.1063/1.3272686.

[181] Liu Y, Tao LQ, Wang DY, Zhang TY, Yang Y, Ren TL. Flexible, highly sensitive pressure sensor with a wide range based on graphene-silk network structure. Appl Phys Lett 2017;110. https://doi.org/10.1063/1.4978374.

[182] Li P, Zhao L, Jiang Z, Yu M, Li Z, Zhou X, et al. A wearable and sensitive graphene-cotton based pressure sensor for human physiological signals monitoring. Sci Rep 2019;9:1–8. https://doi.org/10.1038/s41598-019-50997-1.

[183] Latthe S, Terashima C, Nakata K, Fujishima A. Superhydrophobic Surfaces Developed by Mimicking Hierarchical Surface Morphology of Lotus Leaf. Molecules 2014; 19:4256–83. https://doi.org/10.3390/molecules19044256.

[184] Shathi MA, Chen M, Khoso NA, Rahman MT, Bhattacharjee B. Graphene coated textile based highly flexible and washable sports bra for human health monitoring. Mater Des 2020;193:108792. https://doi.org/10.1016/j.matdes.2020.108792.

[185] Krishnamoorthy K, Navaneethaiyer U, Mohan R, Lee J, Kim SJ. Graphene oxide nanostructures modified multifunctional cotton fabrics. Appl Nanosci 2012;2:119–26. https://doi.org/10.1007/s13204-011-0045-9.

[186] Kuang Y, Shang J, Zhu T. Photoactivated Graphene Oxide to Enhance Photocatalytic Reduction of CO_2. ACS Appl Mater Interfaces 2020;12:3580–91. https://doi.org/10.1021/acsami.9b18899.

[187] Mandal S, Mallapur S, Reddy M, Singh JK, Lee DE, Park T. An Overview on Graphene-Metal Oxide Semiconductor Nanocomposite: A Promising Platform for Visible Light Photocatalytic Activity for the Treatment of Various Pollutants in Aqueous Medium. Molecules 2020;25:8–10. https://doi.org/10.3390/molecules25225380.

[188] Azarang M, Shuhaimi A, Yousefi R, Moradi Golsheikh A, Sookhakian M. Synthesis and characterization of ZnO NPs/reduced graphene oxide nanocomposite prepared in gelatin medium as highly efficient photo-degradation of MB. Ceram Int 2014;40:10217–21. https://doi.org/10.1016/j.ceramint.2014.02.109.

[189] Balandin AA, Ghosh S, Bao W, Calizo I, Teweldebrhan D, Miao F, et al. Superior Thermal Conductivity of Single-Layer Graphene. Nano Lett 2008;8:902–7. https://doi.org/10.1021/nl0731872.

[190] Abbas A, Zhao Y, Zhou J, Wang X, Lin T. Improving thermal conductivity of cotton fabrics using composite coatings containing graphene, multiwall carbon nanotube or boron nitride fine particles. Fibers Polym 2013;14:1641–9. https://doi.org/10.1007/s12221-013-1641-y.

[191] Hu X, Tian M, Qu L, Zhu S, Han G. Multifunctional cotton fabrics with graphene/polyurethane coatings with far-infrared emission, electrical conductivity, and ultraviolet-blocking properties. Carbon N Y 2015;95:625–33. https://doi.org/10.1016/j.carbon.2015.08.099.

[192] Manasoglu G, Celen R, Kanik M, Ulcay Y. An investigation on the thermal and solar properties of graphene-coated polyester fabrics. Coatings 2021;11:1–15. https://doi.org/10.3390/coatings11020125.

[193] Gong J, Liu Z, Yu J, Dai D, Dai W, Du S, et al. Graphene woven fabric-reinforced polyimide films with enhanced and anisotropic thermal conductivity. Compos Part A Appl Sci Manuf 2016;87:290–6. https://doi.org/10.1016/j.compositesa.2016.05.010.

[194] Tang Z, Yao D, Du D, Ouyang J. Highly machine-washable e-textiles with high strain sensitivity and high thermal conduction. J Mater Chem C 2020;8:2741–8. https://doi.org/10.1039/c9tc06155j.

[195] Gupta S, Chang C, Anbalagan AK, Lee CH, Tai NH. Reduced graphene oxide/zinc oxide coated wearable electrically conductive cotton textile for high microwave absorption. Compos Sci Technol 2020;188:107994. https://doi.org/10.1016/j.compscitech.2020.107994.

[196] Maity S, Chatterjee A. Conductive polymer-based electro-conductive textile composites for electromagnetic interference shielding: A review. J Ind Text 2018;47:2228–52. https://doi.org/10.1177/1528083716670310.

[197] Safdar F, Ashraf M, Javid A, Iqbal K. Polymeric textile-based electromagnetic interference shielding materials, their synthesis, mechanism and applications – A review. J Ind Text 2021. https://doi.org/10.1177/15280837211037085.

[198] Tian M, Du M, Qu L, Chen S, Zhu S, Han G. Electromagnetic interference shielding cotton fabrics with high electrical conductivity and electrical heating behavior: Via layer-by-layer self-assembly route. RSC Adv 2017;7:42641–52. https://doi.org/10.1039/c7ra08224j.

[199] Wang C, Guo R, Lan J, Tan L, Jiang S, Xiang C. Preparation of multi-functional fabric via silver/reduced graphene oxide coating with poly(diallyldimethylammonium chloride) modification. J Mater Sci Mater Electron 2018;29:8010–9. https://doi.org/10.1007/s10854-018-8807-8.

[200] Chen KY, Gupta S, Tai NH. Reduced graphene oxide/Fe$_2$O$_3$ hollow microspheres coated sponges for flexible electromagnetic interference shielding composites. Compos Commun 2021;23:100572. https://doi.org/10.1016/j.coco.2020.100572.

[201] Zhou Q, Ye X, Wan Z, Jia C. A three-dimensional flexible supercapacitor with enhanced performance based on lightweight, conductive graphene-cotton fabric electrode. J Power Sources 2015;296:186–96. https://doi.org/10.1016/j.jpowsour.2015.07.012.

[202] Liu Y, Weng B, Razal JM, Xu Q, Zhao C, Hou Y, et al. High-Performance Flexible All-Solid-State Supercapacitor from Large Free-Standing Graphene-PEDOT/PSS Films. Sci Rep 2015;5:1–11. https://doi.org/10.1038/srep17045.

[203] Zang X, Li X, Zhu M, Li X, Zhen Z, He Y, et al. Graphene/polyaniline woven fabric composite films as flexible supercapacitor electrodes. Nanoscale 2015;7:7318–22. https://doi.org/10.1039/c5nr00584a.

[204] Yaghoubidoust F, Wicaksono DHB, Chandren S, Nur H. Effect of graphene oxide on the structural and electrochemical behavior of polypyrrole deposited on cotton fabric. J Mol Struct 2014;1075:486–93. https://doi.org/10.1016/j.molstruc.2014.07.025.

[205] Liu K, Yao Y, Lv T, Li H, Li N, Chen Z, et al. Textile-like electrodes of seamless graphene/nanotubes for wearable and stretchable supercapacitors. J Power Sources 2020;446:227355. https://doi.org/10.1016/j.jpowsour.2019.227355.

[206] Shakir I, Ali Z, Bae J, Park J, Kang DJ. Layer by layer assembly of ultrathin V2O5 anchored MWCNTs and graphene on textile fabrics for fabrication of high energy density flexible supercapacitor electrodes. Nanoscale 2014;6:4125. https://doi.org/10.1039/c3nr06820j.

[207] Yu G, Hu L, Liu N, Wang H, Vosgueritchian M, Yang Y, et al. Enhancing the supercapacitor performance of graphene/MnO$_2$ nanostructured electrodes by conductive wrapping. Nano Lett 2011;11:4438–42. https://doi.org/10.1021/nl2026635.

[208] Yu G, Hu L, Vosgueritchian M, Wang H, Xie X, McDonough JR, et al. Solution-processed graphene/MnO$_2$ nanostructured textiles for high-performance electro chemical capacitors. Nano Lett 2011;11:2905–11. https://doi.org/10.1021/nl2013828.

[209] Cao Y, Zhu M, Li P, Zhang R, Li X, Gong Q, et al. Boosting supercapacitor performance of carbon fibres using electrochemically reduced graphene oxide additives. Phys Chem Chem Phys 2013;15:19550. https://doi.org/10.1039/c3cp54017k.

[210] Zang X, Chen Q, Li P, He Y, Li X, Zhu M, et al. Highly Flexible and Adaptable, All-Solid-State Supercapacitors Based on Graphene Woven-Fabric Film Electrodes. Small 2014;10:2583–8. https://doi.org/10.1002/smll.201303738.

Md Omar Faruk, Md Milon Hossain

8 Smart and functional textiles for personal thermal comfort

Abstract: Textiles are considered the most versatile and cost-effective materials that can maintain personal thermal wellbeing by controlling the indoor and outdoor thermal atmosphere. However, traditional textiles cannot efficiently maintain the thermal comfort of an individual. Therefore, textiles engineered with advanced thermoregulation systems can enhance personal thermal comfort (PTC) by regulating the thermal atmosphere. This chapter summarizes the a progress of novel materials integrated into thermoregulation textiles for PTC. The fundamentals of thermal comfort, including heating and cooling principles and strategies for fabricating thermoregulation textiles, have been highlighted, followed by a discussion of different passive heating and cooling textiles. The subsequent section has presented a brief overview of thermal comfort by responsive textiles follows. The conclusion highlights the challenges and prospects of thermoregulation textiles.

Keywords: Advanced textiles, passive heating, passive cooling, thermoregulation, personal thermal comfort

8.1 Introduction

Thermal well-being is a physiological and/or psychological perception of an individual towards the surrounding environment shaped by various thermal as well as non-thermal aspects, such as humidity, the flow of air, visual and tactile sensation [1]. To ensure thermal comfort, it is essential to understand thermophysiological factors (thermal and non-thermal) for controlling the thermal environment by developing advanced technologies. It is also important to maintain personal thermal comfort (PTC) at the individual level, as thermal comfort plays a major role in maintaining personal well-being [2]. For example, if the human body reaches the condition of hypothermia (i.e. temperature below 35 °C) and/or hyperthermia (i.e. temperature above 37.5–38.3 °C), it may cause fatal effects, even death [2, 3]. Besides, controlling individual's thermal comfort decreases consumption of energy for maintaining comfort using heating, ventilation, and air-conditioning (HVAC) systems that are accountable for approximately two-fifth of the total global energy consumption and greenhouse gas emission [2–4]. Thus, developing advanced technologies can control all the thermophysiological conditions for maintaining personalized thermal comfort.

The generating and releasing of metabolic heat is a continuous physiological event of the human body for maintaining a balanced physiological system. Conduction, convection,

https://doi.org/10.1515/9783110759747-008

radiation, and evaporation are the four heat dissipation pathways that work together for maintaining homeostasis through their dependency on different circumstances (Figure 8.1) [2, 5, 6]. Releasing of body heat through radiation (in mid-IR range) is the primal heat loss route for the human body in the indoor environment, whereas, during intense exercise, the body loses heat through evaporation [2, 5, 7, 8].

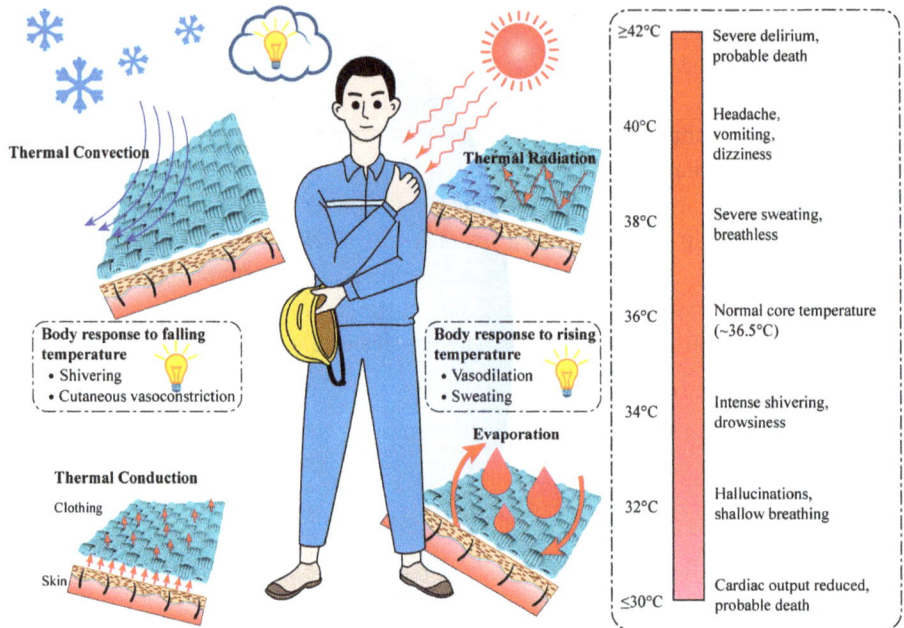

Figure 8.1: The heat transfer interactions among the human body, clothing, and the surrounding environment and their effects on physiological conditions (adapted with kind permission [9], Copyright 2021, Elsevier).

Clothing is considered the most critical factor for maintaining the thermal environment of the human body [1]. Therefore, considering the importance of personal thermal well-being, the development of advanced thermoregulation systems using textiles is an efficient and cost-effective way of maintaining PTC. Effective controlling of the interactions between the incoming and outgoing heat with clothing could result in advanced thermo-regulation for PTC [4, 10]. Such PTC can be achieved by exploiting the excellent thermal conductivity of different natural and synthetic fibres, as well as some polymers (Table 8.1). More importantly, textiles designed with novel thermoregulation systems can optimize body heat in extreme environments [3]. Therefore, instead of facilitating thermal comfort using external devices, these advanced textiles ensure favourable comfort in the microclimate near the human body [4]. As a result, designing thermoregulation textiles by incorporating functional materials, followed by passive and active strategies have shown promise, in recent years. Although there is a huge advancement

Table 8.1: Different fibres, polymers, nanomaterials, and their composites for thermal management applications (adapted with permission from [9], copyright 2021, Elsevier) [9].

Materials	Features	Applications
Cotton	Empty cavity and natural convolutions	Garments, quilt cover
Wool	Porous and voluminous	Sweater, blanket, quilt, and so on
Down fibres	Sub-branches with certain crotches	Down jacket
Ceramic fibres	High chemical stability	Suits for extreme environment
Polyamide fibres and composites	Have strong molecular chains	Multifunctional clothing
Graphene and graphene-based composite incorporated textiles	Highly flexible with excellent thermal conductivity	Advanced textiles for thermoregulation
BNNS/PVA nanocomposites	High thermal conductivity and well-oriented fibres	Cooling textiles
Porous silk fibroin-aerogel fibres	High porosity	Thermal insulating textiles
Biomimetic fibres	Aligned hollow shell	Thermal insulating textiles
PE film and fibres	High IR transmittance and reflectance	Heating and cooling textiles
UHMWPE fibres	Ultra-oriented fibres	Cooling textiles

Note: BNNS/PVA, boron nitride nanosheets/poly(vinyl alcohol); PE, polyethylene; UHMWPE, ultra-high-molecular-weight polyethylene.

in thermoregulation textiles, understanding the fundamentals of textiles for personal comfort, especially using passive heating and cooling textiles are critical. Therefore, this chapter has summarized the fundamentals of advanced heating and cooling textiles using passive thermoregulation.

8.2 Mechanism of thermal comfort through textiles

Thermal comfort through textiles is a process in which clothing acts as a heat exchanger within the body and the adjacent atmosphere by balancing heat generation and heat loss [3]. The factors like temperature, airflow, humidity, and the design of the textiles could influence the heat transfer (HT) between the skin and the environment to balance the thermal comfort [3, 11]. Generally, conduction, convection, and radiation are the core HT systems that are responsible for transferring heat or thermal energy from/to a body to maintain thermal comfort through textiles [3, 11]. Thermal conduction is an impulsive HT phenomenon, where heat is transferred to the lower temperature region from the higher state within two contacted substrates. In contrast, thermal convection results from the air movement because of the temperature gradient of the adjacent environment [3, 11]. The substrates that have a temperature greater than absolute zero is the electromagnetic thermal radiation, which radiation can be reflected by developing a smooth surface on textiles using metallic substrates with high refractive indexes [3, 11].

Recently, six mechanisms of thermal comfort using advanced textiles have been proposed (Figure 8.2) [3, 11]. Compared to the surrounding atmosphere the human body has a much higher temperature in winter. Therefore, the textiles designed with materials having lower thermal conductivity and/or the textiles manufactured with porous fibrous materials are an effective way to keep the body warm (Figure 8.2a). Textiles fabricated with a densely woven or knitted fabric can maintain warmth by reducing the heat convection through suppressing the airflow between the textiles and surrounding skin (Figure 8.2b). Besides, the textiles incorporated with metallic nanomaterials with a high reflectivity could be used for reflecting human body-generated IR (Infrared) thermal radiation (Figure 8.2c). So, these strategies could be feasible to design textiles for the warming effect in a cold environment. In contrast, the mean radiant temperature (≥ 35°C) is comparatively higher than the human body in summer [11]. Therefore, to keep the body cool, that excessive ambient temperature must be controlled. Such cooling effect could be achieved by suppressing thermal conduction by utilizing the materials that have lower thermal conductivity (Figure 8.2d) and strong convection (Figure 8.2e), where thermal convection reduces the heat stress as well local humidity. As a result, convection can control the temperature and humidity of the microclimate. Besides, the humidity is increased in hot conditions because of the release of sweat from the human body. Therefore, proper controlling of convection would release the heat and moisture to maintain cooling effect in summer. In this regard, textiles designed with tunable porosities (i.e., incorporation of textiles with smart materials like shape memory and stimuli responsive materials) could enhance thermal convection and thereby providing cooling effect by improving the flow of air transport [11]. Furthermore, textiles designed with materials having high IR reflective properties could reflect the radiative heat to the surrounding environment to enhance the cooling effect. (Figure 8.2f). In general, textile fibres are more thermally conductive than air, and hence, higher thermal insulation could be achieved by manufacturing fabrics using highly porous materials that can trapped more air content [3]. Alternatively, adequate heat exchange for the cooling effect during summer could be achieved by reducing the thickness of fabrics.

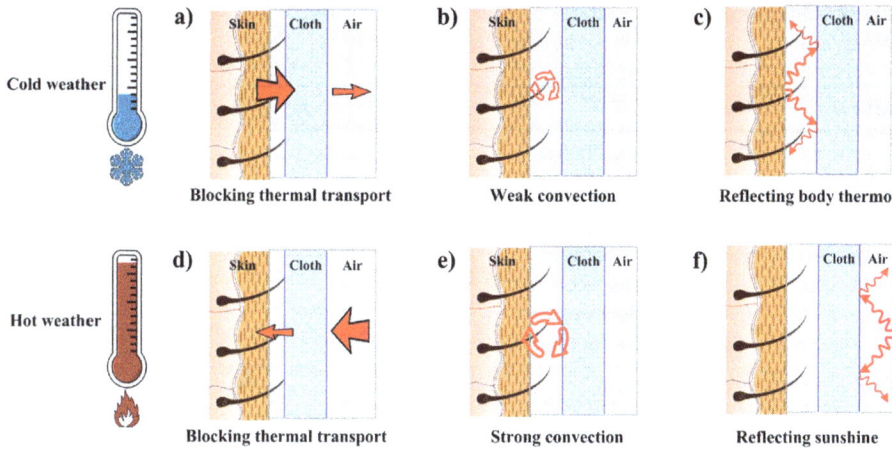

Figure 8.2: The mechanism of maintaining thermal comfort in the winter season is by – (a) blocking thermal transport from the body to the environment, (b) facilitating weak convection near the skin, and (c) reflecting the body heat. Clothing in the summer can maintain thermal comfort by - (d) blocking thermal transport from the outer environment to the body, (e) enhancing strong heat convection near the skin, and (f) reflecting radiative heat coming from the sun (adapted with kind permission [11], Copyright 2019, Springer Nature).

8.3 Strategy of fabricating thermoregulation textiles

8.3.1 Passive heating and cooling

Generally, textiles have played an essential role in balancing the human thermoregulation system for personal comfort; however, traditional textiles are not effective in regulating the thermal system. These conventional textiles are responsible for around 50% of the indoor heating loss through the four primal HT routes (radiation, evaporation, convection, and conduction) [3]. In this regard, novel and cutting-edge strategies have been adapted to design textiles with advanced thermoregulation systems to provide control heating and cooling effect for PTC, by enabling and suppressing radiative HT (Table 8.2) [4, 12]. Such an advanced concept emphasizes the warming or cooling of the human body by enhancing or reducing thermal insulation [3]. Controlling the emissivity level of the exterior side of the textiles as well as the heat dissipation rate within the textiles are two common approaches to designing novel thermoregulation systems for passive heating and cooling textiles [4]. In these concepts, the increasing surface emissivity can enhance the radiative heat dissipation rate of the fabricated textiles, and thereby, the cooling effect can be realized, whereas lowering the surface emissivity would result in a heating effect [4]. Such passive heating and cooling strategies provide satisfactory personal comfort by maintaining inherent features of the textiles and could save up huge indoor and outdoor heat loss [3].

Table 8.2: Advantages and disadvantages of some passive heating and cooling devices (adapted with kind permission from [21], copyright 2021, Elsevier) [13].

Materials	Advantages	Disadvantages	Mode
Polyethylene	– Water-wicking capable – Strong – IR transmittance	– Uncomfortable	Cooling
Polyamide 6	– Resistance to abrasion – Washable and air permeable – Moisture-wicking capable – IR transmittance	–	Cooling
Glass–polymer hybrid metamaterial	– IR transmittance – Excellent cooling power	– Not air permeable – Not suitable for moisture transport	Cooling
Nano-PE textile	– Dual-mode (radiative heating and cooling)	– Complex fabrication	Cooling/heating
TiO_2-coated cotton	– Air-permeable – Washable – Solar reflectance	– Complex fabrication	Cooling
Bilayer (nano-PE-Nylon 6) textile	– Solar reflectance – Moisture permeable – Mechanically strong	– Complex fabrication	Cooling
Photonic structures	– Better thermal insulation.	– Prolonged fabrication – Possibility of uneven cracks	Cooling
Tri-layered (PE-PVDF-Nylon) textile	– Scalable and lightweight – Wearable – High solar reflectivity	– Complex fabrication	Cooling
Porous PVDF fibre	– Scalable – Solar reflectivity	–	Cooling
VO_2 and Ag strip-integrated polyester	– Anisotropic and thermo-responsive – Electrically conductive	– Uncomfortable and heavy – Low air permeable – Complex fabrication	Cooling
Metallic fibres	– Thermal reflectivity	– Brittle, rigid, and heavy	Heating
Nanowire embedded textile	– Vapour permeable – Excellent thermal insulations	– Metal oxidation – Expensive	Heating

Table 8.2 (continued)

Materials	Advantages	Disadvantages	Mode
Ag nanowire-coated cotton	– Mid-IR reflective	– Poor substrate adhesion	Heating
Nanofibres	– Can protect from pollutants – High mid-IR reflectivity	– Complex fabrication	Heating

8.3.2 Active heating and cooling

Another advanced strategy has been developed to design textiles aiming to deliver adequate heating/cooling comfort, known as active heating and cooling. In this active strategy, the Joule heating technique has been exploited for additional warming effects, which can be achieved by using the electrothermal conversion systems powered by external energy [3]. In the Joule heating technology, energy dissipates throughout the whole system as heat, when the electron passes (i.e. movement of current) through an electrode [14]. In this technique, the produced heating energy (Q) is dependent on the flow of current (I) through the electrodes having resistance (R) per unit of time (t), and the total power consumption (P) could be calculated using the following equations [3, 14]:

$$Q = I^2 R t \tag{8.1}$$

$$P = \frac{Q}{t} = I^2 R \tag{8.2}$$

8.4 Textiles for passive heating and cooling

8.4.1 Thermal radiation regulating textiles

On the Earth's surface, solar radiation is primarily distributed in the ultraviolet (UV), visible, and near-IR (NIR) regions, ranging from the wavelength of 0.1–4 μm (100–4,000 nm), having a total power density of about 1,000 W/m² 2, [3, 4, 15, 16]. Based on the average solar reflectivity of skin, the bare human body skin absorbed most of the solar irradiance (>60%) that emits as IR [15]. As an excellent IR emitter, human skin (with an emissivity of 0.98), can emit thermal radiation in the range of 7–14 μm (mid-IR), with the peak emission at 9.5 μm and a net radiation power density of around 100 W/m² [15, 17]. Such radiative heating is crucial for body heat dissipation, as this IR range coincides with the IR transparency window of the earth's atmosphere, and more than 50% of the total heat is dissipated through this atmospheric window of the earth

[2, 15]. Because of the high emissivity (0.75–0.9), traditional textiles cannot control the thermal radiation of the body. Besides, the inferior structural design of traditional clothes is also accountable for most of the body heat dissipation of IR radiation [2, 3]. Such radiative heat losses could be reduced by controlling the thermal radiation using thermoregulated textiles for maintaining thermal comfort [3]. The radiative heat exchanged by using specially designed clothes can be expressed by Kirchhoff's radiation principle, as follows:

$$\varepsilon + t + \rho = 1 \tag{8.3}$$

where ε, t, and ρ denote the emissivity, spectral transmission, and spectral reflectance component, respectively [2, 3]. In the case of cooling effect, it is important to emit a great portion of human body radiation by mid-IR transparent fabric ($t = 1$) and/or highly emissive textile surface ($\varepsilon = 1$), while mid-IR reflective textiles ($\rho = 1$) are essential for warming effect [2]. However, for practical cooling and warming effects, textiles need to be opaque in the visible light wavelength [2].

8.4.1.1 IR-emissive radiative textiles for warming

In a cold environment, the cloth required to reflect the radiative heat to the skin to keep the body warm [3]. The refractive index is higher for particles two times larger than the wavelength of incident radiation. The high refractive index allows better IR reflection, and, thus, affects the warming characteristics of materials [3, 11]. This phenomenon can be explained by Fresnel's law, where the optimal reflection would occur if the particle has an equal diameter (D) to half of the wavelength (i.e. $D = \lambda/2$) [3, 11]. Therefore, IR reflective warming textiles are generally designed by modifying textile structure with IR reflective materials like metallic (e.g., zinc (Zn), copper (Cu), silver (Ag)) nanowires (NW) [3] Besides, the hybrid yarns made of metal-polymer composites, including core spun and blended yarns, are also being exploited to manufacture advanced thermoregulation textiles for thermal warming [2]. Such radiative heating can be reflected in the body by combining the IR-reflective NWs with the traditional textiles to form a reflecting layer in the heating cloth [4]. This concept of passive warming textiles can be exploited by modifying the surface of cotton fabric coated with mesh-like silver nanowires (AgNWs) [18, 19]. To get a better warming effect by reflecting IR radiation to the human skin, the gaps within the NWs networks (200–300 nm) should be kept lower than the IR radiation that is radiated from human skin, with a wavelength of 9.5 µm and size larger than the water vapour molecules. The coating layer containing the network of NWs has the potential to maintain radiative heating by trapping comparatively more radiative heat than the traditional textiles, with a reflectance of around 97% [18]. The controlled spacing within the porous networks of the AgNWs-coated advanced textiles has been found to exhibit improved moisture management. These specially designed textiles not only provide passive heating but also achieve a surface temperature of 38 °C with a small power supply (0.9 V). Besides the warming effect, the AgNW-coated textiles also have the potential to save huge energy consumption (354.7 W per person), while lowering indoor heating from 187 to 96 W [18]

Sometimes, radiative heating textiles cannot demonstrate optimal heating performance, as they are fabricated with low IR reflectance and transmittance materials [3, 20]. Besides, metallic film-based composites having a comparatively higher density compromise their comfortability[3, 20]. For example, Omni-Heat and Mylar blanket are two advanced thermoregulating textiles fabricated using reflective materials to keep the body warm, despite inferior radiative heating performance [20, 21]. However, the Mylar blanket is unpleasant for daily use, as it is fabricated by coating dense metallic film into the solid plastic sheet (polyethylene terephthalate (PET) film) [20, 22]. On the contrary, the Omni-Heat technology has unsatisfactory radiative heating performance, with low reflectivity due to the integration of scattered metallic dots into the inside of the clothes to reflect human body heat [20, 21]. Therefore, it is important to optimize radiative heating and comfort by focusing on controlling IR radiation effectively, like textiles designed with advanced photonic structure [20]. Besides, lowering the surface emissivity of the textile is another promising method of minimizing radiative heat loss for passive warming textiles (Figure 8.3a and b) [4, 20]. In this regard, novel nanophotonic heating clothes were fabricated by dispositioning IR reflective layers of nanoporous metal (Ag nanoparticles) on an IR transparent polydopamine (PDA)-treated nanoporous polyethylene (nano-PE) layer [20]. The fabricated film was then laminated with the cotton fabric (Figure 8.3c). The nanopore size (50–1,000 nm) in IR transparent nano-PE layers was comparable with the wavelength of visible light (400–700 nm) was comparable with the wavelength of visible light (400–700 nm) (Figure 8.3d) that made PE opaque to human eyes by strongly scattering the visible light. The porous Ag layer with a pore size of 50–300 nm (Figure 8.3d) and the interconnected mesh-like porous structure in the nano-PE was responsible for the low emissivity and ensuring desired breathability [20]. The integration of extremely reflective (98.5%) film made of nanoporous Ag particles on the remarkably IR transparent (96.0%) nano-PE has lowered the IR emissivity of the exterior surface of the textile up to 10.1%, compared to the Mylar Blanket (60.6%), Omni-Heat technology (85.4%), and cotton (89.5%) (Figure 8.3e and f). The advanced photonic textile (nano-Ag/PE textile) reduced the ambient set-point temperature to 15 °C, which is comparatively lower (7.1 °C lower) compared to the pure cotton fabric (22.1 °C) and showed a better heating performance. As shown from the Figure 8.3g and h, the person wearing the nano-Ag/PE textile emits less heat than in the traditional cotton fabric, attributed to both the IR reflective and transparent textiles with a low emissivity, which suppresses the heat dissipation to the environment.

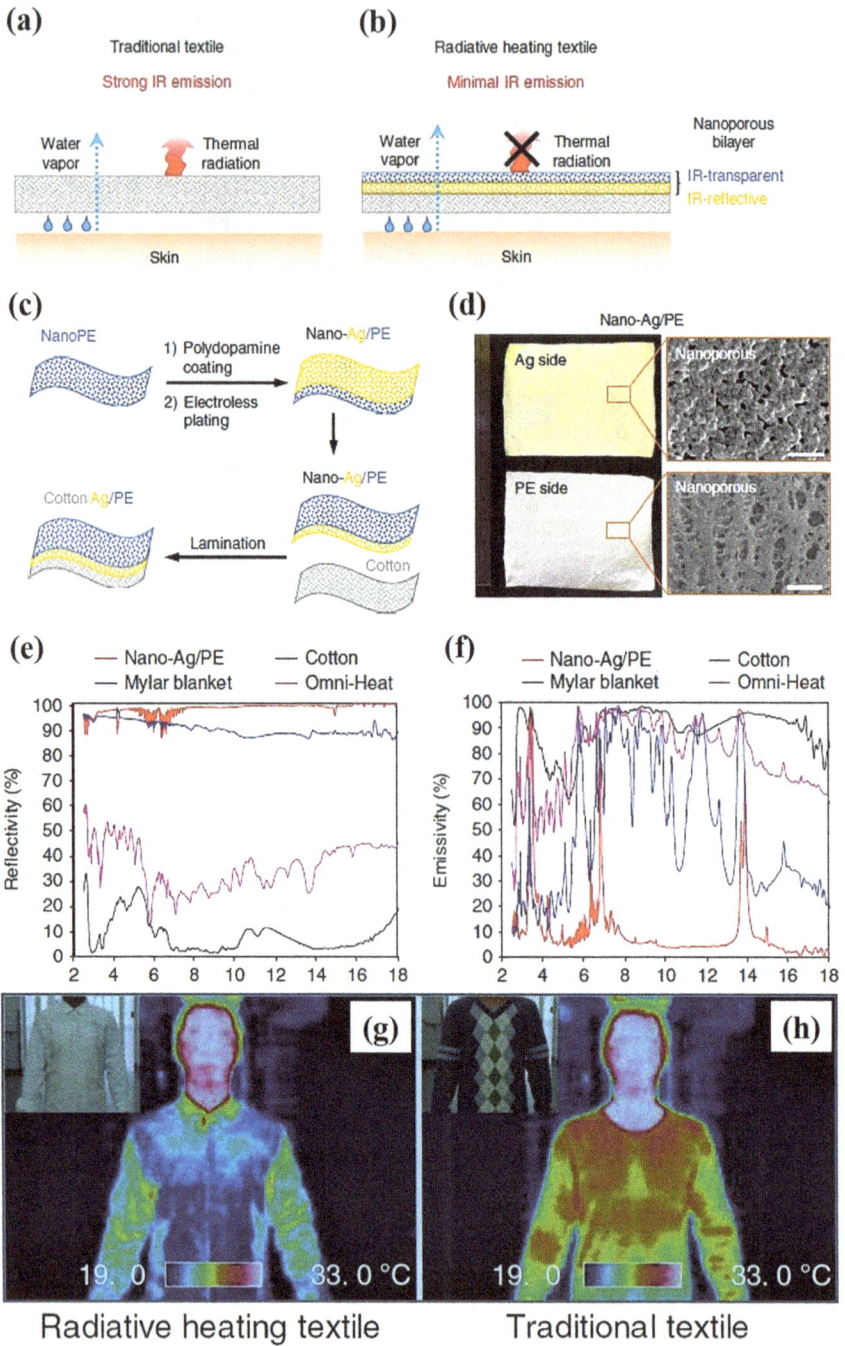

Figure 8.3: (a, b) Comparison of the IR emission in conventional and advanced warming clothes, (c) fabrication techniques of nano-PE textile laminated with the cotton fabric, (d) photos and SEM images of

8.4.1.2 Textiles for radiative cooling

8.4.1.2.1 IR-transparent radiative textiles

Dissipation of black body radiation of a warm object (in the range of 8–13 µm) through the atmospheric window to the surrounding space with a lower temperature is defined as radiative cooling [4, 23]. Such a phenomenon is exploited to fabricate textiles for personal cooling using advanced thermoregulation techniques [4, 23–25]. Generally, textiles are modified to make them transparent to the heat radiation generated by the human body [4]. Traditional clothes made up of natural and synthetic fibres (e.g. wool, cotton, and polyester) are mostly opaque to IR rays and absorb the heat (in the range of mid-IR) that emits from the human body. Such phenomenon results in thermal discomfort by heat accumulation near the human skin [4]. The existence of vibration of the molecular bonds (e.g. C–H, C–N, C–O, and S_2O) of natural fibres is responsible for the opacity of the IR radiation [4]. Like the IR absorption peaks (9.5 µm) of the human body, these chemical bonds also have IR absorption peaks in the same region [4]. However, the fibres, for example, polyethylene and nylon, which are IR transparent do not have any comparable chemical bonds mentioned above that have overlapping absorption peaks in their structure [4]. Considering the above-mentioned characteristics, it is possible to modify the physical and chemical structures of textiles to enhance the cooling effect by facilitating IR radiation transmittance to ensure the comfort of the fabricated textiles [4]. Besides, these IR-transparent textiles also have good wearability, comparable to conventional textiles. The interconnected nanopores with the size distribution of 50–1,000 nm allow the nano-PE to be transparent to mid-IR human body radiation but opaque to visible light (Figure 8.4a) [26]. The comparable pore size in the visible wavelength (400–700 nm) makes the nano-PE opaque to the eyes, by strong scattering of visible light. However, it maintains intrinsic mid-IR transparency attributed to their smaller pore sizes, compared to the IR wavelength. The fabricated nano-PE exhibited more than 90% IR transmittance for wavelength more than 2 µm. However, strong nanopores scattering was responsible for low visible light specular transmittance (Figure 8.4b). Besides, the IR scattering effect of the nano-PE also intensified by increasing the average pore size in the range of 200 nm to 4.2 µm, as the scattering induced transmittance dip was shifted from visible to NIR and then, to the mid-IR (Figure 8.4c). Therefore, the device (simulated skin) covered with nano-PE showed a 2.7 °C lower temperature than the device covered with cotton (Figure 8.4d and e).

Figure 8.3 (continued)
the Ag side and PE side of the fabricated nano-Ag/PE film, with a scale bar of 1 µm, (e, f) FTIR reflectance and emittance of different textiles, respectively, and (g, h) Thermal images of advanced radiative heating textiles and conventional textiles, respectively (adapted under Creative Commons Attribution 4.0 International License from [20].

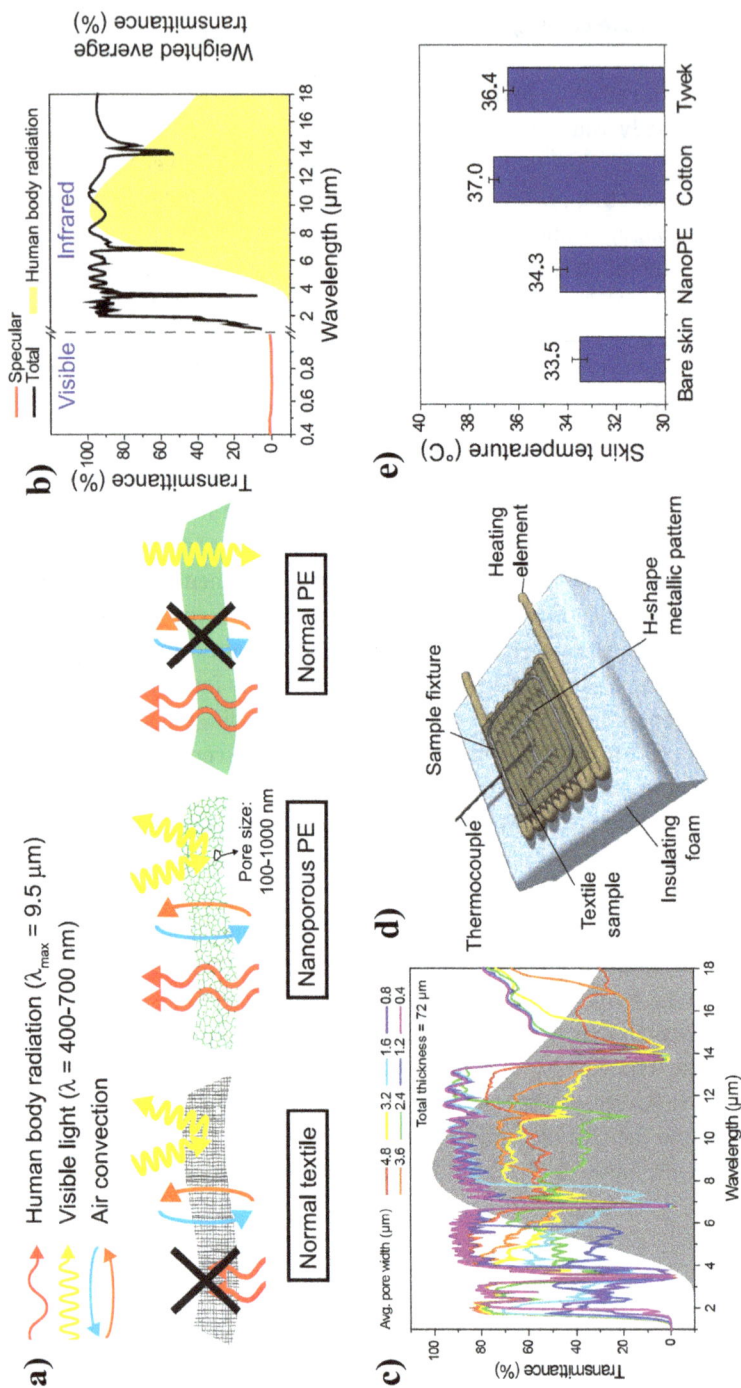

Figure 8.4: (a) Illustration of the comparative thermal properties of nano-PE, normal PE, and cotton; (b) IR and visible light transmittance of nano-PE, where the thickness and average pore size of nano-PE is 12 nm and 400 nm, respectively; (c) The relationship between the pore size increase and transmittance dip movement; (d) the thermal experiment set-up for the thermoregulation textiles, and (e) the comparison of thermal performance of different textiles (adapted with kind permission from [26], Copyright 2016, American Association for the Advancement of Science).

8.4.1.2.2 IR-emissive radiative textiles

Apart from modifying the structures of the fibres for developing IR-transparent radiative cooling textiles, regulating the emissivity of textiles is another way for thermal regulation in a hot environment to get cooling effect [2, 4, 27]. Herein, controlling the surface emissivity and surface temperature plays a crucial role in achieving radiative cooling effects, according to the equation 8.4 [4, 27]:

$$q_{rad} = \sigma \in_{tex} \left(T_{tex}^4 - T_{amb}^4 \right) \tag{8.4}$$

where q_{rad} represents radiation heat flux, $\in_{tex}$ is the emissivity of the textile materials, σ denotes the Stefan–Boltzmann constant, and the term T_{tex} and T_{amb} represents the temperatures of the textile surface and the surrounding environment, respectively. The equation 8.4 indicates that the radiation heat flux will increase while increasing the temperature and emissivity of the textile, and thereby, radiative cooling will be occurred as body heat dissipation rate will be more efficient with high emissivity of textiles [4, 27]. In these IR-emissive radiative cooling, the heat dissipation rate can be enhanced by controlling the emissivity of the outer part of textiles rather than the inner counterpart [2, 4]. This is because the radiation is prevalent while exchanging heat between the outer face of the clothes and the surrounding environment, compared to the heat exchange between the inner surface of textiles and the human body [2, 20, 27]. Recently, a dual-mode textile was fabricated by embedding a bilayer emitter (having different emissivities on each side) inside the nano-PE textiles (having different thicknesses on each side) for IR-emissive radiative heating and cooling [27]. Herein, the emissivity of the textile was controlled by the bilayer emitter, while the thickness of nano-PE regulates the temperature of the emitter from being closer or farther from the hot side (human skin). This specially designed textile achieved a cooling effect, when the thickness of nano-PE between the emitter and the skin was small and the high-emissive layer (high ε_{tex}) was facing towards the outer environment. The effectiveness of thermal conduction between the warm skin and the emitter is dependent on such a short distance between the skin and emitter, resulting in increased emitter temperature (high T_{tex}). Such a combination of the smaller emitter and skin thickness (high T_{tex}) and high emissive (high ε_{tex}) outer layer was responsible for a higher heat transfer coefficient, and, thereby, ensures cooling effect. It was reported that a strong cooling effect (as strong as wearing only nano-PE cooling textiles) would be achieved when there is no thermal resistance between the skin and emitter ($T_{tex} = T_{skin}$) and the $\varepsilon_{tex} = \varepsilon_{skin} = 0.98$ [17, 27]. However, the fabricated textiles would work as heating textiles, when there is a high emitter-to-skin distance and a high-emissivity (high ε_{tex}) surface is facing inside (low-emissivity layer is facing outside (low ε_{tex})), by reducing the thermal conductance (low T_{tex}). The fabricated dual-mode textiles achieved a radiative cooling effect of about 3.1 °C, when the high-emissivity layer (high ε_{tex}) was facing outside, while about 3.4 °C warming effect was achieved when the fabric was flipped inside out (low-emissivity (low ε_{tex}) layer facing outside).

Solar-reflecting radiative textiles

IR thermal radiation is considered the major heat exchange path for the human body. Therefore, regulating the thermal radiation in the mid-IR range, generated from the human body is vital for maintaining the cooling effect in the indoor environment [2, 27]. However, in outdoor environments, circumventing the thermal radiation from the solar systems is essential for the radiative cooling effect in case of PTC [2, 28–30]. Generally, the solar system is composed of light both in the visible and NIR in the range of 0.4–2.5 µm, which is responsible for around 93.4% of solar irradiance (~1,000 W/m^2) [2, 31]. Therefore, it is important to evade the energy from the solar spectrum to maintain personal cooling effect in the outdoor environment. It can be done by designing textiles coated with different reflective materials. In this regard, researchers have exploited different reflective nanomaterials, based on transition metals (e.g. Ag, Ti, and Al), inorganic and/or organic compounds (e.g. TiO$_2$, Fe$_2$CO$_3$, and AZO pigments), and natural compounds like chlorophylls coating materials for developing advanced textiles that enhance the reflection of sunlight sending back to the environment, for personal cooling [2, 32–34]. For example, Wong et al. [33] have reported that the anatase and rutile mixed irregular-shaped TiO$_2$-coated (having a particle size of 293–618 nm) cotton textiles could achieve a maximum temperature reduction of 3.9 °C by maximizing the solar reflectance.

The low IR transmissivity and the lower mean solar reflectivity (around 60%) of cotton textile have been responsible for inhibiting effective heat dissipation [15]. Therefore, manipulating both solar radiation and human body radiation is another efficient way of realizing the optimum cooling effect [2]. Hence, developing clothing that can effectively reflect solar radiation both in the visible and NIR light could be promising in maintaining mid-IR transparent human body radiation for radiative cooling. The IR-transparent aliphatic C–C and C–H bond of PE is effective for fully transmitting out human body radiation for indoor cooling. However, solar reflectance of PE is not sufficient for outdoor cooling because of relatively low refractive index ($n \sim 1.5$) [15–17, 26, 35]. Alternatively, the solid and inorganic ZnO has a relatively higher refractive index ($n \sim 2$) along with the absorption capability from visible (0.4 µm) to mid-IR wavelengths (16 µm) [15, 36, 37]. Such inherent properties influence consideration for the combination of ZnO and PE as the desired radiation selectivity substrate for outdoor radiative cooling textiles [15]. ZnO with particle size 0.1–1 µm is an essential parameter for maximizing reflection in visible and NIR region, as well as high diffusion in mid-IR wavelength. Therefore, strong Mie scattering drastically enhances the scattering cross-sections both in the visible and NIR ranges, with a small scattering in the mid-IR range [15]. It was also found that the increase of the size and density of ZnO nanoparticles increases the solar reflection, but decreases the transmission in mid-IR wavelengths. It was reported that the thickness of the ZnO-based nanocomposite layer in the range of 80–160 µm is optimal for both the high solar reflection and mid-IR transmission for indoor cooling [15]. Considering these possibilities recently, a novel spectrally selective nanoengineered textile has been proposed, which was designed by combining metallic nanomaterials (ZnO) with nanoporous polyethylene

(ZnO–nano-PE) at a weight ratio of ZnO:PE to 2:5 in paraffin oil (PE-to-oil ratio 1:5), for outdoor radiative cooling (Figure 8.5a) [15]. A strong scattering of light in the visible range in all directions under the sun was observed, with a film thickness of 150 μm. The reflectivity and transmissivity spectra (Figure 8.5b) exhibited a solar irradiance reflection of >90% with a higher transmissivity of around 80% in the range of human body thermal radiation (between 7 and 14 μm). As seen from Figure 8.5c-d, in a typical outdoor atmosphere (peak solar irradiance over 900 W/m²), this spectrally selective textile managed to evade the overheating of simulated skin by 5–13 °C (corresponding to a cooling power of more than 200 W/m²) showing much better performance compared to conventional cotton textile. Such excellent radiative cooling was achieved because of the superior solar reflection that decreases heat input from the sun, along with its high transmission to human body thermal radiation that expands the radiative heat output. Importantly, the fabricated textile has a potential to prevent overheating of the simulated skin by up to 8 °C than the textiles made of cotton, if the sweat evaporation comes into play. Compared to the existing cooling textiles [25, 29, 38, 39], an entirely different radiation characteristic of such a nano-engineered outdoor cooling textiles will expand the opportunities of radiative cooling depending on the need.

8.4.2 Conduction heat-regulating textiles

Heat conduction is one of the most common HT routes between human skin and clothing as well as the clothing and outside environment. Therefore, controlling heat conduction is an important factor for enhancing thermal comfort. Heat conduction, instead of the heat radiation, plays a leading role in the HT between the skin and clothes of IR-opaque textiles [2]. More importantly, heat conduction is the only pathway for heat dissipation inside the textiles themselves. However, traditional clothes often have a low heat conduction rate, which is not efficient for maintaining thermal comfort through releasing excess heat from the body. Besides, most of the textiles that we use daily are porous and not efficient for thermal comfort, as the trapped air inside the pores precludes to release of heat. Thus, designing conduction heat-regulating textiles with advanced thermoregulation techniques is essential for effective thermal comfort at the individual level [2]. Enhancing the thermal conductivity and the thermal insulation properties are important for thermal comfort, using smart textiles [2].

8.4.2.1 Advanced cooling textiles with enhanced thermal conductivity

Conventional textiles usually lead to an unsatisfactory cooling effect. They have low thermal conductivity (e.g. the thermal conductivity of cotton is 0.007 W/m K), which impedes the escape of heat generated from the body to the environment [4]. Therefore, accelerating the heat dissipation from the skin towards the external atmosphere

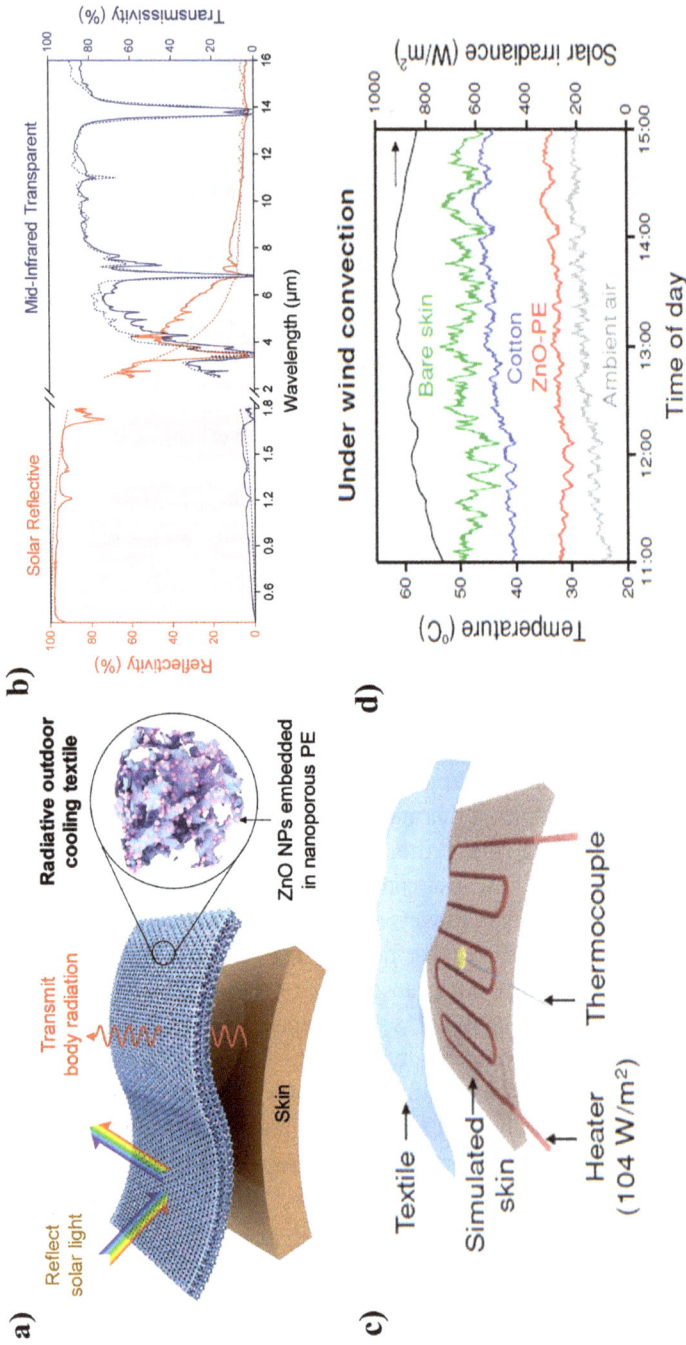

Figure 8.5: (a) Illustration of the ZnO–nano-PE radiative outdoor cooling textile, (b) reflectivity and transmissivity of ZnO–nano-PE from UV to mid-IR range (0.3–16 μm), (c) representation of the thermal management system, where a thermocouple evaluated the simulated skin temperature, and (d) graphical presentation of the temperatures measurement of ZnO–nano-PE-covered, cotton-covered, and bare skin-simulating heaters. Ambient temperature and solar irradiance were measured and plotted for reference (adapted with kind permission from [15], Copyright 2018, John Wiley and Sons).

through increasing thermal conductivity is one of the most effective ways for enhancing the cooling effect [4]. Such a cooling effect could be enhanced by incorporating textiles with highly conductive nanomaterials with good thermal conductivity, without any adverse effect in radiative heat dissipation [2, 4, 40]. There is extensive development in thermally conductive advanced textiles for cooling effect by exploiting highly conductive materials, especially carbonaceous materials including graphene, single-walled carbon nanotubes, and multi-walled carbon nanotubes (MWCNT), due to their high thermal conductivity and emissivity [2, 4, 41–43]. It was reported that the incorporation of resin-coated MWCNT with cotton fabrics resulted in a drastic enhancement of thermal conductivity [41]. The presence of a small amount of MWCNT (11.1%) dramatically improved the thermal conductivity of textiles as high as 78%, which was further boosted by 1.5 times, when the MWCNT content expanded to 50% [41]. Compared to the untreated fabric, the fabric treated with 11.1% and 50% of MWCNT had exhibited 2 and 3.9 °C lower equilibrium surface temperatures, respectively at 50 °C [41]. Although the coating process dramatically enhanced the cooling effect, it affected the permeability and comfortability (e.g. hand feeling) of the fabrics.

Instead of incorporating textiles with conductive nanomaterials, embedding thermally conductive materials into the fibre structures offered more sustainable conductive cooling [2, 4]. Recently, it was reported that highly aligned and thermally conductive boron nitride nanosheets (BNNS)/poly (vinyl alcohol) (PVA) composite fibre (BNNS/PVA) can enhance the thermal transport of the fabricated textiles for tailored cooling [44]. BNNSs have been added with the PVA and dimethyl sulfoxide (DMSO) mixture followed by passing through a solution of methanol (coagulation bath) to fabricate the fibres by 3D printing (Figure 8.6a). Finally, the printed fibres underwent a hot-stretching process (200 °C) for orienting BNNSs properly and uniformly in the fibre structures, which, then, acted as energy routes for heat dissipation. The highly thermally conductive composite fibres achieved an effective cooling effect (55% better than commercial cotton fibres), as they transported the body-generated extra heat to the ambient environment, more efficiently (Figure 8.6b and c).

8.4.2.2 Advanced warming textiles with reduced thermal conductivity

Generally, reducing the thermal conductivity is an important pathway for conduction HT regulated textiles for warming. Therefore, trapping a mass of air inside the textiles is promising for increasing the thermal resistance (for example, down jacket) to achieve a superior thermal-insulated warming textile [2, 45]. Herein, imitating the natural characteristics of down fibres may accumulate air inside the textiles to enhance the thermal insulation by reducing conduction HT [2, 46]. As is evident from the Thinsulate® and PrimaLoft®, commercial products developed by 3M and PrimaLoft, respectively, the much thinner fibres than the regular fibres can effectively reduce the HT, compared to conventional textiles [2, 46]. To enable the recovery of loft after compression, however, such

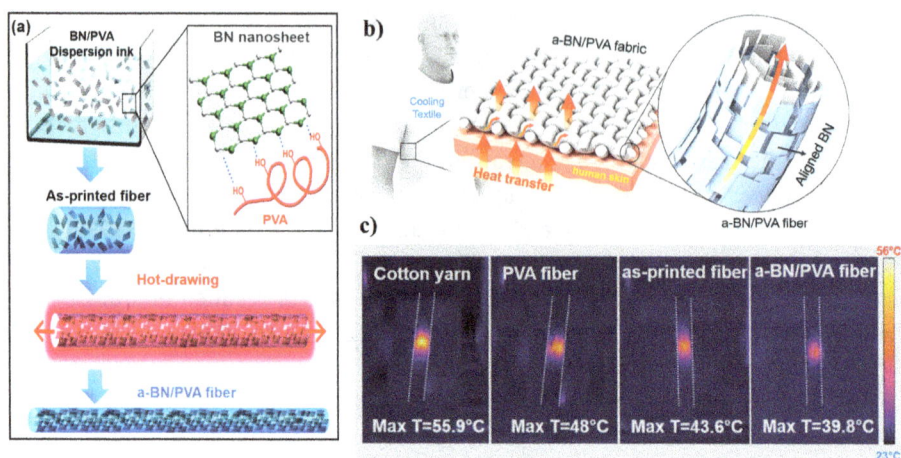

Figure 8.6: (a) Graphical illustration of the synthesis of BNNs/PVA composite fibre, (b) schematic of the thermoregulation principle of as-fabricated conductive cooling textiles, and (c) temperature distribution in as-printed BNSS/PVA fibres composed of cotton yarn, PVA fibre, and as-printed BNNS/PVA fibre without BNNS alignment and aligned BN/PVA (a-BN/PVA) fibre (adapted with kind permission from [44], Copyright 2017, American Chemical Society).

insulating clothes made by extremely thin fibres must have a considerable amount of thick and reinforcing fibres [46]. Incorporating textiles with special cross-sectional-shaped fibres is another alternative to trap and store a huge amount of air inside the textiles, as they cannot be packed as compact as fibres with round shapes [2, 3, 47].

Furthermore, the efficiency of textiles fabricated with hollow fibres in enhancing the thermal insulation to keep the body warm has been proven. These specially designed textiles can prevent HT from the body by storing sufficient air inside the textiles, and thus, boost passive warming by reducing the heat dissipation through heat conduction [2–4]. Focusing on such strategies, recently, bio-mimicked fibrous structures with aligned porous fibres have been developed using the freeze spinning technique, by mimicking the insulating hairs of polar bears to fabricate advanced warming textiles (Figure 8.7a) [48]. When the freezing temperature gradually rose from −196 to 40 °C, the fibres with axially aligned porous microstructures (87% porosity) showed superior mechanical strength than those with random porous structures (Figure 8.7b). The number of layers of fabric and the porosity of fibres are two crucial factors for enhancing thermal insulation using porous fibres. From Figure 8.7c, it can be seen that the single-layered woven textiles with extremely thin (≈0.4 mm thick) and porous (pore size ≈ 30 μm) fibres proved to be superior thermal insulators than the multi-layered textiles. Besides, the high porosity and the aligned micropores of biomimetic fibres had reduced both the heat conduction and convection, as the air was restricted on these axially aligned individual micropores (Figure 8.7d). The multiple reflective effects with the constant incident angle of aligned porous fibres (Figure 8.7d) enhance the overall reflectance

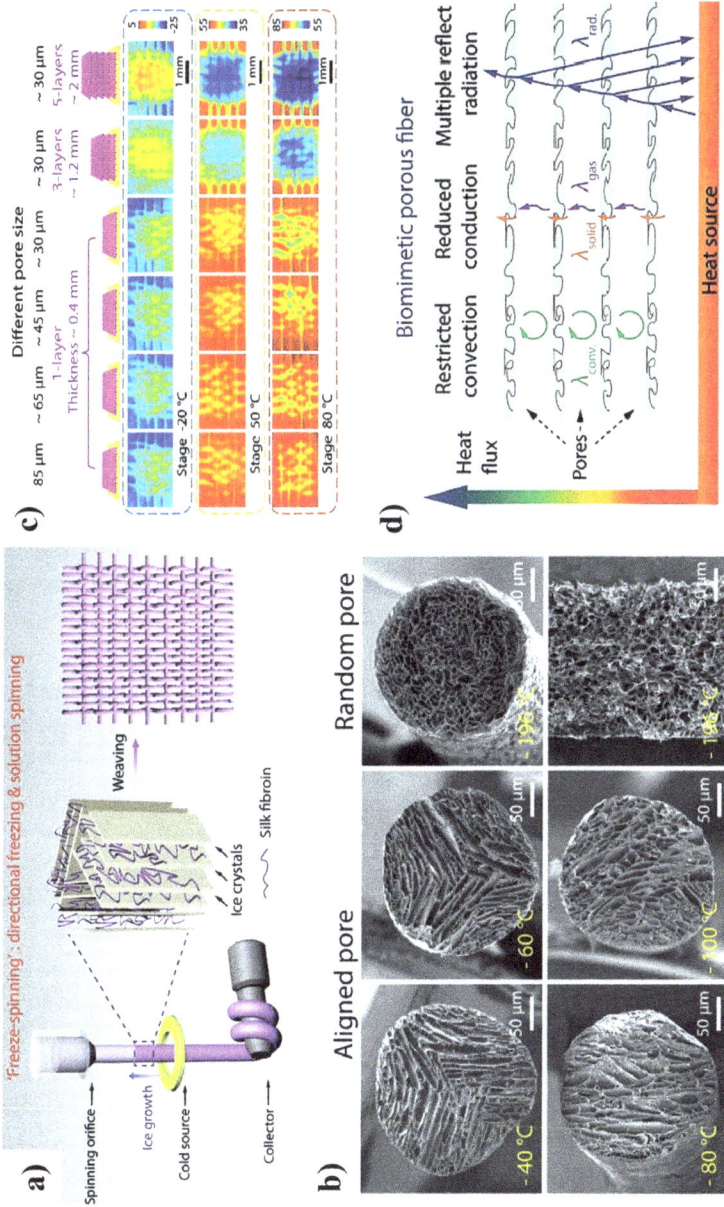

Figure 8.7: (a) Graphical presentation of the fabrication of biomimetic fibres using freeze spinning, (b) radial cross-sectional and longitudinal section (lowermost right corner) of different porous structures of biomimetic fibres, (c) the IR image of woven biomimetic textiles with different pore sizes and layers, and (d) graphical representation of thermal conductivity of the biomimetic porous fibre (adapted with kind permission from [48], Copyright 2018, John Wiley and Sons).

of the fabricated textiles to 70–80%. Therefore, the textiles fabricated with aligned porous biomimetic fibres would be favourable for next-generation warming textiles, as they greatly reduce heat loss through convection, conduction, and radiation.

8.5 Textiles with phase-change materials

Phase-change materials (PCM) can absorb and release thermal energy by changing the phase of the materials over a narrow temperature range, by taking advantage of latent heat enthalpy [2, 4]. Based on the stored latent heat, different organic [e.g. paraffin waxes (PW)] and inorganic (e.g. salt hydrates) PCMs with various melting temperatures are usually employed to control the unexpected temperature variation [9, 49]. Generally, the integration of PCMs into textile materials ensures the thermal comfort of the wearer by acting as a thermal barrier against the differences in the temperature of the surroundings [4]. The PCM-incorporated cooling textiles maintain thermal comfort by absorbing heat from the human body that minimizes the body perspiration [4, 50]. In contrast, heating is maintained by releasing the absorbed heat back to the body, when the temperature falls below the crystallization point of PCMs [50]. For incorporating PCMs into textiles, the melting point of PCM should be maintained between 15 and 35 ° C; also, having a short interval between the melting and solidification temperature is highly important for PCM to be incorporated with textiles[4, 49, 51]. Besides, while incorporating PCM with textiles, it is important to maintain the high thermal conductivity along with high specific heat capacity, and latent heat per volume with a small phase transition volume [4, 49, 51]. As these PCMs-incorporated textiles are exploited for thermal comfort, they should also be chemically stable, safe, eco-friendly, and cost-effective [4, 49, 51]. There are several ways for integrating PCMS with textiles, including adding PCMs in the coating formulation, spinning the PCMs – fibre polymer solution, electrospinning the phase-change fibres (PCFs), and so on [4]. For instance, in a recent study, PW was encapsulated with polyacrylonitrile (PAN) to fabricate core-sheath textiles using co-axial electrospinning and incorporating the heat absorbent $Cs_{0.32}WO_3$ in the sheath (Figure 8.8a) [52]. Herein, incorporating heat-absorbing $Cs_{0.32}WO_3$ has enhanced the heat storage capacity of fabricated textiles, as the NIR transmittance of the composite textiles decreased by the increase of the content $Cs_{0.32}WO_3$ (Figure 8.8b). Besides, the as-spun textiles exhibited high encapsulation efficiency of 54.3% with a latent heat of 60.31 J/ g. Importantly, the fabricated textiles showed excellent stability, as there were almost no changes in latent heat after 500 heating–cooling cycles (Figure 8.8c).

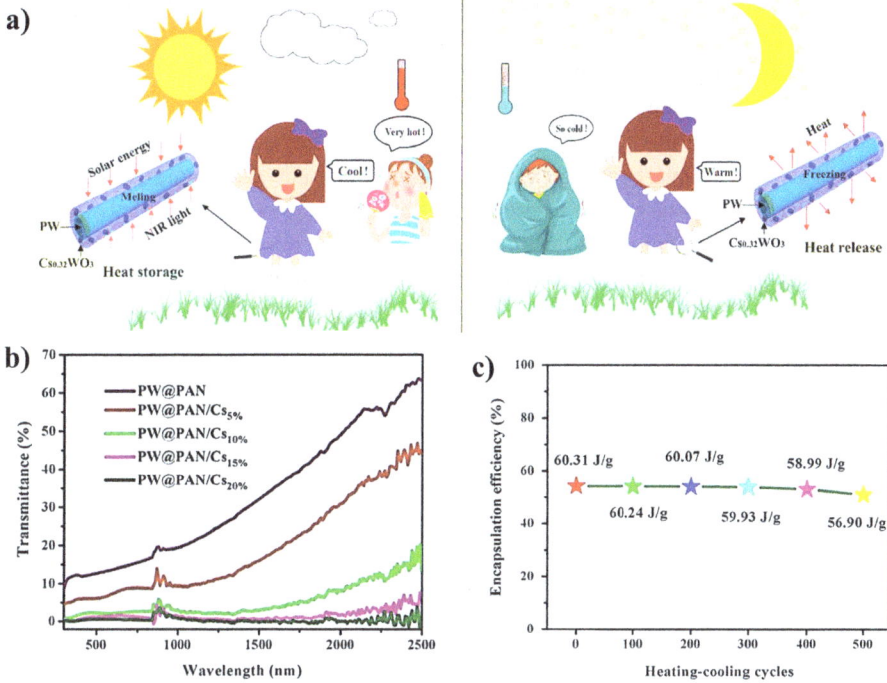

Figure 8.8: (a) Graphical representation of the PCM (PW)-based smart textiles for PTC, (b) the UV–vis–NIR spectroscopy of the PCM-based smart textiles, and (c) the encapsulation efficiency of the as-spun smart textiles at various heating-cooling cycles (adapted with kind permission from [52], Copyright 2018, Elsevier).

8.6 Thermoregulation with dynamic structure modification for extreme environment

Although there is an extensive advancement in thermoregulation textiles for PTC in specific conditions, most of them are not responsive to the changes in the thermal environment. It is believed that textiles with environment-responsive technologies would maintain PTC, as there is evidence that many natural species have evolved to maintain their thermal comfort by adapting to the environmental changes [53]. Evidence showed that *Cataglyphis bombycina*, Saharan silver ants that forage in the extreme temperature of the African desert, can control their body temperature [28]. It was found that the uniquely shaped hair of this desert ant dissipates the body heat back to the desert environment, as these hairs enhance the body reflectivity both in the visible and NIR wavelength, as well enhance the emissivity in the mid-IR range [28]. Such natural incidence has inspired researchers to develop bio-mimicking textiles structure and optical properties for maintaining thermal comfort in extreme environments.

While sweating, the human body is cooled down by absorbing and dissipating body heat to the environment. However, conventional thermal and moisture management textiles compromise the comfortability of the textile. They are fully wetted when the wearers sweat a lot or by getting exposed to external liquids while working in extreme environments like deserts, rain forests, and/or open water bodies. Such challenges can be countered by designing textiles following the liquid collection and release mechanism of the skin of different desert beetles [54, 55]. For example, there are many discrete tiny hydrophilic bumps in the hydrophobic back skin of the Namib Desert beetle [54, 55]. These beetles can condense and adhere to the moisture (from fog) in the tiny hydrophilic bumps, which then grow into sufficiently large droplets and roll off directionally to the beetle's mouth, due to gravity (Figure 8.9a and b) [54, 55]. Such a novel strategy has inspired the development of specially designed cooling and moisture management textiles with enhanced liquid transport [55]. The specialty of such a bio-mimicking textile was to dissipate the sweat in terms of droplets, by transferring the excessive sweat from the skin to the outer surface, as well as completely repelling the external liquids. The fabricated textiles dissipated body heat more quickly, with a rela-

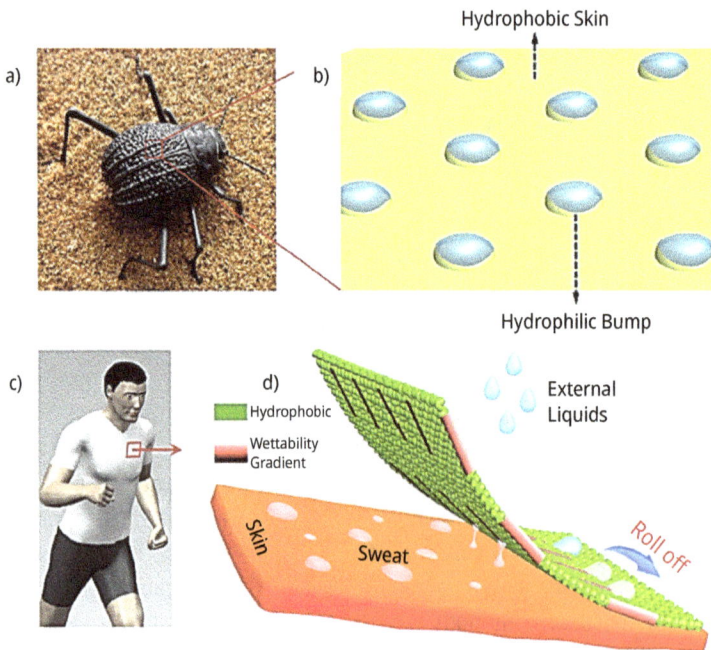

Figure 8.9: (a) Namib desert beetle; (b) schematic of the hydrophobic surface with many tiny hydrophilic bumps on the Namib desert beetle; (c, d) graphics of the human model wearing Namib desert beetle-inspired textiles (Modified according to [55], Copyright 2021, Elsevier).

tively better chill effect than the other moisture management fabric, due to its higher thermal conductivity (about fourfold higher than the traditional cotton fabric). Such excellent thermal and moisture management were achieved by creating gradient wettability across discrete local areas, in the form of loosely dashed lines in a predominantly hydrophobic fabric (Figure 8.9c). Herein, the liquid was continuously transported from the inner to the outer side of the fabric through the localized regions with gradient, and accumulated larger liquid droplets rolled off under the synergic effect of capillary force and gravity (Figure 8.9d). The fabricated cotton fabric facilitated excellent moisture management and directional water transport of 1647.9% at a rate of 91.44 g/min m^2.

8.7 Conclusion

The design of advanced thermoregulating textiles has made a huge breakthrough in maintaining thermal comfort. These advanced textiles are made from highly thermal conductive electroactive fibers and polymers. While advanced thermoregulation has made rapid progress, there are still some limitations and challenges that need to be addressed. It is evident that there is a considerable gap between lab-scale development and practical application. Besides, the comfortability and wearability of these specially designed textiles should be improved. The cost of thermoregulatory textiles designed with advanced materials is higher than that of their counterparts. Consequently, it is necessary to design an inexpensive PTC system. Moreover, most of the literature focuses on specific parameters like thermal conductivity instead of direct cooling or warming of the human body when developing thermoregulation textiles. Thermal comfort is determined by interactions between the body, clothes, and the surrounding thermal environment. Focusing on some specific parameters will not significantly improve thermal comfort. Moreover, thermoregulating textiles must be tested outside of the lab environment by using simulated or real environments.

The bio-compatibility and toxicity of advanced nanomaterials have drawn mass attention as these materials are extensively exploited for fabricating sophisticated devices. Advanced functional textiles that are modified by integrating different nanomaterials directly come into contact with the human body. Therefore, it is imperative to test the biosafety of the fabricated textiles.

References

[1] Tabor J, Chatterjee K, Ghosh TK. Smart Textile-Based Personal Thermal Comfort Systems: Current Status and Potential Solutions. Adv Mater Technol 2020;5:1901155. https://doi.org/10.1002/admt.201901155.

[2] Peng Y, Cui Y. Advanced Textiles for Personal Thermal Management and Energy. Joule 2020;4:724–42. https://doi.org/10.1016/j.joule.2020.02.011.

[3] Faruk MO, Ahmed A, Jalil MA, Islam MT, Shamim AM, Adak B, et al. Functional textiles and composite based wearable thermal devices for Joule heating: progress and perspectives. Appl Mater Today 2021;23:101025. https://doi.org/10.1016/j.apmt.2021.101025.

[4] Pakdel E, Naebe M, Sun L, Wang X. Advanced Functional Fibrous Materials for Enhanced Thermoregulating Performance. ACS Appl Mater Interfaces 2019;11:13039–57. https://doi.org/10.1021/acsami.8b19067.

[5] Hardy JD, DuBois EF. Regulation of Heat Loss from the Human Body. Proc Natl Acad Sci 1937;23:624–31. https://doi.org/10.1073/pnas.23.12.624.

[6] Nakamura K. Central circuitries for body temperature regulation and fever. Am J Physiol – Regul Integr Comp Physiol 2011;301:1207–28. https://doi.org/10.1152/ajpregu.00109.2011.

[7] Wendt D, Van Loon LJC, Van Marken Lichtenbelt WD. Thermoregulation during exercise in the heat: Strategies for maintaining health and performance. Sport Med 2007;37:669–82. https://doi.org/10.2165/00007256-200737080-00002.

[8] Shibasaki M, Wilson TE, Crandall CG. Neural control and mechanisms of eccrine sweating during heat stress and exercise. J Appl Physiol 2006;100:1692–701. https://doi.org/10.1152/japplphysiol.01124.2005.

[9] Zhang X, Chao X, Lou L, Fan J, Chen Q, Li B, et al. Personal thermal management by thermally conductive composites: A review. Compos Commun 2021;23:100595. https://doi.org/10.1016/j.coco.2020.100595.

[10] Ahmed A, Jalil MA, Hossain MM, Moniruzzaman M, Adak B, Islam MT, et al. A PEDOT:PSS and graphene-clad smart textile-based wearable electronic Joule heater with high thermal stability. J Mater Chem C 2020;8:16204–15. https://doi.org/10.1039/d0tc03368e.

[11] Peng L, Su B, Yu A, Jiang X. Review of clothing for thermal management with advanced materials. Cellulose 2019;26:6415–48. https://doi.org/10.1007/s10570-019-02534-6.

[12] Zhao B, Hu M, Ao X, Chen N, Pei G. Radiative cooling: A review of fundamentals, materials, applications, and prospects. Appl Energy 2019;236:489–513. https://doi.org/10.1016/j.apenergy.2018.12.018.

[13] Farooq AS, Zhang P. Fundamentals, materials and strategies for personal thermal management by next-generation textiles. Compos Part A Appl Sci Manuf 2021;142. https://doi.org/10.1016/j.compositesa.2020.106249.

[14] Moreira IP, Sanivada UK, Bessa J, Cunha F, Fangueiro R. A Review of Multiple Scale Fibrous and Composite Systems for Heating Applications. Molecules 2021;26:3686. https://doi.org/10.3390/molecules26123686.

[15] Cai L, Song AY, Li W, Hsu P, Lin D, Catrysse PB, et al. Spectrally Selective Nanocomposite Textile for Outdoor Personal Cooling. Adv Mater 2018;30:1802152. https://doi.org/10.1002/adma.201802152.

[16] ASTM. Standard Tables for Reference Solar Spectral Irradiances: Direct Normal and Hemispherical on 37° Tilted Surface. Astm 2013;03:1–21. http://enterprise.astm.org/SUBSCRIPTION/filtrexx40.cgi?+REDLINE_PAGES/G173.htm (accessed October 15, 2021).

[17] Steketee J. Spectral emissivity of skin and pericardium. Phys Med Biol 1973;18:686–94. https://doi.org/10.1088/0031-9155/18/5/307.

[18] Hsu P-C, Liu X, Liu C, Xie X, Lee HR, Welch AJ, et al. Personal Thermal Management by Metallic Nanowire-Coated Textile. Nano Lett 2015;15:365–71. https://doi.org/10.1021/nl5036572.

[19] Huang GW, Xiao HM, Fu SY. Wearable Electronics of Silver-Nanowire/Poly(dimethylsiloxane) Nanocomposite for Smart Clothing. Sci Rep 2015;5:1–9. https://doi.org/10.1038/srep13971.

[20] Cai L, Song AY, Wu P, Hsu PC, Peng Y, Chen J, et al. Warming up human body by nanoporous metallized polyethylene textile. Nat Commun 2017;8:1–8. https://doi.org/10.1038/s41467-017-00614-4.

[21] Hayes SG, Venkatraman P, editors. Materials and Technology for Sportswear and Performance Apparel. CRC Press; 2018. https://doi.org/10.1201/b19359.

[22] McCann JD. Build the Perfect Survival Kit: Custom Kits for Adventure, Sport, Travel. Krause Publ 2005.

[23] Tso CY, Chan KC, Chao CYH. A field investigation of passive radiative cooling under Hong Kong's climate. Renew Energy 2017;106:52–61. https://doi.org/10.1016/j.renene.2017.01.018.

[24] Catrysse PB, Song AY, Fan S. Photonic Structure Textile Design for Localized Thermal Cooling Based on a Fiber Blending Scheme. ACS Photonics 2016;3:2420–6. https://doi.org/10.1021/acsphotonics.6b00644.

[25] Zhai Y, Ma Y, David SN, Zhao D, Lou R, Tan G, et al. Scalable-manufactured randomized glass-polymer hybrid metamaterial for daytime radiative cooling. Science (80-) 2017;355:1062–6. https://doi.org/10.1126/science.aai7899.

[26] Hsu PC, Song AY, Catrysse PB, Liu C, Peng Y, Xie J, et al. Radiative human body cooling by nanoporous polyethylene textile. Science (80-) 2016;353:1019–23. https://doi.org/10.1126/science.aaf5471.

[27] Hsu PC, Liu C, Song AY, Zhang Z, Peng Y, Xie J, et al. A dual-mode textile for human body radiative heating and cooling. Sci Adv 2017;3:e1700895. https://doi.org/10.1126/sciadv.1700895.

[28] Shi NN, Tsai CC, Camino F, Bernard GD, Yu N, Wehner R. Keeping cool: Enhanced optical reflection and radiative heat dissipation in Saharan silver ants. Science (80-) 2015;349:298–301. https://doi.org/10.1126/science.aab3564.

[29] Raman AP, Anoma MA, Zhu L, Rephaeli E, Fan S. Passive radiative cooling below ambient air temperature under direct sunlight. Nature 2014;515:540–4. https://doi.org/10.1038/nature13883.

[30] Li W, Shi Y, Chen Z, Fan S. Photonic thermal management of coloured objects. Nat Commun 2018;9:4240. https://doi.org/10.1038/s41467-018-06535-0.

[31] Jelle BP, Kalnæs SE, Gao T. Low-emissivity materials for building applications: A state-of-the-art review and future research perspectives. Energy Build 2015;96:329–56. https://doi.org/10.1016/j.enbuild.2015.03.024.

[32] Miao D, Jiang S, Liu J, Ning X, Shang S, Xu J. Fabrication of copper and titanium coated textiles for sunlight management. J Mater Sci Mater Electron 2017;28:9852–8. https://doi.org/10.1007/s10854-017-6739-3.

[33] Wong A, Daoud WA, Liang HH, Szeto YS. Application of rutile and anatase onto cotton fabric and their effect on the NIR reflection/surface temperature of the fabric. Sol Energy Mater Sol Cells 2015;134:425–37. https://doi.org/10.1016/j.solmat.2014.12.011.

[34] Miao D, Li A, Jiang S, Shang S. Fabrication of Ag and AZO/Ag/AZO ceramic films on cotton fabrics for solar control. Ceram Int 2015;41:6312–7. https://doi.org/10.1016/j.ceramint.2015.01.057.

[35] Hamza AA, Sokkar TZN, Mabrouk MA, El-Morsy MA. Refractive index profile of polyethylene fiber using interactive multiple-beam Fizeau fringe analysis. J Appl Polym Sci 2000;77:3099–106. https://doi.org/10.1002/1097-4628(20000929)77:14<3099::AID-APP110>3.0.CO;2-K.

[36] Bond WL. Measurement of the refractive indices of several crystals. J Appl Phys 1965;36:1674–7. https://doi.org/10.1063/1.1703106.

[37] Srikant V, Clarke DR. On the optical band gap of zinc oxide. J Appl Phys 1998;83:5447–51. https://doi.org/10.1063/1.367375.

[38] Li W, Fan S. Nanophotonic control of thermal radiation for energy applications [Invited]. Opt Express 2018;26:15995. https://doi.org/10.1364/oe.26.015995.

[39] Goldstein EA, Raman AP, Fan S. Sub-ambient non-evaporative fluid cooling with the sky. Nat Energy 2017;2. https://doi.org/10.1038/nenergy.2017.143.

[40] Maity S. Optimization of processing parameters of in-situ polymerization of pyrrole on woollen textile to improve its thermal conductivity. Prog Org Coatings 2017;107:48–53. https://doi.org/10.1016/j.porgcoat.2017.03.010.

[41] Abbas A, Zhao Y, Wang X, Lin T. Cooling effect of MWCNT-containing composite coatings on cotton fabrics. J Text Inst 2013;104:798–807. https://doi.org/10.1080/00405000.2012.757007.

[42] Montazer M, Ghayem Asghari MS, Pakdel E. Electrical conductivity of single walled and multiwalled carbon nanotube containing wool fibers. J Appl Polym Sci 2011;121:3353–8. https://doi.org/10.1002/app.33979.

[43] Manasoglu G, Celen R, Kanik M, Ulcay Y. Electrical resistivity and thermal conductivity properties of graphene-coated woven fabrics. J Appl Polym Sci 2019;136:48024. https://doi.org/10.1002/app.48024.

[44] Gao T, Yang Z, Chen C, Li Y, Fu K, Dai J, et al. Three-Dimensional Printed Thermal Regulation Textiles. ACS Nano 2017;11:11513–20. https://doi.org/10.1021/acsnano.7b06295.

[45] Gao J, Pan N, Yu W. Compression behavior evaluation of single down fiber and down fiber assemblies. J Text Inst 2010;101:253–60. https://doi.org/10.1080/00405000802377342.

[46] Gibson P, Lee C. Application of nanofiber technology to nonwoven thermal insulation. Proc 14th Annu Int TANDEC Nonwovens Conf 2004:1–14. https://doi.org/10.1177/155892500700200204.

[47] Wang Z, Zhong Y, Wang S. A new shape factor measure for characterizing the cross-section of profiled fiber. Text Res J 2012;82:454–62. https://doi.org/10.1177/0040517511426614.

[48] Cui Y, Gong H, Wang Y, Li D, Bai H. A Thermally Insulating Textile Inspired by Polar Bear Hair. Adv Mater 2018;30:1706807. https://doi.org/10.1002/adma.201706807.

[49] Cabeza LF, Castell A, Barreneche C, De Gracia A, Fernández AI. Materials used as PCM in thermal energy storage in buildings: A review. Renew Sustain Energy Rev 2011;15:1675–95. https://doi.org/10.1016/j.rser.2010.11.018.

[50] engineering SM-A thermal, 2008 undefined. Phase change materials for smart textiles–An overview. Elsevier n.d.

[51] Mohamed SA, Al-Sulaiman FA, Ibrahim NI, Zahir MH, Al-Ahmed A, Saidur R, et al. A review on current status and challenges of inorganic phase change materials for thermal energy storage systems. Renew Sustain Energy Rev 2017;70:1072–89. https://doi.org/10.1016/j.rser.2016.12.012.

[52] Lu Y, Xiao X, Fu J, Huan C, Qi S, Zhan Y, et al. Novel smart textile with phase change materials encapsulated core-sheath structure fabricated by coaxial electrospinning. Chem Eng J 2019;355:532–9. https://doi.org/10.1016/j.cej.2018.08.189.

[53] Lan X, Wang Y, Peng J, Si Y, Ren J, Ding B, et al. Designing heat transfer pathways for advanced thermoregulatory textiles. Mater Today Phys 2021;17:100342. https://doi.org/10.1016/j.mtphys.2021.100342.

[54] Parker AR, Lawrence CR. Water capture by a desert beetle. Nature 2001;414:33–4. https://doi.org/10.1038/35102108.

[55] Zou C, Lao L, Chen Q, Fan J, Shou D. Nature-inspired moisture management fabric for unidirectional liquid transport and surface repellence and resistance. Energy Build 2021;248:111203. https://doi.org/10.1016/j.enbuild.2021.111203.

Santanu Basak, Animesh Laha

9 Stimuli-responsive smart and functional textiles

Abstract: The definition of stimuli as per biology is an external source of energy or some means that force to react to an organ or part of an organ. Cold or hot sensation, any external mechanical force applied upon, mosquito bites, smell, touch, etc. are a few examples of stimuli, and our body instantly reacts when any of these stimuli are acted upon. Similarly, any smart polymeric materials which can react, sense, and respond to the applied external stimuli such as temperature, light, presence of any chemical agent, moisture, pH, or presence of electric and magnetic fields, etc. are called stimuli-responsive polymeric materials. They are also called smart materials. Any textile structure made using the stimuli-responsive polymer or above-mentioned smart material is called stimuli-responsive textile. As the name suggests, it is the new addition of smart textiles and one of the most emerging areas, and they are gaining much scientific interest among the research community and industry to improve or impart smart functionalities into textiles. Photochromic, solvatochromic, aesthetic appeal, comfort, electrochromic, ionochromic, piezochromic, textile soft display, smart controlled drug release, pH-responsive, chemichromic, mechanochromic, fantasy design with colour changing, wound monitoring, smart wetting properties, and protection against extreme variations in environmental conditions, phase change, etc. are few to name with of various smart functionalities which can be imparted by using stimuli-responsive textiles. Dyeing, printing, finishing (coating or lamination) and dope dyeing (inherent properties), and composite structures are a few methods available for imparting stimuli-responsive properties into textiles.

Keywords: Stimuli-responsive textile, mechanochromic colour, photochromic textile, thermochromic, smart polymer, phase change material

9.1 Introduction

Stimuli-responsive materials are a new addition to material science and are gaining huge popularity and substantial interest among the research community. The beauty of these materials is that they can spontaneously respond against any external stimuli acted upon them, and in most cases, these responses are reversible; that is, the material will come back to its original state when external stimuli are removed. External stimuli could be anything, and magnetic and electric fields, temperature, pH, light, moisture, etc. are the most important types of stimuli normally applied for different applications [1]. The key factors of stimuli-induced effects acting upon the material could break hydrogen bonds and progressive ionization. The presence of functional groups and interactions between

https://doi.org/10.1515/9783110759747-009

them play a key role in stimuli-responsive materials. The ability to respond against any stimuli by any material can be changed or altered by adding suitable filler into the polymer matrix or by incorporating a new functional group [2]. Stimuli-responsive materials are gaining popularity across all sectors of applications such as self-healing coating, diagnostics, soft robots, actuators, sensors, textile-based flexible sensors, etc., and mostly used materials for this purpose are albumin, cellulose, chitosan, gelatin, and graphene [3–9].

Researchers are using various methodologies to impart stimuli-responsive properties to textiles. The stimuli-responsive property could be inherent/permanent, or this could be applied through dyeing, printing, and finishing route where the effect could be subdued over subsequent use or washing if required [10–17]. To obtain inherent stimuli-responsive properties, the stimuli-responsive agent is applied during fibre spinning, such as electro-spinning, wet spinning, etc. [10–12]. The same could be obtained through finishing route, coating or lamination, by treating with hydrogels, application of phase change materials, or application during dyeing or printing [13–17]. The development of polymer composite, nanocomposite, and 4D printed items is another way of imparting stimuli-responsive property into textile [18–21]. After the incorporation of stimuli-responsive properties into textiles, various functional properties improved like aesthetic, comfort, controlled drug release, wound monitoring, protection against extreme weather conditions could easily be achieved which is not possible to get from normal textile substrates. Thus, stimuli-responsive textiles (SRT) are gaining huge popularity among the research community and they are developing various applications like extreme weather clothing, sports clothing, self-cleaning, pH monitoring, shape memory fibre, all types of smart textile applications, and many more [10, 13–14, 17]. This chapter mainly focuses on photochromic textiles, thermochromic textiles, solvatochromic textiles, electrochromic textiles, piezochromic textiles, pH-responsive textiles, chemichromic textiles, mechanochromic textiles, and phase change materials/shape-memory materials.

9.2 Photochromic textile

A group of textile materials that change their colour due to the presence of external stimuli is called chromic textiles. Photochromic textiles are those that change their colours with the change in external environmental conditions as they have turned into different colour in outdoor applications compared to indoor applications. Colour of the textile materials changes due to UV radiation of the environment reacting with the photons of the dyes. These colour materials applied on the textiles are sensitive to UV rays and change their colour under exposure to UV rays. These materials are also known as chameleon textiles. Different classes of dyes like spiropyran, spirooxazine, naphtopyran, etc. behave like photochromic dye. Figure 9.1 shows the reversible changes of colour due to the shift in naphthopyrone structure [14, 15]. When colour change of a textile material happens due to the presence of photons or a source of light, it is known

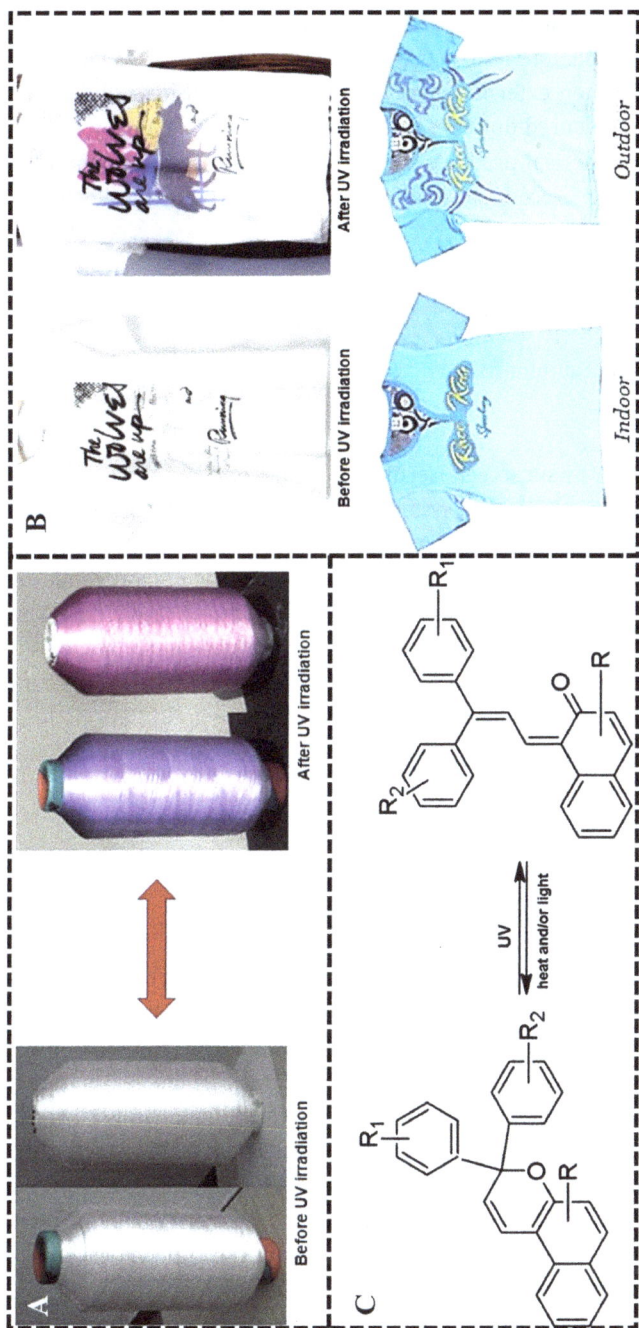

Figure 9.1: (A) Photochromic yarn; [14]; (B) Photochromic fabric [14, 15]; and (C) Mechanism of color change of photochromic textiles under sunlight [14].

as photochromic textiles [22]. This is a reversible phenomenon; colour change will occur when external energy (photon here) acts on the material, and it will again come back to its original colour when external energy is taken away. Positive photochromism, that is, colourless to coloured due to a ring-opening reaction, is mostly used in photochromic effects in commercial products. Based on the mechanism of conformation changes, the photochromic process is classified into the following categories:
a) triplet–triplet photochromism
b) heterolytic cleavage
c) homolytic cleavage
d) trans-cisisomeration
e) photochromism based on tautomerism
f) photodimerization

Apart from applications such as sunscreen, security printing, optical recording, solar energy storage, non-linear optics, biological systems, etc., application of photochromism in the textile sector also got numerous interests among the research community in recent days [23–24]. For textile applications, textile-based sensors using photochromic behaviour are one of the key areas of interest among researchers in the recent past. This technology has secured other application areas such as medical thermography, plastic strip thermometers, photochromic lenses, food packaging, and non-destructive testing of engineered articles and electronic circuitry. Among different application methods available for imparting photochromic property by dyeing, screen printing, finishing, and dope dyeing are the most reliable methods [23–25].

9.3 Thermochromic textiles

Similar to photochromic material, a reversible change in color is observed also in case of thermochromic materials, but under exposure of heat. When thermochromic material is applied onto textiles, a colour change would occur at an elevated temperature, and it comes back to its original colour or colourless condition when the heat source is removed or vice versa. There are many ways to impart thermochromic properties into textiles: the use of microencapsulated dyes, leuco encapsulated dyes, liquid crystal thermochromic dyes, electrothermic yarn utilizing cotton and CNT composite are a few examples [26–28]. The crystallic or molecular structure of the thermochromic pigment reversibly changes after absorbing a certain amount of light or heat. Indeed, it absorbs and emits light at a different wavelength. The history of using microencapsulated dyes as thermochromic materials is quite old. It was developed in the 1970s, and since then textile and novelty industries are major areas where it has been used most [27]. Among other systems discussed, the leuco-dye-type thermochromic system was found to be used mostly. This gives a reversible change from coloured to colourless when it is exposed to an elevated temperature [29]. Recently, the application of thermochromic technology is found in a wide area of applications due to the combined effect of smart textiles and in conjunction with electronic engineering [30–31]. Along with the scientific community, textile designers are also exploiting the special property of thermochromism as one of the most important functional properties along with the aesthetic to create a new definition of look into apparel and various new applications areas. For instance, to develop a colour effect in a woven fabric designers used a carbon fibre in the structure, where change in colour in the pattern was observed when heat is applied utilizing a power supply [32]. Thus, thermochromic textiles created their own space for developing many new areas of application and altogether multidisciplinary research. Few potential application areas are heat-profiling circuitry, charting and mapping the skin, touch-me wallpaper (a temporary hand print that appears with the contact of the human hand), colour changing textile, electric paid combined with thermochromic printed textiles and electronic circuitry etc. and many more [33–34]. It is needless to mention that the thermochromic effect is very specific. It is activated to the very specific points, and thus every individual pixel is addressable and the colour change due to the thermochromic effect can be controlled in real time. Figure 9.2 shows the thermochromic colours that change their hues with

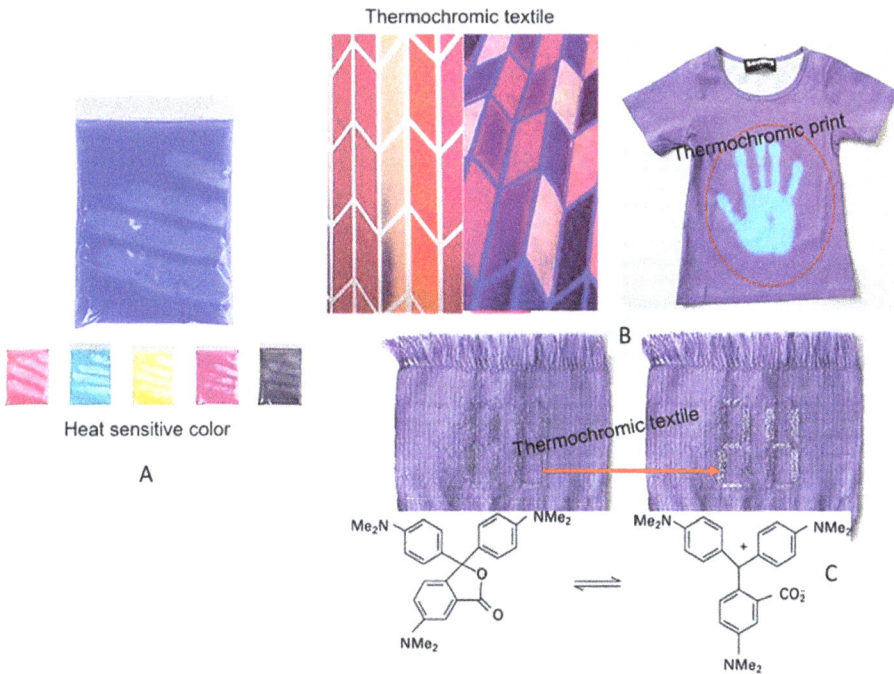

Figure 9.2: (A) Pictorial representation of thermochromic colour [25]; (B) Thermochromic print [27]; (C) Molecular rearrangement (arise form tautomerisation) or structural change of colour in thermochromic textiles under the action of heat as stimuli. Reproduced with kind permission from [28], Copyright 2006, Elsevier.

temperature, and it also represents the picture of the garment composed of thermochromic print in it. This print is invisible in normal conditions, although visible with raising the temperature of the surroundings.

9.4 Solvatochromic textiles

From the point of view of chemistry, the phenomenon of solvatochromic is observed when colour due to solute is different as it is dissolved in different solvents. That is, the spectrum of a specific substance (solute) will change as the substrate is dissolved in a different solvent. Any solvent can be easily differentiated from others by its dielectric constant and hydrogen bonding capacity, which are the most important properties of any solvent. Thus, there would be a different electronic ground state and excited state of a substrate when it is exposed to different solvents. As a result, the size of the energy gap would change as the solvent changes, as well as the absorption or emission spectrum of the substrate with the change in position, intensity, and shape of the spectroscopic bands. Thus, solvatochromism is observed as a change of colour. Solvatochromism has found a huge interest among research scientists in developing smart textile products.

Different solvatochromic dyes and their chemical structures are represented in Figure 9.3. These dyes have alternative double bond and single bond and very much sensitive to response against different kinds of solvents. Colours like 6-propinyl-6-dimethylamino naphthalene, Nile Red are the most common examples of solvatochromic dyes. Properties of the dyes also depend on the polarity of the chemical groups present in it, bond energy, molecular weight, and stereospecific nature of the dyes. Alternative double and single bonds assist to show fluorescent nature of the dyes.

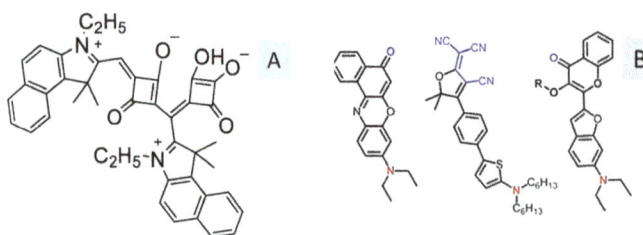

Figure 9.3: (A) Schematic representation of solvatochromic a dye molecule [36], (B) Different solvatochromic dyes. Reproduced with kind permission [37] Copyright 2017, American Chemical Society.

In our daily life, we are using many textile products. Few of them are used for limited period. Personal care products are a very good example of the same. It would be advantageous if the products contain a smart system of visually conveying messages or indication systems during their product realization and more specifically during their use. If the user uses it after a certain period of time, it becomes unhealthy. Indication systems of disposable products like diapers; sanitary pads; small child training paints; patient bed-sheets; wound dressings, towels, and bedsheets in the hospitality sector; etc. are a few examples where such type of indication is advantageous to the user as they will be aware of the time for a replacement. Therefore, a robust and user-friendly massaging system is required to develop that will activate and deactivate as per requirement. There are many approaches are available for imparting the solvatochromic property into textiles. The use of solvatochromic dye is mostly used method and is widely accepted by researchers [35–37].

The working principle of solvatochromic dyes utilizes change of colour due to solvent polarity. This effect is reversible too. The colour change effect is observed when shifting in maximum absorption happened in different solvents. The solvatochromic effect could be of two types: positive solvatochromism and negative solvatochromism. Hypsochromic shift induced due to decrease in solvent polarity is known as positive solvatochromic, and on the other hand, hypsochromic shift induced due to increase in solvent polarity is known as negative solvatochromic. Pyridium, merocyanine, and stilbazolium dyes are a few examples of mostly used solvatochromic dyes. As a textile application technique, microencapsulation is one of the most frequently used. Due to the presence of microspheres on the fabric surface colour changing effect was observed during both the wet and dry state of the fabric. However, the use of solvatochromic dyes still has limited acceptability as they are very specific to reaction conditions. Sometimes these dyes are used to indicate stale or toxic food [38].

9.5 Electrochromic textiles

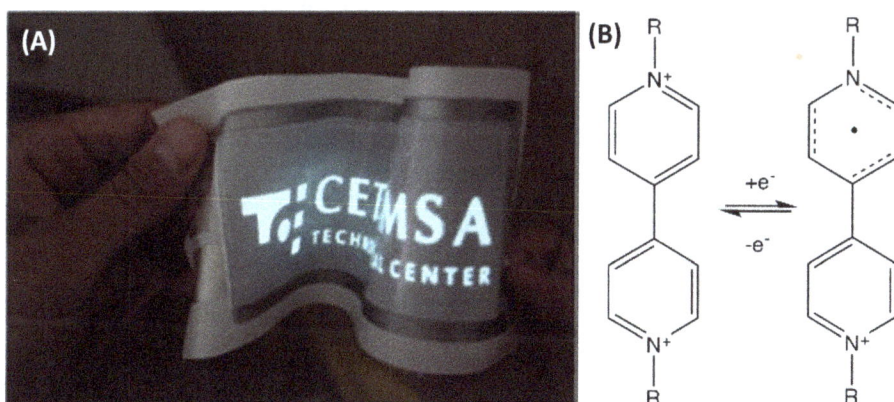

Figure 9.4: (A) Photograph of an electro-chromic textile. Reproduced with kind permission from [38] Copyright 2016, Elsevier; and (B) Redox chemical reaction under the action of electric filed in a elctrochromic textile [39].

Electrochromic effect is also a reversible change in colour or opacity of a material due to the presence of an applied electric field or current, as shown in Figure 9.4. Electricity catalyse chemical reaction and it changes the colour of the polymer and come back to the parent form when the voltage source has been withdrawn. Mostly, ion insertion materials are used as electrochromic materials, and they have the capability of rapidly and reversibly inserted effects. Thus, it is an electricity-induced chemical reaction for colour change. Normally, electrochromic effect occurs in inorganic compounds by dual injection (cathodic) or ejection (anodic) of ions (M) and electrons (e⁻). For a typical cathodic reaction, it turns to blue from colourless for ions like hydrogen, lithium, sodium, silver, etc. On the other hand, for a typical anodic reaction, it turns black from colourless [39].

The electrochromism effects are found to be used in many application areas starting from smart windows or mirrors, anti-glare rear-view mirrors in cars, smart sunglasses, and in devices for optical information and storage [40]. Among other application areas of electrochromism, textile-based electrochromism is found to be in a great way due to its flexibility. Although researchers are facing many challenges and they are working hard to overcome them through continuous research and development, there is a great opportunity for flexible electrochromic devices. Other areas of applications are flexible colour-changing display, visible in natural and ambient light, etc. will offer endless opportunities. Recently, designers are also using colour-changing and light-emitting textile structures using electrochromic materials as flexible displays. This has opened up many applications starting from medical textiles, communicative textiles, and even in art and fashion [41]. Textile composites based on polyaniline-coated metalized conductive fabrics, hybrid textile-film electrochromic devices with poly-3,4-ethylenedioxythiophene: polystyrene sulfonate (PEDOT:PSS), dyeing, etc. are few methods used as application methods for textiles [38, 42–43].

The electrochromic dyes show a reversible colour change in presence of an electric field or loss and gain of an electron. This colour-changing effect was observed due to the transition of metal oxides, that is, ionic transition of metal oxides due to the presence of strongly polarizable metal oxygen bonds in it. This ionic transition of metal oxide also has been used in batteries which is basic in nature. Phthalocyanine dyes are well known for their electrochromism [38].

9.6 Piezochromic textiles

Piezochromism is a property of a special kind of material that changes its colour reversibly when pressure is acted upon the material, as shown in Figure 9.5. The pressure application is done utilizing mechanical force application onto the material; thus, it is also known as mechanochromic. The colour-changing effect of a piezochromic material is closely related to the electronic bandgap change, and this is quite often found in plastics, semiconductors (e.g. hybrid perovskites), and hydrocarbons [44]. Recently, piezochromic luminescent materials with multi-colour switching are gaining much attention among the scientific community due to their attractive applications in areas like advanced photonics, such as optical recording, memory, and sensors [45]. Mechanical forces or pressure acting upon the material could be of different forms like shearing, tensioning, smashing, and compression force or pressing. The applied mechanical force could be anisotropic or iso- tropic. Differences in applied mechanical force would give opportunities in different packing and emission regulations; for example, anisotropic shearing, isotropic pressing, etc. Among all the available options, anisotropic shearing and isotropic pressing show a consistent red-shifted emission. Thus, it confirms the increased interaction, planarization, and charge transfer effect [46]. Material researchers are putting their efforts to get direc- tional change in emission utilizing modifying or altering the molecular structure to get the desired effect. By regulating the molecular dimeric structure of 2,3,4,5-tetra(2- thiazolyl)thiophene molecule, researchers were able to obtain distinct emission responses from anisotropic shearing and isotropic pressing [47]. A similar type of phenomenon was reported by Liu et al. [48], with changing both π–π and charge-transfer interactions. Al- though there has been a great deal of research to develop new materials, there is still a long way to go before the behaviour of these materials can truly be understood. Another point that must be considered is that in most cases crystal to amorphous transition is ob- served and amorphous powder can only offer the explanation of molecular packing and emission changes [49]. On the other hand, crystals are ideal examples as they can provide effective information on critical influencing factors on the emission of the material [50].

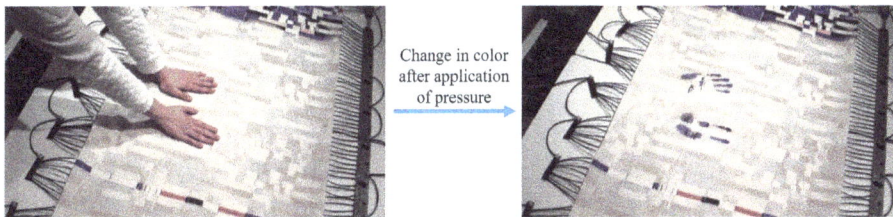

Figure 9.5: Piezochromic textile changing color with application of pressure [47].

9.7 pH-responsive textiles

pH-responsive or pH-sensitive materials are materials that can respond due to the change in the pH of their surrounding medium through changing their dimensions or colour, showed in Figure 9.6. Typically, the material can swell, collapse, change in col- our, etc. due to the change in pH of their surroundings. Due to the presence of a certain functional group in the molecular structure, they can exhibit pH-responsive behaviour. These types of materials are also known as halochromic materials. Mostly, these materials

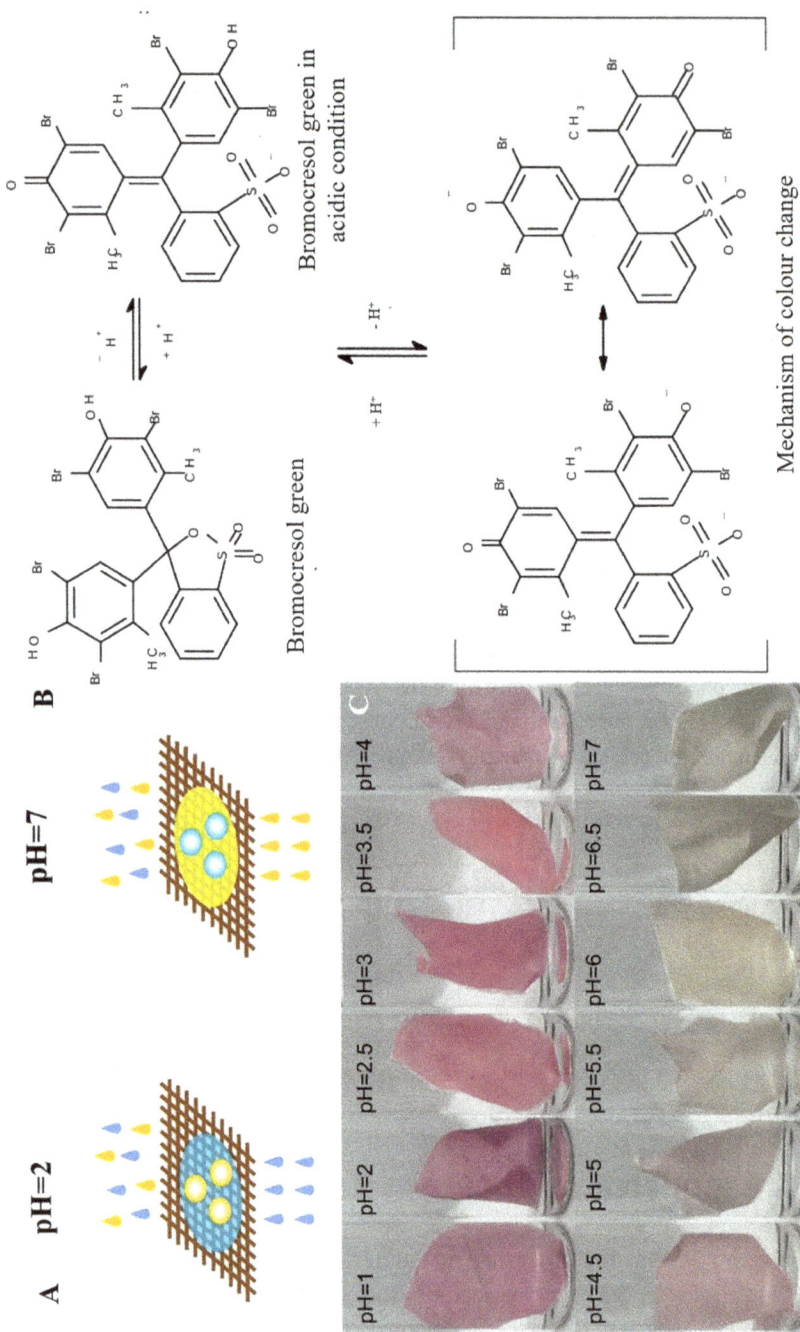

Figure 9.6: (A) Changes color of dyed textile product with change in pH condition [53] Copyright 2019, Elsevier; (B) Mechanism of colour change of pH sensitive textiles; [54] and (C) Change of natural pigment colour with pH [55] Copyright 2022, Elsevier.

are available in two ionic states, namely, acidic or basic, and they can respond to corresponding basic or acidic pH values. The materials have very good potential for a wide range of applications, from controlled drug delivery systems, biomimetics, micromechanical systems, separation processes to surface functionalization [51]. In recent decades, researchers are taking a lot of interest in developing textile-based pH-responsive sensors. The advantages of these sensors are many folds such as they are flexible, they can be washable, and can be used multiple times. These flexible textile-based pH sensors have secured many applications such as personal care products: sweat analysis and real-time information on a person's sweating rate and type of sweat by measuring its pH, sensors for detecting acid rain, universal sensor, pH sensor-based wound dressing, and personal care products [52–54]. Researchers have opted for various methods of imparting pH-responsive properties. Use of halochromic dyes based on sulfonphthaleine, chitosan-carboxymethyl cellulose-indicator dye-based nanocompounds, resorufin-GPTMS hybrid sol–gel, bi-heterocyclic hydrazine dyes, polymeric dye-based on waterborne polyurethane, etc. are a few approaches adopted by the researchers to impart pH-responsive property into textile [55–59]. Halochromic textile is another example of pH-sensitive textile. The fabric coated with sol–gel layer and halochromic pH-sensitive dyes (sulfonphthaleine dyes) shows a clear colour change with changing the pH condition. The response of the sensor depends upon the density of the fabric. Halo chromic shift of the dye also depends on the interaction of dye and fibre. Bromocresol green structure is shown in Figure 9.7. In acidic condition, structure of bromocresol green has been changed, and it also changes its colour. The mechanism of changing colour also has been represented in Figure 9.7. "Congo red," a special class of direct dye, also has shown pH-sensitive behaviour and changes its colour from blue to red with increasing the pH level from 3 to 5 or 6, as shown in Figure 9.7. Azo group present in its structure is mainly responsible for its colour and acts as chromophore of this dye molecule. Indeed, its structure has remained the same, but the position of the chromophoric group has altered and amine groups are protonated due to the acidic condition; that is, due to the

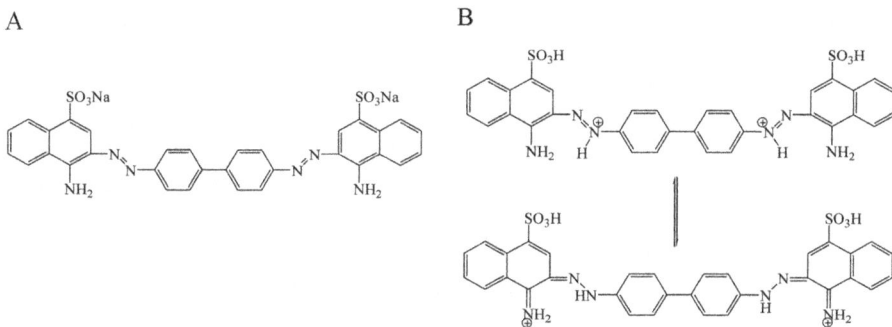

Figure 9.7: Color change of "Congo Red" dye molecule with changing the pH condition: (A) at pH > 5.5 and (B) at pH < 5.5 [59].

change in pH condition. This kind of dye generally has been used for colouration of textile materials and is also very much useful as pH indicator. Like this dye, there are different natural dyes like anthocyanin, betacyanin, coumarin, etc., which act as indicators and change their colour with changing the pH condition.

9.8 Mechanochromic textile

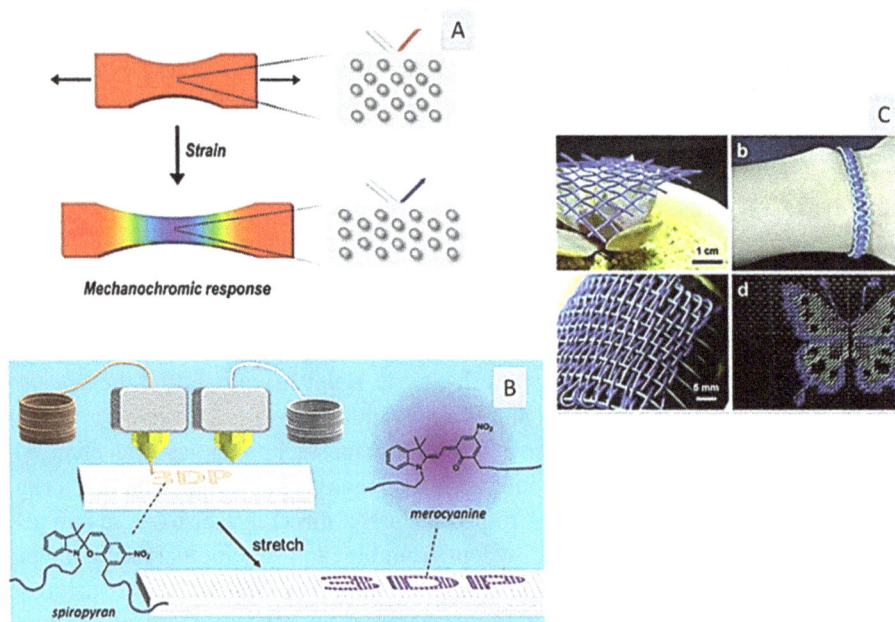

Figure 9.8: (A) Mechanism of mechanochromic textiles that changes color with mechanical action. Reproduced with kind permission from [62] Copyright 2020, John Wiley and Sons; (B) Mechanism of colour change of 3D printed mechanochromic material [64]; and (C) Photographs of chromatic patterns and textiles made from mechanochromic fibers. Reproduced with kind permission from [65] Copyright 2013, Royal Society of Chemistry.

Mechanochromic textile changes colour under different mechanical condition of the textile. Examples of change in colour of the textiles are shown in Figure 9.8. No colour component or dyes have been involved in the total process. Mechanochromic fibres hard core–soft shell microspheres are incorporated inside the commercially available textile fibres. The microspheres assemble to form a photonic crystal structure and displays different structural colours. This method is applicable for a wide range of fibres with their varying diameter and cross-sectional shapes of microspheres. Mechanochromic fibres have changed their colour from green to blue, red to green with stretching, bending, rolling kind of mechanical attributes, and the changing process is repeatable and reversible in nature. The use of hard core-shell microspheres, their shapes, colour, concentration, and structure are important aspects for controlling the mechanochromic properties. Researchers have prepared hard core shell structure by placing polystyrene/polymethyl methacrylate core and polyethylene acrylate shell by following emulsion polymerization process. Thesemechanochromic materials have been incorporated inside the textile or fabric and could be used as dressing products, and show colourful stripes and designs. These kinds of chromic properties may be substituted for the dyeing process in future, and it has smart applications in the wearable industries [60, 61]. Mechanochromic electronic textile (MET) is incorporated in wearable garments, and it is useful for measuring the joint movement of the body. Moreover, electrical response and structural durability, structural colour of MET sensors have remained constant for up to 30,000 stretching/releasing cycles [62, 63]. Figure 9.9 shows the structural changes due to mechanical action. It is observed that the position of the aryl groups is changed due to the mechanical action; however, main chromophoric group remain unchanged. These kinds of changes have occurred during grinding, breaking into small sizes like micro or nano.

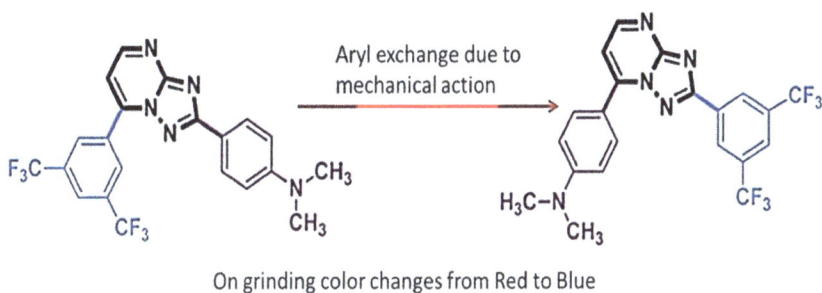

On grinding color changes from Red to Blue

Figure 9.9: Changing of the structure and isomerism of structure due to mechanical action. Reproduced with kind permission from [60], Copyright 2016, American Chemical Society.

Stimuli-responsive textiles also have potential applications in the field of medical textiles. It helps to release drugs in control way and assist in various dermal therapy. It contains various types of aromatic substances: painkillers, hormone therapy, atopic dermatitis, melanoma, etc. Textile materials could be treated with different nano-chemicals, hydrogels (high-molecular-weight material containing hydrophilic groups) by spraying or pad dry cure method. Treatment of textiles with these special polymers also can be possible by microencapsulation methods. Hydrogels absorb biological water from the application area due to the presence of polar groups and assist to release the active ingredient of drugs in control way. Researchers have developed hydrogels from coupling reactions of polycaprolactone and polyethylene glycol with hexamethylene diisocyanate crosslinker. This synthesized hydrogel polymer has been used for coating of fibre-made non-woven textile by dissolving it into water and other copolymers. The concerned hydrogel is adhered to the textile by physical interaction and release bio-active substance to the skin in control way. Different kinds of herbal products like curcumin, aloe extract also can be used in the formulation. Hydrogels could assist to release the active ingredient of the extract in the skin cells. Different kinds of drug carriers are chitosan-modified poly(N-isopropylacrylamide), nanoparticles from poly-e caprolactone with encapsulated active ingredients, chitosan with glutaraldehyde, etc. could be used as hydrogel polymers.

9.9 Phase change material

Phase change materials (PCM) have the potential to absorb and release heat energy with the influence of mechanical strain, thermal condition, and environmental changes depending on solvents, chemical, pH, etc. These materials can store latent heat energy inside their structure and have the capability to use it during changing of phase from solid to liquid or from liquid to solid. During transformation from one phase to another phase, these polymers have the potential to absorb a significant amount of energy. Paraffin wax and bees wax are common natural examples of PCMs, and they have large heat of fusion per unit weight [64]. On heating, wax transforms into liquid by melting phenomenon and again transfers to solid condition when heat stimuli are removed. Phase change micro-capsules of size range 1 to 30 μm have extensive use in the domain of technical textiles for making fabric for high-altitude, fabric for cooling in summer conditions, etc. Microencapsulation is a very effective way of incorporating PCMs in textiles. These capsules are effective at particular condition of temperature, pH, pressure, or any other external conditions. These capsules have the capability of storing energy in a heating and cooling system. Micro-capsules have the advantages of high heat-transfer area, reducing the reactivity and interference of the external environment on the energy stored inside the capsules and also on the changes of volume of content during phase change. However, an effective burst of the micro-capsules of the PCM is still a technical challenge for the researchers. Concerning the mechanism of action when the temperature of the surroundings rise, PCM capsules melt and absorb heat energy, thus providing cooling sensation to the user body [64, 65]. On the other hand, the capsules are also capable to emit heat energy in an extremely cool atmosphere. PCM capsules are incorporated inside the fibrous structure by various application processes such as in-situ polymerization, interfacial polymerization, coacervation phase separation, coating, lamination, spray drying, solvent evaporation, melt spinning, and moulding. As far as the coating process is concerned, PCM-based paraffinic hydrocarbons are dispersed in the water-based solution. This solution contains binder, surfactant, and other polymer mixture. Henceforth, textile products are coated with the mixed formulation containing PCM, subsequently dried and cured. Mainly core of the capsules contain hydrophilic and hydrophobic chain with cover of polymerized shell [65, 66]. The extent of energy stored in the core material depends on the structure and energy of the chemical bond present in it. For example, phosphate encapsulated polyurethane shell works as an effective flame retardant when it is required. For flame-retardant material, poly-ether-polyuretahne and polyester polyurethane coved microcapsule shells are used because these shells are typically heat-resistant. Textile fabrics treated with this kind of capsules are resistant to heat and temperature [67, 68]. PCM capsules have a wide range of applications in the sports sector, as during sports activity, a lot of heat is generated and the extent of emission in the surrounding atmosphere is comparatively less. As a result, people feel uncomfortable due to high thermal stress and low thermal regulation. Using PCM inside the garment assists to absorb excess heat energy and then releasing it slowly into the environment, thus maintaining thermal regulation. PCM also has potential in the field of medical textile. Polyethylene glycol-based capsules could be used for making bandages, gauge, surgical apparel, heat/cool message for pain relief, etc., and it has the capability to act as an antimicrobial, heat/cool agent when it is necessary [65]. Apart from sports and medical textile, PCM has a wide range of applications in different kinds of textile materials starting from gloves, garments, footwear, different underwears, golf shoes, medical sector, automobile textiles, seat cover, belt, ski boots, etc. It is one of the emerging areas of futuristic technical textiles. However, uniform encapsulation of PCM active component, uniform application technique on the textile substrate and integrity, effectivity of capsules in particular environmental condition is still a big challenge for the researchers working in this field. In addition to this, storage durability, washing, and rubbing fastness of the PCM-based treatment are the other applicable issues for wide-scale commercialization. Figure 9.10 shows the basic concept of phase change polymers from solid to liquid and liquid to solid by heat absorption and heat emission. In addition of it, it also shows the transition of heating liquid from solid to liquid state. On the other hand, shape memory polymers have the capability to return their own phase by action of external stimuli [65, 66]. By the action of different kinds of external stimuli, these polymers have potential to change their colour and other characteristics. Methacrylate and polyurethane are the most common examples of shape memory polymers.

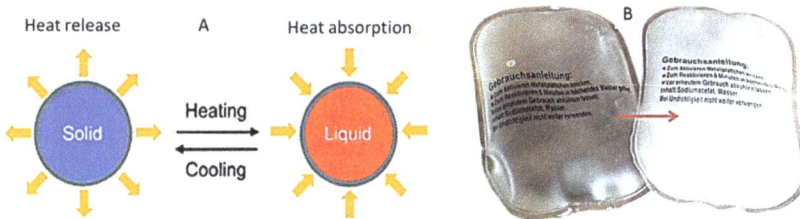

Figure 9.10: (A) Phase change by heat absorption and heat release. Reproduced with permission from [66], Copyright 2014, John Wiley and Sons; (B) Phase changeable heating pack [68].

9.10 Conclusion

In this context, a brief outline has been made on stimuli-responsive textiles, prepared by different physical and chemical means on textile products. Context systematically discussed about the various basic aspects of photochromic, solvatochromic, piezo-chromic, mechanochromic, pH chromic textile materials. Photochromic textiles have good applications in umbrella cloth (different colour in shade and in sunshine, changes colour with particular temperature range, etc). Stimuli-responsive textiles are very much promising and have the potential for use in various sectors of technical textile field. Chemichromic and solvatochromic textiles are developed and mainly used in the field of medical textiles. Mechanochromic textiles are also newer kind of addition in the field of smart textiles and have a lot of scientific potential for use in the areas of sports and defence sectors.

References

[1]   Lu, Q.; Jang, H.S.; Han, W.J.; Lee, J.H.; Choi, H.J. Stimuli-Responsive Graphene Oxide-Polymer Nanocomposites. Macromol. Res. 2019, 27, 1061–1070.
[2]   Patel, D.K.; Seo, Y.R.; Lim, K.T. Stimuli-Responsive Graphene Nanohybrids for Biomedical Applications. Stem Cells Int. 2019, 17, 42–59.
[3]   Tolvanen, J.; Kilpijärvi, J.; Pitkänen, O.; Hannu, J.; Jantunen, H. Stretchable Sensors with Tunability and Single Stimuli- Responsiveness through Resistivity Switching Under Compressive Stress. ACS Appl. Mater. Interfaces 2020, 12, 14433–14442.
[4]   Song, Y.-Y.; Liu, Y.; Jiang, H.-B.; Xue, J.-Z.; Yu, Z.-P.; Li, S.-Y.; Han, Z.-W.; Ren, L.-Q. Janus Soft Actuators with On–Off Switchable Behaviors for Controllable Manipulation Driven by Oil. ACS Appl. Mater. Interfaces 2019, 11, 13742–13751.
[5]   Abu-Thabit, N.Y.; Hamdy, A.S. Stimuli-responsive Polyelectrolyte Multilayers for fabrication of self-healing coatings – A review. Surface and Coatings Technology 2016, 303, 406–424.
[6]   Chatterjee, S.; Chi-leungHui, P. Review of Stimuli-Responsive Polymers in Drug Delivery and Textile Application. Molecules 2019, 24, 2547–2555.
[7]   Shu, T.; Hu, L.; Shen, Q.; Jiang, L.; Zhang, Q.; Serpe, M.J. Stimuli-responsive polymer-based systems for diagnostic applications. J. Mater. Chem. B 2020, 8, 7042–7061.
[8]   Son, H.; Yoon, C. Advances in Stimuli-Responsive Soft Robots with Integrated Hybrid Materials. Actuators 2020, 9, 115–121.
[9]   Li, Z.; Yin, Y. Stimuli-Responsive Optical Nanomaterials. Adv. Mater. 2019, 31, 1807061-1807067
[10]   Niu Z, Qi S, Shuaib SS, Yuan W. Flexible, stimuli-responsive and self-cleaning phase change fiber for thermal energy storage and smart textiles. Composites Part B: Engineering. 2022 Jan 1; 228: 109431–109437.
[11]   Shen X, Hu Q, Ge M. Fabrication and characterization of multi stimuli-responsive fibers via wet-spinning process. SpectrochimicaActa Part A: Molecular and Biomolecular Spectroscopy. 2021 Apr 5; 250: 119245–119252.
[12]   Huang C, Soenen SJ, Rejman J, Lucas B, Braeckmans K, Demeester J, De Smedt SC. Stimuli-responsive electrospun fibers and their applications. Chemical Society Reviews. 2011; 40(5):2417–2434.

[13]   Jahid MA, Hu J, Zhuo H. Stimuli-responsive polymers in coating and laminating for functional textile. In Smart Textile Coatings and Laminates 2019 Jan 1 (pp. 155–173). Woodhead Publishing.

[14]   Morsümbül S, Kumbasar EA. Photochromic textile materials. In IOP Conference Series: Materials Science and Engineering, IOP Publishing. 2018;459(1):012053.

[15]   Photochromic Fabric – changes from one colour to another in response to light; https://www.pinterest.com/pin/431641945508709400/; Accessed on date: 07.02.2023.

[16]   Chatterjee S, Hui PC. Review of applications and future prospects of stimuli-responsive hydrogel based on thermo-responsive biopolymers in drug delivery systems. Polymers. 2021;13 (13):2086–2091.

[17]   Atanasova D, Staneva D, Grabchev I. Textile Materials Modified with Stimuli-Responsive Drug Carrier for Skin Topical and Transdermal Delivery. Materials 2021; 14: 930–936.

[18]   Mohamed AL, Hassabo AG. Cellulosic fabric treated with hyperbranchedpolyethyleneimine derivatives for improving antibacterial, dyeing, pH and thermo-responsive performance. International Journal of Biological Macromolecules. 2021; 170:479–89.

[19]   Gupta D, Basak S, Surface functionalization of wool using 172nm UV Excimer lamp, 2010, 117: 3448–3453.

[20]   McCarthy PC, Zhang Y, Abebe F. Recent applications of dual-stimuli responsive chitosan hydrogel nanocomposites as drug delivery tools. Molecules. 2021; 26(16):4735–4742.

[21]   Patdiya J, Kandasubramanian B. Progress in 4D printing of stimuli responsive materials. Polymer-Plastics Technology and Materials. 2021; 60(17):1845–83.

[22]   Biswas MC, Chakraborty S, Bhattacharjee A, Mohammed Z. 4D Printing of Shape Memory Materials for Textiles: Mechanism, Mathematical Modeling, and Challenges. Advanced Functional Materials. 2021; (19):2100257–2100261.

[23]   Gupta D, Basak S, Surface functionalization of wool using 172nm VUV excimer lamp, Journal of Applied Polymer Science, 2010; 117 (6): 3448–3453.

[24]   Kelly, J. M., Mc Ardle, C. B., Maunder, M. J., editors, Photochemistry and polymeric systems, Royal Society of Chemistry, Cambridge (1993).

[25]   Calder L, Magalhaes J, Ayett R, Louchart S, Padilia S, Chantler M, Vean AM, Kinect based RGB detection for smart costume interaction, 2013, ISEA International, Australian network for art and Technology, University of Sydney; http://www.ses.library.usyd.edu.au/handle/2123/9645; Accessed on date: 1.1.2013.

[26]   Chowdhury MA, Joshi M, Butola BS. Photochromic and thermochromic colorants in textile applications. Journal of Engineered Fibers and Fabrics. 2014 Mar; 9(1):155892501400900113.

[27]   Color-Changing "Smart Thread" Turns Fabric into a Computerized Display, Berkeley school of information, June 6, 2016; https://www.ischool.berkeley.edu/news/2016/color-changing-smart-thread-turns-fabric-computerized-display; Accesses on date: 08.02.2923.

[28]   P Talvenmaa, Introduction to chromic materials in Intelligent textiles and clothing, Woodhead Publishing Series in Textiles, Woodhead Publishing, 2006, 193–205.

[29]   Chen HJ, Huang LH. An investigation of the design potential of thermo-chromic home textiles used with electric heating techniques. Mathematical Problems in Engineering. 2015; 2015: 224–229.

[30]   Bamfield P. Chromic phenomena: technological applications of colour chemistry. Royal Society of Chemistry; 2010.

[31]   Tao X, editor. Smart fibres, fabrics and clothing: fundamentals and applications. Elsevier; 2001 Oct 4.

[32]   Berglin L, Ellwanger M, Hallnäs L, Worbin L, Zetterblom M. Smart Textiles: what for and why?. Nordic Textile Journal. 2005.

[33]   Worbin L. Textile Disobedience. When textile patterns start to interact. Nordic Textile Journal. 2005.

[34]   Robertson S, Taylor S, Christie R, Fletcher J, Rossini L. Designing with a responsive colour palette: The development of colour and pattern changing products. In advances in Science and technology 2008 (Vol. 60, pp. 26–31). Trans Tech Publications Ltd.

[35] Berzina Z. Skin stories: charting and mapping the skin: research using analogies of human skin tissue in relation to my textile practice (Doctoral dissertation, University of the Arts London).

[36] Tang L, Sharma S, Pandey SS, Synthesis and characterization of newly designed and highly solvatochromic double squuarine dye for sensitive and selective recognition towards copper, Molecules, 2022; 27: 6578–6587.

[37] Klymchenko SA, Solvatochromic and Fluorogenic dyes as environment sensitive probes: design and biological applications, Accounts in Chemical Research, 2017; 50: 366–375.

[38] Moretti C, Tao X, Koehl L, Koncar V, Electrochromic textile displays for personal communication in smart textile and their application, Woodhead publishing series in textile, 2016; 539–568.

[39] Electrochromism; https://en.wikipedia.org/wiki/Electrochromism; Accessed on date: 09.02.2023

[40] Moretti C, Tao X, Koehl L, Koncar V. Electrochromic textile displays for personal communication. InSmart Textiles and their Applications 2016 Jan 1 (pp. 539–568). Woodhead Publishing.

[41] Graßmann C, Mann M, Van Langenhove L, Schwarz-Pfeiffer A. Textile Based Electrochromic Cells Prepared with PEDOT: PSS and Gelled Electrolyte. Sensors. 2020; 20(19):5691–5697.

[42] Gicevicius M, Cechanaviciute IA, Ramanavicius A. Electrochromic textile composites based on polyaniline-coated metallized conductive fabrics. Journal of the electrochemical society. 2020; 167(15): 155515–155519.

[43] Zhang R, Cai W, Bi T, Zarifi N, Terpstra T, Zhang C, Verdeny ZV, Zurek E, Deemyad S. Effects of nonhydrostatic stress on structural and optoelectronic properties of methylammonium lead bromide perovskite. The journal of physical chemistry letters. 2017;8(15): 3457–65.

[44] zisah P, Sadoh A, Ravindra NM. 3D Printing of Polymer-Based Gasochromic, Thermochromic and Piezochromic Sensors. In TMS 2019 148th Annual Meeting & Exhibition Supplemental Proceedings, Springer International Publishing. 2019:1545–1561.

[45] Dong Y, Xu B, Zhang J, Tan X, Wang L, Chen J, Lv H, Wen S, Li B, Ye L, Zou B. Piezochromic luminescence based on the molecular aggregation of 9, 10-Bis ((E)-2-(pyrid-2-yl) vinyl) anthracene. AngewandteChemie International Edition. 2012 Oct 22;51(43):10782–5.

[46] Nagura K, Saito S, Yusa H, Yamawaki H, Fujihisa H, Sato H, Shimoikeda Y, Yamaguchi S. Distinct responses to mechanical grinding and hydrostatic pressure in luminescent chromism of tetrathiazolylthiophene. Journal of the American Chemical Society. 2013 17; 135(28): 10322–5.

[47] High-Tech Fabric Changes Color When You Touch It (VIDEO); https://www.thetrentonline.com/watch-high-tech-fabric-changes-color-touch-video/; Accesses on date: 09.02.2023

[48] Sagara Y, Mutai T, Yoshikawa I, Araki K. Material design for piezochromic luminescence: hydrogen-bond-directed assemblies of a pyrene derivative. Journal of the American Chemical Society. 2007;129 (6): 1520–1.

[49] Sato O. Dynamic molecular crystals with switchable physical properties. Nature chemistry. 2016; 8(7): 644–56.

[50] Kocak G, Tuncer CA, Bütün VJ. pH-Responsive polymers. Polymer Chemistry. 2017;8(1): 144–76.

[51] Basak S, Raja ASM, Saxena S, Patil PG, Tannin based polyphenolic bio-macromolecules: creating a new era towards sustainable flame retardancy of polymers, Polymer Degradation and Stability, 2021; 189: 109603–109610.

[52] Kausar A. Review on technological significance of photoactive, electroactive, pH-sensitive, water-active, and thermo-responsive polyurethane materials. Polymer-Plastics Technology and Engineering. 2017; 56(6): 606–616.

[53] Yan T, Zhang T, Zhao G, Zhang C, Li C, Jiao F, Magnetic textile with pH responsive wettability for controllable oil/water separation, Colloids and Surfaces A: Physicochemical and Engineering Aspects, 2019; 575: 235–245.

[54] Stojkoski V, Kert M, Design of pH responsive textiles a sensor material for acid rain, Polymers, 2020; 12: 2251–2259.

[55] Zhou J, Jiang B, Gao C, Zhu K, Xu W, Song D, Stable, reusable, and rapid response smart pH-responsive cotton fabric based on covalently immobilized with naphthalimide-rhodamine probe, Sensors and Actuators B: Chemical, 2022; 355: 457–465.

[56] Schueren LV, Clerck K, Halochromic textile materials as innovative pH sensors in Advances in Science and Technology, Scientific Net, 47–52.

[57] Zhang J, He S, Liu l, Guan G, Lu X, Sun X, Peng H, The continuous fabrication of mechanochromic fibres, Journal of Material Chemistry C, 2016, 4: 2127–2133.

[58] Attia AAM, Shouman MAH, Khedr SAA, Hasan NA, Fixed bed column syudies for the removal of congo red using simmondsia chinensis (jojoba) and coated with chitosan, Indonsia Journal of Chemistry, 2018; 18: 294–305.

[59] Omidi S, Kakanejadifard A. Eco-friendly synthesis of graphene–chitosan composite hydrogel as efficient adsorbent for Congo red. *RSC advances*, 2018; 8: 12179–12189.

[60] Wu J, Cheng Y, Lan J, Wu D, Qian S, Yan L, He Z, Li X, Wang K, Zou B, You J, Molecular Engineering of Mechanochromic materials by programmed C-H Arylation: Making a counterpoint in the chromism trend, Journal of American Chemical Society, 2016; 138: 12803–12812.

[61] Schrettl S, Weder C, Clough J M, Mechanochromism in Structurally Colored Polymeric Materials, Macromolecular Rapid Communications, 2020; 42: 345–355.

[62] Gregory I. Peterson, Michael B. Larsen, Mark A. Ganter, Duane W. Storti, Andrew J. Boydston, 3D-Printed Mechanochromic Materials, ACS Applied Material Interfaces, 2015; 7: 577–583.

[63] Zhang J, He S, Liu L, Guan G, Lu X, Sun X, Peng H, The continuous fabrication of mechanochromic fibers, Journal of Material Chemistry C, 2016; 4: 2127–2133.

[64] Khalil E, Application of phase change materials in textiles: A review, International journal of research and review, 2015; 2: 281–289.

[65] Iqbal K, Khan A, Sun D, Ashraf N, Rahman A, Safdar F, Basid A, Maqsood HF, Phase change materials, their synthesis and application in textiles: A review, The journal of the Textile Institute, 2018; 7: 345–351.

[66] Dong Choon Hyun, Nathanael S. Levinson, Unyong Jeong, Emerging Applications of Phase-Change Materials (PCMs): Teaching an Old Dog New Tricks, Angewandte Chemie International Edition, 2014; 53: 1564–1572.

[67] Biswas MC, Chakroborty S, Bhattacharjee A, Mohammed Z, 4D printing of shape memory materials for textiles: mechanism, mathematical modeling and challenges, Advanced Functional Materials, 2021; 31: 2100257–2100265.

[68] Phase change material; https://en.wikipedia.org/wiki/Phase-change_material; Accessed on date: 06.02.2023.

Unsanhame Mawkhlieng, Abhijit Majumdar

10 Protective smart and functional textiles

Abstract: Protective textiles are a class of functional textiles that serve as a shielding medium to guard people from the dangerous impact, weather, harmful gases, fire, and pathogens. Hence, they are an essential requirement that demands great precision in design and development. The functionality of protective textiles has been improved over the years with the advent of smart materials and nanotechnology. The advantage of textiles is that with judicious designing, it is possible to integrate these supporting elements of smartness without affecting the wearability. Further, the application of nanotechnology has been reported to enhance the functionality. This chapter demonstrates protective textiles that have been developed for ballistic and impact resistance, for extreme cold weather resistance, for protection against hazardous nuclear, biological, and chemical warfare agents, and for flame retardancy. Special attention has been given to the advantages of nanotechnology that has been incorporated into the design and development of the protective smart and functional textiles. While in certain cases such as in liquid body armour, the use of nanotechnology has seen some level of commercial success, in other cases as in the application of nanosilver for extreme cold weather clothing, the success of nanotechnology is still limited to laboratories. However, with intensive research and increasing demand for enhanced performance, nanotechnology will receive more acceptance and recognition from industries. This aspect is discussed as part of challenges that protective smart and functional textiles face today and the opportunities that are present for researchers, academicians, and industries to unfold.

Keywords: Protective textile, smart textile, nanotechnology, body armour, extended cold weather clothing systems, flame retardancy

10.1 Introduction

Smart materials refer to those that can sense, react, and/or respond to the environment around them. There are three kinds of "smartness" that authors have described in literature: passive smart, active smart, and ultra-smart [1]. Passive smart materials can sense the changes that occur around them and hence, behave as sensors. Active smart materials are one step ahead of passive smart materials, in that they, upon sensing, can react to the environmental changes by activating a secondary function to take place in response to the stimulus. Ultra-smart materials are futuristic materials that can even self-regulate, that is, they can reform themselves. These materials when integrated into or with fabrics lead to smart textiles. Hence smart textiles are those that can sense and/or react to the external changes in the environment. Most, if not all, smart textiles are

https://doi.org/10.1515/9783110759747-010

integrated with electronics and are often termed as e-textiles in literature. The most recent example would be the novel digital fibre developed at the Massachusetts Institute of Technology wherein a research team of Prof Fink created a fibre with hundreds of square silicon microchips integrated into the preform that is later used to spin continuous fibres, capable of producing a fabric that can store and process data digitally [2]. Other smart textiles that are not electronic-based are made possible today because of the availability and development of smart materials like phase change materials (PCMs) [3], conductive polymers [4], piezoelectric materials [5, 6], shape-memory polymers [7, 8], photovoltaic materials [9–11], electroactive polymers [12], halochromic materials [13], non-Newtonian fluids [14, 15], and self-healing materials [16, 17].

It must be noted that textile materials with added functionality such as antibacterial fabrics, anti-ultraviolet fabrics, and light guide fabrics, although often categorized as passive smart materials, do not fall under the category of "smart textile" and may be termed as functional textiles instead. Functional textiles are, therefore, those that perform functions above the basic requirement of clothing. Further, textile substrates can be integrated with attachments to enhance functionality, a concept that leads to the development of common enhanced clothing like the spacesuit that has life supporting attachments such as oxygen supply, temperature controls, body activity monitors and regulators, eye shielding and protection, and communication transmitters [18]. Presently, the concept of functional textiles has expanded beyond fabrics that have additional attachments and are now dedicated to inherently functional fabrics also. A detailed classification of functional clothing is provided by Gupta, according to whom, functional clothing can be grouped into six classes: protective, medical, sports, vanity, cross-functional, and clothing for special needs [19].

A class of smart and functional textiles that are intentionally designed to enhance protection to a wearer is called as protective textile or ProTech. This broad class of functional textiles includes those textiles that are: (i) environmental hazard protective (extreme heat or cold, fire, rain, snow, wind, and UV exposure); (ii) biological, chemical, and radiation hazard protective (bacterial and viral infections, penetration and skin contact of hazardous chemicals, toxic gases, particulate matters, and body fluid); (iii) injury protective (impacts, cuts, and slashes, and visibility clothing). Further, textiles that act as sensors to indicate a hazardous level of intoxicants can also be termed as protective textiles. To achieve the optimal functionality of protective textiles, it is necessary to treat them or design them specifically, and in many cases, nanotechnology is utilized. Therefore, the present chapter focuses on protective smart and functional textiles to safeguard a person from impact, flame, nuclear and biological war agents, and extreme weather conditions, and the application of nanotechnology to enhance certain desirable properties thereof. The chapter also highlights the challenges that functional textiles face currently and concludes with futuristic opportunities that will likely present themselves as far as smart and functional protective textiles are concerned.

10.2 Protective textile

Protective textiles are class of technical or functional textiles that not only provides the wearer a ballistic protection but also imparts protective function to the wearer in adverse environmental condition including heat, cold, and chemical. The advent of nanotechnology or nanotechnology-based materials offers a promising approach in manufacturing the line of protective textiles due to their extraordinary properties at the performance level.

10.2.1 The use of nanotechnology and its influence in properties

Nanotechnology has been used for performance enhancement or for development of new functions due to their unique properties. When the dimension of a bulk material is brought down to the nanorange, the surface area increases substantially for a given volume (Figure 10.1). In bulk form, much of the material is "hidden" inside and, therefore, cannot be used advantageously particularly when the interaction with the host is superficially driven. Their application to textile substrates is easy and quick, generally achievable through methods such as coating, printing, spraying, and most commonly, padding as reported by Wong et al. [20]. In addition, the quantity of particles required is generally low. Nanotextiles that are smart and functional have been reported as suitable innovative substitutes in various sectors like medicine, construction, automobiles, communication, sports, fashion, and protection. In this chapter, protective properties that will be discussed include impact resistance, flame retardancy, thermal insulation, and resistance to nuclear, biological, and chemical warfare.

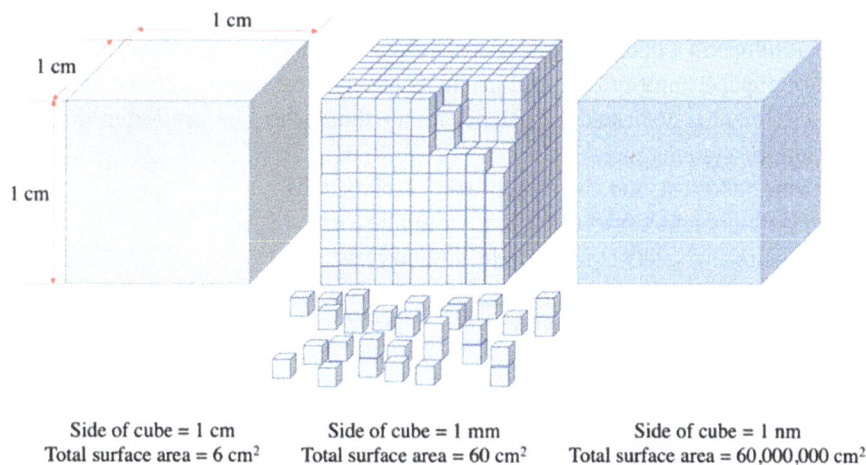

Side of cube = 1 cm	Side of cube = 1 mm	Side of cube = 1 nm
Total surface area = 6 cm^2	Total surface area = 60 cm^2	Total surface area = 60,000,000 cm^2

Figure 10.1: Nanotechnology – the increase in surface area with size reduction.

10.2.2 Projects on smart textiles and functional protective textiles in India

Protective textile is a booming industry in India. As the world becomes more and more health and safety driven, the demand for protective clothing is increasing. The world market is expected to reach US $12.3 billion in 2025 from US $8.8 billion in 2020 [21]. The Indian Protech (protective textile) market is estimated to reach 4,500 crores (US $0.6 billion) by the end of 2022, the major consumer being the defence forces [22]. During the pandemic, the need for personal protective equipment (PPE) increased substantially, with the market touching 7,000 crores (US $0.9 billion) in India alone [23]. At the time of writing this chapter, the Government of India is undertaking several projects with the help of various research institutes in the country to further the industry. Ahmedabad Textile Industry's Research Association is working on nanofibre-based textiles to manufacture facemasks, automotive engine air filter, cigarette filters, water filter media, and treatment devices [24]. Defence Research Development Establishment developed the enhanced chemical protection nuclear biological and chemical (NBC) permeable suit Mk-V that is high adsorbent, water and oil-repellent, fire-retardant, anti-static, and launderable with high level of comfort [25]. Northern India Textile Research Association's Centre of Excellence for Protective Textiles has undertaken various projects such as the development of cut and abrasion-resistant protective textile by using composites metallic yarn, development of stab resistant armour using high modulus polyethylene fibres, development of flame resistant, and chemical resistant laboratory coat/apron, and the use of corn to develop shields for enhanced protection against flame [26]. Defence Research and Development Organisation has been developing a revolutionized bullet-resistant jacket for armed forces that is lightweight, flexible, and relatively inexpensive [27].

To further promote the industry, the Government of India has released a scheme called as Production Linked Incentive Scheme for Textiles in 2021 to promote manmade fibre apparels and fabrics and products of technical textiles. Under technical textiles, a segment is dedicated for defence textile that focuses on protection through several clothing systems such as

i.   Nuclear, biological, and chemical or NBC warfare suits
ii. High visibility and infrared clothing for military use
iii. Head gears and safety equipment for military use
iv. Bullet-resistant clothing
v.   Tents, parachutes, collapsible textiles
vi. Special masks including gas masks
vii. High altitude clothing
viii. Fighter aircraft clothing

Further, a special category of "smart textiles" has also been enlisted as one of the final products that are eligible for this scheme. In brief, the scheme provides incentives to companies that are willing to invest in plant, machinery, equipment, and civil works to produce the notified lines of products and earn a turnover of double the amount of investment in 2-years time. The incentives will be provided for a span of 5 years thereafter up to 2030 [28].

10.2.3 Examples of smart and functional protective textiles

10.2.3.1 Body armour against stab and ballistic impacts

Textile materials have been used for protection against stab and ballistic impact since time immemorial. In earlier times, fabrics from cotton, wool, and linen were used by people to protect themselves against animal attacks and from piercing arms like arrows, spears, and stone tips [29]. With time, the advancement in technology has led to the evolution of weapons and consequently, the armour. The use of textile materials as armour again revived with the advent of high-performance fibres such as nylon (flak jackets) and para-aramid (Kevlar). The subsequent development of ultra-high molecular weight polyethylene has led to the latest version of soft body armour that is lightweight and effective. Multiple layers of such high-performance fabrics are combined to form a panel that is inserted into the pockets of a battle dress uniform. The number of layers varies according to the level of protection the armour is intended to provide. For instance, in the research works of the present authors, 17–18 layers are found to be useful for Level IIIA of the NIJ 0101.06 [30]. It is important to note here that the layers are different from each other – the first 10 layers are unidirectional, and the rest are 2D-woven fabrics. Similarly, a report from Honeywell shows an armour of 21 layers with different compositions broken down into three Spectra Shield, two Gold Flex, seven Kevlar, four Spectra Shield, and five Gold Flex [31]. However, the use of such multi-layered clothing increases the weight and thickness of the garment, which in turn imposes restrictive movements and high strain to the wearer.

In 2003, a new research direction was observed when a group of scientists from the University of Delaware and the Army Research Laboratory, US discovered that the impact resistance of a body armour can be enhanced substantially when treated with a certain non-Newtonian fluid called the shear thickening fluid, regularly abbreviated as STF [15]. The birth of "liquid armour" came into existence hence and is an assembly of high-performance fabrics soaked or padded with STF. The application procedure widely adopted is shown in Figure 10.2. STF is a smart fluid that is uniquely characterized by sudden rise in viscosity when a sudden impact is applied. The fluid temporarily "thickens," and hence the phenomenon is called as shear thickening and is reversible. More specifically, when an applied shear rate reaches a certain value called as critical shear rate, the flow behaviour of the STF is characterized by a steep rise in viscosity. The commonly used STF for protective textile is a combination of silica nanoparticles and polyethylene glycol in certain proportions. The application of STF to fabrics has been consistently proven to be effective even at bullet impact velocities of 430 m/s as demonstrated by Bajya et al. [32]. According to the study, the back-face signature (BFS) of para-aramid fabrics treated with STF was reduced by 2.5–2.8 mm without affecting the areal density of the panels (5 kg/m^2) (Figure 10.3). By judiciously putting STF impregnated fabrics on the rear side of the panel and neat fabrics on the strike face of the panel, the area density of soft armour panels (SAPs) can be reduced by 10%, while maintaining a BFS comparable to or lower than that of an STF-treated homogeneous panel [32]. The latest report from 2017 shows that BAE Systems and Helios Global Technologies will further the technology for its commercialization [33]. Alternatively, another smart fluid called the magnetorheological fluid also has the potential to be used as a performance enhancer [34, 35]. This ferrofluid containing iron particles in oil flows freely under normal circumstances but when influenced by a nearby magnetic field, hardens due to the iron particles coming closer together. Although this technology is still far away from reaching the battlefield, the research at laboratory scale is ongoing and the results are promising [36]. Several articles reviewing the rheological aspect of STFs and their subsequent applications to textiles have been published [37, 38]. A detail progress of the subject focusing on the impact application of STF-treated structures has also been compiled by the present authors. STFs can be single phase or multiphase. In the former, only one type of nanoparticles is used, and in the latter, two or more types are mixed as reinforcements. Those extra additives that are reported in literature include nano and microparticles in various forms such as graphene nanoplatelets, graphene oxide, halloysite nanorods, and ceramics [39–42]. Nanoparticles in the form of zinc oxide nanorods have also been grown on the fabric surface to impact friction to enhance its energy absorption capacity [43, 44]. Researchers from Daresbury Laboratory, Tuskegee University, and Florida Atlantic University investigated how silicon and titanium dioxide nanoparticles, as well as carbon nanotubes, could improve the flexibility of composite-based hard body armour [45]. Gonzalez from Harvard University developed highly oriented and porous nanofibres of para-aramid via a method called immersion Rotary Jet-Spinning for a multi-functional smart armour [46]. High orientation helps in effective redistribution of the forces upon impact whereas the porosity aspect of it adds to insulation causing comfort when worn in cold places. The research team found that the performance of the nanofibres in combination with the woven Twaron fabrics resulted in a performance equivalent to that of 100% woven Twaron sheets while providing 20 times better heat insulation.

10.2.3.2 Extended cold weather clothing system

Extended cold weather clothing system (ECWCS) is a protective ensemble that is developed specifically to assist military personnel combat in extreme cold environments. The system was first developed in the 1980s by the United States Army Natick Soldier Research, Development and Engineering Center, Massachusetts, and has seen tremendous

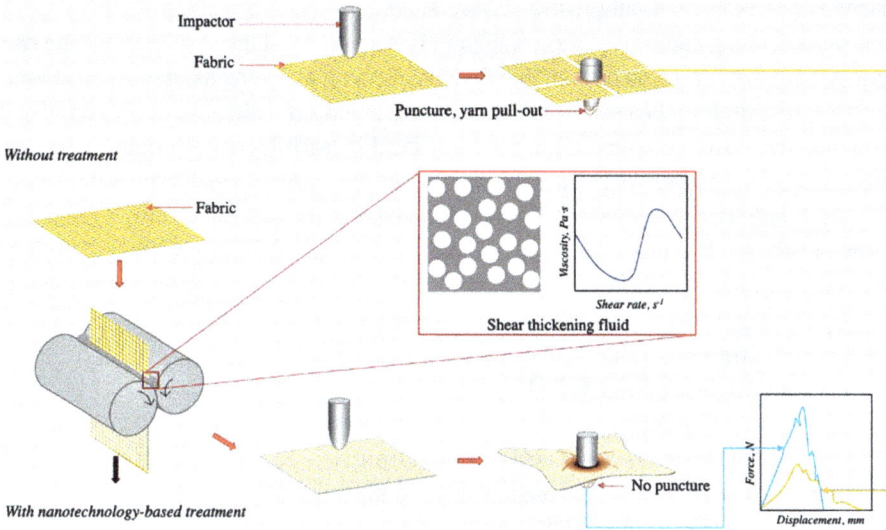

Figure 10.2: Mechanism of energy absorption of nanotechnology-based treated fabric.

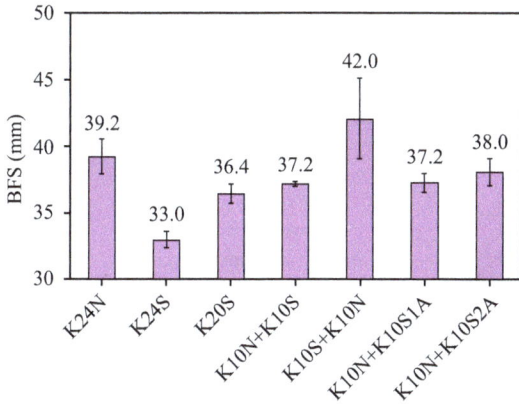

Figure 10.3: BFS of soft armour panels treated with STF-100. Note: K, Kevlar-woven fabric; N, neat; S, STF-treated fabric (the numerical figures represent the number of layers); A, alternate stacking of neat and STF-impregnated fabrics; and the number preceding "A" corresponds to the repeating unit of alternation (Modified according to [47], Elsevier, Copyright 2020).

advancement since then, broadly classified into three generations. Since its inception in the 1980s, the system has witnessed several changes, with the latest generation being the GEN III (third generation), where the "W" that stands typically for "weather" is often replaced with "warfighter". The scope of ECWCS protection expands from −50 °C and +4 °C. ECWCS is a system of clothing which implies that several apparels are used together to obtain the intended protection. These include the undershirt, jacket, trousers, headwear, handwear, overall, liner, and footwear [48]. Multiple layers are required to allow quick and rapid clothing adjustment according to the temperature of the surrounding. The earliest system, GEN I consists of over 20 individual clothing including handwear, headwear, and footwear that can be used in different combinations according to the environmental demand. The system is worn in a three-layer fashion, each layer performing a specific function. The base layer wicks moisture away from the body, the mid-layer is the insulation layer, and the outer layer is the protection layer. In this system, several clothing can form a layer. The latest GEN III, however, consists of 12 components, providing 7 levels of protection as follows:

– Level I: Light-weight undershirt and drawers
– Level II: Mid-weight shirt and drawers
– Level III: High-loft fleece jacket
– Level IV: Wind jacket
– Level V: Soft shell cold weather jacket and trousers
– Level VI: Extreme wet/cold weather jacket and trousers
– Level VII: Extreme cold weather parka and trousers

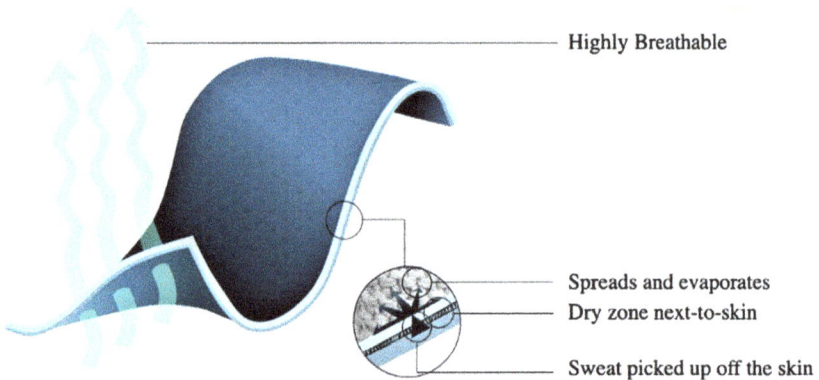

Highly Breathable

Spreads and evaporates
Dry zone next-to-skin

Sweat picked up off the skin

Figure 10.4: Working mechanism of Polartec® Power Dry® (modified after [51]).

A pictorial juxtaposition of the different levels can be readily found online. In fact, a brochure by ADS, the official GEN III ECWCS supplier for the US army, neatly summarizes the entire clothing system [49]. The system functions through (i) insulation; (ii) layering, and (iii) ventilation. Insulation helps in resisting heat dissipation by trapping the air pockets in the system while layering helps to increase the air space and allows easy adjustment to accommodate different activity level. On the other hand, ventilation allows moisture to escape. Each layer of the system is arranged such that the layer that is closest to the body has the least thermal insulation. Subsequent layers generally have more significant thermal insulation. If the layered clothing is to be versatile, each layer should be able to perform as the outside layer, which should be water and windproof. However, they should be breathable as well. There are a few products that are both water and windproof yet water-vapour permeable. One such is Gore-Tex. Other materials are also used, for example, in the case of GEN III that can reach up to seven layers of materials depending on the environment [50]. Some of the layers in the middle intended to provide light insulation are constructed from Polartec® Power Dry™ grid material [51], a structure that can wick and dry fast. The third layer, which is a high-loft fleece jacket providing moderate insulation, is constructed from Polartec® Thermal Pro™ capable of trapping air and retaining body heat (Figure 10.4). GEN III is a huge improvement over the previous versions of the system; however, its substantial bulkiness drives the inventors to explore nanotechnology in their latest designs and the use of nanosilver wires is one such example [53]. It is claimed that the nanowires can generate temperatures up to 110 °C when induced with electric current. The scientists are working on the minimization of weight and bulkiness by using a type of hydrogel coating that helps in quick wicking. Similarly, aerogels have been reported to be effective in such clothing systems because of their low weight (99.8% volume by air) [54]. They are transparent and highly porous and hence have low thermal conductivity. Aerogels considerably reduce the three phenomena of heat transfer: convection, conduction, and radiation. Convection is reduced due to the inability of the air to circulate through the tiny pores, whereas conduction is lowered because silica is a poor conductor of heat. Finally, radiation can be reduced by adding carbon to the aerogel to absorb infrared radiation. Aerogels have seen commercial success; for example, Shiver Shield Extreme Cold Weather Clothing uses this technology [55]. Another nanotechnology based is Outlast® technology that uses PCMs that absorb, store, and release heat [56]. It is a smart technology that can self-regulate the microclimate between the skin and the fabric using special patented Thermocules™ capable of absorbing and releasing heat. Additionally, they can be readily incorporated into the fabric. The working principle of Themocules™ is rather unique. They absorb and store excess heat energy from the environment and release the same when the temperature drops. In contrast to wicking, this technology proactively regulates moisture production in order to manage the heat. The schematic is shown in Figure 10.5. In commercial applications, it is used to keep the body comfortable in Tempex polar coveralls.

10.2.3.3 Nuclear biological and chemical suits/hazmat suits

An NBC suit is also known as chemical or chem suit and is a type of military PPE that protects the wearer from contamination and direct contact of nuclear, biological, and chemical materials. Soldiers can wear NBC suits over their uniforms in military operations and stay protected for several days. Most are made of impermeable material such as rubber, an example of which is a Zodiak [57]. Modern NBC suits incorporate a filter, allowing air, sweat, and condensation to slowly pass through. However, the chances of contamination and penetration of toxic substances exist in such cases. Certain alternative hydrophilic materials such as polyvinyl alcohols, some varieties of polyurethanes, and

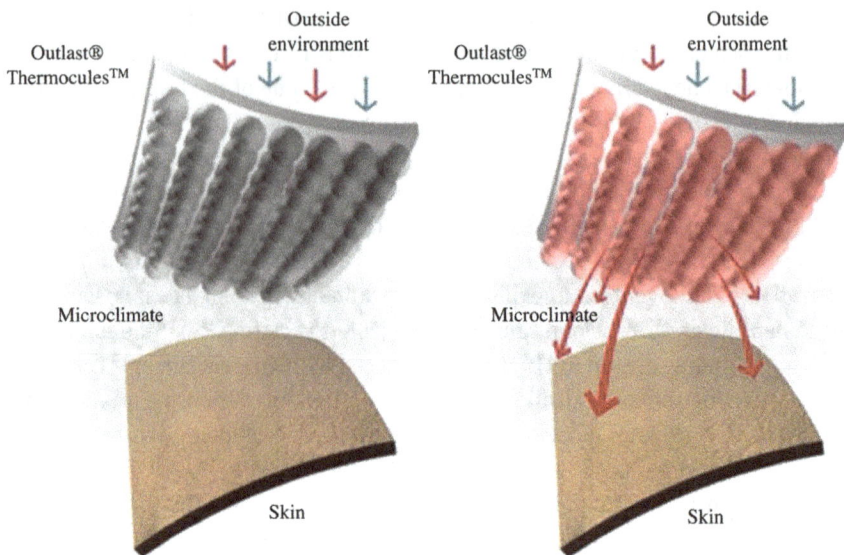

Figure 10.5: (a) Thermocules™ absorbing excess heat and (b) Thermocules™ releasing excess heat (reprinted with permission from [56], ©Outlast Technologies GmbH, Copyright, 2020).

cellophane have been reported to be used as barriers against chemical warfare agents to eliminate the disadvantages of breathable materials. However, these materials work in dry conditions. When wet, they get hydrolyzed and lose their ability to prevent aerosols, solvents, and oils from entering the suit. This associated problem of hydrophilic barriers can be minimized by the use of nanotechnology in the form of activated charcoal in structures made of water permeable porous plastic materials [58]. The entrance of chemical warfare agents is stopped by the activated carbon content that forms a part of the protective layer. For instance, the CWU-66/P* Aircrew Chemical Protective Coverall of GENTEX utilizes this nanotechnology by incorporating spherical activated carbon adsorbents in the protective layer [59]. Furthermore, DEMRON ICE suits from Radiation Shield Technologies incorporate patented carbon nanotechnology that makes them effective against low-level gamma radiation [60]. Likewise, Defence Research and Development Organisation (DRDO), India made Mark-IV-activated carbon-based permeable suit [61]. Additionally, Argonide Corporation's alumina nanofibres are used to protect fabric against biological agents and radiations or as filters against biologically charged contaminated water [62]. Apart from adsorption, another nanotechnology-based approach that is used against chemical agents is the use of nanofibres as sensors.

Nanoparticles have also been used as sensors to detect the presence of chemical warfare agents in air and water. Functionalized fluorescent nanosensors based on carbon nanoparticles have been used by Tuccitto et al. [63] to detect traces of nerve agents (dimethyl methyl phosphonate) both in gaseous and aqueous media. Decontamination of certain war agents including blister agents, insecticide model compounds, and other biological agents has also been reported to be decomposed by certain metal oxides [64]. For instance, Šťastný et al. [65] found that the nerve-agent simulant dimethyl methyl phosphonate can be almost decomposed with the help of a nanocomposite of graphene oxide–manganese oxide. Textiles coated with metal-organic frameworks based on zirconium can help catalyze the degradation of chemical warfare such as soman and VX (an extremely toxic synthetic chemical compound) [66]. Similarly, gold nanoparticles have been explored as sensors in the colourimetric and fluorescence detection of such agents [67]. Oxides of magnesium, calcium, and aluminium have been reported to detoxify mustard gas [68]. Sadeghi et al. [69] found that a nanocomposite of zinc oxide and polyvinyl alcohol can decontaminate an agent named chlorethyl-phenyl sulphide. When applied to fabrics, the nanocomposites help to destroy the lethal chemicals via photocatalytic actions. The nanoparticles are applied to the textiles through coating, spraying, or padding. In other cases, nanoparticles interact chemically with the host. In the former case, the nanoparticles simply adhere to the surface, and hence their application is not durable. Upon laundering or rubbing, there are problems of leaching out and low fastness that prevent the wide acceptance of nanoparticles particularly for commercial textiles. As a result, researchers have explored the use of nanofibres instead [57]. Nanofibres have extremely fine diameters in the nanoscale and can produce fabrics that have high breathability because of enhanced porosity and surface area [70]. Figure 10.6 depicts the nanofibrous-coated webs and activated carbon sphere (ACS) laminated fabric manufacturing process for next-generation NBC suits. With these electrospun nanofibres, other functional compounds such as peroxides, polyoxometalates, and oximes can be incorporated for enhanced detoxication [70]. Nanofibrous webs of polyvinyl alcohol have also been coated on non-woven fabrics of polypropylene along with the ACS fabrics for enhanced protection against the chemical warfare agents and bacteria.

10.2.3.4 Flame-retardant fabrics

Flame-retardant fabrics are protective textiles that are intended to protect humans and things from fire. The textile should not catch fire, and if it does, it should inhibit the propagation of flame. A condition that enables fire to burn, often modelled as the "fire triangle", consists of three necessary ingredients or elements: fuel, oxygen, and heat. Sometimes, a fourth element is also added, a chemical chain reaction that helps sustain certain fire types, in which case, a "fire tetrahedron" model is considered. Therefore, firefighting aims at depriving the fire of at least one these elements [71].

One hazard posed by routinely used textiles is their easy ignitibility that is most prominent in case of natural cellulosic and flax fibres. Hence, fabrics made from such fibres catch fire easily. The increased concern and consequently, legislation and regulation for fire safety have therefore compelled special treatment of fabrics to enhance their resistance to fire. Back in 1735, Wyld described a finishing process of cellulosic textiles using alum, ferrous sulphate, and borax to induce flame retardancy [72]. Alongi

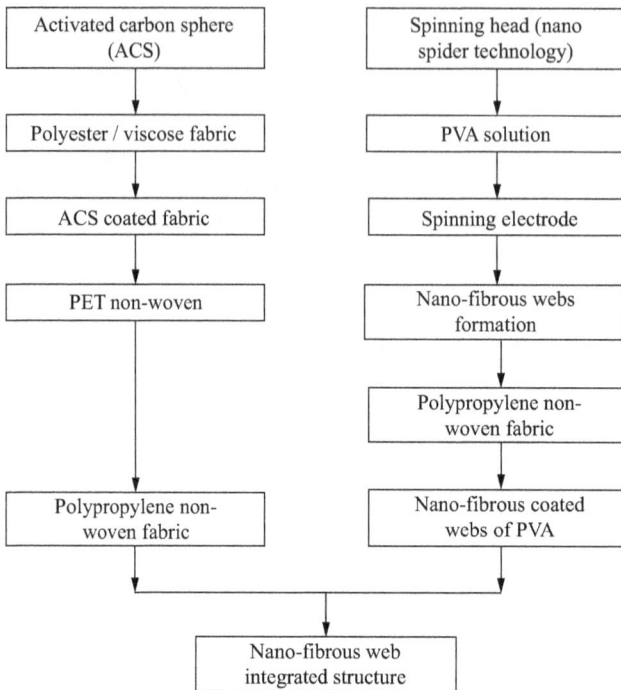

Figure 10.6: Schematic flow diagram of nanofibrous-coated webs and ACS-laminated fabric manufacturing process for NBC suits (reprinted with permission from [70], SCIndeks, Copyright 2018).

et al. [71] have compiled a review on flame-retardant textiles, focusing on the burning hazards, aspects of fame retardancy, regulations and requirements for flame-retardant textiles, environmental issues, and innovative solutions on the subject matter.

The use of nanotechnology for flame retardancy has received attention since the last decade or so. Nanotechnology has been observed to improve the flame retardancy of fabrics provided there exists good compatibility between the polymeric fibre and the nanoparticles. Some of the nanoparticles that has been reported in literature include nanoclay, graphene, carbon nanotubes, and their combinations.

Nanoclay-based flame retardants

The use of nanoclay as an environmentally friendly flame-retardant catalyst or synergist has been reported on many occasions. Kaynak et al. [73] reported hybrid nanocomposites of polyamide-6 and phosphorous compounds reinforced with exfoliated clay that showed reduced peak heat release rate and delayed ignition. It has also been reported that when clay is added to acrylic, the flame retardancy improved through the formation of a ceramic like protective barrier on charring [74]. Further, when nanokaolin white clay is used with nanoparticles of hydroxyl aluminium oxalate, the flame-retardant property of the applied fabric was found to enhance synergistically [75]. Likewise,

Priyanka and Dahiya [76] found that the best V0 (burning combustion is not sustained for more than 10 s) rating in UL-94 burning test was achieved when the phosphorylated epoxy nanocomposites contained 1% nanoclay. Again, Thirumal et al. [77] reported enhanced flame-retardant properties in terms of limiting oxygen index (LOI) and flame spread rate when a small fraction of organically modified clay or organoclay was incorporated into polyurethane foams. Generally, the application of organoclay in small quantity helps to improve both the thermal as well as mechanical properties of the nanocomposites in addition to flame retardancy [78]. In polymer/clay nanocomposites, flame retardancy is determined by the migration of nanoclay particles to the surface during combustion, resulting in a protective barrier that prevents heat transfer, reduce oxygen diffusion, and volatilization of combustible products. A detailed review by Norouzi et al. [79] summarizes the effect of nanoclay on flame retardancy behaviour of many polymer/clay nanocomposite textiles focusing on the filler amount (wt%), preparation techniques, flame retardants used, and optimized state (Figure 10.7).

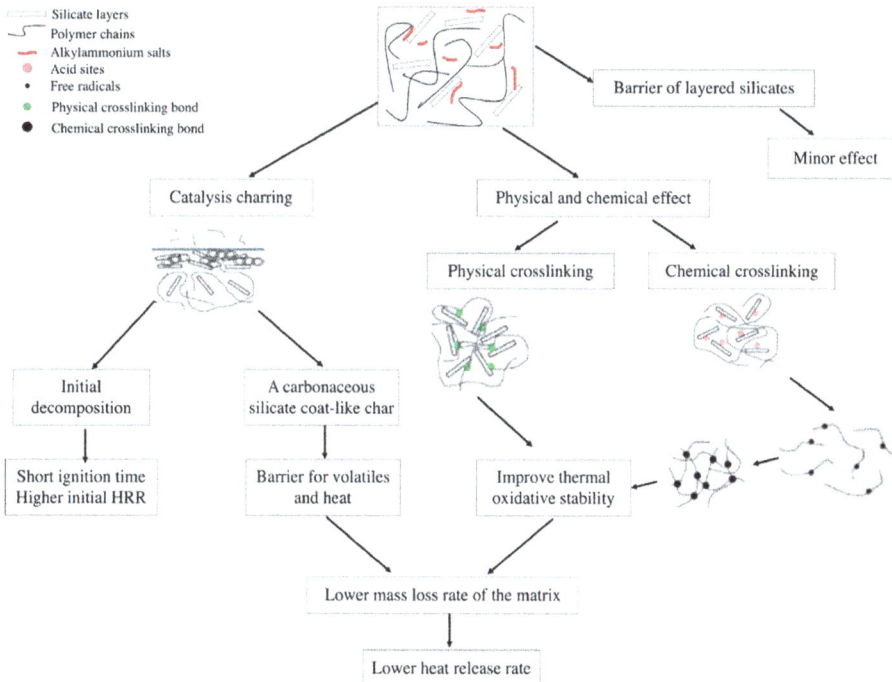

Figure 10.7: Flame-retardant mechanism of polymer/clay nanocomposites during combustion [80] (reprinted with permission from [80], Elsevier, Copyright 2005).

Graphene-based flame retardants

Graphene has been reported as an effective flame-retardant synergist. The char produced when graphene is used is denser and more voluminous as compared to that produced from conventional FR [78, 81]. The mechanism is picturized in Figure 10.8. Graphene-based flame retardants inhibit two of the four key components of the fire tetrahedron required for combustion. These are heat and fuel. Graphene-based flame retardants are postulated to work in three synergistic ways:

i.   The structure of graphene that is two-dimensional helps to develop a thick and dense char-like layer during decomposition. The layer formed through the interlocking carbonaceous structure is anticipated to (a) act as the thermal barrier, (b) slow the pyrolysis of the material, and (c) form tortuous pathways for the flammable gases and vapours, decreasing the flow rate of the combustible fuels reaching the gas phase.

ii. The surface area of graphene and its derivatives, being enormous, they have the ability to adsorb volatile organic compounds and inhibit their release and diffusion during combustion.

iii. Graphene and its derivatives contain reactive oxygen containing groups that may undergo endothermic decomposition. During such times, heat absorption from the surrounding may take place.

Graphene has been explored extensively and has been used in combination with many polymeric matrices such as polyurethane, polypropylene, and polyvinyl acetate. In most cases, graphene is used as a synergist to enhance the flame retardancy of common flame retardants. A summarized review of the related work has been provided by Araby et al. [81] and Sang et al. [82], where a general understanding is that graphene has a huge potential to be a candidate of flame retardant.

Carbon nanotube-based flame retardants

The use of carbon nanotubes (CNTs) has also been explored as a flame-retardant synergist in many textile polymer nanocomposites, and it is generally seen that the flame retardancy improves. It is postulated that in systems reinforced with CNT, the char layer formation acts as thermal insulator and heat barrier, delaying the degradation of the polymer [79]. Their high aspect ratio helps limit fuel, one of the key parameters of the fire tetrahedron, by deviating direct pathways. Hence, the effectiveness of CNT depends upon the parameters of the nanoparticle such as aspect ratio and alignment as well [78]. Additionally, the flame retardancy also depends on the quantity of the CNTs added. CNTs have been incorporated to polymers like epoxies, polylactide, polypropylene, and ethylene-vinyl acetate. In general, when CNTs are added, there is a reduction in peak heat release rate. In fact, when the dispersion forms a jammed network, CNTs can even surpass the performance of nanoclays [83]. CNTs have also been applied to foams by Kim and Davis [84] who reported improvement in flammability. Furthermore, CNTs

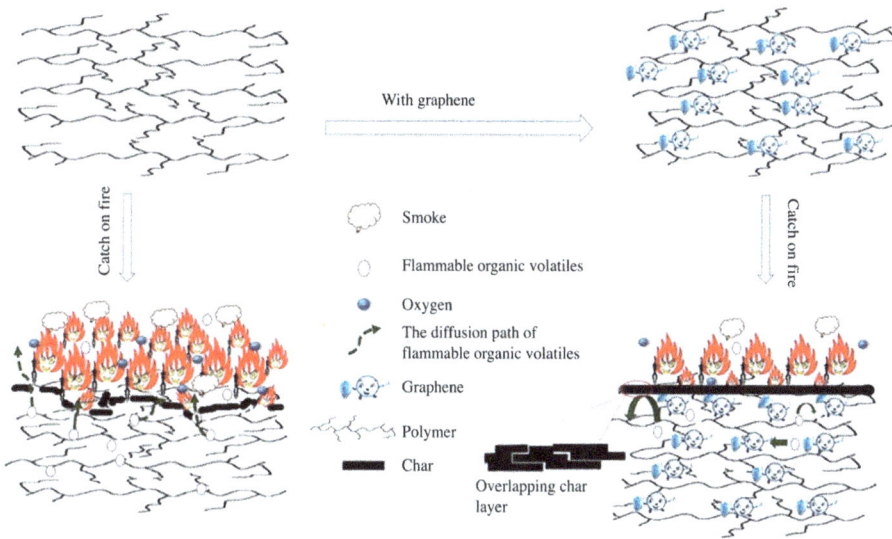

Figure 10.8: Mechanism of graphene-based flame retardant (reprinted with permission from [82], Springer Nature, Copyright 2016).

films that were developed as free-standing sheets have been compared to other flame-resistant polymeric fibres such as Nomex and Twaron [85]. The flame-retardant property of the CNT sheet is found to be superior. In this area of application, again, several publications and reviews have been published [86–89]. The other carbon-based additives that have also been reported in literature include fullerenes [90], graphitic carbon nitride [91], and ACS [92].

10.2.4 Opportunities and challenges of protective smart textiles

An overview of the challenges and opportunities of smart textiles has been provided by Cherenack and van Pieterson [93]. The scope of smart protective textile is expanding as the world becomes more safety and security oriented. The support of governmental bodies through schemes and projects also helps to develop the industry. Further, while some of the smart products have been commercialized, most of them are still developing in laboratories and need further development for commercial purposes. Two of the reasons for this are the practicality aspect and the expensiveness of the inventions. Consequently, the commercial acceptability is low. This shortcoming opens a huge scope and endless opportunities for industries to participate and develop the technologies further, such that the development procedure becomes more adaptable, and the products become more reproducible and cheaper. Smart protective textiles that are based on nanotechnology also need development in terms of applicability and longevity. The application procedure followed in laboratories is often not safe since nanoparticles can be toxic. For instance, when STFs are applied to fabrics for impact resistance improvement in the laboratories, the handlers are often exposed to the silica nanoparticles. This practice puts their health at risk. Hence, opportunities to assess the health- and eco-friendliness of different nanoparticles await. In fact, there is scare literature that explicitly explores the toxicity of different nanoparticles. Most information available is related to nanoparticles that are inhaled and those that are used for drug delivery. Therefore, there is scope to investigate the toxicity of nanoparticles in relation to their use in the smart nanotextiles for protective application. A detailed report has been provided by the Scientific Committee on Emerging and Newly Identified Health Risks, European Commission on the effect of nanoparticles on health and safety [94].

Further, since in many cases, coating is the method of application, longevity is an issue. The effectiveness after washing may be reduced. Therefore, more exploration of binders or fixers is needed to enhance the adhesion between the nanoparticles and the substrate. Sometimes, to alleviate this problem, the nanoparticles are embedded into the fibres during the fibre-spinning process itself. For instance, Khude et al. [95] explored silver-embedded polyester as antibacterial nanocomposites. Here, the silver nanoparticles are mixed with polyester during melt spinning itself. Such developments and approaches increase the efficiency of the nanoparticles and axiomatically, that of the smart textile. Nanoparticles are also generally expensive. However, with mass production and increasing demand, the price is expected to drop eventually. Finally, there is a need for interdisciplinarity, that is, teams from different expertise to work together to form a common platform with the interest and need of the user in mind. Prototypes developed may often be accepted to a lesser extent than anticipated as in the case of Eleksen smart fabric keyboards [93]. Therefore, extensive study of the market and acceptability is inevitable. Finally, there are no standards with respect to the use of nanotechnology for protective textiles, leading to opportunities to research and develop the same.

10.3 Conclusion

In conclusion, smart and functional protective textiles is a growing sector that has a huge potential for mass production and commercialization. The increased concern for safety and health, backed with governmental policies, is a driving factor that will propel the industry to new heights. As far as the specific categories of protective textile elaborated in this chapter are concerned, the functionality aspect is of utmost importance as failure or underperformance may cause serious damage of property and life. As advanced as the products already are, there is always scope for improvement either in terms of performance, design, weight, or cost. In this regard, exploitation of smart materials is a step towards in improving the efficiency of protective textiles. Further, research in various universities and laboratories has shown that nanotechnology has enabled the development of lighter weight alternatives, with even better performance in some cases. However, nanotechnology is yet to develop fully, and research directions in this area are of utmost importance. While working with nanoparticles, special attention should be given to their toxicity and effects on health as well as development of standards.

References

[1] Sarif Ullah Patwary MS. Smart textiles and nano-technology: A general overview. J Text Sci Eng 2015;05. https://doi.org/10.4172/2165-8064.1000181.

[2] Ham B. Engineers create a programmable fiber. MIT News 2021.

[3] Mondal S. Phase change materials for smart textiles- An overview. Appl Therm Eng 2008;28:1536–50. https://doi.org/10.1016/j.applthermaleng.2007.08.009.

[4] Grancarić AM, Jerković I, Koncar V, Cochrane C, Kelly FM, Soulat D, et al. Conductive polymers for smart textile applications. vol. 48. 2018. https://doi.org/10.1177/1528083717699368.

[5] Krajewski AS, Magniez K, Helmer RJN, Schrank V. Piezoelectric force response of novel 2d textile based pvdf sensors. IEEE Sens J 2013;13:4743–8. https://doi.org/10.1109/JSEN.2013.2274151.

[6] Hadimani RL, Bayramol DV, Sion N, Shah T, Qian L, Shi S, et al. Continuous production of piezoelectric PVDF fibre for e-textile applications. Smart Mater Struct 2013;22. https://doi.org/10. 1088/0964-1726/22/7/075017.

[7] Huang Y, Zhu M, Pei Z, Xue Q, Huang Y, Zhi C. A shape memory supercapacitor and its application in smart energy storage textiles. J Mater Chem A 2016;4:1290–7. https://doi.org/10.1039/c5ta09473a.

[8] Hu J, Lu J. Shape memory polymers in textiles. Adv Sci Technol 2013;80:30–8. https://doi.org/10. 4028/www.scientific.net/ast.80.30.

[9] du P, Song L, Xiong J, Wang L, li ni. A photovoltaic smart textile and a photocatalytic functional textile based on co-electrospun TiO$_2$/MgO core–sheath nanorods: Novel textiles of integrating energy and environmental science with textile research. Text Res J 2013;83:1690–702. https://doi. org/10.1177/0040517513490062.

[10] Kumar M. Flexible photovoltaic textiles for smart applications. In: Kosyachenko LA, editor. Sol. Cells – New Asp. Solut., 2011. https://doi.org/10.5772/19950.

[11] Bedeloglu A, Demir A, Bozkurt Y, Sariciftci NS. A photovoltaic fiber design for smart textiles Text Res J 2010;80:1065–74. https://doi.org/10.1177/0040517509352520.

[12] Rossi D De, Carpi F, Lorussi F, Mazzoldi A, Paradiso R, Scilingo EP, et al. Electroactive fabricsandwearble biomonitoring devices. AUTEX Res 2003;3:4–7.

[13] Khattab TA, Abdelrahman MS. From smart materials to chromic textiles. In: Shahid M, Adivarekar R, editors. Adv. Funct. Finish. Text., Springer Singapore; 2020, p. 257–74.

[14] Egres Jr. RG, Lee YS, Kirkwood JE, Kirkwood KM, Wetzel ED, Wagner NJ, et al. "Liquid armor": Protective fabrics utilizing shear thickening fluids. 4th Int. Conf. Saf. Prot. Fabr., 2004, p. 1–8.

[15] Lee YS, Wetzel ED, Wagner NJ. The ballistic impact characteristics of Kevlar woven fabrics impregnated with a colloidal shear thickening fluid. J Mater Sci 2003;38:2825–33. https://doi.org/10. 1023/A:1024424200221.

[16] Langley N. Multi-layered self-healing material system for impact mitigation 2019:2–4.

[17] Varley RJ, van der Zwaag S. Towards an understanding of thermally activated self-healing of an ionomer system during ballistic penetration. Acta Mater 2008;56:5737–50. https://doi.org/10.1016/j. actamat.2008.08.008.

[18] Chang EYW. Fashion styling and design aesthetics in spacesuit: An evolution review in 60 Years from 1960 to 2020. Acta Astronaut 2021;178:117–28. https://doi.org/10.1016/j.actaastro.2020.08.035.

[19] Guptaa D. Functional clothing- definition and classification. Indian J Fibre Text Res 2011;36:312–26.

[20] Wong YWH, Yuen CWM, Leung MYS, Ku SKA, Lam HLI. Selected applications of nanotechnology in textiles. Autex Res J 2006;6:1–8.

[21] Clare. The protective clothing market is estimated to grow from USD 8.8 billion in 2020 to USD 12.3 billion by 2025, at a CAGR of 6.92%. ReportLinker n.d. https://www.globenewswire.com/news-release/2020/07/24/2067349/0/en/The-protective-clothing-market-is-estimated-to-grow-from-USD-8-8-billion-in-2020-to-USD-12-3-billion-by-2025-at-a-CAGR-of-6-92.html (accessed October 8, 2021).

[22] Nagpal K. Protective textile (PROTECH) and equipment. Def ProAc Biz News n.d. https://defproac.in/?p=3583 (accessed December 8, 2021).

[23] Lakshmanan R, Nayyar M. Personal protective equipment in India: An INR 7,000 Cr industry in the making. Invest India n.d. https://www.investindia.gov.in/siru/personal-protective-equipment-india-INR-7000-cr-industry-in-the-making (accessed October 8, 2021).

[24] ATIRA. ATIRA Research and Development activities n.d. https://atira.in/research-development-activities/ (accessed December 8, 2021).

[25] Tomar S. DRDE develop standard for Nuclear, Biological & chemical war protective clothing. Hindustan Times 2021.

[26] NITRA. NITRA Centre of excellence for protective textiles n.d. https://www.nitratextile.org/coepro tech/web/content/index.html (accessed December 8, 2021).

[27] DRDO develops light weight bullet-proof jacket. Econ Times 2021.

[28] Nandi S. Union Cabinet approves Rs 10,683-crore PLI scheme for textile sector. Bus Stand 2021.

[29] Yadav R, Naebe M, Wang X, Kandasubramanian B. Body armour materials: from steel to contemporary biomimetic systems. RSC Adv 2016;6:115145–74. https://doi.org/10.1039/C6RA24016J.

[30] Mawkhlieng U, Majumdar A. Designing of hybrid soft body armour using high-performance unidirectional and woven fabrics impregnated with shear thickening fluid. Compos Struct 2020;253:112776. https://doi.org/10.1016/j.compstruct.2020.112776.

[31] Heinecke J. From fiber to armor. Law Enforc Technol 2007.

[32] Bajya M, Majumdar A, Butola BS, Arora S, Bhattacharjee D. Ballistic performance and failure modes of woven and unidirectional fabric based soft armour panels. Compos Struct 2021;255:112941. https://doi.org/10.1016/j.compstruct.2020.112941.

[33] Liquid armour to become a future choice for protecting soldiers. BAE Syst 2017. https://www.baesys tems.com/en-ca/article/liquid-armour-to-become-a-future-choice-for-protecting-soldiers (accessed December 8, 2021).

[34] Kozłowska J, Leonowicz M. Magnetorheological fluids as a prospective component of composite armours. Compos Theory Pract 2013;13:227–31.

[35] Leonowicz M, Kozłowska J, Wierzbicki Ł, Olszewska K, Zielińska D, Kucińska I, et al. Rheological fluids as a potential component of textile products. Fibres Text East Eur 2014;103:28–33.

[36] Next-generation bullet resistant panels: Magnetic liquid armor. Total Secur Solut 2012. https://www.tssbulletproof.com/blog/next-generation-bullet-resistant-panels-magnetic-liquid-armor/ (accessed December 8, 2021).

[37] Gürgen S, Kushan MC, Li W. Shear thickening fluids in protective applications: A review. Prog Polym Sci 2017;75:48–72. https://doi.org/10.1016/j.progpolymsci.2017.07.003.

[38] Zarei M, Aalaie J. Application of shear thickening fluids in material development. J Mater Res Technol 2020;9:10411–33. https://doi.org/10.1016/j.jmrt.2020.07.049.

[39] Mawkhlieng U, Majumdar A, Bhattacharjee D. Graphene reinforced multiphase shear thickening fluid for augmenting low velocity ballistic resistance. Fibers Polym 2021;22:213–21. https://doi.org/10.1007/s12221-021-0163-2.

[40] Wang Y, Zhu Y, Fu Y. Preparation and properties of HNT-SiO2 compounded shear thickening fluid. Nano Br Reports Rev 2014;9:1450100-1-1450100–7. https://doi.org/10.1142/S1793292014501008.

[41] Huang W, Wu Y, Qiu L, Dong C, Ding J, Li D. Tuning rheological performance of silica concentrated shear thickening fluid by using graphene oxide. Adv Condens Matter Phys 2015;2015:1–5. https://doi.org/10.1155/2015/734250.

[42] Gürgen S, Li W, Kushan MC. The rheology of shear thickening fluids with various ceramic particle additives. Mater Des 2016;104:312–9. https://doi.org/10.1016/j.matdes.2016.05.055.

[43] Ehlert GJ, Sodano HA. Zinc oxide nanowire interphase for enhanced interfacial strength in lightweight polymer fiber composites. ACS Appl Mater Interfaces 2009;1:1827–33. https://doi.org/10.1021/am900376t.

[44] Arora S, Majumdar A, Butola BS. Deciphering the structure-induced impact response of ZnO nanorod grafted UHMWPE woven fabrics. Thin-Walled Struct 2020;156:106991. https://doi.org/10.1016/j.tws.2020.106991.

[45] Kirby D. Super-light body vests to beat bullets. Manchester Evening News 2013.

[46] Gonzalez GM, MacQueen LA, Lind JU, Fitzgibbons SA, Chantre CO, Huggler I, et al. Production of synthetic, para-aramid and biopolymer nanofibers by immersion rotary jet-spinning. Macromol Mater Eng 2017;302:1–11. https://doi.org/10.1002/mame.201600365.

[47] Bajya M, Majumdar A, Butola BS, Verma SK, Bhattacharjee D. Design strategy for optimising weight and ballistic performance of soft body armour reinforced with shear thickening fluid. Compos Part B Eng 2020;183:107721. https://doi.org/10.1016/j.compositeb.2019.107721.

[48] Extended Climate Warfighter Clothing System. MilitaryCom n.d. https://www.military.com/equipment/extended-climate-warfighter-clothing-system-gen-iii (accessed December 8, 2021).

[49] Extended Cold Weather Clothing System. Off GEN III ECWCS n.d. https://docplayer.net/21292912-The-only-generation-iii-ecwcs-authorized-for-u-s-army-issue-800-948-9433-www-adstactical-com.html (accessed December 8, 2021).

[50] Extended climate warfighter clothing system. Military n.d. https://www.military.com/equipment/extended-climate-warfighter-clothing-system-gen-iii (accessed October 8, 2021).

[51] Polartec Power Dry. Polartec® n.d. https://www.polartec.com/fabrics/base/power-grid (accessed October 8, 2021).

[52] Polartec Thermal Pro. Polartec® n.d. https://www.polartec.com/fabrics/insulation/thermal-pro (accessed October 8, 2021).

[53] Morris W. Army develops cloth that could drastically improve cold-weather uniforms. Task Purp 2017.

[54] Shaid A, Bhuiyan MAR, Wang L. Aerogel incorporated flexible nonwoven fabric for thermal protective clothing. Fire Saf J 2021;125:103444. https://doi.org/10.1016/j.firesaf.2021.103444.

[55] What is Aerogel? Shiver Shield n.d. https://shivershield.com/aerogel-information/ (accessed December 8, 2021).

[56] Outlast®. Outlast® Technol n.d. https://www.outlast.com/en/technology/ (accessed October 8, 2021).

[57] Boopathi M, Singh B, Vijayaraghavan R. A review on NBC body protective clothing. Open Text J 2008;1:1–8. https://doi.org/10.2174/1876520300801010001.

[58] Turaga U, Singh V, Lalagiri M, Kiekens P, Ramkumar SS. Nanomaterials for defense applications. NATO Sci. Peace Secur. Ser. B Phys. Biophys., 2012, p. 197–218. https://doi.org/10.1007/978-94-007-0576-0_10.

[59] GENTEX®. Gentex Corp n.d. https://www.yumpu.com/en/document/read/27876173/cwu-66-p-gentex-corporation (accessed October 8, 2021).

[60] Demron ice multi use suit. Radiat Shield Technol n.d. https://cdn.thomasnet.com/ccp/30870223/301216.pdf (accessed October 8, 2021).

[61] NBC Suit Permeable Mk IV. Def Res Dev Organ n.d. https://www.drdo.gov.in/nbc-suit-permeable-mk-iv (accessed October 8, 2021).

[62] Nanofiber filters eliminate contaminants. NASA 2009:92–3. https://www.nasa.gov/pdf/413408main_Nanofiber.pdf (accessed October 8, 2021).

[63] Tuccitto N, Riela L, Zammataro A, Spitaleri L, Li-Destri G, Sfuncia G, et al. Functionalized carbon nanoparticle-based sensors for chemical warfare agents. ACS Appl Nano Mater 2020;3:8182–91. https://doi.org/10.1021/acsanm.0c01593.

[64] Bhuiyan MAR, Wang L, Shaid A, Shanks RA, Ding J. Advances and applications of chemical protective clothing system. J Ind Text 2019;49:97–138. https://doi.org/10.1177/1528083718779426.

[65] Šťastný M, Tolasz J, Štengl V, Henych J, Žižka D. Graphene oxide/MnO 2 nanocomposite as destructive adsorbent of nerve-agent simulants in aqueous media. Appl Surf Sci 2017;412:19–28. https://doi.org/10.1016/j.apsusc.2017.03.228.

[66] Dumé I. Nanomaterial-coated fabric destroys chemical warfare agents. Nanomaterials 2020. https://physicsworld.com/a/nanomaterial-coated-fabric-destroys-chemical-warfare-agents/ (accessed December 8, 2021).

[67] Yue G, Su S, Li N, Shuai M, Lai X, Astruc D, et al. Gold nanoparticles as sensors in the colorimetric and fluorescence detection of chemical warfare agents. Coord Chem Rev 2016;311:75–84. https://doi.org/10.1016/j.ccr.2015.11.009.

[68] Biological and chemical weapon decontamination by nanoparticles (nanotechnology). What-When-How n.d. http://what-when-how.com/nanoscience-and-nanotechnology/biological-and-chemical-weapon-decontamination-by-nanoparticlesnanotechnology/ (accessed December 8, 2021).

[69] Sadeghi M, Yekta S, Ghaedi H. Decontamination of chemical warfare sulfur mustard agent simulant by ZnO nanoparticles. Int Nano Lett 2016;6:161–71. https://doi.org/10.1007/s40089-016-0183-x.

[70] Sinha M, Das B, Prasad N, Kishore B, Kumar K. Exploration of nanofibrous coated webs for chemical and biological protection. Zast Mater 2018;59:189–98. https://doi.org/10.5937/zasmat1802189k.

[71] Alongi J, Horrocks AR, Carosio F, Malucelli G. Update on flame retardant textiles: State of the art. 2013.

[72] Weil ED, Levchik S V. Flame retardants in commercial use or development for textiles. J Fire Sci 2008;26:243–81. https://doi.org/10.1177/0734904108089485.

[73] Kaynak C, Gunduz HO, Isitman NA. Use of nanoclay as an environmentally friendly flame retardant synergist in polyamide-6. J Nanosci Nanotechnol 2010;10:7374–7. https://doi.org/10.1166/jnn.2010.2768.

[74] Wang Z, Han E, Ke W. Fire-resistant effect of nanoclay on intumescent nanocomposite coatings. J Appl Polym Sci 2006;103:1681–9. https://doi.org/10.1002/app.

[75] Chang ZH, Guo F, Chen JF, Yu JH, Wang GQ. Synergistic flame retardant effects of nano-kaolin and nano-HAO on LDPE/EPDM composites. Polym Degrad Stab 2007;92:1204–12. https://doi.org/10.1016/j.polymdegradstab.2007.04.001.

[76] Priyanka, Dahiya JB. Effect of nanoclay on thermal and flame retardant properties of phosphorylated epoxy nanocomposites. Int J Appl Chem 2017;13:515–23.

[77] Thirumal M, Khastgir D, Singha NK, Manjunath BS, Naik YP. Effect of a nanoclay on the mechanical, thermal and flame retardant properties of rigid polyurethane foam. J Macromol Sci Part A Pure Appl Chem 2009;46:704–12. https://doi.org/10.1080/10601320902939101.

[78] Joshi M, Bhattacharyya A. Nanotechnology – A new route to high-performance functional textiles. Text Prog 2011;43:155–233. https://doi.org/10.1080/00405167.2011.570027.

[79] Norouzi M, Zare Y, Kiany P. Nanoparticles as effective flame retardants for natural and synthetic textile polymers: Application, mechanism, and optimization. Polym Rev 2015;55:531–60. https://doi.org/10.1080/15583724.2014.980427.

[80] Qin H, Zhang S, Zhao C, Hu G, Yang M. Flame retardant mechanism of polymer/clay nanocomposites based on polypropylene. Polymer (Guildf) 2005;46:8386–95. https://doi.org/10.1016/j.polymer.2005.07.019.

[81] Araby S, Philips B, Meng Q, Ma J, Laoui T, Wang CH. Recent advances in carbon-based nanomaterials for flame retardant polymers and composites. Compos Part B Eng 2021;212:108675. https://doi.org/10.1016/j.compositesb.2021.108675.

[82] Sang B, Li Z wei, Li X hong, Yu L gui, Zhang Z jun. Graphene-based flame retardants: a review. J Mater Sci 2016;51:8271–95. https://doi.org/10.1007/s10853-016-0124-0.

[83] Kashiwagi T, Du F, Douglas JF, Winey KI, Harris RH, Shields JR. Nanoparticle networks reduce the flammability of polymer nanocomposites. Nat Mater 2005;4:928–33. https://doi.org/10.1038/nmat1502.

[84] Kim YS, Davis R. Multi-walled carbon nanotube layer-by-layer coatings with a trilayer structure to reduce foam flammability. Thin Solid Films 2014;550:184–9. https://doi.org/10.1016/j.tsf.2013.10.167.

[85] Janas D, Rdest M, Koziol KKK. Flame-retardant carbon nanotube films. Appl Surf Sci 2017;411:177–81. https://doi.org/10.1016/j.apsusc.2017.03.144.

[86] Idumah CI. Emerging advancements in flame retardancy of polypropylene nanocomposites. J Thermoplast Compos Mater 2020:089270572093078. https://doi.org/10.1177/0892705720930782.

[87] Kausar A, Rafique I, Muhammad B. Significance of Carbon Nanotube in Flame-Retardant Polymer/ CNT Composite: A Review. Polym – Plast Technol Eng 2017;56:470–87. https://doi.org/10.1080/ 03602559.2016.1233267.

[88] Idumah CI, Hassan A, Affam AC. A review of recent developments in flammability of polymer nanocomposites. Rev Chem Eng 2015;31:149–77. https://doi.org/10.1515/revce-2014-0038.

[89] Arao Y. Flame Retardancy of Polymer Nanocomposite 2015:15–44. https://doi.org/10.1007/978-3- 319-03467-6_2.

[90] Pan Y, Guo Z, Ran S, Fang Z. Influence of fullerenes on the thermal and flame-retardant properties of polymeric materials. J Appl Polym Sci 2020;137:1–18. https://doi.org/10.1002/app.47538.

[91] Vasiljević J, Jerman I, Simončič B. Graphitic carbon nitride as a new sustainable photocatalyst for textile functionalization. Polymers (Basel) 2021;13. https://doi.org/10.3390/polym13152568.

[92] Liu H, Wu W, He S, Song Q, Li W, Zhang J, et al. Activated carbon spheres (ACS)@SnO 2 @NiO with a 3D nanospherical structure and its synergistic effect with AHP on improving the flame retardancy of epoxy resin. Polym Adv Technol 2019;30:951–62. https://doi.org/10.1002/pat.4529.

[93] Cherenack K, Pieterson van L. Smart textiles: Challenegs and opportunities. J Appl Phys 2012;112:091301.

[94] Scientific Committee on Emerging andn Newly Identified Health Risks (SCENIHR). 2006.

[95] Khude P, Majumdar A, Butola BS. Leveraging the antibacterial properties of knitted fabrics by admixture of polyester-silver nanocomposite fibres. Fibers Polym 2018;19:1403–10. https://doi.org/ 10.1007/s12221-018-7889-5.

Sudipta Mondal, Bapan Adak, Samrat Mukhopadhyay

11 Functional and smart textiles for military and defence applications

Abstract: Nowadays, the textiles which are used in military applications have become more functional and capable due to constant technical innovations in this field. Military smart textiles not only improve the performance of the soldiers but also provide more mobility and fast connectivity. Moreover, with the advancement in the field of nanotechnology, wearable sensors/actuators, wireless networking, personal thermal regulation, energy storage, and harvesting a huge revolution has been observed in the developments of military and defence textiles with various smart features. This chapter deals with the advancements in the field of textiles for military and defence applications. The first section discusses about smart body amours for armed forces, their materials, and technologies. The second section covers smart military garments such as uniform with programmable fibres, microwave/radar absorptive textiles, camouflages, multi-spectral camouflage net, and stealth-coated textiles. The third section discusses the different functional military textiles such as extended cold weather clothing system clothing, military sleeping bags, high-altitude pulmonary oedema chambers, military parachutes, and paragliding. The fourth section covers the applications of sensors in various defence applications, application of aerostat/airships in military surveillance, and other intelligent textiles for military and defence applications. The final section highlights the current changes and future perspectives of functional and smart textiles for military and defence applications.

Keywords: Camouflage, sleeping bag, stealth, ECWCS, parachute, military shelters, sensor, smart body amours

11.1 Introduction

Textiles used in military and defence application are very different from the textiles in general usage. These textiles' main purpose is to shield the user from environmental hazards and offer safety features specific to the application field they are used in. They also aim to make the wearer comfortable. All military textiles serve several functions. While certain fabrics are made to ward off the cold, others are made to ward off projectiles (such as bullets and sharp projectiles from bomb explosions), while yet others are made for fire protection. The applications of textiles in the military sector are unlimited as long as comfort factors remain intact [1].

Advancements in textile sciences and engineering have led access to new material, manufacturing technologies resulting to have significant improvements on to textile

https://doi.org/10.1515/9783110759747-011

products. All the when we use the term military textiles, it does not signify only clothing items used by military personnel. Textiles have vivid application in this sector: clothing articles, shelters, accessories, and so on are being manufactured using textiles. Textiles may offer both products and solutions if the demands of the military and defence are properly understood.

Recent advances in chemical technology have allowed us to introduce superior textile auxiliaries that can enhance textile performance. Nanotechnology has made it simpler than ever before to incorporate several functional property characteristics onto a single textile product. Various nanomaterials of different forms (nanoparticles, nanorods, nanocrystals, nanoplatelets, nanofibres, etc.) and having different chemical composition (carbon-based, metal-based, polymer-based, etc.), nanocomposites, nanogel, nanocolloid, nanofoam, and so on, as well as different nanotechnology-based advanced processes (in situ polymerization, sol–gel processing, plasma treatment, nanocoating, etc.) are gaining importance in making functional and smart textiles for protective, military, and defence applications. Especially, different electronic component (flexible sensors, actuators, batteries, solar panels, etc.) incorporated smart wearable electronics, or e-textiles have huge value nowadays in various applications related to defence and security, communications, health monitoring of military personals, energy harvesting, and so on [1, 2].

In military some of the most commonly used textiles are general uniforms, combat suits, cold protection suits, camouflage fabrics, bullet proof jackets, nuclear biological and chemical (NBC) protective suits and NBC shelters, rucksacks, modular lightweight load-carrying equipment (MOLLE), integrated combat kits, rain ponchos, sleeping bags, extended cold weather clothing system (ECWCS), anti-stealth tents and nets, parachute, and many more. With cutting-edge manufacturing technology, products perform effectively and efficiently. While some new products are introduced for military applications, scientists and engineers continue to work on improving existing products – improving performance, making products lighter while maintaining the same or better performance, and extending the life of these products.

11.2 Nanotechnology-empowered multi-functional and smart body armour

Body armours are used to provide safety against mechanical hazards, such as stab by knife or sharp objects, high speed projectile (bullet), and sharp particle from grenade explosion. Although, nowadays a notable research is going on in the field of ballistic armour and other penetration resistance devices and accessories for military and other safety purposes, the current body armour still depends on a stiff and relatively heavy layer of ceramic materials for absorbing ballistic impact. As a result, the body armour becomes heavy and unhandy. Military nanotechnology research has recently received a lot of attention because it can help body armours perform better while weighing less. In addition to the use of different high performance fibre [Kevlar, Twaron®, ultra-high-molecular-weight polyurethane (UHMWPE), polybenzoxazine, polyhydroquinone–diimidazopyridine (M5), etc.]-based textiles, various nanotechnology-based approaches have been explored by textile and material science researchers for making nanoenhanced ballistic textiles with improved performance [3], few of which has been summarized:

i) Treatment of high-performance fibre-based fabrics with shear thickening fluids (STF) consists of liquid polymers containing SiO_2 nanoparticles that harden on ballistic impact [4, 5]. Using STF technology, fluid becomes stiff in milliseconds when mechanical forces are applied, facilitating the movement of body armour with pockets containing liquid bags that react to mechanical stresses [6].

ii) Nanoclay and nanocalcite-filled epoxy matrix and Kevlar-based composite structures improve the impact resistance significantly [7].

iii) Iron nanoparticles in innocuous oil harden when electrically triggered (magnetorheological fluid). It hardens as iron particles are brought closer together [8].

iv) An epoxy or plastic matrix containing spherical nanoparticles of silicon dioxide or titanium dioxide offers improved ballistic resistance while also greatly improving flexibility [9].

A primary focus of current research on military textiles is the improvement of ballistic protection (which is the most pressing need) and the development of new designs that incorporate sensors and embedded sensors into backpacks, clothing, or tents for additional functionality. Acellent Technologies Inc., a US-based company, created a SmartArmor system for monitoring the health of body armour structures in service in collaboration with the US Army [10]. A SMART layer and software are integrated into system to monitor personal body armour in combat zones to increase soldier safety.

In a very recent study, Zhou et al. [11] reported about a Kevlar and epoxy-based triboelectric nanogenerator (KE–TENG) as wearable protective clothing, as schematically represented in Figure 11.1. This intelligent body armour can use impact signals as an alarm when they are transmitted remotely via Bluetooth modules. By using share stiffening gel (SSG), KE–TENG dissipated 72.4% of the force under the impact of a drop hammer. The epoxy coating also enhances the robustness of KE–TENG under impacts, in addition to improving its triboelectric properties. This study also shows that a fibre-shaped supercapacitor (F-SC)'s energy storage capacity is unaffected by the angle of bending or the number of charging and discharging cycles it experiences. Furthermore, it is compatible with KE–TENG via rectifiers, enabling the integration of energy storage and collection systems.

In another very recent study Fan et al. [12] developed a body armour having lightweight, flexible, and able to monitor body temperature as well as be non-flammable in addition to being resistant to stabs and impacts (Figure 11.2). A multi-functional soft body armour composite with three coaxial layers was developed with Kevlar fabric as a

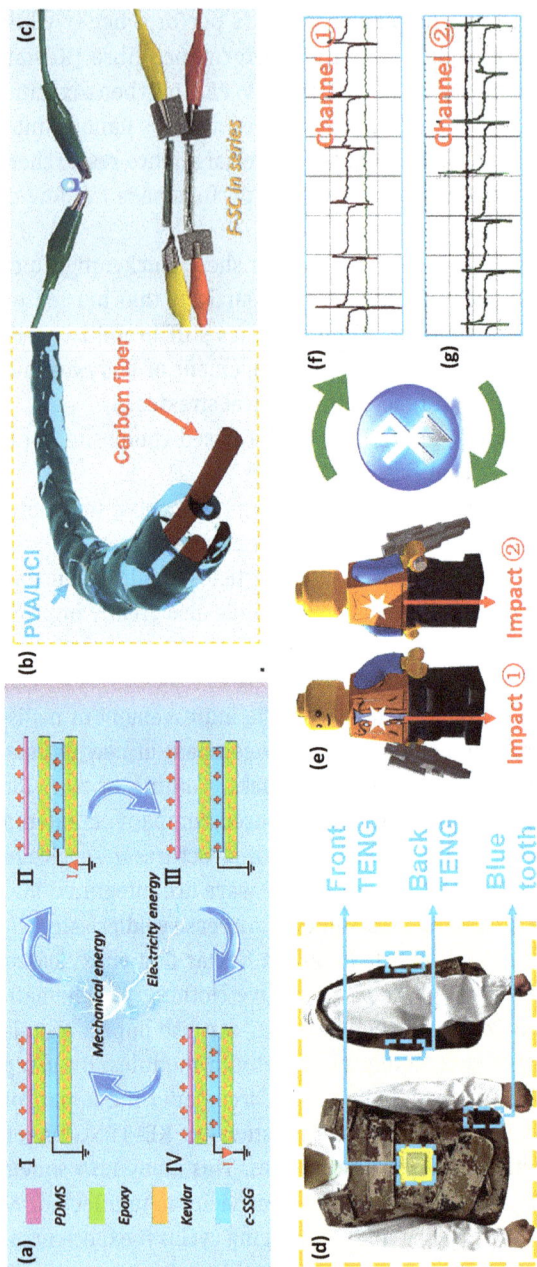

Figure 11.1: (a) Working mechanism of KE–TENG; (b) schematic of fibre shape super capacitor (F-SC); (c) LED lighted by two F-SCs in series; (d) schematic of KE–Kevlar-based body armour; (e) front/back KE–TENG was impacted, signal could be transmitted to (f) Channel 1 and (g) Channel 2 by Bluetooth (reproduced with kind permission from [11]. Copyright 2022, Elsevier).

core protection layer, reduced graphene oxide (RGO), and nano-SiO_2/shear-thickening gel (STG) as functional layers. With a low work voltage of 5 V, the SiO_2/STG/RGO@Kevlar composite fabric (SSG@Kevlar) performs excellently electro-heating, has exceptional strain sensing capability (gauge factor over 41, outstanding linearity ($R^2 > 0.99$), high strain resolution (0.05%), fast response (69 ms)) and a high impact energy sensor. SSG@Kevlar fabric composite has also been shown to have improved knife-stab resistance (a 12 mm reduction in penetration depth) and anti-impact performance (a 40% reduction in peak force) as a result of the introduction of the SiO_2/STG layer. In addition to shear thickening, SiO_2 nanoparticles enhanced yarn interaction forces. SSG@Kevlar fabric composites were also fire-resistant, with no open flame and a glowing appearance under continuous ignition owing to the dense char they produce. Consequently, the developed SSG@Kevlar fabric composite has the potential to provide multi-functional soft body armours that provide protection in a number of different ways to wearers [12]. More details of nanoenhanced smart textiles as soft body armour has been covered in detail in Chapter 10.

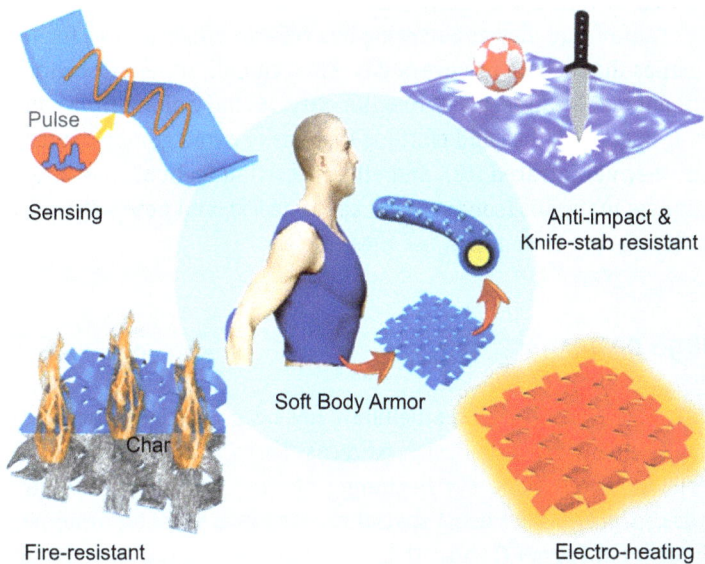

Figure 11.2: Schematics of a multi-functional soft body armour (Modified according to [12]. Copyright 2022, Elsevier).

11.3 Radar/microwave absorbent textiles: clothing for defence personnel

Textiles used in clothing, especially for military personnel, require multi-functional properties. These textiles require to cover a wide array of property parameters to provide protection against several threats, for example, from protection against cold weather, harsh sunlight, protection against thrown projectiles (bullets, bomb fragments, etc.), protection from radiation, and protection against enemy detection technologies.

The nature of their attire can be changed to serve different purposes. The possibilities are endless, thanks to developments in textile technology. On the other hand, as engineering has advanced, so has modern combat technology. Applications of new technologies have been made in the military. Enemy detecting technologies have advanced significantly, necessitating increased anti-detection or concealing efforts against many detection methods simultaneously. Therefore, modern textile components had to be upgraded to tackle such problems and provide solutions against such threats.

The utilization of electromagnetic waves in defence, communication via radar, and advancements in radio frequency engineering has created opportunities for improvement in the textiles that people use every day. Modern military textiles widely differ from conventional textiles in terms of functionality. Military of multiple counties have started using special clothes and clothing systems to avoid getting detected in microwave, radar, thermal infrared (IR), and other detection technologies. Use of camouflage patterns, new materials (conductive textile fabrics), and new technology (anti-thermal and coolants) are in common use these days.

11.3.1 Camouflage patterns

Use of visual camouflage patterns onto a textile fabric has been practiced since decades. Every nation has their own unique design patterns. Through digitization evolution of print patterns occurred since the beginning of early nineteenth century. Military clothing fabrics were printed using special printing inks to incorporate new age materials on the printed fabrics (Figure 11.3). To achieve a reduced IR signature that matches the area where the military operates, iron oxide-based colourants and pigments have been used. Metal oxides show excellent magnetic properties, making them excellent microwave absorbers. The addition of these materials also provides microwave absorbency. Use of carbon particle inhibits the same property as metal oxides.

Figure 11.3: Camouflage patterns of four NATO members during the 1990s and early 2000s: (a) DPM (United Kingdom); (b) Flecktarnmuster (Germany); (c) Centre Europe (France); (d) M81 Woodland (USA) [13]; and (e) US Army's Operation Enduring Freedom (OEF) Camouflage Pattern (OCP) is Crye Precision's MultiCam pattern, which is used in clothing and equipment for troops preparing for Afghanistan (US Army Program Executive Office (PEO) – Soldier, Fort Belvoir, Virginia, USA) (reprinted with kind permission from [14]. Copyright 2012, Elsevier).

11.3.2 Advanced textile materials for microwave absorption

Although conventional textiles materials show good amount of absorption of micro-waves, new developments in material science have open possibilities for the use of metal-based textile fibres and yarns for fabrication of non-woven, knitted, and woven fabrics.

The high electrical conductivity of metallic yarns also leads to excellent electro-magnetic shielding characteristics. Yarn composed of 73% polyester fibre and 27% stainless fibre can provide up to 95–98% shielding against electromagnetic waves [15]. Aluminium-coated filament yarn or ribbons also demonstrate good electromagnetic shielding; these metalloplastic yarns, also known as metallic Zaris, are utilized widely because they are more technologically and commercially feasible than pure metallic yarn or bi-component metal-synthetic yarns. Utilizing conductive yarn will produce fabrics with conductivity and prevent electrostatic discharge, which can naturally ab-sorb microwaves emitted onto the fabric surface.

For broadband frequency absorption, carbon-based composites exhibit excellent microwave absorption property. For instance, graphene-based composites also exhibit excellent microwave absorption in X-band and Ku-band frequencies as reported by

Song et al. [16]. A specially designed silicas textile and freeze-drying are used to fabricate reduced GO in situ using a 3D conductive framework (Figure 11.4). These integrated bi-matrices are capable of enhancing the mechanical properties and thermal stability of composites by coexisting with silica textiles and thermoset polymers. The as-fabricated lightweight composites (~1 g/cm^3) achieved excellent thermal stability beyond 225 °C (95% mass retention in air) and high tensile strength (40 MPa). They also exhibited effective absorption in the full X-band at a low filler loading (4.1 wt% RGO) and a microwave absorption peak at 36 dB. The results suggest that such thermostable polymeric graphene/silica textile composites can be used for practical microwave absorption applications, given the excellent overall performance [16]. Wearable electro conductive cotton fabric coated with RGO in combination with zinc oxide also exhibits excellent microwave absorption in X-band frequencies [17].

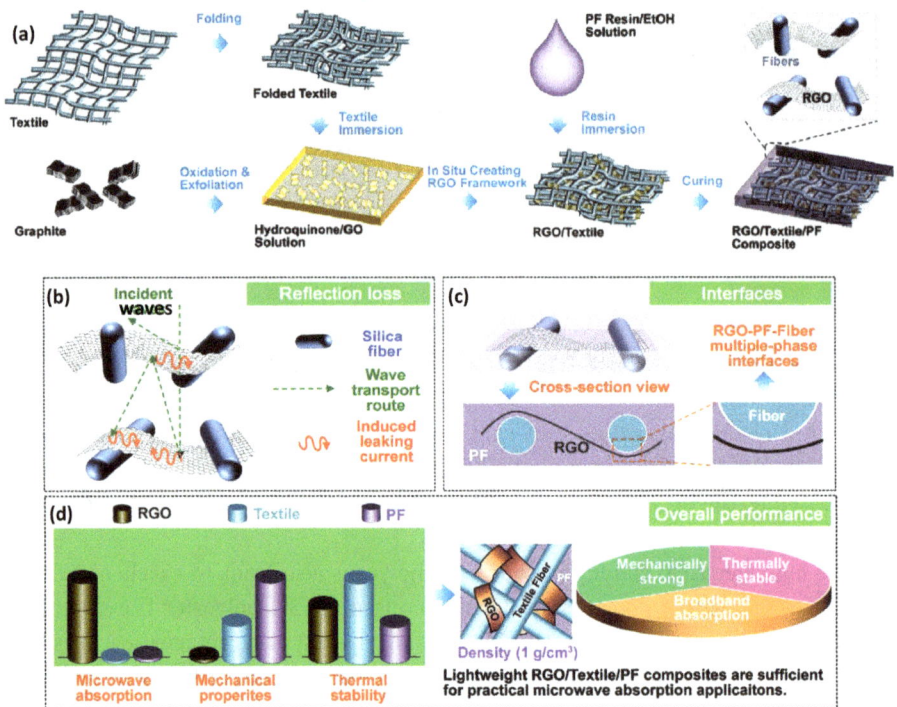

Figure 11.4: Scheme of the procedures for RGO/silica textile/PF composites; schemes of (b) reflection loss, (c) interfaces, and (d) overall performance of the RGO/silica textile/PF composite showing potential for practical microwave absorption applications (reprinted with kind permission from [16]. Copyright 2016, Elsevier).

11.3.3 Special coating for microwave absorption

Conventional textiles sometimes being coated with several substrates, for example, polyurethanes, polyvinyl chlorides, acrylic, and silicone to enhance their required functional property parameters. In solution-based coating system, multiple additives can be mixed in different ratios. Such special recipe formulation can increase the targeted functional property to several folds. UV absorbers, heat absorbers, heat reflectors, conductive metal particles, light blocking particles, and so on can be used to attain desired properties.

Polyurethane mixed with metal-oxide pigment provides excellent absorbency to micro and radio waves to the textile substrate. Whereas polyurethane–aluminium particle-coated fabrics provides electromagnetic shielding and can also reflect higher percentages of radar waves; coating substrate mixed with nano carbon particle also exhibits high degree of microwave absorbency. Most functional textiles fabrics used for military clothing purposes have coated inner surface. With suitable addition in microwave absorbent filler materials military apparel fabrics can be modified to a microwave absorbent one.

11.3.4 Applications of microwave or radar absorbing textiles

In military apparel fabrics, microwave absorbency plays a crucial role. Fabrics made with microwave absorbent materials can avoid getting detected by the equipment's transmitting electromagnetic waves. Fabric incorporated with smaller radar cross section materials cannot be easily identified by radar waves. These clothes absorb certain percentage of EM waves and scatter certain percentages, and reflectance of radar waves can be adjusted by the quantity of absorbers present on the fabric itself.

Further advancements in stealth technologies opted by military, the use of radar absorbing materials, radar absorbing structures have drastically increased. Use of IRR signature control has also advanced the clothing systems used by military personnel.

11.4 Colorants and nanomaterial treated textiles for adaptive camouflage application

An active or adaptive camouflage involves the use of panels or coatings that can alter the appearance, colour, brightness, and reflection of an object to blend in with its surroundings. This means that active camouflage is completely invisible to the naked eye, a technology that makes this possible is organic light-emitting diodes. It would not become invisible with the aid of a camera but would appear as if it were part of its surrounding environment.

Radiation falling on an experimental metamaterial with negative microwave refraction index curves the radiation around the object rather than being reflected or refracted. An optical camouflage technique called "phased array optics" involves projecting a 3D hologram of a background in front of an object so that it becomes invisible.

As modern detection technologies advance, multi-spectral camouflage becomes more and more necessary. A number of IR camouflage materials have been developed previously, including metamaterials and photonic crystals, but their multi-spectral compatibility, cost-effectiveness, and ease of fabrication remained challenging. Ding et al. [18] reported about an oxalate-rich porous alumina (OPA) nanostructured composite film for simultaneous thermal and visible-to-IR camouflage. The nanostructured composite film is made of three layers: a visible-transparent OPA layer, an OPA/metal oxide-based composite layer, and an aluminium substrate (Figure 11.5(a, b)). Every functional layer performed a desired reflection/emission property for both IR and visible camouflage. Infrared camouflage was achieved by low emission (high reflection) of the metal substrate in both IR-detected bands (3–5 and 8–14 μm). Moreover, the intrinsic absorption of oxalate in the undetected band (5–8 μm) increased the surface heat dissipation due to radiative cooling. Additionally, metal oxides in the composite layer can be used to tune the colour of the camouflage to match the background, such as green for forests and brown for deserts. This study presented a simple strategy for modulating multi-spectral absorption/emission properties with great flexibility and therefore has great potential for energy conversion and stealth technology.

In a study reported by Bhattacharya and co-workers [19], different thermochromic colorants have been used to develop colour-changing (chameleon-type) printing on cotton fabrics for defence applications. The use of blue and orange thermochromic colorants combined with turmeric (a natural dye) and graphite has resulted in the development of light green, dark green, black, brown, and sandal-coloured coatings. The chameleon-like camouflage printing technique described in this article is very promising because it allows the same fabric to be used for camouflaging at different terrains, thus speeding up troop movement.

Nanotechnology can be used to achieve electrochromic camouflage, which allows soldiers to disappear without being seen. In order to blend with the environment, material made from electrochromic camouflage can change colour instantly [20]. However, a number of major problems at the device level have prevented recent advancements in active electrochromic camouflage from being applied to practical applications. This technology's complex structure and electrode materials make it challenging for it to be useful and achieve its objective of chameleon-like camouflage. In a study reported by Yu et al. [21] an electrochromic camouflage fabric was constructed by using electrodes with a lateral configuration and inexpensive and readily available ITO film as electrodes, using an all-solution process. A liquid precursor (consisting of ITO nanosticks, polyvinyl alcohol, and isopropanol) was sprayed through a mask to lay down strips of commercial ITO film on a synthetic fabric. Laterally configured polymer-based electrochromic fabric devices (PECFDs) were then produced by sequentially adding electrochromic polymer layers and

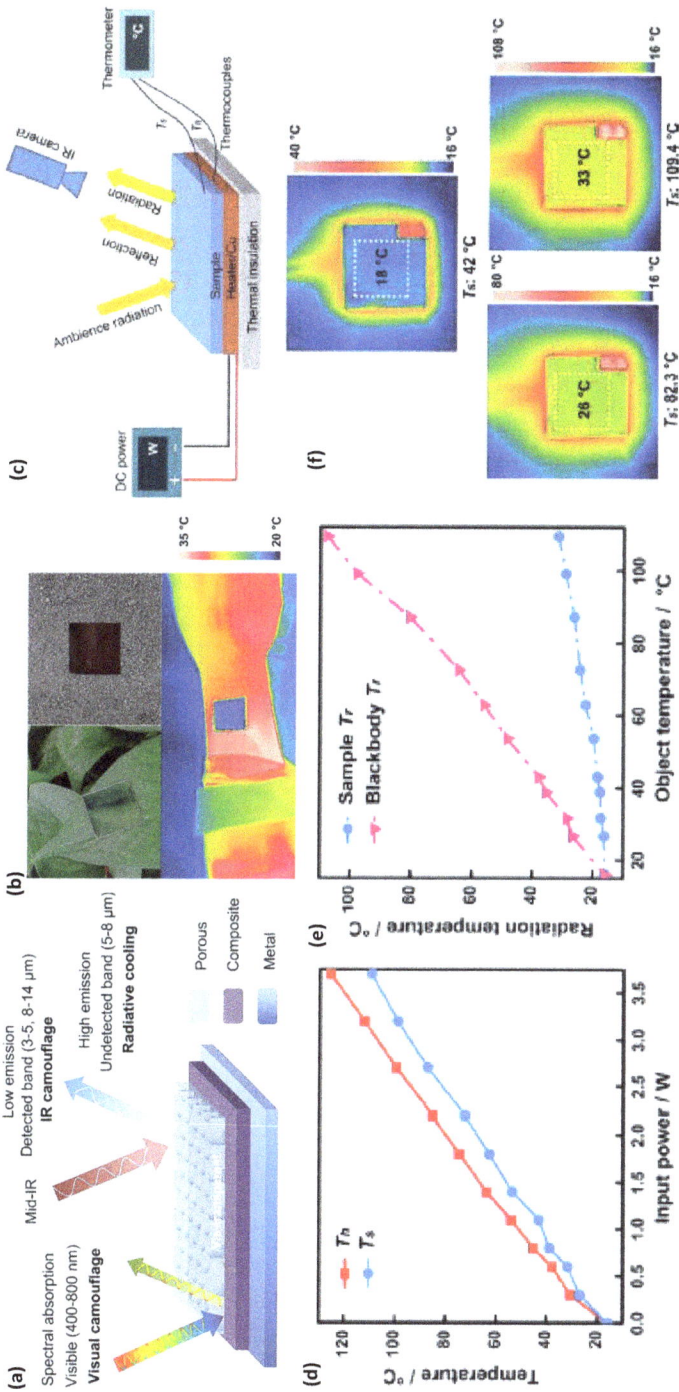

Figure 11.5: Camouflage with multi-band compatibility and performance demonstrations: (a) specifications for multi-band camouflage for respective ranges and different layers; (b) camouflage illustrated in both visible and infrared bands; (c) schematic of the experimental setup for IR camouflage demonstration; (d) the heater/surface temperatures (Th/Ts) were measured for the OPA/Cu sample at different input power levels; (e) temperatures of radiation (Tr) for tape and OPA/Cu (blackbodies) at their real surface temperatures; (f) demonstration of MIR camouflage on the OPA/Cu sample at different surface temperatures (reprinted with kind permission from [18]. Copyright 2022, American Chemical Society).

solid electrolytes onto the surfaces of ITO films and fabrics (Figure 11.6). These PECFDs can change from vegetable-green to soil-brown in colour and are also lightweight, thin, highly flexible, and stable. Using this strategy, transparent electrochromic devices can be made on plastic surfaces and human skin to generate "chameleon camouflage."

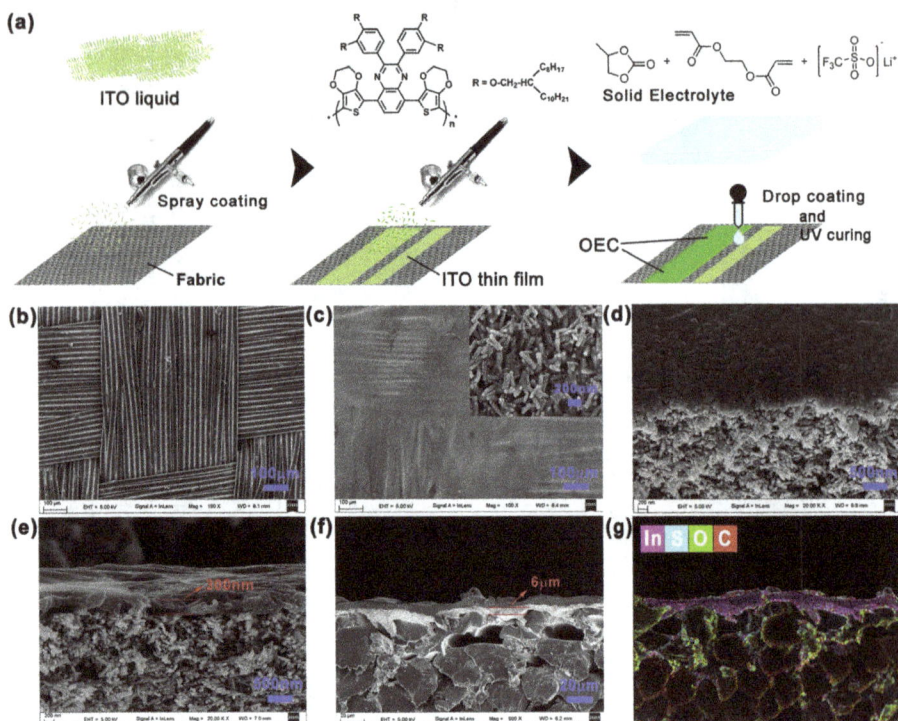

Figure 11.6: (a) The process of PECFD fabrication. Top views of (b) the synthetic fabric; (c) fabric/ITO (inset is an enlarged image); (d, e) cross-sectional views of ITO/EC-polymer at a high magnification; (f) fabric/ITO/EC-polymer at a low magnification; and (g) cross-sectional EDX spectrum of fabric/ITO/EC-polymer (reprinted with kind permission from [21]. Copyright 2019, Elsevier).

11.5 Uniforms with programmable fibres

In the near future, "programmable fibres" may redefine the term "digital camouflage." U.S. Army forces in collaboration with the researchers from Institute for Soldier Nanotechnologies of Massachusetts Institute of Technology (MIT) have developed smart fibres with digital capabilities. When woven into a garment, these fibres can sense/record, store, analyse, and transmit data [22]. The uniforms can both protect as well as perform computations by incorporating microchips into their fibres (Figure 11.7).

It may result in a uniform which generates power, provide physiological data, provide their location to their team, and even alert others if the wearer becomes injured [23].

In order to create a digital electronic containing polymer fibre, Loke et al. [24] placed hundreds of square silicon digital microchips and memory devices (memory density of $\sim 7.6 \times 10^5$ bits/m) in a preform (Figure 11.7). The fibre was drawn to create a continuous electrical connection between the chips over the entire length (10 m) of the fibre. The thin and flexible properties of the fibre enable it to pass through needles, be sewn into fabrics, and wash up to ten times without deterioration. The digital fibre's trained neural network, which is kept inside of it, enables it to predict wearer behaviour with 96% accuracy in real time using data gathered over many days. Digital devices that are capable of not only storing and measuring physiological parameters within a fibre strand but also incorporating neural networks to infer sensory data present intriguing possibilities for fabrics which can sense, memorize, learn, and infer situational context within a fibre strand. These digital fibres can be used to monitor performance, make medical inferences, and detect early diseases in the human body [24].

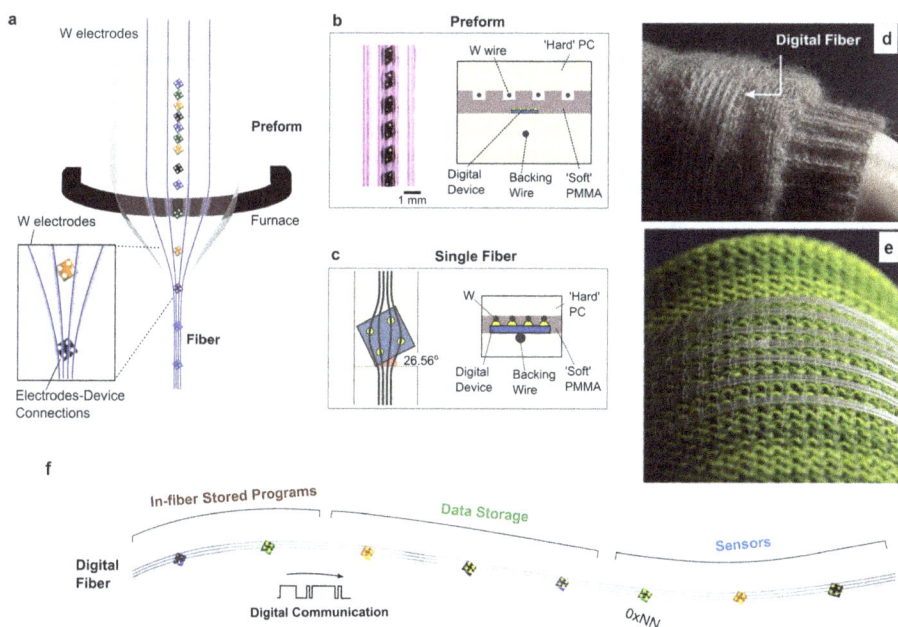

Figure 11.7: (a) Thermal drawing of the programmable fibre developed by MIT and US Army; (b) <left>: an optical micrograph of the rotated devices embedded within the preforms, <right>: diagram of a three-layer preform (hard PC/soft PMMA/hard PC) with cross sections; integration of the digital fibre (d) in the sleeve of a sweater and (e) in a cotton-based fabric; (e) illustrations of various chips encapsulated within thermally drawn fibres with different functions such as sensors and memory (reprinted with kind permission from [24]. Copyright 2021, Springer Nature).

11.6 Stealth-coated textiles

In the fields of military and defence applications textiles cover a multitude smart and functional properties to provide better protection against vivid threats. Stealth is one of the most vital properties for the application in military and defence domain.

Attaining stealth can be a bit tricky, but it is a smart move to stay one step ahead of the enemy. The use of stealth and counter stealth on textiles can be attained by several smart application methodologies. Generally, the conventional textile materials or polymers do not have inherent stealth property. However, there are various ways to incorporate smart stealth property by integrating special features in textiles and polymer-based materials.

Smart coating is the most practical method of achieving stealth properties in textiles. However, there are possibility for incorporation of special materials in yarns to induce stealth in the yarn stage as well as by means of functional finishing. Among different methods for preparation of smart stealth materials, coating provides more sustainable and efficient results with comparatively less effort.

11.6.1 What is stealth and how is it important?

In a broader sense, stealth refers to concealment, that is, hiding the presence or making difficulty in detection. Applications of stealth in military and defence domain cover both offensive and defensive approaches depending upon the requirements. Stealth and counter stealth (counter measures for detection such as radar detection, sonar detection, and thermal infrared (TIR) by enemy armed forces) both technologies are crucial parts for military/defence applications. Stealth required in such application is multi-spectral, which covers protection against several types against detection technologies to make the object almost invisible. Stealth is obtained not only by a single technology but also using a combination of such technologies.

A serious research and development work was started on radar in late 1930s for advanced detection/tracking. Considering that military and defence are under major threat, stealth technologies have also been developed to avoid enemy detection of military vehicles (such as aircraft and trucks). In the year 1989, two aircrafts were equipped with stealth technology in Panama, which were later used in Iraq in the year 1991 for combat purpose [25].

The primary requirement of stealth-coated textiles for such applications is:

i) Multi-spectral shielding from energy signature (e.g. radar, thermal IR, and visual), creating difficulties in detection technologies;

ii) Controlled emission of energy signatures (e.g. thermal and electromagnetic) by incorporating smart materials to reduce emission of thermal energy and by controlling the conductivity.

11.6.2 Working mechanisms stealth-coated textiles

The stealth-coated fabrics should function well in the UV, IR, and visible spectrums while simultaneously blocking radar and the entire electromagnetic spectrum. By doing this, the textile system becomes multi-spectral and offers defence against a variety of various detecting technologies utilized in defence and military applications.

11.6.2.1 Detection technology of radar systems

Radio detection and ranging (Radar) system uses radio or microwaves that emit in the form of transverse electromagnetic waves from a distant source (transmitter) and after hitting the target energy reflects in all directions, while certain portion of that energy is directed to the source antenna (receiver) bringing the location of the target [26]. In a fraction of second, thousands of wave energy are emitted. The target location is measured by the time interval between the transmission and reception of the pulse [25]. While assessing the efficacy of radar protection, textiles are tested in J-band (5.85–8.4 GHz), X-band (8.4–12.4 GHz), and Ku-band (12.4–18 GHz) to study the transmission loss through the textiles.

11.6.2.2 Radar absorbing material coating

Special class of polymer-based material know as RAMs can be applied onto the textile surface, which reduces the radar cross section and therefore makes it much more difficult to detect the object. The material itself absorbs the electromagnetic wave energy and reflects no or very less signal for the receiver to detect the target. One of the most common RAM applied on military aircraft consists of specially developed polymer matrix where high dielectric constant type ferromagnetic nanoparticles [barium ferrite ($BaFe_{12}O_{19}$), strontium ferrite ($SrFe_{12}O_{19}$), epsilon phase iron oxide (ε-Fe_2O_3), substituted epsilon phase iron oxide (ε-$Ga_x Fe_2$-xO_3, ε-$Al_x Fe_2$-xO_3, etc.] are grafted. After absorbing the radar waves, the particles begin to vibrate at the molecular level, producing heat that is later expelled. Another type of RAM is also used in military where ferrite and carbon particles grafted on neoprene sheet. Nanosized ferromagnetic particles, which are superparamagnetic in nature, and strongly polarized by electromagnetic radiation, are used to create a radar absorbent coat, which corrugates the coated surface, reducing or eliminating radar wave energy significantly. Such materials are used for coating on military vehicles, aircrafts, tents, and nets. However, for different types of energy bands of radar, the RAMs should be selected carefully as any single RAM cannot absorb every single energy band transmitted by radar system [27].

11.6.2.3 Working mechanism of thermal infrared stealth

Every material emits radiation in atomic level in a continuous manner in the form of electromagnetic wave at an IR wavelength which is corresponding to its temperature. Infrared stealth is widely used in military applications such as fighter planes, tanks, military personal vehicles, military tents, and nets to make the user invisible to IR-detection equipment. By inclusion of novel materials onto the textile substrate the release of thermal energy to the atmosphere (covering 3–5 and 8–14 µm waveband windows) can be suppressed. These two wavebands represent the main transmission channels for electromagnetic waves in the atmosphere. The best way to tailor the IR stealth is to reduce the emissivity and hence, wavelength selective thermal radiation is needed from IR stealth material.

Primary requirements for an IR stealth material are as follows:
i)   a low emissivity at certain atmospheric windows and
ii) high broadband emissivity outside atmospheric window.

The electromagnetic waves emitted outside the transparency windows cannot be detected due to attenuation and absorption phenomena in the atmosphere [28].

11.6.2.4 Materials used in thermal IR stealth

Energy signature reduced textile fabrics are widely used in military tents, multi-spectral camouflage net (MSCN) nets, MSCN uniforms, and other military products. The reflectance signature is controlled with the help of different material via coating by employing several components in the coating solution. Fabric coated with organic binders of different coating thicknesses can significantly improve the performance of the final product. Different coating thickness has different absorption effects in the IR band. Most of the organic binders do not show strong absorption in the near infrared (NIR) band but due to vibration of functional group they show strong absorption in the thermal IR region. Mixing of metal fillers (copper, aluminium, bronze, etc.) in coating solution results in lower absorption and higher reflection due to their close-packed structure. Metal oxide-based colour pigments (chrome oxide, titanium dioxide, iron oxides, etc.) with compatible binder can also provide the potential of controlling the IR signature [29].

11.6.2.5 Working mechanism of visual/NIR stealth

Visual stealth commonly termed as camouflage is simply to match the spectral reflectance of the object with that of its surrounding terrain. Military personals operate in different terrains such as jungle, desert, semi-desert, snow, mountain, and barren land. Thus, the requirements change depending upon the nature of the terrain. The reflectance of the coatings is tailor-made to provide better stealth effect within visual to near IR region (400–1,200 nm).

Fabrics used for military operations have colour as per the terrain profile and suitably designed pattern with matching energy reflectance signature. In case of land-based operations, fabrics surfaces are coated to match energy signature of terrains foliage. The reflectance of artificial green colour should be matching with that of surrounding terrains chlorophyll, for example, vegetation and plantation. In case of desert/semi-desert-type terrains, the colour schemes are selected as the profile of local sand colour and reflectance tuned with their energy signature.

For coating on textile surface, combination of metal oxide pigments such as chromium trioxide green, iron oxide red, and chrome oxide black together with titanium dioxide and nano carbon black particles are generally used for colour shade development required for particular terrain.

11.6.3 Applications of stealth technology in military

11.6.3.1 Multi-spectral camouflage net

Camouflage net which also known as Camo net is one of most important equipment used by military to provide concealment irrespective of terrain. Camo nets are available in two forms: two-dimensional (2D) and three-dimensional (3D) net. A specially developed coated and printed knitted or woven structure made of synthetic fibres (e.g. polyester, nylon, and polypropylene) or natural fibres (e.g. jute and cotton) provides stealth protection against TIR, Vis–NIR, and radar-based detection technologies. Recently, synthetic polymer-based MSCN nets are preferred over natural fibre ones. In addition to its waterproofing capabilities, synthetic MSCN nets' small weight and flexibility make them simple to roll up, deploy, and transport [30].

Different types of nets such as open mesh fish nets and shrimp nets are manufactured from different types of woven or knitted fabrics which are used as base where the net is being garnished as per pattern requirements. Woven fabrics are mainly used for 3D-type camo nets where fabric is required to incised and garnished on top of the base net structure to fabricate final net, while in 2D-type camo nets, knitted fabric with holes are used which provides required air circulation within the net. The major goal is to prevent internal heat build-up using knitted holes or fabric that has been etched to avoid being detected by TIR technology (most focused TIR frequency ranges are 3–5 μm and 8–12 μm). In terms of radar absorbency, the fabrics preferred to show 3 dB transmission loss in one way and 6 dB total losses in two-way transmission in the radar frequency range of 6–18 GHz. In case of visual/NIR camouflage, fabrics are coated with colour pigments according to the terrain of its use. The most used colour for plain or jungle-type terrains are olive green, light olive green, beige, and brown, while for desert or semi-desert terrains light stone, dark stone, leaf brown, and light olive green are preferable. For snow or mountain-type terrains grey, slate, snow white, and brown shades are preferable. There is no absolute requirement of colours reflectance value, while the requirements for reflectance percentages change based on the geographical locations.

11.6.3.2 Sniper suit

The British Lovat Scouts snipers which were developed during World War I, as a counter measure against enemy snipers, had become a specialist sniper unit. The best sniping camouflage developed till date is called Ghillie suit. Ghillie suits are usually made of one piece or two-piece uniforms, with multiple fabric pieces attached to the base uniform to match the terrain. The diffused coloration and textured appearance make the suit easy to blend with the terrain. Mostly used G-suit is green and forest-oriented, and the second most used G-suit is brown and desert or semi-desert terrain-oriented [31].

11.6.4 Few recent developments in stealth-coated textiles

The use of stealth materials with significant IR properties is important for the camouflage of military targets and managing temperature in special environments. Gu et al. [32] have designed an IR stealth fabric with a dual-working module. The steps involved in preparation of this IR stealth-coated fabric are (i) electroless plating of silver and (ii) coating of high heat latent phase change material (PCM) of eicosane which dispersed in waterborne polyurethane on the one side of electroless-plated fabric (Figure 11.8(a)). Heat absorption via phase change controls the surface temperature of a material. Due to the electroless silver plating, the PCM coating provides a phase change latent heat of 128.5 J/g to further reduce the temperature as a result of reducing the IR emissivity to 0.692 (1–22 m), 0.687 (8–14 m), and 0.655 (3–5 m). When a stealth fabric is prepared in this manner, it can significantly shield or interfere with the target's IR heat signature compared to an untreated fabric, as characterized by using an infrared camera (FLIRONEPRO). As shown in Figure 11.8(b), when the heat plate heats the IR stealth fabric, two parts of the stealth mechanism can be analysed, heat conduction, and heat radiation. After heat is transferred from the hot plate to the coating, the PCM absorbs heat and maintains the temperature relative to the temperature of the coating when it reaches near the phase change point. IR stealth fabrics exhibit consistent and low temperatures on the top layer of electroless-plated fabric. Their low emissivity makes them perfect for obscuring IR features in an IR camera. It can effectively camouflage the target's thermal signature under the detection of IR cameras by effectively suppressing IR radiation. These IR

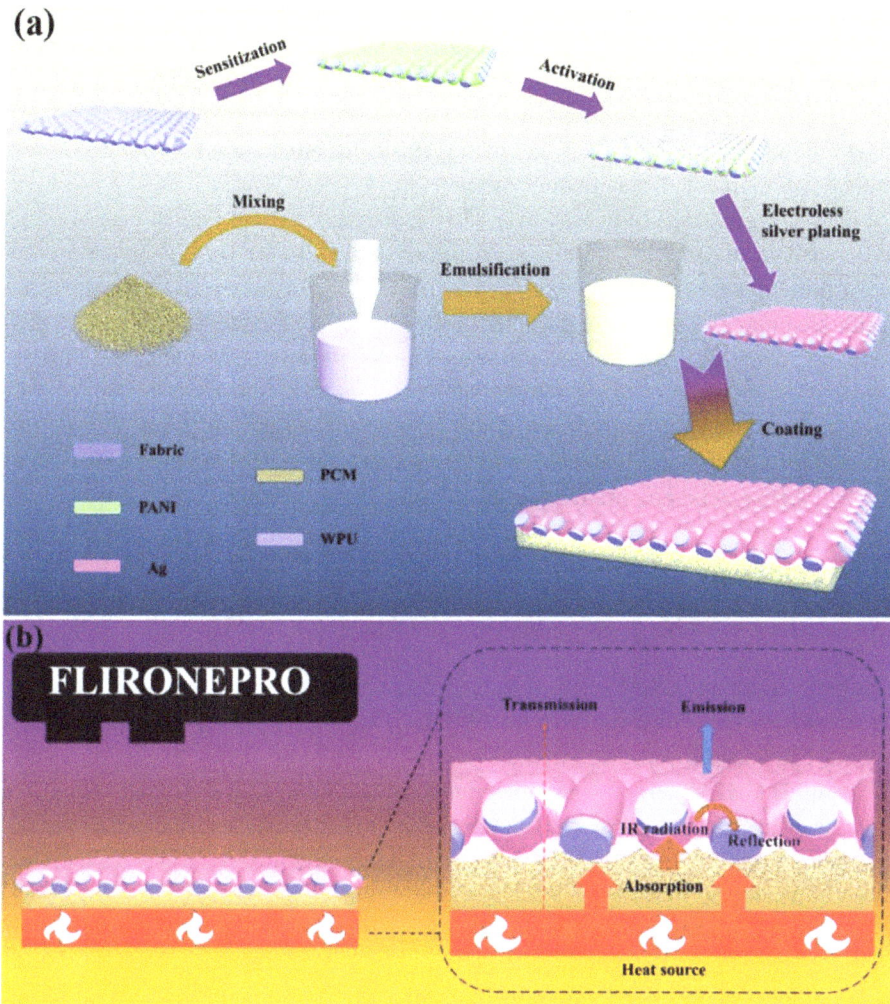

Figure 11.8: (a) Preparation of dual-working module IR stealth fabric; (b) schematic explain the working principle of dual-working module IR stealth fabric (reprinted with kind permission from [32]. Copyright 2022, Elsevier).

stealth fabrics have great potential for practical applications due to the modulation of the dual working modules and good fastness property [32].

In another study, Gu et al. [33] obtained high IR emissivity and temperature control using a double-shell microcapsule. In the first step, eicosane was used as the PCM while melamine urea formaldehyde was used as the shell material to form microcapsules. DSMs are formed by depositing polyaniline (PANI) on the surface of these microcapsules. As measured, DSMs had a particle size of 1.484 m, a core content of 63.25%, a latent heat of 155.1 J/g, and an IR emissivity of 0.722. When DSMs were coated onto polyester fabrics, the coating's IR emissivity increased slightly to 0.794, but impressively, it was possible to cool the fabric by a maximum of 11.2 °C compared to the untreated fabric, and the temperature control process lasted 27 min, which is superior to similar products. Powered by PCM and PANI for low emissivity, the IR camera image shows that the fabric has an obvious IR stealth effect, proving that it is an excellent candidate for IR stealth application [33].

In an interesting study Mao et al. [11] used a sol–gel method to prepare ZnO nanoparticles doped with Al and La. With the coating of ZnO: (Al, La) on cotton fabric, a thin film was formed, which reduced the cotton fabric's IR emissivity. As a result of the addition of Al and La, the ZnO: (Al, La)-coated fabric was able to provide more free electrons in 8–14 μm wavelengths, improving IR scattering and decreasing IR emissivity. The increased doping content of Al improved the samples' IR stealth effect. Thermal radiation was still high on the cotton hand covered by ZnO-coated fabric, but it was lower on the ZnO: (Al, La)-coated cotton hand. This type of coated fabric can be potentially used in stealth applications.

11.7 Terahertz wave technology in smart textiles

Terahertz frequencies lie in between microwave and IR frequencies. An electromagnetic spectrum within a range of 0.1–10 THz is considered to be the terahertz frequency band, as shown in Figure 11.9. The wavelength of this band ranges from 3 mm to 30 μm. These frequencies emit a non-ionizing radiation, which is able to penetrate clothing, synthetic materials, for example, polyethylene, polyester, and other types of shrouds, covers, and enclosures, made of various opaque materials. The radiation is also absorbed selectively by water and organic substances. These unique properties make the terahertz frequency band much more attractive and informative as well as useful in scanning and imaging sector [34].

The two most commonly used methods of T-ray generation are based on photo-conductive antenna and optical rectification. These two methods are used widely in the field of T-ray spectroscopy [35].

11.7.1 Application of terahertz technology

Having a wide array of unique properties, terahertz technology provides variety of applications and opportunities in different fields. The main fields of applications are material science and engineering, biomedical engineering, spectroscopy and imaging technology, and communication and detection technology. Although it is a newly emerged technology, terahertz technology has been taking its place in military applications as well.

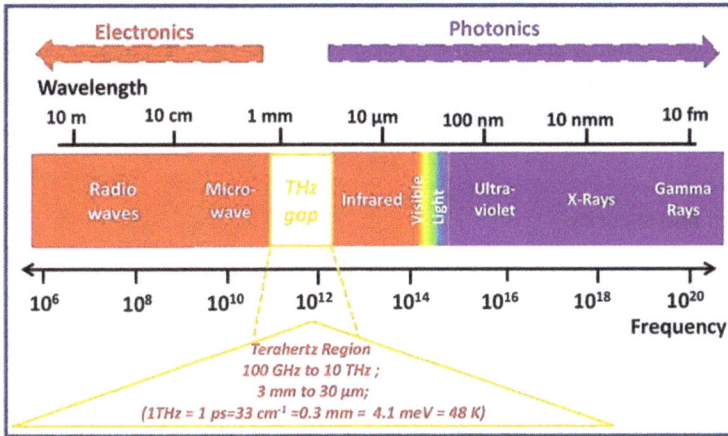

Figure 11.9: Frequency range of different radiations [36].

11.7.1.1 Textiles and composites

Variety of textile fibres and utilization of mixed-fibre yarns in high value textile products can be easily traced through time-domain spectroscopy via terahertz frequency. Terahertz spectroscopy can be reliable, rapid, non-destructive testing for textile identification and for identifying fraud [37]. In textile-based fibre-reinforced polymer (FRP) composites, terahertz spectroscopy is used to study fibre orientation, as the orientation directly relates to the mechanical properties.

11.7.1.2 Biomedical engineering

Terahertz frequency, being a short wavelength, can easily penetrate through soft tissues. Although penetration depth is low, it is being used to examine tissues near the surface; particularly skin and teeth. Being absorbed by water makes it suitable for investigation of tissue hydration. It is also being used for skin cancer detection and detection of burn depth [38].

11.7.1.3 Communication

Terahertz radiation has a higher frequency and bandwidth, which can carry more information making it suitable in the short-distance high-capacity wireless communications.

In outer space, the transmission is lossless, so it can achieve long-range space communications with very little power. Terahertz band having low scattering and much greater penetration through aerosols and clouds make it suitable for communications and radar system used in the stratosphere [39].

11.7.1.4 Detection and security

Through terahertz spectroscopy and image, detecting and characterizing concealed materials through their characteristic transmission or reflectivity spectra in the range of 0.5–10 THz. Many explosives (like C-4, HMX, RDX, and TNT) and illegal drugs (e.g. methamphetamine) have characteristic transmission/reflection spectra in the terahertz range that could be distinguishable from other materials [39].

11.7.1.5 Anti-stealth radar

Terahertz pulse contains a wealth of frequency which enables stealth aircraft to lose the role of narrow-band radar absorbing coating. Furthermore, THz radar has strong anti-stealth ability to shape stealthy and material stealthy [40].

11.8 ECWCS for soldiers

Extended cold weather clothing system (ECWCS) is a protective clothing system developed in 1980 by the United States Army Natick Soldier Research, Development and Engineering Centre, Natick, Massachusetts. The ECWCS is a multi-layered insulating system adjustable to personal preference, metabolism, and prevailing weather conditions. Initially developed to maintain adequate environmental protection in temperatures ranging between +40 °F (+4 °C) and –60 °F (–51 °C) [41].

11.8.1 Extreme cold climate

Although every county has their own standard defining cold weather and extreme cold weather, the U.S. National Weather Service defines extreme cold as –35 °F (–37 °C) with winds less than 5 miles per hour (2.2 m/s) [42].

11.8.1.1 Wind chill factor

Wind chill is a measure of the rate of heat loss from human skin that is under exposure to cold air. Depending upon the increase of speed of wind flow, the cold might start feeling colder [43]:

$$\text{Wind chill} = 13.12 + 0.6215T - 11.37(V^{.16}) + 0.3965T(V^{.16})$$

where T is the temperature in degrees Celsius and V is the wind velocity in kilometre per hour.

With proper understanding of wind chill factor, estimated time can be concluded for occurrence of frostbite to the skin exposed to cold or extreme cold climatic conditions. Figure 11.10 represents plot of temperature against wind speed, and the colour highlights the frostbite time.

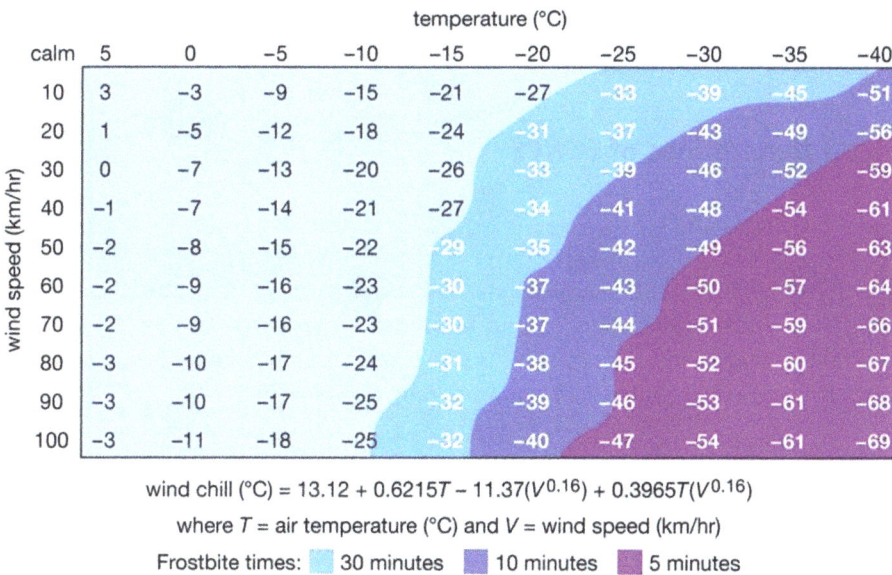

					temperature (°C)					
calm	5	0	−5	−10	−15	−20	−25	−30	−35	−40
10	3	−3	−9	−15	−21	−27	−33	−39	−45	−51
20	1	−5	−12	−18	−24	−31	−37	−43	−49	−56
30	0	−7	−13	−20	−26	−33	−39	−46	−52	−59
40	−1	−7	−14	−21	−27	−34	−41	−48	−54	−61
50	−2	−8	−15	−22	−29	−35	−42	−49	−56	−63
60	−2	−9	−16	−23	−30	−37	−43	−50	−57	−64
70	−2	−9	−16	−23	−30	−37	−44	−51	−59	−66
80	−3	−10	−17	−24	−31	−38	−45	−52	−60	−67
90	−3	−10	−17	−25	−32	−39	−46	−53	−61	−68
100	−3	−11	−18	−25	−32	−40	−47	−54	−61	−69

(wind speed (km/hr) shown in left column)

wind chill (°C) = 13.12 + 0.6215T − 11.37($V^{0.16}$) + 0.3965T($V^{0.16}$)

where T = air temperature (°C) and V = wind speed (km/hr)

Frostbite times: 30 minutes   10 minutes   5 minutes

Figure 11.10: Wind chill chart in degree Celsius [44].

11.8.2 Requirements for ECWCS

For better understanding of the requirements for any extreme cold climate (ECC) clothing system, it is mandatory to know the problems that might occur without having proper protection. The prolonged exposure to temperature, categorized as cold to extreme cold weather, primarily causes frostbite which leads to hypothermia. Frostbite-exposed organs start freezing rapidly and organ movement becomes tough. It also causes loss of sensation, poor blood circulation, redness due to excessive pressure build-up, and then blueness due to lack of oxygen flow on the affected organs, leading to hypothermia. The core temperature gets abnormally low which may cause organ failure finally resulting in death.

Thus, any clothing system for protection against cold to extreme cold climatic conditions should prevent loss of core temperature while keeping the comfort levels intact. Clothing should permit transmission of moisture produced by perspiration. The system should provide insulation and maintain warmth with least amount of bulkiness. It should also resist water penetration, but permeable to wind flow at the same time (depending on protection level). Repelling moisture helps in maintaining the thermal insulation within the microclimate created by layers of clothing while the air flow resistance restricts the heat loss within the clothing system [41].

Factors that influence protection against cold environment assuming that there is no recourse to warm shelter or auxiliary heating are (i) metabolic heat output; (ii) wind chill; (iii) thermal insulation; and (iv) air and moisture vapour permeability [45].

11.8.3 Generations of ECWCS

11.8.3.1 Generation I ECWCS

The first-generation clothing system was a tri-layer garment. A synthetic base layer which was worn next to skin, second layer was an insulated and water or wind-resistant mid-layer made of synthetic or cotton blends, and the outer layer was Gore-Tex and designed for protection from wind, rain, and snow [46].

11.8.3.2 Generation II ECWCS

The second-generation ECWCS was a four-layer system.

Layer 1: Underwear layer
Layer 1 consists of two parts: drawer and undershirt made of polypropylene. It was 100% textured or sometimes non-textured multi-filament polypropylene yarn-based circular or warp-knitted terry loop fabric. The back of the fabric was brushed and worn-brushed aside as inside. The fabric weighed above 5.9 oz/yd^2. The cuff of the garment was made of separate rib-knitted PP fabric with a weight around 5.3 oz/yd^2.

The design features of the garment included pants with elastic polyester webbing for waistband, barracking at the top and bottom of fly, knit cuffs; and undershirt was with raglan sleeves, one piece collar that converts to a turtle neck when the front zipper is closed, knit cuffs [47].

Layer 2: Pile knit
The second layer was the fibrepile shirt and bib overalls. It is a polyester staple fibre pile knit with a polyester backing (brown). It has a minimum weight of 11.0 oz/yd^2. Its back side was a non-piled side which was coated with acrylic resin to act as a binder or anti-curl agent [47].

Layer 3: Field coat liner and trouser liner
The third layer was the field liner trouser and coat liners. In this layer, the fabric weigh was around 4.4 oz/yd^2, where the polyester batting quilted to a ripstop nylon fabric using a dumbbell quilt pattern [47].

Layer 4: Waterproof, moisture vapour-permeable laminated cloth
The fourth layer was a waterproof, moisture vapour-permeable laminated fabric. The base fabric was made of nylon, with a weight of 2.8 ± 0.2 oz/yd^2. The fabric was printed in a woodland camouflage pattern having specified spectral reflectance, laminated with polytetrafluorethylene membrane (weight of the membrane was around 0.5 ± 0.2 oz/yd^2) which was further laminated with a nylon tricot knit 1.5 ± 0.3 oz/yd^2. The final maximum weight of this three-layer laminated fabric was around 5.9 oz/yd^2 [47].

11.8.3.3 Generation III ECWCS

The third generation of clothing system is most recent system in use. The system consists of seven levels of protection, each level having its own set of garments. The seven layers can be worn all together as a part of the system and as per the weather requirements top layers can be used on their own, as shown in Figure 11.11. The seven levels are:

Level 1: Lightweight base layers – This layer acts as a base layer or next to skin layer. It is worn by itself or in conjunction with other levels for added insulation and to aid in the transfer of moisture. This layer consists of light-weight undershirt, which is highly breathable, wicks moisture away from the skin, and dries fast, providing evaporative cooling in warmer weather and insulating in cool weather. All of these are achieved with less weight and bulkiness than previous systems [48].

Level 2: Midweight base layers – This is a second base layer, worn on top of next to skin layer for added insulation and to aid in the transfer of moisture. This is also a two-part garment, undershirt, and drawers provide light insulation in mild climates and serve as a base layer in cold climates. This layer provides extra warmth in cooler conditions but still wicks moisture away and dries fast [48].

Level 3: Fleece cold weather jacket – This level acts as primary insulation layer and worn on top of level 1 and level 2 garments. This jack can also be worn as an outer garment in cool conditions. The jacket is made of polyester fleece serving as an insulation layer, which creates air pockets that entraps air and prevents body heat loss, providing outstanding warmth without weight. The jacket comes in three colours: green, brown, and Special Forces black [48].

Level 4: Wind cold weather jacket – The garment constructed with stretchable nylon acting as base and insulative levels in transitional environments to provide adequate protection against wind and sand.

Level 5: Soft shell jacket and trousers – It is a two-part garment system, consisting of a jacket and a trouser, acting as an insulation layer, basically intended for use in moderate to cold conditions. The fabrics were treated with water-repellent finishes to keep the outer surface dry [48].

Level 6: Extreme cold and/or wet weather jacket and trousers – This set of garments acts as a shell layer, to be worn over other levels in moderate to cold wet conditions alternating between freezing and thawing. The lightweight Jacket and trousers provide an outstanding protection against water and wind. The fabric is printed with NIR signature reduction technology further enhancing soldier survivability.

Level 7: Extreme cold weather (ECW) parka and trousers – this is also a two-part garment layer, especially designed for use during static operations in extreme cold and dry conditions. Constructed with a nylon fabric that has a water-repellent finish, in between the outer and inner layer of the garment, insulative panels are incorporated to obtain excellent thermal insulation [48].

Level 1   Level 2   Level 3   Level 4   Level 5   Level 6   Level 7

Figure 11.11: Seven levels of Generation III ECWCS [49].

11.8.4 Specification of latest generation ECWCS

Fabric used to manufacture garments for ECWCS might change its requirements based on the nation and based on the climatic conditions. Every nation has its own unique requirements. As well as for printed layers, design will also change as per the application terrain and the nation using the said clothing system. The two most commonly seen patterns are universal camouflage patterns also known as ACU of U.S. Army and the other one is multi-terrain pattern of British Armed Forces. Based on different army requirements, IR signature on level 6 garments also required to be tailored. General specifications of Generation III ECWCS have been presented in Tables 11.1 and 11.2.

Table 11.1: Knitted fabric layers specifications for Generation III ECWCS [48], [50–52].

Parameters	Level 1 fabric	Level 2 fabric	Level 3 fabric
Material	100% polyester	Polyester–Spandex	100% polyester
Fabric type	Jersey knitted	Jersey face, shearling grid back	Velour face and back
Base fabric weight (g/m^2)	Approx. 130	Approx. 225	Approx. 240
Thermal insulation, clo	–	0.8 min	1.3 min
Air permeability, CFM	300	150–500	350–500
Main properties needed	High breathability, wicking property	Wicking property, high stretch, breathability, thermal insulation	Excellent breathability, thermal insulation

Table 11.2: Woven fabric layer specifications for Generation III ECWCS [48, 50–52].

Parameters	Level 4 fabric	Level 5 fabric	Level 6 fabric	Level 7 fabric
Layer	Mid-layer	Outer layer	Outer layer	Outer fabric
Material	Nylon–Spandex	Nylon–Spandex	Nylon–polyester tricot-laminated	Down proof Nylon fabric
Base fabric weight (g/m^2)	Approx. 140	Approx. 120	Around 190	Approx. 70
Water vapour resistance	20 m^2/Pa W	20 m^2/Pa W	20 m^2/Pa W	20 m^2/Pa W
Water repellent	Yes	Yes	Yes	Yes
Water proofness, bar	–	–	1.0 min	–
Main properties needed	High stretch, breathability, water resistance, wind, and sand protection	High stretch, breathability, and water resistance	Water proofing, wind proofing, and breathability	High thermal insulation, breathability, and water resistance

11.9 Sleeping bag

No of the region of use, a sleeping bag is without a doubt one of the most important pieces of survival kit. Materials are specifically designed for sleeping bags to provide optimum comfort depending upon the altitude of its use and climatic conditions of that altitude. Most basic principles that work in a sleeping bag are breathability and insulation. Certain components of the system are designed in such a way to prevent the loss of natural heat generated by human body. Different components of sleeping bags have been shown in Figure 11.12.

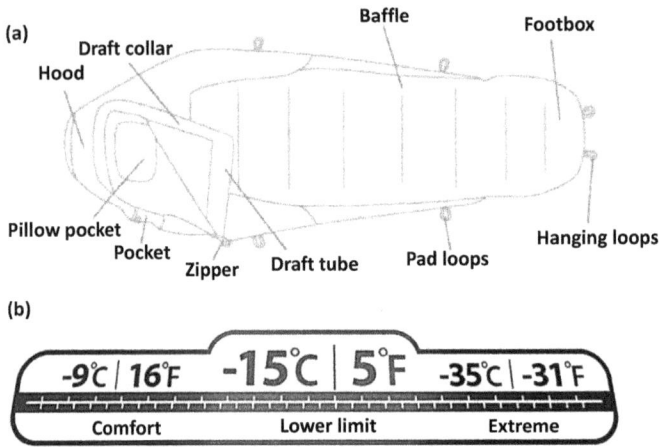

Figure 11.12: (a) Anatomy of a sleeping bag and (b) EN13537 temperature rating of a sleeping bag.

11.9.1 Types of sleeping bags

Based on the application of the product, sleeping bags can be categorized in different types. According to Table 11.3, sleeping bags can be categorized according to their application: sports and recreational sleeping bags, military-grade sleeping bags designed for high-altitude environments, field training and exercises, relief and rescue work, and by their season and weather conditions.

Table 11.3: Temperature ranges of different sleeping bags [53].

Type of sleeping bag	Low temp. limit (°F)	High temp. limit (°F)
Summer	35	Over 35
3 Season	10	35
Winter	Under 10	10

11.9.1.1 Sleeping bag temperature rating

It is very crucial to understand how sleeping bag temperature rating works in order to understand the suitability of the product in different climatic conditions. The activity of human or personal using the sleeping bag should also be taken into consideration. It is also important to understand the different ways how a sleeping bag is given its ratings (Figure 11.12(b)) [54].

Comfort rating: It is the temperature at which an adult female can expect to sleep comfortably.

Lower limit rating: It is the temperature at which an adult male can expect to sleep comfortably.

Extreme rating: It is the lowest temperature at which the sleeping bag will keep an adult female alive. This temperature also poses significant risk of hypothermia, frostbite, and many other cold temperature problems.

11.9.2 Sleeping bag systems

11.9.2.1 ECW or ECC sleeping bag

ECW sleeping bag or ECC sleeping bag, only one sleeping bag made of inner lining and outer shell fabric. This kind of sleeping bag is excellent for protection till -13 °F (-25 °C). The most common structure of this type of sleeping is quilted construction with a front centre opening and equipped with adjustable face closure by means of draw strings.

11.9.2.2 Modular sleeping bag system

Modular sleeping bag provides great insulation from cold to extreme cold climatic conditions. The whole system consists of two to three independent sleeping bag which can be used solely or altogether. The system can provide protection again temperature ranging from -30 to -40 °F. All the individual sleeping bags are made of nylon fabric ribstop-woven, with special features like water-repellent surface, and the bags are reversible in nature as well. The first bag is called a patrol bag, and this can provide protection up to $+30$ °F (-1 °C). This bag weights approximately 3 lbs. and the next bag is called an intermediate bag, for extreme cold conditions, the patrol bag can be inserted inside the intermediate bag for better thermal protection. This double system can provide protection up to a temperature of 14 °F (-10 °C). The weight of the individual bag is approximately 4 lbs. The third and last bag system is called Bivy Cover, manufactured with special water vapour permeable, wind proof three-layered laminated fabric. The individual bag weighs around approximately 2 lbs. The whole system can provide up to 4 h of comfortable sleep at temperature of −40 °F (−40 °C).

11.9.3 General requirements for sleeping bag

The most basic requirement for any sleeping is thermal comfort. For cold weathers, the sleeping bag must have that property to prevent heat loss – heat generated due to basic human metabolism should be prevented from falling down to the ambient temperature. The property parameters of all sleeping bags are solely based on the application landscape showing different requirements. Some sleeping bag must have high air permeability to circulate air along with sweat to provide the user dry and comfort thus leading to sensorial comfort.

For most cold weather sleeping bags, the thermal comfort plays a vital role. The main goal of using the product is to protect against cold and extremely cold weather conditions. Preventing the loss of body temperature becomes crucial in such applications. Any military-grade sleeping bag is designed to provide a comfortable night's rest for longer than four hours in arctic temperatures. Sweating occurs when you move and perform basic activities. Sweating on a fabric may cause heat loss since water is a good conductor of heat, spreading the water rapidly so it will evaporate more quickly. A sufficient air flow in the bag results in holding the warmth as air is a bad conductor of heat.

For most non-military sleeping bags, the comfort plays importance than the protection. Most sports grade sleeping bags are designed in such a way that it can provide optimum conditions for a sound sleep. Here air flow or permeability or some might call the breathability is crucial if the sleeping bag is not being used in high altitudes. Sensorial comfort becomes vital in such cases, where the product is manufactured for extended or prolonged usage times.

The technicality and requirements of all sleeping bags vary based on the terrain of its application, altitude of its application, and the climatic conditions of the terrain of altitude – making these factors as a primary requirement for manufacturing.

11.9.4 Components of a standard sleeping bag

Sleeping bags are manufactured with textile fabrics, insulating fills, and panels within the fabric layers. The most common sleeping bags has two textile fabrics – inner fabric also known as inner lining fabric and outer fabric or sometimes termed as outer shell fabric. Apart from those two fabric layers, all standard sleeping bags come with a carry bag, which is also a textile fabric. Within the two layers of fabrics (e.g. inner and outer fabric) filling materials are inserted for protection and environmental threats. The bag is tailored with zippers on the one side or both sides for easy opening and closing. Some special grades of sleeping bag attached with loops and fasteners in the area where head or face to tighten those areas to prevent air flow. Sleeping bags for hiking and mountaineering are equipped with metal hooks inside loops to keep them secure and steady at one place.

11.9.4.1 Textile fabric components

Sleeping bag composes of basically fine denier nylon-based ripstop-woven fabrics, in some cases, microfibre-based polyester fabrics as well. The whole sleeping bag is made of three fabrics, one that forms inner layer of the bag which comes in direct contact with the user, next is the outer side of the fabric which comes in contact with atmosphere, and the whole bag is protected with a carry bag, and this bag provides easy transportability to the sleeping bag when it is not being used. For longer life nylon 66 is preferred; however when cost-effectiveness comes into play, some prefer to use nylon 6 fibres. In most sports and recreational purpose sleeping bag is manufactured with polyester fibre to reduce the cost extremely when compared to nylon fabric-based sleeping bags.

Inner layer fabric
The inner layer fabric of any sleeping bag is made from 30 to 50 denier nylon or polyester fibre, ribstop or sometimes plain-woven fabric. This fabric is smooth and light weight in nature. The inner fabric of the majority of military-grade sleeping bags weighs between 50 and 55 gsm, has controlled air permeability, and has no protective covering on the back. Even though it is thin, this fabric offers high tearing resistance, excellent sweat absorption, water distribution through the fill, good heat retention for maximum heat dissipation, and comfort while worn.

Outer layer fabric
In the outer layer the sleeping bag is somewhat sturdy as compared to the inner fabric. This fabric layer is also made of similar or same material as inner fabric. Almost in every instance, it is preferably made with ribstop-woven fabric. As the layer comes in direct contact with the atmosphere, it becomes very crucial that this layer of fabric has high strength in terms of tensile and tearing behaviour. This layer is generally coated with breathable grade polyurethane to provide water proofness in the sleeping bag. The fabric is generally a bit heavier than the inner layer. This fabric ranges from 65 to 75 gsm. This fabric layer is woven with 60–75 denier finer multi-filament yarns. Fabric is treated with water-repellent finishing agents to accommodate in water proofness and not letting the fabric surface getting wet while it comes in contact with any liquid.

Carry bag fabric

Each and every sleeping bag comes with its own carry bag. Some carry bags are detachable, and some are inbuilt with the sleeping bag itself. However, the carry is manufactured with coarser denier fibre, sturdy, high abrasion resistance in nature along with high strength features. Manufactured with 150–200 denier filament yarn, with a tighter construction with plain-woven fabric, this fabric has to have a non-breathable coating on its back side to provide higher water proofness as compared to the outer layer fabric.

11.9.4.2 Insulation material

Insulation is one of the most important parameters of any sleeping bag. Between inner and outer layers of sleeping bag, filling materials are inserted to achieve desired insulation required for different weather conditions. Till date, natural filling and synthetic filling materials are in use for insulation purposes.

Fill power

Fill power is a characteristic measure that indicates the volume of a one ounce of down that can cover how many cubic inches of loft or fluffiness. Higher the fill power number higher the insulation can provide. A 650 Fill Power denotes that one ounce of down fill can cover 650 cubic inches. The fill power rating can range from 300 to 900 and above [55].

Down insulation

Down is the naturally obtained duck or geese undercoat that grown under their wings. This material can provide greater degree of thermal insulation all by itself. Down is very light weight in nature as compared to synthetic materials.

For down sleeping bags a lofting power of 600–650 is considered good quality, 700–750 is considered very good, and 800–950+ is considered excellent [56].

Synthetic filling materials for insulation

The most common synthetic filling material used is polyester fibre fillings. Continuous strands of polyester filaments are cut into staple length and these stable fibres can be used in free form of thermally bonded to give them a sheet like appearance. In free form, the filling provides bulk or a very high loft; however keeping the filling in the right place has been always a problem. This free movement of fills can also be restricted by stitching the filling along with both the fabrics in different stitching pattern.

In today's advanced technology, different shapes of polyester filaments are being in use, for example, hollow-shaped filament which entraps air within its hollow tubes to provide higher degree of thermal insulation as compared to same mass of normal solid polyester filaments. PrimaLoft is a brand of patented synthetic microfibre thermal insulation material that was developed for the United States Army in the 1980s. PrimaLoft is a registered trademark of PrimaLoft, Inc. [57]. Climashield Apex is a high-performance insulation material that can be used for garments and for sleeping bags and quilts. It is made out of very fine continuous filament fibres. The fibres do not migrate through fabrics and provide durable insulation. Climashield apex has very good insulating properties, dries fast, and is very breathable [58].

With further improvements in technology, filling materials treated with water-repellent finishes – Therm-a-Rest's eraLoft™ insulation is made of water-resistant polyester that retains warmth when wet and dries fast [59]. Moreover, PCMs have a greater impact in providing improved thermal insulation.

11.9.5 Application of sleeping bags in military

During on-going mission an army personal gets 4–6 h of time to rest. Within this short period of time the comfort and thermoregulation within the sleeping bag play the major role to obtain optimum conditions for a sound sleep. During rescue operations and during warfare, the average sleep time of a soldier is around 5.8 h. During training exercises, it is less than 5 h which is split up in multiple episodes usually lasting less than 2 h each. Without a comfortable night's sleep, combat effectiveness may drop by 15%–20%. In order to function at their best both physically and psychologically, soldiers must get enough sleep.

Military activities like trekking, high altitude missions, and hiking required equipment like sleeping bag to provide the personal comfort while having rest. During such activities, protection against rain, snow, and wind becomes an essential feature provided by the equipment, keeping the user warm and comfortable while the minimum resting time becomes crucial.

11.9.6 Recent developments

Many studies reported that people who used sleeping bags under defined comfort and temperature limits had low skin temperatures, cold, and pain sensations in their feet. Song et al. [60] developed an innovative heating sleeping bag (i.e. MAR_{HT}) by embedding two heating pads in a traditional sleeping bag (i.e. MAR_{CON}) in the region of the feet to improve wearers' local thermal comfort. Two tests were administered on different days to seven female volunteers and seven male volunteers. In each test, the setting temperature was determined from EN 13537 (2012) and tests lasted three hours. Both sexes were able to maintain thermoneutral toe and foot temperatures with MAR_{HT}, whereas MAR_{CON} linearly decreased temperatures during the 3-h exposure. Additionally, all males and most females experienced significantly increased toe blood flow while wearing MAR_{HT}. In MAR_{HT}, the comfort and thermal sensations experienced by both sexes were greatly improved in the feet, while improvement in the whole body was small to moderate, proofing high effectiveness of MAR_{HT} to local thermal heating. Another recent study [61] investigated the thermal comfort properties (relative water vapour permeability, thermal absorptivity, thermal resistance, thermal conductivity, and air permeability) of tri-layered composite fabrics to utilize them in sleeping bags. A total of 12 samples were prepared varying materials in different layers, while the inner layer consists of wool-knitted single jersey fabric, the middle layer made of polyester needle punched non-woven fabric, and the outmost layer nylon-based Gore-Tex branded waterproof breathable fabric. The results of this study support the development of sleeping bags with enhanced comfort levels.

An et al. [62] proposed a CFD simulation model to study the heat transmission and air movement inside and on the surface of four types of sleeping bags (sewn-through, box, trapezoidal, and V-tube) [Figure 11.13(a)]. A thermal resistance tester was used to quantify the thermal properties of four samples. Experimental and simulation results of thermal conductivity and thermal resistance were in good agreement with a maximum difference of 11.22% (Figure 11.13(b)). The sewn-through structure was found to have the worst thermal insulation under tiny wind speeds, while the V-tube shared the highest thermal insulation of all samples with similar porosities. There was mainly convection and conduction of heat inside the sleeping bag, while in a ratio of 30%:70% (Figure 11.13(c)). In terms of heat transmission by the sleeping bag's surface, convection was most responsible, followed by radiation.

11.10 High-altitude pulmonary oedema chamber

11.10.1 What is high-altitude pulmonary oedema?

With the increase in altitude, due to the loss of air pressure and reduction of oxygen in the atmosphere, physical activities gradually become difficult. High altitude refers to an altitude above 2,700 m from the sea level. To perform any physical activity in high altitude, human body wants more oxygen to compensate the reduced barometric pressure. High-altitude pulmonary oedema (HAPO) is a deadly and fatal condition that can occur to mountaineers, archaeologists, scientific explorers, or military personnel working at a high altitude [63].

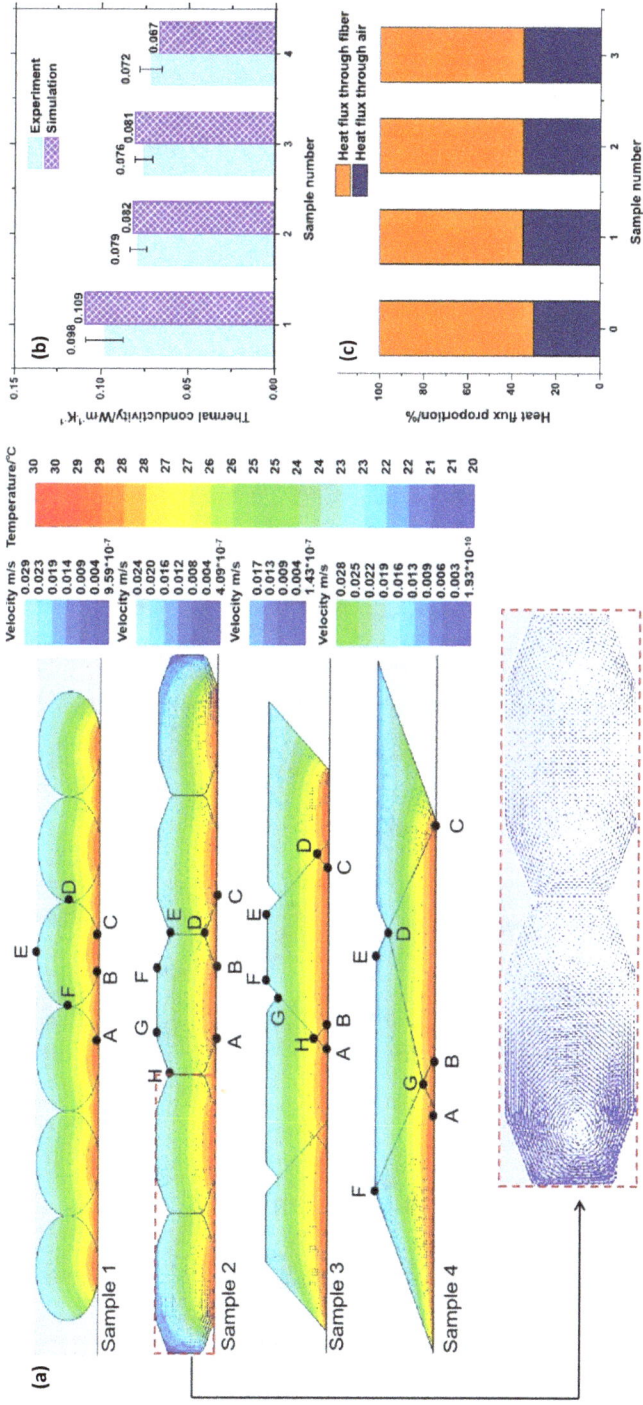

Figure 11.13: (a) The temperature contours and the velocity vectors inside four different sleeping bag samples; (b) experiment results and simulation data of thermal conductivity; and (c) the proportion of heat fluxes through air (convection) and fibre (conduction) (reprinted with kind permission from [62]. Copyright 2021, Elsevier).

11.10.2 Application of HAPO chamber in military and defence

As altitude increases, the atmospheric pressure gradient decreases, making military personnel's required motions uncomfortable. Moreover, the temperature decreases by 1 °C for every 150 metres of height gain. Such life-threatening conditions reduce the performance of the military personals and combat effectiveness. Often getting a patient evacuated to a lower altitude or to a medical facility becomes difficult due to weather conditions and nature of the terrain. HAPO chambers are basically foldable devices (as shown in Figure 11.14), which can be inflated manually to a pressure generally above 2 psi to simulate the atmospheric conditions of below 2,700 metres depending upon the working altitude, and each chamber has its own pressure limitation. Each HAPO chamber comes with an altitude metre inside to indicate the "virtual altitude" achieved within the chamber. The application of HAPO chambers as "life saving devices" in remote and high altitudes, where moving the soldiers immediately from bunkers to a lower altitude, is not possible, and/or supplemental oxygen is not available [64].

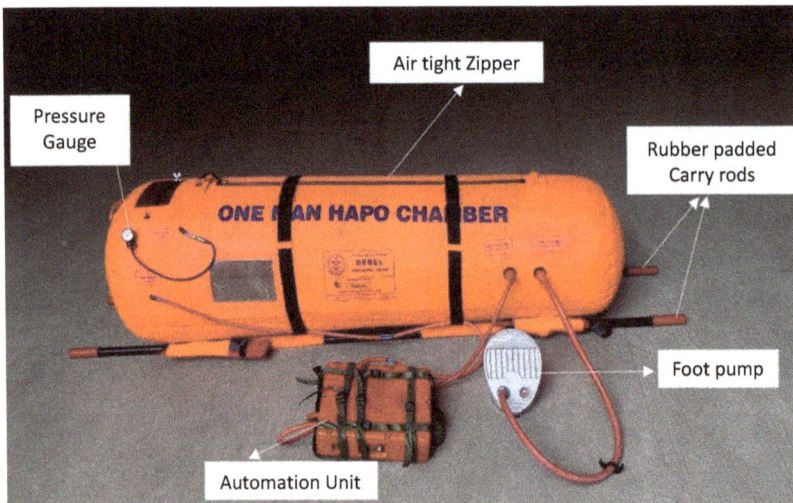

Figure 11.14: Different components of a typical HAPO chamber [65].

11.10.3 Material requirements for HAPO chambers

Material selection for manufacturing a HAPO chambers is very critical. The material should be able to withstand extreme cold operation (temperature up to −40 °C) and undulated rough terrain as well as it should be non-degrading under the exposure of UV radiation at the higher altitude. Generally, polymer-coated fabrics are used for manufacturing the chambers. The glass transition temperature of the coated fabric system determines the vulnerability to sub-zero temperature. Furthermore, the coated fabric should not crack upon repeated folding that occurs during its lifetime. The valves, hoses, and pumps are required to withstand extreme low temperature and exhibit robustness, as these items are considered as "pressure sensitive and critical for life Saving" in difficult times in the absence of medical facilities. The suitable material must be chosen with sufficient consideration since if one component fails, and the entire bag will be rendered useless [66].

11.10.4 Different types of HAPO chambers and their development history

11.10.4.1 Portable altitude chamber

The first portable hyperbaric chamber was patented in America in 1919 by Hermann Stelzner (from Germany). The structure was collapsible suited for single patient and used for the treatment of decompression sickness. It was fabricated from waterproof, airtight material. An electric pump connected by a pipe to the chamber was used to create pressure, which was measured using a manometer and development of excess pressure was prevented by a predetermined security valve. The recompression chamber can keep folded while not in use, taking small storage space thus making it portable to carry. This chamber was basically developed for divers [67].

A German patent was filed in the year 1929 concerning a similar portable device for treatment of allergic diseases by means of isolating the patient from allergic elements present in ambient atmosphere. Another similar patent was obtained by two French inventors G. Perron and A. Hoff in 1940, where a portable pressure chamber designed to protect children from toxic gas in wartime. In 1946, The Russell S. Colley (from the USA) patented an inflatable pressure bag, primarily for patients suffering from insufficient pressurization of the interior of the plane when flying at high altitude. The developed structure was foldable and light in weight with the ease for single man operation, having flexible container with a collapsible central part and two-rounded reinforced ends. It had a window for observation, and it had two handles for carrying. In 1973, an American inventor Miller patented a portable inflatable chamber made of a waterproof, pressure-resistant material for divers suffering from decompression sickness. The unit was divided into two compartments: a fitted mattress with handles for carrying and a chamber for the patient. The size of the whole unit was 71 cm × 230 cm when inflated. The first portable hyperbaric chamber designed for the treatment of high-altitude illness was in 1979 by a German inventor H. Becker. The structure was conical in shape, made of two bag one inside another. The inside bag was of PVC-coated fabric, and the outside bag was made of reinforced waterproof nylon fabric fitted with a zipper, while the size of unit was 200 cm × 37–60 cm. Hyperbaric chambers were forgotten when Gamow presented its new version for specifically to treat high altitude illness [67].

11.10.4.2 Gamow bag

The Gamow bag is a lightweight portable hyperbaric chamber that can produce an environment inside the chamber equivalent to altitudes descending more than 1,500 metre. This portable device is mainly used for treatments of high-altitude sickness such as high-altitude pulmonary edema and high-altitude cerebral edema even at higher altitudes exceeding 3,500 metres. The Gamow bag is a windowed, cylindrical-shaped, portable hyperbaric chamber constructed of non-permeable nylon with polymer coating that requires constant pressurization with a foot pump attached to the bag. The complete device is 7 feet (213 cm) long, with a diameter of 21 inches (53 cm), weighs 14.9 pounds (6.8 kg), and inflates to a volume of 17 cubic feet (0.48 m³). The increased ambient pressure produced inside the bag helps counteract pulmonary hypertension and increases the partial pressure of oxygen, thereby improving ventilation and reducing hypoxemia [68].

11.10.4.3 CERTEC bag

In the year 1989, a French firm named CERTEC developed a hyperbaric chamber like the Becker and Gamow bags (ref10). CERTEC bag made of coated polyamide fabric, conical to cylindrical in shape, having size around 220 cm in length and diameter is 65 cm at widest, weighs approximately 4.5 kg including bag and the pump. The system is attached with long airtight zipper for ease of access, fitted four circular straps along with a fast release buckle, and has a large central window for observation. Additionally, it has eight handles for transportation. The system comes with a pump and silicone hose connection. The bag also contains a monometer in the lining and a window for altimeter. There are two relief valves which allow inflation and deflation [67].

The patient is usually placed into the bag, inflated to the point where the pop-off valves hiss and held under pressure for an hour. You need to keep pumping multiple times per minute to flush fresh air through the system if your bag does not have a CO_2 scrubber system. The patient is taken out of the bag and given another evaluation at the end of the hour.

11.11 Textiles in military shelters

Irrespective of the terrain, level of altitude of applications, military has always been reliant in textiles whether woven textile, knitted or non-woven whatever form it may be. The wide range of properties offered by textile materials had broadened the scope of its use into military application. Shelters of all kinds had been built of practical textile materials in order to migrate from one place to another to withstand the difficulties or obstructions thrown by adverse weather conditions or even during combat.

Being light weight in terms of weight to area ratio, covering wider spectrum in terms of camouflaging, providing multiple types of protection, for example, protection against heat, cold, rain, and snow can be obtained by functional textiles keeping the cost effectiveness in mind. Different applications require different kinds of property parameters which are easily met by different functional textiles.

11.11.1 General requirements for the fabrics used in military shelters

The required fabric properties may change based on the geographical location, altitude of application, and its climatic conditions. The various environmental or functional factors which are considered in military shelters mainly are (i) anti-fungal or anti-microbial, (ii) UV stability, (iii) water repellency and proofing, (iv) flame retardancy, (v) mosquito or insect-repellent, and (vi) temperature blocking (reduced heat entrapment), (vii) IR/visual camouflage effect, (viii) light weight, (ix) high strength to weight ratio, (x) tear strength, (xi) abrasion resistance, (xii) durable properties, (Xiii) durable colour, and (xiv) service temperature.

11.11.2 Different components of military shelters

Shelters used in military application are primarily used for providing protection against weather challenges of any and all kinds of potential threats. These shelters are capable of providing protection to multiple personals at a time and have been opted in military for single personal as well as multiple personals.

Almost every shelter must have the below mentioned components altogether or in combination:

a) Roofing fabric: This kind of fabric covers the top part of the shelter, providing roof over the head, also known as outer fabric. It protects from direct sunlight, rainfall, snowfall, and other factors. From natural canvas to synthetic fabric all are being used for such purposes. In some cases, special property parameters are incorporated while manufacturing these kinds of fabrics. Sturdy, high resistance to wear and tear, water repellency, fire retardancy, and UV reflective finishes are the common requirements for these fabrics.

b) Ground fabric: It is also known as flooring fabric. This category of fabrics is spread on the ground of the shelter area. Numerous manufacturers have recently suggested nylon/polyester fabrics with excellent levels of water resistance, flame retardancy, and fungus resistance for this use. Commonly used fabrics, however, are constructed of polypropylene that has been laminated together and coated, giving them a larger mass and stronger abrasion resistance. Such fabrics are highly waterproof, provides excellent protection against fungus, micros, and mildew. The fabrics are supposedly anti-skid in nature and water repellency is incorporated while manufacturing.

c) Inner linings: This layer of fabric is generally of low mass per unit area and forms the side walls of the shelter. Generally, they are treated with water-repellent and fire-retardant chemicals.

d) Ropes or cordages: The edges of roofing fabrics are secured with ropes or cordages. Each of the corners of the fabrics is tightly tied with pegs and ropes knotted on those.

e) Support systems: This covers different equipment for whole structure (e.g. poles or telescopic poles, anchors, locking mechanisms, ground stakes, or pegs or twisted pins). With the advancements in technology, poles of FRP have replaced the metallic poles which are heavier in weight and less flexible as compared to the FRP composites. Whereas, the ground stakes are manufactured with zinc-chromium-plated iron making them less prone to corrosion.

11.11.3 Different types of military shelters

11.11.3.1 Tents and domes

Tents are the most basic shelter seen to be used in any military establishments. Most tents are used as temporary mode of shelter. A fabric is covered on the top of one or more poles or support structures. The cloth is tightly spread, stretched over the ground support structure, and tightly secured by cords. Most tent cloths have loops on its edges and sides where cordages are attached to pegs into the ground. The base cloth is made of sturdy materials, for example, nylon, polyester, cotton, and cotton/polyester. The inner sides are generally coated with polyurethane or PVC and made to withstand robust climatic conditions such as heavy rainfall, high winds, and to some extent also from mild to moderate snowfall.

Most commonly identified tents are in triangular-shaped roof with and without sidewalls. The test setup basically made upon requirements such as the duration of the establishment and the purpose of the establishment. The tents for human are rectangular in structure with triangular shed on top, whereas the tents for storage purposes do not require much height, although more length is needed as compared with personal shelters.

Domes are shelters with much more structured ground support systems. These shelters are spherical in shape and established while the duration of stay is longer. These structures have its own specially made outer shell, which is tailored based on

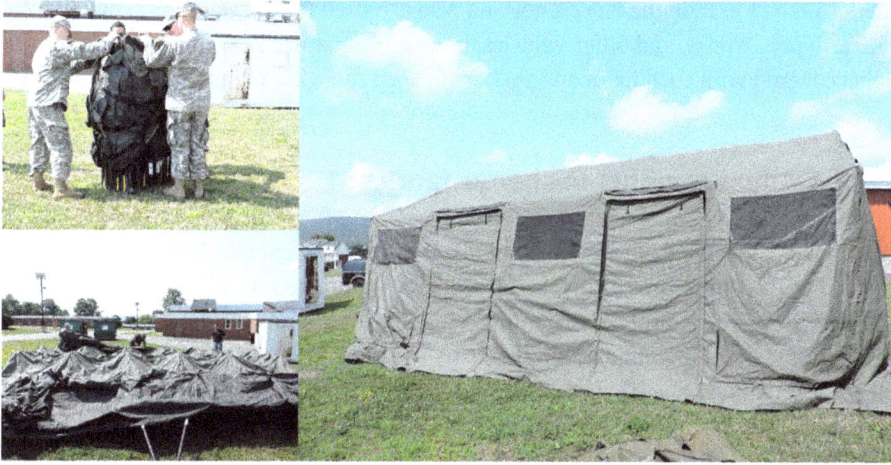

Figure 11.15: Deployed HDT Base-X 305 military shelter (reprinted with kind permission from [69]. Copyright 2021, Elsevier).

the required area to be covered. These fabrics are coated on either face side or sometimes both sides with polyurethane or PVC-based coating formulations with additives required to impart special functional properties. Fabrics are super flexible, light weight, and most commonly made of polyester. The durable and sturdy support systems are manufactured with high quality anti-corrosive steel or similar-type metal or hybrid metals. In the field of military, these types of shelters are established for many military applications such as field research, medical, and isolation compartments. In higher altitudes, the inside walls are lined with thermal insulating panels to keep the ambient temperature warmer.

11.11.3.2 Specialty shelters

Synthetic camouflage nets is one of such special kind of fabric that is being used for the outer cover of tents – either replacing the outer fabric or just placed on top of the outer fabric of the tent. These nets come in both 2D and 3D structures. The 2D kind of net is multiple visual spectrums matched coated fabric with same coverage of objects of its application terrains. Whereas the 3D kind is incised with special patterns which allows them to be placed flexibly on top of any structure providing concealment. Every landscape found in different part of the world having different ratio of foliage, ground, rock, sand, and snow, which can be replicated into nets just by maintaining the area coverage. Green belt area, snow mountain area, desert area, and so on have different ratios of colours, and hence, nowadays the nets are manufactured as per the custom requirement. Military shelters in jungle, deserts, and so on use such special shelter fabrics to hide their presence and also are used to conceal their armoured vehicles, deploy trucks, and other vehicles. In country boarder lines, tanks and arms are being covered with such net-made shelters to keep them out of enemy sites.

11.11.4 Advancements in shelter fabrics

With the advancements in the world of textiles and allied chemicals, it becomes much easier to incorporate multi-functional property parameters onto one single piece of fabric. Whether it is protection against heat or protection against cold – same cloth can perform both functions simultaneously. The surface of the fabrics is treated with blend of functional chemicals that can blackout light while some fabrics can also absorb heat from the atmosphere and keep the inside warm. The development in insulation fields has blessed the world with materials that can prevent heat loss from the inside. These insulating materials are highly suitable for tents that are being established in high altitudes with cold to extreme cold climatic conditions.

Other special function properties such as stealth and deception can also be in application in high-end military tent or shelter fabrics. Such fabrics having similar IR signature of its surrounding terrain and capable of blending itself completely to naked eye through radar detection and TIR detection technology as well.

11.12 Parachute and paragliding

A parachute is a device which enables a person to jump from an aircraft and float safely before reaching to the ground. This device slows down the vertical descent of a body falling through the atmosphere or the velocity of a body moving horizontally. It increases the surface area of body; thus, the increased air resistance slows the body in motion. Employment of parachutes has been found widely in military warfare for safely dropping supplies and equipment as well as personnel, and they are deployed for slowing a returning space capsule after re-entry into Earth's atmosphere. Parachutes are also used in recreational sports like paragliding, parasailing, and skydiving [70].

11.12.1 Glide ratio

The ratio of the forward distance travelled to the vertical distance that an aircraft descends when gliding without any power called a gliding ratio [71]. Glide ratio is the number of feet a glider travels horizontally in still air for every foot of altitude lost:

$$\text{Glide ratio}(E) = \frac{\text{Lift}}{\text{Drag}}$$

The glide ratio (E) is numerically equal to the lift-to-drag ratio under these conditions; but is not necessarily equal during other manoeuvres, especially if speed is not constant. A glider's glide ratio varies with airspeed. Glide ratio usually varies little with carried load; a heavier parachute glides faster but maintains nearly similar glide ratio [72].

11.12.2 Parachute components

The most important part of any parachute is no doubt the canopy also known as the main parachute. It is the part which provides sufficient air resistance to help in gliding the system. It is made of strong and robust material to withstand the air pressure of the altitude when it is being opened. The size of the component varies depending upon the altitude, air pressure of the altitude as well as size and weight of the load it is carrying.

The main parachute canopy is used in conjunction with a reserve parachute assembly, just in case of emergency if the main canopy is malfunctioned during opening, the reserve parachute to act in. The other main components of a parachute are (i) slider, (ii) stabilizers, (iii) Hackie, ripcord, pilot chute, (iv) harness/containers, (v) risers and associated steering toggles, (vi) reserve static line, (vii) altimeter, (viii) three ring system, and (ix) hook knife. The readers can read about all these parachute components in detail in the reference [73].

11.12.3 Properties of parachute fabric

Property requirements of parachute fabrics vary based on its application. The general properties are described further:

i) **High tensile strength**
Fabrics which are suitable to use in a parachute canopy should have high tensile strength. The most important criteria a parachute fabric should have been the strength to weight ratios to be high. For one square feet of canopy can withstand one pound of payload.

ii) **High tearing resistance**
Parachute fabric must have sufficient resistance to tearing while it is being deployed. For evaluating the performance of fabrics suitable for parachute application, different types of tearing resistance are checked (e.g. cut slit tear resistance and nail puncture method) as per the application or working altitude.

iii) **Porosity**
Porosity in parachute fabrics is one of the most critical property parameters. Based on the porosity of the canopy fabric, the chute opening altitude is decided or vice versa. The main canopy should be able to withstand the air pressure to float upon opening. On higher altitude where there is less pressure of air present, the canopy should be able to hold the air to successfully glide to reach the surface. For obtaining "zero porosity" fabrics are generally coated with silicon in both the sides.

iv) **Resistance to wear and tear**
The canopy fabric used is any parachute is intended for multiple usages. With proper maintenance and care of the parachute, the fabric can last up to 150–200 times in its whole lifespan. Generally, these fabrics have very high tear resistance and soft in feel, thus making it low packing in its nature. Even though these fabrics are stored in small pack covers in properly folded conditions, fabrics can withstand longer period without degradation.

11.12.4 Materials and specifications for making parachutes

There are certain materials that are commonly used on almost each and every parachute system.

11.12.4.1 Fibres and fabrics used in making parachutes

The most predominant fibre used in manufacturing of parachute fabric is Nylon and to be a bit more specific Nylon 6,6. However, the other fibres which have the potential and also used for making parachutes are Kevlar, Vectran, polyester, and silk. As mentioned in different specifications, the major differences include the weave, weight, and finish for different parachute fabrics in use. The various types of fabric materials in use of a parachute system include canopy fabric, pack cloth, tapes, webbings, mesh, elastic fabrics, stiffener materials, and foams. Parachute canopies are mostly ripstop structure-woven nylon fabrics. Ripstop weave is a derivative of plain weave with heavier threads woven into the material at right angles resulting in a boxlike pattern. This weave increases tear strength and results in stronger and sturdy fabric. The parachute containers are made of duck nylon fabrics or cordura fabrics. Table 11.4 summarizes the different fibres and their advantages which can be potentially used for making military parachutes.

Table 11.4: Potential fibres and their advantages for making parachute fabrics.

Material	Advantages
Nylon	– Most common material in use, specially Nylon 6,6 – High strength, multi-filament, heat, and UV-stabilized material – Good elongation
Kevlar	– Very high tensile strength and modulus – High resistance to abrasion – Does not degrade at high temperature – High wear and tear
Vectran	– Very high strength – High resistance to degradation under heat and weathering
Polyester	– Good resistance to weather – Moderate to high strength, multi-filament, and heat resistant – Easily available and less costly

11.12.4.2 Specifications of parachute fabrics

It is mandatory that all parachute systems are built under government approval programs, and if not all, then material used for constructing the parachute system must have some form of specification approval. The most common of these systems is the military specification (MIL-SPEC) system. However, in commercial uses, other than MIL-SPECs such as Federal Standards, PIA Standards, and Federal Aviation Administration (FAA) specifications are also acceptable. Table 11.5 represents the specifications of different types of parachute fabrics.

The manufactures can use any of specification for manufacturing parachute components, if the manufacturer can prove compliance with this specification, and that the specification is acceptable to the FAA for use in the parachute system. Amongst the specifying authority, the MIL-SPEC systems are readily available and worldwide accepted.

11.12.5 Paragliding

Paragliding is aerial sports meant for recreational purposes, where a parachute is attached to the sportsmen's body by harness which allows the sportsmen to glide in air by running off from slopes, snow mountains, and so on. This adventure sports can be performed in conjugation with many other water sports (rafting, surfing, etc.), land sports (mountaineering, trekking, etc.), and air sports (paragliding, jumping, etc.) as well.

Table 11.5: Specifications – military parachute/canopy with parachute support fabrics [74].

S. no.	Specification	Yarn (Denier)	Construction (threads/inch)	Weight (Oz/Yd²)	Breaking strength (lbs)	Tearing strength (DaN)	Air permeability (cfm)	Thickness (inch)	Weave
Personnel chutes									
1	PIA C 44378 Type 1	30	126 × 132	1.20	45/45	5/5	0–5	0.003	Ribstop
2	PIA C 44378 Type 2	30	120 × 120	1.12	42/42	5/5	30–50	0.003	Ribstop
3	PIA C 44378 Type 3	30	114 × 132	1.17	45/45	5/5	30–50	0.003	Ribstop
4	PIA C 44378 Type 4	30	126 × 132	1.20	45/45	5/5	0–3	0.003	Ribstop
5	PIA C 44378 Type 5	30	126 × 132	1.17	45/45	5/5	0–5	0.003	Ribstop
6	PIA C 44378 Type 6	40 × 50	115 × 116	1.50	50/50	7/7	0–3	0.003	Ribstop
Personnel and cargo chutes									
1	PIA C 7020 Type 1	30	120 × 120	1.1	42/42	5/5	100 ± 20	0.003	Ribstop
2	PIA C 7020 Type 2	40 × 70	120 × 76	1.6	50/50	5/5	130 ± 30	0.004	Ribstop
3	PIA C 7020 Type 3	40 × 70	120 × 76	1.1	50/50	5/5	130 ± 30	0.004	Ribstop
Cargo chutes									
1	PIA C 7350 Type 1	100	68 × 68	2.25	85/85	10/10	90–140	0.007	–
2	PIA C 7350 Type 2	210	52 × 52	3.5	130/130	30/30	150–200	0.014	–

S. no.	Specification	Yarn (Denier)	Construction (threads/inch)	Weight (Oz/Yd²)	Breaking strength (lbs)	Tearing strength (DaN)	Air permeability (cfm)	Thickness (inch)	Weave
Pack cloth									
1	PIA C 7219 Type 3 Class 1 and Class 3	420	60 × 45	7.25	325/275	–	–	–	Plain
Zero porosity									
1	Zero porosity	30	113 × 132	1.3	47/47	12/12	0	0.003	Ribstop
Low porosity									
1	Low porosity	30	113 × 132	1.1	47/47	5/5	0–3	0.003	Ribstop

Technically speaking, paragliding fabrics are similar to low weight parachute fabrics. Generally, the fabrics used in paragliding have slightly stiffer hand feel in comparison to the parachute fabrics which are very soft, smooth, and limpy. The fabrics for paragliding are manufactured with nylon 66 yarns having lower linear density with high strength. Fabrics are ribstop pattern woven with compact structures to attain high tearing strength on the final fabric. The fabrics are generally coated to obtain zero air permeability. All manufacturing industries have their own secret coating formulation. Very low coating add-on is applied to make the fabrics impermeable to air. The most common coating substrates are polyurethane and silicone. Table 11.6 represents the specification of typical paragliding fabrics.

Table 11.6: Typical specifications for PU-coated high-tenacity Nylon 6,6 fabric-based paragliding fabrics.

S. no.	Parameters	Unit of measurement	Values				
1	Mass	g/m^2	26 ± 2	29 ± 2	32 ± 2	38 ± 1.5	40 ± 2
2	Tear strength	kg	≥1.5	≥1.5	≥1.5	≥1.5	≥1.5
3	Breaking strength	kg/5 cm	≥22	≥22	≥25	≥33	≥33
4	Elongation on bias direction						
	Under 3 lbs		≤10	≤10	≤10	≤10	≤2
	Under 5 lbs	%	≤18	≤18	≤18	≤18	≤3
	Under 10 lbs		≤30	≤30	≤30	≤30	≤15
5	Air permeability	L/m^2/s	≤20	≤20	≤20	≤20	≤100

11.12.6 Military application of parachutes and paragliders

Primarily, objects or humans are discharged from a moving air vehicle (aircraft, jet plane, etc.) at a higher altitude to an intended or pre-calculated landing co-ordinate for safe delivery to the ground. Occasionally, parachutes are used as an emergency landing or evacuations. Whether it is carrying an object or a human passenger, after a certain height free-fall under gravitational force the speed slows down by using a parachute which is opened either by means of pulling ripcord either by manually, mechanically, or by electronically.

Parachutes in use for aircraft extraction for heavy loaded cargo, a drogue parachute ejected in the airstream from the cargo door of an aircraft. For heavy loaded cargo, multi-parachute systems are used for the supply of military equipment and for safe recovery of spacecrafts after its re-entry.

Some parachutes are used for landing high speed vehicles, aircraft, and space shuttles by reducing their speed significantly. These kinds of parachute are termed as

"Break Parachute." These are basically ribbon-type parachute structures. The structure varies based on the requirements of its application as well.

For military aircrafts at an inoperable situation, parachutes are used as a safe ejection and emergency landing device. Almost every military aircraft has its pilot seats assembled with emergency parachute system for unfavourable situation, for example, during engine malfunction or during battle, the pilot, and the cabin crew can easily eject themselves through the cockpit.

11.13 Coated and laminated textiles for aerostat/ airship used for military surveillance

Different lighter-than-air (LTA) aircrafts such as aerostats and airships are used in a variety of applications in defence, such as surveillance and detection of aerial threats [75]. LTA aircrafts generally work in an altitude of 2–5 km from the sea level, and they are generally filled with an LTA gas (generally helium or non-flammable hydrogen) for lifting purpose. LTA systems, especially stratospheric airships, face significant challenges in terms of material performance balancing all the requirements. Ideally, it should be strong, lightweight, flexible even at very low temperatures, able to retain helium for a long duration, and weather-resistant (i.e. protect against ultraviolet light, and ozone). The most common fabric for these types of applications is multi-layer-coated or laminated fabrics, in which each layer performs a particular function [76].

An aerostat or airship is made based on "balloon-within-a-balloon" concept containing an inner balloon and an outer envelope. The inner balloon and outer envelope are called "ballonets" and "hulls," respectively. The hull structure is filled with an LTA gas (lifting gas) such as helium to keep the airship aloft, whereas the ballonets are filled with ambient air to expand and contract by releasing air in opposition to the lifting gas. Figure 11.16 schematically represents three different layers of hull including function of each layer. In the external protective layer high weather-resistant polymers [such as polyvinylidene fluoride, Tedlar® or polyvinyl fluoride, etc.] are used. It is very common to use an aluminized topcoat for better weather resistance [77]. A fabric in the middle of the structure acts as strength layer which are generally made of woven fabric based on different conventional fibres (polyester or nylon) or high performance fibres (Vectran, Zylon, Kevlar, UHMWPE, etc.). The internal functional layer (barrier later) is made of a polymer having excellent helium gas barrier property. Generally, biaxially oriented polyester film (Mylar®), polyvinylidene chloride, thermoplastic polyurethane, and so on are used in internal functional layer.

Recently, researchers are finding different polymer nanocomposite-based options for outer weather-resistant and internal gas barrier layers to overcome the issues with conventional coated or laminated fabrics used in LTA envelope. Different polymer nanocomposites especially polyurethane nanocomposites have huge potential for making

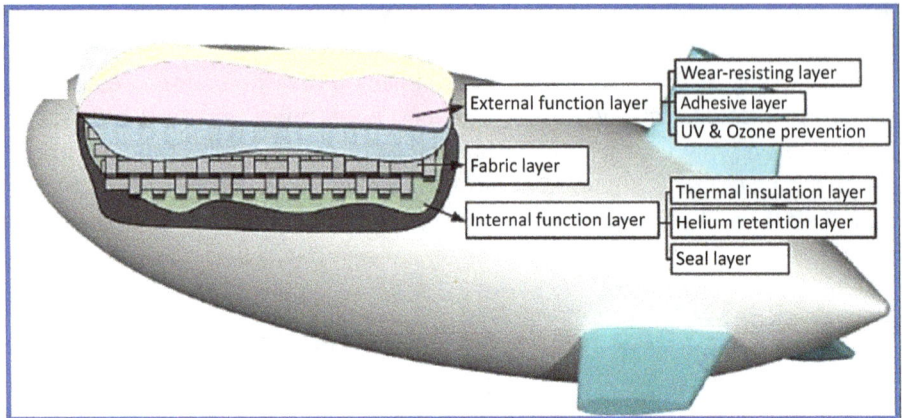

Figure 11.16: A typical structure of LTA envelope showing different layers and their functions (Modified according to [78]. Copyright 2018, Elsevier).

coated and laminated textiles with improved weather resistance and gas barrier properties [76, 77, 79–81]. Layered structured nanomaterials or nanoplatelets such as nanoclay, grapheme, and its derivatives show strong potential in improving helium gas barrier properties of polymer nanocomposites by increasing tortuous path length of the gas molecules [78, 82–84]. Similarly, polyurethane nanocomposites reinforced with UV shielding nanomaterials such as rutile-TiO_2, ZnO, CeO_2, Al_2O_3, and graphene derivatives [85–87] show good weather resistance property and can be potentially used in the protective outer layer of LTA aircrafts.

11.14 Nanoenhanced textile-based sensors and actuators in security/defence application

With the advancement in technologies, electronics are miniaturized, allowing them to be integrated into textiles and used by soldiers or civilians. Introducing different electronics such as sensor and actuators into military textiles could enable soldiers to achieve high level of performance and capabilities that have never been achieved before. In e-textile, a conductive metal or polymer fibre must be embedded to carry signals that are created by sensors that react to a wide variety of input parameters, including movement, light, sound, chemicals, liquid vapours, and gases in the environment. A sensor is classified as either a chemical sensor, light sensor, heat sensor, acoustic sensor, activity recognition sensor, and motion sensor or a location detection sensor. Table 11.7 summarizes different types of sensors used in military and defence applications with their potential and applications.

Table 11.7: Different types of sensor for military and defence applications [88].

Sensor type	Input parameter	Input device	Output signal	Output device	Application
Biometric	Heart rate, body temperature	ECG, EEG	Electrical, mechanical	Digital display	Soldiers' health monitoring
Acoustic	Sound	Microphone, audio recording, speech recognition, ultrasonic detectors	Electrical	Headphones, loudspeaker, piezospeaker, speech synthesis, ultrasonic transmitter	Detection of approaching vehicles, enemies or aircraft
Temperature	Heat and cold	Resistance temperature detectors (RTDs), thermistors	Heat	Thermal devices	Detection of body as well as environmental temperature
Location	X, Y, Z, and T collected by satellite	Wi-Fi, Bluetooth, Cell ID, ultrasonic, radio frequency identification (RFID), GPS, ultra-wideband radio	Electrical	Computer screen, digital display	GPS system can be used for detection of the location
Buttons/touch input	Textile switch, fabric keyboard	Keypad, wristband	Electrical	Digital display	Sending information and biometric data
Optical	Infrared (IR) camera, image recognition, laser	Cameras, light sensors	Electrical	Digital display, position display	Detecting the location of gunshots

Even though some of these sensors are already used by the military for environmental sensing, research and development can improve existing technology and help integrate the remaining types into military applications. Nanomaterials can be used to create smaller and more sensitive sensors and actuators than conventional technology by exploiting their unique properties. Military field operatives will highly benefit from portable and efficient sensors especially in the following applications:

i) Position and motion sensors based on small, lightweight accelerometers, and GPS

ii) Thermal sensors with high sensitivity to IR radiation

iii) Biochemical sensors

iv) Sensors and devices for monitoring health and delivering drugs/nutrients

v) Environmental sensors

In addition to detecting enemies, environmental sensing can also detect potential biochemical threats. A right sensor can detect blast situations and provide health risk information in these situations. Embedded button-sized microphones in conductive-woven fabrics have been reported to detect remote objects such as approaching vehicles [89]. By comparing and analysing the sound from each microphone, a microcontroller determines the sound's direction.

In order to provide remote wireless health diagnostics using nano-biosensor systems based on textiles, woven or printed connections can be used to connect nanobiosensor systems to compact textile integrated wireless electronics [90]. In a study reported by Sempionatto et al. [91] described about an electrochemical monitoring platform using wireless wearable rings for rapid detection of explosives and nerve agents in vapours and liquids. In the ring-based sensor system which consisted of two parts, electrochemical sensors were printed on plastic rings and a battery-powered, stamp-size potentiostat was used as an electronic interface for signal processing and transmission (Figure 11.17). The 3D-printed compact ring structure integrates a wide range of electrochemical capabilities, enabling fast square-wave voltammetry and chronoamperometry analyses as well as interchangeable screen-printed sensing electrodes that can be used to detect different chemical threats rapidly. The ring system was remarkably miniaturized and integrated despite its high analytical performance. Using this wearable sensor ring system, nitroaromatic and peroxide explosives can be monitored voltammetrically and amperometrically, and organophosphate nerve agents can be detected amperometrically. Detecting and alerting the wearer of multiple chemical threats in both liquid and vapour phases simultaneously is a significant achievement for a miniaturized wearable sensor ring platform which a great potential for meeting the demands of diverse defence and security scenarios [91].

11.15 Wearable motherboard: the smart shirt for military soldiers

Wearable motherboards or smart shirts allow electronics and textiles to be seamlessly integrated. To create a flexible and wearable smart shirt, sensors and interconnection technology are integrated onto a basic fabric (Figure 11.18). A flexible data bus that integrated into the structure transports the signals from the sensors to the multifunction controller/processor. Using appropriate communications protocols, the controller then transmits the signals wirelessly to desired locations such as to a hospital, doctor's office, and battlefield triage station. Furthermore, smart shirts can be used as a bidirectional information infrastructure by transmitting information from external

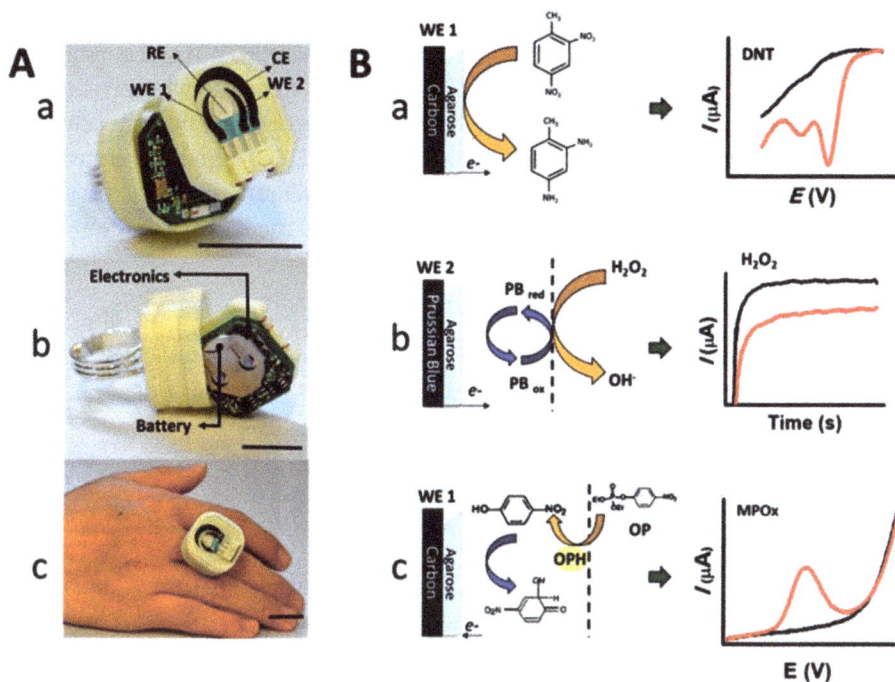

Figure 11.17: (A) The ring-based sensor platform detects vapour and liquid explosives as well as nerve agents. Images showing (a) the screen-printed electrodes integrated into the polymeric ring case based on (from left to right) a carbon working electrode (WE 1), an Ag/AgCl reference electrode (RE), a carbon/Prussian-blue working electrode (WE 2), and a carbon counter electrode (CE); (b) bottom view of integrated circuit board and coin battery compartment capable of performing SWV and CA; (c) ring sensor worn on wearer's middle finger. (B) The ring sensor platform is shown here in a schematic format showing the redox detection process of the different chemical threats as well as the multiplexed vapour phase detection of (a) DNT (red) on carbon WE 1 and corresponding background SWV (black) in vapour phase; (b) peroxides detection on the carbon-Prussian blue WE 2 (red) along with the background CA (black); (c) MPOx vapour detection on carbon WE 1 (red) along with the background SWV response (black) [91].

sources to sensors and wearer. The controllers provide power (energy) to wearable motherboards. As smartphones becoming more popular, the controller can be obviated because all the communication and processing can be shifted on them.

Georgia Tech has developed a wearable Motherboard or an "intelligent" garment for the twenty-first century through research on sensate liners for combat casualty care. With this Georgia Tech Wearable Motherboard (GTWM), sensors, monitors, and information processing devices can be incorporated into an extremely versatile platform [92]. GTWM can monitor humans' vital signs unobtrusively and in a systematic way. A flexible wearable monitoring device has been created by attaching appropriate sensors to this motherboard via the developed Interconnection Technology. It transmits the information to monitoring devices such as electrocardiogram machines, temperature recorders, and voice recorders via the flexible data bus (based on fibre optic and specialty fibres) integrated into the structure. Additionally, GTWM serves as a valuable information infrastructure by transmitting information from external sources to sensors. The GTWM is lightweight, so anyone can wear it easily, whether they are new-born baby or senior citizen. There are many applications for GTWM, including telemedicine, post-operative recovery monitoring, sudden infant death syndrome prevention, monitoring athletes and astronauts, law enforcement personnel, and combat soldiers.

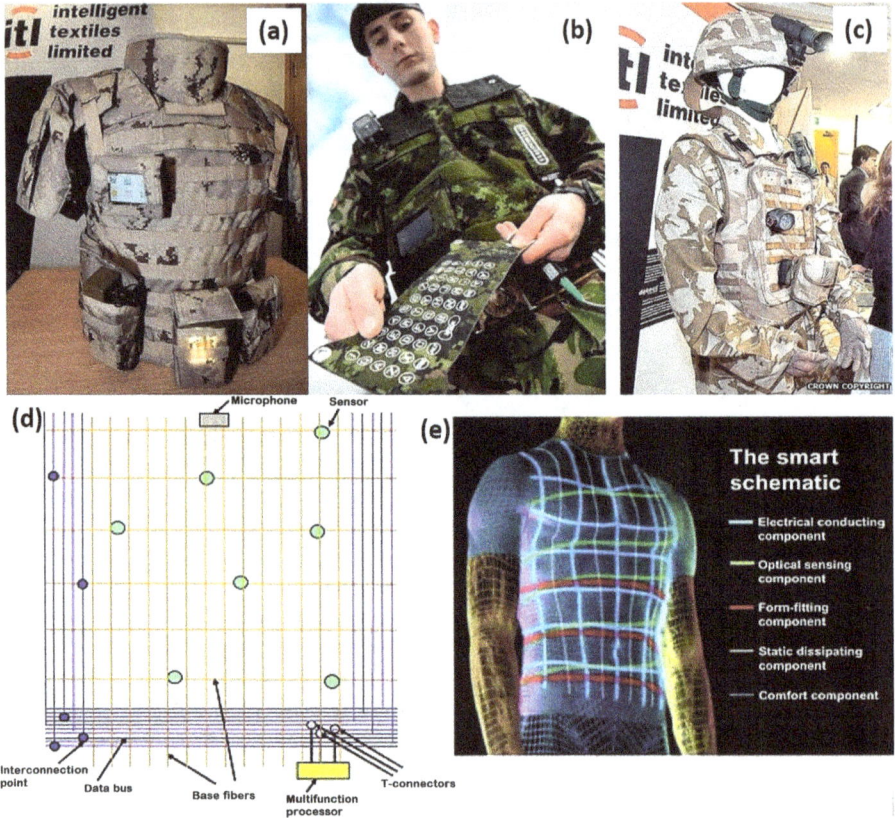

Figure 11.18: (a–c) Wearable keyboard and computer for military uniforms (reprinted with kind permission from [88]. Copyright 2015, Elsevier); (d, e) wearable motherboard architecture (reprinted with kind permission from [93]. Copyright 2021, Elsevier).

In addition, Sensatex has developed a smart shirt for the US Navy that uses different sensing elements for diagnosing and treating wounded soldiers on the battlefield [94]. Soldiers can become part of the digital battlefield by wearing wearable computers or motherboards. The wearable computer is a device attached to or integrated into an individual's clothing, and although it was once considered mostly useful as bulky

maintenance equipment, it has now become the device that will power the future soldier's electronic heart.

In the past, wearable computers were often used in the military to monitor equipment malfunctions and diagnose soldiers' wounds at a distance. Today, with the development of state-of-the-art technology and inventive minds in government laboratories
and industry, computers are being wired into clothing and capable of tracking enemy
targets, connecting soldiers to air, sea, and land forces, monitoring health, and translat-
ing native languages. Additionally, technology is improving anti-terrorist efforts and
law enforcement [92].

11.16 Electrospun nanofibres in security/defence application

The process of electrospinning involves applying an electric field to an extruded polymer solution to produce nanofibres. A 3D porous network of nanofibrous mat can be
created using electrospinning under appropriate conditions, with a consistent pore
size distribution and a tunable interconnected porosity. Researchers have tried electrospinning of several synthetic and natural polymers, polymers composites, and
polymer nanocomposites (reinforced by graphene, carbon nanotubes (CNTs), MXene,
etc.) for several mechanical, electrical, and biological applications.

When dealing with chemical and biological threats (which include chemicals like
nerve agents, mustard gas, blood agents like cyanides, and biological toxins like bacterial
spores, viruses, and rickettsiae) in a variety of environments like combat, urban, agricultural, and industrial, military, firefighter, law enforcement, and medical personnel need
high-level protection. Permeable adsorptive protective over garments like those worn by
the US military or hazardous materials suits are examples of current protective gear
that is based on full barrier protection. These suits have two obvious drawbacks: weight
and moisture retention, which make it impossible to wear them continuously. Polymer
nanofibres are regarded as suitable membrane materials for this application due to
their low weight, large surface area, and porous (breathable) nature. Nanofibres may be
good candidates for detecting interfaces for chemical and biological poisons at concentration levels of parts per billion due to their high sensitivity to warfare chemicals.

In one of the researches by Pandey et al. [95], an electrospun polyacrylonitrile
(PAN) nanofibre decorated with zeolitic imidazolate framework (ZIF-8) has been proposed as an effectual triboelectric positive material for a high-performance TENG. The
proposed material (TENG) can also serve as a self-powered visible light communication
system (VLC) for wireless human–machine communication. PAN@ZIF-8 nanofibres
have significantly higher surface roughness, which results in more charges being induced upon contact electrification. Therefore, PZ-TENG has a significant improvement

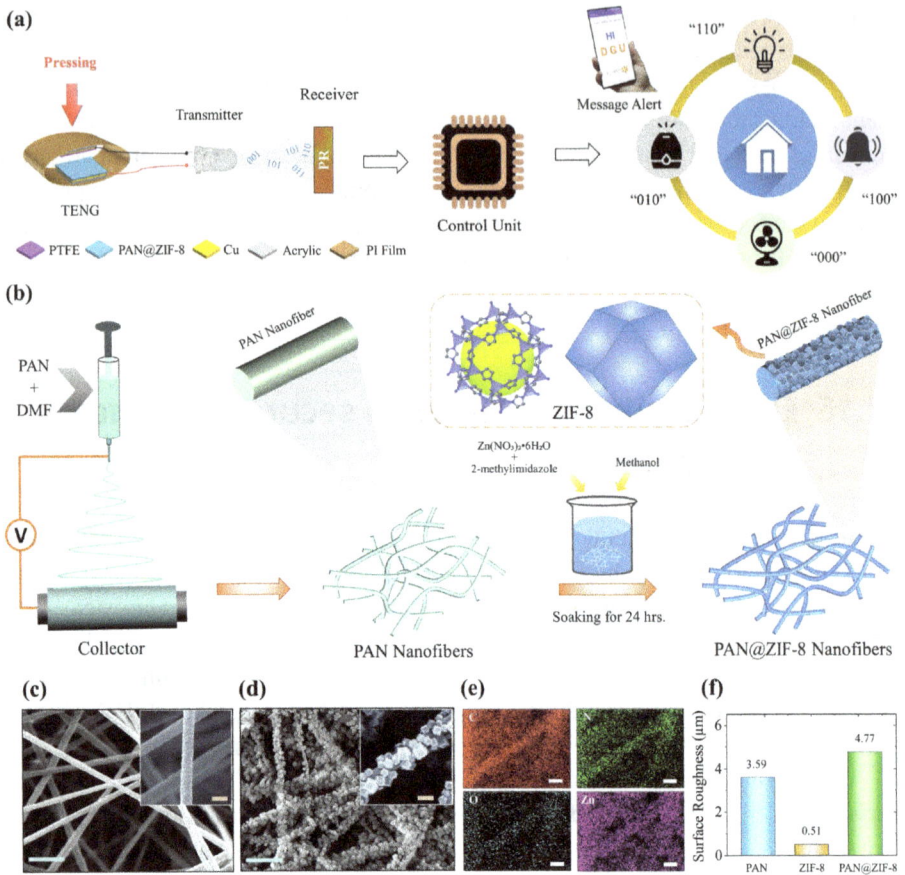

Figure 11.19: (a) Schematic of the PZ-TENG driven self-powered VLC system for wireless and secure human–machine interactions for multi-functional applications; (b) a schematic showing how ZIF-8 nanocrystals are synthesized on PAN nanofibres. FESEM images of (c) PAN nanofibres and (d) PAN@ZIF-8 nanofibres, with insets of magnified SEM images; (e) EDS mapping of PAN@ZIF-8 nanofibre, C (red), O (blue), N (green), and Zn (pink); (f) surface roughness (Ra) is plotted for three different triboelectric materials (reprinted with kind permission from [95]. Copyright 2023, Elsevier).

in energy harvesting performance compared to PAN nanofibres and ZIF-8 crystal-built TENG devices (Figure 11.19).

In another work by Teo et al. [96] a system with five distinct levels of organization was constructed using electrospun fibres with composite nanofibres at the first level. The core composite nanofibre was covered with a second layer of composite material at the second level. The third level was provided by the nanofibre's surface modification. The nanofibres were arranged into an assembly to provide the fourth level of organization. The nanofibre assembly was finally enclosed within a matrix or formed into a bulk structure with a predetermined shape at the final level. Individual strands of nanofibres need a distinct addition because chemicals are typically more sensitive to or reactive to specific stimulants. It is possible to assemble the composite nanofibres, each with a unique level of reactivity, to create a single, multi-functional yarn. The researchers propose that the woven fabric will offer protection against a number of threats and might communicate with the user if the fabric becomes obsolete.

In another article by Deeraj et al. [97], electrospun TiC@TiO$_2$ core–shell carbon fibre mats are described that are synthesized by electrospinning nano titanium carbide embedded PAN fibres directly. The details of the manufacturing process are shown in the schematic (Figure 11.20).

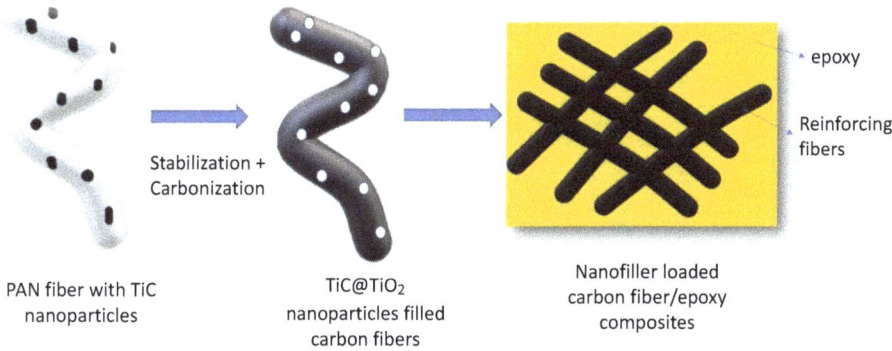

PAN fiber with TiC nanoparticles

Stabilization + Carbonization

TiC@TiO$_2$ nanoparticles filled carbon fibers

epoxy

Reinforcing fibers

Nanofiller loaded carbon fiber/epoxy composites

Figure 11.20: Manufacturing process of electrospun TiC@TiO$_2$ core–shell carbon fibre mats (Modified according to [97]. Copyright 2022, Elsevier).

The findings demonstrate that during the conversion of the PAN fibre to carbon fibre, the surface oxidation of TiC to TiO$_2$ occurred. These TiC@TiO$_2$ carbon fibre mats were tested for electromagnetic interference shielding effectiveness over the Ku-band (12–18 GHz) and achieved a shielding effectiveness of –33.31 dB for a single layer (thickness: 0.21 mm).

In another interesting research by Guo et al. [98] a series of spatial-scale conductive frameworks containing tantalum carbide (TaC) nanoparticles were fabricated through electrospinning and high-temperature pyrolysis. The method presented in this work is simple and effective for preparing lightweight, thin composite fabrics with high EMI shielding performance, and the TaC/C electrospun non-woven composite fabrics obtained from this research have potential for use in aerospace, defence, and electronic devices. The idea of the proposed mechanism is represented in Figure 11.21.

Due to the impedance mismatch between the air and fabrics at the interface induced by the plentiful free electrons creating surface plasmon resonance, some of the externally incident electromagnetic waves that hit the surface of the composite fabrics would be directly reflected. The electromagnetic energy would swiftly attenuate into thermal energy or other types of energy while the residual electromagnetic waves

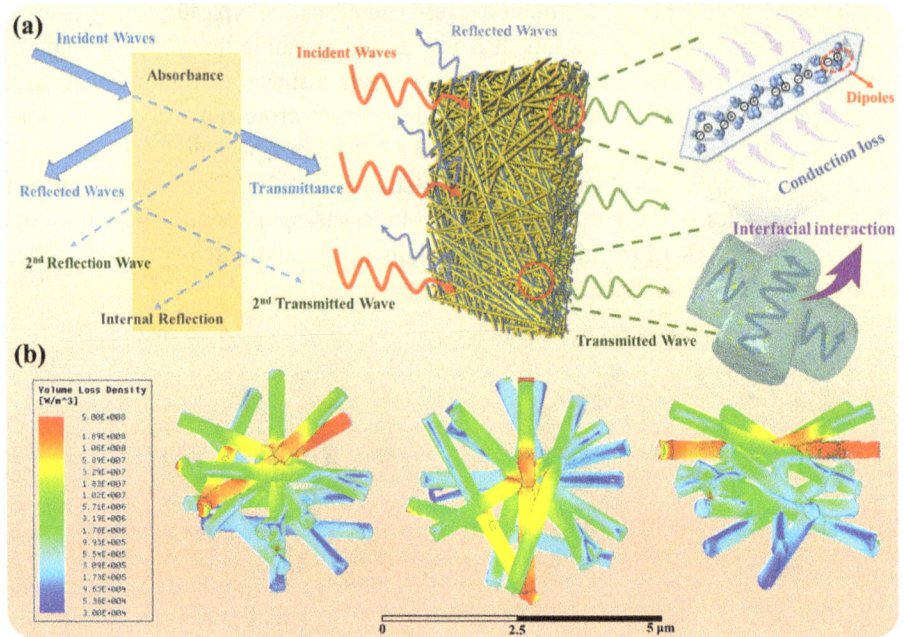

Figure 11.21: (a) Schematic description of the EMI shielding mechanism in the TaC/C electrospun non-woven fabrics; (b) the finite element simulation of TaC/C electrospun non-woven fabrics (reprinted with kind permission from [98]. Copyright 2021, Elsevier).

continued to travel through the fabrics. This was due to a mass of multiple internal reflection that existed inside porous fibre architectures.

11.17 Smart skin material for military applications

Protective gear is essential for maintaining employee safety, but it leaves a lot to be desired. For instance, when soldiers are involved in missions in contaminated settings, breathability – that is, the passage of water vapour from the wearer's body to the outside – is essential in protective military uniforms to minimize heat-stress and weariness. Breathability is negatively hampered by the same materials (adsorbents or barrier layers) that offer protection in modern clothing. This is however, an emerging area. Since it is very sensitive, there are not many publications which report the results. However, some scientific publications report some interesting results.

In one of the works [99, 100], researchers led by Lawrence Livermore National Laboratory scientist Francesco Fornasiero have created a smart, breathable fabric that can repel biological and chemical warfare agents. "We demonstrated a smart material that is both breathable and protective by successfully combining two key elements: a base membrane layer comprising trillions of aligned carbon nanotube pores and a threat-responsive polymer layer grafted onto the membrane surface," the lead scientist Fornasiero commented. Here, it is shown how a biostable membrane can quickly, arbitrarily, and irreversibly switch from a highly permeable state in a safe environment to a chemically protective one when threatened by organophosphates like sarin. The physical collapse of an ultrathin copolymer layer on the membrane surface, which effectively blocks transport through membrane pores made of single-walled carbon nanotubes (SWNTs), allows for a dynamic response to chemical stimuli. By using nanometre-wide SWNTs for ultrafast moisture conduction, it is possible to simultaneously increase size-sieving selectivity and water-vapour permeability while overcoming the limitation of conventional membrane materials' breathability/ protection trade-off.

In another article by Bui et al. [101] membranes with high degree of bioprotection has been reported. As a consequence of a concentration-gradient driving force, CNTs show exceptional fast water vapour transport, enabling membranes with sub-5 nm CNTs as conductive pores to feature outstanding breathability while maintaining a high degree of bioprotection by size exclusion. In order to achieve this new paradigm of protective yet breathable adaptive clothing, the researchers recently proposed a chemical threat-responsive membrane made of two parts: a highly breathable CNT-membrane that offers an effective barrier against biological threats, and a thin responsive functional layer grafted or coated on the membrane surface that, upon contact with a chemical warfare agent, either closes the CNT pore entrance or self-exfoliates in the area.

For the first time, the researchers demonstrated that CNT nanochannels can sustain gas-transport rates that are more than an order of magnitude faster than those predicted by Knudsen diffusion theory when a concentration gradient is used as the driving force. The researchers achieved this by quantifying the single-pore permeability to water vapour.

In another very interesting report [102] the idea of smart skin which is being developed by BAE Systems' UK division has been disclosed. With the goal of enabling robots to "feel" their surroundings has been perceive and process information similarly to animals and transmit that information to a machine's "brain." Thousands or even hundreds of thousands of sensors are incorporated into the skin, giving the machines the ability to "sense." All machines could use this incredibly sophisticated skin in future warfare to detect heat, damage, and stress. Armed drones, tanks, other land vehicles, and naval ships might all have smart skin applied to them in the future.

Aircraft smart skin technology can enhance an aircraft's structural performance and enable self-perception, self-diagnosis, self-adaptation, self-learning, and self-repair in addition to other capabilities. An important sort of aircraft smart skin that has recently attracted a lot of attention is aircraft smart skin for structural health monitoring (SHM). Three major issues are preventing aircraft smart skin for SHM from being realized and used in engineering: size, weight, and power consumption. In order to create aircraft smart skin, the primary idea behind it is to integrate a significant number of sensor arrays, actuator arrays, microprocessors, and large-scale signal wires into large-scale skin structures with or without curved surface during the production or assembly process.

Aircraft smart skin for SHM can continuously monitor structural health status and environment parameters online such as strain, damage, impact, corrosion, vibration, acceleration, pressure, temperature, humidity, and pH. Thanks to a large number of sensors and microprocessors integrated with skin structures, advanced signal processing methods, mechanical modelling methods, and onboard SHM systems. There is an interesting review [102] which reports the progress in research on aircraft smart skin. The different kinds of aircraft smart skin have been appraised in detail in Figure 11.22.

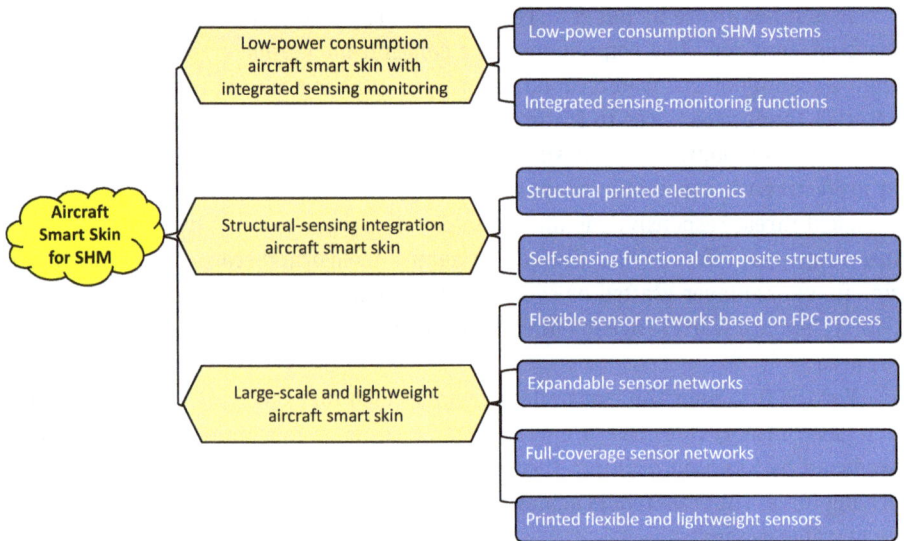

Figure 11.22: Different kinds of smart aircraft smart skin [102].

In a very interesting work reported by Nasa Langley research lab, a smart skin [103] developed for aircrafts can be immensely useful. The Langley Research Center of NASA has created sensor technologies for composite aircraft surface SHM. The outcome of a lightning strike on a conventional aircraft can range from little damage to severe damage requiring costly repairs that can keep the plane out of service for a long time. Before the airplane can actually be struck by lightning when a lightning leader passes through the atmosphere near it, the electromagnetic emissions from the moving electrical charge will radiate the aircraft's surface. The radiated emissions at the aeroplane will intensify as the lightning leader moves closer to it. By design, the lightning's frequency bandwidth falls inside the region of SansEC resonance. As a result, the external oscillating magnetic field of the lightning-radiated emission will passively power the SansEC coil. The coil will produce its own oscillating magnetic and electric fields through resonance. These fields produce what are known as Lorentz forces, which affect the direction and momentum of the lightning attachment and, as a result, affect the location of the aircraft's striking entrance and exit sites and damage. Figure 11.23 illustrates the working of a smart skin aircraft.

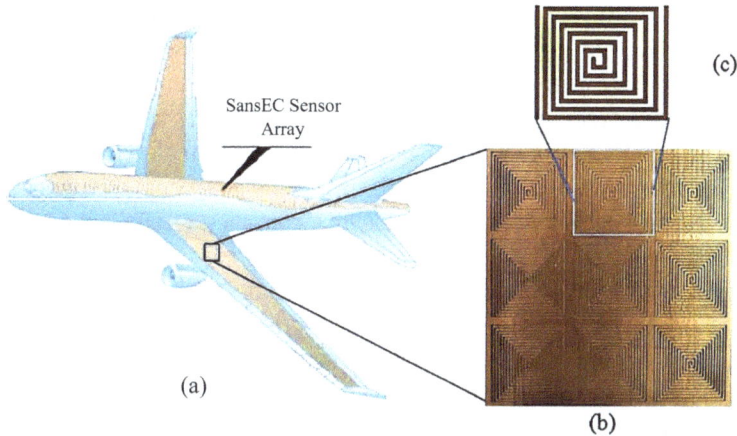

Figure 11.23: Schematic of smart skin for composite aircraft: (a) an aircraft with smart skin; (b) a sensor array including a plurality of sensors; and (c) a sensor [103].

11.18 All other developments in this field

Developments in military textiles sector are a continuous process. With each new emerging products of solution provided by textiles the user feedback keeps the developments fuelled. All over the world, many military-owned research and development organization and establishments are working independently or jointly with industry partners targeting the new product developments and existing product performance improvements.

From the late twentieth century, textiles have started becoming multi-disciplinary field as other fields of science and technology had started to fuse with conventional textile making the invention of smart textiles. One of such revolutions was e-textile or electronic textiles by attaching electrical components directly onto the fabric and interconnect them through the fabric over an arbitrary wiring structure [102]. Using electrically conductive materials in fibrous form, such as metals or carbon, or even conductive polymers such as PANI that can be used as wiring within a piece of fabric. Today the inherently conductive polymers such as PANI, polypyrrole (PPy), and polytiophene (PT) have attained a level where many industrial applications have become reality [104]. Based on the adopted approach, electronic textiles are also known as smart textiles, intelligent textiles, and wearable electronics.

Today, the use of conductive polymers, conductive fibres [copper (Cu) and silver-plated copper (Cu/Ag) filaments, brass (Ms) and silver-plated brass (Ms/Ag) filaments, and aluminium (Al) filaments to copper-clad aluminium (CCA) filaments], conductive yarns, conductive inks for printing, and conductive coating are widely used for making textiles electrically conductive. Some popular fields of application of conductive fabrics are in medical sector – monitoring vital signals of human body, in healthcare sector – special garments for cold weather (WarmX® undershirt which is a knitted sleeveless vest including two heated areas around the kidneys at the front and back, they used silver yarn coated polyamide, powered by a mobile phone sized 12 V battery) [105], military application includes in protective measure against electromagnetic interference, shielding systems [copper-coated nylon fabric – shielding efficiency 40–80 dB at 4 MHz to 1 GHz frequency range] [106].

Chromic textiles, also known as chameleon textiles, are the fabrics or fibres that can change its colour when influenced by external stimuli from the environment. A change in colour presents an opportunity in many textile application domains because colour is so important to our daily life. Chromic textiles are potential to become a powerful tool for monitoring, while also supplying a high degree of safety and comfort to the user [107]. Types of chromism include photochromism (induced by sunlight or UV rays), thermochromism (induced by changes in temperature), electrochromism (induced by electricity), solvatochromism (induced by polarity of the solvent), halochromism (induced by changes in pH value), acidochromism (induced by acids), and ionochromism (induced by ions).

Photochromic textiles change colour reversibly as a result of stimulation by electromagnetic radiation. These structures have been prepared via conventional dyeing techniques to insert photochromic molecules into fibres or via applying microcapsules containing photochromic dyes onto textile surfaces [108]. The best-known classes of photochromic dyes are spiropyrans, spirooxazines, naphthopyrans, diartyltenenes, fulgids, and azobenzenes.

Thermochromic textiles change their molecular or supramolecular structure and absorption spectrum as a result of the variation in environmental temperature. In addition to photochromic compounds, thermochromic compounds can also be used in the development of camouflage patterns for military protective clothing [19].

Electrochromic textiles change colour upon application of an electrical potential. Having a frequency response, high precision, and excellent discoloration ability, meanwhile without changing the normal fabric properties [109]. Thin films of vanadium oxide in polyethylene terephthalate nanofibres coated with poly(3,4-ethylenedioxythiophene)-polystyrene sulfonate (PEDOT:PSS) exhibits electrochromism [110].

When compared to other materials, textiles offer hardness, strength, ductility, flexibility, and biocompatibility properties that enable them to be easily modified to a variety of end-user requirements. Modern-day wearable system becomes more versatile, and with the use of smart chromic material applied textiles, the user can change its look depending on environmental changes and individual preference. This can also be beneficial for a wide range of specific applications such as medical, diagnostic, health-care, fitness, wellness, and environmental.

Since the dawn of the twenty-first century, requirements for protection against hazardous agents like nuclear and chemical radiation along with protection against biological threats have emerged due to the intentional release of chemicals or infectious agents as well as biological weapons made up of living organisms like bacteria, toxins, and viruses. A lot of research focus has shifted towards the protection against radiation and biohazard threats in areas such as industrial, agricultural, and medical work, during military operations and in terrorism attacks. The protective clothing system normally includes respirator, hooded jacket, and trousers or one-piece coverall, gloves, and overboots, individually or together in an ensemble. A wide variety of materials and design of garments are being developed for a particular situation or background. Weight, comfort, level, and duration of protection required are important parameters to be considered [1].

Todays' load carriage system used by soldiers have different designs based on necessity to carry specialist military equipment, such as different types of weaponry, communication equipment, sensor systems, ammunition, and personal survival equipment. The most commonly used load carriage systems are rucksacks, integrated combat kits, cartridge holders and ammo pouch/belts, and hydration packs.

MOLLE is one such latest generation carriage system developed at the U.S. Army Soldier Systems Center, seen to be used by NATO – US army and British armed forces. The whole system is divided into several component packs – assault pack is a backpack with 32 L of storage capacity, one large capacity rucksack (65 L) with attached hydration pouch, one medium capacity rucksack (50 L), and several modular pouches for utility purposes [111]. The whole system is manufactured with four to five different textile fabrics of different specifications. Cordura fabrics and heavy weight nylon fabrics are being used recently. Research is going on to incorporate camouflage effects on to this accessory, weight reduction, and improvement of the quick release mechanism and rapid re-attachment of all the component bags/pouches. Most recent designs of this system come with tactical assault panel, which allows the system to be attached and used along with the outer tactical vest.

11.19 Current challenges and future scope

Military sector has always kept its reliability on the textile sector. With advancements in technical textiles fields, functional fabrics can cater a wide array of problems. Greater scope of opportunities has opened technical textiles with improvements in properties critical to military applications. These advancements not only beneficial to military sector but it has also opened new doors in aviation and aerospace, marine defence sectors as well as making the wholistic improvements in all sector related to defence.

With drastic changes in climatic conditions, the performance of the military uniform and protective clothing should be upgraded. New warfare explosives and use of radiation (nuclear), biological weapons have made new requirement for improving the level of protection. Thus, protection against such threats requires specially tailored solutions. In combination with, other materials – nanomaterials, meta materials, and flexible electronics – have made significant contribution in leading the developments of technical textiles. Whereas improvements in stealth and anti-stealth technology have made it easy to seek out enemy while keeping themselves camouflaged in the surrounding. With such high-tech improvements, carrying out difficult missions becomes easier than before and more research is needed on this.

The deployment of innovative solutions has always been difficult despite all the on-going developments. Despite the improvements occurring at research centres, institutions with government funding for research, and private businesses, the capability of such new technical developments has always been constrained. In addition to capability issues, fund management and distribution problems frequently obstruct the progress of development. In addition, the manufacturing of new materials is still restricted to laboratory or small-scale settings for testing their performance, making it highly challenging to put them into practice or produce the necessary number in a limited amount of time.

Technologies developed by government research facilities are transferred to many private industries after successful trial runs. It eliminates the capacity constrains to some extent as one single technology or solution is being separately developed at multiple industry setups. Thus, producing large volume is not so difficult. The technology developed by private industries due to competitive marketing, business rivalry, earning more profit, and so on holds the potential of the newly developed product or solution reach its full glory. No matter how much unfair it may seem, from a business point of view, it is fair to some extent. A private industry spending time, resources, money, and so on to reach a final solution of any problem may try to have its full profits to itself only, although some industries are taking patent rights restricting its usage rights.

Nanotechnology has huge potential for manufacturing smart textiles for military and defence applications. In fact, there are a variety of military nanotechnology programs being proposed and actively pursued to improve existing systems and materials and to enable the development of new ones but it also has some limitations. The high cost of the majority of nanomaterials, their propensity to aggregate, their toxicity, poor wash durability, and the lack of bulk scale machines for nanotechnology-based processing are a few general problems with nanotechnology that continue to limit the commercialization of nanoenhanced textiles for military and defence applications. The use of nanotechnology in general medicine might also require careful regulation since some of its applications were developed to enhance soliders' endurance and performance.

The suitable power source is one of the main challenges which are being faced by electronic textiles. We frequently take our electrical infrastructure for granted and fail to consider the need for portable power for electronic textiles. Batteries, on the other hand, are pricey and cumbersome, and rechargeable batteries must be taken out and plugged in at night, which is not a real answer. One alternative to collecting the lost energy that we produce in our daily lives is the usage of parasitic power. As the power needs for microelectronics drop, some wearable subsystems may start to take the place of batteries. A project by Joe Paradiso called Parasitic Power Shoes uses excess energy from walking steps to generate electricity [112]. However, energy harvesting does not come without sacrificing comfort, which is why the Parasitic Shoes sacrifice comfort.

11.20 Conclusion

The use of smart textiles within the military is growing enormously due to their advanced capabilities and functionality as well as their ability to provide additional security to soldiers who wear and use them. There are multiple types of smart textiles, some of which have advanced insulation capabilities, ballistic protection, and are made from waterproof fabrics; others have GPS capabilities, sensors, and motion-tracking capabilities. Today's advanced military clothing system is able to endure physical, environmental, physiological, and battlefield stress. A key method of developing technologically advanced uniforms and materials for the military is to use textile-based materials integrated with nanotechnology and electronics. Active and intelligent textiles combined with electronics can improve a soldier's performance by detecting and adjusting to predefined states and responding to situations. Textile and material researchers are working with great effort to find out new applications for functional and smart textiles as well as enhance the performances of existing materials using advanced processing techniques, nanotechnology, and multi-functional materials (chemicals, polymers, and electronics). Currently, the key drivers for the global smart textile market for military applications are (i) increased demand for advanced military textiles, (ii) high level of protection and safety, (iii) use of lightweight textiles, and (iv) miniaturization of electronic materials.

References

[1]   Steffens F, Gralha SE, Ferreira ILS, Oliveira FR. Military Textiles – An Overview of New Developments. Key Eng Mater 2019;812:120–6. https://doi.org/10.4028/www.scientific.net/KEM.812.120.

[2]   Wilusz E, editor. Military Textiles. Elsevier Science; 2008.

[3]   Scott RA, editor. Textiles for Protection. Woodhead Publishing Ltd., Cambridge, England; 2005.

[4]   Wei M, Lin K, Sun L. Shear thickening fluids and their applications. Mater Des 2022;216:110570. https://doi.org/10.1016/j.matdes.2022.110570.

[5]   Li X, Cao HL, Gao S, Pan FY, Weng LQ, Song SH, et al. Preparation of body armour material of Kevlar fabric treated with colloidal silica nanocomposite. Plast Rubber Compos 2008;37:223–6. https://doi.org/10.1179/174328908X309439.

[6]   Decker MJ, Halbach CJ, Nam CH, Wagner NJ, Wetzel ED. Stab resistance of shear thickening fluid (STF)-treated fabrics. Compos Sci Technol 2007;67:565–78. https://doi.org/10.1016/j.compscitech.2006.08.007.

[7]   Pekbey Y, Aslantaş K, Yumak N. Ballistic impact response of Kevlar Composites with filled epoxy matrix. Steel Compos Struct 2017;24:191–200. https://doi.org/https://doi.org/10.12989/scs.2017.24.2.191.

[8]   Sadiku ER, Agboola O, Mochane MJ, Fasiku VO, Owonubi SJ, Ibrahim ID, et al. The Use of Polymer Nanocomposites in the Aerospace and the Military/Defence Industries. Res. Anthol. Mil. Def. Appl. Util. Educ. Ethics, IGI Global; 2021, p. 323–56. https://doi.org/10.4018/978-1-7998-9029-4.ch018.

[9]   Study shows new body armour can benefit from nanotechnology. Phys Org n.d. https://phys.org/news/2006-08-body-armour-benefit-nanotechnology.html%0A.

[10] Acellent's SmartArmor system. Acellent n.d. https://www.acellent.com/industries-2/military%0A.

[11] Zhou J, Wang S, Zhang J, Liu S, Wang W, Yuan F, Gong X. Intelligent body armor: Advanced Kevlar based integrated energy devices for personal safeguarding. Compos Part A: App Sci Manuf 2022;161:107083. https://doi.org/10.1016/j.compositesa.2022.107083.

[12] Fan T, Sun Z, Zhang Y, Li Y, Chen Z, Huang P, et al. Novel Kevlar fabric composite for multifunctional soft body armor. Compos Part B Eng 2022;242:110106. https://doi.org/10.1016/j.compositesb.2022.110106.

[13] Talas L, Baddeley RJ, Cuthill IC. Cultural evolution of military camouflage. Philos Trans R Soc B Biol Sci 2017;372:20160351. https://doi.org/10.1098/rstb.2016.0351.

[14] Krueger GP. Psychological issues in military uniform design. Adv. Mil. Text. Pers. Equip., Elsevier; 2012, p. 64–82e. https://doi.org/10.1533/9780857095572.1.64.

[15] Das A, Alagirusamy R, editors. Technical Textile Yarns. Woodhead Publishing Series in Textiles; 210AD.

[16] Song W-L, Guan X-T, Fan L-Z, Zhao Y-B, Cao W-Q, Wang C-Y, et al. Strong and thermostable polymeric graphene/silica textile for lightweight practical microwave absorption composites. Carbon N Y 2016;100:109–17. https://doi.org/10.1016/j.carbon.2016.01.002.

[17] Gupta S, Chang C, Anbalagan AK, Lee C-H, Tai N-H. Reduced graphene oxide/zinc oxide coated wearable electrically conductive cotton textile for high microwave absorption. Compos Sci Technol 2020;188:107994. https://doi.org/10.1016/j.compscitech.2020.107994.

[18] Ding D, He X, Liang S, Wei W, Ding S. Porous Nanostructured Composite Film for Visible-to-Infrared Camouflage with Thermal Management. ACS Appl Mater Interfaces 2022;14:24690–6. https://doi.org/10.1021/acsami.2c03509.

[19] Karpagam KR, Saranya KS, Gopinathan J, Bhattacharyya A. Development of smart clothing for military applications using thermochromic colorants. J Text Inst 2016:1–6. https://doi.org/10.1080/00405000.2016.1220818.

[20] Sharon M, Rodriguez ASL, Sharon C, Gallardo PS, editors. Nanotechnology in the Defense Industry. Wiley; 2019. https://doi.org/10.1002/9781119460503.

[21]   Yu H, Qi M, Wang J, Yin Y, He Y, Meng H, et al. A feasible strategy for the fabrication of camouflage electrochromic fabric and unconventional devices. Electrochem Commun 2019;102:31–6. https://doi.org/10.1016/j.elecom.2019.03.006.

[22]   Uniforms with programmable fiber could transmit data and more. US Army DEVCOM Army Res Lab Public Aff 2021. https://www.army.mil/article/247472/uniforms_with_programmable_fiber_could_transmit_data_and_more%0A.

[23]   Mizokami K. Smart Fibers Could Turn Army Uniforms Into Wearable Computers. Hear Mag Media, Inc 2021. https://www.popularmechanics.com/military/research/a36732071/army-uniform-fibers-create-wearable-computers/%0A.

[24]   Loke G, Khudiyev T, Wang B, Fu S, Payra S, Shaoul Y, et al. Digital electronics in fibres enable fabric-based machine-learning inference. Nat Commun 2021;12:3317. https://doi.org/10.1038/s41467-021-23628-5.

[25]   Sharon M, Rodriguez AS, Sharon C, Gallardo PS. Nanotechnology in the Defense Industry Advances, Innovation, and Practical Applications. John Wiley & Sons; 2019.

[26]   Zohuri B. Radar Energy Warfare and the Challenges of Stealth Technology. Berlin: Springer; 2020.

[27]   Mouritz AP. Introduction to aerospace materials. Elsevier; 2012.

[28]   Peng L, Liu D, Cheng H, Zhou S, Zu M. A Multilayer Film Based Selective Thermal Emitter for Infrared Stealth Technology. Adv Opt Mater 2018;6:1801006. https://doi.org/10.1002/adom.201801006.

[29]   Zhou X, Xin B, Liu Y. Research progress on infrared stealth fabric. J Phys Conf Ser 2021;1790:012058. https://doi.org/10.1088/1742-6596/1790/1/012058.

[30]   Kumar N, Dixit A. Nanotechnology for defence applications. Springer International Publishing; 2019.

[31]   Spicer M. Illustrated manual of sniper skills. Pen and Sword; 2016.

[32]   Gu J, Wang W, Yu D. Temperature-control and low emissivity dual-working modular infrared stealth fabric. Colloids Surfaces A Physicochem Eng Asp 2022;653:129966. https://doi.org/10.1016/j.colsurfa.2022.129966.

[33]   Gu J, Wang W, Yu D. Temperature control and low infrared emissivity double-shell phase change microcapsules and their application in infrared stealth fabric. Prog Org Coatings 2021;159:106439. https://doi.org/10.1016/j.porgcoat.2021.106439.

[34]   Pawar AY, Sonawane DD, Erande KB, Derle DV. Terahertz technology and its applications. Drug Invent Today 2013;5:157–63. https://doi.org/10.1016/j.dit.2013.03.009.

[35]   Shumyatsky P, Alfano RR. Terahertz sources. J Biomed Opt 2011;16:033001. https://doi.org/10.1117/1.3554742.

[36]   Ghann W, Uddin J. Terahertz (THz) Spectroscopy: A Cutting-Edge Technology. Terahertz Spectrosc. – A Cut. Edge Technol., InTech; 2017. https://doi.org/10.5772/67031.

[37]   Kurabayashi T, Saitoh F, Watanabe N, Tanno T. Identification of textile fiber by terahertz spectroscopy. 35th Int. Conf. Infrared, Millimeter, Terahertz Waves, 2010, p. 1–2.

[38]   Sharma V, Arya D, Jhildiyal M. Terahertz technology and its applications. 5th Int. Conf. Adv. Comput. Commun. Technol., 2011, p. 175–8.

[39]   Ergün S, Sönmez S. Terahertz technology for military applications. J Manag Inf Sci 2015;3:13–6.

[40]   Sethy PK, Mishra PR, Behera S. An introduction to terahertz technology, its history, properties and application. Int. Conf. Comput. Commun., 2015.

[41]   Hu J. The use of smart materials in cold weather apparel. In Textiles for cold weather apparel. Woodhead Publishing; 2009.

[42]   Extreme cold weather clothing. Wikipedia, Free Encycl n.d. https://www.wikiwand.com/en/Extreme_cold_weather_clothing.

[43]   wind chill. Britannica n.d. https://www.britannica.com/science/wind-chill.

[44]   Wind Chill Chart. US Dept Commer Natl Ocean Atmos Adm Natl Weather Serv n.d. https://www.weather.gov/safety/cold-wind-chill-chart.

[45] DEVELOPMENTS IN COLD WEATHER CLOTHING. Ann Occup Hyg 1975. https://doi.org/10.1093/ann hyg/17.3-4.279.

[46] Extended "extreme" cold weather clothing system. Saf One Train Int Inc n.d. https://safetyoneinc. com/2017/01/24/extended-extreme-cold-weather-clothing-system/%0A.

[47] Auerbach MA, Jugueta RD. Candidate Fabrics for the 2nd Generation Extended Cold Weather Clothing System. 1997.

[48] Extended Climate Warfighter Clothing System. MilitaryCom n.d. https://www.military.com/equip ment/extended-climate-warfighter-clothing-system-gen-iii%.

[49] Extended Cold Weather Clothing System. TacticalGearCom n.d. https://tacticalgear.com/experts/ex tended-cold-weather-clothing-system.

[50] Generation III Extended Cold Weather Clothing System. CIE Hub n.d. https://ciehub.info/clothing/ CW/ECWCS/GEN3.html.

[51] Base Layers: UnderCover Comfort. Tru-Spec n.d. https://www.truspec.com/base-layers/gen-iii-ecwcs

[52] Candidate fabrics for the 2nd Generation Extended Cold Weather Clothing System n.d. https://apps. dtic.mil/sti/pdfs/ADA336776.pdf.

[53] How to Choose Sleeping Bags for Camping. Recreat Equipment, Inc n.d. https://www.rei.com/learn/ expert-advice/sleeping-bag.html%0A.

[54] Leslie D. What Temperature Sleeping Bag do I need? 2015. https://www.snowys.com.au/blog/what-temperature-sleeping-bag-do-i-need/%0A.

[55] A Guide to Down Jacket Warmth: Down Fill Power vs Down Weight. Triple FAT Goose 2021. https://triplefatgoose.com/blogs/down-time/a-guide-to-down-jacket-warmth-down-fill-power-vs-down-weight.

[56] What Is Down Fill Power? Recreat Equipment, Inc n.d. https://www.rei.com/learn/expert-advice/ what-is-down-fill-power.html%0A.

[57] Seek the unseen. Primaloft n.d. https://www.primaloft.com/%0A.

[58] Climashield Apex continuous filament insulation 133g/sqm, 4oz/sqyd. Extremtextil n.d. https://www.extremtextil.de/en/climashield-apex-continuous-filament-insulation-133g-sqm-4oz-sqyd.html.

[59] Meyers J. Down vs. Synthetic sleeping bags & quilts: how to choose insulation 2021. https://www. thermarest.com/blog/synthetic-vs-down-sleeping-bag/.

[60] Song WF, Zhang CJ, Lai DD, Wang FM, Kuklane K. Use of a novel smart heating sleeping bag to improve wearers' local thermal comfort in the feet. Sci Rep 2016;6:19326. https://doi.org/10.1038/ srep19326.

[61] Kumar B. S, T. M. Thermal property analysis of tri layer composite fabric towards the utility of sleeping bag. Res J Text Appar 2022;26:124–37. https://doi.org/10.1108/RJTA-04-2020-0032.

[62] An Y-Y, Tu L-X, Shen H, Xu G-B, Zhang G-R, Zhu H-Q, et al. Numerical simulation and validation on heat transfer of four structures of sleeping bag. Int Commun Heat Mass Transf 2021;129:105707. https://doi.org/10.1016/j.icheatmasstransfer.2021.105707.

[63] Revaiah RG, Kotresh TM, Kandasubrmanian B. Military customized fabric hyperbaric chamber – design and safety aspects. J Text Inst 2022;113:869–81. https://doi.org/10.1080/00405000.2021. 1907988.

[64] Revaiah RG, Kotresh TM, Kandasubramanian B. Technical textiles for military applications. J Text Inst 2020;111:273–308. https://doi.org/10.1080/00405000.2019.1627987.

[65] HAPO chamber. DRDO, Minist Defence, Gov India n.d. https://www.drdo.gov.in/hapo-chamber%0A.

[66] Luks AM. Physiology in Medicine: A physiologic approach to prevention and treatment of acute high-altitude illnesses. J Appl Physiol 2015;118:509–19. https://doi.org/10.1152/japplphysiol.00955. 2014.

[67] Dubois C, Herry J-P, Kayser B. Portable hyperbaric medicine, some history. J Wilderness Med 1994;5:190–8. https://doi.org/10.1580/0953-9859-5.2.190.

[68]   Freeman K, Shalit M, Stroh G. Use of the Gamow Bag by EMT-Basic Park Rangers for Treatment of High-Altitude Pulmonary Edema and High-Altitude Cerebral Edema. Wilderness Environ Med 2004;15.

[69]   Lee DS, Iacocca M, Joshi YK. Energy usage modeling for heating and cooling of off-grid shelters. J Build Eng 2021;35:102054. https://doi.org/10.1016/j.jobe.2020.102054.

[70]   Parachute. Britannica n.d. https://doi.org/https://www.britannica.com/technology/parachute.

[71]   Glide ratio. An Illus Dict Aviat by McGraw-Hill Companies, Inc 2005. https://encyclopedia2.thefreedictionary.com/glide+ratio.

[72]   Glider flying handbook. Aviation Supplies & Academics; 2004.

[73]   Reeves E. What's a Baffle? Sleeping bag technology explained 2017. https://www.sierra.com/blog/camping/whats-baffle-parts-sleeping-bags-explained/.

[74]   Specifications – Military Parachute / Canopy with Parachute Support Fabrics n.d. https://perftex.com/wp-content/uploads/2022/01/military-specifications.pdf.

[75]   Joshi M, Adak B. Introduction to LTA Systems. Coat. Laminated Text. Aerostats Airships, Boca Raton: CRC Press; 2022, p. 1–32. https://doi.org/10.1201/9780429432996-1.

[76]   Joshi M, Adak B, Varshney S, Tiwari R. Laminated Textiles for the Envelope of LTA Vehicles. Coat. Laminated Text. Aerostats Airships, Boca Raton: CRC Press; 2022, p. 141–64. https://doi.org/10.1201/9780429432996-5.

[77]   Adak B, Joshi M. Coated or Laminated Textiles for Aerostat and Stratospheric Airship. Adv. Text. Eng. Mater., Hoboken, NJ, USA: John Wiley & Sons, Inc.; 2018, p. 257–87. https://doi.org/10.1002/9781119488101.ch7.

[78]   Joshi M, Adak B, Butola BS. Polyurethane nanocomposite based gas barrier films, membranes and coatings: A review on synthesis, characterization and potential applications. Prog Mater Sci 2018;97:230–82. https://doi.org/10.1016/j.pmatsci.2018.05.001.

[79]   Adak B, Joshi M. Recent Developments in Gas Barrier Polymer Nanocomposite Coatings. Nanotechnol. Text., Jenny Stanford Publishing; 2020, p. 661–94. https://doi.org/10.1201/9781003055815-22.

[80]   Adak B, Parasuram S, Joshi M, Butola BS. Modelling for Performance Analysis of Aerostats/Airships. Coat. Laminated Text. Aerostats Airships, Boca Raton: CRC Press; 2022, p. 299–316. https://doi.org/10.1201/9780429432996-10.

[81]   Joshi M, Adak B, Chatterjee U. Polyurethane Nanocomposite-Based Advanced Materials for Aerostat/Airship Envelopes. Coat. Laminated Text. Aerostats Airships, Boca Raton: CRC Press; 2022, p. 165–97. https://doi.org/10.1201/9780429432996-6.

[82]   Adak B, Butola BS, Joshi M. Effect of organoclay-type and clay-polyurethane interaction chemistry for tuning the morphology, gas barrier and mechanical properties of clay/polyurethane nanocomposites. Appl Clay Sci 2018;161:343–53. https://doi.org/10.1016/j.clay.2018.04.030.

[83]   Adak B, Joshi M, Butola BS. Polyurethane/functionalized-graphene nanocomposite films with enhanced weather resistance and gas barrier properties. Compos Part B Eng 2019;176:107303. https://doi.org/10.1016/j.compositesb.2019.107303.

[84]   Adak B, Joshi M, Butola BS. Polyurethane/clay nanocomposites with improved helium gas barrier and mechanical properties: Direct versus master-batch melt mixing route. J Appl Polym Sci 2018;135:46422. https://doi.org/10.1002/app.46422.

[85]   Adak B, Butola BS, Joshi M. Calcination of UV shielding nanopowder and its effect on weather resistance property of polyurethane nanocomposite films. J Mater Sci 2019;54:12698–712. https://doi.org/10.1007/s10853-019-03739-7.

[86]   Ahmed A, Adak B, Bansala T, Mukhopadhyay S. Green Solvent Processed Cellulose/Graphene Oxide Nanocomposite Films with Superior Mechanical, Thermal, and Ultraviolet Shielding Properties. ACS Appl Mater Interfaces 2020;12:1687–97. https://doi.org/10.1021/acsami.9b19686.

[87]   Joshi M, Sandhoo R, Adak B. Nano-ceria and nano-zirconia reinforced polyurethane nanocomposite-based coated textiles with enhanced weather resistance. Prog Org Coatings 2022;165:106744. https://doi.org/10.1016/j.porgcoat.2022.106744.

[88]   Nayak R, Wang L, Padhye R. Electronic textiles for military personnel. Electron. Text., Elsevier; 2015, p. 239–56. https://doi.org/10.1016/B978-0-08-100201-8.00012-6.

[89]   Berzowska J. Electronic Textiles: Wearable Computers, Reactive Fashion, and Soft Computation. Textile 2005;3:58–75. https://doi.org/10.2752/147597505778052639.

[90]   Mishra RK, Martín A, Nakagawa T, Barfidokht A, Lu X, Sempionatto JR, et al. Detection of vapor-phase organophosphate threats using wearable conformable integrated epidermal and textile wireless biosensor systems. Biosens Bioelectron 2018;101:227–34. https://doi.org/10.1016/j.bios. 2017.10.044.

[91]   Sempionatto JR, Mishra RK, Martín A, Tang G, Nakagawa T, Lu X, et al. Wearable Ring-Based Sensing Platform for Detecting Chemical Threats. ACS Sensors 2017;2:1531–8. https://doi.org/10. 1021/acssensors.7b00603.

[92]   Wearable computers help make individual soldiers part of the digital battlefield. Mil Aerosp Electron ENewsletters 2002. https://www.militaryaerospace.com/computers/article/16710872/wear able-computers-help-make-individual-soldiers-part-of-the-digital-battlefield%0A.

[93]   Park S, Jayaraman S. Wearables: Fundamentals, advancements, and a roadmap for the future. Wearable Sensors, Elsevier; 2021, p. 3–27. https://doi.org/10.1016/B978-0-12-819246-7.00001-2.

[94]   McLoughlin J, Sabir T, editors. High-Performance apparel: Materials, development, and applications. Woodhead Publishing; 2017.

[95]   Pandey P, Thapa K, Ojha GP, Seo M-K, Shin KH, Kim S-W, et al. Metal-organic frameworks-based triboelectric nanogenerator powered visible light communication system for wireless human-machine interactions. Chem Eng J 2023;452:139209. https://doi.org/10.1016/j.cej.2022.139209.

[96]   Teo W-E, Ramakrishna S. Electrospun nanofibers as a platform for multifunctional, hierarchically organized nanocomposite. Compos Sci Technol 2009;69:1804–17. https://doi.org/10.1016/j.compsci tech.2009.04.015.

[97]   Deeraj BDS, Shebin KJ, Bose S, Sampath S, Joseph K. Electrospun carbon fibers embedded with core–shell TiC@TiO nanostructures and their epoxy composites for potential EMI shielding in the Ku band. Nano-Structures & Nano-Objects 2022;32:100912. https://doi.org/10.1016/j.nanoso.2022.100912.

[98]   Guo H, Wang F, Luo H, Li Y, Lou Z, Ji Y, et al. Flexible TaC/C electrospun non–woven fabrics with multiple spatial-scale conductive frameworks for efficient electromagnetic interference shielding. Compos Part A Appl Sci Manuf 2021;151:106662. https://doi.org/10.1016/j.compositesa.2021.106662.

[99]   Second Skin Protects Against Chemical Weapons, Biological Warfare Agents. Lawrence Livermore Natl Lab 2020. https://scitechdaily.com/second-skin-protects-against-chemical-weapons-biological-warfare-agents%0A.

[100] Li Y, Chen C, Meshot ER, Buchsbaum SF, Herbert M, Zhu R, et al. Autonomously Responsive Membranes for Chemical Warfare Protection. Adv Funct Mater 2020;30:2000258. https://doi.org/10. 1002/adfm.202000258.

[101] Bui N, Meshot ER, Kim S, Peña J, Gibson PW, Wu KJ, et al. Ultrabreathable and Protective Membranes with Sub-5 nm Carbon Nanotube Pores. Adv Mater 2016;28:5871–7. https://doi.org/10. 1002/adma.201600740.

[102] Wang Y, Hu S, Xiong T, Huang Y, Qiu L. Recent progress in aircraft smart skin for structural health monitoring. Struct Heal Monit 2022;21:2453–80. https://doi.org/10.1177/14759217211056831.

[103] Smart Skin for Composite Aircraft (LAR-TOPS-129): For lightning strike protection and damage sensing on aircraft. NASA's Technol Transf Progr n.d. https://doi.org/https://technology.nasa.gov/ patent/LAR-TOPS-129.

[104] Ulrich H. Introduction to industrial polymers. Munich: Hanser Publishers; 1982.

[105] Van Langenhove L. Smart textiles for medicine and healthcare: materials, systems and applications. Woodhead Publishing Limited; 2007.

[106] Albrecht W, Fuchs H, Kittelmann W, editors. Nonwoven fabrics: raw materials, manufacture, applications, characteristics, testing processes. John Wiley & Sons; 2006.

[107] Chowdhury MA, Joshi M, Butola BS. Photochromic and Thermochromic Colorants in Textile Applications. J Eng Fiber Fabr 2014;9:155892501400900. https://doi.org/10.1177/155892501400900113.

[108] Cheng T, Lin T, Brady R, Wang X. Fast response photochromic textiles from hybrid silica surface coating. Fibers Polym 2008;9:301–6. https://doi.org/10.1007/s12221-008-0048-7.

[109] Sheng M, Zhang L, Wang D, Li M, Li L, West JL, et al. Fabrication of dye-doped liquid crystal microcapsules for electro-stimulated responsive smart textiles. Dye Pigment 2018;158:1–11. https://doi.org/10.1016/j.dyepig.2018.05.025.

[110] Eren E, Karaca GY, Alver C, Oksuz AU. Fast electrochromic response for RF-magnetron sputtered electrospun V2O5 mat. Eur Polym J 2016;84:345–54. https://doi.org/10.1016/j.eurpolymj.2016.09.027.

[111] Physiological, Biomechanical, and Maximal Performance Comparisons of Female Soldiers Carrying Loads Using Prototype U.S. Marine Corps Modular Lightweight Load-Carrying Equipment (MOLLE) with Interceptor Body Armor and U.S. Army All-Purpose Lightweight Ind. 1999.

[112] Kymissis J, Kendall C, Paradiso J, Gershenfeld N. Parasitic power harvesting in shoes. Dig. Pap. Second Int. Symp. Wearable Comput. (Cat. No.98EX215), IEEE Comput. Soc; n.d., p. 132–9. https://doi.org/10.1109/ISWC.1998.729539.

Akanksha Pragya, Kony Chatterjee, Tushar K. Ghosh

12 Sensors and actuators for textiles: from materials to applications

Abstract: The integration of electrical functionalities into textiles offers exponentially expanding opportunities for new smart and connected products with capabilities that cut across traditional product boundaries. The hierarchical structure of textiles, from fibres and yarn to fabrics, and the myriad interlacement patterns attainable through weaving, knitting, and non-woven technologies provide numerous possibilities for ubiquitous integration of electrical devices such as sensors and actuators. While electronic textiles (e-textiles) offer new disruptive opportunities in applications such as healthcare, it also poses certain challenges that arise from the need to preserve the inherent desirable qualities of textiles such as comfort, flexibility, conformability, strength, and breathability. Since textiles take up the intimate space around the human body, sensors deployed therein can sense numerous useful parameters like motion, stress, strain, moisture, chemicals, and bio-signals corresponding to the wearer's body. To implement different sensing modalities (e.g., capacitive and resistive) into textiles, various techniques such as geometric microengineering (e.g., multimaterial layers) and surface functionalization (e.g., in situ polymerization) can be used. While sensors provide useful physiological and environmental information, fibre- and yarn-based actuators offer the capabilities to create responsive textiles that can react to environmental or applied stimuli such as moisture, heat, and electricity. Sensors and actuators complement each other, and a truly "smart" textile system of the future should include these two functionalities together to trigger a response within the textiles using integrated sensors, thereby creating a closed-loop holistic system. This chapter presents the research advances on sensors and actuators for e-textiles with their significance in a functional system. The discussion extends into relevant measurement principles, materials, and methods potentially compatible with textile products and processes. Finally, a reflection on the current challenges and possible future directions in sensors and actuators for e-textiles is made.

Keywords: e-textiles, sensors, actuators, dielectric polymers, conducting polymers, elastomers

12.1 Introduction

Textiles have been used for comfort and protection against the environment since the earliest of human civilization. Humans have used it to convey their personal identity and societal status. Arguably, fibres have been used to create textile structures even

https://doi.org/10.1515/9783110759747-012

before anatomically modern humans began using fibres to make textiles [1]. Over the thousands of years, fibrous structures or textiles have expanded into numerous technical products because of their unique combination of strength, flexibility, porosity, and so on [2, 3]. In most applications, however, textiles have been used as "passive" components that did not need any external power source to perform their inherent functions. However, since the late 1980s, there has been a growing interest in developing textiles with "active" functional capabilities by incorporating electronic devices. The promise of the convergence of textiles and electronics has driven efforts towards seamless and massive integration of sensors and actuators into textiles. These textiles with electrical functionalities or e-textiles are capable of adapting and/or responding to the environment around them [4]. They can be used for a myriad of applications, including healthcare [5], communication [6], safety [7], energy harvesting [8], and personalized heating and cooling [2].

Sensors and actuators are essential devices for any functional electrical system of today and in the future. Sensor and actuator technologies enable every smart and automated system, from interconnected home devices such as door locks, televisions, thermostats, lights, and even appliances to smart buildings, autonomous vehicles, power grids, and so on. The smart and interconnected technologies evolving in the forms of cyber-physical systems [9, 10] and the Internet of things (IoT) [11, 12] depend critically on the fast and efficient sensor and actuator networks. These technologies have the potential to revolutionize every aspect of human society, including education, agriculture, healthcare, and energy efficiency, and thereby improve the quality of life for billions of people around the globe.

In a typically smart and connected system, an embedded sensor network is used to collect relevant information and share it with other devices and systems over the Internet for analysis using appropriate algorithms and sending alerts to users, and even triggering an automated response from an actuator in proportion to that sensed input. Thus, sensors and actuators in a functional system are inextricably linked and work together from opposite ends. Working in a similar manner, e-textiles can be part of the IoT ecosystem to collect, send, and respond to data they acquire from the human environments. For instance, e-textiles or smart textiles that can communicate physiological information such as heart rate, temperature, breathing, stress, movement, and posture can be transformative. Figure 12.1 presents a textile-based smart personal thermal comfort system (PCTS) proposed recently [2]. As part of an interconnected system, the autonomous PCTS should also allow the users to regulate their intimate environment's conditions (temperature, humidity, etc.), independent of the building's temperature. The system is designed to sense (using temperature and moisture sensors) and respond to the changes in the environment around the wearer's body. Thermoregulation in such a system can be achieved by incorporating devices like moisture or temperature-responsive actuators and heating elements into them. Similar systems are envisioned for healthcare, communications, and many other applications.

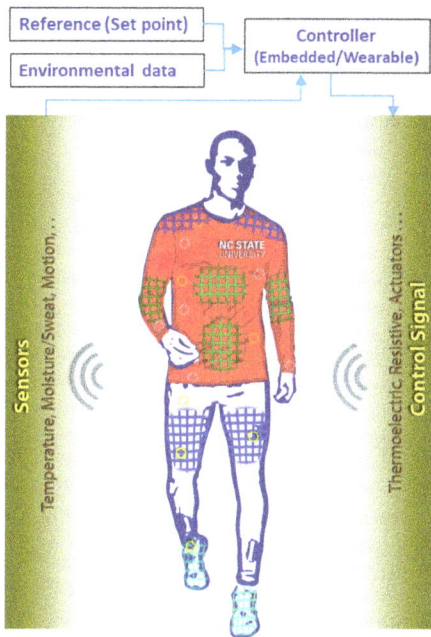

Figure 12.1: Conceptual depiction of an ideal smart textile system that combines responsive actuators with sensors to provide a personal thermal comfort system with minimal user intervention (reproduced with permission from [7]. Copyright 2020 WILEY-VCH Verlag GmbH & Co. KGaA, Weinheim).

Although significant progress has been made in the field, commercially successful products have been elusive so far. There are many challenges in designing textile-based smart systems. In an ideal e-textile system, smart functionalities should be integrated unobtrusively to preserve the unique and essential "textile" characteristics of strength, flexibility, texture, softness, porosity, and so on. The process should begin with a fundamental understanding of the hierarchical nature of textile structures and their unique methods of assembly. For the most part, the desirable properties of textiles (e.g., flexibility) arise from the hierarchical structure of the fibre assembly through the complex interaction of the inherent fibre material properties and the characteristic fibre structure at multiple length scales. Electronic functionalities can be engineered into textiles at one or more of the hierarchical levels of polymers, fibres, yarns, or fabrics, using processes compatible with textiles. The choice here is critical to the product's performance and, ultimately, the practical utility of the technology.

A summary of the research advances on sensors and actuators for e-textiles is presented in this chapter. It begins with a brief introduction to sensors, and their significance in a functional system, followed by their classification and working principles. A similar discussion on actuators is presented next. In all cases, the focus here is primarily on materials, methods, and devices that are potentially compatible with textiles; therefore, the discussion is limited to soft and flexible fibre/textile-based devices. Finally, a discussion on challenges and future directions in sensors and actuators for e-textiles is presented.

12.2 Textile-based sensors

Sensors are all around us, anywhere we live, eat, work, and play. Among the numerous things they do, they regulate the temperature around us, turn on the lights when we enter a dark room, trigger the fire alarm in case of smoke or fire, or help park our car. Sensors make every automated system work. A sensor is a device that converts a physical (light, proximity, touch, etc.) or chemical (gases, humidity, etc.) stimulus to a meaningful electrical or optical signal. Most textile-based sensors are designed to be in our intimate space, typically in our clothing, to measure physiological parameters such as temperature, blood pressure, and heart rate from the wearer's body. With the recent rise in wearable sensors, especially in healthcare applications, the role of textile-based sensors in the healthcare delivery of the future has become even more important. The potential applications include the monitoring of simple human body motions (breathing, speaking, blinking, and joint motions), to the more complex tasks of monitoring metabolic, cardiovascular, sleep, neurologic, and muscular activities. Similarly, textile sensors can also be integrated with any household textile product, such as mattress covers in our beds, or technical textile products such as airbags in our cars. In all applications, its role is to monitor one or more application-specific stimuli.

The conventional semiconductor-based sensors are rigid and not readily compatible with textile products and processes. In contrast, the textile-based sensors must be lightweight, flexible, conformable, and relatively easy to include in the textile fabrication process to ensure seamless integration with textile products. These requirements present complex fundamental challenges due to low electrical conductivity, mechanical nonlinearity, viscoelastic losses, soft/hard interfaces, safety, environmental stability, and scalable fabrication. Despite these and other formidable challenges, a significant progress is being made in this interdisciplinary field.

Successful development of functional e-textile devices and systems within the constraints of a given application requires an understanding of the application as well as the underlying challenges. In the following sections, an overview of textile-based sensors, classified on the basis of their transduction mechanism, is presented. The discussion comprises a detailed survey of the current sensing concepts, methodologies, and useful metrics commonly used to compare sensor performance, highlighting their advantages and disadvantages.

12.2.1 Materials for sensor fabrication and integration

In an electronic textile system, sensor(s) can be incorporated at any level of the hierarchy of the structure: fibre, yarn, fabric, or product. In all cases, however, the primary material for fabrication is polymeric. Common fibre-forming polymers (cellulose, polyester, polyamide, etc.) are electrically insulating and otherwise (optically or chemically) mostly functionally inert. Hence, sensor fabrication often involves imparting electrical, optical, or chemical functionalities to textile materials [13, 14].

For most e-textile sensor fabrication, electrical conductivity is one of the necessary attributes to perform its sensory function. Electrical conductivity can be imparted to a textile structure using metals [15–17], inherently conducting polymers (CPs) [18–20], conductive polymer composites (CPCs) [21, 22], among other materials such as metal-organic frameworks (MOFs) [21, 23], carbonaceous materials [such as carbon black (CB), carbon nanotubes (CNTs), and graphene] [24–26], and MXenes [13, 27]. These materials could be used in the shape of fibres or in any other suitable form such as a printed layer or interconnect.

The use of metals (gold, silver, copper, nickel, etc.) in the fabrication of e-textiles is limited by their relatively high stiffness and low extensibility compared to fibre-forming polymers. These shortcomings can be mitigated to a certain extent by using geometrical shapes or structural features to develop flexibility. Besides conductive coatings [28, 29], buckled structures [30] and serpentine lines [31] have been used in the fabrication of stretchable and flexible metallic e-textile components, including sensors.

Intrinsically conductive polymers (ICPs) constitute an important class of organic functional materials. Shirakawa, Louis, and MacDiarmid first discovered the electrically CPs in 1977 via doping the semiconducting polymer, *trans*-polyacetylene, with various halogens [32]. Since the discovery, ICPs ushered in a new era of polymeric materials that combine the electrical, magnetic, and optical properties of metals or semiconductors and the mechanical and facile processing properties of polymers. The first known stable metallic polymer, processible in the metallic form, was polyaniline (PANI) [33].

In addition to excellent electrical and optical properties [34, 35], ICPs present many attractive attributes such as lightweight, tunable electrical conductivity and environmental stability, ease of processing, and economic viability. Besides PANI, other notable ICPs for sensor applications include polypyrrole (PPy), poly(3,4-ethylenedioxythiophene):polyelectrolyte poly(styrenesulphonate) (PEDOT:PSS), and polythiophene. In general, these materials may be either insulators or semiconductors in their undoped form; upon doping, charges are injected into the polymer macromolecule through the redox reaction to form delocalized charge carriers [36, 37].

Carbon and carbonaceous materials have been widely used in textile sensor fabrication as well as in other e-textile devices for their relatively lightweight, excellent mechanical and electrical properties, and specific surface area for functionalization.

Among the notable carbon materials are the traditional carbon fibres and CB [38, 39], the nanostructured carbon nanofibres (CNFs) [40], single-walled CNTs (SWCNTs) and multiwalled CNTs (MWCNTs) [41, 42], and graphene [43, 44]. For sensing applications, carbonaceous materials are most commonly used in polymer composites with a wide range of polymer matrices, often applied in the form of paste or conductive ink as a coating on textile substrates [45, 46], or spun into fibres [43, 47]. Alternatively, carbonaceous materials can be directly coated onto textile substrates [42].

Historically, CB is the most commonly used conductive particle to induce conductivity in polymers. The single CB particle, with its diameter in hundreds of nanometres, fuses together in three-dimensional aggregates. The aggregate structure, morphology, and microporosity of CB particles for the CB-filled composites greatly affect the percolation threshold (see the discussion on CPCs later in the section). Similarly, the high aspect ratio CNFs, having a diameter in the range of 10–500 nm and a length between 0.5 and 200 μm with good electrical properties, are considered an important filler to impart electrical conductivity to otherwise insulating polymers [48].

Amongst the nanostructured carbon materials, two forms of CNTs are widely used: the SWCNTs having a diameter of 1–3 nm [49] and the MWCNTs that are composed of a series of nested individual tubes with a net diameter ranging from ~10 nm to over 70 nm [50, 51]. Continuous yarns made of graphene and CNTs have been prepared via wet spinning, dry spinning from arrays (or forest), spinning from dispersion, mechanical twisting of ribbons, in situ spinning, and so on [43, 47, 52–57]. CNT films (or sheets) are also prepared by direct growth techniques, solution-based deposition, dispersion of CNTs into films, and so on [58–61]. These materials show promising advantages as electrode materials (specifically in capacitors) due to high specific area, mechanical reliability under mechanical deformation, ionic mobility, and other things. Graphene films can be prepared via thermal annealing, casting, foaming, and so on [62–64].

Electromechanical properties of carbon-based materials can be further improved by hybridization with ICPs (e.g., PANI, PEDOT:PSS, PPy, etc.) [18–20], metals (e.g., iron, copper, etc.) [15–17], or other carbonaceous materials (e.g., carbon/graphene, graphene/CNTs, and graphene quantum dot/carbon nanofibre) [24, 25, 44]. For example, graphite with a laminar structure of tens of microns in length, combined with submicron spherical CB particles, is more conductive than anyone used as a single filler. Smaller CB particles are more likely to bridge the interstices between the larger graphite particles to form conductive pathways [65].

As a class of functional material, flexible and soft polymer composites have been a subject of investigation for many years because they can effectively combine the respective advantages of each component, often synergistically, thus creating versatile functionalities with outstanding processability [66, 67]. Depending on the choice of filler particles, one or more target attributes (e.g., electrical, thermal, or chemical) of the polymer can be enhanced.

One of the most important groups of polymer composites, the CPCs, is formed by embedding conductive particles (e.g., carbonaceous, metallic, or ICPs) into a polymer matrix. The electrical conductivity of CPCs depends critically on the concentration of the conductive particles. Above a threshold concentration, described as the percolation threshold, conductive networks are formed throughout the CPCs that allow electrical charge transfer between the conducting segments that are dispersed throughout the polymeric matrix [68]. Percolation represents a standard model to describe the distribution of species within disordered systems. Percolation is defined as the development of long-range connectivity (network) in a random system [69, 70].

The electrical conductivity of a CPC is determined by both the filler and the matrix polymer properties as well as its structure. The intrinsic electrical properties of the filler material [71], its size [72], its shape [73], as well as its level of dispersion in the matrix are all critical to the performance of the CPC [74]. Since the electrical charge transfer with the CPC depends on the filler properties, spatial distribution of the filler, and so on, a tiny change in its external environment may result in a corresponding change of the conductive network, causing a remarkable change in the electrical resistance of the CPCs. Therefore, in sensor applications, the CPCs have been widely used to measure stress/strain [75], temperature [76], vapour [77], biofluids [78], and so on.

Certain sensors (such as capacitive and piezoelectric) use dielectrics in their structure which acts as the responsive element to stimuli. A dielectric is an electrical insulator that can be polarized by applying an electric field or mechanical stress. Commonly used dielectric materials include polydimethylsiloxane (PDMS), polytetrafluoroethylene (PTFE), polyvinylidene fluoride (PVDF), poly(vinylidene fluoride–trifluoroethylene) copolymer, polylactic acid (PLA), polyurethane (PU), polyimide (PI), and polyamides. The dielectrics maybe used in the form of yarn, film, foam, or fabric (woven, knitted, nonwoven, or 3D spacer) [79–82].

Hydrogels containing conductive and chemically or optically active materials are being increasingly used in textile-based sensor due to their good dispersion properties, eco-friendliness, and ability to preserve functional agents over long time [83, 84].

Methods of fabrication used for textile sensors depend primarily on the type of sensors as well as the materials being used. While the focus of this section is limited to materials and their forms relevant to textile sensors, the sensor fabrication process may include any combinations of the well-developed conventional methods for organic microelectronics such as coating [29], chemical/physical vapour deposition [85, 86], printing [48], in situ polymerization [87], or traditional textile processes such as electrospinning [26, 88], braiding [56], twisting/plying [89, 90], core-spinning [91, 92], weaving [93–95], and knitting [91, 96, 97].

Individual sensor or a sensor array can be formed in a fibre or a yarn configuration for integration in the fabric [98] or only upon assembly of the yarns in a fabric [99]. Common methods to integrate sensors into textiles involve stitching [98], embroidery [100], weaving [101], and knitting [102]. Embroidery is preferred as a more precise, simple, and flexible process [100].

12.2.2 Capacitive sensors

Capacitive sensor is one of the most common in currently used systems. The same is true for the emerging area of sensors in e-textiles. The simplest implementation of capacitive sensing is in the form of the parallel plate capacitor in which two conductive plates (or electrodes) of equal area A are separated by a distance d, as shown in Figure 12.2a. Usually, a dielectric layer is placed between the electrodes to maintain the separation. Upon application of an electric potential to the initially uncharged capacitor, equal and opposite charges accumulate on the electrodes. The amount of charge a capacitor can store depends on the applied potential, the capacitor's physical characteristics (parameters A and d), and the permittivity (ε) of the dielectric material. The capacitance C of the capacitor is the amount of charge stored per unit potential (volt) and can be expressed as follows:

$$C = \varepsilon \frac{A}{d} \qquad (12.1)$$

In a capacitive sensor, any perturbation in the environment causes one or more of the parameters, A, d, or ε to change, which in turn alters the net capacitance. Thus, a capacitive sensor can detect proximity, position, pressure, and displacement due to the corresponding geometrical changes (i.e. A and d), while moisture, pH, and so on can be monitored based on permittivity (ε) changes [103–105].

The geometry of the capacitive sensor has been implemented in e-textile sensors both in the parallel plate and cylindrical configurations. In general, the electrodes are prepared by treating the textile substrates with conductive materials. On the other hand, the dielectric is often made of soft textile materials such as PDMS, PU, and PI [89, 95, 103, 104]. In linear, cylindrical capacitive sensors, the conductive and dielectric layers are arranged coaxially in a multicore–shell structure [104, 108]. Planar, fabric-like sensors are prepared by vertically stacking separate laminated layers of films and fabrics in a multilayered assembly [106, 109]. In other implementations of the necessary electrode-dielectric arrangement, functional fibres or yarns are coupled via twisting, sewing, weaving, and so on, to complete the sensor assembly [96, 99, 110].

The fibre-based capacitive humidity sensor in Figure 12.2b used copper and sputtered silver as electrodes and humidity-sensitive PI as the dielectric [104]. The moisture-induced change in permittivity of the PI was measured as a change in capacitance to monitor human breath [104]. Fabric-based sensors based on the simple layering of electrodes and dielectric in a planar structure have also been reported [106, 111]. Recently, an ultrathin, breathable fabric sensor was prepared by using thermoplastic PU (TPU) as the dielectric layer, sandwiched between two TPU–silver nanowires layers on either side (see Figure 12.2c) [106]. The micropatterned dielectric scaffold was introduced to enable a greater change in the electrode spacing upon applying pressure. Moreover, the air voids introduced due to the micropattern improve its effective dielectric constant. As a result, the sensor showed enhanced sensitivity of ~8.31–1 kPa (<1 kPa), ultralow detection limit (0.5 Pa), and durability in repeated loadings [106].

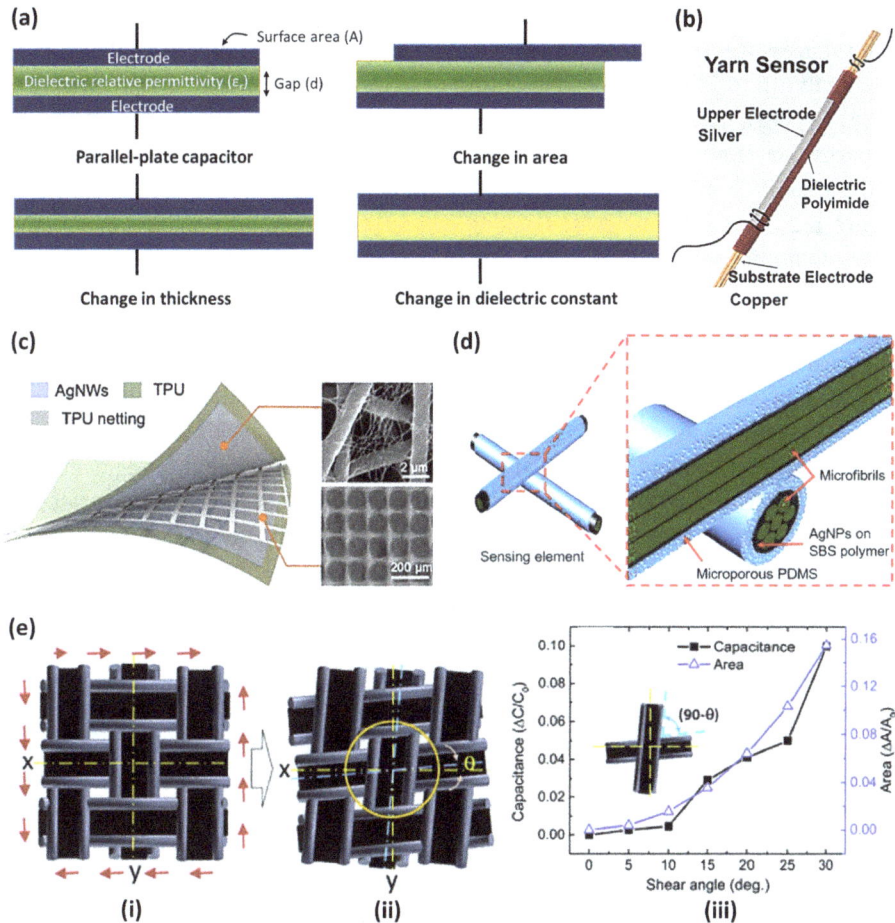

Figure 12.2: (a) Schematic of a parallel-plate capacitor and different modes of capacitance change; (b) all-in-one linear yarn-type coaxial capacitive humidity sensor based on humidity-sensitive polyimide dielectric, and copper and silver as electrodes (reproduced with permission from [104]. Copyright 2021, SAGE Publications). (c) Multilayer capacitive sensor assembly made with TPU netting as dielectric and AgNWs/TPU nanofibre as electrodes. Inset: (top) SEM image of the AgNWs/TPU nanofibre electrode; (bottom) SEM image of the patterned TPU netting (reproduced with permission from [106]. Copyright 2022, American Chemical Society). (d) Schematic of cross-sectional view of fibre-based capacitive pressure sensor (reproduced from [107] with permission from the Royal Society of Chemistry). (e) (i and ii) Schematic representations of the deformation of a woven sensor array subjected to shear, (iii) capacitive response and resulting change in the area (reproduced with permission from [95]. Copyright 2018 WILEY-VCH Verlag GmbH & Co. KGaA, Weinheim).

Another seamless method to prepare planar fabric sensors is by converting the sensing yarns into woven [95, 112] or knitted [96, 97] fabric. Based on this approach, a plied yarn sensor was prepared by plying two core-spun yarns fabricated by wrapping silver-coated nylon fibres with cotton fibres and a PU layer [89]. The plied yarn can act as a pressure sensor due to the deformation of the sheath layers caused by applied pressure. The plied yarns were woven into a sensor array that displayed good sensitivity up to 35 kPa and stable output over loading 500 cycles. Cross-stacking of fibres has been used to produce capacitive sensors for pressure monitoring. Weaving, in particular, is a facile and scalable method to prepare an array of sensors [89, 112, 113] for spatially resolved mechanical stimuli detection [113]. The sensor shown in Figure 12.2d was made by cross-stacking two fibres prepared by coating conductive fibres with microporous PDMS [107]. The conductive fibres were prepared by depositing silver nanoparticles onto poly(styrene-block-butadiene-styrene) polymer-coated aramid microfibres. The micropores of the PDMS layer helped to increase the effective permittivity of the dielectric, thereby enhancing the sensor's sensitivity [107]. In a similar approach towards the fabrication of fibre-based multimodal and multifunctional sensors, Kapoor et al. [95] reported a woven sensor array of bicomponent fibres made of ordered insulating PDMS and conducting PDMS/CB composite segments (see Figure 12.2e). The multifunctional characteristics of the sensors were demonstrated by measuring tactile, tensile, and shear deformations, as well as wetness and biopotential. The same sensor structure produced by waving melt–spun bicomponent fibres was later demonstrated by Tabor et al. [99]. The bicomponent fibres were made by a commercial tricomponent extrusion method using polyamide-6, polyamide-6 with CB fillers, and PLA, as the insulating, conducting, and sacrificial polymers, respectively. The sensitivity of the sensor array could be tuned by changing the fabric design [127]. In another approach, core-spun yarn made of nickel-coated cotton yarn wrapped by graphene oxide (GO)-doped PU nanofibres was woven into a fabric-based sensor array. The application for the sensor array was demonstrated for voice recognition, airflow monitoring, and muscle movements [113].

Capacitive sensors have often been proposed as pressure sensors capable of sensing a wide range of pressure for applications in health diagnostics to monitor breathing, speaking, blinking, joint motions, respiration, perspiration, and other bio-signals [79, 114–117]. Ideally, a capacitive sensor design with a large surface area and a small thickness is desirable such that the sensitivity and accuracy are not compromised. Highly compressible dielectric materials are essential to achieve high sensitivity. A lower Young's modulus of the dielectric leads to greater deformation when pressure is applied to the sensor, resulting in a larger change in capacitance [118]. To achieve such performance, air as dielectric has been proposed [118]. Common approaches to introduce air gaps between electrodes are microstructures [80, 119], porous dielectric [120–122], or 3D spacer [123, 124]. Introducing microfeatures via twisting or buckling imparts additional stretchability to the sensors [125, 126]. It is important to note that the sensing performance of a capacitive (or any other) sensor can be tuned by an interplay of its structure and the property of the constituent materials. Significant progress has been made in the field of textile-based capacitive sensing. However, scalability, integration, and lifetime problems remain active areas of research.

12.2.3 Resistive sensors

Resistive sensors have been in use for over a century. The development of resistive sensors dates back to the discovery of the change in electrical resistance in copper and iron wires when subjected to a mechanical strain by Lord Kelvin in 1856 [127]. In a resistive sensor, the electrical resistance of a resistive component within the sensor changes due to changes in environmental stimuli such as pressure, moisture, and temperature. Based on the stimuli type, a resistive-type sensor may be described as piezoresistive, photoresistive, thermoresistive, and so on. The most common type, however, is the piezoresistive type. Piezoresistivity refers to the phenomenon in which the electrical resistance (R) of a conductive material changes with an imposed strain or stress. The change in resistance (ΔR) of the resistor of length L, under a normal strain ($\Delta L/L$), can be expressed in terms of its resistivity (ρ) and Poisson's ratio (μ) [128]:

$$\frac{\Delta R}{R} = (1 + 2\mu)\frac{\Delta L}{L} + \frac{\Delta \rho}{\rho} \tag{12.2}$$

Thus, the fractional change in resistance ($\Delta R/R$) of a resistive sensor depends on the dimensional changes described by $\Delta L/L$ and μ, as well as bulk resistivity ($\Delta \rho/\rho$) of the resistor [128]. Depending on the material, either of these parameters can be dominant. While the bulk resistivity change is dominant in semiconductors (such as silicone) [129], the change in geometric components dominates in the case of metals and many elastomeric composites [130].

In the most common approach in fabricating piezoresistive sensors, fibres or fibrous structures that respond to stimuli by changing their electrical properties are used. In general, the conductive network in the textile structure is disrupted at the material or higher levels of hierarchy [91, 131]. The textile structures in the form of fibres or fabrics (knitted, woven, etc.) could be entirely made of or coated with CPCs, ICPs, or other conductive materials. CPCs are obtained by incorporating conductive carbonaceous or metallic micro/nanoparticles into an insulating, viscoelastic polymer matrix [16–21]. The insulator–conductor transition and the piezoresistive effect in these composites are well explained through the strain-induced disruption of percolation pathways (see Figure 12.3a) [132].

Linear piezoresistive sensors can be obtained by spinning various ICPs and CPCs into yarn-like structures using dry, jet, wet, and electrospinning techniques [135]. They can also be prepared by simply coating pristine textile fibre/yarn with conductive materials [136, 137]. A composite fibre spun from poly(styrene–butadiene–styrene)/CNT solution demonstrated good conductivity, excellent sensitivity over a large strain (<50%), and durability [10]. For the composite fibres, the net sensitivity and dynamic range of CPC-based sensors can be tuned by manipulating the filler loading, shape dimension, distribution, or by a hybrid filler system [16–21]. In another approach, a linear resistive sensor was prepared by electrospinning TPU solution into yarns and subsequently applying CNTs to the surface of the TPU fibres. The stretchable conductive fibres were

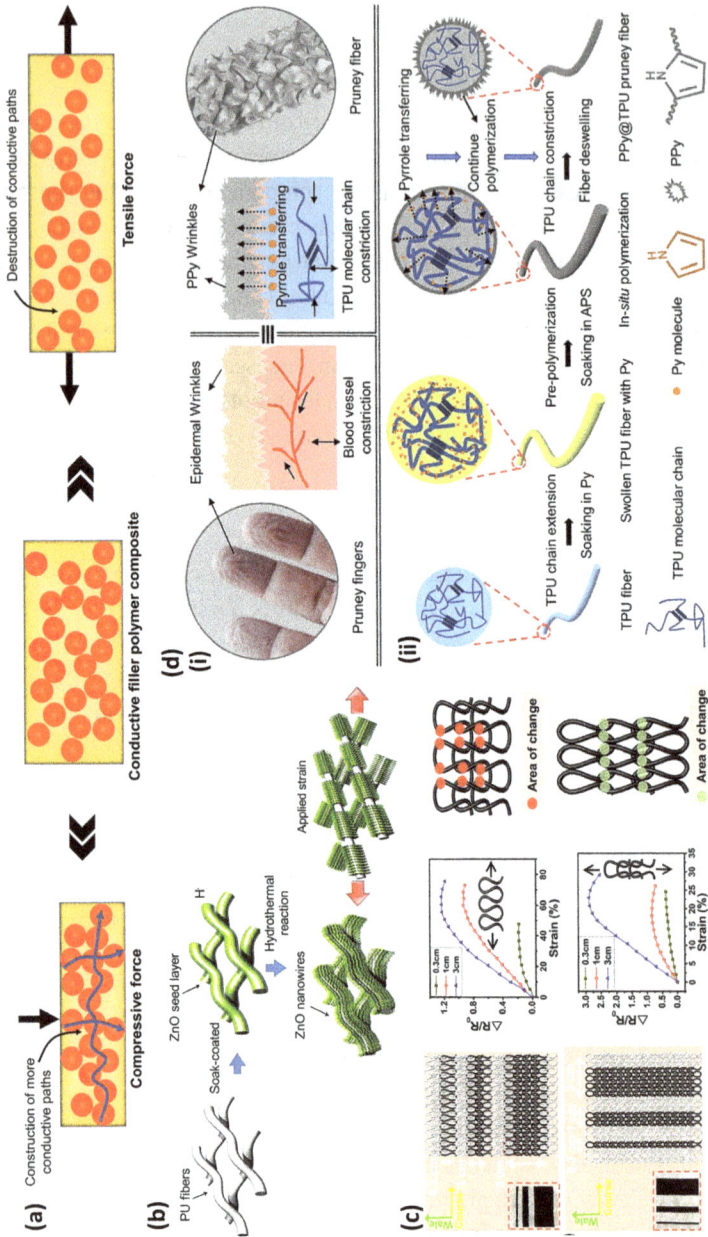

Figure 12.3: (a) Mechanism of disruption and formation of the conductive path in a conductive polymer composite under axial strain; (b) fabrication of the zinc oxide NWs@PU stretchable fabric-like strain sensor (reproduced with permission from [133]. Copyright 2016 WILEY-VCH Verlag GmbH & Co. KGaA, Weinheim). (c) Weft and warp-knitted strain sensors and corresponding changes in resistance due to change in the area of contact between the loops upon application of strain (reproduced with permission from [91]. Copyright 2021, American Chemical Society); and (d) (i) microstructure on a linear strain sensor inspired by the pruney fingers in humans, and (ii) fabrication of the sensor to induce pruney-finger-like microstructure on its surface (reproduced with permission from [134]. Copyright 2021 Elsevier B.V).

used to monitor body motion [26]. Stretchable piezoresistive CNT-embedded sensing yarns were also obtained by wrapping an elastic core fibre with electrospun CNT-embedded PU nanofibres. The yarns made into a woven fabric could detect multiple mechanical stimuli of pressure, stretching, and bending [93]. Graphene nanoplate and poly(vinyl alcohol) coating on twisted woollen yarns through layer-by-layer assembly technique have also been used to fabricate highly stretchable, sensitive piezoresistive yarns [22]. The sensitivity in such sensors increases as the coating thickness is decreased [138].

Printing of conductive sensory layers on woven or knitted fabric substrates has been used to fabricate textile sensors [139, 140]. Topracki et al. demonstrated reversible positive and negative piezoresistive sensory behaviour of screen-printed carbon nanofibres in a plasticized poly(vinyl chloride) or silicone matrix on knitted fabrics. The printed fabrics showed a high sensitivity over a large range of strain. Planar sensors obtained by applying CPCs or other conductive layers on textile materials may suffer from surface decay or delamination over time due to external factors such as temperature, humidity, and abrasion. Additionally, the impermeable coating also compromises the breathability of the textile structure by blocking its pores. One of the methods to apply a conductive layer on a woven fabric while maintaining its breathability was proposed by Liao et al. in the hydrothermal growth of zinc oxide (ZnO) nanowires over a PU fabric substrate (see Figure 12.3b) [133]. Here the piezoresistive response reportedly originated from the microcracks between the conductive particles upon stretching while the substrate remained conductive [133]. In some instances, resistance variation occurs due to changes in the contact area of the conductive elements [91, 141]. For example, Wang et al. used knitted graphene-based conductive core-spun yarns for strain sensing to detect human motion [91]. While the knitted loops are together with low initial electrical resistance, under strain, the resistance increases due to the separation of the conductive paths (see Figure 12.3c) [91].

Geometric microengineering is an effective strategy to enhance the sensing performance of piezoresistive sensors. Recently, Chen et al. prepared linear pressure-sensing piezoresistive fibres which had pruney finger-like wrinkles [134]. The sensor was prepared by in situ polymerization of pyrrole on prestretched TPU fibres, as shown in Figure 12.3d. The resultant wrinkles on the yarn surface exhibited high sensitivity (0.15 kPa^{-1}), fast response time (47 ms), low detection limit (0.2 g), and high stability. In another example, a planar CNT/PDMS-based piezoelectric sensor with the interlocked microdome-patterned film was prepared to impart high specificity and sensitivity to the pressure sensor [142].

Piezoresistive sensors are one of the most common types due to their simplicity, reliability, tunable resolution, and availability of a wide range of materials [143–145]. Sensors made from CPCs are usually stretchable and conformable, and can detect subtle loadings over a wide strain range. However, repeated mechanical deformation may lead to severe mechanical and electrical hysteresis [146, 147], post-yield resistance fluctuations under high strain [148], and nonlinearity [149]. These are some of the issues that need to be addressed in order to fabricate durable and high-performance textile-based piezoresistive sensors.

12.2.4 Piezoelectric sensors

The prefix piezo- in Greek represents "press" or "squeeze". Piezoelectric materials have lattice crystals with non-centrosymmetry[1] such that they can undergo reversible contraction or elongation under an external electric field [150]. The distortion leads to a reorientation of electrical charges within the material, resulting in a polarization of positive and negative charges [150]. Thus, the piezoelectric effect is the ability of a material to generate an electric charge in response to applied mechanical stress, and conversely, these materials deform under an applied electric field (Figure 12.4a). In a textile-based piezoelectric sensor, mechanical deformation causes a voltage to develop across the electrodes. The piezoelectric layer is the most important component of the sensor. Flexible piezoelectric layers are made of organic dielectric materials, for example, PVDF and poly(L-lactic acid) [151–153]. For a material to be useful as a piezoelectric sensor or generator, the fabrication process should maximize the formation of crystalline piezoelectric phases, such as β-phase in PVDF and δ´-crystalline phase in odd-numbered nylons (e.g. Nylon-11) [154, 155]. Additionally, to render the material piezoelectric, a process called poling is necessary. For example, in the case of PVDF, the dipoles are aligned in the poling process by subjecting the polymer in a suitable form (fibre or film) to a sufficiently high electric field [156]. Bicomponent melt spinning has been used to extrude PVDF fibres with the integrated inner electrode CB/high-density polyethylene inner electrode [157]. After applying the outer electrodes and poling, the fibres were shown to have comparable sensitivity to axial tension and compression with commercial piezoelectric films. Electrospinning has emerged as a useful route to obtain the piezoelectric phase in dielectric materials through a precise interplay of process parameters [155, 158]. The spinning process allows for simultaneous control of the piezoelectric β-phase of PVDF, thus combining fabrication and field-induced poling into one step.

Organic dielectrics are often blended with their inorganic counterparts (e.g. ZnO and barium titanate) to improve piezoelectric properties [159–161]. However, sensors made of organic/inorganic composite piezoelectric materials may suffer from delamination [162] and crack formation [163]. Surface modifiers such as polydopamine have been used during the fabrication process to improve the piezoelectric properties, interfacial adhesion, and mechanical strength [164].

---

1 A centrosymmetric material has points of inversion symmetry throughout its volume. A material that does not have points of inversion symmetry is said to be non-centrosymmetric, e.g. polar or chiral molecules.

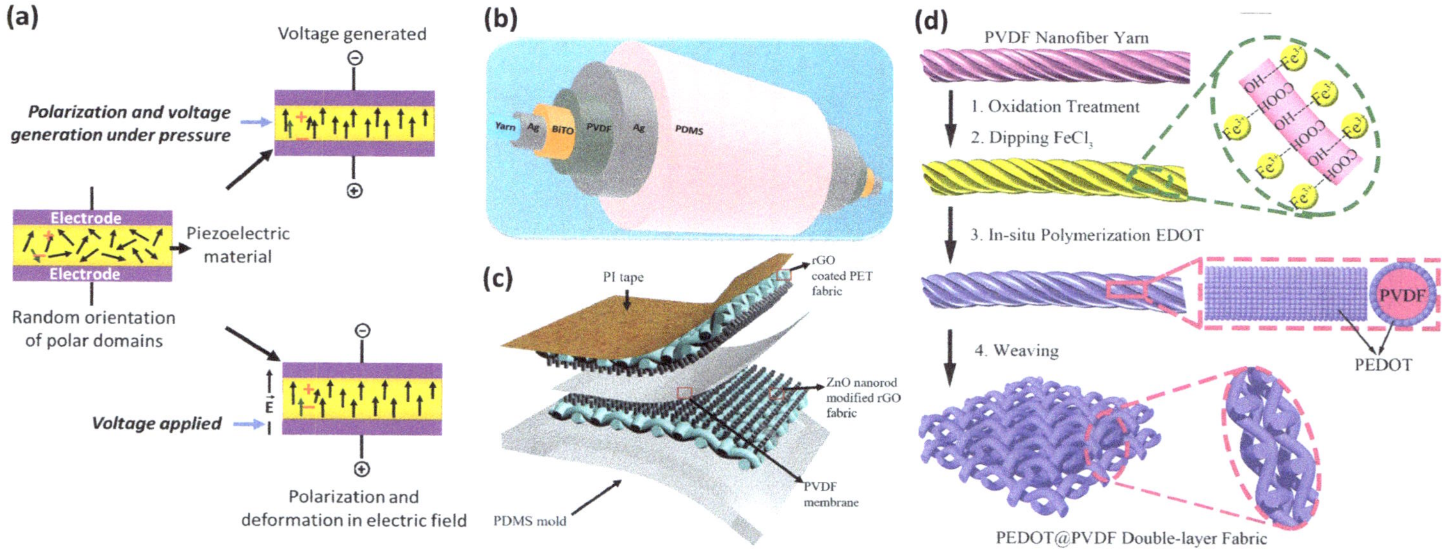

Figure 12.4: (a) Working principle of a piezoelectric sensor and (b) a coaxial piezoelectric yarn sensor (reproduced with permission from [165]. Copyright 2018 Elsevier Ltd.). (c) Multilayer piezoelectric fabric pressure sensor (reproduced with permission from [166]. Copyright 2021, Tsinghua University Press and Springer-Verlag GmbH Germany, part of Springer Nature). (d) Preparation of PEDOT@PVDF nanofibre-based woven fabric sensor (reproduced under a Creative Commons Attribution 4.0 (CC BY 4.0) from [141]. Copyright 2017, The Author(s)).

Piezoelectric textile-based sensors can have linear or planar geometry. Linear sensors are based on coaxial yarn structures that can be prepared by sequentially applying the functional layers to a pristine textile yarn [165, 167, 168]. As shown in Figure 12.4b, a battery-free piezoelectric yarn sensor was prepared for breath sensing using the brush coating technique. The sensor demonstrated a good variation in output (i.e. 0.2–0.4 V) for inhalation/exhalation. It could distinctly differentiate between slow/fast and constant breathing conditions [165]. Planar piezoelectric sensors are based on stacked assemblies of piezoelectric layers and electrodes [169, 170]. Tan et al. prepared a textile pressure sensor to detect wrist or finger movements [166]. The piezoelectric sensing mainly relied on the synergistic piezoelectric effect of single-crystalline ZnO nanorods and β-phase PVDF membrane arranged in a multilayered assembly, as shown in Figure 12.4c. The resulting sensitivity was 0.62 V/kPa in the pressure range of 0–2.25 kPa, with a low limit of detection of 8.71 kP, high output voltage, and mechanical stability over 2,000 cycles [166]. Other reported sensor configurations include the arrangement of the piezoelectric and electrode components in a twisted or interlaced/cross-stacked structure [90, 141, 171]. Twist optimization is desirable for a twisted assembly because a higher twist increases the charge generation due to increased inter-yarn contact area but collaterally decreases the mechanical strength [90]. Fabric-based piezoelectric fabric sensors have also been prepared by cross-stacking functional yarns in a woven or knitted assembly [101, 141]. Zhou et al. reported a self-powered fabric pressure sensor using PVDF yarn made from electrospun nanofibres and coated with PEDOT, woven into double layers [141]. The woven structure was capable of undergoing significant change in contact area at very ultra-low loads of ~ 2–5 Pa, and exhibited high sensitivity (18.38 kPa^{-1}, at ~ 100 Pa), wide pressure range (0.002–10 kPa), fast response time (15 ms), and good durability (7,500 cycles); see Figure 12.4d [141].

Textiles embedded with a piezoelectric sensor have attracted widespread attention because they can reliably and continuously monitor human movement, and physiological information such as respiration, pulse rate, and heartbeat [141, 164]. They have great potential in practical application due to their large dynamic range, high-frequency response, and low-power consumption. However, piezoelectric sensors can only be used for measuring dynamic mechanical stimuli because the output voltage generated by the piezoelectric materials is an impulsive signal and can be detected only when the movement is in the transition. Therefore, the major challenge for piezoelectric-driven pressure sensors is to achieve a stable static signal measurement with low-cost and easy fabrication.

12.2.5 Triboelectric sensors

A triboelectric sensor is an electromechanical device based on the triboelectric effect, a form of contact electrification wherein materials with different electronegativity in physical contact become electrically charged. As a result, any relative motion between the material surfaces through separation, sliding, and so on caused by mechanical forces is measured in the form of an electrical potential between electrodes attached to the materials [172]. This phenomenon is used to sense momentary mechanical stimuli such as touch, pressure, vibration, displacement, and rotation [172, 173]. Due to their inherent capability to generate (tribo-)electricity, triboelectric sensors are almost always self-powered (referred to as triboelectric nanogenerators) and can be used as battery-less, self-powered sensing devices [172, 173].

Structurally, textile-based triboelectric sensors are composed of a triboelectric material (the sensing element) where the charges develop and electrode(s) to collect the charges. The tribonegative component is usually made of PDMS, PVDF, polyester, PTFE, and so on, while nylon, cellulosics, PMMA, and so on form the tribopositive part [92, 174–177]. The components of the sensor can be arranged in various configurations based on the mode of contact electrification: contact-separation mode (CSM) [178, 179], single-electrode mode (SEM) [8, 180], linear sliding mode [181, 182], and free-standing mode [183]; see Figure 12.5a. The sensor's performance is guided by the relative position of the sensory materials in the tribo-series, a list that ranks materials according to their tendency to gain or lose electrons [184]. A large difference in the tribo-series produces a higher number of electrified charges, resulting in a higher triboelectric output [183, 185].

Textile-based triboelectric sensors are widely used as pressure and strain sensors; and most often based on SEM [187–189] or CSM [176, 185, 190] modes. Commonly, triboelectric sensors are prepared by stacking layers of tribopositive and tribonegative materials and electrodes atop each other [191–193]. Cao et al. demonstrated a simple SEM sensor assembly using nylon and silk fabric as triboelectric materials and screen-printed CNT ink patches as the electrode (Figure 12.5b) [179]. The resulting touch/gesture sensor exhibited excellent durability over 10,000 cycles of contact and separation and a high washability and linearity [179]. The high sensitivity was possible due to the rough surface produced by the numerous microfibres of textile and CNTs. In an alternative approach to fabricate sensors, functionalized yarns with different triboelectric behaviour were interlaced by weaving [94, 194], knitting [176, 185, 186], or stitching [186, 191] (Figure 12.5c). Common deposition and spinning techniques can be employed to prepare functional yarns, as discussed in Section 12.2.1.

Stand-alone linear, yarn-like triboelectric sensors are not commonplace, and the associated fabrication is not as straightforward [187, 195]. Lin et al. fabricated a self-powered, flexible force and bend sensor with a timbo-like core morphology to enable SEM working, as shown in Figure 12.5d [187]. Due to the novel structure and sensing mechanism, the linear sensor achieved high sensitivities (5.20 V/N under pressing; 1.61 V/rad under bending), fast response time (<6 ms), and long-term stability (>40,000 cycles).

As pressure sensors, the triboelectric output signal depends on both the contact area and the relative change in the contact area. The effect of various geometries of woven, knitted, and stitched textile sensors on the sensitivity, working range, linearity,

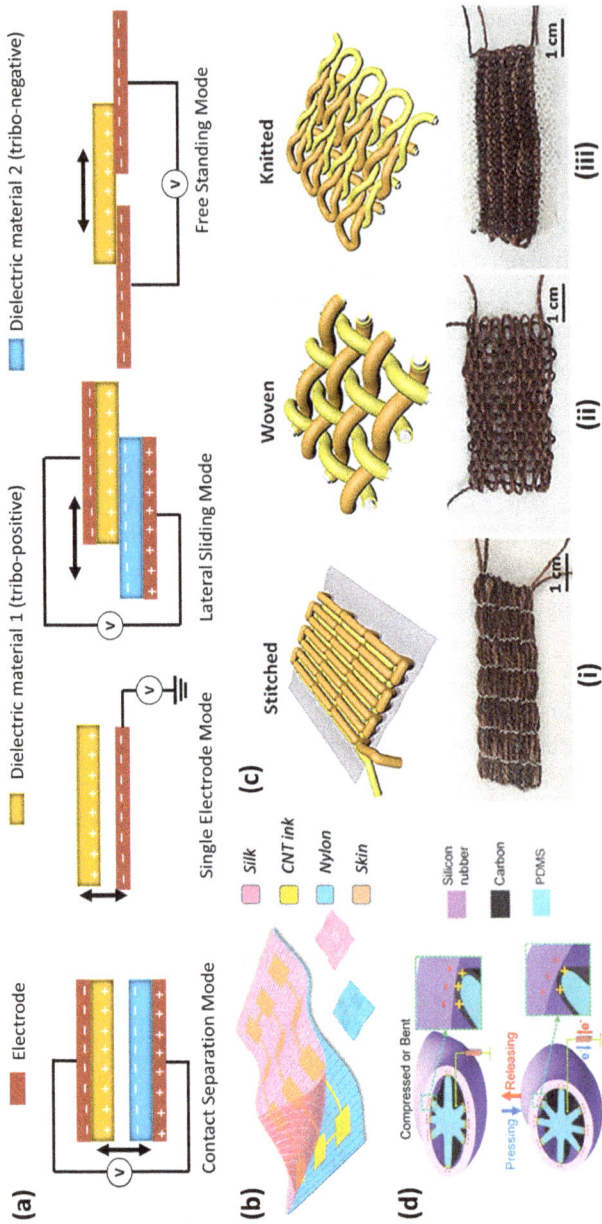

Figure 12.5: (a) Four operational modes of tribo-electrification; (b) structure design and sensing mechanism of a multilayer fabric sensor (reproduced with permission from [179]. Copyright 2022 American Chemical Society); (c) triboelectric fabric-like sensors prepared by the arrangement of functional yarns into (i) stitched, (ii) woven, and (iii) knitted structure (reproduced with permission from [186], Copyright 2020 Elsevier Ltd.); and (d) stand-alone yarn-like triboelectric pressure sensor based on single-electrode mode (reproduced with permission from [187]. Copyright 2018 WILEY-VCH Verlag GmbH & Co. KGaA, Weinheim).

and saturation of pressure sensors was studied by Zhao et al. [186]. In addition to the high sensing performance, the ease of integration of sensors using processes that are compatible with traditional textile manufacturing is desirable [196, 197]. Accordingly, a pressure sensor for physiological signal monitoring was prepared via machine-knitting. This process allowed seamless integration of the sensor without any compromise on its electromechanical durability, wearability, or washability [198]. The knitted sensor produced high pressure sensitivity (7.84 mV/Pa; <4 kPa), fast response time (20 ms), high stability (>100,000 cycles), wide working frequency bandwidth (up to 20 Hz), and good wash resistance (>40 washes) [198]. However, in the higher pressure region (>4 kPa), a drastic decrease in linearity and sensitivity (~96%) was observed due to the saturation of the effective contact area [198]. Such an issue highlights the limitation of interlaced fabric structures for triboelectric sensing, especially in high-pressure regions.

With various available working modes, geometry, and structure, textile-based triboelectric sensors are one of the most versatile sensors. They are widely used to design self-powered devices for physiological sensing, gait analysis, and human–machine interface [176, 179, 199]. These sensors allow high output signal, stable sensing, and good linearity (in low-pressure regions), and can be practically made with any common textile material. Their self-powered sensing ability help address the limited lifetime, high replacement cost, and environmental pollution of Li-battery-driven sensors. However, their drawbacks include the drop in linearity in the high-pressure region [198] and moisture sensitivity [179].

12.2.6 Inductive sensors

An inductive sensor uses the principle of electromagnetic induction to detect or measure external stimuli. Whenever a current passes through an inductor coil, it develops a magnetic flux (Φ_i). The electromotive force or electric potential (V) induced in the electrical circuit is proportional to the temporal variation of the magnetic flux through the N turns of the inductor coil [200]:

$$V = -N\frac{d\Phi(i)}{dt} \tag{12.3}$$

In a typical inductive sensor, when a non-ferromagnetic (e.g. living tissue) or ferromagnetic object enters the magnetic field, the inductance of the coil changes due to the induced eddy current. The change in inductance is measured as a perturbation caused by the object. For instance, the object's position with respect to the sensing coil can be determined by measuring the inductance. The working mechanism of an inductive sensor for proximity detection and sensing mechanical stimuli can be seen in Figure 12.6a.

This principle of magnetic field-induced conductivity has been used to monitor the heart rate, and the results were in sync with those measured with ECG [201]. On the

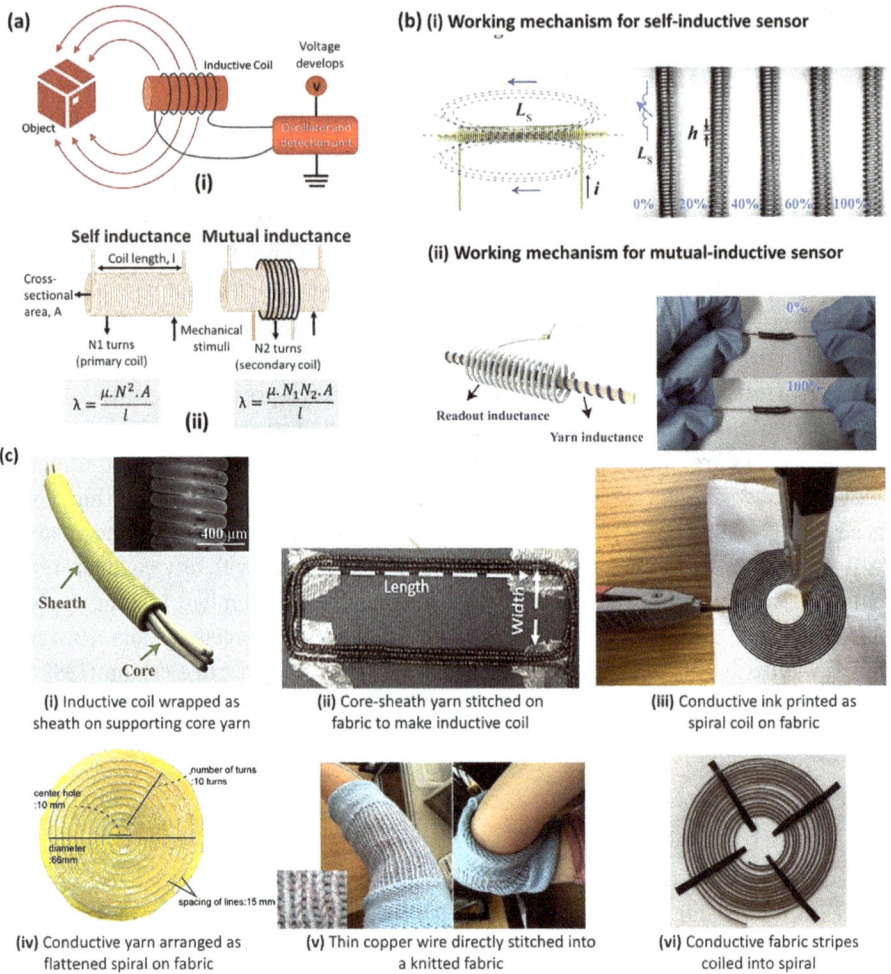

(a) (i) Mechanism of an inductive proximity sensor, and (ii) inductive sensor based on self-inductance and mutual inductance.

(b) (i) Working mechanism for self-inductive sensor

(ii) Working mechanism for mutual-inductive sensor

$$\lambda = \frac{\mu . N^2 . A}{l}$$

$$\lambda = \frac{\mu . N_1 N_2 . A}{l}$$

(c)

(i) Inductive coil wrapped as sheath on supporting core yarn

(ii) Core-sheath yarn stitched on fabric to make inductive coil

(iii) Conductive ink printed as spiral coil on fabric

(iv) Conductive yarn arranged as flattened spiral on fabric

(v) Thin copper wire directly stitched into a knitted fabric

(vi) Conductive fabric stripes coiled into spiral

Figure 12.6: (a) (i) Mechanism of an inductive proximity sensor, and (ii) inductive sensor based on self- inductance and mutual inductance. (b) Mechanism of self- and mutual inductance for active and passive sensing (reproduced with permission from [203]. Copyright 2020 Elsevier Ltd.). (c) The different types of textile-based inductive coils for sensing application: (i) compound yarn sensor: conductive yarn wrapped around supporting textile yarn. Inset: SEM image of the yarn sensor (reproduced with permission from [203]. Copyright 2020 Elsevier Ltd.). (ii)–(vi) Planar sensor: (ii) rectangular coil prepared by stitching the compound yarn on a fabric (reproduced under Common Creatives Attribution 4.0 International (CC BY 4.0) from [205]. Copyright 2020 The Authors. Published by WILEY-VCH Verlag GmbH & Co. KGaA, Weinheim); (iii) spiral coil prepared from conductive ink screen printed on fabric (reproduced with permission from [206]. Copyright 2020, SAGE Publications); (iv) spiral coil from conductive yarn (reproduced with permission from [201]. Copyright 2014, Springer Science Business Media New York); (v) inductive coil prepared by stitching copper wire into a knitted fabric (reproduced under Common Creatives Attribution 4.0 International (CC BY 4.0) from [207]); and (vi) laser-cut conductive fabric coiled as a spiral inductive coil (reproduced with permission from [206]. Copyright 2020, SAGE Publications).

other hand, measurements for mechanical stimuli (strain, pressure, torque, etc.) are carried out by using the dimensional variations of the inductive coil [202]. The inductance is a function of the geometry and electrical conductivity of the coil. For an inductor coil of a fixed conductivity, any deformation in the coil will change the inductance and provide information about the mechanical stimuli. As shown in Figure 12.6b, textile sensors also adopt the self-inductance or mutual inductance mechanism for active or passive monitoring of mechanical stimuli, respectively [203, 204].

The inductive coil can be prepared in different ways in inductive textile sensors, as summarized in Figure 12.6c. Simple one-dimensional coils can be made by wrapping metallic wire over an insulating textile core (e.g. yarn) [203]. A linear stretchable strain sensor prepared by wrapping copper wire in a spiral manner around a PU core (shown in Figure 12.6b(i)) was used to detect human motion [203]. The resulting sensor showed high sensitivity (0.4 mH/%), fast response (<0.1 s), and stable performance without a baseline drift after 4,000 cycles of deformation [203]. Two-dimensional coils resembling a flat spiral structure can be made by stitching a compound or metallic yarn [201, 205] on a fabric or printing conductive ink [206]. Conductive fabric cut into ribbons may also be coiled to prepare an inductive coil [206]. In a comparative analysis of different methods of coil fabrication, Sun et al. reported embroidery as a preferred way to make fabric-integrable flat coils; laser-cut yarns have the least resistance, but they are also highly unstable, while printed coils adhere well to the fabric but have high electrical resistance [206]. In contrast to attaching an inductive coil to the fabric, the intermeshing of an inductive coil within the fabric structure is an excellent way to fabricate a robust and unobtrusive sensor [202, 207]. Electroconductive fibres (polymeric and metallic) were knitted as circular fabric and were used to measure strain and displacement during body motion using self-inductance and mutual inductance, respectively [202]. The sensor assembly was prepared using a highly scalable flat bed-knitting technology.

Two-dimensional inductive coils are prepared in a variety of different configurations that affect their inductive properties [206, 208–212]. In general, the inductance and sensitivity improve as the length, width, and pitch of the coil increase [208–210]. In addition, patterning planar coils such as zigzag geometry or plying the conductive yarn before stitching into a coil are ways to improve inductance further [206, 212]. Thus, high-performance textile sensors are fabricated through an interplay of material and geometrical parameters [205, 212].

Inductive textile sensors allow both active and passive monitoring of body motion, gesture, and bio-signals such as heartbeat, respiration, and displacement [201, 203, 205, 207]. In general, inductive sensors are better due to their reliable output, stability during high-frequency movements, facile and cheap fabrication [213–215]. In addition, the inductive coils can also function as an antenna for wireless sensing and data transmission [206, 215]. Some of the issues related to inductive coil sensor include cross-talk and parasitic capacitance [216] that needs to be tackled to ensure accurate sensing. The literature in the field of textile-based inductive sensors is somewhat limited and needs further exploration.

12.2.7 Optical sensors

Optical sensors use visible or UV light to detect, interrogate, and quantify objects or environments. Fibre-optic sensors function on the principle that the light transmitted through the core of the optical fibre can be affected by the stimuli in its environment [217]. Optical fibres usually have two concentric layers; the core and the cladding illustrated in Figure 12.7a(i). The refractive index of the cladding is designed to be slightly lower than that of the core to facilitate total internal reflection/waveguiding property in the fibre core [218]. Optical sensors in the form of fibres have specific waveguiding properties to carry out various sensing operations [219, 220]. As shown in Figure 12.7a (ii), an optical sensor comprises a wavelength-selectable light source, an optical fibre (sensing material) that interacts with the stimuli, and a light detector that detects variation in photometric parameters upon interaction with stimuli [217, 221]. In general, optical fibre sensors can be either intrinsic or extrinsic. Intrinsic sensors modulate light while it is still propagating in the fibre, and the fibre itself acts as a sensor. In contrast, the extrinsic sensors involve optics to extract the light from the fibre and perform the modulation process outside. Here the fibre is used as a transmission channel for an extrinsic optic sensor [222, 223].

Conventionally, rigid and brittle glass optical fibres (GOFs) have been used for sensing. Eventually, the focus was shifted to polymeric optical fibres (POF) due to their flexibility, lightweight, and textile-integrability [223, 224]. POFs are also safer as they do not produce sharp edges upon breaking [225]. They are available in single/ multimode [226], step-index/graded-index [226], and single/multicore types [227, 228]. Commonly used polymers for POFs are PMMA, polystyrene, amorphous fluorinated polymer (CYTOP), low water-absorbing cyclic olefin polymers (TOPAS and ZEONAX), high-strength plastic polycarbonate, and so on. The relatively lower Young's modulus of POF provides good sensitivity for strain [229], force [230], temperature [231], pressure [232], humidity [233], and so on.

In general, wearable textile-based POF sensors work on the principles of, but not limited to, intensity and wavelength modulation. These modulations occur via mechanisms such as micro/macro-bending, interferometry, fibre Bragg's grating, and scattering. Optical fibre sensors may contain a single POF or multiple POFs assembled via twisting or cross-stacking, which are eventually integrated into the woven or knitted fabric via sewing or embroidery [230, 235, 237, 239]. POFs may be wrapped with textile yarns prior to interlacement (e.g., weaving) [240], or may be inserted as in-lays vis-a-vis direct knitting [239] to avoid unnecessary bending of the optical fibre.

Intensity modulation based on a micro- or macro-bend loss phenomenon can be used to detect variations in physical stimuli such as strain, pressure, and temperature with good sensitivity (see Figure 12.7b) [241–243]. This is a preferred mode of sensing due to the simplicity in implementation and signal processing [244]. A simple way of measurement is to wrap the POF or POF-integrated fabric around the deforming body, such as the human chest, and measure the intensity change due to bend loss [235, 236, 245, 246].

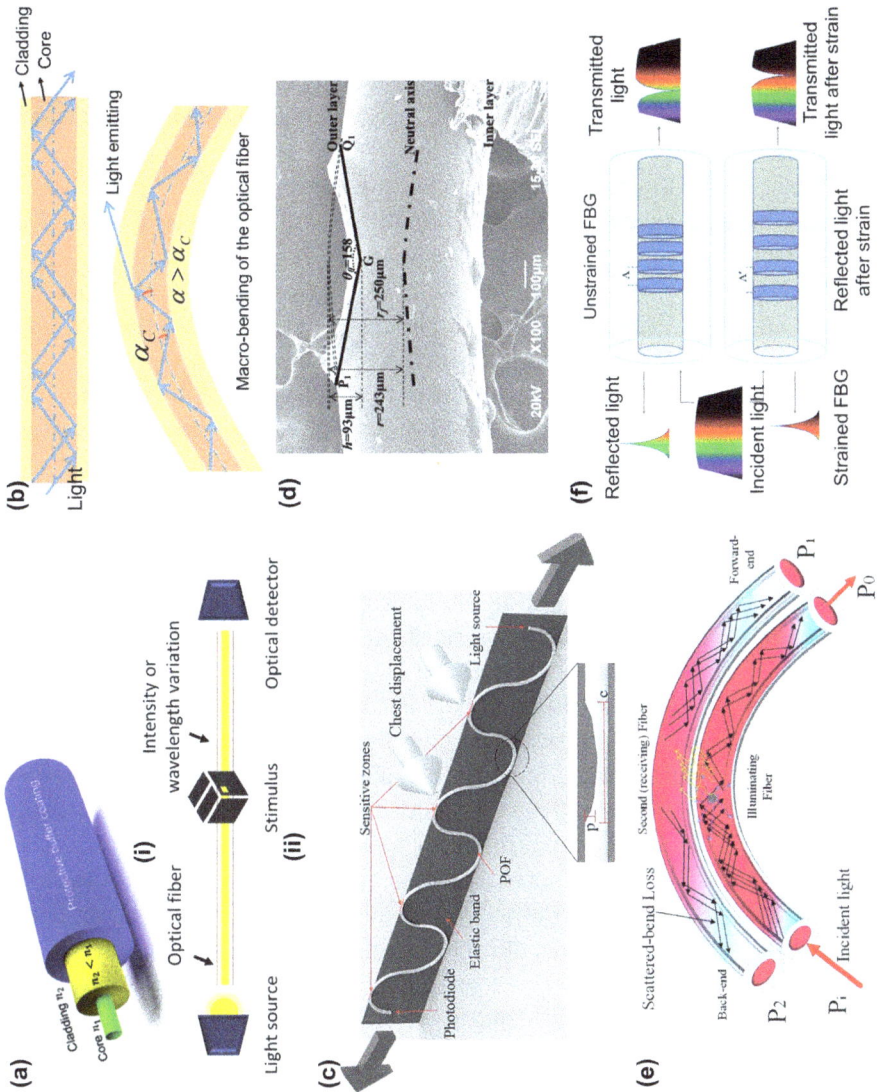

Figure 12.7: (a) (i) Schematic of an optical fibre cable (reproduced under Creative Commons Attribution 3.0 International (CC BY 3.0) from [218]) and (ii) working of an optical fibre sensor; (b) mechanism of intensity attenuation due to bending of optical fibre (reproduced under Creative Commons Attribution 4.0 International (CC BY 4.0) from [234]); (c) overview of POF-based sensor for breath and heart rate monitoring. Inset shows the magnified view of the fibre longitudinal section, where the sensitive zone depth (*p*) and length (*c*) are shown (reproduced with permission from [235], Copyright 2018 Elsevier Ltd.). (d) An etched notched on POF surface (reproduced with permission from [236]. Copyright IOP Publishing). (e) Illustration of coupled POF sensors wherein the bending of fibres causes the scattering and macro-bend loss (reproduced under Creative Commons Attribution 4.0 International (CC BY 4.0) from [237]). (f) Schematic of the working principle of fibre Bragg grating (FBG) sensors and its response to strain (reproduced under Creative Commons Attribution 4.0 International (CC BY 4.0) from [238]).

Figure 12.7c shows a POF sensor for simultaneous breathing and heart rate measurement by detecting the curvature variation on the POF embedded in an elastic band [235]. An appropriate signal-processing technique was used to eliminate body movement influence and enable an e-health solution for continuous and dynamic monitoring of patients during their daily activities [235]. Another way to obtain POF bending is by pre-packaging POFs between two plates with saw-shaped edges to prepare micro-bend modulators [242, 247]. Here, the intensity attenuation depends on the tooth geometry and the gap between the plates [247]. The modulator–POF assembly can be integrated into beds or pillows [243] to measure breathing, heart rate, and body movement simultaneously and non-invasively [242, 247]. Notched POFs have been used to improve the sensitivity of the macro-bends (Figure 12.7d), and the effect of its geometry on the sensitivity has been studied [236, 245]. While intensity modulation based on a micro- or macro-bend loss is effective, interferometry has also been proposed for intensity modulation. Dass et al. proposed a Mach–Zehnder interferometric sensor-based curvature sensor to measure temperature and showed a sensitivity of -11.92 dB/m and a resolution of 8.4×10^{-5} m^{-1} in the curvature range of 0–1 m^{-1}, with highly stable negligible hysteresis [241]. In certain POF sensors, intensity variation may occur due to an interplay of macro-bending, surface scattering, and variation of the fibre refractive index due to the stress-optic effect [248].

Another facile method of optical sensing is based on the combination of side emittance and scattered-bend loss between driving and driver POF, as shown in Figure 12.7e. This phenomenon occurs when sensors are made by coupling POFs via twisting or cross-stacking and are useful for measuring force or displacement [230, 232, 237]. The light coupling at the intersection of two POFs can be utilized for quantitative sensing [232]. Bunge et al. prepared a 3 × 3 textile-integrated array of force sensors by assembling a stiff driving fibre and a flexible sensing fibre [230]. Upon application of a normal force at the cross-over point, light leaks from the driving into the sensing fibre. This happens, firstly, due to the increased side-emission of light caused by a larger contact area under increased force. The second reason is the scattering at the cross-over point, which enhances the reception of the side-emitted light into the sensing fibre by changing the propagation direction of the light. The resulting sensor was highly sensitive and could support multi-touch, a difficult challenge with GOF-based sensor arrays [232]. In addition, by using a non-circular trilobal-shaped driving fibre and combining it with a sensor fibre such that the latter aligned with the direction of higher light emission of the former, the light coupling power was improved by 5 dBs [230]. POFs based on bend-scattering reportedly provide the highest dynamic range for displacement sensor [237].

In the other mode of sensing using wavelength modulation, POF sensors can be used to measure strain, temperature, pressure, humidity, and so on [233, 249, 250]. FGB is an easy way to obtain wavelength modulation without any attenuation [251, 252]. FBG is constructed by introducing periodic modulations within a short segment of POF that act as a back-reflector for specific wavelengths of light (see Figure 12.7f). FGBs are encapsulated in polymers such as PDMS to fabricate sensors to measure body posture [252] or vital signs such as body temperature, respiratory rate, and heart rate [251, 253]. PDMS is chosen due to its non-reactivity to the human skin and resistance to EM waves, UV absorption, and radiation. PDMS-embedded FBG POF can be easily integrated with textiles for physiological measurements [253, 254]. PDMS can also substantially increase the temperature sensitivity of the encapsulated FBG-based sensor [253]. However, a general issue with the FBG sensors is cross-sensitivity to undesirable stimuli [255]; for example, the temperature–strain cross-sensitivity issue is a well-known one [256]. The problem can be addressed by having two gratings that respond differently to temperature and strain. The simplest version of this approach is to co-locate two gratings but shield one of them from the effects of strain.

Optical sensors have garnered increased attention over electrical sensors due to their passive sensing, immunity to EM interference, electric isolation, small size, lightweight, high accuracy, easy multiplexing, and so on [221, 230]. Immunity to EM is essential for physiological monitoring during MRI where electromagnetic sensors cannot be used [243, 257]. However, POFs are viscoelastic materials that do not have a constant response to stress or strain and present hysteresis during dynamic measurements. POF sensors are a versatile class of sensors with various modes and mechanisms of sensing, but they still rely on large, distinguishable, and rigid devices. As a result, the degree of freedom and imperceptibility that is expected of a typical textile-based sensor are compromised. Future work in optical sensors must focus on these two aspects to create devices that allow seamless integration with textiles/clothing and enable large-scale fabrication.

12.2.8 Chemical sensors

A chemical sensor is designed to interact with chemical species (analytes) in the environment and generate electrical or optical signals that provide specific information such as concentration, composition, and activity about the analyte (Figure 12.8a). The analytes can be pH, ions, glucose, lactate, and so on, present in the bio-fluids like sweat, breath, wounds, tears, and saliva, or pollens, gases, pH, temperature, and so on, in the environment [258]. The sensors can be classified as electrochemical (e-chem: amperometric, potentiometric, conductometric, etc.) [258], or optical (colorimetry, fluorescence, luminescence, etc.) based on their transduction mechanism [259–263]. In e-chem sensors, information about the chemical interaction between the analyte and the sensing material (i.e. the receptor) is obtained as variations in electrical signals (current, voltage, conductivity, etc.) [258]. Such sensors provide highly quantitative information about the analytes. On the other hand, an optical-type chemical sensor reacts to a specific chemical interaction between the analyte and the receptor by changing the sensing component's colour or degree of fluorescence/luminescence [259–263]. Here, the output signal is visually perceived and is qualitative. In textile-based chemical sensors, the surface of the fibre, yarn, or fabric is functionalized by immobilizing a specific

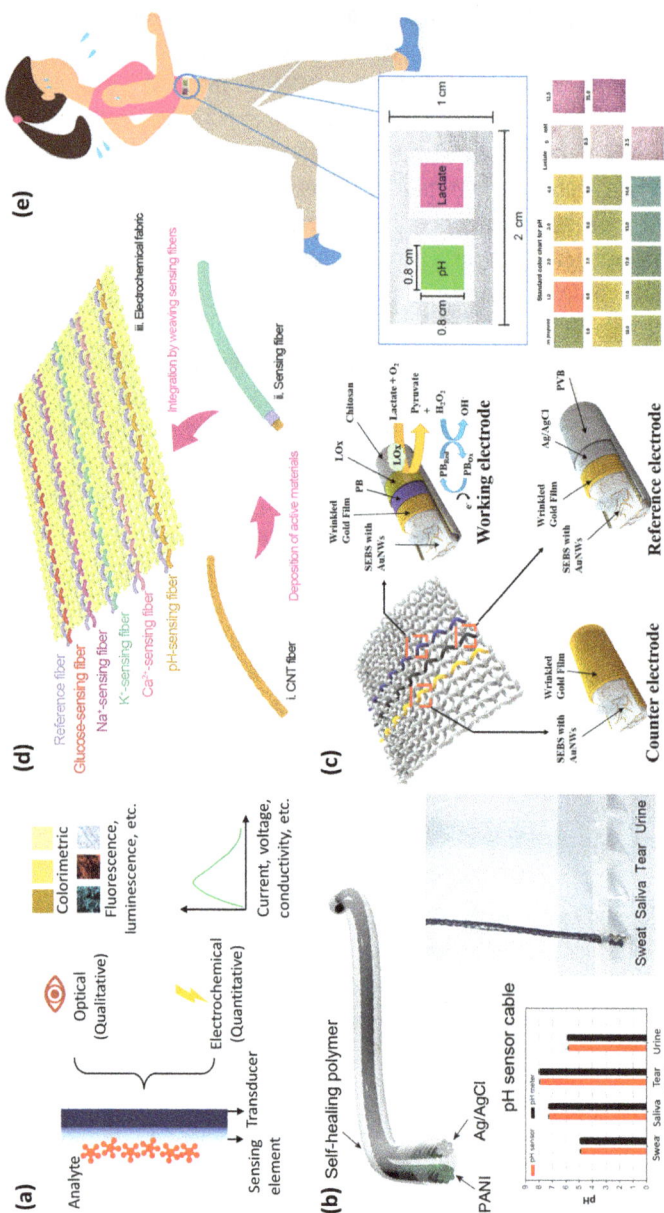

Figure 12.8: (a) Working principle of a chemical sensor, (b) schematic of linear pH sensor prepared by coating individual electrodes with self-healing polymer and side-by-side comparison of pH reading of bio-fluids using textile sensor and pH meter (reproduced with permission from [271]. Copyright 2019 Elsevier B.V.). (c) Schematic of the working, counter, and reference electrodes woven into the fabric for lactate sensing (reproduced with permission from [268]. Copyright 2020 Elsevier B.V.). (d) Multi-functional sweat sensor for glucose, ions, and pH monitoring (reproduced with permission from [272]. Copyright 2018 WILEY-VCH Verlag GmbH & Co. KGaA, Weinheim). (e) Colorimetric fabric sensor: the colour of the pH indicator gradually shifts from red to blue as the pH increases, and the purple colour intensifies with increased lactate concentration in the sweat (reproduced with permission from [259]. Copyright 2018 Elsevier B.V.).

chemical (the receptor) which may act as the sensing component [261, 263]. In this section, we will discuss the fabrication and working of e-chem sensors followed by optical chemical (o-chem) sensors.

In general, textile-based e-chem sensors consist of three electrodes: sensing (or working) electrode (SE), counter (or auxiliary) electrode (CE), and reference electrode (RE). The electrodes are prepared by functionalizing pristine textile substrates with conductive materials and/or biorecognition elements (BRE) using techniques summarized in Section 12.2.1. Ligand-functionalized metallic, a particular type of carbonaceous, and polymeric nanoparticles are used as a sensing material in chemical sensors due to their high surface-to-volume ratio and excellent biocompatibility [264, 265]. The receptor of an e-chem sensor is an analyte-specific BRE[44] coated on the SE where the reaction of interest occurs [266, 267]. In measurements based on amperometric sensing, the current between the SE and the CE at a fixed potential is measured to detect information about the analyte [259, 268]. In the case of potentiometric sensors, the electrical potential between the SE and the RE is measured to assess the analyte of interest [269, 270].

Wearable textile e-chem sensors in a yarn-like structure prepared by coupling the RE or CE with BRE-coated SE have been reported. For instance, a potentiometric pH sensor cable with self-healing property is fabricated using PANI as a pH-sensitive material and used to measure the pH of various bioanalytes and fruits [271]. PANI was used as a pH-sensitive material due to its redox equilibrium between H_3O^+ and PANI phase transitions. The RE was prepared by coating carbon yarns with Ag/AgCl and dielectric inks. The sensor was used to measure the pH of small volumes of bio-fluids, and showed results that matched those from a commercial pH meter (Figure 12.8b). Additionally, the sensor showed high repeatability, negligible hysteresis, and long-term stability due to low baseline drift [271]. Textile-based planar e-chem sensors, on the other hand, are generally prepared by integrating individual electrodes within the fabric structure via weaving, knitting, and sewing/embroidery to complete the sensor assembly [268, 273, 274]. Wang et al. proposed a gold fibre-based wearable lactate sensor by coating conductive and BRE components on a pre-strained gold/SEBS CPC-based yarn to ensure the highest intrinsic strain (see Figure 12.8c) [268]. The sensor components were woven into a fabric and showed high sensitivity (19.13 μA/mM cm^2) up to 100% strain [268]. In general, ions and pH sensors are based on potentiometric sensing [271, 275], whereas sensing analytes such as glucose and lactate are based on amperometric measurements [268, 276]. Multi-functional textile sensors can be fabricated by combining these two modes, as shown in Figure 12.8d. Each unit efficiently detects glucose (via chronoamperometry) and Na^+, K^+, Ca^{2+}, and pH (via open-circuit voltage measurement) for real-time health monitoring [272].

The other category of textile-based chemical sensors is the o-chem sensors. Such sensors have a very simple design where the sensing element – a linear yarn or a textile patch integrated with fabric/clothing – undergoes intensity variation in colour, fluorescence, luminescence, and so on [259–261, 277]. Commonly used receptors are reactive dyes, fluorescent dyes, highly luminescent quantum dots, and so on. Promphet et al.

reported a textile-based colorimetric sensor for pH and lactate detection in human sweat by depositing an indicator dye (mixture of methyl orange and bromocresol green) and lactate assay on a cotton substrate via padding [259]. Usually, for such sensors, a dye fixator (here, chitosan) is used for efficient coating. Figure 12.8e shows the colour variation with varying pH and lactate concentrations. While o-chem devices mostly perform qualitative sensing, for more precise and complex colour signatures, detection via smartphones is used to monitor changes in the acquired images of the sensing element that undergoes colorimetric change [278]. The straightforward transduction mechanism of o-chem sensors allows them to be highly user-compatible and suitable for non-expert operators.

Textile-based chemical sensors are extensively used for bio-monitoring [259, 268, 279]. Here, the fabric picks the bio-fluid directly from the skin. The wetting and wicking properties of textile substrates facilitate the efficient analyte transportation to the transducer [275, 280]. Recently, the concept of Janus wettability has been employed to improve further the self-pumping ability of bio-fluids through textile [281].

Although optical assays offer advantages like lightweight, portability, and a wide range of analyte detection [282], they pose challenges such as the limit of detection, tedious data collection, and analytics. In contrast, the e-chem sensors have advantages like a fast response, device miniaturization, portability, sensitivity, selectivity, cost-efficiency, and minimal power consumption [282]. Generally, coated textile surfaces are prone to surface decay due to delamination or leaching, affecting the textile-based chemical sensor's longevity. On the other hand, hydrogel-based textile sensors have shown remarkable stability for an extended period without suffering from substrate surface delamination [277, 283]. Another significant issue is associated with the irreversibility of the reaction between the receptor and the analyte that renders chemical sensors useless after the first or initial few cycles of sensing. To ensure multiple use cycles, full reversibility of the sensing element without the need for a separate cleaning process is required. The MOFs have emerged as suitable candidates due to their excellent reversibility for the adsorption and desorption of the analyte molecule [284]. Besides, high temperature, moisture stability [285, 286], and highly crystalline structure of MOFs enable precise sensor–analyte interaction. These attributes highlight the significance of MOFs in fabricating chemical sensors with high response, selectivity, stability, and washability [284].

12.3 Textile-based actuators

An actuator is a device that is capable of converting an external source of input energy such as thermal, electric, magnetic, or hydraulic into mechanical motion. Comparable to muscles in the human body that converts energy to motion, actuators work to perform a mechanical action. Traditional actuators are ubiquitously used in many applications; however, they are rigid, bulky, and noisy, making them unsuitable for wearable electronic applications. An important method of benchmarking actuator performance is to compare them to the robust performance of biological muscles, which work for billions of cycles, generating 25% strain and 150–300 kPa stress within a few milliseconds [287–289].

Fibre and yarn-based actuators provide a method of imperceptibly incorporating actuation functionalities into fabrics without sacrificing their inherent qualities. For this reason, we will focus exclusively on fibre and yarn-based actuators. It is also important to note the difference in textiles between fibres and yarns; a multitude of fibres are twisted together to be held by friction, resulting in the creation of yarns [290]. However, in this section, the terms "fibre" and "yarn" will be used interchangeably, as commonly observed in published works about textile-based actuators.

12.3.1 Ionic actuators

Actuators that contract and expand due to the diffusion of ions into and out of the polymer from a surrounding electrolyte, thereby converting electrical energy into mechanical energy, are classified as ionic actuators (IA). IAs can be further classified into those composed of CNTs, CPs, electrorheological fluids, ionic polymer gels, and ionic polymer-metal composites [291]. Out of these, we will discuss CNTs and CPs in detail due to their extensive applications in fibre/yarn-based actuators. If the polymer matrix with the mobile ions is electrically conducting (such as CPs or CNTs), these ions can balance the charge created from the conductive matrix as the externally applied voltage is changed [292]. This results in strong local electric fields at low overall voltages, as well as a higher amount of charge transfer [292]. First proposed by Baughman in 1991, these actuators are attractive due to their high work densities per cycle of actuation, high force generation capabilities combined with high power densities, and low operating voltages [293]. However, they do suffer from low cycle life and energy conversion efficiencies [293]. The subsequent sections further discuss the performance of these IAs and their application in fibre/yarn-based actuators.

12.3.1.1 Carbon nanotube actuators

CNT-based yarn actuators have been explored extensively due to their flexibility and ease of processability using conventional textile processes such as knitting and weaving [294, 295]. CNTs are cylindrical structures composed of one (SWCNTs) or multiple (MWCNTs) graphene layers with either open or closed ends [29]. The high strength and flexibility of CNT yarns enable them to be used for various actuation mechanisms while being integrated into textile structures unobtrusively. Foroughi et al. first demonstrated torsional CNT actuators fabricated via twist spinning CNT yarns from

MWCNT forests, partially immersing this twisted MWCNT yarn and a CE in an electrolyte and applying a voltage between the two electrodes, creating a partial untwist in the MWCNT yarn. The mechanism of actuation is attributed to the fact that the internal pressure caused by ion insertion from the electrolyte causes the yarn contraction, resulting in torque in the immersed part of the yarn to be higher than in the unimmersed part. This results in a torque imbalance, causing the paddle in the middle of the yarn to rotate. However, this device is limited because it requires a large quantity of liquid electrolyte into which the CNT yarn needs to be submerged. To combat this shortcoming, Lee et al. demonstrated a solid-state CNT actuator capable of torsional and tensile actuation, wherein the CNT anode and cathode yarns are coated with solid gel electrolyte and then plied together in opposite twist directions to create a torque-balanced system, as shown in Figure 12.9 (a–i) [296].

Actuation resulted in the twist in each yarn to decrease, whereas the twist of the plying between the two yarns increased, being able to achieve a torsional muscle stroke of 53°/mm at a low voltage of 5 V [296]. The torsional actuation stroke of a material can be understood as the angular rotation when a torque is applied to a slender beam, with the rotation occurring around the beam's longer axis [298]. Such a solid-state actuator may also be more amenable to textile integration. Moving away from e-chem actuation methods, a new class of emerging twisted CNT actuators includes infiltrating the CNT yarn host with shape or volume changing guest materials to create thermal, vapour, and water-driven actuation, as shown in Figure 12.9(j–l) [299]. Impregnating CNT yarns with various guest materials such as silicones to enable torsional actuation when exposed to hexane and ether due to swelling [297], poly(diallyldimethylammonium chloride) (PDDA) to enable water-driven swelling actuation [300], PU resin enabling electrothermal actuation wherein the PU resin expands due to applied electric current generating heat [301], and polystyrene–poly(ethylene–butylene)–polystyrene (SEBS) polymer and paraffin wax enabling thermal actuation of CNT actuators have also been demonstrated [302, 303]. These are advantageous over e-chem methods because, unlike the latter, these do not need an electrolyte or CE that adds to the actuator's weight and volume.

12.3.1.2 Electrically conducting polymer actuators

ICPs are classified as either cationic or anionic salts of highly conjugated, π-stacked planar polymers, which are semiconductors in their undoped state but can be made conducting using positive (p-type) or negative (n-type) dopants via chemical or e-chem methods [29]. Unlike conventional polymers, ICPs have high electron affinities and redox activities. Their charge transfer mechanism depends on the orientation of their backbone chain, the degree of order of the chain, and their impurities and structural defects [304]. ICP actuators undergo a change in shape due to e-chem reduction and oxidation, arising from the insertion of bulky ions and resultant conformation changes

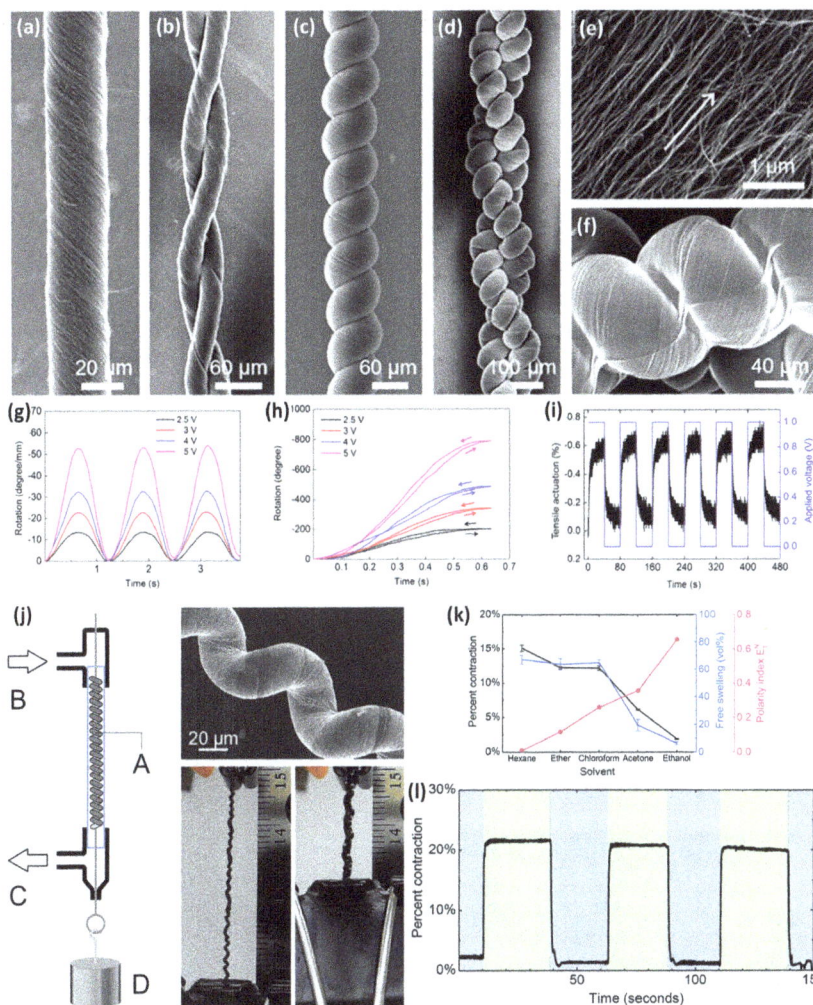

Figure 12.9: Electrochemically driven solid-state torsional CNT actuator yarns [296]: SEM images of (a, b) neat S-twist MWCNT yarn and neat Z-S twist two-ply yarn, respectively; (c, d) neat Z coiled yarn and neat S-Z plied yarn, respectively; (e) surface of the single-ply non-coiled yarn; (f) solid gel electrolyte PVA/H$_2$SO$_4$ coated on two-ply coiled yarns used for tensile actuation. Actuation performance of MWCNT yarns with solid gel electrolyte [296]: (g) torsional rotation per yarn length at applied voltage for a two-ply, twist-spun, Z-S yarn; (h) reversible actuation of the same yarn with untwisting and retwisting of a paddle is illustrated with arrows; (i) tensile actuation strain for a plied MWCNT yarn as it raises an 11 MPa load (reprinted with permission from [296]. Copyright 2014 American Chemical Society). Solvent-driven actuation in helically twisted CNT yarns with cross-linked elastomer guest material [297]. (j) Set-up for measuring actuation with the yarn: (A) inside a capillary tube with an inlet (B) and an outlet (C) allowing for solvent and gas flow, respectively, with a load (D) applied to the yarn; (k) actuation by different solvents for a coiled CNT yarn with less polar solvents causing greater actuation; (l) actuation of the coiled hybrid CNT yarn when exposed to ethanol (blue) and ethyl ether (green) showing rapid contraction (<0.25 s) and expansion (reproduced with permission [297]. Copyright 2015 WILEY-VCH Verlag GmbH & Co. KGaA, Weinheim).

within the polymer due to the delocalization of the π electrons in their conjugated backbone chain [295]. Thus, this electro-chemo-mechanical deformation can result in a greater than 30% strain in PPy – one of the most commonly used ICP actuator materials, followed by PANI [292, 305]. ICP actuators provide the advantage of having large bending displacements at low operating voltages, as well as ease of synthesis and tunability [291]. There have been two main methods of incorporating ICPs into textile-based actuators: (i) synthesis of ICP yarns or fibres using various methods such as solution spinning and electrospinning [306, 307], and (ii) the easier and more explored method of coating commercially available yarns with ICPs [29, 308], such as lyocell/PPy [309], chitosan/PPy [310], PANI on PET yarn [311], and PEDOT:PSS on silk yarn [312]. ICP fibre actuators are a growing field of interest, but these are usually microfibres that are not suitable for textile-based applications due to their complex processing requirements, poor mechanical properties compared to textile fibres and yarns, and their limitation to being processed from a solution rather than a melt, thereby limiting their applicability in bulk processes [29, 313]. Hence, the second method of combining ICPs with textiles using coating methods is a more promising approach for the fabrication of textile-based actuators.

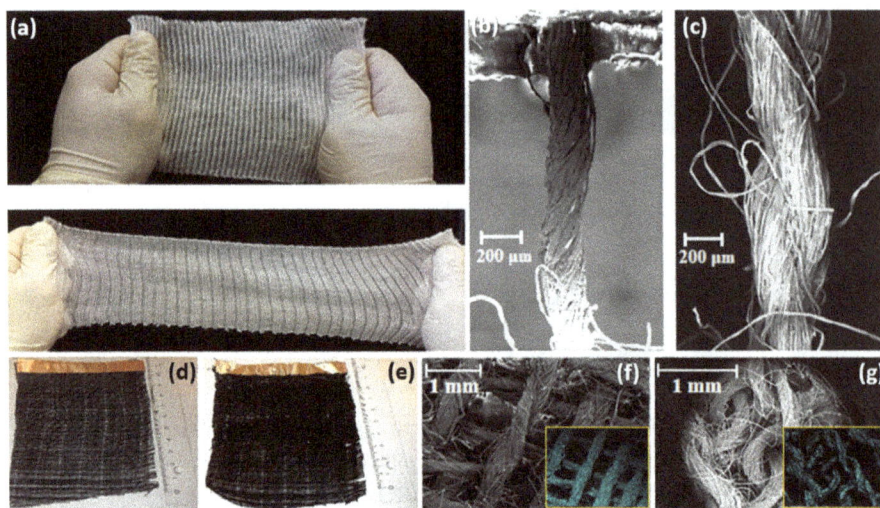

Figure 12.10: Lyocell yarns with conducting polymers PEDOT and PPy deposited onto them to create textile actuators: (a) knitted lyocell fabric in unstretched and stretched states, (b) PPy-coated single lyocell yarn with (c) showing the twisted and coated yarn, (d) vapour phase processing (VPP) PEDOT-coated and (e) PEDOT-PPy-coated lyocell woven fabric, and (f) and (g) show SEM images of PEDOT-PPy-coated woven and knitted fabrics, respectively, with insets showing the EDX sulphur map over the conductive fabrics (reproduced under Creative Commons Attribution-NonCommercial license (CC BY-NC 4.0) from [309]. Copyright 2017, The Authors).

Maziz et al. knitted a fabric using cellulose-based lyocell yarns as core with PEDOT as an electrically conductive layer deposited onto this yarn, followed by the deposition of PPy as the actuating layer (see Figure 12.10) [309]. PPy was used due to its ability to deliver high actuation stresses at low operating voltages, and cellulose provided the hydroxyl groups required as attachment points for the PEDOT oligomers. PEDOT provided an electrically conductive layer that allows for even e-chem deposition of PPy. By applying an alternating potential of −1.0 and 0.5 V, they achieved an individual yarn radial thickness change of 14% with actuating stress of approximately 0.5 MPa. Maziz et al. knitted these functional yarns in a rib structure that showed an isometric actuation strain increase from 0.075% for a single yarn to 3% for the knitted fabric. While the results were promising, Maziz et al. noted that predicting the effect of complex textile architectures on actuator behaviour and performance needs further exploration due to the interaction of electromechanical behaviour of the actuator yarns with the mechanical and frictional forces between the yarns in the fabric, as well as the fibres within the yarns themselves [309].

As mentioned earlier, the limitation of IAs such as ICPs and CNT actuators is once again their need for an ionic liquid that enables actuation. This adds to the complexity and weight of the actuators; however, it does make them viable candidates for moisture or sweat-based actuators, as is the case for CNT actuators. In the subsequent sections, we will explore more such actuators as well as fibre and yarn actuators that do not need a medium for ion exchange.

12.3.2 Electric field-driven actuation

Electroactive polymer (EAP)-based actuators undergo a dimensional change in response to an externally applied electric field. Examples of EAPs, reported in the literature, include dielectric elastomers (DE), relaxor ferroelectric polymers, electrostrictive polymers, and liquid crystal elastomers [292]. EAPs typically exhibit large stress and/or strain upon electrical stimulation [314, 315]. Because of their generally superior electromechanical attributes, EAPs have drawn significant attention in the development of fully organic actuators. In general, EAPs can produce actuation strains as high as two orders of magnitude greater than their rigid and fragile alternatives [287, 314–318].

12.3.2.1 Dielectric elastomer actuators

Lightweight and conformable EAPs are considered suitable candidates for integration in textile structures due to their general polymeric attributes of resilience, toughness, low cost, facile processability, and scalability. Of all EAPs, DEs exhibit the largest actuation strain upon exposure to an electric field, efficiently coupling input electrical energy and output mechanical energy [319]. Hence, they have been extensively explored for various applications such as refreshable Braille [320], textile actuators [321], and biomimicking robots [322, 323].

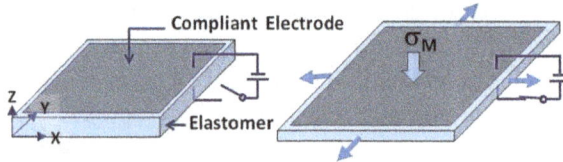

Figure 12.11: DEA configuration with compliant electrodes applied to the dielectric. The electric field-induced charges on the electrodes generate a compressive (Maxwell) normal stress.

The electromechanical response of DEs is due to the development of a "Maxwell stress" upon application of an external electrical field [324]. While the stress originating from the electrostatic attraction between conductive layers applied to opposite surfaces of the DE film serves to compress it, like charges along each film surface repel each other, resulting in further stretching of the film in the lateral (x and y) directions (see Figure 12.11). The effective electrostatic (Maxwell) stress can be expressed as $p = \varepsilon \varepsilon_0 E^2$ (where ε_0 and ε represent the permittivity of free space and the dielectric constant of the elastomer, respectively, and E is the magnitude of the applied electric field) [321].

In general, DEs are attractive for soft actuators because of their low elastic modulus, high dielectric strengths (>100 MV/m), and actuation strains greater than 300% [287]. A wide range of DEs have been explored for their actuation behaviour. These include silicones, acrylic elastomers, PUs, isoprene, interpenetrating networks, fluoroelastomers, and block copolymers, with the first two (silicones and acrylics) being the most extensively explored [322, 325]. Among these, silicones demonstrate a faster electromechanical response with more reproducible actuation cycles, and acrylic elastomers show larger actuation strains [326]. When it comes to integration into textiles, researchers have focused on using electrically conductive textile fabrics as the compliant electrodes for integration with dielectric materials due to their low electrical resistance combined with skin conformability [327–329]. Textile electrodes used in DE actuators (DEA) exhibit some desirable qualities such as being stretchable in their plane, being mechanically compliant such that they do not restrict the actuation of the DEA, low electrical resistance, and ease of integration into the DEA without complex processing steps [328]. Guo et al. manually attached a stretchable conductive textile to a pre-stretched VHB 3905 (3M, USA) acrylic adhesive film creating a square-shaped planar DEA capable of a relative areal expansion of 16.4% when charged to 9 kV, as well as developing a crawling robot with the conductive textile electrode capable of traveling 18 mm in 3 min [329]. The safety of the textile electrode compared to potentially carcinogenic materials such as carbon grease enabled them to be more skin-safe alternatives [329]. In an interesting investigation of DEAs for therapeutic applications, Pourazadi et al. simulated both analytically and numerically the design of an active compression bandage that can help with venous dysfunctions, taking into account the compressibility of the calf and varying compliancy in different regions [330]. They reported the pressures in different regions of the calf for varying stretch ratios of the active compression bandage as it actuates [330].

In the case of fabrication of actuating fibres/yarns, Arora et al. developed an early prototype of a DE-based fibre actuator with commercially available thin-walled silicone and PU hollow tubes used as the DE material, with CB-filled silicone used as the outer electrode and conductive silver grease or calcium chloride solution used as the inner electrode [321]. The hollow tubes were used since they had good elastomeric properties and could be commercially extruded into fibre form. While the prototype fibre actuators made of silicone were able to generate actuation strains of 7% and 18% in the axial and radial directions, respectively, the same for the PU-based actuators were much lower. However, the PU-based fibre actuators demonstrate higher blocking stress, measured under isometric conditions, compared to those made with silicone. The difference is attributed to the complex interaction between the higher elastic modulus and dielectric constant of PU and the experimental conditions [321]. Recently, Chortos et al. demonstrated DE fibres made via multimaterial, multicore–shell 3D printing, with the dielectric matrix ink composed of silicone and fumed silica and the conductive electrodes made of silicone and CB [331]. These fibres, bundled together, produced 3% actuation strain at 8 kV of applied external voltage.

Although the DE-based actuators provide the advantage of producing large actuation strains while simultaneously being soft and compliant, they require large actuation voltages (3–10 kV), which may ultimately render them as bulky devices or unsafe for wearable [332].

12.3.3 Environmental stimuli-based actuators

The human body can produce both moisture and heat, especially during physical exertion. Therefore, the ability of stimuli-responsive materials to use heat and moisture as actuation triggers can be beneficial in terms of harvesting the heat and moisture released from the body to create smart textiles that can either contract or expand appropriately to cool down the body.

12.3.3.1 Thermally responsive actuators

Fibre/yarn-type thermally responsive actuators, when integrated into fabrics, can provide next-to-skin thermal comfort by altering their permeability or bulk characteristics, thereby either cooling or heating the body [333–335]. Thermally induced shape change materials include shape-memory polymers (SMPs) that can change their shape when exposed to heat – transitioning from an initial, temporary state to a permanent state above and below their glass transition temperature (T_g) [336]. SMP materials include chemically cross-linked thermosets such as epoxy-based polymers and polystyrene and thermoplastic polymers such as polyether-esters and PU [336, 337]. While conventionally, SMPs are limited to demonstrating a one-way effect wherein only one permanent shape can be achieved above T_g, which is then lost when the temperature is lowered and they revert to their temporary shape, a new class of temperature memory effect materials can show reversible deformation because they can be programmed to remember a deformation temperature rather than a deformation shape [338, 339].

Efforts towards the integration of SMPs into textiles at the fibre and yarn levels include self-assembled melt-spun SM fibres composed of triblock copolymers that produced uniaxial length changes of 500% when exposed to cyclic temperature stimuli ranging from 20 to 120 °C [340], biodegradable thermoplastic SMPs composed of suture monofilaments capable of 200% elongation when the temperature is raised to 41 °C [341], and melt-spun PU fibres wherein their shape recovery could be controlled using the draw ratio and heat setting during melt-spinning to achieve a more complete shape recovery compared to as-spun fibres [342]. While thermally induced actuation of SMPs is one of the most popular implementations of shape-memory actuation [343, 344], they require large temperature gradients, which may not be possible to implement on the human body. Moreover, they have low work capacities, limiting the amount of actuation they can achieve.

To better integrate the thermally responsive actuators into textiles without the drawbacks of SMPs, Haines et al. demonstrated low-cost and high-strength fishing line actuators composed of Nylon 6,6 monofilament with large tensile actuation stroke made possible due to inserting excessive twist and coil into the filament yarn [345]. These twisted and coiled Nylon 6,6 monofilaments were able to reversibly thermally contract from 4% to 34% while the temperature was increased from 20 to 240 °C. By adjusting the coil spring index (ratio of the mean coil diameter to the fibre diameter), Haines et al. were able to vary the tensile stroke and load-bearing capabilities of these actuators. Such actuators have the capability to change their permeability in response to temperature changes while working in conjunction with textile-based temperature sensors.

12.3.3.2 Moisture-responsive actuators

Since moisture is so readily available in the environment and produced by the human body through sweat, humidity-driven actuators provide an excellent mechanism for the integration of actuation behaviour into textiles that form conformal contact with the human body. Moreover, most natural and a few manmade fibres can absorb and desorb environmental moisture, thereby showing reversible dimensional changes, which can be used to create humidity-driven actuators. A variety of humidity-driven textile actuators have been proposed using materials such as Nafion [346–348], graphene [349, 350], CNTs [351, 352], cotton [353], and silk [354]. Many twisted and coiled yarn structures have been explored for creating such actuators because polymeric fibres and yarns can be twisted at two hierarchical levels (twist level and coil level), resulting in large actuation strokes when exposed to moisture [352]. He et al. used helically twisted MWCNT fibres to produce hydrophilic primary fibres (HPFs), which were able to generate 10.8 MPa of contractive actuation stress when exposed to moisture for 400 ms [351]. The oxygen-plasma-treated HPFs showed easy wettability to water vapour, with microchannels formed in the twisted HPFs providing space for water vapour to infiltrate and expand the HPFs. Such actuators, including the CNT-twisted and -coiled actuators mentioned in Section 12.3.1.1, and the Nylon 6,6 filament actuators mentioned in Section 12.3.3.1, represent an emerging class of twisted and coiled artificial muscle yarns that hold great potential for integration into textiles due to their textile processing-compatible construction, flexibility, comfort, and large actuation strokes [299].

Moisture-triggered actuation using a commercial cotton-polyamide fabric treated with a polymer/MWCNT dispersion has been reported by Gong et al. [355]. The nanoporous network fabric is capable of undergoing dimensional changes when exposed to drying and wetting cycles. Similarly, GO fibres have also been explored as moisture-responsive actuators due to their ability to absorb and desorb molecules on their surface, including NO_2, OH, K, and NH_3 [356, 357]. Cheng et al. used GO fibres to create bilayer graphene/GO fibre structures which were able to rapidly (8°/s) bend in response to moisture exposure, recovering their shape when the moisture is removed [352]. While carbon-based fibres such as CNTs and graphene have numerous actuator applications, they are still limited by their higher costs, complex manufacturing steps, and lack of large-scale commercialization. In addition, few studies have been undertaken to illustrate whether such actuators are truly wearable and can provide the wearer comfort, especially when placed next to the skin. Conventional textile yarns such as silk fibres, capable of contracting 70% when the relative humidity is changed from 20% to 80 %, have also been investigated for actuation, as shown in Figure 12.12 [354]. Jia et al. proposed the actuation mechanism caused by the formation of hydrogen bonds with the hydrophilic groups in silk fibroin, leading to a loss of the original inter- and intrachain hydrogen bonds within silk, causing an expansion of the system. Hence, water was found to enhance the mobility of the silk fibroin [354]. Upon absorption of water into the twisted and plied filament, the individual fibre segments will untwist in the S twist direction due to expansion, producing a torque that increases the plying of the two fibres in the Z twist direction. Liu et al. also demonstrated spider dragline silk filaments capable of shrinking in length by up to 50% with a corresponding increase in dimensions in the radial direction when exposed to high humidity [358]. The hygroscopic nature of silk due to the presence of hydrophilic groups in its backbone chain, combined with the presence of nanopores, enables these filaments to

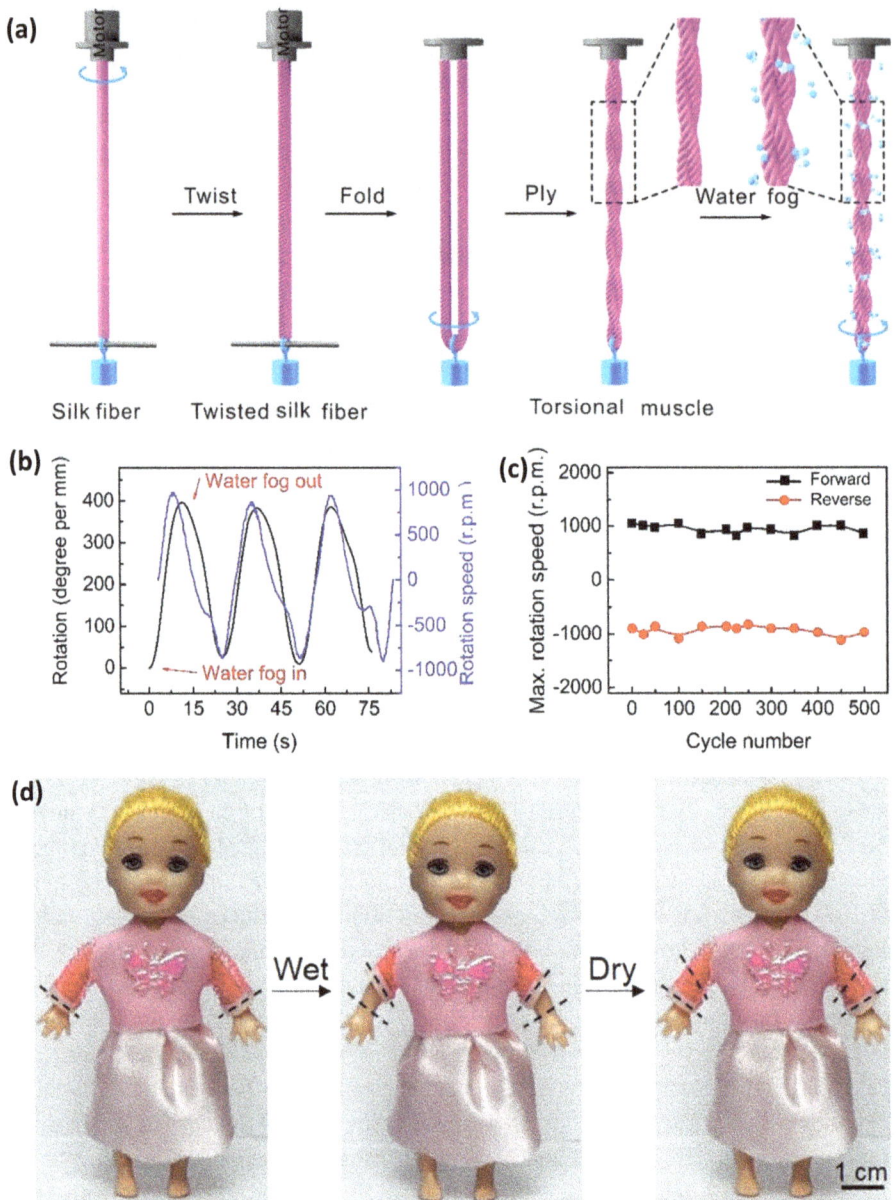

Figure 12.12: Moisture-responsive silk filament actuators: (a) fabrication of a two-ply torsional silk filament actuator with the twisted silk fibre folded and plied in half to create a large stroke actuator; (b) torsional actuation and rotation speed for the silk actuator when moisture was delivered for 15 s and removed for 18 s; (c) forward and reverse actuation speed of the actuator with a ratio of 0.8 for the forward to reverse rotation speed; and (d) photos showing the woven silk actuator contracting when exposed to moisture and recovering when the moisture is removed (reproduced with permissions from [354]. Copyright 2019 WILEY-VCH Verlag GmbH & Co. KGaA, Weinheim).

have moisture regain of 30%, enabling them to behave as torsional actuators when exposed to moisture-rich environments [359].

While humidity-driven actuators are promising due to the abundance of moisture in the atmosphere, as well as the human body's ability to generate such a stimulus during sweating, it is essential to note that many of these actuators require high levels of humidity (>60%) to show appreciable actuation. This can be a limiting factor when the wearer is in a sedentary state where the human body does not produce a large amount of moisture gradient. Thus, these actuators would still need some external stimulus which could detract from the comfort and wearability of these fibres and yarns. Nevertheless, the idea of exploring humidity-triggered actuators made of conventional textile materials with high moisture-regain remains an exciting area of exploration for textile-based actuators [360, 361].

12.4 Conclusion and future outlook

Textile-based electronics enabled by appropriately designed sensors and actuators offer a way of producing low-cost, lightweight, conformable, stretchable, and mechanically robust large-area electronic systems for a diversity of applications. The integration of sensors and actuators together provides a symbiotic interaction between the two. Such an instance has been demonstrated in a thermally responsive Ni–Ti (nickel–titanium) yarn-reinforced shape-memory alloy (SMA) actuator with integrated silver-plated thermal sensor yarns to monitor the deflection of the Ni–Ti yarn [362]. While the signal from the sensor was nonlinear and influenced by the heating of SMAs, it is an interesting attempt to integrate both sensors and actuators into the same textile system. To date, remarkable progress has been made in the development of textile-compatible sensors and actuators. Important among these are textile-based multimodal sensors [95], multifunctional sensors [42, 95], self-powered sensors [126, 170], environmentally driven actuators [363, 364], multistimuli-responsive actuators [365, 366], self-actuating and self-sensing systems [367, 368], and so on.

While CNT and graphene-based fibres and yarns are promising in the design of fibre-based actuators due to their large response, ease of modification, and processing (i.e., twisting, knitting, and weaving) [301, 350], many current textile yarns, such as cellulosic and silk yarns, are also capable of responding to external stimuli such as moisture due to their high moisture regain [358, 369]. In addition to sensors and actuators, the e-textile devices proposed in the literature span the whole spectrum that includes batteries and supercapacitors for energy storage [88, 370], transistors [371, 372], antennas [373, 374], and so on for use in functional e-textile systems. Over the last few decades, the research in e-textiles has resulted in innovative ideas increasingly visible in the marketplace, albeit with limited capabilities. Progress in the IoT and other enabling technologies (such as augmented reality, artificial intelligence, and blockchain) have opened new avenues of innovation in the field of wearable sensors to create assistive garments that can be trained using machine learning tools to predict potential health risks, such as fall and stroke reoccurrence [375–377]. Although a variety of applications are being explored, the most significant driving force behind the ongoing research in textile-compatible sensors and actuators is their vast potential in healthcare; more specifically, in physiological monitoring [378], therapeutic [379], assistive technology [380], and so on. One of the ultimate healthcare product goals is a wearable point-of-care textile platform for remote and continuous monitoring of vital signs for implementing round-the-clock treatment protocols.

While the emerging applications in e-textiles offer many potentially transformative opportunities, most research innovations are still far from realization into commercial products. There remains a host of challenges and critical technology gaps. Among the most pressing challenges is developing flexible polymer-based materials that are sufficiently electrically conductive, safe and biocompatible, structurally and environmentally stable, and textile compatible. CNT and other nanoparticle-based polymer composites hold promise in this context, but the long-term safety of using these materials next to the human skin is yet to be ascertained. Other material-related issues of concern are the viscoelastic nature of polymeric materials manifested in nonlinear deformation, hysteresis, creep, and stress relaxation. The fibre-forming polymers, their composites, and the resulting textile structures (e.g. twisted yarns and knitted/woven fabrics) used in e-textile fabrication exhibit these characteristics. Consequently, the electronic devices built from these are likely to behave similarly [146, 147]. For example, the nonlinearity of fibre and textile-based capacitive and piezoresistive sensors for pressure sensing is attributed largely to the material behaviour [95, 146, 147]. The viscoelastic behaviour can also manifest in the baseline drift and electromechanical instability over many deformation cycles of electrical devices. Research in new material development and optimization of the fibrous structure using appropriate design and fabrication routes and possibly encapsulation can prove very useful in mitigating these issues.

Yet another serious challenge is the scalability of manufacturing. Academic research in e-textiles often results in insights, ideas, and hypotheses that are not fully grounded in manufacturability or tested with physical products or prototypes to facilitate the next step into commercial manufacturing. The manufacturing of textiles is fundamentally different from that of conventional silicon-based or thin-film-based devices and systems. The methods of assembly of textile structures involve continuous processes like extrusion, twisting, and interlacement of materials from nano-, micro-, meso-scales into larger hierarchical structures. While at each step of integration, the length scale increases by orders of magnitude, the increase of a corresponding level of inconsistency is also likely within the resulting textile structure. The length matters because of the proportional increase in electrical resistance. These well-developed processes are necessary to obtain flexibility, porosity, and so on, valued in textile products. Careful adaptation of these processes for the integration of sensors and actuators into textile structures is necessary.

Further issues arise during the use of the products based on the physiochemical interaction between the inherently insulating textiles and the electrically conductive moieties through deformation, abrasion, and exposure to intimate and peri-personal environmental factors (e.g., sweat, humidity, and microbial attack). Consequently, the performance of the e-textile devices (sensitivity, limit of detection, dynamic range, etc.) gradually depletes. The unique issues related to the manufacturing processes and subsequent use must be considered when designing, developing, and integrating e-textile devices like sensors and actuators. Unfortunately, there are too many published ideas for the fabrication and integration of sensors into e-textiles that fail this test. Lastly, at the system level, challenges arise because of the presence of hard and soft interfaces. While the e-textile devices are soft and flexible, they need to be connected to rigid components such as printed circuit boards (PCBs). The interfacial mechanical mismatch may lead to fatigue failure of the interconnect.

Despite all the challenges, the future of e-textiles is remarkably promising because textile substrates having varied structural (porosity, texture, etc.) and mechanical (strength, extensibility, etc.) attributes are ideal for deploying electronic systems on unconventional shapes and interfaces, such as curved, soft, and deformable surfaces, where traditional silicon-based electronics would pose difficult challenges. E-textiles can revolutionize almost every aspect of our daily lives and how we interact with our environment. The development of textile-compatible sensors and actuators based on the fundamentals of textile materials and their processing offers the most promising avenue towards achieving the goal of fully functional e-textile systems.

References

[1]   Adovasio JM, Lynch TF. Preceramic Textiles and Cordage from Guitarrero Cave, Peru. Am Antiq 1973;38:84–90. https://doi.org/10.2307/279313.
[2]   Tabor J, Chatterjee K, Ghosh TK. Smart Textile-Based Personal Thermal Comfort Systems: Current Status and Potential Solutions. Adv Mater Technol 2020;5:1901155. https://doi.org/10.1002/admt.201901155.
[3]   Velusamy S, Roy A, Sundaram S, Kumar Mallick T. A Review on Heavy Metal Ions and Containing Dyes Removal Through Graphene Oxide-Based Adsorption Strategies for Textile Wastewater Treatment. Chem Rec 2021;21:1570–610. https://doi.org/10.1002/tcr.202000153.
[4]   Koncar V. 1 – Introduction to smart textiles and their applications. In: Koncar V, editor. Smart Text. Their Appl., Oxford: Woodhead Publishing; 2016, p. 1–8. https://doi.org/10.1016/B978-0-08-100574-3.00001-1.
[5]   Basu A. Chapter 7 – Smart electronic yarns and wearable fabrics for human biomonitoring. In: Ehrmann A, Nguyen TA, Nguyen Tri P, editors. Nanosensors Nanodevices Smart Multifunct. Text., Elsevier; 2021, p. 109–23. https://doi.org/10.1016/B978-0-12-820777-2.00007-8.

[6]   Gorgutsa S, Bachus K, LaRochelle S, Oleschuk RD, Messaddeq Y. Washable hydrophobic smart
       textiles and multi-material fibers for wireless communication. Smart Mater Struct 2016;25:115027.
       https://doi.org/10.1088/0964-1726/25/11/115027.

[7]   Dolez PI, Decaens J, Buns T, Lachapelle D, Vermeersch O. Applications of smart textiles in
       occupational health and safety. IOP Conf Ser Mater Sci Eng 2020;827:012014. https://doi.org/
       10.1088/1757-899X/827/1/012014.

[8]   Shuai L, Guo ZH, Zhang P, Wan J, Pu X, Wang ZL. Stretchable, self-healing, conductive hydrogel
       fibers for strain sensing and triboelectric energy-harvesting smart textiles. Nano Energy
       2020;78:105389. https://doi.org/10.1016/j.nanoen.2020.105389.

[9]   Bullón Pérez J, González Arrieta A, Hernández Encinas A, Queiruga-Dios A. Industrial Cyber-Physical
       Systems in Textile Engineering. In: Graña M, López-Guede JM, Etxaniz O, Herrero Á, Quintián H,
       Corchado E, editors. Int. Jt. Conf. SOCO'16-CISIS'16-ICEUTE'16, Cham: Springer International
       Publishing; 2017, p. 126–35. https://doi.org/10.1007/978-3-319-47364-2_13.

[10]   Wang W, Fang Y, Nagai Y, Xu D, Fujinami T. Integrating Interactive Clothing and Cyber-Physical
       Systems: A Humanistic Design Perspective. Sensors 2020;20:127. https://doi.org/10.3390/s20010127.

[11]   Avellar L, Stefano Filho C, Delgado G, Frizera A, Rocon E, Leal-Junior A. AI-enabled photonic smart
       garment for movement analysis. Sci Rep 2022;12:4067. https://doi.org/10.1038/s41598-022-08048-9.

[12]   Ramaiah GB. Theoretical analysis on applications aspects of smart materials and Internet of Things
       (IoT) in textile technology. Mater Today Proc 2021;45:4633–8. https://doi.org/10.1016/
       j.matpr.2021.01.023.

[13]   Zheng Y, Yin R, Zhao Y, Liu H, Zhang D, Shi X, et al. Conductive MXene/cotton fabric based pressure
       sensor with both high sensitivity and wide sensing range for human motion detection and E-skin.
       Chem Eng J 2021;420:127720. https://doi.org/10.1016/j.cej.2020.127720.

[14]   Chen R, Ma X, Chai Y, Hua K, Gui Q, He Y, et al. A Polyester/Polypyrrole Textile-Based Ultrasensitive
       Wearable Microdistance Sensor. Macromol Mater Eng 2021;306:2100478. https://doi.org/10.1002/
       mame.202100478.

[15]   Li BM, Yildiz O, Mills AC, Flewwellin TJ, Bradford PD, Jur JS. Iron-on carbon nanotube (CNT) thin films
       for biosensing E-Textile applications. Carbon 2020;168:673–83. https://doi.org/10.1016/
       j.carbon.2020.06.057.

[16]   Subramaniam C, Yamada T, Kobashi K, Sekiguchi A, Futaba DN, Yumura M, et al. One hundred fold
       increase in current carrying capacity in a carbon nanotube–copper composite. Nat Commun
       2013;4:2202. https://doi.org/10.1038/ncomms3202.

[17]   Ammara S, Shamaila S, zafar N, Bokhari A, Sabah A. Nonenzymatic glucose sensor with high
       performance electrodeposited nickel/copper/carbon nanotubes nanocomposite electrode. J Phys
       Chem Solids 2018;120:12–9. https://doi.org/10.1016/j.jpcs.2018.04.015.

[18]   Mottaghitalab V, Spinks GM, Wallace GG. The development and characterisation of polyaniline –
       single walled carbon nanotube composite fibres using 2-acrylamido-2 methyl-1-propane sulfonic
       acid (AMPSA) through one step wet spinning process. Polymer 2006;47:4996–5002. https://doi.org/
       10.1016/j.polymer.2006.05.037.

[19]   Li S, Ping Guo Z, Yun Wang C, G. Wallace G, Kun Liu H. Flexible cellulose based
       polypyrrole–multiwalled carbon nanotube films for bio-compatible zinc batteries activated by
       simulated body fluids. J Mater Chem A 2013;1:14300–5. https://doi.org/10.1039/C3TA13137H.

[20]   Meng C, Qian Y, He J, Dong X. Wet-spinning fabrication of multi-walled carbon nanotubes
       reinforced poly(3,4-ethylenedioxythiophene)-poly(styrenesulfonate) hybrid fibers for high-
       performance fiber-shaped supercapacitor. J Mater Sci Mater Electron 2020;31:19293–308. https://
       doi.org/10.1007/s10854-020-04464-7.

[21]   Moghadam BH, Hasanzadeh M, Simchi A. Self-Powered Wearable Piezoelectric Sensors Based on
       Polymer Nanofiber–Metal–Organic Framework Nanoparticle Composites for Arterial Pulse
       Monitoring. ACS Appl Nano Mater 2020;3:8742–52. https://doi.org/10.1021/acsanm.0c01551.

[22] Dhakal KN, Khanal S, Krause B, Lach R, Grellmann W, Le HH, et al. Electrically conductive and piezoresistive polymer nanocomposites using multiwalled carbon nanotubes in a flexible copolyester: Spectroscopic, morphological, mechanical and electrical properties. Nano-Struct Nano-Objects 2022;29:100806. https://doi.org/10.1016/j.nanoso.2021.100806.

[23] Rauf S, Vijjapu MT, Andrés MA, Gascón I, Roubeau O, Eddaoudi M, et al. Highly Selective Metal–Organic Framework Textile Humidity Sensor. ACS Appl Mater Interfaces 2020;12:29999–30006. https://doi.org/10.1021/acsami.0c07532.

[24] Zhao J, Zhu J, Li Y, Wang L, Dong Y, Jiang Z, et al. Graphene Quantum Dot Reinforced Electrospun Carbon Nanofiber Fabrics with High Surface Area for Ultrahigh Rate Supercapacitors. ACS Appl Mater Interfaces 2020;12:11669–78. https://doi.org/10.1021/acsami.9b22408.

[25] Li L, Li H, Guo Y, Yang L, Fang Y. Direct synthesis of graphene/carbon nanotube hybrid films from multiwalled carbon nanotubes on copper. Carbon 2017;118:675–9. https://doi.org/10.1016/j.carbon.2017.03.078.

[26] Li Y, Zhou B, Zheng G, Liu X, Li T, Yan C, et al. Continuously prepared highly conductive and stretchable SWNT/MWNT synergistically composited electrospun thermoplastic polyurethane yarns for wearable sensing. J Mater Chem C 2018;6:2258–69. https://doi.org/10.1039/C7TC04959E.

[27] Zheng X, Wang P, Zhang X, Hu Q, Wang Z, Nie W, et al. Breathable, durable and bark-shaped MXene/textiles for high-performance wearable pressure sensors, EMI shielding and heat physiotherapy. Compos Part Appl Sci Manuf 2022;152:106700. https://doi.org/10.1016/j.compositesa.2021.106700.

[28] Kim H, Lee S, Kim H. Electrical Heating Performance of Electro-Conductive Para-aramid Knit Manufactured by Dip-Coating in a Graphene/Waterborne Polyurethane Composite. Sci Rep 2019;9:1511. https://doi.org/10.1038/s41598-018-37455-0.

[29] Chatterjee, Tabor, Ghosh. Electrically Conductive Coatings for Fiber-Based E-Textiles. Fibers 2019;7:51. https://doi.org/10.3390/fib7060051.

[30] Lacour SP, Jones J, Wagner S, Li T, Suo Z. Stretchable Interconnects for Elastic Electronic Surfaces. Proc IEEE 2005;93:1459–67. https://doi.org/10.1109/JPROC.2005.851502.

[31] Li T, Suo Z, Lacour SP, Wagner S. Compliant thin film patterns of stiff materials as platforms for stretchable electronics. J Mater Res 2005;20:3274–7. https://doi.org/10.1557/jmr.2005.0422.

[32] Shirakawa H, Louis EJ, MacDiarmid AG, Chiang CK, Heeger AJ. Synthesis of electrically conducting organic polymers: halogen derivatives of polyacetylene, (CH)x. J Chem Soc Chem Commun 1977:578–80. https://doi.org/10.1039/C39770000578.

[33] Salaneck WR, Lundström I, Huang W-S, Macdiarmid AG. A two-dimensional-surface 'state diagram' for polyaniline. Synth Met 1986;13:291–7. https://doi.org/10.1016/0379-6779(86)90190-6.

[34] Pal T, Banerjee S, Manna PK, Kar KK. Characteristics of Conducting Polymers. In: Kar KK, editor. Handb. Nanocomposite Supercapacitor Mater. Charact., Cham: Springer International Publishing; 2020, p. 247–68. https://doi.org/10.1007/978-3-030-43009-2_8.

[35] Ghomi M, Zare EN, Varma RS. Properties of Conducting Polymers. Conduct. Polym. Anal. Chem., vol. 1405, American Chemical Society; 2022, p. 39–65. https://doi.org/10.1021/bk-2022-1405.ch002.

[36] Heeger AJ. Semiconducting and Metallic Polymers: The Fourth Generation of Polymeric Materials. J Phys Chem B 2001;105:8475–91. https://doi.org/10.1021/jp011611w.

[37] Patil AO, Ikenoue Y, Basescu N, Colaneri N, Chen J, Wudl F, et al. Self-doped conducting polymers. Synth Met 1987;20:151–9. https://doi.org/10.1016/0379-6779(87)90554-6.

[38] Barroso Bogeat A. Understanding and Tuning the Electrical Conductivity of Activated Carbon: A State-of-the-Art Review. Crit Rev Solid State Mater Sci 2021;46:1–37. https://doi.org/10.1080/10408436.2019.1671800.

[39]   Spahr ME, Gilardi R, Bonacchi D. Carbon Black for Electrically Conductive Polymer Applications. In:
        Rothon R, editor. Fill. Polym. Appl., Cham: Springer International Publishing; 2017, p. 375–400.
        https://doi.org/10.1007/978-3-319-28117-9_32.

[40]   Rodriguez NM. A review of catalytically grown carbon nanofibers. J Mater Res 1993;8:3233–50.
        https://doi.org/10.1557/JMR.1993.3233.

[41]   Robert C, Feller JF, Castro M. Sensing Skin for Strain Monitoring Made of PC–CNT Conductive
        Polymer Nanocomposite Sprayed Layer by Layer. ACS Appl Mater Interfaces 2012;4:3508–16.
        https://doi.org/10.1021/am300594t.

[42]   Kim SJ, Song W, Yi Y, Min BK, Mondal S, An K-S, et al. High Durability and Waterproofing rGO/
        SWCNT-Fabric-Based Multifunctional Sensors for Human-Motion Detection. ACS Appl Mater
        Interfaces 2018;10:3921–8. https://doi.org/10.1021/acsami.7b15386.

[43]   Sugimoto Y, Irisawa T, Hatori H, Inagaki M. Yarns of carbon nanotubes and reduced graphene
        oxides. Carbon 2020;165:358–77. https://doi.org/10.1016/j.carbon.2020.04.087.

[44]   Rajan G, Morgan JJ, Murphy C, Torres Alonso E, Wade J, Ott AK, et al. Low Operating Voltage
        Carbon–Graphene Hybrid E-textile for Temperature Sensing. ACS Appl Mater Interfaces
        2020;12:29861–7. https://doi.org/10.1021/acsami.0c08397.

[45]   Zhao Q, Zhang K, Zhu S, Xu H, Cao D, Zhao L, et al. Review on the Electrical Resistance/Conductivity
        of Carbon Fiber Reinforced Polymer. Appl Sci 2019;9:2390. https://doi.org/10.3390/app9112390.

[46]   Castellino M, Chiolerio A, Shahzad MI, Jagdale PV, Tagliaferro A. Electrical conductivity phenomena
        in an epoxy resin–carbon-based materials composite. Compos Part Appl Sci Manuf 2014;61:108–14.
        https://doi.org/10.1016/j.compositesa.2014.02.012.

[47]   Jalili R, Aboutalebi SH, Esrafilzadeh D, Shepherd RL, Chen J, Aminorroaya-Yamini S, et al. Scalable
        One-Step Wet-Spinning of Graphene Fibers and Yarns from Liquid Crystalline Dispersions of
        Graphene Oxide: Towards Multifunctional Textiles. Adv Funct Mater 2013;23:5345–54. https://doi.
        org/10.1002/adfm.201300765.

[48]   Toprakci HAK, Kalanadhabhatla SK, Spontak RJ, Ghosh TK. Polymer Nanocomposites Containing
        Carbon Nanofibers as Soft Printable Sensors Exhibiting Strain-Reversible Piezoresistivity. Adv Funct
        Mater 2013;23:5536–42. https://doi.org/10.1002/adfm.201300034.

[49]   Raval JP, Joshi P, Chejara DR. 9 – Carbon nanotube for targeted drug delivery. In: Inamuddin, Asiri
        AM, Mohammad A, editors. Appl. Nanocomposite Mater. Drug Deliv., Woodhead Publishing; 2018,
        p. 203–16. https://doi.org/10.1016/B978-0-12-813741-3.00009-1.

[50]   Masyutin AG, Bagrov DV, Vlasova II, Nikishin II, Klinov DV, Sychevskaya KA, et al. Wall Thickness of
        Industrial Multi-Walled Carbon Nanotubes Is Not a Crucial Factor for Their Degradation by Sodium
        Hypochlorite. Nanomaterials 2018;8:715. https://doi.org/10.3390/nano8090715.

[51]   Egbosiuba TC, Abdulkareem AS, Tijani JO, Ani JI, Krikstolaityte V, Srinivasan M, et al. Taguchi
        optimization design of diameter-controlled synthesis of multi walled carbon nanotubes for the
        adsorption of Pb(II) and Ni(II) from chemical industry wastewater. Chemosphere 2021;266:128937.
        https://doi.org/10.1016/j.chemosphere.2020.128937.

[52]   Mirfakhrai T, Oh J, Kozlov M, Fok ECW, Zhang M, Fang S, et al. Electrochemical actuation of carbon
        nanotube yarns. Smart Mater Struct 2007;16:S243–S249. https://doi.org/10.1088/0964-1726/16/2/
        S07.

[53]   Vigolo B, Pénicaud A, Coulon C, Sauder C, Pailler R, Journet C, et al. Macroscopic Fibers and Ribbons
        of Oriented Carbon Nanotubes. Science 2000;290:1331–4. https://doi.org/10.1126/
        science.290.5495.1331.

[54]   Jiang K, Li Q, Fan S. Spinning continuous carbon nanotube yarns. Nature 2002;419:801–801. https://
        doi.org/10.1038/419801a.

[55]   Zhang X, Jiang K, Feng C, Liu P, Zhang L, Kong J, et al. Spinning and Processing Continuous Yarns
        from 4-Inch Wafer Scale Super-Aligned Carbon Nanotube Arrays. Adv Mater 2006;18:1505–10.
        https://doi.org/10.1002/adma.200502528.

[56]   Bradford PD, Bogdanovich AE. Electrical Conductivity Study of Carbon Nanotube Yarns, 3-D Hybrid Braids and their Composites. J Compos Mater 2008;42:1533–45. https://doi.org/10.1177/0021998308092206.

[57]   Aboutalebi SH, Jalili R, Esrafilzadeh D, Salari M, Gholamvand Z, Aminorroaya Yamini S, et al. High-Performance Multifunctional Graphene Yarns: Toward Wearable All-Carbon Energy Storage Textiles. ACS Nano 2014;8:2456–66. https://doi.org/10.1021/nn406026z.

[58]   Ma W, Song L, Yang R, Zhang T, Zhao Y, Sun L, et al. Directly Synthesized Strong, Highly Conducting, Transparent Single-Walled Carbon Nanotube Films. Nano Lett 2007;7:2307–11. https://doi.org/10.1021/nl070915c.

[59]   Hu L, Hecht DS, Grüner G. Carbon Nanotube Thin Films: Fabrication, Properties, and Applications. Chem Rev 2010;110:5790–844. https://doi.org/10.1021/cr9002962.

[60]   Song L, Ci L, Lv L, Zhou Z, Yan X, Liu D, et al. Direct Synthesis of a Macroscale Single-Walled Carbon Nanotube Non-Woven Material. Adv Mater 2004;16:1529–34. https://doi.org/10.1002/adma.200306393.

[61]   Endo M, Muramatsu H, Hayashi T, Kim YA, Terrones M, Dresselhaus MS. 'Buckypaper' from coaxial nanotubes. Nature 2005;433:476–476. https://doi.org/10.1038/433476a.

[62]   Wu Z-S, Tan Y-Z, Zheng S, Wang S, Parvez K, Qin J, et al. Bottom-Up Fabrication of Sulfur-Doped Graphene Films Derived from Sulfur-Annulated Nanographene for Ultrahigh Volumetric Capacitance Micro-Supercapacitors. J Am Chem Soc 2017;139:4506–12. https://doi.org/10.1021/jacs.7b00805.

[63]   Shao Y, El-Kady MF, Lin C-W, Zhu G, Marsh KL, Hwang JY, et al. 3D Freeze-Casting of Cellular Graphene Films for Ultrahigh-Power-Density Supercapacitors. Adv Mater 2016;28:6719–26. https://doi.org/10.1002/adma.201506157.

[64]   Xiong Z, Liao C, Han W, Wang X. Mechanically Tough Large-Area Hierarchical Porous Graphene Films for High-Performance Flexible Supercapacitor Applications. Adv Mater 2015;27:4469–75. https://doi.org/10.1002/adma.201501983.

[65]   Phillips C, Al-Ahmadi A, Potts S-J, Claypole T, Deganello D. The effect of graphite and carbon black ratios on conductive ink performance. J Mater Sci 2017;52:9520–30. https://doi.org/10.1007/s10853-017-1114-6.

[66]   Sarkar B, Alexandridis P. Block copolymer–nanoparticle composites: Structure, functional properties, and processing. Prog Polym Sci 2015;40:33–62. https://doi.org/10.1016/j.progpolymsci.2014.10.009.

[67]   Yan N, Liu H, Zhu Y, Jiang W, Dong Z. Entropy-Driven Hierarchical Nanostructures from Cooperative Self-Assembly of Gold Nanoparticles/Block Copolymers under Three-Dimensional Confinement. Macromolecules 2015;48:5980–7. https://doi.org/10.1021/acs.macromol.5b01219.

[68]   Stauffer D, Aharony A. Introduction To Percolation Theory: Second Edition. 2nd ed. London: Taylor & Francis; 2017. https://doi.org/10.1201/9781315274386.

[69]   McCall C, Dimitrov N, Sieradzki K. Underpotential Deposition on Alloys. J Electrochem Soc 2001;148:E290. https://doi.org/10.1149/1.1371801.

[70]   Uemura YJ, Birgeneau RJ. Crossover from Spin Waves to Quasilocalized Excitations in the Diluted Antiferromagnet ($Mn_{0.5}Zn_{0.5}F_2$). Phys Rev Lett 1986;57:1947–50. https://doi.org/10.1103/PhysRevLett.57.1947.

[71]   Dal Lago E, Cagnin E, Boaretti C, Roso M, Lorenzetti A, Modesti M. Influence of Different Carbon-Based Fillers on Electrical and Mechanical Properties of a PC/ABS Blend. Polymers 2019;12:29. https://doi.org/10.3390/polym12010029.

[72]   Ruschau GR, Yoshikawa S, Newnham RE. Resistivities of conductive composites. J Appl Phys 1992;72:953–9. https://doi.org/10.1063/1.352350.

[73]   Bigg DM, Stutz DE. Plastic composites for electromagnetic interference shielding applications. Polym Compos 1983;4:40–6. https://doi.org/10.1002/pc.750040107.

[74] Matsuhisa N, Kaltenbrunner M, Yokota T, Jinno H, Kuribara K, Sekitani T, et al. Printable elastic conductors with a high conductivity for electronic textile applications. Nat Commun 2015;6:7461. https://doi.org/10.1038/ncomms8461.

[75] Taya M, Kim WJ, Ono K. Piezoresistivity of a short fiber/elastomer matrix composite. Mech Mater 1998;28:53–9. https://doi.org/10.1016/S0167-6636(97)00064-1.

[76] Dan L, Elias AL. Flexible and Stretchable Temperature Sensors Fabricated Using Solution-Processable Conductive Polymer Composites. Adv Healthc Mater 2020;9:2000380. https://doi.org/10.1002/adhm.202000380.

[77] Chen J, Tsubokawa N. Novel gas sensor from polymer-grafted carbon black: Vapor response of electric resistance of conducting composites prepared from poly(ethylene-block-ethylene oxide)-grafted carbon black. J Appl Polym Sci 2000;77:2437–47. https://doi.org/10.1002/1097-4628 (20000912)77:11<2437::AID-APP12>3.0.CO;2-F.

[78] Mozammal Hossain MD, Moon J-M, Gurudatt NG, Park D-S, Choi CS, Shim Y-B. Separation detection of hemoglobin and glycated hemoglobin fractions in blood using the electrochemical microfluidic channel with a conductive polymer composite sensor. Biosens Bioelectron 2019;142:111515. https://doi.org/10.1016/j.bios.2019.111515.

[79] Wang Y, Zhang L, Zhang Z, Sun P, Chen H. High-Sensitivity Wearable and Flexible Humidity Sensor Based on Graphene Oxide/Non-Woven Fabric for Respiration Monitoring. Langmuir 2020;36:9443–8. https://doi.org/10.1021/acs.langmuir.0c01315.

[80] He X, Liu Z, Shen G, He X, Liang J, Zhong Y, et al. Microstructured capacitive sensor with broad detection range and long-term stability for human activity detection. Npj Flex Electron 2021;5:1–9. https://doi.org/10.1038/s41528-021-00114-y.

[81] Zhu P, Du H, Hou X, Lu P, Wang L, Huang J, et al. Skin-electrode iontronic interface for mechanosensing. Nat Commun 2021;12:4731. https://doi.org/10.1038/s41467-021-24946-4.

[82] Wu R, Ma L, Patil A, Hou C, Zhu S, Fan X, et al. All-Textile Electronic Skin Enabled by Highly Elastic Spacer Fabric and Conductive Fibers. ACS Appl Mater Interfaces 2019;11:33336–46. https://doi.org/10.1021/acsami.9b10928.

[83] Wang F, Jiang J, Sun F, Sun L, Wang T, Liu Y, et al. Flexible wearable graphene/alginate composite non-woven fabric temperature sensor with high sensitivity and anti-interference. Cellulose 2020;27:2369–80. https://doi.org/10.1007/s10570-019-02951-7.

[84] Sun X, Agate S, Salem KS, Lucia L, Pal L. Hydrogel-Based Sensor Networks: Compositions, Properties, and Applications – A Review. ACS Appl Bio Mater 2021;4:140–62. https://doi.org/10.1021/acsabm.0c01011.

[85] Bashir T, Naeem J, Skrifvars M, Persson N-K. Synthesis of electro-active membranes by chemical vapor deposition (CVD) process. Polym Adv Technol 2014;25:1501–8. https://doi.org/10.1002/pat.3392.

[86] Wei Q, Xu Y, Wang Y. 3 – Textile surface functionalization by physical vapor deposition (PVD). In: Wei Q, editor. Surf. Modif. Text., Woodhead Publishing; 2009, p. 58–90. https://doi.org/10.1533/9781845696689.58.

[87] Pragya A, Deogaonkar-Baride S. Effect of yarn interlacement pattern on the surface electrical conductivity of intrinsically conductive fabrics. Synth Met 2020;268:116512. https://doi.org/10.1016/j.synthmet.2020.116512.

[88] Sun B, Long Y-Z, Chen Z-J, Liu S-L, Zhang H-D, Zhang J-C, et al. Recent advances in flexible and stretchable electronic devices via electrospinning. J Mater Chem C 2014;2:1209–19. https://doi.org/10.1039/C3TC31680G.

[89] Zhang Q, Wang YL, Xia Y, Zhang PF, Kirk TV, Chen XD. Textile-Only Capacitive Sensors for Facile Fabric Integration without Compromise of Wearability. Adv Mater Technol 2019;4:1900485. https://doi.org/10.1002/admt.201900485.

[90] Ryu CH, Cho JY, Jeong SY, Eom W, Shin H, Hwang W, et al. Wearable Piezoelectric Yarns with Inner Electrodes for Energy Harvesting and Signal Sensing. Adv Mater Technol n.d.;n/a:2101138. https://doi.org/10.1002/admt.202101138.

[91] Wang L, Tian M, Qi X, Sun X, Xu T, Liu X, et al. Customizable Textile Sensors Based on Helical Core–Spun Yarns for Seamless Smart Garments. Langmuir 2021;37:3122–9. https://doi.org/10.1021/acs.langmuir.0c03595.

[92] Gao Y, Li Z, Xu B, Li M, Jiang C, Guan X, et al. Scalable core–spun coating yarn-based triboelectric nanogenerators with hierarchical structure for wearable energy harvesting and sensing via continuous manufacturing. Nano Energy 2022;91:106672. https://doi.org/10.1016/j.nanoen.2021.106672.

[93] Qi K, Zhou Y, Ou K, Dai Y, You X, Wang H, et al. Weavable and stretchable piezoresistive carbon nanotubes-embedded nanofiber sensing yarns for highly sensitive and multimodal wearable textile sensor. Carbon 2020;170:464–76. https://doi.org/10.1016/j.carbon.2020.07.042.

[94] Dong K, Deng J, Zi Y, Wang Y-C, Xu C, Zou H, et al. 3D Orthogonal Woven Triboelectric Nanogenerator for Effective Biomechanical Energy Harvesting and as Self-Powered Active Motion Sensors. Adv Mater 2017;29:1702648. https://doi.org/10.1002/adma.201702648.

[95] Kapoor A, McKnight M, Chatterjee K, Agcayazi T, Kausche H, Bozkurt A, et al. Toward Fully Manufacturable, Fiber Assembly–Based Concurrent Multimodal and Multifunctional Sensors for e-Textiles. Adv Mater Technol 2019;4:1800281. https://doi.org/10.1002/admt.201800281.

[96] Tabor J, Agcayazi T, Fleming A, Thompson B, Kapoor A, Liu M, et al. Textile-Based Pressure Sensors for Monitoring Prosthetic-Socket Interfaces. IEEE Sens J 2021;21:9413–22. https://doi.org/10.1109/JSEN.2021.3053434.

[97] Ou J, Oran D, Haddad DD, Paradiso J, Ishii H. SensorKnit: Architecting Textile Sensors with Machine Knitting. 3D Print Addit Manuf 2019. https://doi.org/10.1089/3DP.2018.0122.

[98] Agcayazi T, Tabor J, McKnight M, Martin I, Ghosh TK, Bozkurt A. Fully-Textile Seam-Line Sensors for Facile Textile Integration and Tunable Multi-Modal Sensing of Pressure, Humidity, and Wetness. Adv Mater Technol 2020;5:2000155. https://doi.org/10.1002/admt.202000155.

[99] Tabor J, Thompson B, Agcayazi T, Bozkurt A, Ghosh TK. Melt-Extruded Sensory Fibers for Electronic Textiles. Macromol Mater Eng 2022;307:2100737. https://doi.org/10.1002/mame.202100737.

[100] Qureshi S, Stojanović GM, Simić M, Jeoti V, Lashari N, Sher F. Silver Conductive Threads-Based Embroidered Electrodes on Textiles as Moisture Sensors for Fluid Detection in Biomedical Applications. Materials 2021;14:7813. https://doi.org/10.3390/ma14247813.

[101] Ahn Y, Song S, Yun K-S. Woven flexible textile structure for wearable power-generating tactile sensor array. Smart Mater Struct 2015;24:075002. https://doi.org/10.1088/0964-1726/24/7/075002.

[102] Seyedin S, Razal JM, Innis PC, Jeiranikhameneh A, Beirne S, Wallace GG. Knitted Strain Sensor Textiles of Highly Conductive All-Polymeric Fibers. ACS Appl Mater Interfaces 2015;7:21150–8. https://doi.org/10.1021/acsami.5b04892.

[103] Lee J, Kwon H, Seo J, Shin S, Koo JH, Pang C, et al. Conductive Fiber-Based Ultrasensitive Textile Pressure Sensor for Wearable Electronics. Adv Mater 2015;27:2433–9. https://doi.org/10.1002/adma.201500009.

[104] Ma L, Wu R, Miao H, Fan X, Kong L, Patil A, et al. All-in-one fibrous capacitive humidity sensor for human breath monitoring. Text Res J 2021;91:398–405. https://doi.org/10.1177/0040517520944495.

[105] Nocke A, Schröter A, Cherif C, Gerlach G. Miniaturized textile-based multi-layer ph-sensor for wound monitoring applications. Autex Res J 2012;12:20–2. https://doi.org/10.2478/v10304-012-0004-x.

[106] Yu P, Li X, Li H, Fan Y, Cao J, Wang H, et al. All-Fabric Ultrathin Capacitive Sensor with High Pressure Sensitivity and Broad Detection Range for Electronic Skin. ACS Appl Mater Interfaces 2021;13:24062–9. https://doi.org/10.1021/acsami.1c05478.

[107] Chhetry A, Yoon H, Park JY. A flexible and highly sensitive capacitive pressure sensor based on conductive fibers with a microporous dielectric for wearable electronics. J Mater Chem C 2017;5:10068–76. https://doi.org/10.1039/C7TC02926H.

[108] Yang Z, Jia Y, Niu Y, Yong Z, Wu K, Zhang C, et al. Wet-spun PVDF nanofiber separator for direct fabrication of coaxial fiber-shaped supercapacitors. Chem Eng J 2020;400:125835. https://doi.org/10.1016/j.cej.2020.125835.

[109] Zhou M, Zhang H, Qiao Y, Li CM, Lu Z. A flexible sandwich-structured supercapacitor with poly(vinyl alcohol)/H3PO4-soaked cotton fabric as solid electrolyte, separator and supporting layer. Cellulose 2018;25:3459–69. https://doi.org/10.1007/s10570-018-1786-3.

[110] Kapoor A, McKnight M, Chatterjee K, Agcayazi T, Kausche H, Bozkurt A, et al. Toward Fully Manufacturable, Fiber Assembly–Based Concurrent Multimodal and Multifunctional Sensors for e-Textiles. Adv Mater Technol 2019;4:1800281. https://doi.org/10.1002/admt.201800281.

[111] Vu CC, Kim J. Highly elastic capacitive pressure sensor based on smart textiles for full-range human motion monitoring. Sens Actuators Phys 2020;314:112029. https://doi.org/10.1016/j.sna.2020.112029.

[112] Kapoor A, McKnight M, Chatterjee K, Agcayazi T, Kausche H, Ghosh T, et al. Soft, flexible 3D printed fibers for capacitive tactile sensing. 2016 IEEE Sens., 2016, p. 1–3. https://doi.org/10.1109/ICSENS.2016.7808918.

[113] You X, He J, Nan N, Sun X, Qi K, Zhou Y, et al. Stretchable capacitive fabric electronic skin woven by electrospun nanofiber coated yarns for detecting tactile and multimodal mechanical stimuli. J Mater Chem C 2018;6:12981–91. https://doi.org/10.1039/C8TC03631D.

[114] Milne SD, Seoudi I, Al Hamad H, Talal TK, Anoop AA, Allahverdi N, et al. A wearable wound moisture sensor as an indicator for wound dressing change: an observational study of wound moisture and status. Int Wound J 2016;13:1309–14. https://doi.org/10.1111/iwj.12521.

[115] Zheng Y-N, Yu Z, Mao G, Li Y, Pravarthana D, Asghar W, et al. A Wearable Capacitive Sensor Based on Ring/Disk-Shaped Electrode and Porous Dielectric for Noncontact Healthcare Monitoring. Glob Chall 2020;4:1900079. https://doi.org/10.1002/gch2.201900079.

[116] Hassan G, Sajid M, Choi C. Highly Sensitive and Full Range Detectable Humidity Sensor using PEDOT:PSS, Methyl Red and Graphene Oxide Materials. Sci Rep 2019;9:15227. https://doi.org/10.1038/s41598-019-51712-w.

[117] Wang X, Deng Y, Chen X, Jiang P, Cheung YK, Yu H. An ultrafast-response and flexible humidity sensor for human respiration monitoring and noncontact safety warning. Microsyst Nanoeng 2021;7:1–11. https://doi.org/10.1038/s41378-021-00324-4.

[118] Pyo S, Choi J, Kim J. Flexible, Transparent, Sensitive, and Crosstalk-Free Capacitive Tactile Sensor Array Based on Graphene Electrodes and Air Dielectric. Adv Electron Mater 2018;4:1700427. https://doi.org/10.1002/aelm.201700427.

[119] Luo Z, Chen J, Zhu Z, Li L, Su Y, Tang W, et al. High-Resolution and High-Sensitivity Flexible Capacitive Pressure Sensors Enhanced by a Transferable Electrode Array and a Micropillar–PVDF Film. ACS Appl Mater Interfaces 2021;13:7635–49. https://doi.org/10.1021/acsami.0c23042.

[120] Ko Y, Vu CC, Kim J. Carbonized Cotton Fabric-Based Flexible Capacitive Pressure Sensor Using a Porous Dielectric Layer with Tilted Air Gaps. Sensors 2021;21:3895. https://doi.org/10.3390/s21113895.

[121] Li R, Zhou Q, Bi Y, Cao S, Xia X, Yang A, et al. Research progress of flexible capacitive pressure sensor for sensitivity enhancement approaches. Sens Actuators Phys 2021;321:112425. https://doi.org/10.1016/j.sna.2020.112425.

[122] Chou H-H, Lee W-Y. 3 – Tactile sensor based on capacitive structure. In: Zhou Y, Chou H-H, editors. Funct. Tactile Sens., Woodhead Publishing; 2021, p. 31–52. https://doi.org/10.1016/B978-0-12-820633-1.00004-8.

[123] Vu CC, Kim J. Simultaneous Sensing of Touch and Pressure by Using Highly Elastic e-Fabrics. Appl Sci 2020;10:989. https://doi.org/10.3390/app10030989.

[124] Nie B, Huang R, Yao T, Zhang Y, Miao Y, Liu C, et al. Textile-Based Wireless Pressure Sensor Array for Human-Interactive Sensing. Adv Funct Mater 2019;29:1808786. https://doi.org/10.1002/adfm.201808786.

[125] Li L, Xiang H, Xiong Y, Zhao H, Bai Y, Wang S, et al. Ultrastretchable Fiber Sensor with High Sensitivity in Whole Workable Range for Wearable Electronics and Implantable Medicine. Adv Sci 2018;5:1800558. https://doi.org/10.1002/advs.201800558.

[126] Lou M, Abdalla I, Zhu M, Yu J, Li Z, Ding B. Hierarchically Rough Structured and Self-Powered Pressure Sensor Textile for Motion Sensing and Pulse Monitoring. ACS Appl Mater Interfaces 2020;12:1597–605. https://doi.org/10.1021/acsami.9b19238.

[127] Thomson W. XIX. On the electro-dynamic qualities of metals: – Effects of magnetization on the electric conductivity of nickel and of iron. Proc R Soc Lond 1857;8:546–50. https://doi.org/10.1098/rspl.1856.0144.

[128] Bao M. Chapter 1 – Introduction to MEMS Devices. In: Bao M, editor. Anal. Des. Princ. MEMS Devices, Amsterdam: Elsevier Science; 2005, p. 1–32. https://doi.org/10.1016/B978-044451616-9/50002-3.

[129] Rowe ACH. Piezoresistance in silicon and its nanostructures. J Mater Res 2014;29:731–44. https://doi.org/10.1557/jmr.2014.52.

[130] Zhang JXJ, Hoshino K. Chapter 6 – Mechanical transducers: Cantilevers, acoustic wave sensors, and thermal sensors. In: Zhang JXJ, Hoshino K, editors. Mol. Sens. Nanodevices Second Ed., Academic Press; 2019, p. 311–412. https://doi.org/10.1016/B978-0-12-814862-4.00006-5.

[131] Kanoun O, Bouhamed A, Ramalingame R, Bautista-Quijano JR, Rajendran D, Al-Hamry A. Review on Conductive Polymer/CNTs Nanocomposites Based Flexible and Stretchable Strain and Pressure Sensors. Sensors 2021;21:341. https://doi.org/10.3390/s21020341.

[132] Wang M, Gurunathan R, Imasato K, Geisendorfer NR, Jakus AE, Peng J, et al. A Percolation Model for Piezoresistivity in Conductor–Polymer Composites. Adv Theory Simul 2019;2:1800125. https://doi.org/10.1002/adts.201800125.

[133] Liao X, Liao Q, Zhang Z, Yan X, Liang Q, Wang Q, et al. A Highly Stretchable ZnO@Fiber-Based Multifunctional Nanosensor for Strain/Temperature/UV Detection. Adv Funct Mater 2016;26:3074–81. https://doi.org/10.1002/adfm.201505223.

[134] Chen S, Li J, Liu H, Shi W, Peng Z, Liu L. Pruney fingers-inspired highly stretchable and sensitive piezoresistive fibers with isotropic wrinkles and robust interfaces. Chem Eng J 2022;430:133005. https://doi.org/10.1016/j.cej.2021.133005.

[135] Mirabedini A, Foroughi J, G. Wallace G. Developments in conducting polymer fibres: from established spinning methods toward advanced applications. RSC Adv 2016;6:44687–716. https://doi.org/10.1039/C6RA05626A.

[136] Chatterjee K, Tabor J, Ghosh TK. Electrically Conductive Coatings for Fiber-Based E-Textiles. Fibers 2019;7:51. https://doi.org/10.3390/fib7060051.

[137] Shi B, Wang T, Shi L, Li J, Wang R, Sun J. Highly stretchable and strain sensitive fibers based on braid-like structure and sliver nanowires. Appl Mater Today 2020;19:100610. https://doi.org/10.1016/j.apmt.2020.100610.

[138] Park JJ, Hyun WJ, Mun SC, Park YT, Park OO. Highly Stretchable and Wearable Graphene Strain Sensors with Controllable Sensitivity for Human Motion Monitoring. ACS Appl Mater Interfaces 2015;7:6317–24. https://doi.org/10.1021/acsami.5b00695.

[139] Toprakci HAK, Kalanadhabhatla SK, Spontak RJ, Ghosh TK. Polymer Nanocomposites Containing Carbon Nanofibers as Soft Printable Sensors Exhibiting Strain-Reversible Piezoresistivity. Adv Funct Mater 2013;23:5536–42. https://doi.org/10.1002/adfm.201300034.

[140] Cochrane C, Koncar V, Lewandowski M, Dufour C. Design and Development of a Flexible Strain Sensor for Textile Structures Based on a Conductive Polymer Composite. Sensors 2007;7:473–92. https://doi.org/10.3390/s7040473.

[141] Zhou Y, He J, Wang H, Qi K, Nan N, You X, et al. Highly sensitive, self-powered and wearable electronic skin based on pressure-sensitive nanofiber woven fabric sensor. Sci Rep 2017;7:12949. https://doi.org/10.1038/s41598-017-13281-8.

[142] Park J, Lee Y, Hong J, Ha M, Jung Y-D, Lim H, et al. Giant Tunneling Piezoresistance of Composite Elastomers with Interlocked Microdome Arrays for Ultrasensitive and Multimodal Electronic Skins. ACS Nano 2014;8:4689–97. https://doi.org/10.1021/nn500441k.

[143] Ma Y, Liu N, Li L, Hu X, Zou Z, Wang J, et al. A highly flexible and sensitive piezoresistive sensor based on MXene with greatly changed interlayer distances. Nat Commun 2017;8:1207. https://doi.org/10.1038/s41467-017-01136-9.

[144] Luo Y, Li Y, Sharma P, Shou W, Wu K, Foshey M, et al. Learning human–environment interactions using conformal tactile textiles. Nat Electron 2021;4:193–201. https://doi.org/10.1038/s41928-021-00558-0.

[145] Sengupta D, Romano J, Kottapalli AGP. Electrospun bundled carbon nanofibers for skin-inspired tactile sensing, proprioception and gesture tracking applications. Npj Flex Electron 2021;5:1–14. https://doi.org/10.1038/s41528-021-00126-8.

[146] Yu S, Wang X, Xiang H, Zhu L, Tebyetekerwa M, Zhu M. Superior piezoresistive strain sensing behaviors of carbon nanotubes in one-dimensional polymer fiber structure. Carbon 2018;140:1–9. https://doi.org/10.1016/j.carbon.2018.08.028.

[147] Costa P, Ribeiro S, Lanceros-Mendez S. Mechanical vs. electrical hysteresis of carbon nanotube/styrene–butadiene–styrene composites and their influence in the electromechanical response. Compos Sci Technol 2015;109:1–5. https://doi.org/10.1016/j.compscitech.2015.01.006.

[148] Montazerian H, Dalili A, Milani AS, Hoorfar M. Piezoresistive sensing in chopped carbon fiber embedded PDMS yarns. Compos Part B Eng 2019;164:648–58. https://doi.org/10.1016/j.compositesb.2019.01.090.

[149] Georgousis G, Roumpos K, Kontou E, Kyritsis A, Pissis P, Koutsoumpis S, et al. Strain and damage monitoring in SBR nanocomposites under cyclic loading. Compos Part B Eng 2017;131:50–61. https://doi.org/10.1016/j.compositesb.2017.08.006.

[150] Dineva P, Gross D, Müller R, Rangelov T. Piezoelectric Materials. In: Dineva P, Gross D, Müller R, Rangelov T, editors. Dyn. Fract. Piezoelectric Mater. Solut. Time-Harmon. Probl. BIEM, Cham: Springer International Publishing; 2014, p. 7–32. https://doi.org/10.1007/978-3-319-03961-9_2.

[151] Udovč L, Spreitzer M, Vukomanović M. Towards hydrophilic piezoelectric poly-L-lactide films: optimal processing, post-heat treatment and alkaline etching. Polym J 2020;52:299–311. https://doi.org/10.1038/s41428-019-0281-5.

[152] Katsouras I, Asadi K, Li M, van Driel TB, Kjær KS, Zhao D, et al. The negative piezoelectric effect of the ferroelectric polymer poly(vinylidene fluoride). Nat Mater 2016;15:78–84. https://doi.org/10.1038/nmat4423.

[153] Yang J, Chen Q, Xu F, Jiang H, Liu W, Zhang X, et al. Epitaxy Enhancement of Piezoelectric Properties in P(VDF-TrFE) Copolymer Films and Applications in Sensing and Energy Harvesting. Adv Electron Mater 2020;6:2000578. https://doi.org/10.1002/aelm.202000578.

[154] Soin N, Boyer D, Prashanthi K, Sharma S, Narasimulu AA, Luo J, et al. Exclusive self-aligned β-phase PVDF films with abnormal piezoelectric coefficient prepared via phase inversion. Chem Commun 2015;51:8257–60. https://doi.org/10.1039/C5CC01688F.

[155] Anwar S, Hassanpour Amiri M, Jiang S, Abolhasani MM, Rocha PRF, Asadi K. Piezoelectric Nylon-11 Fibers for Electronic Textiles, Energy Harvesting and Sensing. Adv Funct Mater 2021;31:2004326. https://doi.org/10.1002/adfm.202004326.

[156] Nilsson E, Lund A, Jonasson C, Johansson C, Hagström B. Poling and characterization of piezoelectric polymer fibers for use in textile sensors. Sens Actuators Phys 2013;201:477–86. https://doi.org/10.1016/j.sna.2013.08.011.

[157] Lund A, Jonasson C, Johansson C, Haagensen D, Hagström B. Piezoelectric polymeric bicomponent fibers produced by melt spinning. J Appl Polym Sci 2012;126:490–500. https://doi.org/10.1002/app.36760.

[158] Szewczyk PK, Gradys A, Kim SK, Persano L, Marzec M, Kryshtal A, et al. Enhanced Piezoelectricity of Electrospun Polyvinylidene Fluoride Fibers for Energy Harvesting. ACS Appl Mater Interfaces 2020;12:13575–83. https://doi.org/10.1021/acsami.0c02578.

[159] Wang S, Shao H-Q, Liu Y, Tang C-Y, Zhao X, Ke K, et al. Boosting piezoelectric response of PVDF-TrFE via MXene for self-powered linear pressure sensor. Compos Sci Technol 2021;202:108600. https://doi.org/10.1016/j.compscitech.2020.108600.

[160] Chen C, Bai Z, Cao Y, Dong M, Jiang K, Zhou Y, et al. Enhanced piezoelectric performance of BiCl3/PVDF nanofibers-based nanogenerators. Compos Sci Technol 2020;192:108100. https://doi.org/10.1016/j.compscitech.2020.108100.

[161] Zhang X, Le M-Q, Zahhaf O, Capsal J-F, Cottinet P-J, Petit L. Enhancing dielectric and piezoelectric properties of micro-ZnO/PDMS composite-based dielectrophoresis. Mater Des 2020;192:108783. https://doi.org/10.1016/j.matdes.2020.108783.

[162] Sundar U, Lao Z, Cook-Chennault K. Investigation of Piezoelectricity and Resistivity of Surface Modified Barium Titanate Nanocomposites. Polymers 2019;11:2123. https://doi.org/10.3390/polym11122123.

[163] Kim H, Torres F, Villagran D, Stewart C, Lin Y, Tseng T-LB. 3D Printing of BaTiO3/PVDF Composites with Electric In Situ Poling for Pressure Sensor Applications. Macromol Mater Eng 2017;302:1700229. https://doi.org/10.1002/mame.201700229.

[164] Su Y, Chen C, Pan H, Yang Y, Chen G, Zhao X, et al. Muscle Fibers Inspired High-Performance Piezoelectric Textiles for Wearable Physiological Monitoring. Adv Funct Mater 2021;31:2010962. https://doi.org/10.1002/adfm.202010962.

[165] Maria Joseph Raj NP, Alluri NR, Vivekananthan V, Chandrasekhar A, Khandelwal G, Kim S-J. Sustainable yarn type-piezoelectric energy harvester as an eco-friendly, cost-effective battery-free breath sensor. Appl Energy 2018;228:1767–76. https://doi.org/10.1016/j.apenergy.2018.07.016.

[166] Tan Y, Yang K, Wang B, Li H, Wang L, Wang C. High-performance textile piezoelectric pressure sensor with novel structural hierarchy based on ZnO nanorods array for wearable application. Nano Res 2021;14:3969–76. https://doi.org/10.1007/s12274-021-3322-2.

[167] Qin Y, Wang X, Wang ZL. Microfibre–nanowire hybrid structure for energy scavenging. Nature 2008;451:809–13. https://doi.org/10.1038/nature06601.

[168] Mokhtari F, Foroughi J, Zheng T, Cheng Z, M. Spinks G. Triaxial braided piezo fiber energy harvesters for self-powered wearable technologies. J Mater Chem A 2019;7:8245–57. https://doi.org/10.1039/C8TA10964H.

[169] Kim M, Wu YS, Kan EC, Fan J. Breathable and Flexible Piezoelectric ZnO@PVDF Fibrous Nanogenerator for Wearable Applications. Polymers 2018;10:745. https://doi.org/10.3390/polym10070745.

[170] Maity K, Mandal D. All-Organic High-Performance Piezoelectric Nanogenerator with Multilayer Assembled Electrospun Nanofiber Mats for Self-Powered Multifunctional Sensors. ACS Appl Mater Interfaces 2018;10:18257–69. https://doi.org/10.1021/acsami.8b01862.

[171] Ji SH, Cho Y-S, Yun JS. Wearable core-shell piezoelectric nanofiber yarns for body movement energy harvesting. Nanomaterials 2019;9:555. https://doi.org/10.3390/nano9040555.

[172] Wang ZL. Triboelectric Nanogenerators as New Energy Technology for Self-Powered Systems and as Active Mechanical and Chemical Sensors. ACS Nano 2013;7:9533–57. https://doi.org/10.1021/nn404614z.

[173] Kim DW, Lee JH, Kim JK, Jeong U. Material aspects of triboelectric energy generation and sensors. NPG Asia Mater 2020;12:1–17. https://doi.org/10.1038/s41427-019-0176-0.

[174] Lou M, Abdalla I, Zhu M, Yu J, Li Z, Ding B. Hierarchically Rough Structured and Self-Powered Pressure Sensor Textile for Motion Sensing and Pulse Monitoring. ACS Appl Mater Interfaces 2020;12:1597–605. https://doi.org/10.1021/acsami.9b19238.

[175] Lee J-E, Shin Y-E, Lee G-H, Kim J, Ko H, Chae HG. Polyvinylidene fluoride (PVDF)/cellulose nanocrystal (CNC) nanocomposite fiber and triboelectric textile sensors. Compos Part B Eng 2021;223:109098. https://doi.org/10.1016/j.compositesb.2021.109098.

[176] He Q, Wu Y, Feng Z, Fan W, Lin Z, Sun C, et al. An all-textile triboelectric sensor for wearable teleoperated human–machine interaction. J Mater Chem A 2019;7:26804–11. https://doi.org/10.1039/C9TA11652D.

[177] Wang W, Zhou J, Wang S, Yuan F, Liu S, Zhang J, et al. Enhanced Kevlar-based triboelectric nanogenerator with anti-impact and sensing performance towards wireless alarm system. Nano Energy 2022;91:106657. https://doi.org/10.1016/j.nanoen.2021.106657.

[178] Busolo T, Szewczyk PK, Nair M, Stachewicz U, Kar-Narayan S. Triboelectric Yarns with Electrospun Functional Polymer Coatings for Highly Durable and Washable Smart Textile Applications. ACS Appl Mater Interfaces 2021;13:16876–86. https://doi.org/10.1021/acsami.1c00983.

[179] Cao R, Pu X, Du X, Yang W, Wang J, Guo H, et al. Screen-Printed Washable Electronic Textiles as Self-Powered Touch/Gesture Tribo-Sensors for Intelligent Human–Machine Interaction. ACS Nano 2018;12:5190–6. https://doi.org/10.1021/acsnano.8b02477.

[180] Yang C-R, Ko C-T, Chang S-F, Huang M-J. Study on fabric-based triboelectric nanogenerator using graphene oxide/porous PDMS as a compound friction layer. Nano Energy 2022;92:106791. https://doi.org/10.1016/j.nanoen.2021.106791.

[181] Chen C, Guo H, Chen L, Wang Y-C, Pu X, Yu W, et al. Direct Current Fabric Triboelectric Nanogenerator for Biomotion Energy Harvesting. ACS Nano 2020;14:4585–94. https://doi.org/10.1021/acsnano.0c00138.

[182] Liu J, Cui N, Du T, Li G, Liu S, Xu Q, et al. Coaxial double helix structured fiber-based triboelectric nanogenerator for effectively harvesting mechanical energy. Nanoscale Adv 2020;2:4482–90. https://doi.org/10.1039/D0NA00536C.

[183] Yu A, Pu X, Wen R, Liu M, Zhou T, Zhang K, et al. Core-Shell-Yarn-Based Triboelectric Nanogenerator Textiles as Power Cloths. ACS Nano 2017;11:12764–71. https://doi.org/10.1021/acsnano.7b07534.

[184] Zou H, Zhang Y, Guo L, Wang P, He X, Dai G, et al. Quantifying the triboelectric series. Nat Commun 2019;10:1427. https://doi.org/10.1038/s41467-019-09461-x.

[185] Chen C, Zhang L, Ding W, Chen L, Liu J, Du Z, et al. Woven Fabric Triboelectric Nanogenerator for Biomotion Energy Harvesting and as Self-Powered Gait-Recognizing Socks. Energies 2020;13:4119. https://doi.org/10.3390/en13164119.

[186] Zhao Z, Huang Q, Yan C, Liu Y, Zeng X, Wei X, et al. Machine-washable and breathable pressure sensors based on triboelectric nanogenerators enabled by textile technologies. Nano Energy 2020;70:104528. https://doi.org/10.1016/j.nanoen.2020.104528.

[187] Lin Z, He Q, Xiao Y, Zhu T, Yang J, Sun C, et al. Flexible Timbo-Like Triboelectric Nanogenerator as Self-Powered Force and Bend Sensor for Wireless and Distributed Landslide Monitoring. Adv Mater Technol 2018;3:1800144. https://doi.org/10.1002/admt.201800144.

[188] Dudem B, Mule AR, Patnam HR, Yu JS. Wearable and durable triboelectric nanogenerators via polyaniline coated cotton textiles as a movement sensor and self-powered system. Nano Energy 2019;55:305–15. https://doi.org/10.1016/j.nanoen.2018.10.074.

[189] Tian X, Hua T. Antibacterial, Scalable Manufacturing, Skin-Attachable, and Eco-Friendly Fabric Triboelectric Nanogenerators for Self-Powered Sensing. ACS Sustain Chem Eng 2021;9:13356–66. https://doi.org/10.1021/acssuschemeng.1c04804.

[190] Chen C, Chen L, Wu Z, Guo H, Yu W, Du Z, et al. 3D double-faced interlock fabric triboelectric nanogenerator for bio-motion energy harvesting and as self-powered stretching and 3D tactile sensors. Mater Today 2020;32:84–93. https://doi.org/10.1016/j.mattod.2019.10.025.

[191] Shin Y-E, Lee J-E, Park Y, Hwang S-H, Chae HG, Ko H. Sewing machine stitching of polyvinylidene fluoride fibers: programmable textile patterns for wearable triboelectric sensors. J Mater Chem A 2018;6:22879–88. https://doi.org/10.1039/C8TA08485H.

[192] Xiong J, Lee PS. Progress on wearable triboelectric nanogenerators in shapes of fiber, yarn, and textile. Sci Technol Adv Mater 2019;20:837–57. https://doi.org/10.1080/14686996.2019.1650396.

[193] Xu F, Dong S, Liu G, Pan C, Guo ZH, Guo W, et al. Scalable fabrication of stretchable and washable textile triboelectric nanogenerators as constant power sources for wearable electronics. Nano Energy 2021;88:106247. https://doi.org/10.1016/j.nanoen.2021.106247.

[194] Zhao Z, Yan C, Liu Z, Fu X, Peng L-M, Hu Y, et al. Machine-Washable Textile Triboelectric Nanogenerators for Effective Human Respiratory Monitoring through Loom Weaving of Metallic Yarns. Adv Mater 2016;28:10267–74. https://doi.org/10.1002/adma.201603679.

[195] Dong K, Deng J, Ding W, Wang AC, Wang P, Cheng C, et al. Versatile Core–Sheath Yarn for Sustainable Biomechanical Energy Harvesting and Real-Time Human-Interactive Sensing. Adv Energy Mater 2018;8:1801114. https://doi.org/10.1002/aenm.201801114.

[196] Chen L, Chen C, Jin L, Guo H, Wang AC, Ning F, et al. Stretchable negative Poisson's ratio yarn for triboelectric nanogenerator for environmental energy harvesting and self-powered sensor. Energy Environ Sci 2021;14:955–64. https://doi.org/10.1039/D0EE02777D.

[197] Dong K, Peng X, An J, Wang AC, Luo J, Sun B, et al. Shape adaptable and highly resilient 3D braided triboelectric nanogenerators as e-textiles for power and sensing. Nat Commun 2020;11:2868. https://doi.org/10.1038/s41467-020-16642-6.

[198] Fan W, He Q, Meng K, Tan X, Zhou Z, Zhang G, et al. Machine-knitted washable sensor array textile for precise epidermal physiological signal monitoring. Sci Adv n.d.;6:eaay2840. https://doi.org/10.1126/sciadv.aay2840.

[199] Zhang Z, He T, Zhu M, Sun Z, Shi Q, Zhu J, et al. Deep learning-enabled triboelectric smart socks for IoT-based gait analysis and VR applications. Npj Flex Electron 2020;4:1–12. https://doi.org/10.1038/s41528-020-00092-7.

[200] Saslow WM. Chapter 12 – Faraday's Law of Electromagnetic Induction. In: Saslow WM, editor. Electr. Magn. Light, San Diego: Academic Press; 2002, p. 505–58. https://doi.org/10.1016/B978-012619455-5.50012-7.

[201] Koo HR, Lee Y-J, Gi S, Khang S, Lee JH, Lee J-H, et al. The Effect of Textile-Based Inductive Coil Sensor Positions for Heart Rate Monitoring. J Med Syst 2014;38:2. https://doi.org/10.1007/s10916-013-0002-0.

[202] Wijesiriwardana R. Inductive fiber-meshed strain and displacement transducers for respiratory measuring systems and motion capturing systems. IEEE Sens J 2006;6:571–9. https://doi.org/10.1109/JSEN.2006.874488.

[203] Wu R, Ma L, Liu S, Patil AB, Hou C, Zhang Y, et al. Fibrous inductance strain sensors for passive inductance textile sensing. Mater Today Phys 2020;15:100243. https://doi.org/10.1016/j.mtphys.2020.100243.

[204] Sardini E, Serpelloni M, Pasqui V. Wireless Wearable T-Shirt for Posture Monitoring During Rehabilitation Exercises. IEEE Trans Instrum Meas 2015;64:439–48. https://doi.org/10.1109/TIM.2014.2343411.

[205] Tavassolian M, Cuthbert TJ, Napier C, Peng J, Menon C. Textile-Based Inductive Soft Strain Sensors for Fast Frequency Movement and Their Application in Wearable Devices Measuring Multiaxial Hip Joint Angles during Running. Adv Intell Syst 2020;2:1900165. https://doi.org/10.1002/aisy.201900165.

[206] Sun D, Chen M, Podilchak S, Georgiadis A, Abdullahi QS, Joshi R, et al. Investigating flexible textile-based coils for wireless charging wearable electronics. J Ind Text 2020;50:333–45. https://doi.org/10.1177/1528083719831086.

[207] Fobelets K, Panteli C. Ambulatory Monitoring Using Knitted 3D Helical Coils. Eng Proc 2022;15:6. https://doi.org/10.3390/engproc2022015006.

[208] Fava JO, Lanzani L, Ruch MC. Multilayer planar rectangular coils for eddy current testing: Design considerations. NDT E Int 2009;42:713–20. https://doi.org/10.1016/j.ndteint.2009.06.005.

[209] Kiziroglou ME, Wright SW, Yeatman EM. Coil and core design for inductive energy receivers. Sens Actuators Phys 2020;313:112206. https://doi.org/10.1016/j.sna.2020.112206.

[210] Liu X, Liu J, Wang J, Wang C, Yuan X. Design Method for the Coil-System and the Soft Switching Technology for High-Frequency and High-Efficiency Wireless Power Transfer Systems. Energies 2018;11:7. https://doi.org/10.3390/en11010007.

[211] Lee, Jeong-Han. Smart Wear: Heart activity monitoring system using inductive sensors. Korea Sci 2013;62:21–7.

[212] García Patiño A, Khoshnam M, Menon C. Wearable Device to Monitor Back Movements Using an Inductive Textile Sensor. Sensors 2020;20:905. https://doi.org/10.3390/s20030905.

[213] Seyedin S, Zhang P, Naebe M, Qin S, Chen J, Wang X, et al. Textile strain sensors: a review of the fabrication technologies, performance evaluation and applications. Mater Horiz 2019;6:219–49. https://doi.org/10.1039/C8MH01062E.

[214] Patiño AG, Menon C. Inductive Textile Sensor Design and Validation for a Wearable Monitoring Device. Sensors 2021;21:225. https://doi.org/10.3390/s21010225.

[215] Chen H-Y, Conn AT. A Stretchable Inductor With Integrated Strain Sensing and Wireless Signal Transfer. IEEE Sens J 2020;20:7384–91. https://doi.org/10.1109/JSEN.2020.2979076.

[216] Zhou S, Xu L, Lu J, Yang Y. Simulation and Optimization of High Performance On-Chip Solenoid MEMS Inductor. 2018 19th Int. Conf. Electron. Packag. Technol. ICEPT, 2018, p. 710–5. https://doi.org/10.1109/ICEPT.2018.8480563.

[217] Zubia J, Arrue J. Plastic Optical Fibers: An Introduction to Their Technological Processes and Applications. Opt Fiber Technol 2001;7:101–40. https://doi.org/10.1006/ofte.2000.0355.

[218] Correia R, James S, Lee S-W, Morgan SP, Korposh S. Biomedical application of optical fibre sensors. J Opt 2018;20:073003. https://doi.org/10.1088/2040-8986/aac68d.

[219] Selm B, Gürel EA, Rothmaier M, Rossi RM, Scherer LJ. Polymeric Optical Fiber Fabrics for Illumination and Sensorial Applications in Textiles. J Intell Mater Syst Struct 2010;21:1061–71. https://doi.org/10.1177/1045389X10377676.

[220] Quandt BM, Scherer LJ, Boesel LF, Wolf M, Bona G-L, Rossi RM. Body-Monitoring and Health Supervision by Means of Optical Fiber-Based Sensing Systems in Medical Textiles. Adv Healthc Mater 2015;4:330–55. https://doi.org/10.1002/adhm.201400463.

[221] Li X, Yang C, Yang S, Li G. Fiber-Optical Sensors: Basics and Applications in Multiphase Reactors. Sensors 2012;12:12519–44. https://doi.org/10.3390/s120912519.

[222] Culshaw B. CHAPTER 24 – Principles of Fiber Optic Sensors. In: Pal BP, editor. Guid. Wave Opt. Compon. Devices, Burlington: Academic Press; 2006, p. 371–87. https://doi.org/10.1016/B978-012088481-0/50025-5.

[223] Cennamo N, Zeni L. Polymer Optical Fibers for Sensing. Macromol Symp 2020;389:1900074. https://doi.org/10.1002/masy.201900074.

[224] Dhawan A, Muth JF, Kekas DJ, Ghosh TK. Optical nano-textile sensors based on the incorporation of semiconducting and metallic nanoparticles into optical fibers. MRS Online Proc Libr OPL 2006;920. https://doi.org/10.1557/PROC-0920-S05-06.

[225] Najafi B, Mohseni H, Grewal GS, Talal TK, Menzies RA, Armstrong DG. An Optical-Fiber-Based Smart Textile (Smart Socks) to Manage Biomechanical Risk Factors Associated With Diabetic Foot Amputation. J Diabetes Sci Technol 2017;11:668–77. https://doi.org/10.1177/1932296817709022.

[226] Tricker R. 37 – Optical Fibres in Power Systems. In: Laughton MA, Warne DJ, editors. Electr. Eng. Ref. Book Sixt. Ed., Oxford: Newnes; 2003, p. 37–1. https://doi.org/10.1016/B978-075064637-6/50037-X.

[227] Bunge C-A, Beckers M, Lustermann B. 3 – Basic principles of optical fibres. In: Bunge C-A, Gries T, Beckers M, editors. Polym. Opt. Fibres, Woodhead Publishing; 2017, p. 47–118. https://doi.org/10.1016/B978-0-08-100039-7.00003-8.

[228] Bunge C-A, Bremer K, Lustermann B, Woyessa G. 4 – Special fibres and components. In: Bunge C-A, Gries T, Beckers M, editors. Polym. Opt. Fibres, Woodhead Publishing; 2017, p. 119–51. https://doi.org/10.1016/B978-0-08-100039-7.00004-X.
[229] Kiesel S, Vickle PV, Peters KJ, Hassan T, Kowalsky M. Intrinsic polymer optical fiber sensors for high-strain applications. Smart Struct. Mater. 2006 Smart Sens. Monit. Syst. Appl., vol. 6167, SPIE; 2006, p. 285–95. https://doi.org/10.1117/12.657436.
[230] Bunge C-A, Kallweit JP, Houri MA, Mohr B, Bērziòš A, Grauberger C, et al. Textile Multitouch Force-Sensor Array Based on Circular and Non-Circular Polymer Optical Fibers. IEEE Sens J 2020;20:7548–55. https://doi.org/10.1109/JSEN.2020.2985328.
[231] Fasano A, Woyessa G, Stajanca P, Markos C, Stefani A, Nielsen K, et al. Fabrication and characterization of polycarbonate microstructured polymer optical fibers for high-temperature-resistant fiber Bragg grating strain sensors. Opt Mater Express 2016;6:649–59. https://doi.org/10.1364/OME.6.000649.
[232] Rothmaier M, Luong MP, Clemens F. Textile Pressure Sensor Made of Flexible Plastic Optical Fibers. Sensors 2008;8:4318–29. https://doi.org/10.3390/s8074318.
[233] Woyessa G, Fasano A, Markos C, Rasmussen HK, Bang O. Low Loss Polycarbonate Polymer Optical Fiber for High Temperature FBG Humidity Sensing. IEEE Photonics Technol Lett 2017;29:575–8. https://doi.org/10.1109/LPT.2017.2668524.
[234] Gong Z, Xiang Z, OuYang X, Zhang J, Lau N, Zhou J, et al. Wearable Fiber Optic Technology Based on Smart Textile: A Review. Materials 2019;12:3311. https://doi.org/10.3390/ma12203311.
[235] Leal-Junior AG, Díaz CR, Leitão C, Pontes MJ, Marques C, Frizera A. Polymer optical fiber-based sensor for simultaneous measurement of breath and heart rate under dynamic movements. Opt Laser Technol 2019;109:429–36. https://doi.org/10.1016/j.optlastec.2018.08.036.
[236] Ying DQ, Tao XM, Zheng W, Wang GF. Fabric strain sensor integrated with looped polymeric optical fiber with large angled V-shaped notches. Smart Mater Struct 2012;22:015004. https://doi.org/10.1088/0964-1726/22/1/015004.
[237] Ghaffar A, Mehdi M, Hussain S, Hussian N, Ali S, JianHui S, et al. The coupling of scattered-bend loss in POF based the displacement measurement sensor. Sens Bio-Sens Res 2020;29:100351. https://doi.org/10.1016/j.sbsr.2020.100351.
[238] Massaroni C, Saccomandi P, Schena E. Medical Smart Textiles Based on Fiber Optic Technology: An Overview. J Funct Biomater 2015;6:204–21. https://doi.org/10.3390/jfb6020204.
[239] Guignier C, Camillieri B, Schmid M, Rossi RM, Bueno M-A. E-Knitted Textile with Polymer Optical Fibers for Friction and Pressure Monitoring in Socks. Sensors 2019;19:3011. https://doi.org/10.3390/s19133011.
[240] Koyama Y, Nishiyama M, Watanabe K. Smart Textile Using Hetero-Core Optical Fiber for Heartbeat and Respiration Monitoring. IEEE Sens J 2018;18:6175–80. https://doi.org/10.1109/JSEN.2018.2847333.
[241] Dass S, Jha R. Micrometer Wire Assisted Inline Mach–Zehnder Interferometric Curvature Sensor. IEEE Photonics Technol Lett 2016;28:31–4. https://doi.org/10.1109/LPT.2015.2478957.
[242] Yang X, Chen Z, Elvin CSM, Janice LHY, Ng SH, Teo JT, et al. Textile Fiber Optic Microbend Sensor Used for Heartbeat and Respiration Monitoring. IEEE Sens J 2015;15:757–61. https://doi.org/10.1109/JSEN.2014.2353640.
[243] Lau D, Chen Z, Teo JT, Ng SH, Rumpel H, Lian Y, et al. Intensity-Modulated Microbend Fiber Optic Sensor for Respiratory Monitoring and Gating During MRI. IEEE Trans Biomed Eng 2013;60:2655–62. https://doi.org/10.1109/TBME.2013.2262150.
[244] Marques CAF, Webb DJ, Andre P. Polymer optical fiber sensors in human life safety. Opt Fiber Technol 2017;36:144–54. https://doi.org/10.1016/j.yofte.2017.03.010.

[245] Zheng W, Tao X, Zhu B, Wang G, Hui C. Fabrication and evaluation of a notched polymer optical fiber fabric strain sensor and its application in human respiration monitoring. Text Res J 2014;84:1791–802. https://doi.org/10.1177/0040517514528560.

[246] Aitkulov A, Tosi D. Design of an All-POF-Fiber Smartphone Multichannel Breathing Sensor With Camera-Division Multiplexing. IEEE Sens Lett 2019;3:1–4. https://doi.org/10.1109/LSENS.2019.2912982.

[247] Hu H, Sun S, Lv R, Zhao Y. Design and experiment of an optical fiber micro bend sensor for respiration monitoring. Sens Actuators Phys 2016;251:126–33. https://doi.org/10.1016/j.sna.2016.10.013.

[248] Leal-Junior AG, Frizera A, Vargas-Valencia L, dos Santos WM, Bó APL, Siqueira AAG, et al. Polymer Optical Fiber Sensors in Wearable Devices: Toward Novel Instrumentation Approaches for Gait Assistance Devices. IEEE Sens J 2018;18:7085–92. https://doi.org/10.1109/JSEN.2018.2852363.

[249] Liu HB, Liu HY, Peng GD, Chu PL. Strain and temperature sensor using a combination of polymer and silica fibre Bragg gratings. Opt Commun 2003;219:139–42. https://doi.org/10.1016/S0030-4018(03)01313-0.

[250] Sheng H-J, Fu M-Y, Chen T-C, Liu W-F, Bor S-S. A lateral pressure sensor using a fiber Bragg grating. IEEE Photonics Technol Lett 2004;16:1146–8. https://doi.org/10.1109/LPT.2004.824998.

[251] Lo Presti D, Massaroni C, D'Abbraccio J, Massari L, Caponero M, Longo UG, et al. Wearable System Based on Flexible FBG for Respiratory and Cardiac Monitoring. IEEE Sens J 2019;19:7391–8. https://doi.org/10.1109/JSEN.2019.2916320.

[252] Zaltieri M, Lo Presti D, Massaroni C, Sabbadini R, Schena E, Bravi M, et al. An FBG-based Smart Wearable Device for Monitoring Seated Posture in Video Terminal Workers. 2020 IEEE Int. Workshop Metrol. Ind. 40 IoT, 2020, p. 713–7. https://doi.org/10.1109/MetroInd4.0IoT48571.2020.9138213.

[253] Fajkus M, Nedoma J, Martinek R, Vasinek V, Nazeran H, Siska P. A Non-Invasive Multichannel Hybrid Fiber-Optic Sensor System for Vital Sign Monitoring. Sensors 2017;17:111. https://doi.org/10.3390/s17010111.

[254] Ciocchetti M, Massaroni C, Saccomandi P, Caponero MA, Polimadei A, Formica D, et al. Smart Textile Based on Fiber Bragg Grating Sensors for Respiratory Monitoring: Design and Preliminary Trials. Biosensors 2015;5:602–15. https://doi.org/10.3390/bios5030602.

[255] Webb DJ. Polymer Fiber Bragg Grating Sensors and Their Applications. Opt. Fiber Sens., CRC Press; 2015.

[256] Munster P, Horvath T. Intelligent Technical Textiles Based on Fiber Bragg Gratings for Strain Monitoring. Sensors 2020;20:2951. https://doi.org/10.3390/s20102951.

[257] Grillet A, Kinet D, Witt J, Schukar M, Krebber K, Pirotte F, et al. Optical Fiber Sensors Embedded Into Medical Textiles for Healthcare Monitoring. IEEE Sens J 2008;8:1215–22. https://doi.org/10.1109/JSEN.2008.926518.

[258] Bandodkar AJ, Wang J. Non-invasive wearable electrochemical sensors: a review. Trends Biotechnol 2014;32:363–71. https://doi.org/10.1016/j.tibtech.2014.04.005.

[259] Promphet N, Rattanawaleedirojn P, Siralertmukul K, Soatthiyanon N, Potiyaraj P, Thanawattano C, et al. Non-invasive textile based colorimetric sensor for the simultaneous detection of sweat pH and lactate. Talanta 2019;192:424–30. https://doi.org/10.1016/j.talanta.2018.09.086.

[260] Park YK, Oh HJ, Bae JH, Lim JY, Lee HD, Hong SI, et al. Colorimetric Textile Sensor for the Simultaneous Detection of NH_3 and HCl Gases. Polymers 2020;12:2595. https://doi.org/10.3390/polym12112595.

[261] Staneva D, Betcheva R, Chovelon J-M. Optical sensor for aliphatic amines based on the simultaneous colorimetric and fluorescence responses of smart textile. J Appl Polym Sci 2007;106:1950–6. https://doi.org/10.1002/app.26724.

[262] Boukhriss A, Messoudi ME, Roblin J-P, Aaboub T, Boyer D, Gmouh S. Luminescent hybrid coatings prepared by a sol–gel process for a textile-based pH sensor. Mater Adv 2020;1:918–25. https://doi.org/10.1039/D0MA00211A.

[263] Yao J, Ji P, Wang B, Wang H, Chen S. Color-tunable luminescent macrofibers based on CdTe QDs-loaded bacterial cellulose nanofibers for pH and glucose sensing. Sens Actuators B Chem 2018;254:110–9. https://doi.org/10.1016/j.snb.2017.07.071.

[264] Mandal R, Baranwal A, Srivastava A, Chandra P. Evolving trends in bio/chemical sensor fabrication incorporating bimetallic nanoparticles. Biosens Bioelectron 2018;117:546–61. https://doi.org/10.1016/j.bios.2018.06.039.

[265] Suárez PL, García-Cortés M, Fernández-Argüelles MT, Encinar JR, Valledor M, Ferrero FJ, et al. Functionalized phosphorescent nanoparticles in (bio)chemical sensing and imaging – A review. Anal Chim Acta 2019;1046:16–31. https://doi.org/10.1016/j.aca.2018.08.018.

[266] Machini WBS, Teixeira MFS. Electrochemical Properties of the Oxo-Manganese-Phenanthroline Complex Immobilized on Ion-Exchange Polymeric Film and Its Application as Biomimetic Sensor for Sulfite Ions. Electroanalysis 2014;26:2182–90. https://doi.org/10.1002/elan.201400289.

[267] Zhi M, Xiang C, Li J, Li M, Wu N. Nanostructured carbon–metal oxide composite electrodes for supercapacitors: a review. Nanoscale 2013;5:72–88. https://doi.org/10.1039/C2NR32040A.

[268] Wang R, Zhai Q, An T, Gong S, Cheng W. Stretchable gold fiber-based wearable textile electrochemical biosensor for lactate monitoring in sweat. Talanta 2021;222:121484. https://doi.org/10.1016/j.talanta.2020.121484.

[269] Parrilla M, Cánovas R, Jeerapan I, Andrade FJ, Wang J. A Textile-Based Stretchable Multi-Ion Potentiometric Sensor. Adv Healthc Mater 2016;5:996–1001. https://doi.org/10.1002/adhm.201600092.

[270] Manjakkal L, Dang W, Yogeswaran N, Dahiya R. Textile-Based Potentiometric Electrochemical pH Sensor for Wearable Applications. Biosensors 2019. https://doi.org/10.3390/bios9010014.

[271] Yoon JH, Kim S-M, Park HJ, Kim YK, Oh DX, Cho H-W, et al. Highly self-healable and flexible cable-type pH sensors for real-time monitoring of human fluids. Biosens Bioelectron 2020;150:111946. https://doi.org/10.1016/j.bios.2019.111946.

[272] Wang L, Wang L, Zhang Y, Pan J, Li S, Sun X, et al. Weaving Sensing Fibers into Electrochemical Fabric for Real-Time Health Monitoring. Adv Funct Mater 2018;28:1804456. https://doi.org/10.1002/adfm.201804456.

[273] Wang R, Zhai Q, Zhao Y, An T, Gong S, Guo Z, et al. Stretchable gold fiber-based wearable electrochemical sensor toward pH monitoring. J Mater Chem B 2020;8:3655–60. https://doi.org/10.1039/C9TB02477H.

[274] Farajikhah S, Choi J, Esrafilzadeh D, Underwood J, Innis PC, Paull B, et al. 3D textile structures with integrated electroactive electrodes for wearable electrochemical sensors. J Text Inst 2020;111:1587–95. https://doi.org/10.1080/00405000.2020.1720968.

[275] Herrero EJ, Bühlmann P. Potentiometric Sensors with Polymeric Sensing and Reference Membranes Fully Integrated into a Sample-Wicking Polyester Textile. Anal Sens 2021;1:188–95. https://doi.org/10.1002/anse.202100027.

[276] Liu X, Lillehoj PB. Embroidered electrochemical sensors on gauze for rapid quantification of wound biomarkers. Biosens Bioelectron 2017;98:189–94. https://doi.org/10.1016/j.bios.2017.06.053.

[277] Staneva D, Grabchev I, Bosch P. Fluorescent Hydrogel–Textile Composite Material Synthesized by Photopolymerization. Int J Polym Mater Polym Biomater 2015;64:838–47. https://doi.org/10.1080/00914037.2015.1030654.

[278] Owyeung RE, Panzer MJ, Sonkusale SR. Colorimetric Gas Sensing Washable Threads for Smart Textiles. Sci Rep 2019;9:5607. https://doi.org/10.1038/s41598-019-42054-8.

[279] Zhao Y, Zhai Q, Dong D, An T, Gong S, Shi Q, et al. Highly Stretchable and Strain-Insensitive Fiber-Based Wearable Electrochemical Biosensor to Monitor Glucose in the Sweat. Anal Chem 2019;91:6569–76. https://doi.org/10.1021/acs.analchem.9b00152.

[280] He J, Xiao G, Chen X, Qiao Y, Xu D, Lu Z. A thermoresponsive microfluidic system integrating a shape memory polymer-modified textile and a paper-based colorimetric sensor for the detection of glucose in human sweat. RSC Adv 2019;9:23957–63. https://doi.org/10.1039/C9RA02831E.

[281] He X, Yang S, Pei Q, Song Y, Liu C, Xu T, et al. Integrated Smart Janus Textile Bands for Self-Pumping Sweat Sampling and Analysis. ACS Sens 2020;5:1548–54. https://doi.org/10.1021/acssensors.0c00563.

[282] Mohan AMV, Rajendran V, Mishra RK, Jayaraman M. Recent advances and perspectives in sweat based wearable electrochemical sensors. TrAC Trends Anal Chem 2020;131:116024. https://doi.org/10.1016/j.trac.2020.116024.

[283] Zhang D, Zhang Y, Lu W, Le X, Li P, Huang L, et al. Fluorescent Hydrogel-Coated Paper/Textile as Flexible Chemosensor for Visual and Wearable Mercury(II) Detection. Adv Mater Technol 2019;4:1800201. https://doi.org/10.1002/admt.201800201.

[284] Li H-Y, Zhao S-N, Zang S-Q, Li J. Functional metal–organic frameworks as effective sensors of gases and volatile compounds. Chem Soc Rev 2020;49:6364–401. https://doi.org/10.1039/C9CS00778D.

[285] Yuan S, Feng L, Wang K, Pang J, Bosch M, Lollar C, et al. Stable Metal–Organic Frameworks: Design, Synthesis, and Applications. Adv Mater 2018;30:1704303. https://doi.org/10.1002/adma.201704303.

[286] Ding M, Cai X, Jiang H-L. Improving MOF stability: approaches and applications. Chem Sci 2019;10:10209–30. https://doi.org/10.1039/C9SC03916C.

[287] Bar-Cohen Y. Electroactive Polymer (EAP) Actuators as Artificial Muscles: Reality, Potential, and Challenges, Second Edition. 1000 20th Street, Bellingham, WA 98227-0010 USA: SPIE; 2004.

[288] Fink JK. Sensor Types and Polymers, John Wiley & Sons, Inc.; 2012, p. 1–42. https://doi.org/10.1002/9781118547663.ch1.

[289] Gregorio CC, Granzier H, Sorimachi H, Labeit S. Muscle assembly: a titanic achievement? Curr Opin Cell Biol 1999;11:18–25. https://doi.org/10.1016/S0955-0674(99)80003-9.

[290] Lord PR. Handbook of Yarn Production: Technology, Science and Economics. Elsevier; 2003.

[291] Melling D, Martinez JG, Jager EWH. Conjugated Polymer Actuators and Devices: Progress and Opportunities. Adv Mater 2019;31:1808210. https://doi.org/10.1002/adma.201808210.

[292] Mirfakhrai T, Madden JDW, Baughman RH. Polymer artificial muscles. Mater Today 2007;10:30–8. https://doi.org///dx.doi.org/10.1016/S1369-7021(07)70048-2.

[293] Baughman RH. Conducting polymer artificial muscles. Synth Met 1996;78:339–53. https://doi.org/10.1016/0379-6779(96)80158-5.

[294] Di J, Zhang X, Yong Z, Zhang Y, Li D, Li R, et al. Carbon-Nanotube Fibers for Wearable Devices and Smart Textiles. Adv Mater 2016;28:10529–38. https://doi.org/10.1002/adma.201601186.

[295] Fang X, Chatterjee K, Kapoor A, Ghosh T. Fiber-Based Sensors and Actuators. Handb. Fibrous Mater., vol. Volume 2, John Wiley & Sons, Ltd; 2020, p. 681–720. https://doi.org/10.1002/9783527342587.ch25.

[296] Lee JA, Kim YT, Spinks GM, Suh D, Lepró X, Lima MD, et al. All-Solid-State Carbon Nanotube Torsional and Tensile Artificial Muscles. Nano Lett 2014;14:2664–9. https://doi.org/10.1021/nl500526r.

[297] Lima MD, Hussain MW, Spinks GM, Naficy S, Hagenasr D, Bykova JS, et al. Efficient, Absorption-Powered Artificial Muscles Based on Carbon Nanotube Hybrid Yarns. Small Weinh Bergstr Ger 2015;11:3113–8. https://doi.org/10.1002/smll.201500424.

[298] Aziz S, Spinks GM. Torsional artificial muscles. Mater Horiz 2020;7:667–93. https://doi.org/10.1039/C9MH01441A.

[299] Haines CS, Li N, Spinks GM, Aliev AE, Di J, Baughman RH. New twist on artificial muscles. Proc Natl Acad Sci 2016;113:11709–16. https://doi.org/10.1073/pnas.1605273113.

[300] Kim SH, Kwon CH, Park K, Mun TJ, Lepró X, Baughman RH, et al. Bio-inspired, Moisture-Powered Hybrid Carbon Nanotube Yarn Muscles. Sci Rep 2016;6:23016. https://doi.org/10.1038/srep23016.

[301] Song Y, Zhou S, Jin K, Qiao J, Li D, Xu C, et al. Hierarchical carbon nanotube composite yarn muscles. Nanoscale 2018;10:4077–84. https://doi.org/10.1039/C7NR08595H.

[302] Chun K-Y, Hyeong Kim S, Kyoon Shin M, Hoon Kwon C, Park J, Tae Kim Y, et al. Hybrid carbon nanotube yarn artificial muscle inspired by spider dragline silk. Nat Commun 2014;5:3322. https://doi.org/10.1038/ncomms4322.

[303] Lima MD, Li N, Andrade M, Fang S, Oh J, Spinks GM, et al. Electrically, Chemically, and Photonically Powered Torsional and Tensile Actuation of Hybrid Carbon Nanotube Yarn Muscles. Science 2012;338:928–32. https://doi.org/10.1126/science.1226762.

[304] Naveen MH, Gurudatt NG, Shim Y-B. Applications of conducting polymer composites to electrochemical sensors: A review. vol. 9. 2017. https://doi.org/10.1016/j.apmt.2017.09.001.

[305] Hara S, Zama T, Takashima W, Kaneto K. Gel-like Polypyrrole Based Artificial Muscles with Extremely Large Strain. Polym J 2004;36:933–6. https://doi.org/10.1295/polymj.36.933.

[306] Zhou J, Mulle M, Zhang Y, Xu X, Li EQ, Han F, et al. High-ampacity conductive polymer microfibers as fast response wearable heaters and electromechanical actuators. J Mater Chem C 2016;4:1238–49. https://doi.org/10.1039/C5TC03380B.

[307] Uh K, Yoon B, Lee CW, Kim J-M. An Electrolyte-Free Conducting Polymer Actuator that Displays Electrothermal Bending and Flapping Wing Motions under a Magnetic Field. ACS Appl Mater Interfaces 2016;8:1289–96. https://doi.org/10.1021/acsami.5b09981.

[308] Martinez JG, Richter K, Persson N-K, Jager EWH. Investigation of electrically conducting yarns for use in textile actuators. Smart Mater Struct 2018;27:074004. https://doi.org/10.1088/1361-665X/aabab5.

[309] Maziz A, Concas A, Khaldi A, Stålhand J, Persson N-K, Jager EWH. Knitting and weaving artificial muscles. Sci Adv n.d.;3:e1600327. https://doi.org/10.1126/sciadv.1600327.

[310] Ismail YA, Martínez JG, Al Harrasi AS, Kim SJ, Otero TF. Sensing characteristics of a conducting polymer/hydrogel hybrid microfiber artificial muscle. Sens Actuators B Chem 2011;160:1180–90. https://doi.org/10.1016/j.snb.2011.09.044.

[311] Kim B, Koncar V, Dufour C. Polyaniline-coated PET conductive yarns: Study of electrical, mechanical, and electro-mechanical properties. J Appl Polym Sci 2006;101:1252–6. https://doi.org/10.1002/app.22799.

[312] Ryan JD, Mengistie DA, Gabrielsson R, Lund A, Müller C. Machine-Washable PEDOT:PSS Dyed Silk Yarns for Electronic Textiles. ACS Appl Mater Interfaces 2017;9:9045–50. https://doi.org/10.1021/acsami.7b00530.

[313] Hofmann AI, Östergren I, Kim Y, Fauth S, Craighero M, Yoon M-H, et al. All-Polymer Conducting Fibers and 3D Prints via Melt Processing and Templated Polymerization. ACS Appl Mater Interfaces 2020;12:8713–21. https://doi.org/10.1021/acsami.9b20615.

[314] Pelrine null, Kornbluh null, Pei null, Joseph null. High-speed electrically actuated elastomers with strain greater than 100%. Science 2000;287:836–9. https://doi.org/10.1126/science.287.5454.836.

[315] Madden JDW, Vandesteeg NA, Anquetil PA, Madden PGA, Takshi A, Pytel RZ, et al. Artificial muscle technology: physical principles and naval prospects. IEEE J Ocean Eng 2004;29:706–28. https://doi.org/10.1109/JOE.2004.833135.

[316] Zupan M, Ashby M f., Fleck N a. Actuator Classification and Selection – The Development of a Database. Adv Eng Mater 2002;4:933–40. https://doi.org/10.1002/adem.200290009.

[317] Zhang QM, Li H, Poh M, Xia F, Cheng Z-Y, Xu H, et al. An all-organic composite actuator material with a high dielectric constant. Nature 2002;419:284–7. https://doi.org/10.1038/nature01021.

[318] Ashley S. Artificial Muscles. Sci Am 2003;289:52–9.

[319] Pelrine R, Kornbluh R, Kofod G. High-Strain Actuator Materials Based on Dielectric Elastomers. Adv Mater 2000;12:1223–5. https://doi.org/10.1002/1521-4095(200008)12:16<1223::AID-ADMA1223>3.0.CO;2-2.

[320] Chakraborti P, Toprakci HAK, Yang P, Di Spigna N, Franzon P, Ghosh T. A compact dielectric elastomer tubular actuator for refreshable Braille displays. Sens Actuators Phys 2012;179:151–7. https://doi.org/10.1016/j.sna.2012.02.004.

[321] Arora S, Ghosh T, Muth J. Dielectric elastomer based prototype fiber actuators. Sens Actuators Phys 2007;136:321–8. https://doi.org/10.1016/j.sna.2006.10.044.

[322] Shao H, Wei S, Jiang X, Holmes DP, Ghosh TK. Bioinspired Electrically Activated Soft Bistable Actuators. Adv Funct Mater 2018;28:180299-n/a. https://doi.org/10.1002/adfm.201802999.

[323] Wei S, Ghosh TK. Bioinspired Bistable Dielectric Elastomer Actuators: Programmable Shapes and Application as Binary Valves. Soft Robot 2021. https://doi.org/10.1089/soro.2020.0214.

[324] Shankar R, Ghosh TK, Spontak RJ. Electroactive Nanostructured Polymers as Tunable Actuators. Adv Mater 2007;19:2218–23. https://doi.org/10.1002/adma.200602644.

[325] Fang X. Anisotropic D-EAP Electrodes and their Application in Spring Roll Actuators. Dissertation. North Carolina State University, 2017.

[326] Michel S, Zhang XQ, Wissler M, Löwe C, Kovacs G. A comparison between silicone and acrylic elastomers as dielectric materials in electroactive polymer actuators. Polym Int 2010;59:391–9. https://doi.org/10.1002/pi.2751.

[327] Rogers JA, Someya T, Huang Y. Materials and Mechanics for Stretchable Electronics. Science 2010;327:1603–7. https://doi.org/10.1126/science.1182383.

[328] Allen D, Farmer S, Gregg R, Voit W. Stretchable conductive fabric simplifies manufacturing of low-resistance dielectric-elastomer-system electrodes. In: Bar-Cohen Y, editor. Electroact. Polym. Actuators Devices EAPAD XX, Denver, United States: SPIE; 2018, p. 35. https://doi.org/10.1117/12.2292108.

[329] Guo J, Xiang C, Helps T, Taghavi M, Rossiter J. Electroactive textile actuators for wearable and soft robots. 2018 IEEE Int. Conf. Soft Robot. RoboSoft, 2018, p. 339–43. https://doi.org/10.1109/ROBOSOFT.2018.8404942.

[330] Pourazadi S, Ahmadi S, Menon C. On the design of a DEA-based device to potentially assist lower leg disorders: an analytical and FEM investigation accounting for nonlinearities of the leg and device deformations. Biomed Eng OnLine 2015;14:103. https://doi.org/10.1186/s12938-015-0088-3.

[331] Chortos A, Mao J, Mueller J, Hajiesmaili E, Lewis JA, Clarke DR. Printing Reconfigurable Bundles of Dielectric Elastomer Fibers. Adv Funct Mater 2021;31:2010643. https://doi.org/10.1002/adfm.202010643.

[332] Yang Y, Wu Y, Li C, Yang X, Chen W. Flexible actuators for soft robotics. Adv Intell Syst 2020;2:1900077. https://doi.org/10.1002/aisy.201900077.

[333] Mondal S, Hu J, Yang Z, Liu Y, Szeto Y. Shape memory polyurethane for smart garment. Res J Text Appar 2002;6:75–83. https://doi.org/10.1108/RJTA-06-02-2002-B007.

[334] Mondal S, Hu JL. Water vapor permeability of cotton fabrics coated with shape memory polyurethane. Carbohydr Polym 2007;67:282–7. https://doi.org/10.1016/j.carbpol.2006.05.030.

[335] Mondal S. Phase change materials for smart textiles: An overview. Appl Therm Eng 2008;28:1536–50. https://doi.org/10.1016/j.applthermaleng.2007.08.009.

[336] Behl M, Lendlein A. Shape-memory polymers. Mater Today 2007;10:20–8. https://doi.org/10.1016/S1369-7021(07)70047-0.

[337] Lendlein A, Kelch S. Shape-Memory Polymers. Angew Chem Int Ed 2002;41:2034–57. https://doi.org/10.1002/1521-3773(20020617)41:12<2034::AID-ANIE2034>3.0.CO;2-M.

[338] Behl M, Kratz K, Noechel U, Sauter T, Lendlein A. Temperature-memory polymer actuators. Proc Natl Acad Sci U S A 2013;110:12555–9. https://doi.org/10.1073/pnas.1301895110.

[339] Xie T, Page KA, Eastman SA. Strain-Based Temperature Memory Effect for Nafion and Its Molecular Origins. Adv Funct Mater 2011;21:2057–66. https://doi.org/10.1002/adfm.201002579.

[340] Ahir SV, Tajbakhsh AR, Terentjev EM. Self-Assembled Shape-Memory Fibers of Triblock Liquid-Crystal Polymers. Adv Funct Mater 2006;16:556–60. https://doi.org/10.1002/adfm.200500692.

[341] Lendlein A, Langer R. Biodegradable, Elastic Shape-Memory Polymers for Potential Biomedical Applications. Science 2002;296:1673–6. https://doi.org/10.1126/science.1066102.

[342] Kaursoin J, Agrawal AK. Melt spun thermoresponsive shape memory fibers based on polyurethanes: Effect of drawing and heat-setting on fiber morphology and properties. J Appl Polym Sci 2007;103:2172–82. https://doi.org/10.1002/app.25124.

[343] Amirkiai A, Abrisham M, Panahi-Sarmad M, Xiao X, Alimardani A, Sadri M. Tracing evolutions of elastomeric composites in shape memory actuators: A comprehensive review. Mater Today Commun 2021;28:102658. https://doi.org/10.1016/j.mtcomm.2021.102658.

[344] Hager MD, Bode S, Weber C, Schubert US. Shape memory polymers: Past, present and future developments. Prog Polym Sci 2015;49–50:3–33. https://doi.org/10.1016/j.progpolymsci.2015.04.002.

[345] Haines CS, Lima MD, Li N, Spinks GM, Foroughi J, Madden JDW, et al. Artificial Muscles from Fishing Line and Sewing Thread. Science 2014;343:868.

[346] Kusoglu A, Weber AZ. Water Transport and Sorption in Nafion Membrane. In: Page KA, Soles CL, Runt J, editors., Washington, DC: ACS Symposium Series, Vol. 1096; 2012, p. 175–99.

[347] Jung J-H, Jeon J-H, Sridhar V, Oh I-K. Electro-active graphene–Nafion actuators. Carbon 2011;49:1279–89. https://doi.org/10.1016/j.carbon.2010.11.047.

[348] Zhong Y, Zhang F, Wang M, Gardner CJ, Kim G, Liu Y, et al. Reversible Humidity Sensitive Clothing for Personal Thermoregulation. Sci Rep 2017;7:44208. https://doi.org/10.1038/srep44208.

[349] Cheng H, Hu Y, Zhao F, Dong Z, Wang Y, Chen N, et al. Moisture-Activated Torsional Graphene-Fiber Motor. Adv Mater 2014;26:2909–13. https://doi.org/10.1002/adma.201305708.

[350] Cheng H, Liu J, Zhao Y, Hu C, Zhang Z, Chen N, et al. Graphene Fibers with Predetermined Deformation as Moisture-Triggered Actuators and Robots. Angew Chem Int Ed 2013;52:10482–6. https://doi.org/10.1002/anie.201304358.

[351] He S, Chen P, Qiu L, Wang B, Sun X, Xu Y, et al. A Mechanically Actuating Carbon-Nanotube Fiber in Response to Water and Moisture. Angew Chem 2015;54:14880–4. https://doi.org/10.1002/anie.201507108.

[352] Chen P, Xu Y, He S, Sun X, Pan S, Deng J, et al. Hierarchically arranged helical fibre actuators driven by solvents and vapours. Nat Nanotechnol 2015;10:1077–83. https://doi.org/10.1038/nnano.2015.198.

[353] Merati AA. Application of Stimuli-Sensitive Materials in Smart Textiles. In: ul-Islam S, Butola BS, editors., Hoboken, NJ, USA: John Wiley & Sons, Inc; 2018, p. 1–29. https://doi.org/10.1002/9781119488101.ch1.

[354] Jia T, Wang Y, Dou Y, Li Y, Andrade MJ de, Wang R, et al. Moisture Sensitive Smart Yarns and Textiles from Self-Balanced Silk Fiber Muscles. Adv Funct Mater 2019:1808241. https://doi.org/10.1002/adfm.201808241.

[355] Gong J, Lin H, Dunlop JWC, Yuan J. Hierarchically Arranged Helical Fiber Actuators Derived from Commercial Cloth. Adv Mater 2017;29:1605103. https://doi.org/10.1002/adma.201605103.

[356] Geim AK. Graphene: Status and Prospects. Science 2009;324:1530–4. https://doi.org/10.1126/science.1158877.

[357] Geim AK, Novoselov KS. The rise of graphene. Nat Mater 2007;6:183–91. https://doi.org/10.1038/nmat1849.

[358] Liu D, Tarakanova A, Hsu CC, Yu M, Zheng S, Yu L, et al. Spider dragline silk as torsional actuator driven by humidity. Sci Adv 2019;5:eaau9183. https://doi.org/10.1126/sciadv.aau9183.

[359] Padaki NV, Das B, Basu A. Advances in understanding the properties of silk. In: Basu A, editor. vol. 163, Amsterdam: Elsevier Woodhead; 2015, p. 3–16.

[360] Li W, Xu F, Sun L, Liu W, Qiu Y. A novel flexible humidity switch material based on multi-walled carbon nanotube/polyvinyl alcohol composite yarn. Sens Actuators B Chem 2016;230:528–35. https://doi.org/10.1016/j.snb.2016.02.108.

[361] Sakai Y, Sadaoka Y, Matsuguchi M. Humidity sensors based on polymer thin films. Sens Actuators B Chem 1996;35:85–90. https://doi.org/10.1016/S0925-4005(96)02019-9.

[362] Mersch J, Bruns M, Nocke A, Cherif C, Gerlach G. High-Displacement, Fiber-Reinforced Shape Memory Alloy Soft Actuator with Integrated Sensors and Its Equivalent Network Model. Adv Intell Syst 2021;3:2000221. https://doi.org/10.1002/aisy.202000221.

[363] Suh D, Truong TK, Suh DG, Lim SC. Torsional Actuator Powered by Environmental Energy Harvesting from Diurnal Temperature Variation. ACS Sustain Chem Eng 2016;4:6647–52. https://doi.org/10.1021/acssuschemeng.6b01502.

[364] Xue J, Ge Y, Liu Z, Liu Z, Jiang J, Li G. Photoprogrammable Moisture-Responsive Actuation of a Shape Memory Polymer Film. ACS Appl Mater Interfaces 2022;14:10836–43. https://doi.org/10.1021/acsami.1c24018.

[365] Li X, Liu J, Li D, Huang S, Huang K, Zhang X. Bioinspired multi-stimuli responsive actuators with synergistic color- and morphing-change abilities. Adv Sci 2021;8:2101295. https://doi.org/10.1002/advs.202101295.

[366] Chen Z, Liu J, Chen Y, Zheng X, Liu H, Li H. Multiple-stimuli-responsive and cellulose conductive ionic hydrogel for smart wearable devices and thermal actuators. ACS Appl Mater Interfaces 2021;13:1353–66. https://doi.org/10.1021/acsami.0c16719.

[367] Liu Y, Zhong Y, Wang C. Recent advances in self-actuation and self-sensing materials: State of the art and future perspectives. Talanta 2020;212:120808. https://doi.org/10.1016/j.talanta.2020.120808.

[368] Kruusamäe K, Punning A, Aabloo A, Asaka K. Self-sensing ionic polymer actuators: a review. Actuators 2015;4:17–38. https://doi.org/10.3390/act4010017.

[369] Ducéré V, Bernès A, Lacabanne C. A capacitive humidity sensor using cross-linked cellulose acetate butyrate. Sens Actuators B Chem 2005;106:331–4. https://doi.org/10.1016/j.snb.2004.08.028.

[370] Hu L, Cui Y. Energy and environmental nanotechnology in conductive paper and textiles. Energy Environ Sci 2012;5:6423–35. https://doi.org/10.1039/C2EE02414D.

[371] Hamedi M, Herlogsson L, Crispin X, Marcilla R, Berggren M, Inganäs O. Fiber-embedded electrolyte-gated field-effect transistors for e-textiles. Adv Mater 2009;21:573–7. https://doi.org/10.1002/adma.200802681.

[372] Qing X, Wang Y, Zhang Y, Ding X, Zhong W, Wang D, et al. Wearable fiber-based organic electrochemical transistors as a platform for highly sensitive dopamine monitoring. ACS Appl Mater Interfaces 2019;11:13105–13. https://doi.org/10.1021/acsami.9b00115.

[373] Huang J-S, Jiang T-Y, Wang Z-X, Wu S-W, Chen Y-S. A novel textile antenna using composite multifilament conductive threads for smart clothing applications. Microw Opt Technol Lett 2016;58:1232–6. https://doi.org/10.1002/mop.29771.

[374] El Gharbi M, Martinez-Estrada M, Fernández-García R, Ahyoud S, Gil I. A novel ultra-wide band wearable antenna under different bending conditions for electronic-textile applications. J Text Inst 2021;112:437–43. https://doi.org/10.1080/00405000.2020.1762326.

[375] Abedi V, Avula V, Chaudhary D, Shahjouei S, Khan A, Griessenauer CJ, et al. Prediction of long-term stroke recurrence using machine learning models. J Clin Med 2021;10:1286. https://doi.org/10.3390/jcm10061286.

[376] Lockhart TE, Soangra R, Yoon H, Wu T, Frames CW, Weaver R, et al. Prediction of fall risk among community-dwelling older adults using a wearable system. Sci Rep 2021;11:20976. https://doi.org/10.1038/s41598-021-00458-5.

[377] Karthikeyan S, Sankar T, Vijayakarthick M, Ravi T, Rajasekar B. Role of IOT in healthcare using smart textiles. 2020 Int. Conf. Power Energy Control Transm. Syst. ICPECTS, 2020, p. 1–6. https://doi.org/10.1109/ICPECTS49113.2020.9337054.

[378] Liu M, Pu X, Jiang C, Liu T, Huang X, Chen L, et al. Large-area all-textile pressure sensors for monitoring human motion and physiological signals. Adv Mater 2017;29:1703700. https://doi.org/10.1002/adma.201703700.

[379] Liu M, Ward T, Young D, Matos H, Wei Y, Adams J, et al. Electronic textiles based wearable electrotherapy for pain relief. Sens Actuators Phys 2020;303:111701. https://doi.org/10.1016/j.sna.2019.111701.

[380] Sanchez V, Walsh CJ, Wood RJ. Textile Technology for Soft Robotic and Autonomous Garments. Adv Funct Mater 2021;31:2008278. https://doi.org/10.1002/adfm.202008278.

R. Rathinamoorthy

13 Medical textiles: materials, applications, features, and recent advancements

Abstract: The textile industry is one of the largest industries in the world that provides the basic need of every individual. Apart from clothing, the application areas of textiles are widespread under the aegis of technical textiles. Out of several industrial applications, the use of textiles in the medical field has significant importance. In this view, this chapter aims in consolidating the fundamentals and latest advancements in medical textiles. The first part of the chapter discusses the essentials of medical textiles with various application areas and their requirements. In this section, the general classifications of medical textiles, fibres used, fabric structures preferred, and special requirements of those materials have been elaborated. The second part of the chapter details the recent advancement in the textiles that are used in medical applications, namely smart medical textiles, wound dressing, compression garments, and special applications like sutures.

Keywords: Medical textile, medical fibres, woven structure, knitted fabrics, nonwovens, wound dressing, compression garments, sutures

13.1 Introduction

Medical textile is one of the fast-growing sectors in technical textiles. The sector is expected to grow with a compound annual growth rate of 4% between 2021 and 2026. The industry will reach approximately USD 17 billion at the end of 2026 [1]. Research works in this segment are also growing faster and updated every year. A recent analysis result revealed that out of the total published research works, more than half of the works were published in the last 5 years [2]. The product groups addressed in this category are the different textile materials that are generally used in medical and life science-related applications. The product range in this sector largely varies from disposable healthcare and hygiene products like napkins and diapers to speciality products like operation room textiles, sutures, scaffolds, and sensors. The selection of fibres and fabrics for this application largely varies based on the requirements of the end use. Various property requirements of medical textile raw materials based on their application are listed in Figure 13.1, as reported by Yimin [3].

https://doi.org/10.1515/9783110759747-013

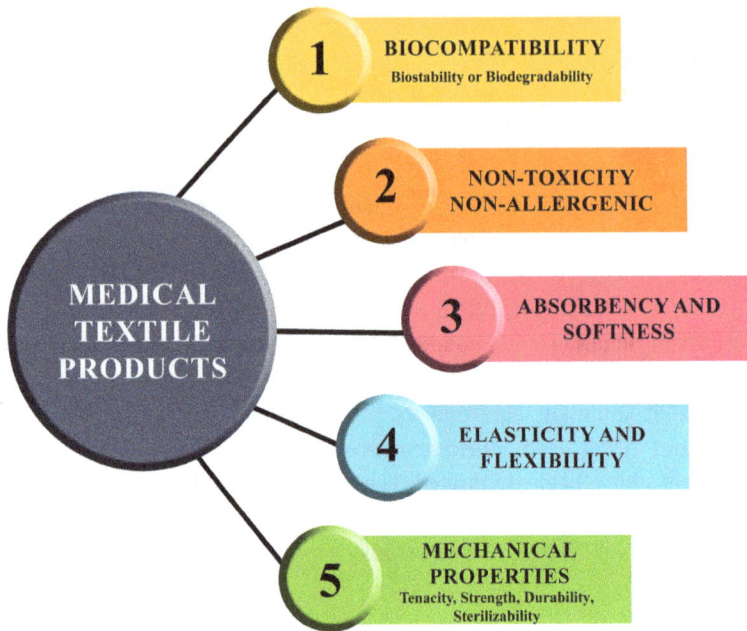

Figure 13.1: Required properties of medical textile products.

13.2 Classification of medical textile

Medical textile products are generally classified into the following categories [4–7].

13.2.1 Non-implantable materials

These are the materials that are used only for external application on the human body. The main function of such medical textiles is to protect the skin from external infections or absorb the body fluids or apply medication to the damaged skin. The non-implantable materials can be listed as wound dressings, plasters, bandages, gauze, compression bandages, and so on. These products are generally manufactured from all types of textile fibres based on the application requirements and end uses [7].

13.2.2 Implantable material

Implantable medical textiles are the materials that are used inside the human body or skin. The main requirement for such an application is biocompatibility. These materials are generally used to repair the damaged internal organs, or damaged skin, wounds, during surgery. These materials can also be categorized as soft tissue implants, orthopaedic implants, and cardiovascular implants. Soft tissue implants are used in the application areas like ligaments and cartilage. Orthopaedic implants are used as artificial bones and cardiovascular implants are used in the heart valves and vascular grafts. Surgical sutures are also one of the most commonly used implants in the human body for the closure of skin wounds. In these applications, several fibres are used ranging from natural (silk) and biodegradable polymers (chitin, collagen, chitosan, etc.) to synthetic polymers (polyester, polyamide, polyethylene, polytetrafluoroethylene, etc.) [8].

13.2.3 Extra corporeal devices

Similar to implantable textiles, extra corporeal devices are implantable artificial organs that are made of textile materials. These are artificial organs used inside the human body to support vital organs like artificial kidneys, liver, and lungs. The main purpose of the artificial organ is to purify the blood during the body's function through dialysis, filtration, and adsorption processes [9].

i) **Artificial kidney**: The organ is manufactured using polyethylene glycol–polyethylene terephthalate (PET) block copolymer membrane. The filtration and purification process of the circulating blood is performed by either a flat sheet or a bundle of hollow regenerated cellulosic fibres in the form of cellophane [9].

ii) **Artificial lung**: This artificial part functions by providing oxygen to the blood and removing carbon dioxide from the blood. The micromembrane made of hollow polypropylene fibre and hollow silicone fibres helps in this process by enabling higher permeability to gases and very lower permeability to liquids.

iii) **Artificial liver**: This organ is chiefly made of hollow viscose fibres to filter the patients' blood and remove waste. The complex metabolism of the liver can be solved with the use of double lumen structured hollow fibre within a hollow fibre. In this structure, the blood runs outside and in contact with the liver cell and after purification it runs inside the fibre [9].

iv) **Textile device to support heart**: The cardiac support device (CSD) is developed from preformed multifilament polyester fibres that are used to wrap the cardiac ventricle to confirm with the surface of the heart closely. The CSD is placed on the epicardial surface by stay sutures. The support device can be implanted on a live heart directly through cardiopulmonary bypass. Clinical studies on dogs showed consistent improvement with CSD usage through significant reduction in left ventricular end-diastolic dimension and a significant improvement in left ventricular ejection fraction. Application of CSD in different methods based on requirements is provided in Figure 13.2 [10].

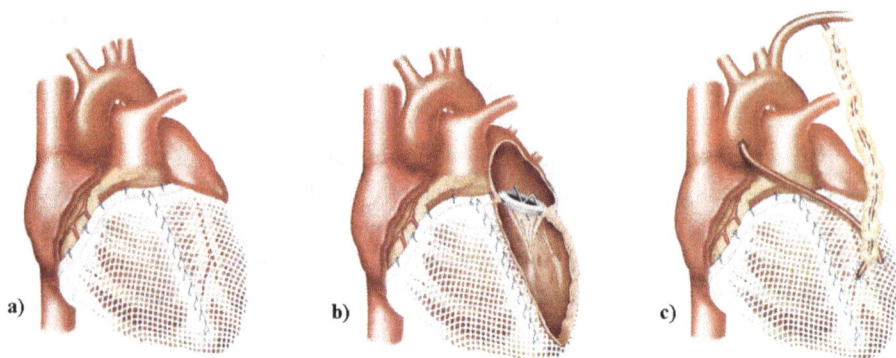

Figure 13.2: (a) Implantation of the cardiac support device (CSD) as the sole surgical procedure, (b) surgical implantation of the CSD with concomitant mitral valve replacement, and (c) surgical implantation of the CSD along with coronary artery bypass surgery (reproduced with kind permission from [10], Copyright 2003, Elsevier B.V.).

13.2.4 Healthcare and hygiene products

These are the products that are used to protect the medical staff and patients from infection. These products are either washable or disposable after use. The products in this range are from personal protective cloths to disposable masks. Another category that is covered under this section is personal hygiene products like napkins, tissues, hospital bed linen, uniforms, and diapers. The main requirements of these textiles are that they should be non-allergic, non-toxic, and non-carcinogenic in nature [11].

13.3 Fibres used in medical textiles

Several textile fibres from natural sources like cotton, silk, and several synthetic biodegradable and non-biodegradable fibres are used in medical applications based on the requirements. The fibres used in the medical textile application are classified in Figure 13.3.

As far as the natural fibres are concerned, cellulose-based (plant) materials such as cotton are most commonly used in medical applications due to their high moisture absorbency and comfort characteristics. Due to this comfortable nature, cellulose-based natural fibres are commonly used in hygienic and healthcare clothing and non-implantable textile products. The porous structure of the cotton fibres shows excellent moisture-wicking properties with handles and soft drapes. Similar to plant-based textiles, animal-based textiles are protein-based, namely silk and wool. They also have unique characteristics due to their surface morphology and structure. Silk fibres are directly used as a suture in wound closure applications. The other most common applications of natural fibres in

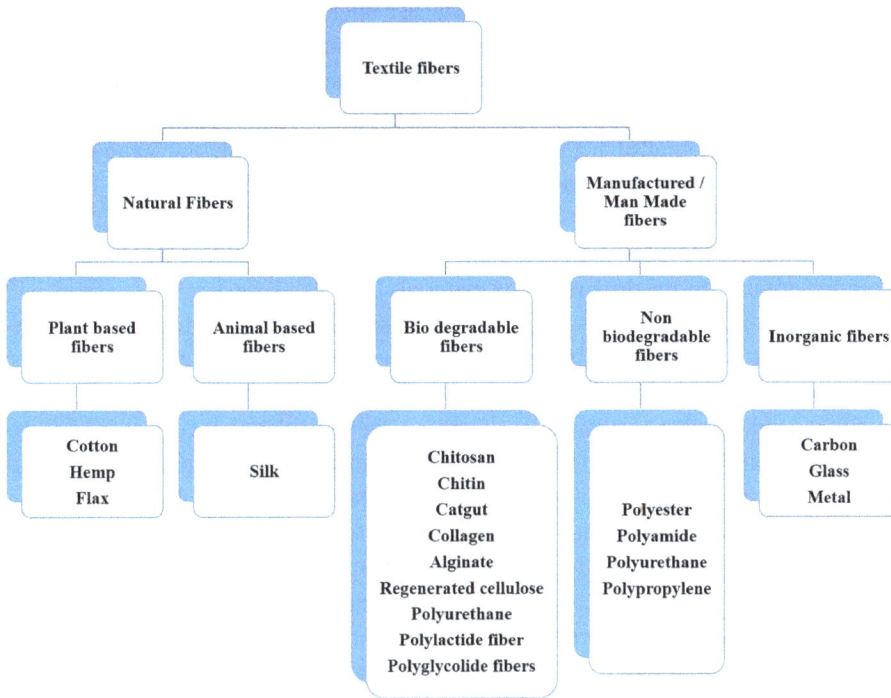

Figure 13.3: Fibres used in medical textile applications.

healthcare materials like bedding, uniforms, pillow covers, sheets, surgical gowns, and surgical hosiery. Concerning the bedding applications, thermal and tactile properties are the major concern. In addition, the natural fibres are also used as gauze, wadding, and bandages. They are also used as absorbent swabs in non-implantable sections [12]. Bast fibres are obtained from the bark of the plant stem and they are commonly used in hygiene products. They also have their traditional applications as sutures. In general, studies reported the following advantages of natural fibres [13]:

– Unique structural characteristics (both surface structure and molecular)
– Have higher tensile and elastic properties
– Fibres like silk and wool show inherent bacterial resistance
– Have been found to be resistant to body fluids
– Help in improving the healing ability of skin infection
– Cost-effective

When synthetic fibres are considered, they are cheaper, durable, and stronger, and have low maintenance. Further, they are also resistant to water and stain [14]. For instance, polypropylene fibres are the most commonly used synthetic material in healthcare products like hygiene products, non-absorbable sutures, surgical gowns, and protective clothing in the medical field. These fibres are also often used in implantable materials. The only disadvantage of polypropylene fibre is its tendency to degrade in sunlight. Polyethylene is the next commercialized synthetic material in the medical textile field. They are used from sanitary napkins to multicomponent wound dressings [15]. High strength, flexibility, emulsifying, excellent film-forming, water solubility, and adhesive properties are the main advantages of polyvinyl alcohol polymers. They can be either used for making sutures or as a wound dressing. They are reported as one of the best raw materials for biomedical applications. They can also be clubbed with other polymers and are used as the drug-releasing materials [16]

Similarly, acrylic fibres are used as superabsorbent materials due to their structural characteristics. The small fibre diameters and longitudinal grooves with a higher surface area of these fibres offer higher moisture absorption. Hence, they are commonly used in wound dressing and bandages, where a higher absorbency is required [17]. Polyester fibre is another important fibre that is used in many medical applications. Braided structures of these fibres are often used for implantable medical applications like tendons, arteries, heart valves, vascular prostheses, vascular grafts, surgical sutures, and artificial ligaments [18]. Other than this, for normal applications, gowns, masks, surgical covers, drapes, bedding, protective cloth, surgical hosiery, blankets, and cover stock are also developed by polyester fibres. The use of polyamide fibres also established their application in sutures, tendons, and lumens. Polyamide materials are also used in the wound dressing application due to their flexible and anti-adherent wound dressing [19]. In structural scaffold preparation, polyurethane nanofibres are commonly used. This material is also used in ligament reconstruction, wound dressing, and bandages. The main advantage of polyurethane fibres is their hydrophilic and hydrophobic nature (dual behaviour) [20]. However, the prominent disadvantages of synthetic fibres is their lower melting temperature, poor insulation, and prone to heat damage. Compared to the natural fibre, they lack moisture absorption characteristics. A detailed literature survey on various other specialty fibres and their application in medical textiles can be found elsewhere [21]. A detailed classification of various fibre types, fabrics, and their various application areas was detailed in Table 13.1 as reported in the literature [8].

Table 13.1: Fibres used in different applications (reproduced with kind permission from [8], Copyright 2000, CRC Press LLC).

Fibres used	Fabric types used	Areas for application
For non-implantable materials		
Cotton, viscose, lyocell	Nonwoven	Absorbent pad
Alginate fibre, chitosan, silk, viscose, lyocell, cotton	Woven, nonwoven, and knitted	Wound-contact layer
Viscose, lyocell, plastic film	Woven and nonwoven	Base material for pads and bandages

Table 13.1 (continued)

Fibres used	Fabric types used	Areas for application
Cotton, viscose, lyocell, and polyamide fibre	Woven, nonwoven, and knitted	Simple bandages
Cotton, viscose, lyocell, and elastomeric-fibre yarns	Woven, nonwoven, and knitted	High-support bandages
Cotton, viscose, lyocell, and elastomeric-fibre yarns	Woven and knitted	Compression bandages
Cotton, viscose, lyocell, polyester fibre, polypropylene fibre, and polyurethane foam	Woven and nonwoven	Orthopaedic bandages
Cotton, viscose, plastic film, polyester, glass, and PP fibre	Woven, nonwoven, and knitted	Plasters
Cotton, viscose, lyocell, alginate fibre, and chitosan	Woven, nonwoven, and knitted	Gauze dressing
Cotton	Woven	Lint
Viscose, cotton linters, and wood pulp	Nonwoven	Wadding
Polylactide fibre, polyglycolide fibre, and carbon	Needle-punched nonwoven	Scaffold
Implantable materials		
Collagen, catgut, polyglycolide, and polylactide fibre	Monofilament and braided	Biodegradable sutures
Polyester, polyamide, PTFE, PP, and PE fibre	Monofilament and braided	Non-biodegradable sutures
PTFE, polyester, silk, collagen, PE, and polyamide fibre	Woven and braided	Artificial tendon
Polyester fibre, carbon fibre, and collagen	Braided	Artificial ligament
Low-density polyethylene fibre	–	Artificial cartilage
Chitin	Nonwoven	Artificial skin
Poly (methyl methacrylate) fibre, silicon fibre, and collagen	–	Contact lens and artificial cornea
Silicone, polyacetyl fibre, and polyethylene fibre	–	Artificial joints/bones
PTFE fibre and polyester fibre	Woven and knitted	Vascular grafts
Polyester fibre	Woven and knitted	Heart valves
Healthcare/hygiene products		
Cotton, polyester fibre, and polypropylene fibre	Woven, nonwoven	Surgical gowns

Table 13.1 (continued)

Fibres used	Fabric types used	Areas for application
Viscose	Nonwoven	Surgical caps
Viscose, polyester fibre, and glass fibre	Nonwoven	Surgical masks
Polyester fibre and polyethylene fibre	Woven and nonwoven	Surgical drapes and cloths
Cotton, polyester fibre, and polyamide fibre	Knitted	Surgical hosiery
Cotton and polyester fibre	Knitted and woven	Blanket
Cotton	Woven	Sheets and pillowcases
Cotton and polyester fibre	Woven	Uniform
Polyester fibre and polypropylene fibre	Nonwoven	Protective clothing
Superabsorbent fibres and wood fluff	Nonwoven	Absorbent layer
Polyethylene fibre	Nonwoven	Outer layer
Viscose and lyocell	Nonwoven	Cloths/wipes

13.4 Fabrics used in medical textiles

A wide range of fabrication techniques including weaving, knitting, braiding, and nonwoven has been adopted in the manufacturing of medical textiles for different applications [22]. Like the role of fibre composition on the potential characteristics of textiles, the fabrication methods also play a crucial role. The structural as well as functional characteristics of the fabrics can alter the requirements of a medical textile. This section details different fabrication methods and their specific application in the field of medical textiles. The common fabric types used in medical textile products are provided in Figure 13.4.

13.4.1 Knitted fabrics

Knitting is the fabrication method where the yarns are inter-looped or intra-looped to form the fabric structure. In general, knitted fabrics were noted to have better advantage over other fabrics like woven and nonwoven structures in terms of higher elasticity, conformability, absorbency, and so on. Knitted fabrics have notable applications in the field of medical textiles which includes bandages, surgical stockings, and orthopaedic equipment like braces and supports, and also knitted fabrics were found in implants

Non-woven

Applications include Surgical masks, gowns, drapes, pads medical filters, **wound dressings** and hygiene products

Braided

Applications include sutures, prostatic stents, braided composite bone plate, artificial ligament or tendon, artificial cartilage, implants, dental floss, surgical cables

Woven

Applications include hospital bedding, wound dressing, gauzes, artificial blood vessels, vascular grafts & also in hygiene products

Knitted

Applications include bandages surgical stockings, orthopedic equipment like braces and supports and also in implants

Types of Fabrics used in Medical Textiles

Figure 13.4: Fabric types used in medical textile application.

[24]. This section elaborates on different applications of medical textiles where knitted fabric structures are effectively being used.

13.4.1.1 Pressure/compression bandages and garments

Compression garments are the type of medical textiles that are mainly developed from the elastane yarns. These garments are developed in the shape of the body and are lesser than the actual measurement. Hence, in the applications, the garments can impart certain pressure on the applied area [25]. The applications of such pressure on the affected area can increase the pressure on the inflammation along with higher blood flow. For the compression garment application, warp and weft-knitted fabrics are most commonly used. The sports and medical compression garments mainly requires compression in both wales and course directions and hence the weft-knitted single and double jersey fabrics are commonly used. However, based on the structural aspect, rib and interlock structures were also used, where a higher extension was required in the wales direction. In the case of weft knit structures, both wales and course direction elongations can be obtained in the tricot structure. These structures are commonly used in swimwear and tightfitting garments, whereas the stretch requirements are used only in the course direction, and the Raschel structures are used from the warp-knitting method [26]. The stiffness of the material and the elasticity of the compression garment are the most important properties.

Pressure bandages/garments are the materials that should be able to maintain pressure on the scar area as per the requirements. These are very effective in the treatment of different problems including chronic venous ulcers [27] and varicose vein disorders. The main role of these bandages/garments is that they apply pressure on the skin which in turn can help in the healing of the wound by increasing the blood pressure. The performance of these fabrics is highly dependent on the stress relaxation behaviour and interfacial pressure of the fabrics. Over some time, the fabrics which are stretched under a constant strain will get relaxed and this in turn can affect the functionality [28]. Various researchers have analysed the pressure and stress behaviour of different knitted structures for their applicability as compression/ pressure garments. Also, different parameters of the knitted structures were analysed to understand their impact on the functionality of the end product. A study analysed the effect of stitch length and strain percentage on the interfacial pressure values and pressure reduction percentages. They have examined two different weft-knitted structures (plain and interlock). The increase in the stitch length led to a decrease in the pressure values, whereas in the case of strain percentage, a higher strain percentage reduced pressure values in plain fabrics and interlock structures, and strain percentage has not shown a significant effect [29]. In a similar research, Bandari et al. analysed the stress relaxation behaviour of different weft-knitted structures which varied with the number of back bed tuck stitches. The analysis of the effect of fabric structure on stress relaxation behaviour revealed that increasing the number of tuck stitches in the structure reduced the initial and residual stresses and increased the stress relaxation. These findings revealed that the inclusion of tuck stitches is not favourable for the application of pressure garments/bandages. As the number of tuck stitches increases, the stretch relaxation behaviour of the fabric increases and thus the study suggested that the structure is not suitable for pressure garments, varicose stockings, and compression bandages.

In addition to weft-knitted structures, warp-knitted techniques were also used in medical applications. Pressure bandages were made with the multi-axial warp knitting technique with the combination of soybean, polyester, and elastomeric yarns and bamboo, polypropylene, and elastomeric yarns with different degrees of yarn directions (+45°, 90°, −45°). The developed bandages were evaluated for tensile strength, elongation, and elasticity properties. Fabrics that are made of bamboo, polypropylene, and elastomeric yarns provided better tensile strength and elongation than the other structure. Moreover, having a yarn direction of 45° yielded better tensile strength. When considering the elasticity percentage, the difference in the fibre types did not show a significant effect. All the samples showed good elasticity but it got decreased with the increase in applied load [30]. Figure 13.5 represents the structure of warp and weft-knitted fabric as reported in the literature [31].

Knitted spacer fabrics are one of the key players in medical textiles. These fabrics show lesser bulk density with more volume. Researchers evaluated the effectiveness of these spacer fabrics as knee braces by comparing their properties with the commercially available knee braces which are made of weft-knitted structure and neoprene-laminated materials. Spacer-knitted fabrics showed better performance in terms of dimensional, mechanical and, especially comfort properties, than the commercially available neoprene materials. In the case of knee braces, which are being worn for a long

Loop formation direction

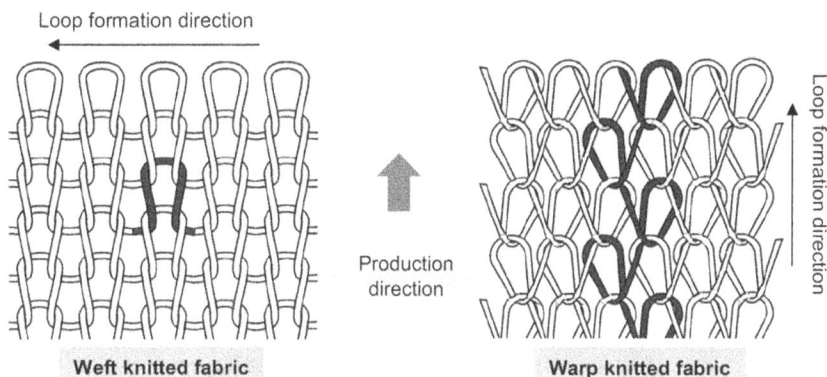

Weft knitted fabric

Production direction

Loop formation direction

Warp knitted fabric

Figure 13.5: Warp and weft-knitted fabric structure and the loop formation directions (reproduced with kind permission from [31] Copyright (2017), Elsevier B.V.).

time, comfort properties are very important. Dry thermal insulation values between 50 and 60 W^{-1} km^2 × 10^3 are favourable for knee braces, and the spacer-knitted fabrics yielded such values. Moreover, to support the comfort behaviour of spacer fabrics, absorbency and water vapour permeability properties of the spacer fabrics were also noted to be better than the commercial structures analysed [32].

13.4.1.2 Wound dressings

The knitted structures hold a predetermined position in the wound dressing applications because of their greater extensibility, elasticity, and flexibility [33]. For wound dressing applications, spacer-knitted fabrics are getting more popular. Spacer fabrics are well suited for replacing the absorbent layer of the wound dressing due to their better permeability, absorbency, thermal conductivity, and cushioning properties. Warp-knitted spacer fabrics were developed with 100% polyester, and their potential to serve as an absorbent layer of wound dressing was analysed against commercially available wound dressings which were mainly focused on the management of exuding wounds like burns, ulcers, and surgical wounds [34]. The findings have revealed that the spacer fabrics show better air permeability; however, variations can be noted between different samples which are highly dependent on the fabric areal and bulk density. The next important requirement concerning wound dressing is thermal conductivity. It implies how the heat is transferred from the wound surface and provides a cooling effect. Since the fibres being used in the fabrics can influence the thermal conductivity, researchers analysed fabrics made of 100% polyester with only varying fabric structural properties like densities, thickness, and spacer yarn angle. The increased areal density reduces the space that can trap the air, and the increased spacer yarn angle provides a gap between fabric surfaces. Both in turn result in better thermal ventilation. Water vapour permeability of warp-knitted spacer fabrics is found to be better than that of commercial wound dressings. This shows that the spacer fabrics can work better in transporting extrudes or sweat from the wound area and maintaining ventilation. In the case of absorbency, though spacer fabrics showed lesser absorbency than that of dressing materials which use specific absorbent material, the polyester spacer fabrics can work equally better due to their better wicking characteristics [34].

Similarly, weft-knitted spacer fabrics of different structures (varied spacer yarn length and spacer yarn inclination) made of cotton yarns are analysed [35]. To overcome the bacterial growth in cotton structures, antibacterial modifications have been made. The angle of spacer yarn inclination is noted to have a significant impact on the properties such as air permeability, water vapour permeability, absorbency, thermal conductivity, and compressibility of the fabric. Increasing the inclination angle led to an increase in air and water vapour permeability and absorbency while decreasing thermal conductivity and compressibility. Moreover, the variation in the stitches in the outer layer of the fabric also imparts variations in the property, that is, implementing alternate tuck and knit stitch in the structure increased air and water vapour permeability, absorbency, and thermal conductivity. While compared with the commercially available wound dressings, weft-knitted spacer fabrics are better in terms of permeability and cushioning effect which provides a better atmosphere for the wound to heal. However, absorbency is needed to be improved. Thus, weft-knitted spacer fabrics are well suitable for wounds that exudate lesser [35].

Apart from spacer fabrics, knitted gauze is currently replacing the traditional woven gauzes which are being used in wound dressing. The better absorbency of knitted structure than woven structure paved the path for using knitted gauze in the place of woven gauze. Al Faruque et al. [36] developed knitted gauze with half and full needle technology with 100% cotton yarns. The developed knitted gauze (Figure 13.6) is compared with the woven gauze for performance. The comparison analysis revealed that knitted gauze fabrics are better than woven gauze fabrics in terms of absorbency. Moreover, gauze knitted by full needle techniques showed better absorbency than the half needle. Being at the initial stage, detailed research on different structures of knitted gauze is essential that can bring a great shift in the wound dressings. In addition to this, knitting technology is replacing nonwoven applications in wound dressings where better mechanical properties and degradation stabilities are required. Biocompatible alginate fibres were knitted to overcome the drawbacks of the nonwoven fabrication process [37].

13.4.1.3 Implants

Artificial blood vessels

Knitted fabrics are often used as artificial blood vessels to replace damaged ones. Though both woven and knitted structures are being used, the knitted structure has

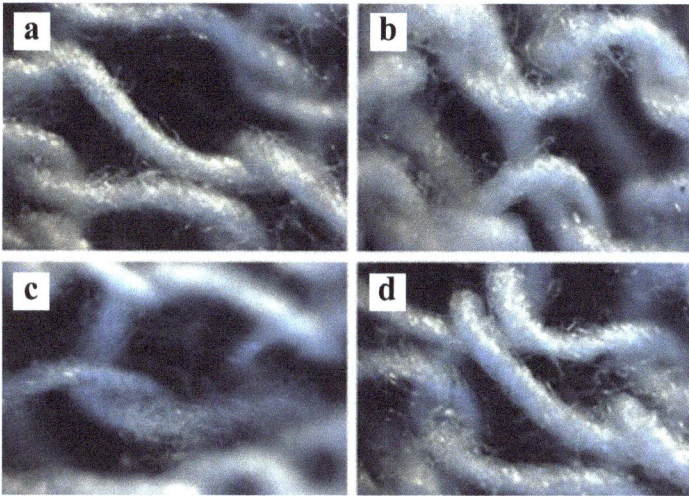

Figure 13.6: Microscopic view of knitted gauze developed by Al Faruque et al. [36] (reproduced under Creative commons license [36], Copyright (2018), Austin Publishing Group).

the advantage of better elasticity than the woven. These knitted structures are capable of replacing the larger diameter vessels (diameter of more than 6 mm) [38].

Vascular implants

Vascular grafts/stents are the substitutes for damaged or diseased arteries. The main requirements of a vascular graft are mechanical and biological mimicking of natural arteries. Knitted fabric structures made of 100% polyester yarns were used as reinforcement of polyurethane vascular grafts. The elasticity and strength of the grafts were noted to improve with the knitted fabrics [39]. Similarly, fabrics made out of polyester/spandex blend also showed better properties [40].

In summary, knitted fabric structures were found to have greater applications in the medical field. Literature works have revealed that the knitted structures have their own unique characteristics that can find a wide range of purposes that can satisfy the need for medical treatments. Knitted structures are inevitable in external medical needs like bandages, pressure garments, and wound dressings. However, they are also being used in different implants including artificial blood vessels, vascular implants, cardiac stents, and urinary bladder reconstructions [41]. The analysis of structural parameters has revealed that different structures can serve different purposes which can be modified to yield better results. Table 13.2 summarizes different knit structures that have been developed to provide different end uses in the field of medical science.

Table 13.2: Applications of knit structures in different areas of medical textiles.

S. no	Proposed application	Properties analysed	Fibres used	Knit structure	Reference
1	Bandages	Tensile strength Elongation Elastic properties	Soybean/polyester/ elastomeric blend Bamboo/polypropylene/ elastomeric blend	Multiaxial warp-knitted fabrics	[30]
2	Knee braces	Dimensional properties Mechanical properties Comfort properties	80/20 polyester/Lycra; spacer yarn – polyester monofilament 60/40 polyester/Lycra; spacer yarn – polyester monofilament Skinlife and polyester yarn with Lycra; spacer yarn – polyamide monofilament Polyester/spandex; spacer yarn – polyester	Warp-knitted and weft-knitted spacer fabrics	[32]
3	Pressure garments	Interfacial pressure values Pressure reduction percentages	Textured polyester yarns	Tubular weft-knitted (plain and interlock structures)	[29]
4	Pressure garments/ varicose stockings/ compression bandages	Stress relaxation	Textured polyester yarns	Weft-knitted rib structures with varied number of tuck stitches	[28]
5	Wound dressing	Air and water vapour permeability Absorbency Thermal conductivity	100% polyester monofilaments	Warp-knitted spacer fabrics	[34]
6	Wound dressing	Air and water vapour permeability Absorbency Thermal conductivity Compression	Cotton modified for antibacterial performance	Weft-knitted spacer fabrics	[35]
7	Wound dressing	Absorption	100% cotton	Gauze	[36]

Table 13.2 (continued)

S. no	Proposed application	Properties analysed	Fibres used	Knit structure	Reference
8	Wound dressing	Tensile strength Phosphate-buffered saline (PBS) storage and absorbency Water storage and absorbency	Alginate fibres	1 × 1 rib structure	[37]
9	Vascular grafts – reinforcement	Tensile strength	Polyester filament yarns	Tubular weft-knitted structure	[39]
10	Vascular grafts – reinforcement	Tensile strength Radial compliance	100/0, 75/25, 50/50, 25/75 Polyester/spandex blend ratios	Tubular weft-knitted structure	[40]
11	External vein graft	Radial compliance Flexibility Stiffness	Nitinol, polyester, polyurethane filaments	Plain knit mesh, segmented knit mesh	[42]

13.4.2 Woven fabrics

Weaving is a technology that produces fabrics by means of the interlacement of two sets of yarns (warp and weft). These woven structures are better in stability when compared to the knitted structure. In the field of medical textiles, woven fabrics find application in the areas of hospital bedding, wound dressing, gauzes, and also in hygiene products to some extent. In tissue engineering processes, woven fabric structures are notably used as scaffolds and also as a reinforcing material for hydrogels.

13.4.2.1 Artificial blood vessels

Artificial blood vessels of larger diameter are the well-known application of woven structure in medical textiles. Researchers have analysed the effect of structural parameters of PET fabrics to alter the biocompatibility to serve as artificial blood vessels. The cell adhesion of the developed blood vessels depends on the coarseness and pores of the surface. Woven structures were developed with different weft yarn densities to alter the surface characteristics, thereby altering the cell adhesion and proliferation. The detailed analysis showed that the endothelialization can be improved in the woven structure with a flatter surface and pore size among yarns should be less than 20 μm [43].

13.4.2.2 Vascular grafts

Woven structures are used in grafts because of their unique characteristics of higher stability, higher strength, and lower porosity. This makes it suitable for high flow arteries, and high-stress locations like the thoracic aorta [44]. Seamless woven polyester grafts were developed with texturized polyester yarns by means of tubular weaving. The clinical trials revealed better results for the developed vascular prosthesis [45].

13.4.2.3 Wound dressings

In terms of wound dressings, woven gauzes are very popular. A wide range of commercial dressings is developed with woven structures. Table 13.3 summarizes different commercially available wound dressings and their specifications [44].

Table 13.3: Commercially available wound dressings with woven structure.

Product	Company	Fibre	Weave structure	Application
Meditull Jelonet dressing	Fleming Medical	100% cotton thread	Leno weave	Primary contact layer for granulating wounds
Euronet paraffin	Pharmacy Line	Cotton	Woven gauze	Burns, ulcers, skin grafts, and traumatic injuries
Paranet	Synergy Health	Cotton	Open mesh leno weave	Superficial wounds, burns, skin grafts, and traumatic injuries
Bactigras	Smith and Nephew	Not specified	Leno weave	Minor burns and scalds
Pharmatull Plus	Biological Pharmaceutical	100% cotton	Leno gauze	Not specified
UniTulle	ZENTA	Not specified	Leno weave	Not specified
JELONET	Smith and Nephew	Not specified	Leno weave	Minor burns, lacerations, abrasions, and leg ulcers
Sofra-Tulle	Hoechst Marion Roussel Ltd	Not specified	Leno weave	Primary wound contact layer for infected wounds

13.4.3 Nonwoven structures

In the field of medical textiles, nonwoven fabrics find a major role, especially as disposable products. Surgical masks, gowns, drapes, pads, medical filters, wound dressings, and so on are well-known medical applications that largely use nonwoven structures. The major advantages of nonwoven fabric structures that facilitate the potential of nonwoven in medical textiles include [46]:
– Higher flexibility
– Easy and quick production process
– Low production cost
– Absorbency
– Breathability
– Excellent barrier properties
– Lightweight

Out of several nonwoven manufacturing methods, spunlace, spunbond, and melt-blown are the most desirable techniques for the manufacturing of nonwovens in the area of medical textiles [47]. The primary drawback of nonwoven over woven and knit structures is the absence of interlooping or interlacing of yarns which ends up in lower mechanical strength [46]. However, all the nonwoven manufactured through different methods are provided in this section.

13.4.3.1 Spunlace nonwovens

These are one of the most efficient methods of nonwoven production, where high-speed water jets are used to bond the fibres inside the structure. The lightweight and chemical-free production methods are the major advantages of spunlace nonwoven. Bulkiness, softness, drape, and higher tensile strength are the added benefits of spunlaced products [48]. Spunlaced nonwoven materials are totally free from binders and chemical additives, and hence it permits easy sterilization of the fabrics. The major reasons that make the spunlace or hydroentangled nonwoven fabrics in the field of medical textiles are their higher absorption characteristics and also the absence of binders [47]. Spunlace nonwovens are found to be applied in wound dressings, surgical fabrics, baby diapers, and other sanitary products. The only disadvantage that is reported in the case of spunlace nonwoven production is higher water consumption.

13.4.3.2 Spun-laid nonwovens

Spun-laid nonwovens are known for their structural uniformity and mechanical property. These products are usually manufactured using calendaring process. Hence, the main disadvantage of the nonwoven is the lack of bulkiness and softness compared to other production methods. Instead, the calendaring process might melt the Structure and reduce the permeability characteristics [49].

13.4.3.3 Spunbond nonwovens

Spun-bonded fabrics play a vital role in health and hygiene-related products like baby care, feminine hygiene, and medicine. Due to their higher porosity and other structural characteristics, spun-bonded fabric helps in keeping the wearer's skin dry and comfortable. Further, spun-bonded fabrics work better with hygiene products due to their smooth and soft nature. Spun-bonded fabrics had better air permeability properties [50]. These fabrics are cheaper to produce and hence widely used in disposable products like surgical masks and surgical gloves. Spun-bonded structures are also used in wound dressing material due to their breathability and absorbency. Out of all, spun-bonded nonwovens are mostly used in the applications like baby and feminine hygiene products.

13.4.3.4 Meltblown nonwovens

Meltblown layers are very much beneficial in the filtration process. These layers are used in medical protective wear like masks to filter bacteria and other microorganisms. In general, meltblown fabrics are comprised of more number of lower diameter fibres which ends up in reduced pore size and pore volume. This leads to reduced air permeability in the case of melt-blown than spun-bonded fabrics [50]. However, as far as the applications are considered, the meltblown nonwoven fabrics are also highly used in hygiene products along with spun-bonded nonwoven. Due to their higher softness and lower thickness, they are highly used in baby and adult diapers, napkins, and other hygiene products.

13.4.3.5 Airlaid nonwovens

Compared to other nonwovens, Airlaid nonwovens have higher absorption and moisture transfer capacity. These nonwovens are one of the main materials for the diapers and feminine hygiene products as a core absorbent material. Due to their higher absorbency, they help the product to remain thin with higher absorption [51]. Researchers reported that the control of pore size is the main advantage of air-laid nonwovens. A smaller pore size helps in gaining better distribution of moisture and rewetting characteristics. Instead, a larger pore size gives a higher water absorption and less distribution [52]

13.4.3.6 Thermal-bonded nonwovens

These are the structures developed via heat and pressure directly applied on the fibre web, to produce nonwoven by passing through the metal calendaring rollers. Thermal-bonded nonwoven is mainly used in the disposable products as a top and bottom layer of the disposable products. The fabrics are also used in wet wipes, breast pads, diapers, napkins, and also in surgical apparel.

13.4.4 Nonwoven composites

Nonwoven composites were developed using elasticated spun-bonded and meltblown fabrics incorporated with cotton fibres. The hydroentanglement process has been adopted to develop the composite. The combined action of synthetic webs and incorporated cotton fibres in the composite resulted in the achievement of desirable properties for application in medical textiles such as absorbency, air permeability, hydrostatic head, tensile strength, stretchability, and barrier properties [50]. Other than this, different structures of the nonwovens can also be bonded together to get the composites with different properties. The most popular structure is spunbond–meltblown–spunbond (SMS), spunbond–meltblown–meltblown–spunbond, and spunbond–spunbond–meltblown–meltblown–spunbond in different weight ranges. Out of these, the SMS structure is the most successful combination as it showed a higher barrier against microorganisms due to the microporous structure of meltblown layer. SMS fabrics are also used as cover stock for diaper and sanitary napkins. Further, to add, the higher tensile strength and hydrophobic nature of the spun-bonded layer act as a liquid barrier. Based on the requirements, all the above-mentioned structures are used in both hygiene and medical products.

13.5 Functionalization of textile materials with antibacterial finish

Though different fibres and fabric structures have their own contribution to Achieve the requirements of medical applications, medical textiles can be functionalized by different physical, chemical, or biological processes. These surface-processing techniques can functionalize the material and make it more suitable for a specific application. Different additives can be either coated on the surface of the textile materials or can be impregnated which is often decided based on the nature of the textile material, the nature of the additive, and the required applications. The most common finish that is used in medical textiles is antibacterial finishing. Various types of antibacterial finishes used in the textiles for functionalization are provided in Figure 13.7.

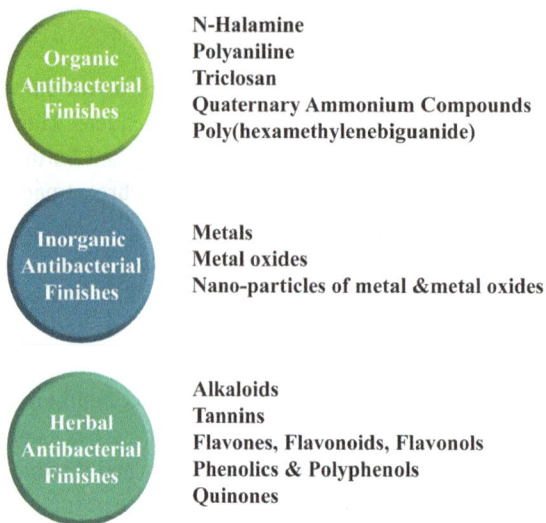

Organic Antibacterial Finishes
N-Halamine
Polyaniline
Triclosan
Quaternary Ammonium Compounds
Poly(hexamethylenebiguanide)

Inorganic Antibacterial Finishes
Metals
Metal oxides
Nano-particles of metal & metal oxides

Herbal Antibacterial Finishes
Alkaloids
Tannins
Flavones, Flavonoids, Flavonols
Phenolics & Polyphenols
Quinones

Figure 13.7: Different types of antibacterial agents used in medical textile functionalization.

The antibacterial-treated fabrics work by leaching the finished substance into the required applications. The leached substance will act as a barrier to the growth of the microorganism or kill the microorganism. The general applications of antibacterial finishing in the textile are used to control the growth of microorganisms, reduce odour, reduce cross-infections, restrict the spread of disease, and also control the degradation of textiles by different bacteria, fungus, and mildews. Antibacterial agents that restrict or reduce bacterial growth are referred to as bacteriostatic and the agents that kill the bacteria are known as bactericidal [53]. Similarly, every antibacterial agent must have some durability with textiles during the laundry or washing process. The durability of the antibacterial treatment is majorly based on the type of interaction. Simple and direct application of antibacterial finishes results in a temporary finish, where a finishing with the aid of chemical binders helps in a durable finish that lasts very long for multiple laundries. Further, the durability of the finishes can also be increased for a prolonged time with the help of micro- or nanoencapsulation techniques. Inorganic antibacterial agents are generally derived from metal-based elements like silver, copper, gold, zinc, and titanium [54]. Whereas the organic antibacterial agents are mostly obtained from synthetic agents and also from natural sources. Examples of such organic antimicrobial agents are quaternary ammonium (QA) compounds (QACs), *N*-halamines, polyhexamethylenebiguanide, triclosan, silicon-based quaternary agent, iodophors, phenols, thiophenols, heterocyclics, nitro compounds, urea, amines, and formaldehyde derivatives [55]. The following section details the important antibacterial agents in each category and their interaction mechanism with bacteria during the antibacterial reaction.

13.5.1 Inorganic antibacterial agents

As far as the metal oxides (inorganic) antibacterial agents are considered, different metals have different interaction mechanisms with bacteria. In the case of silver (Ag), when it is in contact with an aqueous environment, the emission of Ag^+ and Ag^{2+} ions makes them bacterial resistant [56]. The interaction or penetration of Ag^+ ions on the bacterial cell wall damages the DNA and proteins inside the bacteria, which ultimately damages the metabolism of the bacteria. Hence, the treatment leads to cell death and provides antibacterial ability [57]. Studies reported that the reduction in the Ag particle size leads to an increase in the ratio of surface area to volume and provides better inhibition properties against bacteria [58]. The other common metal oxide used in medical textiles is zinc (Zn) and nano-sized zinc oxide (ZnO). Similar to silver, ZnO particles also interact with bacteria's outer wall and penetrate inside the cell to inhibit the growth. The only difference compared to silver is that the Zn particle once penetrated into bacteria restricts enzyme distribution and amino acid metabolism. Nano-Zn and ZnO showed significant antibacterial properties against a wide spectrum of bacteria [59]. But in the case of copper, the antibacterial activity is obtained by using their affinity towards the amine and carboxylic groups. When these metals are treated, the nanoparticle penetrates into the cell structure and disturbs the helical structure of DNA and hence altering the biochemical process. Through a different pathway, namely ATPase activity reduction and inhibition with subunits, the interaction with DNA and reduction or inhibition in protein synthesis were reported as the main cause of antibacterial activity [60].

Concerning the inorganic antibacterial agents, the inhibition of bacterial growth happens via radical generation, particle ionization, and acid formation in the structure. These chemical changes significantly affect the biological process of the cell and inhibit its growth. In the case of gram-negative bacteria, a higher retardancy was noted with silver, copper, and gold particles. However, with respect to the copper nanoparticle and Zn, a higher retardancy was noted with gram-positive types. In the case of silver, gold, and copper, the higher positive charges of these metals are easily attracted by negatively charged cell walls of the gram-negative species and hence cause easier penetration of metals into cells. In the case of zinc and copper nanoparticles, reactive oxygen formation was reported to have an effect on bacteria. The negative charge of these reactive oxygen substances will not attach to the surface of the cell; instead, these metals perform organelle modifications like disturbs in protein generation, DNA structure, oxidation, and depletion of sulphhydryls [54]. The detailed mechanism of the metal ions and nanoparticles is provided in Figure 13.8. The sequence of the mechanism is numbered in this figure as follows: (1) as a first step, the metal ion releases from the source; (2) represents the direct interaction of the metal ions; (3) indicates the electrostatic interaction between the metal nanoparticle and the cell wall and thus leading to impaired membrane function and impaired nutrient assimilation; in the next step (4), the formation of extracellular and intracellular reactive oxygen species damage the lipids and proteins inside the microorganism. It is also true that the oxidative stress process restricts

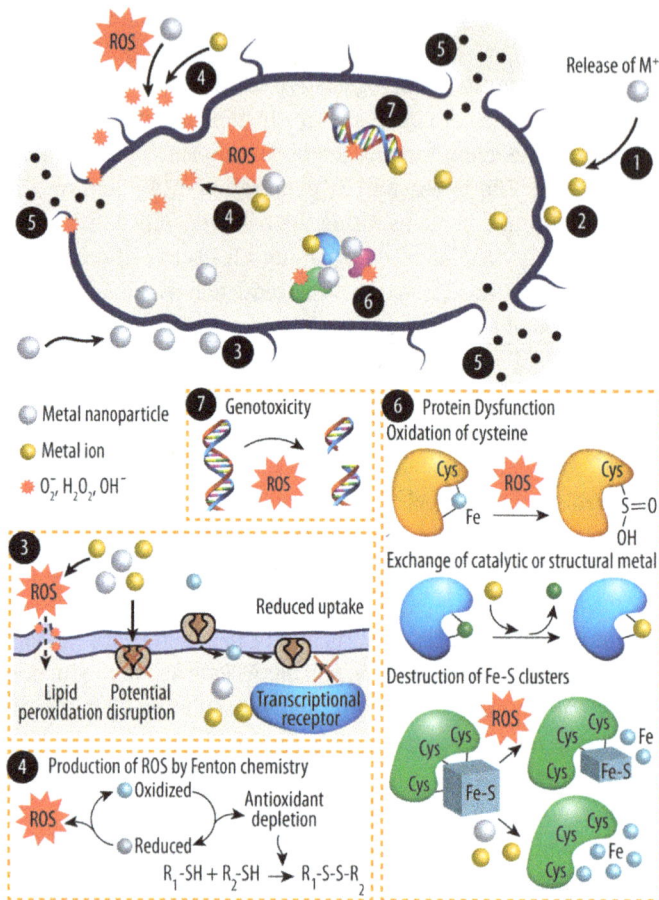

Figure 13.8: Antibacterial mechanisms of metal ions and nanoparticles (reproduced under Creative commons license [61], Copyright 2021, Elsevier B.V.).

the DNA replication as mentioned earlier. All these damages cause the cells to develop a higher level of oxidative stress and plasma damage and hence the cell content is leaked (5). This process ultimately kills the cell. Steps (6) and (7) represent the interference of metal ions and nanoparticle with both proteins and DNA. This process impairs the metabolic activity of the cell and leads to destruction [61].

13.5.2 Organic antibacterial agents

Organic antibacterial agents are generally strong against gram-positive bacteria than the negative strains. Unlike inorganic substances, the antibacterial interaction of the organic antibacterial substances mainly depends upon the disturbance in the biochemical pathway and organelle modification in the bacterial cell. No electrostatic interaction with organic substances and bacterial cells was reported. For instance, triclosan, one of the widely used organic antibacterial agents, acts on the bacteria's biosynthesis pathway and reduces some essential component production called FabI. Subsequent analysis confirmed the inhibition of FabI development against triclosan and resulted in the formation of FabI–NAD1–triclosan ternary complex [62]. In the case of QACs, the antibacterial activity of the substance mainly depends on structural factors like alkyl chain length, perfluorinated groups availability, and the number of cationic ammonium groups in the molecule. The inhibition action of the QA mainly depends upon the attraction of the QA's cationic ammonium group and negatively charged cell wall. This interaction restricts cell growth (protein activity) by developing a complex between QA and bacteria [63]. This interaction by QA containing cationic ammonium moiety activated at different conditions, namely polar and nonpolar interaction. However, after penetration, the alkylammonium group physically restricts/interrupts the function of the cell. Studies also reported that the QA can affect bacterial DNA by diminishing its multiplication ability [64].

Poly(hexamethylene biguanide) (PHMB) is a polycationic amine with cationic biguanide as a repeating unit. The antibacterial activity of such organic substance is based on the interaction which changes the physical properties of the bacterial cell wall membrane. The interaction changes the sold outer membrane into liquid through a phase transition. Further, the magnesium and calcium cations released from these PHMB substances create affinity towards both the gram-positive and gram-negative strains [65]. Likewise, *N*-halamines are another such popular antibacterial agents with organic compounds containing one or more nitrogen–halogen compounds. These bonds are usually of amide, imide, or amine types. Due to these structural differences, *N*-halamines are known for their wide spectrum of antibacterial activity. These compounds release chlorine during the electrophilic substitution with H in the N–Cl bond. These changes will occur in the presence of water, and hence these free chlorine molecules react and hinder the enzymatic metabolic process of microorganism. It is also important that during that process, the halamines regenerate their antibacterial activity by reversing their N–H bond [66]. Polyaniline (PANI) is another important polymer that has antibacterial activity. The antibacterial mechanism of PANI is based on reactive oxygen formation. The H^+ ions in the PANI structure are responsible for creating dissolved oxygen molecules that are later converted as hydroperoxy radicals. During the interaction reaction, the organic matters in the bacterial cells are oxidized by these hydroxyl groups and decompose the structure. Further, the electrostatic adherence between the PANI and bacteria also demolishes the cell wall and helps in bacteria death [67]. Several natural polymers with antibacterial agents are commercially used in medical textile applications. Due to their inherent properties and biodegradable nature, these antimicrobial agents are widely used compared to other organic and inorganic agents.

Chitosan is a natural polymeric material derived from chitin. The polymer is naturally non-toxic and has antibacterial ability inherently. Due to the polycationic nature of the chemical structure, they interact with the negatively charged cell walls of the bacteria to obstruct the metabolism [68]. In the case of the textile application considered, chitosan is also used as an anti-static agent, as deodorant finish, and also to enhance the dye absorption in fabrics. Despite their wide spectrum of activity against varieties of bacteria and viruses, their durability is poor [69]. The effectiveness of chitosan as an antibacterial agent can be altered using factors like its molecular weight, pH of the application, and types of solvent used. Structurally, chitosan contains three reactive sites with a primary amine and two hydroxyl groups per glucosamine unit. Hence, these structures are capable of frequent chemical modifications and mimic glycosaminoglycan components. This offers a higher versatility with biocompatibility, biodegradability, antibacterial, haemostatic, and antioxidant activities [70]. One of the common disadvantages of chitosan is its poor solubility in aqueous media; hence, researchers reported that these structural modifications will help in reducing disadvantages and are more suitable for textile applications [71].

13.5.3 Herbal antibacterial agents

Despite the existence of several synthetic antibacterial agents in the market, the use of natural herbal antibacterial agents has its own segment of customers. Natural antimicrobial compounds are usually derived from microbes, animals, and plants, which were explored by several researchers. However, plant-based materials showed very good potential against a wide spectrum of bacterial species. Plants usually consist of chemical compounds that demonstrate benefits in terms of antibacterial, antioxidants, and antifungal activities. Alkaloids, sulphur-containing compounds, terpenoids, and polyphenols are the common chemical groups found in plants based on their types. Table 13.4 provides a list of broad chemical compounds derived from plant and various subchemicals which are responsible for antibacterial properties of different herbal antimicrobial materials.

13.5.3.1 Alkaloids

Alkaloids have wide structural diversity in their structure. The structure of alkaloids contains a single nitrogen atom with unshared pair of electrons. The presence of basic nitrogen is the unique feature of alkaloids. Generally, alkaloids are mostly basic and hence possess the name alkaloid. Alkaloids are mostly solid but sometimes they are liquid when they lack oxygen. They are soluble in nonpolar solvents and insoluble in water. Though alkaloids possess one single nitrogen, they may possess up to five nitrogens in the form of either primary, secondary, or tertiary amine [72]. As far as the antibacterial action of the alkaloids is concerned, though several mechanisms are unknown, few are detailed already. Alkaloids often affect the cell division of bacterial strains by inhibiting the nucleic acid synthesis inside the cell. The other mechanism in which alkaloid reacts are respiratory inhibition, enzyme inhibition, bacterial membrane disruption, and also by affecting the virulence genes [73]

13.5.3.2 Tannins

Tannins are another important antimicrobial agents that occur in plant species. Tannins are polymeric phenols that can be either hydrolysable or condensed. In the case of hydrolysable tannins, gallic acids form the basic structure, whereas the condensed tannins are obtained from flavonoids [74]. The antimicrobial activity of tannins is based on oxidation and polymerization activities. Tannins form both covalent and non-covalent bonds with proteins in the bacterial cell. And hence they restrict the growth of microorganisms by inhibiting their biological processes [73, 74].

13.5.3.3 Flavones, flavonoids, and flavonols

Flavones are chemicals developed with an aromatic ring in their structure, which yields flavonol after hydroxylation. Similarly, flavonoids also occur after further hydroxylation. The main antibacterial mechanism reported for this group is a complex formation with extracellular and soluble proteins. As these components are developed in the plant to protect it from external microorganisms attaching, it is obvious that these components have antibacterial properties [74].

13.5.3.4 Phenolics and polyphenols

Phenols can have a simple or a complex structure based on the chemical compounds they are getting substitutions and hydroxylations. The antibacterial inhibition mechanism of phenols mainly inhibits the enzymes in the bacteria. Studies reported that the inhibition takes place in the sulphhydryl groups on proteins [74, 75].

13.5.3.5 Quinones

Quinones are effective antimicrobial agents due to their interaction mechanism with bacterial cell. They typically contain two ketones in the aromatic ring whereas, phenolic compounds have one. The antibacterial properties of quinones arise from their ability to donate their free radicals. Hence, they can possibly surface attack the polypeptides in the bacteria cell walls or they can also reach the amino acids in protein

Table 13.4: Broad chemical compounds from plants and various subchemicals that provide antibacterial properties.

General chemical group	Subchemicals that provide functional properties
Alkaloids	– Piperine – Berberine – Quinoline alkaloid – Reserpine – Sanguinarine – Tomatidine – Sanguinarine – Chanoclavine – Steroidal alkaloid conessine – Squalamine
Sulphur-containing compounds	– Allicin – Ajoene – Dialkenyl – Dialkyl sulphides, S-allyl cysteine – S-Ally-mercapto cysteine – Isothiocyanates – Sulphoraphane – Phenethyl isothiocyanate – Berteroin
Phenolic compounds	– Resveratrol – Baicalein – Biochanin A hrysosplenol-D – Chrysoplenetin – Kaempferol – Quercetin – 4′,6′-Dihydroxy-3′,5′-dimethyl-2′-methoxychalcone – Catechin gallate – 3-*p-trans*-Coumaroyl-2-hydroxyquinic acid – Naringenin – Eriodictyol – Taxifolin – Sakuranetin – Quercetin and apigenin – Curcumin – Sophoraflavanone B – Morin – Urease and 4′,7,8-trihydroxyl-2-isoflavene

Table 13.4 (continued)

General chemical group	Subchemicals that provide functional properties
Coumarins	– Aegelinol – Agasyllin – 4′-Senecioiloxyosthol – Osthole – Asphodelin A 4′-*O*-β-D-glucoside – Asphodelin A – Lorobiocin – Novobiocin – Coumermycin A1 – Ergamottin epoxide – 6-Geranyl coumarin gallbanic acid
Terpenes	– Linalool – Geraniol – Nerolidol – Plaunotol – Farnesol – Geranylgeraniol – Phytol – Arnesol – Nerolido – Dehydroabietic acid – (4*R*)-(−)-Carvone – (4*S*)-(+)-Carvone – Thymol – Carvacrol – Eugenol – Menthol – Ursolic acid – A-Amyrin – Eugenol and cinnamaldehyde

and deactivate them. These properties inhibit the growth of bacteria and show potential antibacterial ability [74].

13.6 Textile material as sutures

Sutures are the materials that hold the cuts, wounds, and skin damages until the natural healing process initiates. The use of sutures helps in increasing the wound closure and stops bleeding. These sutures are classified as absorbable and non-absorbable based on their chemical characteristics, in which non-absorbable sutures will not be dissolved or decomposed by the body's action and required a removal after use. Absorbable sutures will be decomposed inside the wound or body and will be excreted through different modes like urine and faeces [76]. The suture ideally should have certain properties. Knot strength is one of the necessary parameters of the material that decides the slippage ability of the suture. The breaking strength of the suture is the next parameter that must be evaluated. Along with that, the capillarity, elasticity, and pliability of the material are the important characteristics of sutures.

13.6.1 Natural fibres for sutures

Several materials of natural origin are used in both absorbable and non-absorbable sutures. For instance, the gut is a natural monofilament strand taken from the serosal layer of healthy cattle, sheep, or goats. These materials are normally used for soft tissue wounds. These materials should not be used for cardio- or neuro-related surgeries as the body has a strong reaction to this material [77]. These filaments are manufactured in a uniform shape or twisted strand form. Similarly, these catgut materials were treated with chromium salt to enhance their strength and resistance to absorption. In the case of non-absorbable sutures, silk, cotton, and linens are the materials commonly used. While considering the silk fibres, they require a special degumming process to remove the gummy substances from the silk fibres. In this process, up to 30% of the original volume of the silk fibre is removed to obtain the denier range of 20–22. The basic material of these sutures is (keratin) protein. They are commonly used as fine filament after twisting it, in commercial applications. They possess higher knotting strength and security in the suturing process. However, they possess the least tensile strength compared to other suture materials [78].

Cotton and flax materials are generally used after the bleaching process. Based on the requirement, these materials were also finished with starch or wax to ensure the performance. Both the materials are made of cellulose and used in the form of twisted strands. Out of these materials, linen-based materials are noted to be stronger and 10% higher strength when they are in the wet stage, whereas the cotton sutures are weaker than the linen material. Out of these, linen materials had better handling and excellent knotting properties than cotton sutures [78].

13.6.2 Synthetic fibres for suture

Polyglactin, polyglyconate, and polyglycolic acids are few examples of synthetic textile-absorbable sutures. Out of these mateials, polyglycolic acids are made into either braided structures or twisted structures, whereas polyglactin and polyglyconate are considered and are developed as monofilaments. All of these materials are absorbed through the hydrolysis process, and the rate of absorbance varies from 60 to 210 days.

The lowest degradation time was reported for polyglactin, and the maximum time was estimated for polyglyconate. Several commercial sutures that are biodegradable are made of polymers of glycolides, lactide, and polydioxanones. The main advantages of these polymers are their lower tissue reactions. They degrade through a hydrolysis

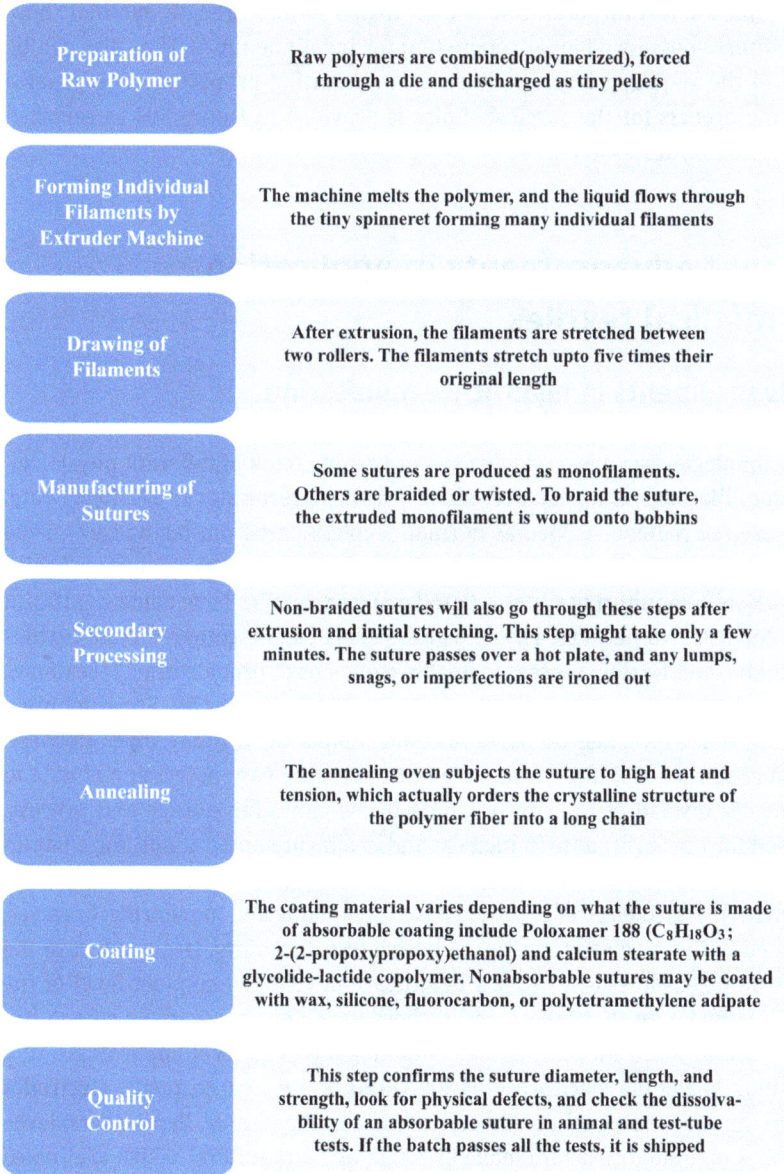

Preparation of Raw Polymer	Raw polymers are combined(polymerized), forced through a die and discharged as tiny pellets
Forming Individual Filaments by Extruder Machine	The machine melts the polymer, and the liquid flows through the tiny spinneret forming many individual filaments
Drawing of Filaments	After extrusion, the filaments are stretched between two rollers. The filaments stretch upto five times their original length
Manufacturing of Sutures	Some sutures are produced as monofilaments. Others are braided or twisted. To braid the suture, the extruded monofilament is wound onto bobbins
Secondary Processing	Non-braided sutures will also go through these steps after extrusion and initial stretching. This step might take only a few minutes. The suture passes over a hot plate, and any lumps, snags, or imperfections are ironed out
Annealing	The annealing oven subjects the suture to high heat and tension, which actually orders the crystalline structure of the polymer fiber into a long chain
Coating	The coating material varies depending on what the suture is made of absorbable coating include Poloxamer 188 ($C_8H_{18}O_3$; 2-(2-propoxypropoxy)ethanol) and calcium stearate with a glycolide-lactide copolymer. Nonabsorbable sutures may be coated with wax, silicone, fluorocarbon, or polytetramethylene adipate
Quality Control	This step confirms the suture diameter, length, and strength, look for physical defects, and check the dissolvability of an absorbable suture in animal and test-tube tests. If the batch passes all the tests, it is shipped

Figure 13.9: Manufacturing sequence of the sutures.

process from 90- to 180-day duration [78]. Several other polymeric materials like poly-esters, polyamides, polypropylene, polybutylene terephthalate, and polyurethane pol-yether are commonly used. These fibres are extruded into monofilaments through a chemical reaction. The extruded monofilaments were either used as such or treated or coated with chemicals with functional properties. The monofilaments were also twisted or braided to form sutures. Due to their higher tensile strength, synthetic non-absorbable sutures possess a higher retainment for indefinite times. They also can be engineered to the required levels of handle and knotting properties. The general manufacturing process for the surgical suture is provided in Figure 13.9 as reported by Shahjalal and Maruf Hasan [77].

13.7 Recent advancements in application of medical textiles

13.7.1 Advancements in healthcare monitoring

Wearable technologies are being widely used in keeping track of different physiologi-cal parameters like body temperature, blood pressure, respiratory rate, heart rate, sweat, and sleep of patients as well as normal people. Apart from taking care of the health status of an individual, these devices help make a database of human health information. Health monitoring systems based on smart textiles have made significant progress in the last two decades. Physiological signals can be gathered from textiles using stretchable and flexible sensors to enable clinicians to provide timely treatment of patients based on their vital motion and physiological signals [79]. Fog computing is an approach that overcomes the issue of cloud computing. It is developed between the cloud and Internet of things devices which enables the easy processing of data to the near devices instead of transmitting data to the cloud for storage and process. With the need for quick response in medical and healthcare applications, fog comput-ing has a remarkable purpose in healthcare monitoring [80]. Ag/AgCl adhesives are the most commonly used electrodes for fog-based computing and monitoring; however, they are noted with few disadvantages including less durability, less comfort to the wearer, and non-reusable [81]. Being a healthcare device, the comfort level of the device is very crucial. In that aspect, researchers have developed textile composites that can be used as electrodes in the fog computing-based medical applications. In a system for monitoring the electrocardiogram (ECG) of the wearer, textile electrodes were used to improve the signal quality and comfort of the wearer. Researchers devel-oped electrodes of knitted structure made of cotton and Nylon fibres which are coated with silver and copper. The different knit structures including Jacquard, Alan, Full Mi-lano, and Single Jersey were analysed to optimize the comfortability and signal quality.

With Ag/AgCl as a reference electrode, the performance of the textile electrodes was analysed. Silver-coated yarns were noted to have better conductivity than copper-coated yarns. While comparing the different textile electrodes developed, better performance in terms of both signal quality and wearer comfort was shown by Alan fabric structure made of 30% cotton and 70% silver-coated nylon [81]. However, researchers have also reported a few drawbacks of textile electrodes which can be effectively improved in the future, which includes the flexible nature of textile materials that leads to more abrasion which in turn causes more noise in the signal. Moreover, metal coating over textile materials can limit their long-term usage [81].

Near-field communication (NFC) is being employed in healthcare monitoring because of the effective short-range wireless communication. Researchers developed a textile antenna to be integrated with garments that can communicate body temperature and sweat to the NFC readers. Textile antennas are developed by incorporating conductive yarns in the textile substrate (Figure 13.10). The silver-coated nylon threads were incorporated into cotton substrates to develop a textile antenna. Moreover, mobile applications were developed to read the data. These antennas can be incorporated into daily wear and can be effectively used to monitor the temperature and sweat of the wearer. These wireless-sensing systems provide more accurate results even when it is bent up to 150° [82].

Figure 13.10: Textile NFC antennae (reproduced with kind permission from [83], Copyright 2019, John Wiley & Sons, Inc.).

With a step ahead, the monitored healthcare parameters are connected to the cloud, where the data will be analysed and any abnormalities will give an alert to the healthcare provider. Researchers have developed smart garments that sense ECG, EMG, temperature, and muscle activity, and the data will be transmitted to the cloud as well as the mobile phones of the wearer. The wearer can visualize their body activities. Also the data which are transmitted with the cloud will be analysed for any abnormalities, and an alert message will be sent to the medical personnel [84]. Other researchers developed a sliver thread-sewn electrode (using single jersey-knitted fabric) to monitor the ECG data collection and compared it with the traditional electrode system. The study analysed the durability of the developed sensors for stretching, bending, and washing, and evaluated their conductivity. Figure 13.11 represents the developed textile-based ECG monitoring sensor [85]. The results of the study reported no significant difference in the developed textile ECG monitoring device compared to the commercial sensors. The statistical analysis also revealed no significant difference in heart rate and R–R interval (time between heartbeats) between the commercial and developed electrodes. The washing test results showed very less changes in the conductivity, and there was a slight increase in the resistance after eight washes. In the case of stretch and bend tests, the stretch or bending process did not affect the conductivity of the developed sensor [85].

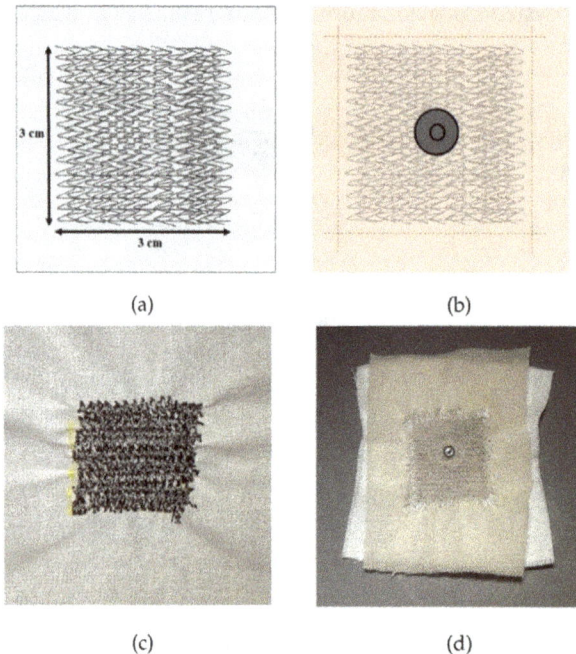

(a) (b)

(c) (d)

Figure 13.11: (a) Sewn electrode stitch design; (b) design of fabric protective cover and snap connector; (c) stitched electrode; (d) complete electrode with protective cover and snap connector (reproduced under creative commons license from [85], Copyright 2020, MDPI, Basel, Switzerland).

Fang et al. [86] developed mechanically durable textile triboelectric sensors to measure the skin temperature and skin pressure. The sensor converts the skin deformation into electric signals and helps in monitoring patients. Poly(dimethylsiloxane), a common biocompatible and waterproof material used in bioelectronics is used in the core and that was covered by protective textile outer layers. Other than these materials, fluorinated ethylene propylene (FEP) nonwoven textile and single-walled carbon nanotube (CNT) conductive networks are also used as an electrode. These conductive materials were deposited on the cotton through a spray-coating method. The developed sensor acted in combination of triboelectric effect with electrostatic induction. This sensor converts the pressure developed on the surfaces into electric signals. The rise and fall of the heartbeat can create a deductible pulse on the surface of the sensor and hence the relative motion created in the CNTs and FEPs. The study showed that the developed sensor can hold a signal-to-noise ratio of 23.3 dB. The measured systolic and diastolic pressure results of the developed sensor were found to be on par with the hospital pressure measurement cuff [86]. Another study developed a dual-sensing device made of textiles for skin temperature and pressure measurement. The researchers developed electrospun piezoelectric polyvinylidene fluoride nanofibrous membrane doped with ZnO nanoparticles (PVDF/ZnO NFM) and flexible thermal-resistant carbon nanofibres, in which the PVDF/ZnO NFM serves as a pressure-sensing and carbon nanofibre used for temperature-sensing purposes. The product also uses a conductive fabric with PU film used as an electrode and insulating material. The product has a pressure sensitivity of 15.75 and 52.09 mV/kPa in the range of 4.9–25 and 25–45 kPa. It can detect the temperature in the range of 25–100 °C. Researchers demonstrated the application of the developed sensing device in the human carotid pulse capture. The study also reported the temperature monitoring capability inside a mask. A detailed outline of the dual-sensing device is provided in Figure 13.12 as reported in the literature [87].

A double-knit cotton fabric was coated with PANI via an in situ polymerization deposition process. The developed samples were analysed for their strain-resisting performance along with other textile parameters. The results of the study reported that the resistance of the developed conductive fabric initially increased and then reduced. At the maximum extension, the lowest resistance was reported. In the case of repeated stretch and recovery cycle, the resistance of the developed fabric underwent a regular change of "increase–decrease–increase–decrease". The extension of the loop length of the conductive fabric was the main reason for the increment in the resistance during stretching. This resistance also increases during the adjacent layer contact during the maximum extension. Overall, the study showed that the developed PANI-coated cotton knitted fabric's performance was satisfactory in terms of sensing linearity and sensitivity for smaller strain levels. For larger strains, though the sensing repeatability has reduced, the changes in resistance were noted to be stable. Based on these findings, the study proposed that the developed PANI/cotton-knitted conductive sensors can be used

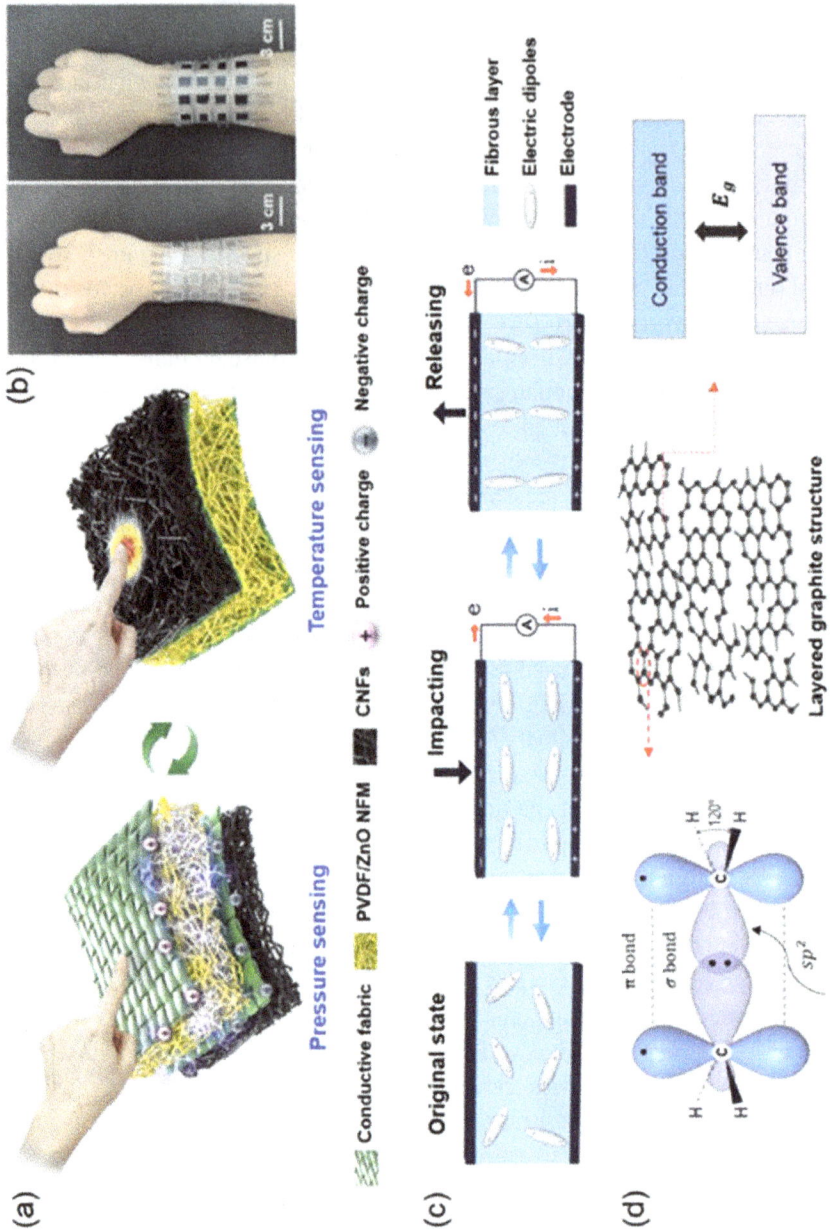

Figure 13.12: Schematic diagram and sensing principle of dual-model electronic skin textile. (a) Schematic diagram of a dual-model textile for pressure and temperature sensing. (b) Optical photographs of the front side and reverse side of electronic skin textile (4 × 4 pixels) conformably attached on an arm. (c) Working principle of piezoelectric response of PVDF/ZnO NFM during impacting/releasing process in a pressure-sensing model. (d) Schematic diagram of electrical conductivity and temperature response principle of CNFs (reproduced with kind permission from [87], Copyright 2021, Elsevier B.V.).

for human body movement monitoring purposes [88]. Though several initiatives were made to produce the monitoring devices using textile materials in the medical sector, the commercialization of such products is still a stack. To achieve such commercial importance, the following points need to be addressed essentially [89]:

– The durability of the products against washing and long-term performance by considering physical activities including laundry
– Power management efficiency, easy handling, including the ability to wear/use clothes without any assistance
– Aesthetic and comfort characteristics include handling, breathability, absorption, and draping
– The product cost versus product life

13.7.2 Advancements in wound dressings

As the complexity of wound management is still a challenging task in the medical and healthcare sector, several smart wound dressings have been developed to address this issue. Any wound dressing that can sense and interact with the wounds or wound environment through smart materials incorporated in the wound dressing is known as smart wound dressings. These wound dressings can able to respond differently at different stages of the wound healing process [90]. There are several types of such wound dressing developed to address the wound management, namely stimuli-responsive wound dressings [91,92], self-healing wound dressing [93], biomechanical wound dressing [94,95], self-removable wound dressing [96], and monitoring wound dressings [97]. A detailed scheme of the different properties of the mentioned smart wound dressing is provided in Figure 13.13 as reported by Dong and Guo [98].

13.7.2.1 Stimuli-responsive wound dressing

These are all dressing materials that respond to the changes in the wound environment through physical and chemical modes known as a stimuli-responsive wound dressing. These dressings can react to the different changes in the wound including pH, temperature, oxygen level, and glucose level [99, 100]. PVA-based hydrogels were developed as a thermal-responsive type of wound dressing. At normal body temperature, the dressing can able to release the drug at a constant level, when the body temperature rises in the wound site due to some infection or inflammation, immediately the temperature-sensitive material triggers a secondary drug release into the site to handle the situation. The temperature-sensitive N-isopropylacrylamide was the responsible material for this action; at normal conditions due to the hydrogen bonds between the drug and this material, slow leaching of the drug was observed. However, a rise in temperature breaks

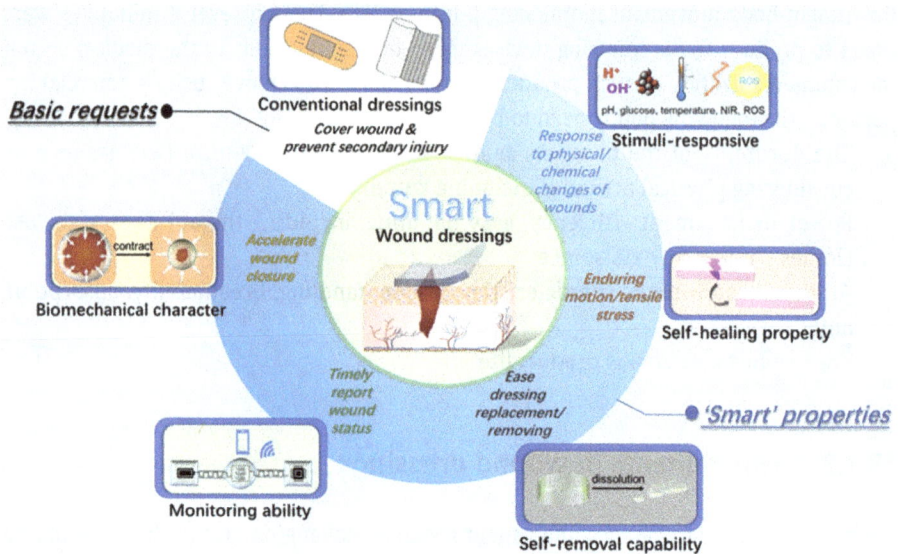

Figure 13.13: Scheme of the properties for smart wound dressings and the smart wound dressings emerged nowadays to fit those requests. Smart wound dressings that can meet the requests that emerge during the wound healing process are desired in future wound management (reproduced with kind permission from [98], Copyright 2021, Elsevier B.V.).

the hydrogen bonds due to the shrinkage of the sensitive material and thus it leaches more drugs into the site [100].

13.7.2.2 Self-healing wound dressings

Self-healing wound dressings are commonly applied to the emotional wounds. These are all the wounds that are situated in the regions where muscles stretch frequently and hence delay the healing process. Self-healing wound dressings are a type of hydrogel dressing that has higher tensile strength and elongation to sustain emotional wounds. The main advantage of these drug-loaded self-healing dressing is that they are biodegradable and hence the necessity of dressing change is reduced significantly [101]. A research group developed chitosan-based hydrogel wound dressing for the wounds in the elbow joints, where it required a high amount of stretch properties. The wound dressing showed a higher potential to withstand the stretch requirements and also due to its biodegradable nature, the wound dressing does not need to be removed from the wound [101].

13.7.2.3 Biomechanical wound dressings

Biomaterials like alginate, collagen, and chitosan are used as a hydrogel to treat wounds. Compared to conventional dressing, these dressings will promote wound healing due to the inherent nature (antibacterial activity) of the biomaterials used. A study developed a hydrogel wound dressing with contraction ability. On application, the wound dressing adheres to the skin firmly, and later changes in the environment (temperature or pH) enables the wound dressing to contract hence a higher wound closure speed can be expected. The in vivo analysis showed an accelerated wound healing with higher vascularization and collagen deposition with higher formation of granulation tissues on-site [102].

13.7.2.4 Self-removable wound dressing

Changing the dressing is one of the painful processes and it also damages the wound closure significantly by tearing the newly formed epithelial cells. Hence, researchers reported several developments in the wound dressing structure that facilitates easy removal of dressing from the wound site. In this aspect, researchers developed wound dressing with the capacity of on-demand dissolution [96, 103]. Liang et al. used deferoxamine mesylate to dissolve the wound dressing developed with the catechol group. When the reagent was added to the hydrogel at the wound site, the chelating agent dissolves the hydrogel partially and reduces its adhesive capacity. Hence, this helps in the easy removal of wound dressing without damaging the wound closure [104]. Similarly, UV light triggered the dissolution of wound dressing, and metallic ion triggered dissolution was also developed by researchers to enhance the easy removal of wound dressing [98].

13.7.2.5 Wound dressing with monitoring

Wound monitoring is one of the important aspects of speeding up of wound healing. Hence, continuous monitoring will reduce the cross-infection of wounds and aid in faster healing. In these aspects, wound dressings are designed to monitor the various changes in the wound site. Wound treatment can be more optimised when the wound characteristics are closely monitored. Bio sensors are being developed and incorporated into the wound dressing to sense different characteristics including pH and protease content. To monitor the pH of the wound, pH-sensitive biosensors have been incorporated into the wound dressing. The optical fibres were fabricated to sense pH by dip-coating in a sol–gel matrix encapsulated with pH dyes. These fibres are then inwoven into wound bandages (Figure 13.14). The fibre ends which are connected to the photodetector will communicate the signal to the computer through a wireless

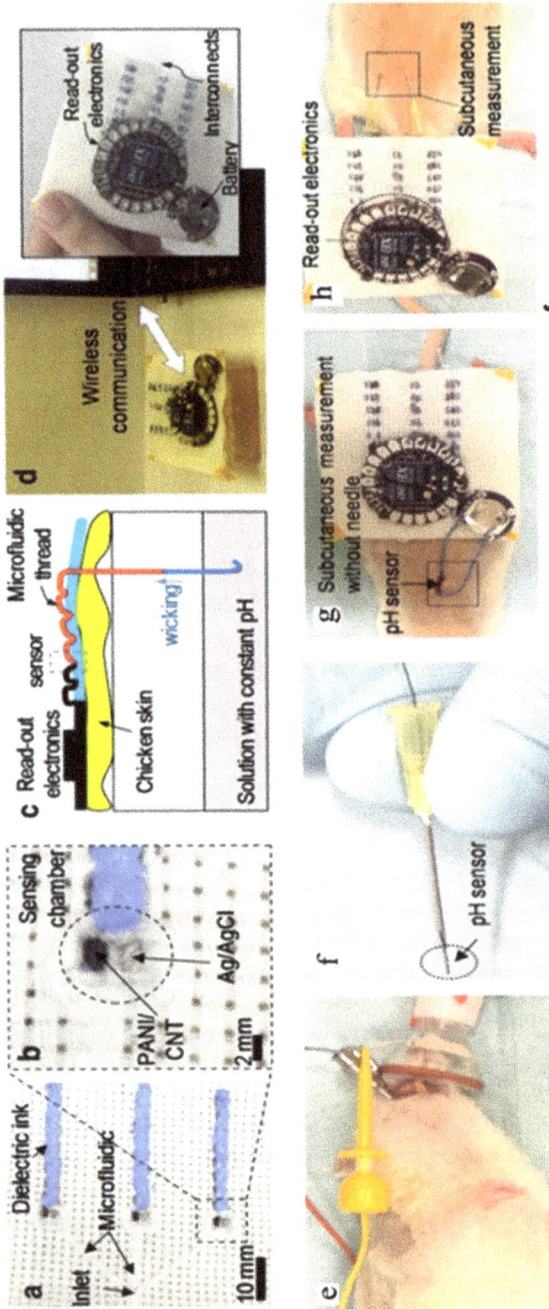

Figure 13.14: (a and b) Optical image of a multiplexed microfluidic pH sensors assay. (c) Schematic illustration of measuring pH in an in vitro skin model. (d) Sensing system communicates with an external computer via a wireless system. (e) The view of a thread-based sensor inserted into the stomach of a rat via the mouth using an oral gavage needle as a guide. (f) pH sensor passed through a needle before subcutaneous implantation. (g and h) Implanted sensors are connected to the patch (reproduced under creative commons license from [107], Copyright 2016, Springer Nature).

connection [105]. Similarly, Nocke et al. [106] developed a smart wound dressing that can monitor the pH of the wound with the help of a fibre sensor that is coated with a pH-responsive hydrogel which is integrated into the wound dressing.

Other researchers developed wound dressing with cationic dye. The dye turns out colourless when it is exposed to bacteria. Hence, when the wound gets infected, the bacteria growth will be triggering the colour change in the wound dressing. Usually, methylene blue is used in the wound dressing, and the colour of the dressing fads as an indicator of wound infection [108]. Similarly, pH changes in the wounds were also monitored with the help of pH-sensitive dyes. The wound dressing was already embedded with sensors to detect the colour changes of the dyes and it will directly communicate with the doctor through mobile applications. The main advantage of such dressings is that the patient can be at the home and at the same time continuously monitored by the doctors [109]. Similarly, several other researchers detected different systems to monitor the wound temperatures, glucose levels, and other important parameters as the requirements demand. Detailed analysis of the latest developments in wound dressing can be found in the literature [98, 110, 111].

13.7.3 Advancements in compression garments

As compression bandages are mainly developed with knit structures, recently several researchers focused on the use of woven fabrics in this application. Studies evaluated the effect of weave type, fibres on the compression properties, and also recent researchers focused on comfort aspects of compression garments. Though the research works were performed since the beginning, the focus on the comfort aspect, influence of the fabric parameters, and properties on the compression is the recent area.

Recently, a study analysed the use of plain-woven fabric with different fibre content as a compression bandage, namely 100% bleached cotton, viscose-Lycra, viscose polyamide, and cotton/polyamide/polyurethane bandages. The study compared the influence of fabric parameters on the compression characteristics of woven bandages. They have reported that an increase in the yarn twist from 300 to 900 increases the elongation up to 43.4%. Further increment in the twist reduces the pressure generated in the compression bandage. Hence, the selection of optimum yarn twist is one of the essential parameters in providing compression to the wearer. Similarly, the yarn tenacity is also largely influenced by the yarn twist. It increases to a certain level and decreases later. While comparing both cotton and viscose woven compression bandages, a higher stretch was noted with viscose-Lycra fabric than cotton fabric. When the selected samples were applied to the real-time patients and the pressures analysed while walking, the results showed a higher pressure with cotton bandages followed by cotton/polyamide/polyurethane, viscose-Lycra, and viscose polyamide bandages. The differences are mainly associated with the fibre types and yarn twists used in this structure [112]. Another study evaluated the moisture and comfort characteristics of compression bandages made of cotton, viscose/polyamide, viscose/Lycra, and cotton/polyamide/polyurethane fibres. The results reported that an increase in the garment stretch increases the porosity of the fabric and hence it reduces the thermal resistance of the fabric. However, when the applied pressure increased from 0.5 to 10 N, the thermal resistance value decreased. A multiple layer bandaging process with higher pressure reduces the thermal resistance value of the compression bandages.

The study also reported that an increase in extension increases the air permeability and water vapour permeability. However, at maximum extension, the water vapour permeability of the compression bandages starts reducing, due to the reduction in pore size [113]. To evaluate the comfort characteristics of compression bandages, a study compared the five different commercial compression bandages and cotton bandages for their frictional properties. The results of the study exhibited that with increased pressure on the compression garment, very less differences were noted in the friction coefficient. However, increasing the normal load step by step reduces the friction coefficient of the compression bandages. In the case of fabric stretch was considered, the fabric stretching process significantly alters the thickness of the fabric. However, the stretch happens only in the direction of warp direction alone. This stretching of fabric reduces the density and hence subsequent reduction in the friction [114]. Figure 13.15 represents the influence of various fabric parameters on the thermal insulation value of compression garments.

Another study evaluated the effect of woven compression bandages on the muscle activation of flexor carpi, soleus (SO), and medial gastrocnemius (MG) muscles. Results showed a significant improvement in muscle activity. The use of woven compression bandages reduced the muscle activity up to 18.24% and 10.66%, respectively, for the SO and MG muscles during the walking action. However, there is a higher reduction in muscle movement up to 22.68% and 25.56% in the SO and MG muscle while using cotton–polyacrylic–polyurethane compression bandage [115]. One of the other sectors in which the application of compression garments emerging is scar management. As compression therapy is always known as a method to manage burn scar, the use of compression garments has also been evaluated recently. A study compared the effect of the use of compression garments on burn wound scar management of patients in vivo. A trial was performed by providing compression garments to 49 patients, and 51 patients were given silicon garments. Fifty-three subjects were provided with both silicon and compression garments for a 6-month study. The results of the study did not show any positive impact of usage of compression garments in scar management. Instead, some adverse effects were noted in the subjects who used both silicon and compression garments [116]. Though it was believed that applying pressure on a burn wound will heal the process faster, the use of a compression garment in the wound healing process did not prove clinically, for better healing [117]. Studies also measured the comfort characteristics of the compression garments with inlays, which are commonly used as orthopaedic supports. Researchers measured the effect of laundry on the comfort characteristics of such compression garments and reported

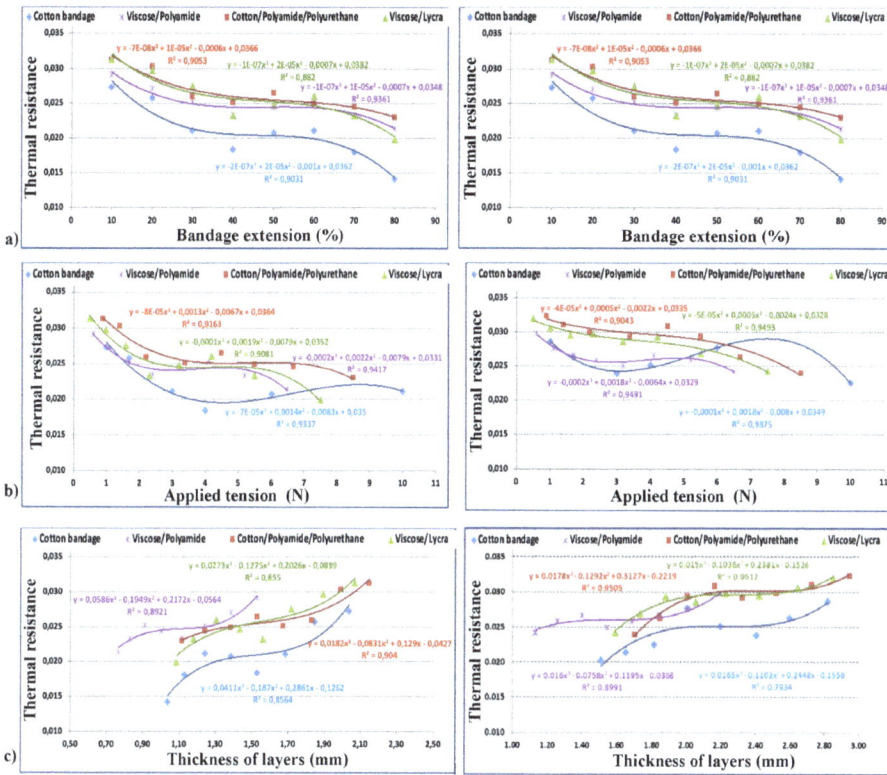

Figure 13.15: Effect of various fabric parameters on the thermal resistance behaviour of compression garments: (a) bandage extension, (b) applied tension, and (c) thickness of layers (reproduced under creative commons license from [113], Copyright 2020, AUTEX).

that the material shrinks, and loop density increases upon laundry; however, the shrinkage was noted to be reduced with repeated laundry. A study reported that the increased shrinkage changes the internal structure of the fabric, and hence, the level of pressure applied to the body varies significantly. A 3.6% increment in the applied pressure was noted as an effect of the first laundry. However, the study did not show this much increment in the pressure in the successive wash cycles [118].

A study reviewed 48,945 research articles in the area of compression garment application and selected 20 studies based on a set of inclusion parameters defined by authors. The results of various experimental studies showed mixed results. Though several researchers reported the athletic recovery ability of the compression garments after exercise, the results still need much corroboration. The authors found mixed results among different researchers with a higher level of heterogeneity. The findings of the review suggested that the use of compression garments increased the recovery rate of swelling, power, muscle soreness, and strength. Further, the use of a compression garment did not affect the heart rate of the user in any study. The results were inconsistent among researchers regarding the changes in exercise-induced muscle damages like clearance of blood lactate, creatine kinase, and lactate dehydrogenase (LDH-5) after the compression garment usage [119]. Though mixed effects of compression garments are reported in the literature, compression garments exist in the market for a longer time, and the research direction shifted to the comfort side recently, compared to the pressure measurement and control-oriented researches in the past. Studies analysed several fabric parameters and fibre properties and their influence on comfort properties. As a result, the recent shift in research mainly concentrates on the wearer or patients comfort more than the pressure applied and other technical parameters.

13.7.4 Advancements in sutures

Sutures are one of the much-researched segments in the medical textile due to their importance in wound care. The increasing requirements and new needs make the researchers engaged in developing new functional aspects with the existing sutures. Sutures can be bioabsorbable or removable based on the polymer types used in their construction. The following section details the various advancements in the suture application in Figure 13.16, as reported by Dennis et al. [120].

13.7.4.1 Antimicrobial sutures

Recently, higher attention was provided to antimicrobial sutures as it heals the wound effectively. Due to their nature of the structure, sutures are highly susceptible to contamination irrespective of their natural or synthetic origin. This will lead to wound contamination and bacterial colonization. This kind of localized surgical site infection is quite coming and occurs in 5% of total procedures performed [121]. Studies evaluated the use of antibacterial materials like chitin in the development of sutures to incorporate their inherent antibacterial ability. Sutures developed from chitin-based material showed higher mechanical strength (63%) for up to 14 days after the application. The in vivo results on rat models reported that a complete absorption of biodegradable sutures was noted in 42 days [122]. Another research proposed an approach to modify the surface of the synthetic sutures through plasma treatment and created nanotopographies on the surface. The study claims that these surface modifications restrict the adhesion of bacteria on the surface of the suture and thus it reduces the chances for cross-infection. Researchers also suggested the potentiality of this method for wide application in different materials [123].

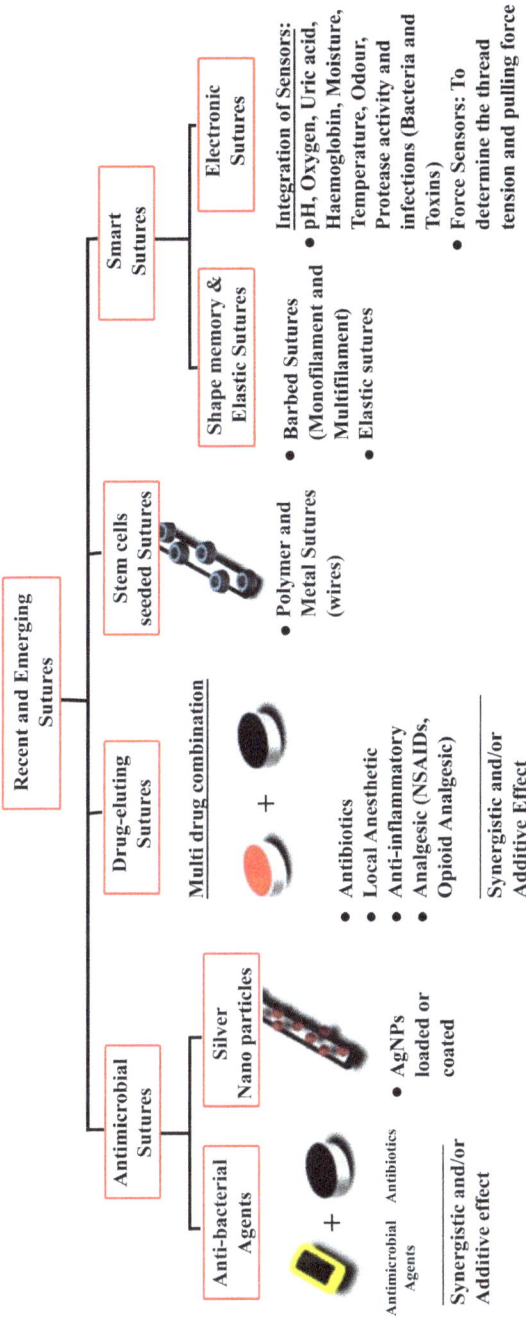

Figure 13.16: Schematic representation of recent and emerging suture technologies (reproduced with kind permission from [120]) Copyright (2016), John Wiley & Sons, Inc.).

13.7.4.2 Antimicrobial coating for suture

As discussed earlier, to avoid these cross-infections, these sutures are either dipped or coated with functional finishing agents. This will help sutures to retain their antibacterial activities from few days to a few weeks. However, studies also reported that the coating or finishing process has a potential impact on the physical and handling properties of the suture [124, 125]. Other than these methods, antibacterial agents were also imparted by a simple evaporation method. In this method, sutures are functionalized with functional agents at a specified temperature, pressure, and time. This process is simple and effective in developing antimicrobial sutures without any impact on physical properties [126] Triclosan is one of such well-known antibacterial agents used in several commercial products, and was also evaluated for their effectiveness in the suture applications. The results of the research reported that the polyglactin suture that was coated with triclosan showed a wide spectrum of antibacterial activity. Both the *in-vivo* and *in-vitro* analyses showed higher effectiveness against *Staphylococcus aureus* and *Staphylococcus epidermidis* [127].

Poly[(aminoethyl methacrylate)-co-(butyl methacrylate)] (PAMBM) is another such potential material evaluated by researchers in recent times. The PAMBM-coated sutures were proven to have higher effectiveness in killing the *S. aureus* rapidly. The findings of the study also showcased that the effect of PAMBM is noted higher than the triclosan-treated sutures [128, 129]. Studies also reported the use of octenidine-palmitate in the suture to incorporate antibacterial effectiveness. The results of the study reported that the octenidine-palmitate-coated sutures had a higher release profile for up to 96 h due to the lower solubility of the palmitic acid in octenidine [130]. Other than these materials, the use of nanoparticles on the suture is also widely adopted for the incorporation of antibacterial properties. The silver nanoparticle is one of such most commonly used materials in the suture coating process. Previous studies performed with silver nanoparticles reported having significant antibacterial activity on the suture [131]. Silver nanoparticle-coated poly(glycolic acid) sutures reported higher antibacterial effectiveness of 99.5% against *S. aureus*, without any cytotoxicity to fibroblast cells [132]. The use of nanoparticles was also found to be effective by several researchers against the common skin-borne bacteria and also they found to show a considerable amount of anti-inflammatory effects [133].

13.7.4.3 Drug-releasing suture

As in wound dressing, researchers also incorporated required medicines into the structure of the suture to alleviate the post-operative issues like wound site inflammation and cross-infection. The use of such medicines in the suture itself will reduce the cost and risk of administrating medicine in the later stages. The loading of medicine or drug on the suture can be either performed via coating by dipping or by grafting or by electrospinning method as discussed elsewhere. [134, 135]. Researchers reported that the incorporation of the required concentration of drug into the suture and the sustainable release for the stipulated period are the major challenges in this field [120]. Additionally, maintaining the mechanical properties of the suture after all this drug incorporation was also found to be challenging. As far as the suture structure is concerned, the braided structure is mostly analysed by the researcher over monofilaments as they are frequently infected. Research analysed the use of levofloxacin hydrochloride and poly(e-caprolactone) in the braided silk suture and analysed their drug-releasing capacity. The sutures were found to be very effective in inhibiting bacterial growth and also showed an average knot pull strength with no cytotoxicity [136].

Studies reported that the use of tetracyclin coating on sutures is more effective in controlling the growth of E. coli than S. aureus due to their higher drug concentration [137]. Another study performed an *in-vivo* analysis on the rat wound model with PLGA-based suture loaded with an anaesthetic (bupivacaine). The results showed that at different concentrations of the drug the suture material had a desirable mechanical property. The results showed a successful diffusion of the drug into the wound closure area of the rats tested. The drug-release profile of the suture is directly proportional to the concentration of the drug. The results confirmed the usefulness of such sutures in the post-operative treatment and the results did not show any adverse effects [138]. The use of drug-loaded suture coated with tacrolimus in the porcine model was analysed and reported in different research. The study reported a controlled release of drug from suture and thus it restricted the neo-intimal hyperplasia at the anastomotic suture site [139]. Lee et al. reported the effectiveness of surgical sutures with ibuprofen on the pain relief ability in post-operative wounds successfully. Researchers also mentioned that the sutures loaded with ibuprofen were maintained with similar mechanical properties to that of unloaded sutures. Hence, a prolonged release with better mechanical properties was achieved [140].

Another advancement in the drug-releasing suture is that these sutures were also treated with growth factors and stem cells to increase the rapid tissue regeneration of the wound sites. These stem cell-loaded sutures increase the number of cells developed at the wound site rapidly and enhance wound healing [141]. The major challenge reported by the researchers is to maintain the desired mechanical and physical properties of the suture even after loading the stem cells into it. Recent research works showed the performance of adipose-derived stem cells loaded with biodegradable sutures. The results of the study reported an even distribution of cells on the wound site and an improved metabolic activity in the wound site [142], whereas braided sutures coated with pluripotent stem cells benefited the mechanical repair of tendons. The *in-vitro* analysis results showed that the cells were alive in the suture and after application, a higher metabolic activity in the injured tendons was reported [143]. From the results, it can be evident that the sutures with such stem cells with increased durability and mechanical properties can be adopted in several life-saving situations to heal the wound rapidly. However, future research works are still needed to address the different wound requirements and reduce the scar formation at the wound site. Similarly, more clinical studies are also required in this direction to enhance the understandability of different drug, suture, and tissue interactions.

13.7.4.4 Smart sutures

Smart sutures are the type of sutures that act according to the situation. To achieve such an effect, various methods were used. One such method is the use of shape-memory polymers in the suture application. These polymers have the capacity to return to the original or default stage from its temporary stage with the help of external stimuli like heat, light, or pH [144]. These sutures are stretched to the required level as a temporary shape at the critical temperature. During the applications, the suture will be loosely fixed on the wound site, and later either by external heat or by body heat the knots will be tightened, and thus it reduces the complexity for surgeons [145]. This kind of sutures offers great flexibility and also pliability with desired mechanical properties over conventional sutures. Similarly, elastic sutures were also developed recently. The use of such sutures avoids tissue necrosis and slacking of sutures during post-operative periods. The non-elastic sutures may fail in post-operative situations like a continuous cough. Hence, these thermoplastic polyurethane-based elastic sutures will withstand such applied pressure and stress in the post-operative conditions due to their higher elongation [146]. Electronic sutures are the other recent developments that incorporate the ability to monitor or sense the situation and react to the kind of material in the suture. Studies reported the use of smart silicon sensors integrated into the polymeric sutures for wound monitoring. The study reported that these sutures upon application can able to measure the temperature and pH changes in the wound site accurately. This will help the medical practitioners to monitor the wound and avoid complexity such as wound infections [147].

Other researchers proposed a concept of smart robotic suture which is programmed for self-tightening of knot or smart anchoring facility to close the wound [148]. The researchers developed a soft scalable and flexible tendon-like artificial muscle (STAM) that works by hydraulic activity. The provided fluidic pressure helps in the elongation of the muscle and contracts or returns to its original stage when the pressure is released. The proposed advantage of the developed smart suture (S^2 Suture) is that it can automatically tighten the knot or it can also deploy anchor points to stabilize the suture at the wound closures without any external action. The structure of the developed suture has STAM, a pressure lock, and a commercial needle as shown in Figure 13.17 A. The suture is made of a silicone tube that can release its micro-coiled elastic energy while the pressure is applied via the hydraulic pressurization process. The details of the anchoring mechanism are provided in Figure 13.17B. One end of STAM is connected with cone-shaped suture tip (Figure 13.16B) and when the pressure-locking mechanism holds or releases its pressure, the tip facilitates the tissue puncture and helps in suturing. As shown in Figure 13.17C–E,

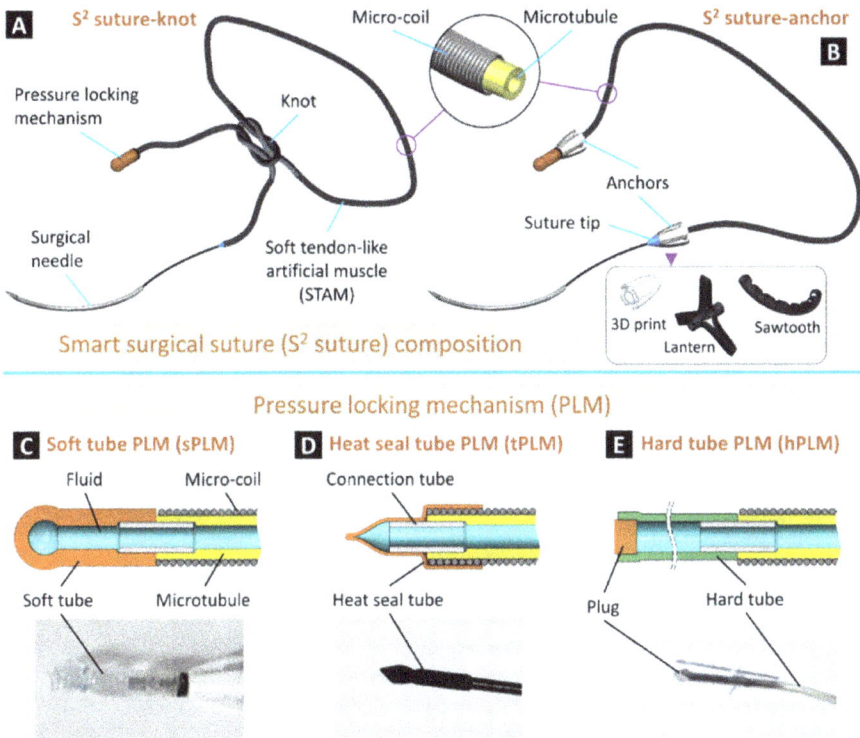

Figure 13.17: Structure of the smart surgical sutures (S^2 sutures). (A) The S^2 suture knot can be knotted as conventional surgical sutures. (B) The S^2 suture anchor is formed by combining the S^2 suture knot with three different types of anchors. (C–E) Design of different pressure-locking mechanisms (PLMs) and their prototypes (reproduced/adapted under creative commons license from [148], Copyright 2021, Springer Nature).

researchers developed three different models of pressure-release mechanism, namely soft tube, heat seal tube made of PET, and hard tube made of polytetrafluoroethylene. Uniform tension distribution along the length of the suture is one of the main advantages of this S^2 suture compared to the conventional sutures [148].

The study also demonstrates the ability of two different prototypes by forming loose knots with the product, and also demonstrated the knot-tightening ability of the S^2 suture by slowly releasing the pressure with a miniature syringe. Though no tension measurements were reported, the self-tightening capacity of the sutures was demonstrated. Figure 13.18A reports the tightening process of the STAM (OD1.49 × L70 mm) and Figure 13.18B shows the tightening in the second type of suture (OD0.8 × L100 mm). The knots were reported to be secured even after a week without any loose and slippage as reported in Figure 13.18C [148].

A recent research reported the development of a battery-free smart suture that can sense and transmit the information to the doctor from deep surgical sites. The suture is

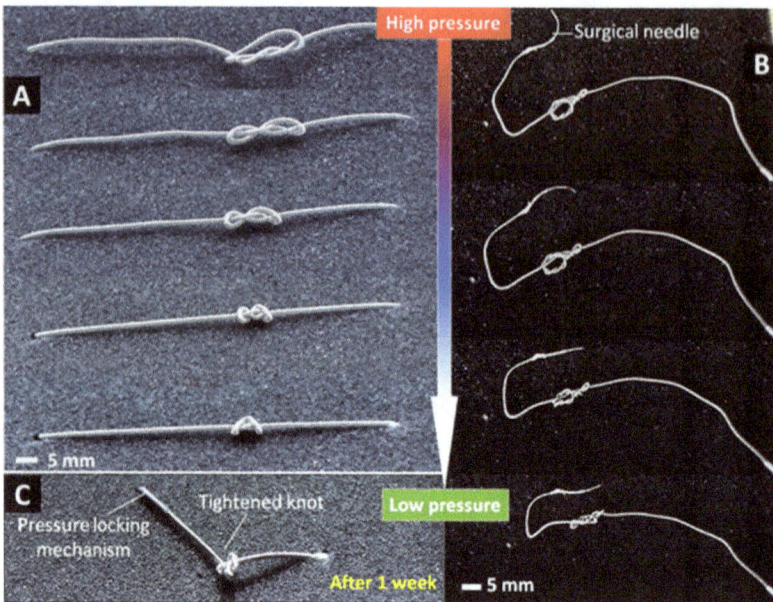

Figure 13.18: Self-tightening capability and knot security of the S^2 suture knot. (A) A prototype (OD1.49 × L70 mm) is pressurized to 100% elongation and tied in a loose knot with both ends fixed. The knot is tightened when reducing input pressure. (B) Similar to (A) but with a prototype OD0.8 × L100 mm and both ends are set free. (C) Stability of the tightened knots after 1 week (reproduced/adapted under creative commons license from [148] Copyright 2021, Springer Nature).

integrated with electronic sensors to measure and monitor wound integrity, gastric leakage, and tissue micro-motions. The developed suture is a conductive polymer coated with the wireless reader so that it can be effectively read from outside the body. The developed suture can be detected up to 50 mm based on the length of the suture used. However, the depth can also be increased by increasing the conductivity of the thread. The main advantage of the suture is that it can also be able to detect broken stitches during wound separation. When the stitch breaks, the receiver reports a reduction in signal due to the reduction in the antenna formed by the suture [149]. As the temperature of the wound ideally represents the healing nature, temperature-sensing sutures are always one of the highly demanded materials. A study developed hybrid smart polycaprolactone and chitosan multifunctional sutures to monitor the wound temperature and aid cell growth and provide antibiotics to control infection. Researchers embedded functionalized nanodiamond and reduced graphene oxide during manufacturing. The temperature sensing was achieved by the nitrogen-vacancy centres of the functionalized nanodiamond along with fluorescent polymers. The developed sutures were able to sense the temperature in the precision of 1 °C for a window of 25–40 °C. Due to the use of reduced graphene oxide coating the sutures possessed higher mechanical properties. The study also reported good biocompatibility with 3-day cell adherence and cytotoxicity analysis [150].

13.8 Conclusions

Medical textile is one of the important technical textile sectors in which the contribution of textiles is significantly higher than the other domains. This chapter summarized the different types of fibres and fabrics used in various medical applications from the wound dressing, compression bandages, and also in healthcare and hygienic application. The review also discussed the different types of organic, inorganic, and natural antibacterial agents used in the medical textile application based on the importance of antibacterial finish in medical textile. As far as the advancements are considered, developments in the smart medical fabrics, wound dressing materials, compression bandages, and sutures were detailed. Smart textile materials in medical applications are generally used for monitoring purposes and also to alert the changes in patients' health. In the case of wound dressing, wound monitoring, self-healing dressings, and stimuli-responsive types are recently commercialized in this domain. In the case of compression garments though studies initially focused on pressure amount and application, most of the recent research aimed to increase the comfort of the wearer. Finally, the chapter details the advancements in sutures including antimicrobial-coated, drug-loaded, and smart sutures.

References

[1]   Global Medical Textiles Market. By Type: Disposable, Non Disposable; By Product: Surgical Gowns, Operating Room Drapes, Sterilization Wraps, Face Masks, Others; By Fabric; By Application; Regional Analysis; Historical Market and Forecast (2017-2027); https://www.expertmarketresearch.com/reports/medical-textiles-market. 2022.
[2]   Morris H, Murray R. Medical textiles. Textile Progress 2020. https://doi.org/10.1080/00405167.2020.1824468.
[3]   Yimin Q. Medical Textile Materials. 1st ed. Elsevier, Textile Institute; 2019.
[4]   Rajendran S, Anand SC. Developments in Medical Textiles. Textile Progress 2002;32:1–42. https://doi.org/http://dx.doi.org/10.1080/00405160208688956,.
[5]   Sabit Adanur. Wellington Sears Handbook of Industrial Textiles. 1995.
[6]   Anand SC. Medical Textile. In: Anand SC, editor. Proceedings of the 2nd international Conference, Bolton Institute, UK: 1999; p. 1–256.
[7]   Zhezhova S, Jordeva S, Golomeova Longurova, S. G, Jovanov S. Application of Technical Textile in Medicine. TEKSTILNA INDUSTRIJA 2021;2:21–29. https://doi.org/10.5937/tekstind2102021Z.
[8]   Alistair JR, Subhash CA. Medical textiles. Handbook of technical textiles, Bolton Institute, 2001, p. 407–24.
[9]   Chaudary SN, Borkar SP. Textiles for extracorporeal devices. Indian Textile Journal 2009:79–84.
[10] Sabbah HN. The cardiac support device and the Myosplint: treating heart failure by targeting left ventricular size and shape. The Annals of Thoracic Surgery 2003;75:S13–9. https://doi.org/10.1016/S0003-4975(03)00463-6.
[11] Parvin F, Islam I, Urmy Z, Ahmed S. A study on the textile materials applied in human medical treatment. European Journal of Physiotherapy and Rehabilitation Studies 2020;1.

[12] Petrulyte S, Petrulis D. Modern textiles and biomaterials for healthcare. In: Bartels VT, editor. Handbook of Medical Textiles, Woodhead Publishing House; 2011, p. 1–35.

[13] Afzal A, Ullah A. Textile fibers. In: Ahmad S et al., editor. Advanced textile testing techniques, CRC Press: Florida, USA; 2017, p. 107–28.

[14] Nony P, Scribner K, Hesterberg T. Synthetic Vitreous Fibers. Encyclopedia of Toxicology: Third Edition 2014:448–53. https://doi.org/10.1016/B978-0-12-386454-3.01172-6.

[15] Kim YK. The use of polyolefins in industrial and medical applications. Polyolefin Fibres: Structure, Properties and Industrial Applications: Second Edition 2017:135–55. https://doi.org/10.1016/B978-0-08-101132-4.00005-9.

[16] Kamoun EA, Kenawy ERS, Chen X. A review on polymeric hydrogel membranes for wound dressing applications: PVA-based hydrogel dressings. Journal of Advanced Research 2017;8:217–33. https://doi.org/10.1016/j.jare.2017.01.005.

[17] Serrano-Aroca Á. Latest Improvements of Acrylic-Based Polymer Properties for Biomedical Applications. In: Reddy BS, editor. Acrylic Polymers in Healthcare, IntechOpen; London; 2017. https://doi.org/10.5772/intechopen.68996.

[18] Gokarneshan N, Anitha Rachel D, Rajendran V, Lavanya B, Ghoshal A. Emerging Research Trends in Medical Textiles. 2015.

[19] Gupta BS. Manufacture, types and properties of biotextiles for medical applications. Biotextiles As Medical Implants 2013:3–47. https://doi.org/10.1533/9780857095602.1.3.

[20] Davis FJ, Mitchell GR. Polyurethane Based Materials with Applications in Medical Devices. In: Bártolo P, Bidanda B, editors. Bio-Materials and Prototyping Applications in Medicine., Springer, Boston, MA.; 2008. https://doi.org/https://doi.org/10.1007/978-0-387-47683-4_3.

[21] Afzal A, Zubair U, Saeed M, Afzal M, Azeem A. Fibres for Medical Textiles. Fibers for Technical Textiles, Springer Nature Switzerland AG.; 2020.

[22] Rohani Shirvan A, Nouri A. Chapter 13 – Medical textiles. The Textile Institute Book Series, 2020.

[23] Rohani Shirvan A, Nouri A. Medical textiles. Advances in Functional and Protective Textiles, Elsevier; 2020, p. 291–333. https://doi.org/10.1016/B978-0-12-820257-9.00013-8.

[24] Lázár K. Application of knitted fabrics in technical and medical textiles. Forty-Fifth International Congress IFKT, 2010.

[25] Krimmel G. The construction and classification of compression garments. Template Pract: Compress Hosiery Upper Body. Lymphoid 2009:2–5.

[26] Ying X, Tao X. Compression Garments for Medical Therapy and Sports. Polymers (Basel) 2018;10:663. https://doi.org/https://doi.org/10.3390/polym10060663.

[27] Brizzio E, Amsler F, Lun B, Blättler W. Comparison of low-strength compression stockings with bandages for the treatment of recalcitrant venous ulcers. Journal of Vascular Surgery 2010;51. https://doi.org/10.1016/j.jvs.2009.08.048.

[28] Bandari SSM, Asayesh A, Latifi M. The Effect of Fabric Structure and Strain Percentage on the Tensile Stress Relaxation of Rib Weft Knitted Fabrics. Fibers and Polymers 2020;21. https://doi.org/10.1007/s12221-020-9450-6.

[29] Maleki H, Aghajani M, Sadeghi AH, Jeddi AAA. On the pressure behavior of tubular weft knitted fabrics constructed from textured polyester yarns. Journal of Engineered Fibers and Fabrics 2011;6. https://doi.org/10.1177/155892501100600204.

[30] Akalin M, Kocak D, Mistik SI, Uzun M. Investigation of Elastic Properties of Multiaxial Warp Knitted Bandages. Medical and Healthcare Textiles, 2010. https://doi.org/10.1533/9780857090348.323.

[31] Schrank V, Beer M, Beckers M, Gries T. Polymer-optical fibre (POF) integration into textile fabric structures. Polymer Optical Fibres: Fibre Types, Materials, Fabrication, Characterisation and Applications 2017:337–48. https://doi.org/10.1016/B978-0-08-100039-7.00010-5.

[32] Pereira S, Anand SC, Rajendran S, Wood C. A study of the structure and properties of novel fabrics for knee braces. Journal of Industrial Textiles 2007;36. https://doi.org/10.1177/1528083707072357.

[33] Zhang X, Ma P. Application of Knitting Structure Textiles in Medical Areas. Autex Research Journal 2018;18:181–91. https://doi.org/10.1515/aut-2017-0019.

[34] Tong SF, Yip J, Yick KL, Yuen CWM. Exploring use of warp-knitted spacer fabric as a substitute for the absorbent layer for advanced wound dressing. Textile Research Journal 2015;85. https://doi.org/10.1177/0040517514561922.

[35] Asayesh A, Ehsanpour S, Latifi M. Prototyping and analyzing physical properties of Weft knitted spacer fabrics as a substitute for wound dressings. Journal of the Textile Institute 2019;110. https://doi.org/10.1080/00405000.2018.1557358.

[36] Al Faruque A, Sarker E, Sowrov K, Alam T. Development of Knitted Gauze Fabric as Wound Dressing for Medical Application Adv Res Text Eng. 2018; 3(1): 1021.

[37] Chen Z, Song J, Xia Y, Jiang Y, Murillo LL, Tsigkou O, et al. High strength and strain alginate fibers by a novel wheel spinning technique for knitting stretchable and biocompatible wound-care materials. Materials Science and Engineering C 2021;127. https://doi.org/10.1016/j.msec.2021.112204.

[38] Zhang J. The Application and Development of Artificial Blood Vessels, 2016. https://doi.org/10.2991/mmme-16.2016.119.

[39] Xu W, Zhou F, Ouyang C, Ye W, Yao M, Xu B. Mechanical properties of small-diameter polyurethane vascular grafts reinforced by weft-knitted tubular fabric. Journal of Biomedical Materials Research – Part A 2010;92. https://doi.org/10.1002/jbm.a.32333.

[40] Yang H, Zhu G, Zhang Z, Wang Z, Fang J, Xu W. Influence of weft-knitted tubular fabric on radial mechanical property of coaxial three-layer small-diameter vascular graft. Journal of Biomedical Materials Research – Part B Applied Biomaterials 2012;100 B. https://doi.org/10.1002/jbm.b.31955.

[41] Gokarneshan N, Dhatchayani U. Mini Review: Advances in Medical Knits. Journal of Textile Engineering & Fashion Technology 2017;3. https://doi.org/10.15406/jteft.2017.03.00095.

[42] Singh C, Wang X. A new design concept for knitted external vein-graft support mesh. Journal of the Mechanical Behavior of Biomedical Materials 2015;48. https://doi.org/10.1016/j.jmbbm.2015.04.001.

[43] Hu X, Hu T, Guan G, Yu S, Wu Y, Wang L. Control of weft yarn or density improves biocompatibility of PET small diameter artificial blood vessels. Journal of Biomedical Materials Research – Part B Applied Biomaterials 2018;106. https://doi.org/10.1002/jbm.b.33909.

[44] Rajendran S, Anand SC. Woven textiles for medical applications. Woven Textiles: Principles, Technologies and Applications, 2019. https://doi.org/10.1016/B978-0-08-102497-3.00011-8.

[45] Unnikrishnan M, Viswanathan S, Balasubramaniam K, Muraleedharan C v., Lal AV, Mohanan P v., et al. The making of indigenous vascular prosthesis. Indian Journal of Medical Research 2016;143. https://doi.org/10.4103/0971-5916.192059.

[46] Rostamitabar M, Abdelgawad AM, Jockenhoevel S, Ghazanfari S. Drug-Eluting Medical Textiles: From Fiber Production and Textile Fabrication to Drug Loading and Delivery. Macromolecular Bioscience 2021;21. https://doi.org/10.1002/mabi.202100021.

[47] Chellamani KP, Vignesh Balaji RS, Veerasubramanian D. Medical Textiles: The Spunlace process and its application possibilities for hygiene textiles. J Acad Indus Res 2013;1:735.

[48] Rajendran S. INFECTION CONTROL AND BARRIER MATERIALS: AN OVERVIEW. Medical Textiles and Biomaterials for Healthcare: Incorporating Proceedings of MEDTEX03 International Conference and Exhibition on Healthcare and Medical Textiles 2006:131–5. https://doi.org/10.1533/9781845694104.3.131.

[49] Ajmeri JR, Ajmeri CJ. Nonwoven materials and technologies for medical applications. Handbook of Medical Textiles 2011:106–31. https://doi.org/10.1533/9780857093691.1.106.

[50] Sikdar P, S Bhat G, Hinchliff D, Islam S, Condon B. Microstructure and physical properties of composite nonwovens produced by incorporating cotton fibers in elastic spunbond and meltblown webs for medical textiles. Journal of Industrial Textiles 2021. https://doi.org/10.1177/15280837211004287.

[51] Brydon AG, Pourmohammadi A. Dry-laid web formation. Handbook of Nonwovens 2007:16–111. https://doi.org/10.1533/9781845691998.16.

[52] Maggio IR, Guichon O. A new spunbond technology for new products. Nonwovens Industrial Textiles 2001;3:68–69.

[53] Hajipour MJ, Fromm KM, Ashkarran AA, De Aberasturi DJ, De Larramendi IR, Rojo T. Antibacterial properties of nanoparticles. Trends Biotechnol 2012;30:1–13. https://doi.org/https://doi.org/10.1016/j.tibtech.2012.06.004.

[54] Saidin S, Jumat MA, Mohd Amin NAA, Saleh Al-Hammadi AS. Organic and inorganic antibacterial approaches in combating bacterial infection for biomedical application. Materials Science and Engineering: C 2021;118:111382. https://doi.org/10.1016/J.MSEC.2020.111382.

[55] Hassabo A, Kamel M. Anti-microbial finishing for natural textile fabrics. Journal of Textiles, Coloration and Polymer Science 2021;0:0–0. https://doi.org/10.21608/jtcps.2021.72333.1054.

[56] Lansdown ABG. A Pharmacological and Toxicological Profile of Silver as an Antimicrobial Agent in Medical Devices. Advances in Pharmacological Sciences 2010;2010:16. https://doi.org/10.1155/2010/910686.

[57] Chaloupka K, Malam Y, Seifalian AM. Nanosilver as a new generation of nanoproduct in biomedical applications. Trends in Biotechnology 2010;28:580–8. https://doi.org/10.1016/J.TIBTECH.2010.07.006.

[58] Jones CF, Grainger DW. In vitro assessments of nanomaterial toxicity. Advanced Drug Delivery Reviews 2009;61:438–56. https://doi.org/10.1016/J.ADDR.2009.03.005.

[59] Seil JT, Webster TJ. Antimicrobial applications of nanotechnology: methods and literature. International Journal of Nanomedicine 2012;7:2767–81. https://doi.org/10.2147/IJN.S24805.

[60] Kim JH, Cho H, Ryu SE, Choi MU. Effects of Metal Ions on the Activity of Protein Tyrosine Phosphatase VHR: Highly Potent and Reversible Oxidative Inactivation by Cu2+ Ion. Archives of Biochemistry and Biophysics 2000;382:72–80. https://doi.org/10.1006/ABBI.2000.1996.

[61] Godoy-Gallardo M, Eckhard U, Delgado LM, de Roo Puente YJD, Hoyos-Nogués M, Gil FJ, et al. Antibacterial approaches in tissue engineering using metal ions and nanoparticles: From mechanisms to applications. Bioactive Materials 2021;6:4470–90. https://doi.org/10.1016/J.BIOACTMAT.2021.04.033.

[62] Parikh SL, Xiao G, Tonge PJ. Inhibition of InhA, the Enoyl Reductase from Mycobacterium tuberculosis, by Triclosan and Isoniazid. Biochemistry 2000;39:7645–7650.

[63] Tiller JC, Liao CJ, Lewis K, Klibanov AM. Designing surfaces that kill bacteria on contact. Proc Natl Acad Sci U S A 2001;98. https://doi.org/10.1073/pnas.111143098.

[64] Marini M, Bondi M, Iseppi R, Toselli M, Pilati F. Preparation and antibacterial activity of hybrid materials containing quaternary ammonium salts via sol-gel process. European Polymer Journal 2007;43. https://doi.org/10.1016/j.eurpolymj.2007.06.002.

[65] Kaehn K. Polihexanide: A safe and highly effective biocide. Skin Pharmacology and Physiology 2010;23. https://doi.org/10.1159/000318237.

[66] Qîan L, Sun G. Durable and regenerable antimicrobial textiles: Chlorine transfer among halamine structures. Industrial and Engineering Chemistry Research 2005;44. https://doi.org/10.1021/ie049493x.

[67] Liang X, Sun M, Li L, Qiao R, Chen K, Xiao Q, et al. Preparation and antibacterial activities of polyaniline/Cu0.05Zn0.95O nanocomposites. Dalt Trans 2015;41:2804–2811. https://doi.org/https:doi.org/10.1039/C2DT11823H.

[68] Joshi M, Ali SW, Purwar R, Rajendran S. Ecofriendly antimicrobial finishing of textiles using bioactive agents based on natural products. Indian Journal of Fibre and Textile Research 2009;34:295–304.

[69] Hassabo AG, Mohamed AL. Multiamine Modified Chitosan for Removal Metal Ions from their Aqueous Solution. BioTechnology: An Indian Journal 2016;12:59–69.

[70] Dragostin OM, Samal SK, Dash M, Lupascu F, Pânzariu A, Tuchilus C, et al. New antimicrobial chitosan derivatives for wound dressing applications. Carbohydrate Polymers 2016;141. https://doi.org/10.1016/j.carbpol.2015.12.078.

[71] Alves NM, Mano JF. Chitosan derivatives obtained by chemical modifications for biomedical and environmental applications. International Journal of Biological Macromolecules 2008;43. https://doi.org/10.1016/j.ijbiomac.2008.09.007.

[72] Cushnie TPT, Cushnie B, Lamb AJ. Alkaloids: An overview of their antibacterial, antibiotic-enhancing and antivirulence activities. International Journal of Antimicrobial Agents 2014;44. https://doi.org/10.1016/j.ijantimicag.2014.06.001.

[73] Othman L, Sleiman A, Abdel-Massih RM. Antimicrobial activity of polyphenols and alkaloids in middle eastern plants. Frontiers in Microbiology 2019;10. https://doi.org/10.3389/fmicb.2019.00911.

[74] Cowan MM. Plant products as antimicrobial agents. Clinical Microbiology Reviews 1999;12. https://doi.org/10.1128/cmr.12.4.564.

[75] Coppo E, Marchese A. Antibacterial Activity of Polyphenols. Current Pharmaceutical Biotechnology 2014;15. https://doi.org/10.2174/1389201015040140825121142.

[76] Chu CC. Types and properties of surgical sutures. Biotextiles As Medical Implants 2013:231–73. https://doi.org/10.1533/9780857095602.2.232.

[77] Shahjalal Md, Hasan SMdM. Surgical sutures, the most common implantable medical textiles. Bangladesh Textile Today 2019;12:106–07.

[78] Ajmeri JR, Ajmeri CJ. Surgical sutures: The largest textile implant material. Medical Textiles and Biomaterials for Healthcare: Incorporating Proceedings of MEDTEX03 International Conference and Exhibition on Healthcare and Medical Textiles 2006:432–40. https://doi.org/10.1533/9781845694104.7.432.

[79] Ahmed A, Adak B, Mukhopadhyay S. Smart Textile-Based Interactive, Stretchable and Wearable Sensors for Healthcare. Nanosensors for Futuristic Smart and Intelligent Healthcare Systems, 2022, p. 112.

[80] Pareek K, Tiwari PK, Bhatnagar V. Fog Computing in Healthcare: A Review. IOP Conference Series: Materials Science and Engineering 2021;1099. https://doi.org/10.1088/1757-899x/1099/1/012025.

[81] Wu W, Pirbhulal S, Sangaiah AK, Mukhopadhyay SC, Li G. Optimization of signal quality over comfortability of textile electrodes for ECG monitoring in fog computing based medical applications. Future Generation Computer Systems 2018;86. https://doi.org/10.1016/j.future.2018.04.024.

[82] Jiang Y, Pan K, Leng T, Hu Z. Smart Textile Integrated Wireless Powered near Field Communication Body Temperature and Sweat Sensing System. IEEE Journal of Electromagnetics, RF and Microwaves in Medicine and Biology 2020;4. https://doi.org/10.1109/JERM.2019.2929676.

[83] Jiang Y, Xu L, Pan K, Leng T, Li Y, Danoon L, et al. e-Textile embroidered wearable near-field communication RFID antennas. IET Microwaves, Antennas and Propagation 2019;13. https://doi.org/10.1049/iet-map.2018.5435.

[84] Sethuraman SC, Kompally P, Mohanty SP, Choppali U. MyWear: A Smart Wear for Continuous Body Vital Monitoring and Emergency Alert 2020.

[85] Arquilla K, Webb AK, Anderson AP, Smead HJ. Textile Electrocardiogram (ECG) Electrodes for Wearable Health Monitoring. Sensors 2020;20:1013. https://doi.org/10.3390/s20041013.

[86] Fang Y, Zou Y, Xu J, Chen G, Zhou Y, Deng W, et al. Ambulatory Cardiovascular Monitoring Via a Machine-Learning-Assisted Textile Triboelectric Sensor. Advanced Materials 2021;33. https://doi.org/10.1002/adma.202104178.

[87] Wang Y, Zhu M, Wei X, Yu J, Li Z, Ding B. A dual-mode electronic skin textile for pressure and temperature sensing. Chemical Engineering Journal 2021;425:130599. https://doi.org/10.1016/J.CEJ.2021.130599.

[88] Zhou X, Hu C, Lin X, Han X, Zhao X, Hong J. Polyaniline-coated cotton knitted fabric for body motion monitoring. Sensors and Actuators, A: Physical 2021;321. https://doi.org/10.1016/j.sna.2021.112591.

[89]   Zahid M, Anwer Rathore H, Tayyab H, Ahmad Rehan Z, Abdul Rashid I, Lodhi M, et al. Recent developments in textile based polymeric smart sensor for human health monitoring: A review. Arabian Journal of Chemistry 2022;15. https://doi.org/10.1016/j.arabjc.2021.103480.

[90]   Rodrigues M, Kosaric N, Bonham CA, Gurtner GC. Wound healing: A cellular perspective. Physiological Reviews 2019;99. https://doi.org/10.1152/physrev.00067.2017.

[91]   Li WP, Su CH, Wang SJ, Tsai FJ, Chang CT, Liao MC, et al. CO2 Delivery to Accelerate Incisional Wound Healing Following Single Irradiation of Near-Infrared Lamp on the Coordinated Colloids. ACS Nano 2017;11. https://doi.org/10.1021/acsnano.7b01442.

[92]   Bhadauriya P, Mamtani H, Ashfaq M, Raghav A, Teotia AK, Kumar A, et al. Synthesis of yeast-immobilized and copper nanoparticle-dispersed carbon nanofiber-based diabetic wound dressing material: Simultaneous control of glucose and bacterial infections. ACS Applied Bio Materials 2018;1. https://doi.org/10.1021/acsabm.8b00018.

[93]   Li S, Wang L, Zheng W, Yang G, Jiang X. Rapid Fabrication of Self-Healing, Conductive, and Injectable Gel as Dressings for Healing Wounds in Stretchable Parts of the Body. Advanced Functional Materials 2020;30. https://doi.org/10.1002/adfm.202002370.

[94]   Li G, Wang Y, Wang S, Liu Z, Liu Z, Jiang J. A Thermo- and Moisture-Responsive Zwitterionic Shape Memory Polymer for Novel Self-Healable Wound Dressing Applications. Macromolecular Materials and Engineering 2019;304. https://doi.org/10.1002/mame.201800603.

[95]   Blacklow SO, Li J, Freedman BR, Zeidi M, Chen C, Mooney DJ. Bioinspired mechanically active adhesive dressings to accelerate wound closure. Science Advances 2019;5. https://doi.org/10.1126/sciadv.aaw3963.

[96]   Ding X, Li G, Zhang P, Jin E, Xiao C, Chen X. Injectable Self-Healing Hydrogel Wound Dressing with Cysteine-Specific On-Demand Dissolution Property Based on Tandem Dynamic Covalent Bonds. Advanced Functional Materials 2021;31. https://doi.org/10.1002/adfm.202011230.

[97]   Gong X, Hou C, Zhang Q, Li Y, Wang H. Thermochromic Hydrogel-Functionalized Textiles for Synchronous Visual Monitoring of On-Demand in Vitro Drug Release. ACS Applied Materials and Interfaces 2020;12. https://doi.org/10.1021/acsami.0c14665.

[98]   Dong R, Guo B. Smart wound dressings for wound healing. Nano Today 2021;41. https://doi.org/10.1016/j.nantod.2021.101290.

[99]   Kiaee G, Mostafalu P, Samandari M, Sonkusale S. A pH-Mediated Electronic Wound Dressing for Controlled Drug Delivery. Advanced Healthcare Materials 2018;7. https://doi.org/10.1002/adhm.201800396.

[100]   Montaser AS, Rehan M, El-Naggar ME. pH-Thermosensitive hydrogel based on polyvinyl alcohol/sodium alginate/N-isopropyl acrylamide composite for treating re-infected wounds. International Journal of Biological Macromolecules 2019;124. https://doi.org/10.1016/j.ijbiomac.2018.11.252.

[101]   Qu J, Zhao X, Liang Y, Zhang T, Ma PX, Guo B. Antibacterial adhesive injectable hydrogels with rapid self-healing, extensibility and compressibility as wound dressing for joints skin wound healing. Biomaterials 2018;183. https://doi.org/10.1016/j.biomaterials.2018.08.044.

[102]   Li Q, Ouyang Y, Lu S, Bai X, Zhang Y, Shi L, et al. Perspective on theoretical methods and modeling relating to electro-catalysis processes. Chemical Communications 2020;56. https://doi.org/10.1039/d0cc02998j.

[103]   Hua Y, Gan Y, Li P, Song L, Shi C, Bao C, et al. Moldable and Removable Wound Dressing Based on Dynamic Covalent Cross-Linking of Thiol-Aldehyde Addition. ACS Biomaterials Science and Engineering 2019;5. https://doi.org/10.1021/acsbiomaterials.9b00459.

[104]   Liang Y, Li Z, Huang Y, Yu R, Guo B. Dual-Dynamic-Bond Cross-Linked Antibacterial Adhesive Hydrogel Sealants with On-Demand Removability for Post-Wound-Closure and Infected Wound Healing. ACS Nano 2021;15. https://doi.org/10.1021/acsnano.1c00204.

[105] Pasche S, Schyrr B, Wenger B, Scolan E, Ischer R, Voirin G. Smart Textiles with Biosensing Capabilities. Smart and Interactive Textiles, vol. 80, 2012. https://doi.org/10.4028/www.scientific. net/ast.80.129.

[106] Nocke A, Schröter A, Cherif C, Gerlach G. Miniaturized textile-based multi-layer pH-sensor for wound monitoring applications. Autex Research Journal 2012;12. https://doi.org/10.2478/v10304-012-0004-x.

[107] Mostafalu P, Akbari M, Alberti KA, Xu Q, Khademhosseini A, Sonkusale SR. A toolkit of thread-based microfluidics, sensors, and electronics for 3D tissue embedding for medical diagnostics 2016;2. https://doi.org/10.1038/micronano.2016.39.

[108] He H, An F, Huang Q, Kong Y, He D, Chen L, et al. Metabolic effect of AOS-iron in rats with iron deficiency anemia using LC-MS/MS based metabolomics. Food Research International 2020;130. https://doi.org/10.1016/j.foodres.2019.108913.

[109] Mirani B, Pagan E, Currie B, Siddiqui MA, Hosseinzadeh R, Mostafalu P, et al. An Advanced Multifunctional Hydrogel-Based Dressing for Wound Monitoring and Drug Delivery. Advanced Healthcare Materials 2017;6. https://doi.org/10.1002/adhm.201700718.

[110] Farahani M, Shafiee A. Wound Healing: From Passive to Smart Dressings. Advanced Healthcare Materials 2021;10. https://doi.org/10.1002/adhm.202100477.

[111] Liang Y, Liang Y, Zhang H, Guo B. Antibacterial biomaterials for skin wound dressing. Asian Journal of Pharmaceutical Sciences 2022. https://doi.org/10.1016/J.AJPS.2022.01.001.

[112] Aboalasaad ARR, Sirková BK. Analysis and prediction of woven compression bandages properties. Journal of the Textile Institute 2019;110. https://doi.org/10.1080/00405000.2018.1540284.

[113] Aboalasaad ARR, Skenderi Z, Brigita Kolčavová S, Khalil AAS. Analysis of Factors Affecting Thermal Comfort Properties of Woven Compression Bandages. Autex Research Journal 2020;20. https://doi.org/10.2478/aut-2019-0028.

[114] Chassagne F, Benoist E, Badel P, Convert R, Schacher L, Molimard J. Characterization of Fabric-to-Fabric Friction: Application to Medical Compression Bandages. Autex Research Journal 2020;20. https://doi.org/10.2478/aut-2019-0050.

[115] Aboalasaad ARR, Sirková BK, Goncu-Berk G. Enhancement of muscle's activity by woven compression bandages. Industria Textila 2021;72. https://doi.org/10.35530/IT.072.04.1789.

[116] Wiseman J, Ware RS, Simons M, McPhail S, Kimble R, Dotta A, et al. Effectiveness of topical silicone gel and pressure garment therapy for burn scar prevention and management in children: a randomized controlled trial. Clinical Rehabilitation 2020;34. https://doi.org/10.1177/0269215519877516.

[117] Ford J. Rapid Review of the efficacy of pressure garment therapy for the treatment of hypertrophic burn scars. https://WwwMedidexCom/Evidence-Based-Procurement-Board-Ebpb/865-Pressure-Garments-in-ScarsHtml 2017.

[118]Ališauskiene D, Mikučioniene D. Investigation on alteration of compression of knitted orthopaedic supports during exploitation. Medziagotyra 2012;18. https://doi.org/10.5755/j01.ms.18.4.3097.

[119] Marqués-Jiménez D, Calleja-González J, Arratibel I, Delextrat A, Terrados N. Are compression garments effective for the recovery of exercise-induced muscle damage? A systematic review with meta-analysis. Physiology & Behavior 2016;153:133–48. https://doi.org/10.1016/J.PHYSBEH.2015.10.027.

[120] Dennis C, Sethu S, Nayak S, Mohan L, Morsi Y, Manivasagam G. Suture materials – Current and emerging trends. Journal of Biomedical Materials Research – Part A 2016;104. https://doi.org/10.1002/jbm.a.35683.

[121] van Niekerk JM, Vos MC, Stein A, Braakman-Jansen LMA, Voor In 't Holt AF, van Gemert-pijnen JEWC. Risk factors for surgical site infections using a data-driven approach. PLoS ONE 2020;15. https://doi.org/10.1371/journal.pone.0240995.

[122] Shao K, Han B, Gao J, Jiang Z, Liu W, Liu W, et al. Fabrication and feasibility study of an absorbable diacetyl chitin surgical suture for wound healing. Journal of Biomedical Materials Research – Part B Applied Biomaterials 2016;104. https://doi.org/10.1002/jbm.b.33307.

[123] Serrano C, García-Fernández L, Fernández-Blázquez JP, Barbeck M, Ghanaati S, Unger R, et al. Nanostructured medical sutures with antibacterial properties. Biomaterials 2015;52. https://doi.org/10.1016/j.biomaterials.2015.02.039.

[124] Fischer J, Scalzo H, Pokropinski Jr H. Method of Making a Packaged Antimicrobial Suture. US8112973B2, 2012.

[125] Chen X, Hou D, Tang X, Wang L. Quantitative physical and handling characteristics of novel antibacterial braided silk suture materials. Journal of the Mechanical Behavior of Biomedical Materials 2015;50. https://doi.org/10.1016/j.jmbbm.2015.06.013.

[126] Mingmalairak C. Antimicrobial sutures: new strategy in surgical site infections. Science Against Microbial Pathogens: Communicating Current Research and Technological Advances 2011.

[127] Rothenburger S, Spangler D, Bhende S, Burkley D. In Vitro Antimicrobial Evaluation of Coated VICRYL* Plus Antibacterial Suture (Coated Polyglactin 910 with Triclosan) using Zone of Inhibition Assays. Surgical Infections 2002;3. https://doi.org/10.1089/sur.2002.3.s1-79.

[128] Melo MN, Ferre R, Castanho MARB. Antimicrobial peptides: Linking partition, activity and high membrane-bound concentrations. Nature Reviews Microbiology 2009;7. https://doi.org/10.1038/nrmicro2095.

[129] Zasloff M. Antimicrobial peptides of multicellular organisms. Nature 2002;415. https://doi.org/10.1038/415389a.

[130] Obermeier A, Schneider J, Föhr P, Wehner S, Kühn KD, Stemberger A, et al. In vitro evaluation of novel antimicrobial coatings for surgical sutures using octenidine. BMC Microbiology 2015;15. https://doi.org/10.1186/s12866-015-0523-4.

[131] Dubas ST, Wacharanad S, Potiyaraj P. Tunning of the antimicrobial activity of surgical sutures coated with silver nanoparticles. Colloids and Surfaces A: Physicochemical and Engineering Aspects 2011;380. https://doi.org/10.1016/j.colsurfa.2011.01.037.

[132] Ho CH, Odermatt EK, Berndt I, Tiller JC. Long-term active antimicrobial coatings for surgical sutures based on silver nanoparticles and hyperbranched polylysine. Journal of Biomaterials Science, Polymer Edition 2013;24. https://doi.org/10.1080/09205063.2013.782803.

[133] Zhang S, Liu X, Wang H, Peng J, Wong KKY. Silver nanoparticle-coated suture effectively reduces inflammation and improves mechanical strength at intestinal anastomosis in mice. Journal of Pediatric Surgery 2014;49. https://doi.org/10.1016/j.jpedsurg.2013.12.012.

[134] Zurita R, Puiggalí J, Rodríguez-Galán A. Loading and release of ibuprofen in multi- and monofilament surgical sutures. Macromolecular Bioscience 2006;6. https://doi.org/10.1002/mabi.200600084.

[135] Gupta B, Jain R, Singh H. Preparation of antimicrobial sutures by preirradiation grafting onto polypropylene monofilament. Polymers for Advanced Technologies 2008;19. https://doi.org/10.1002/pat.1146.

[136] Chen X, Hou D, Wang L, Zhang Q, Zou J, Sun G. Antibacterial Surgical Silk Sutures Using a High-Performance Slow-Release Carrier Coating System. ACS Applied Materials and Interfaces 2015;7. https://doi.org/10.1021/acsami.5b06239.

[137] Viju S, Thilagavathi G. Characterization of tetracycline hydrochloride drug incorporated silk sutures. Journal of the Textile Institute 2013;104. https://doi.org/10.1080/00405000.2012.720758.

[138] Weldon CB, Tsui JH, Shankarappa SA, Nguyen VT, Ma M, Anderson DG, et al. Electrospun drug-eluting sutures for local anesthesia. Journal of Controlled Release 2012;161. https://doi.org/10.1016/j.jconrel.2012.05.021.

[139] Morizumi S, Suematsu Y, Gon S, Shimizu T. Inhibition of Neointimal Hyperplasia With a Novel Tacrolimus-Eluting Suture. J Am Coll Cardiol 2011;58. https://doi.org/10.1016/j.jacc.2011.02.062.

[140] Lee JE, Park S, Park M, Kim MH, Park CG, Lee SH, et al. Surgical suture assembled with polymeric drug-delivery sheet for sustained, local pain relief. Acta Biomaterialia 2013;9. https://doi.org/10.1016/j.actbio.2013.06.003.

[141] Guyette JP, Fakharzadeh M, Burford EJ, Tao ZW, Pins GD, Rolle MW, et al. A novel suture-based method for efficient transplantation of stem cells. Journal of Biomedical Materials Research – Part A 2013;101 A. https://doi.org/10.1002/jbm.a.34386.

[142] Reckhenrich AK, Kirsch BM, Wahl EA, Schenck TL, Rezaeian F, Harder Y, et al. Surgical sutures filled with adipose-derived stem cells promote wound healing. PLoS ONE 2014;9. https://doi.org/10.1371/journal.pone.0091169.

[143] Yao J, Korotkova T, Riboh J, Chong A, Chang J, Smith RL. Bioactive Sutures for Tendon Repair: Assessment of a Method of Delivering Pluripotential Embryonic Cells. Journal of Hand Surgery 2008;33. https://doi.org/10.1016/j.jhsa.2008.06.010.

[144] Xia Y, He Y, Zhang F, Liu Y, Leng J. A Review of Shape Memory Polymers and Composites: Mechanisms, Materials, and Applications. Advanced Materials 2021;33. https://doi.org/10.1002/adma.202000713.

[145] Lendlein A, Langer R. Biodegradable, elastic shape-memory polymers for potential biomedical applications. Science (1979) 2002;296. https://doi.org/10.1126/science.1066102.

[146] Lambertz A, Vogels RRM, Busch D, Schuster P, Övel SJ, Neumann UP, et al. Laparotomy closure using an elastic suture: A promising approach. Journal of Biomedical Materials Research – Part B Applied Biomaterials 2015;103. https://doi.org/10.1002/jbm.b.33222.

[147] Kim DH, Wang S, Keum H, Ghaffari R, Kim YS, Tao H, et al. Thin, flexible sensors and actuators as "instrumented" surgical sutures for targeted wound monitoring and therapy. Small 2012;8. https://doi.org/10.1002/smll.201200933.

[148] Phan PT, Hoang TT, Thai MT, Low H, Davies J, Lovell NH, et al. Smart surgical sutures using soft artificial muscles. Scientific Reports 2021;11. https://doi.org/10.1038/S41598-021-01910-2.

[149] Kalidasan V, Yang X, Xiong Z, Li RR, Yao H, Godaba H, et al. Wirelessly operated bioelectronic sutures for the monitoring of deep surgical wounds. Nature Biomedical Engineering 2021;5. https://doi.org/10.1038/s41551-021-00802-0.

[150] Houshyar S, Bhattacharyya A, Khalid A, Rifai A, Dekiwadia C, Kumar GS, et al. Multifunctional Sutures with Temperature Sensing and Infection Control. Macromolecular Bioscience 2021;21. https://doi.org/10.1002/mabi.202000364.

Chetna Verma, Manali Somani, Ankita Sharma, Pratibha Singh,
Surabhi Singh, Shamayita Patra, Mukesh Kumar Singh,
Samrat Mukhopadhyay, Bhuvanesh Gupta

14 Design and development of chitosan-based textiles for biomedical applications

Abstract: Chitosan (CS) is considered to be one of the most abundant and prominent biopolymers, exhibiting excellent physicochemical properties such as hydrogel nature as well as anti-microbial and haemostatic nature. Moreover, its inherent properties make it suitable for wound healing applications and designing anti-microbial polymeric systems. CS can also be combined with other polymers to enhance its performance in wound care. Further, it may be immobilized on textile support for the development of bioactive surfaces for numerous biomedical applications. The haemostatic nature of the CS-based fabric also helps in designing materials for controlling excessive bleeding in patients. This chapter is focused on utilizing functional CS to yield more effective therapies and vastly extend human health span.

Keywords: Polysaccharides, bioactive surfaces, immobilization, fabrication, infection-resistant dressings, asymmetric membranes

14.1 Introduction

A sequence of junctures to revitalize mis-phased tissue to preserve the coherence is involved in complex biological activities of wound alleviation [1–4]. The prime wound healing needs are genuine exudates management, ideal spreading of gases, moisture-laden wound bed, and minimum bacterial invasion [5, 6]. Delay in the wound healing process is attributed to vascular inefficiency, infection, inflammation for a prolonged period, or excess exudates formation leading to maceration of skin tissue at the wound site [7, 8]. Therefore, the fundamental function of wound management is to promote rapid wound healing to obtain both functional and cosmetic results [9]. However, their use as an infection control device is relatively a new concept [10]. The innovations in wound care dressings have become the focal point of interest [11, 12]. The key function of a wound dressing is to protect against microbial infection, absorb excess wound exudates, and allow proper permeation of gases across the matrix for enhancing wound healing.

Chitosan (CS) is a highly recommended natural polysaccharides, which is found an appropriate carrier of drugs, moisture attracting finishes, and wound treating processes. CS is derived by partial de-acetylation of chitin, a compound formed by 2-acetamido-2-deoxy-β-D glucose by 1,4-beta linkage. Chitin and CS have the same backbone as cellulose. The only difference is that chitin has an acetamido group at

https://doi.org/10.1515/9783110759747-014

the C-2 position, and CS consists amino group at C2 (Figure 14.1). CS has found special recognition in wound care due to its exceptional mucoadhesive character in the swollen state that assists in adherence in epithelial cells. For the last few decades, CS has been extensively exploited for biomedical applications, including wound dressings, for its ability to form a hydrogel. The hydrophilic nature of CS accommodates the effective wound exudates content. The outstanding features of CS, like anti-microbial potential, biocompatibility, environment stimuli-derived degradability, wound healing capability, bring down scare formation, and a soak-up substantial quantity of wound exudates, attracted numerous research organizations of both sides of the Atlantic to work on CS-filled wound dressings [13–15].

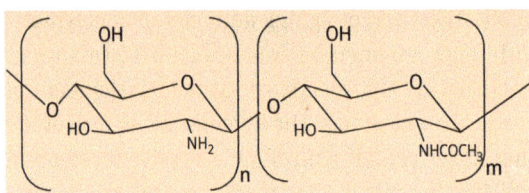

Figure 14.1: Chemical structure of chitosan.

14.2 Chitosan-based wound care systems

CS offers excellent wound healing features. The material, however, is slightly rigid in its dry state, which makes it somewhat uncomfortable in contact with tissues. The blending of an appropriate polymer with CS provides an alternative to introduce desired properties into the material. Several polymers have been incorporated into CS to achieve superabsorbent behaviour in the blended material [16–20].

CS has also been blended with cellulose to fabricate wound dressing and evaluated for its mechanical and anti-bacterial activities [16]. Cellulose and CS may be blended in different proportions and the films evaluated for their characteristics such as hydrophilicity and water vapour transmission rate.

CS blends with different polyethylene glycols (PEGs) have been reported to develop membranes for wound dressing [21]. PEG is known for its protein resistance, low toxicity, immunogenicity, and high biocompatibility, making the PEG an appropriate substance to be used with CS in the form of a binary biological membrane. PEG: CS composites are beneficial for improving the biological advantages during wound curing. The blending of CS with PEG presents an advantageous effect on biocompatibility with the bit of compromising impact on the mechanical strength of the membrane of a few micron pore sizes [22, 23]. CS: PEG membrane's-controlled dissolution is favourable to creating multi-phase morphology. The ratio of PEG with CS, the molecular weight of PEG, and the degree of cross-linking influence the mechanical stability, swelling, and pore size distribution in the composite membrane. The swelling behaviour of these membranes was found to be pH-dependent. A couple of synthetic polymer was blended with CS to design microporous membranes [23]. The composite structure is subsequently made porous by the coagulation process so that a dense layer is formed on top of a highly porous structure. The reinforced fabrics are found suitable to develop strong dressing materials. The tetracycline (TC)-loaded CS dressings offer controlled release for 48 h, which proves the potential of dressings as excellent materials for wound care.

The interaction of PEG with CS may lead to the development of a flexible matrix due to the plasticization effect of PEG. The behaviour of porous morphology with the addition of –20 in the blend reflects the limited interaction of these two components (Figure 14.2).

The porosity of composite membranes is obtained by freeze-drying of CS–PEG-coated fabric. Cotton offers the support layer for the CS hydrogel and provides space for the imbibition of the water in a system. The porosity increases as the PEG-20 content in membranes rises. The porosity varies in the range of 54–70%. The increase in porosity may be ascribed to the low level of interaction of PEG with CS [24].

Thus, the PEG chains tend to segregate within the CS matrix and may push apart the PEG domains from within the matrix and increase the specific volume of the composite membrane. As a result, the PEG chains create segregation inside the CS matrix, which may increase the separation between the PEG domain and CS matrices. Consequently, a high specific volume composite membrane is developed, which is proved

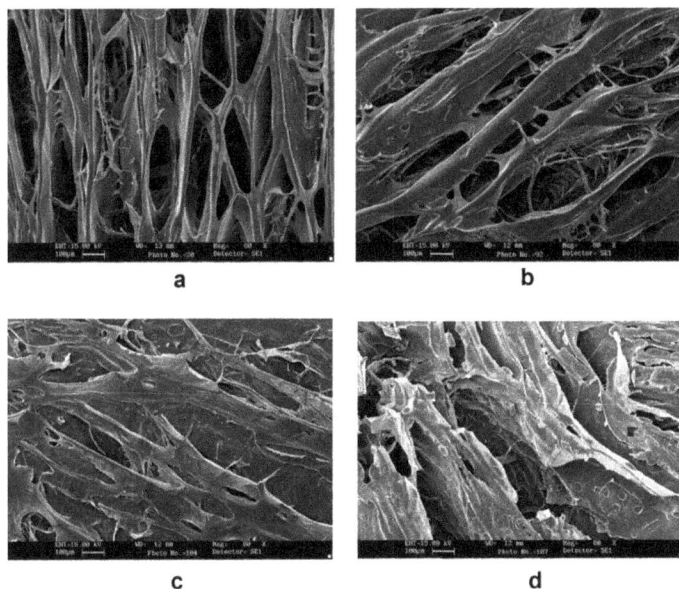

Figure 14.2: Scanning electron microscope photographs coated cotton fabrics: (a) CS-coated membrane; and CPC membranes with (b) 10% PEG-20; (c) 30% PEG-20; (d) 50% PEG-20 (reproduced with kind permission from [24], Copyright 2009, Wiley).

by decreasing membrane density by increasing PEG-20 content [17]. A combination of CS–PEG with polyvinylpyrrolidone (PVP) was searched due to the better biocompatibility of the PVP fraction. However, the PVP addition leads to morphological changes on the hydrogel-coated membranes [18].

Excellent healing has been observed in the CS–PEG–PVP system [18]. The healing of the CS-coated dressing on the mouse was followed for 21 days (Figure 14.3). It was observed that the healing in CPPC–TC was fast and exhibited excellent scar-preventive nature.

The sustained drug release was found up to 504 h, variously treated dressing samples registered remarkable variation in the degree of wound contraction among the selected wounds, and wound healing was significantly considerable by 288 h. On the 21st day, the wounds in all the groups had achieved complete healing. One scar was still observed on the animal's back. The scar dimension measurement data revealed that the animal group that did not receive any wound care treatment had the largest scar size (10.67%) [18].

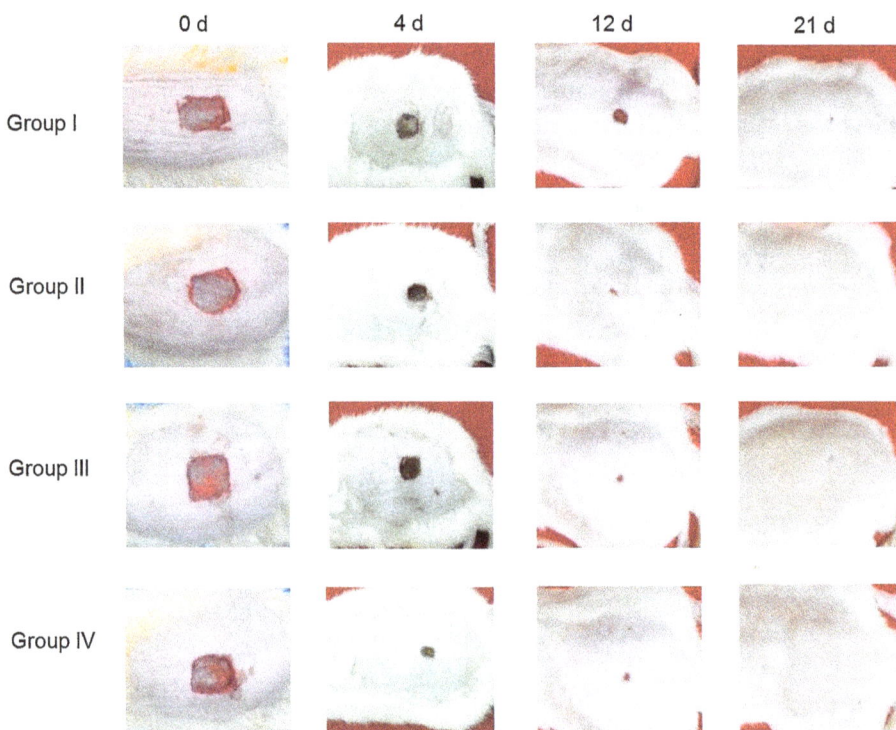

Figure 14.3: Macroscopic surfacing of wound healing wrapped with (1) (Group I): cotton gauze (2) (Group I): CS dressing, (3) (Group III): CPPC dressing, and (4) (Group IV): CPPC–TC dressing at the different healing period: (a) 0 days, (b) 4 days, (c) 12 days, and (d) 21 days (reproduced with kind permission from [18], Copyright 2016, Elsevier).

Several other combinations have also been tried where CS has been used as one of the essential components for wound healing. Dextran-based (DN) composites have been manufactured by blending CS: glycerol (G) in different compositions. Furthermore, aloe vera (AV) and manuka honey (MH) were also found to assist wound healing. Out of these combinations, AV-based dressing displayed the most effective wound healing. The zone of inhibition technique optimized aloe vera MH for anti-bacterial activity at different concentrations [19, 20]. The DNG/Ch/AV and DNG/Ch/MH bio–nano composite dressings were evaluated for their performance against *Staphylococcus aureus* and *Escherichia coli* bacteria. The anti-microbial potential of aloe vera-based wound dressings was assessed by a zone of inhibition as illustrated in Figure 14.4A (a). The zone of inhibition was enhanced 10–15 mm by an increase in aloe vera content from 10% to 20% for Gram-positive (*S. aureus*) 7–10 mm for Gram-negative *E. coli bacteria,* which further increased from 17 to 20 mm by an enhancement in MH application from 10% to 20% for same Gram-positive and Gram-negative bacteria. The zone of inhibition was reduced 5% in the case of MH concentration higher than 20% in the case of *S. aureus* and 40% against *E. coli* which indicates a significant decline in anti-bacterial activity. The optimum quantity of MH was found to be 20% to get the maximum zone of inhibition against both Gram-positive and Gram-negative bacteria [19, 20].

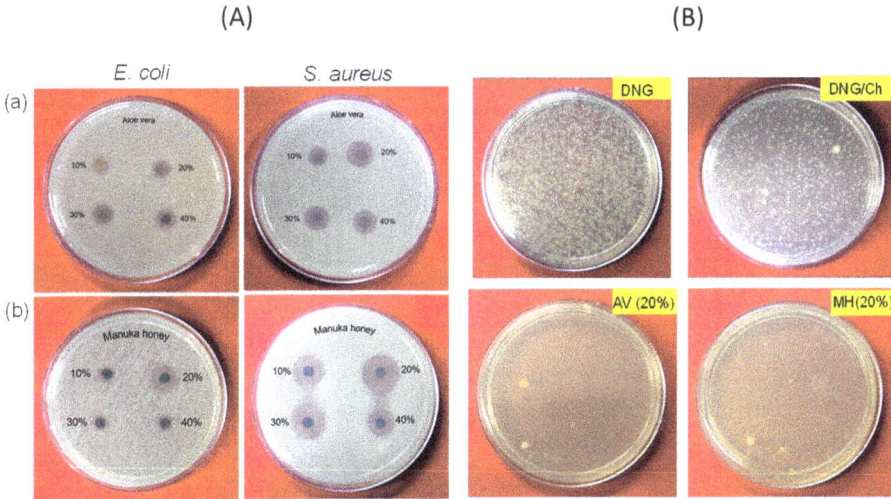

Figure 14.4: A: (a) Anti-bacterial activity of Dextran: nanoclay:glycerol:chitosan:aloe vera (DNGCSAV) dressings by zone of inhibition against Gram-negative and Gram-positive bacteria; (b) anti-bacterial activity of dextran:nanoclay:glycerol:chitosan:manuka honey (DNGCSMH) dressings by zone of inhibition against Gram-negative and Gram-positive bacteria. B: Anti-bacterial activity of DNGCSAV and DNGCSMH dressings to DNG and DNG:CS dressings by CFU approach (reproduced with permission from [19], Copyright 2018, Elsevier).

CO and sandalwood oil (SO) have been explored to develop infection-resistant dressings that offer good wound healing [20]. Vastly promising observations have been reported in these systems compared to systems without CO and SO. The material unveiled excellent bioactivity and did not allow bacterial adhesion on its surface (Figure 14.5).

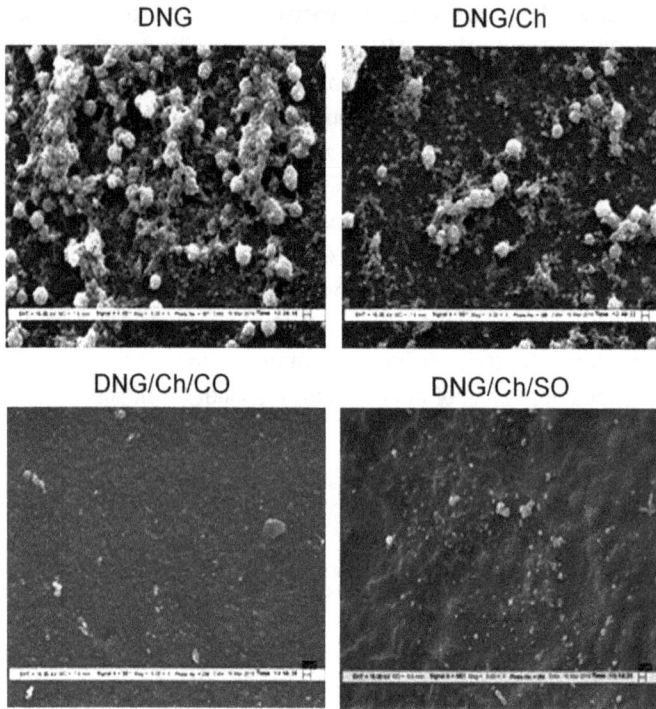

Figure 14.5: SEM photomicrographs exhibiting the intensity of *S. aureus* cohesion against various nano biocomposite wound dressings (reproduced with kind permission from [20], Copyright 2018, Elsevier).

In another study, a pH-sensitive semi-interpenetrating network of CS and PVP blends has been developed for controlled drug delivery [25]. CS and PVP were cross-linked using glutaraldehyde and were incorporated with amoxicillin. The hydrogel was freeze-dried, leading to a porous matrix. The freeze-dried porous samples were evaluated in terms of their pH-dependent swelling and were superior to the air-dried samples. It was found that freeze-dried hydrogel membranes could release around 73% of the drug incorporated into the membrane. In another study, CS and gelatin have been blended to fabricate an artificial skin suitable for skin tissue engineering [26]. Here, asymmetric membrane structure to the scaffold was provided using the freezing and lyophilization method. The dressing was further investigated for its water uptake ability and in vitro fibroblast compared to CS alone. The CS-based systems have been extended to the alginate as the additive for the wound dressings [27]. The calcium alginate shows superiority over other blended components because of its natural haemostatic nature and gel-forming ability, which helps remove the dressing from the wound. A feature like this is essential and can be attributed as a sign of comfortable dressing.

Aloe vera: Curcumin: oxidized pectin gelatin formulation was adhered to the cotton non-woven sheet to develop an anti-microbial wound care dressing.

Rapid wound healing was observed in just eight days by AV-loaded (OP-gel) dressings. Furthermore, decisive anti-inflammatory action and diminished scar formation were exerted due to aloe vera. Histology analysis revealed neovascularization and an ordered collagen deposition along with nuclei migration [28]. Nanocomposite films using castor oil as the matrix material and filled with CS-modified ZnO nanoparticles were prepared as anti-bacterial wound dressings. Films exhibited anti-bacterial activity against *E. coli*, *S. aureus*, and *Micrococcus luteus* (*M. luteus*) bacteria [29].

The physical and anti-microbial properties of *Cassia angustifolia* and *Tamarindus indica* with CS solution treated cotton dressing were found to improve [30]. Some essential oils were mixed with nano CS (40–80 nm particle size) to be treated in the wound care dressing by double-step method. The essential oil release profile was found dependent on the concentration of components [31].

Bioplastics for functional packaging were developed by loading eugenol: CS nanoparticles (particle size less than 100 nm). The active packing was found thermally stable with anti-oxidant functionality. Similarly, essential oil (*Lippia sidiodes) was micro*encapsulated inside the alginate: cashew gum by spray-drying technique to achieve fungicide and bactericide potential [32].

Thymol: zein nanoparticles composite stabilized with CS hydrochloride and caseinate were effective in controlling Gram-positive bacterial growth [33]. Similarly, an anti-microbial and anti-oxidant wound care dressing was developed by embedding the gelatin films with thymol [34]. Nanosilver is an excellent anti-microbial agent in many matrices [35, 36].

Carboxymethyl cellulose (CMC)/Ag/curcumin nanocomposite dressing was fabricated where CMC/Ag film exhibited a synergistic effect in the anti-bacterial action against *E. coli* in the presence of curcumin [37]. Wound dressings based on honey are used for healing and reducing the odour of abscesses, diabetic foot ulcers, and leg ulcers because of their anti-bacterial activity. Honey, PVA, CS nanofibres (HPCS)-based nanofibrous wound dressing was developed for anti-bacterial activity against *S. aureus*, *E. coli*, MRSA, and *P. aeruginosa* and compared with commercial dressing Aquacel Ag. Enhanced wound closure, cell viability, and proliferation compared to control were observed [38].

The essential oils like cinnamon, lemongrass, and peppermint were encapsulated with cellulose-containing nanospun membranes to impart a sufficient zone of inhibition against Gram-negative *E. coli* bacteria. Additionally, the nanofibrous wound care composite was found biocompatible for skin cells [39]. Functional modification of CS by graft polymerization of the specific monomer is another way to develop better performing materials. In one of the studies, acrylic acid (AAc) and HEMA were graft copolymerized onto CS to yield CS-based membranes with efficient wound healing and drug delivery system [40].

CS-*g*-2-hydroxyethyl methacrylate formulation was found suitable for delayed delivery of drugs than CS-*g*-AAc to develop anti-microbial hydrogel. CS-*g*-HEMA was found to be a better matrix for drug delivery than CS-*g*-AA as it possesses hydrogel properties.

The addition of HEMA enhanced the cytocompatible, haemocompatible, and thrombogenic potential of wound care dressings. Gamma-ray irradiation was used to create a polypropylene-gamma irradiate non-woven sheet for functional applications. This non-woven sheet was further modified by grafting *N*-isopropyl acrylamide to enhance its functionalities by opting UV-photo-grafting technique. Eventually, CS-loaded polypropylene-*g*-AA-*g*-NIPAAm biograft non-woven sheet was developed by freeze-drying method. The functionalized PP non-woven sheet was established as a wound care dressing with sufficient moisture vapour permeability and anti-microbial potential, which was quite comparable with commercial wound dressings [41]. Thermo-responsive and hydrophilic CS-containing wound care system was developed by polymerizing AAc on PP non-woven fabric (NWF) sheet by direct current oxygen plasma. The union of poly (*N*-isopropyl acrylamide) with CS is followed by coupling agent carbodiimide to develop PP-*g*-CS-*g*-PNIPAAm composite for wound care management [42]. The effect of graft composition on the surface profile and porous structure of PP NWFs was analysed using techniques such as scanning electron microscopy and electron spectroscopy for chemical analysis. The presence of CS on NWF enhanced the water holding potential (Table 14.1), which is essential for wound dressing and proved by the assistance of *the Sprague–Dawley rat* model. The water-retaining potential of NWF was found better in the case of PP-*g*-CS-*g*-PNIPAAm than CS alone. The wound area filling and wound healing speed were excellent with PP-*g*-CS-*g*-PNIPAAm, and healing was completed at 408 h.

Table 14.1: The water content of NWFs.

Samples	Water content (%)
Original NWF	2.7 ± 0.2
Plasma-modified NWF	376 ± 21
AAc-coated NWF	465 ± 15
PP-*g*-CS NWF (20 °C)	447 ± 20
PP-*g*-CS NWF (40 °C)	454 ± 14
PP-*g*-CS-*g*-PNIPAAm NWF (20 °C)	634 ± 15
PP-*g*-CS-*g*-PNIPAAm NWF (40 °C)	550 ± 24

Data are present with mean ± SD (N-8)

Adopted with kind permission from [49], Copyright 2008, Elsevier.

14.3 Chitosan-based haemostatic systems

Haemostasis is a primary systemic response to the uncontrolled bleeding at the injury site. The haemostasis mechanism starts with the employment of several blood components like platelets to seal the puncture site, which activates the prothrombin activator. Further, the prothrombin activator initiates the conversion of prothrombin to thrombin with the help of calcium ions, which assists in the fibrin clotting. The natural coagulation process is inadequate to manage the haemorrhagic condition, which implies the requirement of excellent haemostatic material. Thus, the fabrication of haemostatic materials has been a topic of research interest since ancient times.

CS has been known as an excellent haemostatic material and has acquired worldwide prominence as a commercial product in excessive bleeding controls [43]. The haemostatic nature of CS accelerates blood clotting by electrostatic interaction between the cationic amine group of chitosan and anion surface of blood cells; consequently, the accumulation of red blood cells (RBCs) reduces the bleeding immediately [44, 45].

The association of haemostatic materials with CS developed smart anti-microbial and scar-free wound dressings [46–48].

Polysaccharide-modified CS foam sponges were manufactured through Schiff cross-linking reaction linking oxidized dialdehyde cellulose and chitosan for haemostatic purposes. The performance of these haemostatic sponges was evaluated on rabbit femoral artery and mouse tail vein, which was found successful and reveals that cellulose-modified chitosan sponge's synergistic effect in water retention potential, mechanical strength, and reduced haemolysis. The endogenous coagulation route mechanism of haemostatic through adhering or activating the RBCs or platelet to promote blood clotting can be utilized for civilian and armed forces in daily life injuries.

The agglomeration of chitosan on PP NWF with a polyphosphate was found favourable to enhance the haemostatic potential of these sponges. The haemostasis performance of these sponges was monitored by contacting it with human blood. The reduced clotting rates were recorded due to the higher absorbance potential of haemoglobin than dressing material. The addition of PP65 or PP45 by 6.7% w/w or 10% w/w, respectively, with chitosan reduces the absorbance values than pure chitosan as shown in Figure 14.6B. Further increase in chitosan loading on PP sponges up to 15% did not enhance blood clotting rates significantly.

The blood clotting rates on wound care gauzes were significantly slower than chitosan ($p < 0.05$). However, the absorbance value in the absence of clotting supporting gauzes was significantly higher than chitosan-based materials ($p < 0.001$) but not gauze [49].

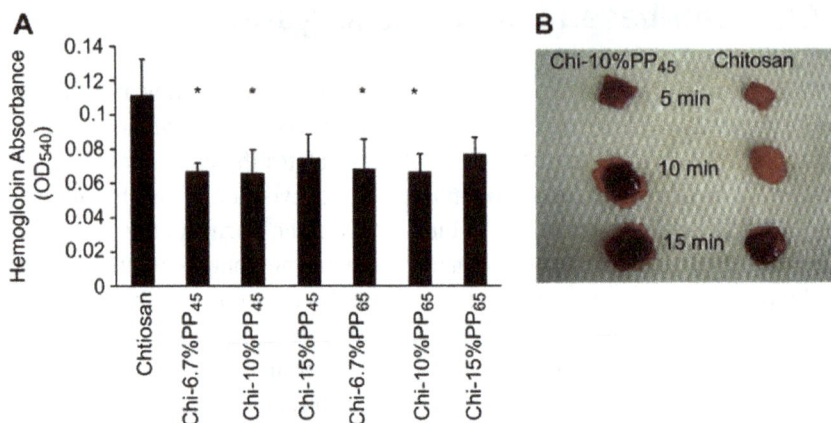

Figure 14.6: (A) Effect of the ratio between PP and chitosan in blood-clotting sponges on blood-clotting rates, evaluated by absorbance of haemoglobin from lysed un-coagulated RBCs. The $p < 0.05$ compared to chitosan is analysed by one-way ANOVA with post hoc Scheffe test, n ¼ 4. (B) Photograph proving rapid blood clotting on Chi-10%PP45 than pure chitosan (reproduced with kind permission from [49], Copyright 2008, Elsevier).

14.4 Chitosan immobilization for anti-microbial systems

The plasma functionalization of PP has been performed to increase surface functionality so that wettability, dyeability, and reactivity with other organic molecules may be enhanced [45]. Although functionality such as hydroxyl, amino, amide, and carboxyl may be created on the PP surface, it is more interesting to have carboxyl functionality on the PP surface so that a covalent bonding with chitosan may be accomplished. Bratskaya et al. [50] carried out oxygen plasma exposure of PP to create functional groups on the surface. The CS was subsequently coated on the surface, and the bonding between carboxyl groups on the PP surface and amino groups from CS was carried out by thermal treatment at more than 80 °C. It was found that around 47% of CS was bonded to the surface and the rest leached out during washing.

Similarly, Elsabee et al. [51] treated the PP films with corona discharge followed by the coating of CS and its derivative to incorporate anti-fungal and anti-bacterial properties to PP films. The chitosan and its derivatives were effective biocidal when becoming part of biofilms.

CS immobilization has a significant contribution in developing anti-microbial sutures. The process involves the grafting of AAc and subsequent CS immobilization [52]. The grafting led to the formation of the PP structure with PAA side chains along its backbone. The carboxyl groups offered sites for the binding of CS molecules by EDC coupling (Figure 14.7). This composite structure was an excellent biomaterial with good biocompatibility and anti-microbial nature.

PVP: CS membrane was electrospun by keeping 6:4 ratios. FTIR and Raman spectroscopy proved the hydrogen bonding between CS and PVP. A tiny shift of 531 eV was detected in the amide of acetylated groups. Fluorouracil was integrated into PVP: CS membrane and the resultant membrane were found effective in damaging A549 alveolar basal epithelial cells in human beings [53]. Chitosan:polyvinyl alcohol (CS:PVA) nanofibres containing chitosan carboxymethyl nanoparticles were effective in healing promotion to treat mouse skin wounds [54].

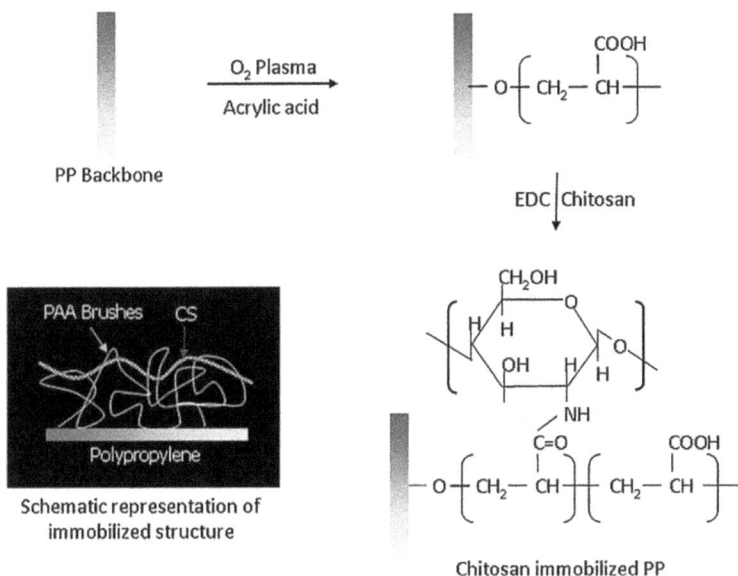

Figure 14.7: Schematic representation of the suture development process by plasma grafting and chitosan immobilization (reproduced with permission from [52], Copyright 2011, Wiley).

14.5 Conclusion

CS is an interesting biopolymer that has significantly benefited the human healthcare sector by fostering the development of advanced biomedical textiles with exciting properties – attracting features of chitosan, such as anti-microbial nature and haemostatic features display tremendous potential in wound care systems. The polymer has been exploited into dressings by meticulously modifying it using different biomaterials with a subsequent coating variety of fabrics. The fabric acts as the support where the coated matrix helps in the healing process. Alternatively, CS may be immobilized onto polymer support followed by the drug incorporation that can sustain a drug's release from its matrix. Nano offers numerous possibilities in fabricating-based dressing and opens up enormous possibilities to design dressings for hastening healing and scar prevention. The herbal composition in a CS-based system is another powerful innovation, and efforts have been directed to combine the therapeutic behaviour of the herbs and the anti-microbial nature of the nanosystems.

References

[1] Mogoşanu GD, Grumezescu AM. Natural and synthetic polymers for wounds and burns dressing. Int J Pharm 2014;463:127–36. https://doi.org/10.1016/j.ijpharm.2013.12.015.

[2] Mayet N, Choonara YE, Kumar P, Tomar LK, Tyagi C, Du Toit LC, et al. A comprehensive review of advanced biopolymeric wound healing systems. J Pharm Sci 2014;103:2211–30. https://doi.org/10.1002/jps.24068.

[3] Babu RP, O'Connor K, Seeram R. Current progress on bio-based polymers and their future trends. Prog Biomater 2013;2:8. https://doi.org/10.1186/2194-0517-2-8.

[4] Vasconcelos A, Pêgo AP, Henriques L, Lamghari M, Cavaco-Paulo A. Protein matrices for improved wound healing: Elastase inhibition by a synthetic peptide model. Biomacromolecules 2010;11:2213–20. https://doi.org/10.1021/bm100537b.

[5] Nardini JT, Chapnick DA, Liu X, Bortz DM. Modeling keratinocyte wound healing dynamics: Cell-cell adhesion promotes sustained collective migration. J Theor Biol 2016;400:103–17. https://doi.org/10.1016/j.jtbi.2016.04.015.

[6] Morgado PI, Aguiar-Ricardo A, Correia IJ. Asymmetric membranes as ideal wound dressings: An overview on production methods, structure, properties, and performance relationship. J Memb Sci 2015;490:139–51. https://doi.org/10.1016/j.memsci.2015.04.064.

[7] Brook I. Microbiology and antimicrobial management of sinusitis. J Laryngol Otol 2005;119:251–8. https://doi.org/10.1258/0022215054020304.

[8] Bowler PG, Davies BJ. The microbiology of infected and noninfected leg ulcers. Int J Dermatol 1999;38:573–8. https://doi.org/10.1046/j.1365-4362.1999.00738.x.

[9] Lin, Shan-Yang; Chen, KO-shao; Run-Chu L. Design and evaluation of drug-loaded wound dressing having thermoresponsive, adhesive, absorptive and easy peeling properties. Biomaterials 2001:2999–3004. https://doi.org/10.1016/S0142-9612(01)00046-1

[10] Walker, M.; Hobot, J. A.; Newman, G.R.; Bowler PG. Scanning electron microscopic examination of bacterial immobilisation in carboxymethyl cellulose (AQUACEL) and alginate dressing. Biomaterials 2003:883–90. https://doi.org/10.1016/S0142-9612(02)00414-3

[11] Hasibuan, Poppy Anjelisa Zaitun; Yuandani; Tanjung, Masitta; Gea, Saharman; Pasaribu, Khatarina Meldawati; Harahap, Mahyuni; Perangin-Angin, Yurika Almanda; Prayoga, Andre; Ginting JG. Antimicrobial and antihemolytic properties of a CNF_AgNP-chitosan film_ A potential wound dressing material. Heliyon 2021:e08197. https://doi.org/10.1016/j.heliyon.2021.e08197

[12] Jayakumar R, Prabaharan M, Sudheesh Kumar PT, Nair S V., Tamura H. Biomaterials based on chitin and chitosan in wound dressing applications. Biotechnol Adv 2011;29:322–37. https://doi.org/10.1016/j.biotechadv.2011.01.005.

[13] Peh K, Khan T, Ch'ng H. Mechanical, bioadhesive strength and biological evaluations of chitosan films for wound dressing. J Pharm Pharm Sci 2000;3:303–11.

[14] Khan TA, Peh KK. A preliminary investigation of chitosan film as dressing for punch biopsy wounds in rats. J Pharm Pharm Sci 2003;6:20–6.

[15] Shahid-ul-Islam, Butola BS. Recent advances in chitosan polysaccharide and its derivatives in antimicrobial modification of textile materials. Int J Biol Macromol 2019;121:905–12. https://doi.org/10.1016/j.ijbiomac.2018.10.102.

[16] Wu, Yu-Bey; Yu, Shu-Huei; Mi, Fwu-Long; Wu, Chung-Wei; Shyu, Shin-Shing; Peng, Chih-Kang; Chao A-C. Preparation and characterization on mechanical and antimicrobial properties of chitosan/cellulose blends Carbohydr. Polym. 2004:435–40. https://doi.org/10.1016/j.carbpol.2004.05.013

[17] Gupta B, Saxena S, Arora A, Alam MS. Chitosan-polyethylene glycol coated cotton membranes for wound dressings. Indian J Fibre Text Res 2011;36:227–80.

[18] Anjum S, Arora A, Alam MS, Gupta B. Development of antimicrobial and scar preventive chitosan hydrogel wound dressings. Int J Pharm 2016;508:92–101. https://doi.org/10.1016/j.ijpharm.2016.05.013.

[19] Singh S, Gupta A, Gupta B. Scar free healing mediated by the release of aloe vera and manuka honey from dextran bionanocomposite wound dressings. Int J Biol Macromol 2018;120:1581–90. https://doi.org/10.1016/j.ijbiomac.2018.09.124.

[20] Singh S, Gupta A, Sharma D, Gupta B. Dextran based herbal nanobiocomposite membranes for scar free wound healing. Int J Biol Macromol 2018;113:227–39. https://doi.org/10.1016/j.ijbiomac.2018.02.097.

[21] Zhang, M.; Li, X.H.; Gong, Y.D.; Zhao, N.M.; Zhang XF. Properties and Biocompatibility of chitosan films modified by blending with PEG. Biomaterials 2002:2641–8. https://doi.org/10.1016/S0142-9612(01)00403-3

[22] Zeng, Minfeng; Fang Z. Preparation of sub-micrometer porous membrane from chitosan/polyethylene glycol semi-IPN. J. Membr. Sci. 2004:95–102. https://doi.org/10.1016/j.memsci.2004.08.004

[23] Zeng, Minfeng; Fang, Zhengping; Xu C. Effect of compatibility on the structure of the microporous membrane prepared by selective dissolution of chitosan/synthetic polymer blend membrane. J. Membr. Sci. 2004:175–81. https://doi.org/10.1016/j.memsci.2003.11.020

[24] Gupta B, Arora A, Saxena S, Alam MS. Preparation of chitosan-polyethylene glycol coated cotton membranes for wound dressings: Preparation and characterization. Polym Adv Technol 2009;20:58–65. https://doi.org/10.1002/pat.1280.

[25] Risbud, Makrand V.; Hardikar, Ananswardhan A.; Bhat, Sujata V.; Bhonde RR. pH-sensitive freeze-dried chitosan-polyvinyl pyrrolidone hydrogel as controlled release system for antibiotic delivery. J. Control. Release 2000:23–30. https://doi.org/10.1016/S0168-3659(00)00208-X

[26] Mao J, Zhao L, De Yao K, Shang Q, Yang G, Cao Y. Study of novel chitosan-gelatin artificial skin in vitro. J Biomed Mater Res – Part A 2003;64:301–8. https://doi.org/10.1002/jbm.a.10223.

[27] Paul W, Sharma CP. Chitosan and Alginate Wound Dressings: A Short Review. Trends Biomaterials Artif Organs 2004;18:18–23.

[28] Tummalapalli M, Berthet M, Verrier B, Deopura BL, Alam MS, Gupta B. Composite wound dressings of pectin and gelatin with aloe vera and curcumin as bioactive agents. Int J Biol Macromol 2016;82:104–13. https://doi.org/10.1016/j.ijbiomac.2015.10.087.

[29] Díez-Pascual AM, Díez-Vicente AL. Wound healing bionanocomposites based on castor oil polymeric films reinforced with chitosan-modified ZnO nanoparticles. Biomacromolecules 2015;16:2631–44. https://doi.org/10.1021/acs.biomac.5b00447.

[30] Chandrasekar S, Vijayakumar S, Rajendran R. Functional finishing of health care cotton for enhanced efficiency of antibacterial activity by chitosan and herbal nanocomposites. Shengtai Xuebao/ Acta Ecol Sin 2020;40:473–7. https://doi.org/10.1016/J.CHNAES.2020.08.004.

[31] Hosseini SF, Zandi M, Rezaei M, Farahmandghavi F. Two-step method for encapsulation of oregano essential oil in chitosan nanoparticles: Preparation, characterization and in vitro release study. Carbohydr Polym 2013;95:50–6. https://doi.org/10.1016/j.carbpol.2013.02.031.

[32] Woranuch S, Yoksan R. Eugenol-loaded chitosan nanoparticles: II. Application in bio-based plastics for active packaging. Carbohydr Polym 2013;96:586–92. https://doi.org/10.1016/j.carbpol.2012.09.099.

[33] Zhang Y, Niu Y, Luo Y, Ge M, Yang T, Yu L, et al. Fabrication, characterization and antimicrobial activities of thymolloaded zein nanoparticles stabilized by sodium caseinate-chitosan hydrochloride double layers. Food Chem 2014;142:269–75. https://doi.org/10.1016/j.foodchem.2013.07.058.

[34] Kavoosi G, Dadfar SMM, Purfard AM. Mechanical, Physical, Antioxidant, and Antimicrobial Properties of Gelatin Films Incorporated with Thymol for Potential Use as Nano Wound Dressing. J Food Sci 2013;78. https://doi.org/10.1111/1750-3841.12015.

[35] Hernández-Rangel, A; Silva-Bermudez; P; Espana-Sanchez, B.L.; Luna-Hernandez E., Almaguer-Flores, A.; Ibarra, C.; Garcia-Perez, V.I.; Velasquillo, C; Luna-Barcenas G. Fabrication and in vitro behavior of dual-function chitosan_silver nanocomposites for potential wound dressing applications Mater. Sci. Eng. 2019:750–65. https://doi.org/10.1016/j.msec.2018.10.012

[36] Yang, Jueying; Chen, YU; Zhao, Lin; Feng, Zhipan; Peng, Kelin; Wei, Ailing; Wang, Yalun; Tong, Zongrui; Cheng B. Preparation of a chitosan/carboxymethyl chitosan/ AgNPs polyelectrolyte composite physical hydrogel with self-healing ability, antimicrobial properties, and good biosafety simultaneously, and its application as a wound dressing Compos. B. Eng. 2020:108139. https://doi.org/10.1016/j.compositesb.2020.108139

[37] Varaprasad K, Vimala K, Ravindra S, Narayana Reddy N, Venkata Subba Reddy G, Mohana Raju K. Fabrication of silver nanocomposite films impregnated with curcumin for superior antibacterial applications. J Mater Sci Mater Med 2011;22:1863–72. https://doi.org/10.1007/s10856-011-4369-5.

[38] Sarhan WA, Azzazy HME. High concentration honey chitosan electrospun nanofibers: Biocompatibility and antibacterial effects. Carbohydr Polym 2015;122:135–43. https://doi.org/10.1016/j.carbpol.2014.12.051.

[39] Liakos I, Rizzello L, Hajiali H, Brunetti V, Carzino R, Pompa PP, et al. Fibrous wound dressings encapsulating essential oils as natural antimicrobial agents. J Mater Chem B 2015;3:1583–9. https://doi.org/10.1039/c4tb01974a.

[40] Santos, K.S.C.R. dos; Coelho, J.F.J.; Ferreira, P; Pinto, I; Lorenzetti, S.G; Ferreira, E.I; Higa, O.Z, Gil M. Synthesis and characterization of membranes obtained by graft copolymerization of 2-hydroxyethyl methacrylate and acrylic acid onto chitosan. Int. J. Pharm. 2006:37–45. https://doi.org/10.1016/j.ijpharm.2005.11.019

[41] Yang, Jen Ming; Lin HaT. Properties of Chitosan containing PP-g-AA-g-NIPAAm bigrafting nonwoven fabric for wound dressing. J. Membr. Sci. 2004:1–7. https://doi.org/10.1016/j.memsci.2004.03.019

[42] Chen JP, Kuo CY, Lee WL. Thermo-responsive wound dressings by grafting chitosan and poly(N-isopropylacrylamide) to plasma-induced graft polymerization modified non-woven fabrics. Appl Surf Sci 2012;262:95–101. https://doi.org/10.1016/j.apsusc.2012.02.106.

[43] Khan MA, Mujahid M. A review on recent advances in chitosan based composite for hemostatic dressings. Int J Biol Macromol 2019;124:138–47. https://doi.org/10.1016/j.ijbiomac.2018.11.045.

[44] Hu Z, Zhang DY, Lu ST, Li PW, Li SD. Chitosan-based composite materials for prospective hemostatic applications. Mar Drugs 2018;16:1–25. https://doi.org/10.3390/md16080273.

[45] Wei X, Ding S, Liu S, Yang K, Cai J, Li F, et al. Polysaccharides-modified chitosan as improved and rapid hemostasis foam sponges. Carbohydr Polym 2021;264. https://doi.org/10.1016/j.carbpol.2021.118028.

[46] Hemamalini T, Vikash N, Brindha P, Abinaya M, Dev VRG. Comparison of acid and water-soluble chitosan doped fibrous cellulose hemostat wet laid nonwoven web for hemorrhage application. Int J Biol Macromol 2020;147:493–8. https://doi.org/10.1016/j.ijbiomac.2020.01.085.

[47] Li J, Wu X, Wu Y, Tang Z, Sun X, Pan M, et al. Porous chitosan microspheres for application as quick in vitro and in vivo hemostat. Mater Sci Eng C 2017;77:411–9. https://doi.org/10.1016/j.msec.2017.03.276.

[48] Ouyang Q, Hou T, Li C, Hu Z, Liang L, Li S, et al. Construction of a composite sponge containing tilapia peptides and chitosan with improved hemostatic performance. Int J Biol Macromol 2019;139:719–29. https://doi.org/10.1016/j.ijbiomac.2019.07.163.

[49] Ong, Shin-Yeu; Wu, Jian; Moochhala, Shabbir M.; Tan, Mui-Hong; Lu J. Development of a chitosan-based wound dressing with improved hemostatic and antimicrobial properties. Biomaterials 2008:4323–32. https://doi.org/10.1016/j.biomaterials.2008.07.034

[50] Bratskaya S, Marinin D, Nitschke M, Pleul D, Schwarz S, Simon F. Polypropylene surface functionalization with chitosan. J Adhes Sci Technol 2004;18:1173–86. https://doi.org/10.1163/1568561041581270.

[51] Elsabee MZ, Abdou ES, Nagy KSA, Eweis M. Surface modification of polypropylene films by chitosan and chitosan/pectin multilayer. Carbohydr Polym 2008;71:187–95. https://doi.org/10.1016/j.carbpol.2007.05.022.

[52] Saxena S, Ray AR, Kapil A, Pavon-Djavid G, Letourneur D, Gupta B, et al. Development of a New Polypropylene-Based Suture: Plasma Grafting, Surface Treatment, Characterization, and Biocompatibility Studies. Macromol Biosci 2011;11:373–82. https://doi.org/10.1002/mabi.201000298.

[53] Grant JJ, Pillai SC, Perova TS, Hehir S, Hinder SJ, McAfee M, Breen A. Electrospun fibres of chitosan/PVP for the effective chemotherapeutic drug delivery of 5-fluorouracil. Chemosensors. 2021; 31;9(4):70. https://doi.org/10.3390/chemosensors9040070.

[54] Dai T, Tanaka M, Huang YY, Hamblin MR. Chitosan preparations for wounds and burns: antimicrobial and wound-healing effects. Expert review of anti-infective therapy. 2011; 1;9(7):857–79. https://doi.org/10.1586/eri.11.59.

Anupam Chowdhury, Srijan Das, Wazed Ali

15 Smart textiles for energy harvesting applications

Abstract: In the yesteryears, textile materials were primarily intended for apparels, home furnishing, and in few technical textile areas. But recently, the unique properties of a textile material owing to its material property, weave patterns, and different fabrication techniques have made it a promising candidate for smart applications. Therefore, smart textile materials can be used to power electronic items demanding energy on a miniscule scale, namely gadgets like electronic watches and smartphones. The drive for renewable and clean energy is the driving wheel behind exploration for areas that do not include energy sources from fossil fuels. Depletion of fossil fuels is an alarming concern, and hence, we should look for such energy sources that can be created from the surrounding environment. It cannot be much better than harvesting the immense potential out of the mechanical energy sources that are generally wasted. This chapter encompasses different concepts underlying the energy harvesting mechanisms of smart textiles. Piezoelectric, thermoelectric, photovoltaic, triboelectric, and piezoelectric nanogenerators are most promising in this segment. Lithium-ion batteries, polymer-based batteries, capacitors, and supercapacitors have catered to growing energy demands for a long time. But issues like portability, life cycle, and performance over a period of time were debatable for a while. Hence, different textile-based supercapacitors, batteries, and solar energy harvesting devices are researched nowadays and are a furore among the material scientists working in these areas. The efficiency and basic working of smart textiles are based on their unique structure and different production methods, which play an important role in upscaling their overall usability in their end use applications. Different potential applications will be one of the highlights of this chapter, along with different challenges and scopes of these segments of smart textiles.

Keywords: Energy harvesting, piezoelectric textiles, triboelectric nanogenerators, photovoltaic textiles, thermoelectric textiles

15.1 Introduction

The energy consumption of today is primarily met by fossil fuels. With ever-increasing energy demand and the environmental damage caused by burning fossil fuels, including the greenhouse effect, air pollution, particulate emission, and the increasing amount of CO_2 released from combustion, the search for alternative energy is ongoing. Harvesting wasted energy can be a viable solution to this problem.

https://doi.org/10.1515/9783110759747-015

Energy harvesting is the conversion of various types of ambient energy present around us into electrical energy. Electrical energy can be used to power and run factories, cities, and large machines, which is generally referred to as mega energy [1]. However, with the advent of the Internet of things which is supposed to be an integral part of social and industrial development, there is also a need to power this system [2]. The problem with powering such a sensor network is that it consists of a large number of sensors and hence would require a large number of batteries to power such sensors. Thus, there is an urgent requirement for energy harvesters that are small in size, can deliver power in a sustainable way, and power such sensors. Such devices are meant to at least partially replace batteries [3]. Batteries can only furnish limited energy, are prone to degeneration, need regular demand replacements, and are not environment friendly.

Energy harvesting for microenergy systems can be done via different routes. Different phenomena and methods have been employed recently for constructing nanogenerators that can harvest energy like thermoelectric (TE), piezoelectric, triboelectric, and photovoltaic (PV) energy. Such generators can harvest wasted kinetic, solar, and thermal energy and enable the sensors or small-scale devices to be self-sufficient. The power density and energy conversion efficiency of such energy harvesting techniques are shown in Figure 15.1.

If we consider the human body, human activities throughout the day, including mechanical motions, bioheat, and exposure to sunlight, can be excellent sources of energy. Such energy can be easily harvested using wearables [4]. However, rigid, bulky structures are unsuitable for textile-based applications that require flexibility and lightweight.

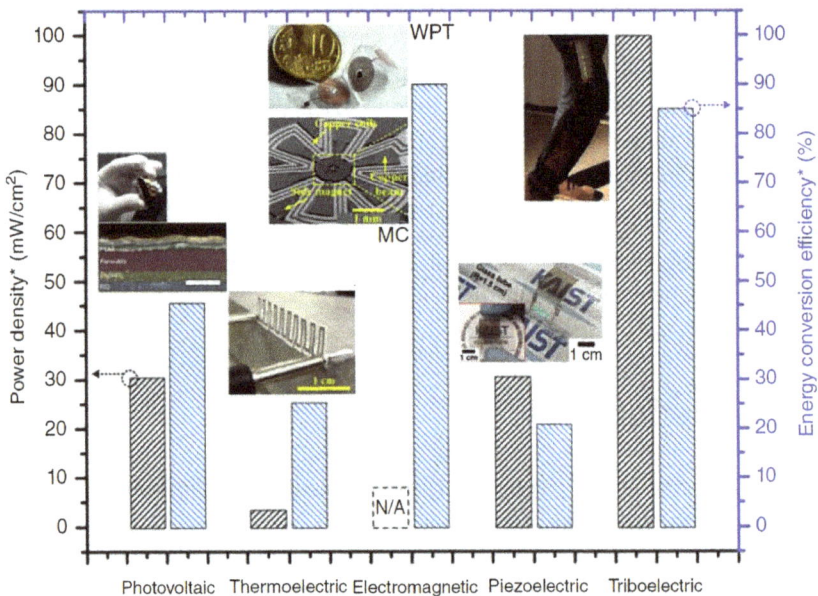

Figure 15.1: Power density and energy conversion efficiency of various energy harvesting techniques (reprinted with permission from [5], Elsevier, Copyright 2018).

Moreover, the structure needs to be breathable for proper use. Textile-based energy harvesters are meant to tackle such problems and, at the same time, provide enough power for various applications.

This chapter discusses primarily the need for energy storage followed by different types of devices and phenomena that can be utilized to harvest energy, including piezoelectric, TE, triboelectric, and PV.

15.2 Energy storage

To date, most wearables are powered by batteries. Batteries, as discussed earlier, are not ideal for textile applications. Also, there needs to be ample opportunity for the energy generated by such energy harvesting technologies to be stored for future use. For such a reason, there is also a requirement for energy storage technologies that can be effortlessly integrated into textiles. Textile batteries and textile-based supercapacitors have been developed for the same reason. Supercapacitors, even though they are promising, still suffer from energy density and high self-discharge rates. Reviews on the same are available in the literature [6, 7].

15.3 Thermoelectric energy harvesters

TE energy generators are meant to convert thermal energy into electrical energy. Research on TE generators has been on the rise, with special emphasis on wearable TE generators that are meant to harvest energy from body heat [8]. TE devices require no movement. Textile-based TE devices are more directed towards flexible and wearable applications which demand mechanical flexibility and electrical stability simultaneously.

15.3.1 Theory

TE devices are based on the Seebeck effect, which states that an electric voltage is generated when there is a difference of temperature formed across the TE device. When a temperature gradient is established between two junctions formed by conductors or semiconductors, the previously uniformly distributed electrons move towards the cold end from the hot end. This uneven distribution of electrons sets up a voltage difference across the material [9].

The TE efficiency of a TE device can be expressed via zT, a dimensionless quantity, via the following equation:

$$zT = \sigma S^2 T / k \qquad (15.1)$$

where T is the absolute temperature, S is the Seebeck coefficient, $S^2\sigma$ is the power factor, k is the total thermal conductivity, and σ is the electrical conductivity. The total thermal conductivity can be further expressed as a sum of lattice thermal conductivity (k_l) and electron thermal conductivity (k_e) [10]:

$$k = k_l + k_e \qquad (15.2)$$

High electrical conductivity, low thermal conductivity, and high Seebeck coefficient are prerequisites of good TE material [9]. The converse of the Seebeck effect is the Peltier effect, which guides the conversion of electrical energy to thermal energy and is applied for refrigeration or cooling effect [9, 11].

The TE power output of generators may again be defined as [11]

$$P = \frac{U^2 R}{(R+r)^2} \qquad (15.3)$$

where U is the open-circuit voltage, R is the loading resistance, and r is the internal resistance.

15.3.2 Device architecture

A TE generator consists of an array of semiconductors, connected thermally in parallel and electrically in series [9]. A schematic of a TE generator is shown in Figure 15.2. They have been broadly classified as per their physical dimensions as in 1D, which includes fibres, filaments, yarns, 2D structures as in fabrics, and 3D structures as in assembled textile products. 1D textiles have been produced using wet spinning, electrospinning, thermal drawing, gelation process, casting, and so on [11]. 2D and 3D architectures are based on woven, nonwoven, knitted fabrics, and fabrics with yarns, fibres, and filaments embroidered onto them [11].

15.3.3 Materials used

Conventional inorganic TE materials are inflexible and brittle, making them difficult to be produced directly into flexible and self-supporting structures. However, such inorganic materials can reach very high zT values, which cannot be achieved using organic TE materials. PbTe (lead telluride), Bi_2Te_3 (bismuth telluride), SiGe (silicon–germanium), SnSe (tin selenide), and skutterudite ($CoAs_2$) are some examples [11, 13]. The incorporation of such materials may also hamper the air permeability of the device to be used.

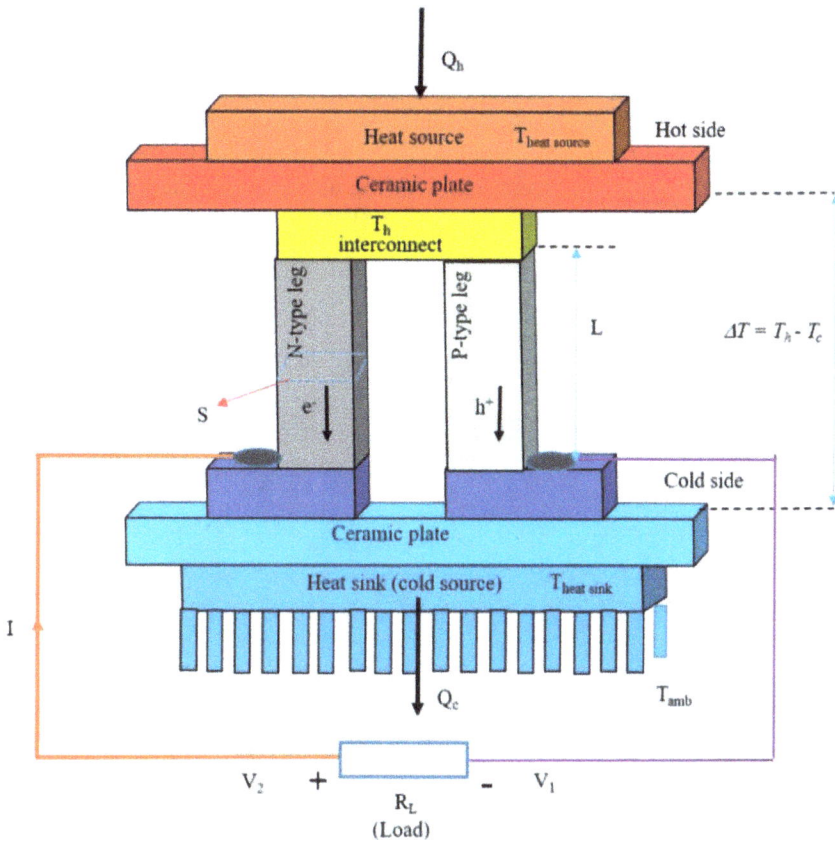

Figure 15.2: Schematic of thermoelectric generator (reprinted with permission from [12], IntechOpen, Copyright 2019).

Organic TE materials are typically solution-processable, which enables them to be spun into filaments or be used as a coating using simple solution-coating procedures [11]. Organic-based materials have lower zT value typically but are easily processable and allow the production of flexible devices that can be integrated into wearables. Polymers like polyphenylenevinylene, poly-3-butylthiophene, polyacetylene, polyaniline (PANi), poly-3-methylthiophene, polythiophenes, polypyrrole, and poly-3,4-ethylenedioxythiophene (PEDOT) have been used in this regard [13]. Other than polymers, carbon compounds like graphene and carbon nanotubes (CNTs) have also been used in making TE textiles.

15.3.4 Thermoelectric devices based on inorganic materials

TE devices were initially predominantly made of inorganics. For example, flexible 3D spacer textiles coated with non-toxic Al:ZnO thin films were used to construct a TE nanogenerator (Figure 15.3). The Al:ZnO films acted as the TE material, whereas the copper thin films were the contact layers. The device could produce a maximum power output of 0.16 µW at 62 K temperature difference [14]. The temperature-dependent Seebeck coefficient ranged from 52.1 to 57.8 µV/K at a temperature range of 14–62 K.

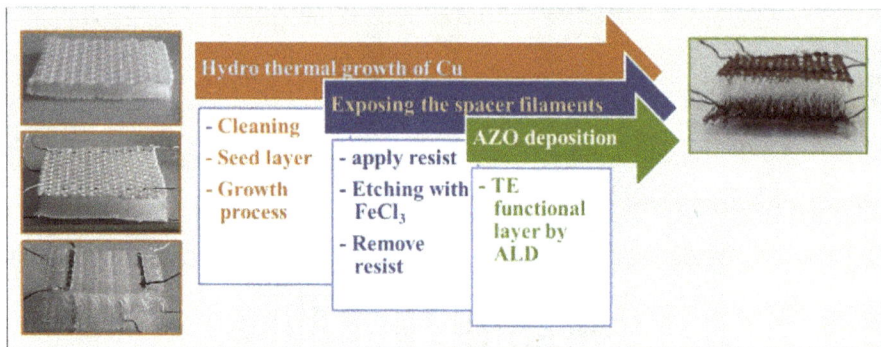

Figure 15.3: Preparation steps and scheme for making thermoelectric textiles using 3D spacer fabrics, uncoated spacer fabrics(left), spacer fabric coated with Cu and Al:ZnO (right) (reprinted with permission from [14], Elsevier, Copyright 2021).

Another study reported on TE textile made from a thin-film hybrid of the biodegradable polymer (nanocellulose) produced from reed and the biocompatible semiconductor copper iodide [15]. At a temperature gradient of 50 K, the planned module's output power density was 44 µW/cm^2. In a different investigation, cotton and polyester were used to deposit copper iodide thin films via sequential ionic layer adsorption and reaction. The films on TE textiles were made up of nanoscale-sized flakes (50 nm). Cotton and polyester with CuI coatings are depicted in Figure 15.4, and exhibit Seebeck coefficients between 120 and 180 µV/K and are constant at a temperature range between 290 and 365 K [16].

Other than the above-mentioned studies, TE generators based on thin-film semiconductors like Bi_2Te_3 [17], Sb_2Te_3 [18], $Bi_{0.5}Sb_{1.5}Te_3$ - Bi_2Se_3, $(Te_{85}Se_{15})_{45}As_{30}Cu_{25}$, and In_4Se_3 [11] have also been reported.

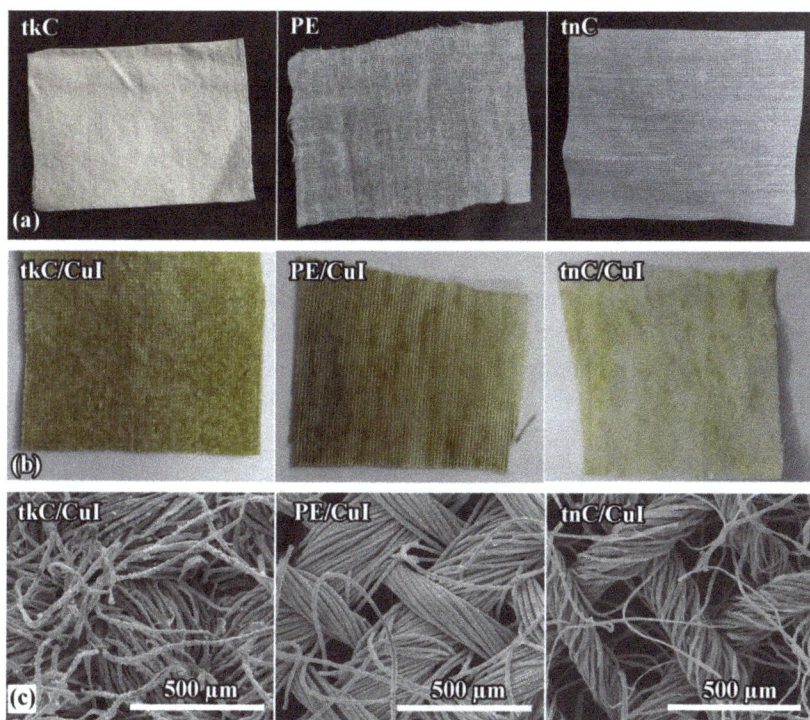

Figure 15.4: Photos displaying (a) thick cotton (tkC), polyester (PE), and thin cotton (tnC) fabrics; (b) CuI-coated thermoelectric textiles tkC/CuI, PE/CuI, and tnC/CuI obtained after deposition of CuI films by successive ionic layer adsorption and reaction (SILAR) method and its (c) SEM images (reprinted with permission from [16], Elsevier, Copyright 2020).

15.3.5 Thermoelectric devices based on conductive polymers and carbon materials

It has been extensively reported that PEDOT and graphene-based polymers can be used to create TE textiles. CNTs have also been used extensively for making TE textiles. The combined electrospinning and spraying technology was used for producing CNT/polyvinylpyrrolidone (PVP)/polyurethane (PU) composite TE fabrics with high air permeability and stretchability (250%) comparable to pure nanofibre fabrics [19]. The use of PVP increased the dispersion of CNTs and acted as interfacial glue between the CNTs and the PU skeleton. As a result, even after 1,000 bends, both the electrical conductivity and the Seebeck coefficient remained intact. An ultrahigh power factor (14 ± 5 mW/mK^2) with high electrical and thermal conductivity was reported recently for aligned CNT macroscopic fibres [20]. A textile TE generator with high TE performance, weavability, and scalability was developed based on these CNT fibres. The observed power factor was a result of the sample's ultrahigh electrical conductivity, owing to its morphology and its improved Seebeck coefficient. Another study reported that the inclusion of CNT in polyester/yarn fabrics coated with polyaniline/CNTs significantly enhances both the electrical conductivity and the Seebeck coefficient [21]. Electrical conductivity improved from 0.011 to 0.1345 S/cm when CNT content grew from 0.5 to 10 wt%. The highest recorded Seebeck coefficient of 11.4 µV/K was found in a sample containing 5% CNT. Ultrasonic induction can also be used to create TE textiles by adding single-walled CNT/PANI composites on polyester [3]. When compared to textiles treated with heating or dip coating, the ultrasonic induction-treated materials displayed improved interfacial stability and maintained their performance even after being bent up to 150 times. In addition, the textiles' highest power factor was 0.28 µW/mK^2. The manufactured device is shown in Figure 15.5.

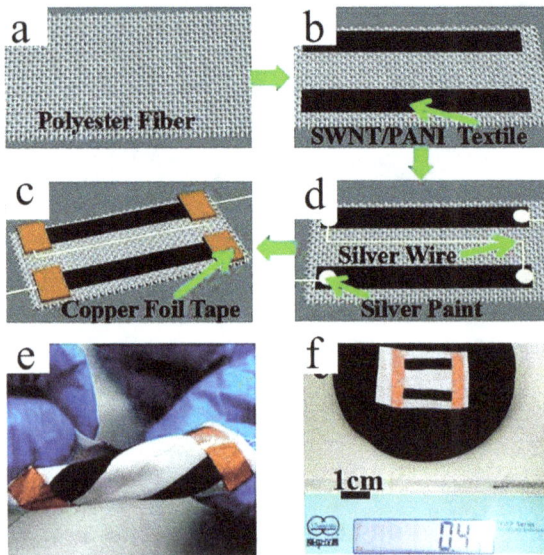

Figure 15.5: Assembly process of a thermoelectric generator based on single-walled carbon nanotube/polyaniline (a)–(d) displaying flexibility of the devices fabricated from two pieces of p-type textiles (e)–(f) displaying weight of the device of length 5 cm and width 2 cm (reprinted with permission from [3], RSC, Copyright 2016).

In another study, TE conductive textiles were fabricated by knife-coating of cellulosic fabrics using graphene nanoplatelets with loading between 0.4% and 2% inside an acrylic coating paste [22]. In a few seconds, the cloth with the greatest graphene concentration experienced a temperature rise of 100 °C.

15.3.6 Organic/inorganic hybrids

A TE device was recently developed using atomic layer deposition (ALD) and molecular layer deposition (MLD) of thin-film methods [23]. The device was composed of n-type ALD-grown ZnO or ALD/MLD-grown ZnO-organic components and p-type spray/immersion-coated PEDOT:PSS components. Thicker coatings in the order of 300–500 nm were found to perform better than the thinner ones. For ZnO, the lower deposition temperature of 100 °C was more advantageous than the higher (200 °C). In addition to being mechanically more flexible, hybrid inorganic–organic materials are better suitable for flexible energy harvesting applications in conjunction with textiles. However, simultaneous and efficient processing of polymers and inorganic materials remains to be a challenge [10]. The manufacturing of such materials is time and energy consuming and costly. Moreover, the blending of rigid inorganic molecules and flexible polymers is difficult.

15.4 Photovoltaic energy harvesters

In PV cells, the production of electrical energy occurs by two different materials in electrical contact when they are subjected to irradiation. The electrical field created by the junction of P-type and N-type semiconductors separates pairs of positive and negative charges that are created when light strikes a PN junction cell. This junction field is the result of introducing minute quantities of certain impurities to either side of the junction, which then creates a potential energy barrier across the junction. This results in electrons migrating from P-type to N-type, while "holes" travel in the other direction, thereby setting up a current.

The power conversion efficiency of a solar cell can be expressed as [24]

$$\eta = \frac{P_{out}}{P_{in}} = \frac{I_{max} * V_{max}}{P_{in}} = \frac{I_{sc} * V_{oc} * FF}{P_{in}} \tag{15.4}$$

where

$$FF = \frac{I_{mpp} * V_{mpp}}{I_{sc} * V_{oc}} \tag{15.5}$$

where FF is the fill factor and I_{max} and V_{max} are the solar cell's maximum values at its peak power (the ratio of the actual maximum obtainable power to the product of the open-circuit voltage and short circuit current), I_{sc} is the current at zero voltage, V_{oc} is the voltage at zero current, I_{mpp} is the current at maximum power, V_{mpp} is the voltage at maximum power, and P_{in} is the total incident irradiance.

15.4.1 Solar cells

Solar cells consist of two electrodes at the extremes and two semiconductor materials sandwiched in between (Figure 15.6). The initial solar cells that still dominate the market are mono/multi-crystalline single junction types made of crystalline silicon. However, the disadvantage of such types of solar cells resides in the fact that they are brittle and not suitable for textile integration. The second generation of solar cells made of gallium arsenide had far lower efficiencies but could be made as thin films. The third generation introduced organic solar cells and hybrid solar cells like dye-sensitized solar cells, which are still not commercial on account of increased cost and problems arising from mass production [25].

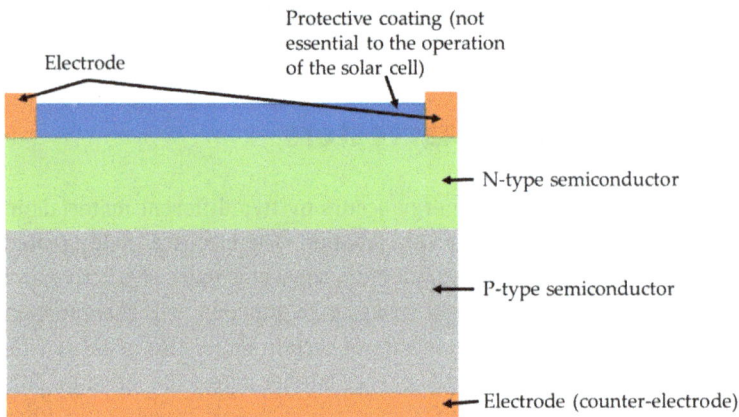

Figure 15.6: Schematic of a solar cell (reprinted with permission from [25], MDPI, Copyright 2020).

15.4.2 Construction of photovoltaic textiles

The simplest method for the construction of PV textiles is to somehow incorporate solar cells or solar cell arrays into the textile substrate. Products employing the same methodology include Maier Sports' prototype of a winter outdoor jacket, Tommy Hilfiger's solar-powered jacket from 2014, The Solar Shirt' from 2014, and Kingstons Beam Backpack [25]. However, the incorporation of solar cells onto textiles hampers the comfort of the wearer. Problems related to adhesion and electrical connection between cells also exist for such types of products.

Another method of producing PV textiles is to produce PV fibres that consist of an anode, a PV material, and a cathode together in the same fibre. A schematic of such a device is shown in Figure 15.7. Usually, a filament is first coated with a conductive material, and then concentric layers of PV materials can be coated, followed by the final electrode [26]. The advantage of this type of structure is that the fibres become inherently

PV and can be assembled into any structure [25]. However, when such fibres or filaments are subjected to further rounds of assembling like weaving and knitting, the
chances of abrasion are higher, which reduce the PV efficiency significantly [26].

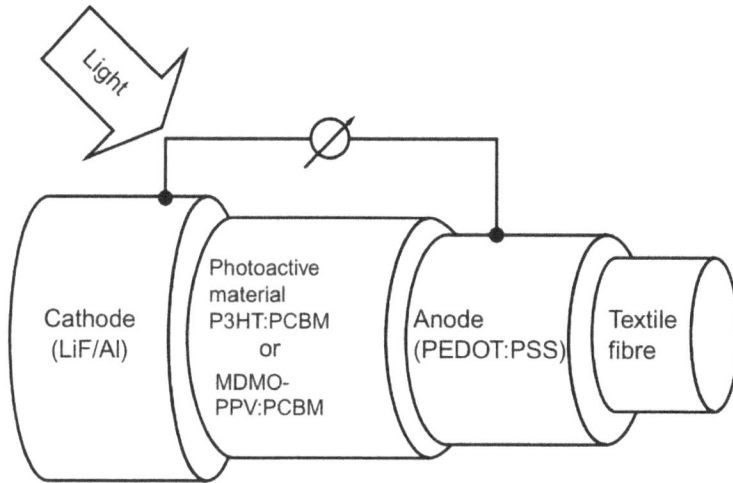

Figure 15.7: Schematic of a photovoltaic fibre (reprinted with permission from [26], Elsevier,
Copyright 2015).

PV textiles can also be produced by depositing thin films of PV materials on textile fabrics. The films may be attached via laminating or sewing onto the fabric. Figure 15.8
shows a PV textile produced directly by depositing silicon films on fabrics. However,
the film deposition should not hamper the mechanical properties of the textile, and the
textile should be able to withstand high temperatures for deposition as well. Glass, poly-

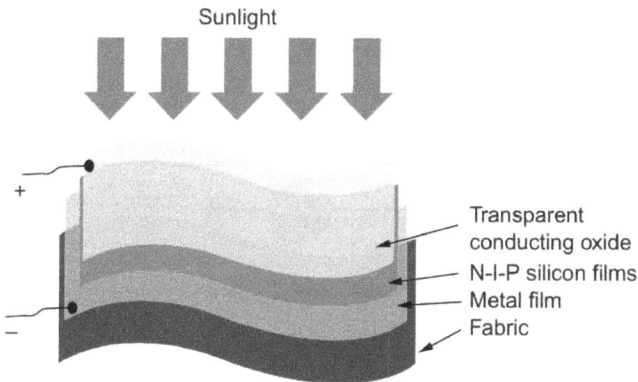

Figure 15.8: Schematic of silicon deposition of photovoltaics on fabric (reprinted
with permission from [26], Elsevier, Copyright 2015).

benzimidazole, polyimide, polyether ether ketone, and poly-(*p*-phenylene benzobisoxa-zole) fibres are found to be suitable for such applications [26].

Recently, a stainless-steel mesh fabric was also utilized as an electrode and sub-strate for constructive PV textiles [24]. PEDOT:PSS and poly(3-hexylthiophene):phenyl-C61-butyric acid methyl ester(P3HT:PCBM) were subsequently placed by dip-coating on steel mesh fabric, and a back metal electrode was afterwards deposited by thermal evaporation. The power conversion efficiency was about 0.69%.

15.4.3 Dye-sensitized solar cells (DSSC)

Dye-sensitized solar cells (DSSCs) utilize organic dyes to generate electricity from sun-light. In the case of DSSCs, the bulk of the semiconductor plays the role of a charge transporter, whereas the photoelectrons are provided by photosensitive dyes. DSSC, in essence, consists of an electrode, a dye, a redox mediator, and a counter electrode [27]. The substrate needs to be essentially transparent and conductive. The working mechanism of a DSSC is shown in Figure 15.9.

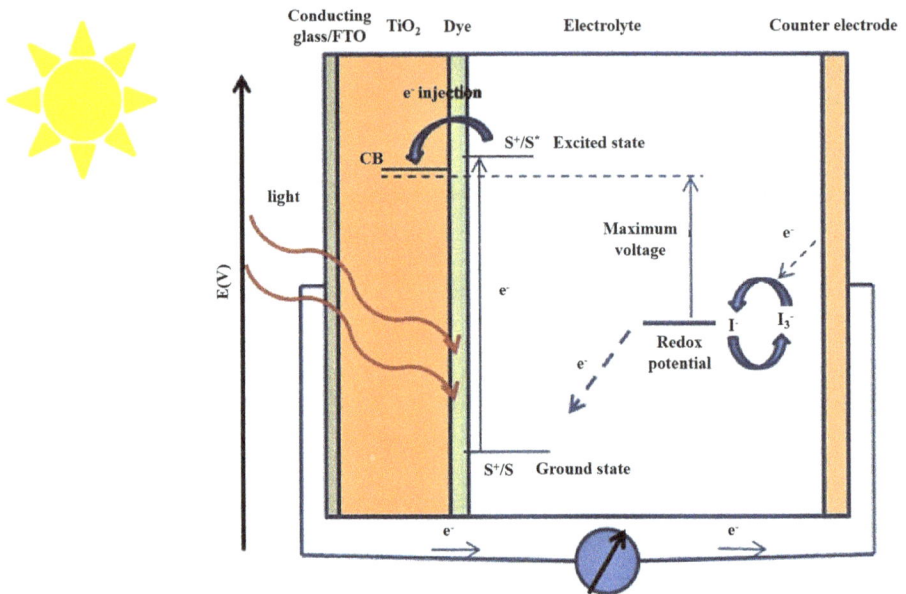

Figure 15.9: Working principle of DSSC (reprinted with permission from [27], Springer Nature, Copyright 2018).

Briefly, the mechanism may be described as follows. When a photon is absorbed by the photosensitizer, electrons of the dye get displaced from the ground state to the excited state. These electrons are then injected into the conduction band of the TiO_2 electrode. The electrons are then again moved towards the counter electrode, which then reduces I_3^- to I^-. The dye accepts electrons from I^- ion, which gets oxidized to I_3^-. The I_3^- ion then again moves back to the counter electrode and gets converted to I^-. The equations may be written as follows [27]:

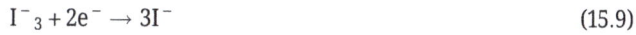

$$S^+/S + h\nu \rightarrow S^+/S^*$$ (15.6)

$$S^+/S^* \rightarrow S^+/S + e^- \ (TiO_2)$$ (15.7)

$$S^+/S^* + e^- \rightarrow S^+/S$$ (15.8)

$$I^-_3 + 2e^- \rightarrow 3I^-$$ (15.9)

DSSC has been incorporated into textile structures in a multitude of ways. In one such study, DSSC was sewn into the textile fabric [28]. The steps included weaving of the electrodes, then depositing TiO_2 and the counter-electrode material, followed by applying heat treatment. Then the core-integrated DSSC device was sewn, and the dye was loaded. The electrodes were woven in a 3/1 twill structure to maintain their flexibility. The device had an energy conversion efficiency of 5.8%.

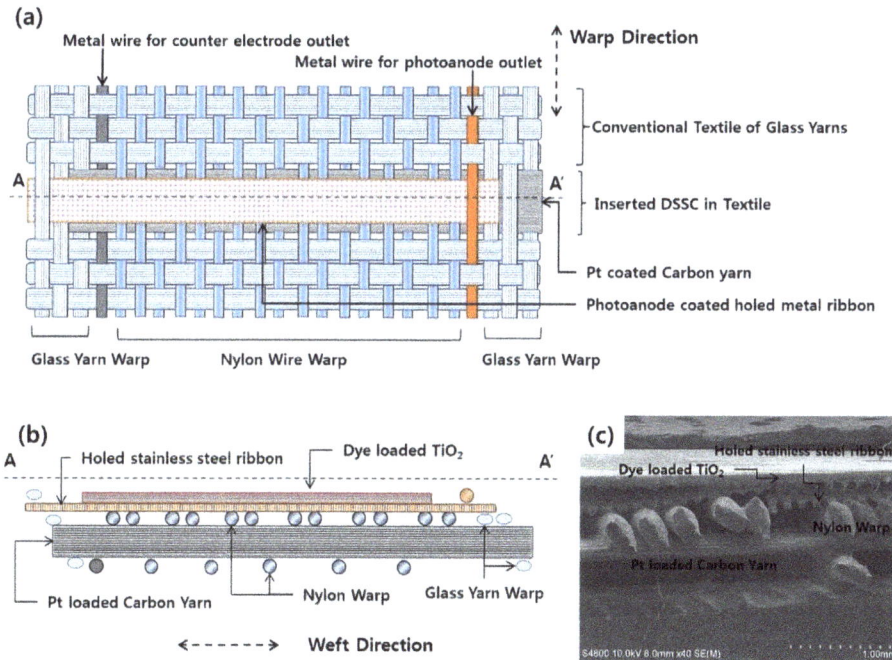

Figure 15.10: (a) Schematic illustration of the structure of the inserted DSSCs, (b) cross-sectional view, and (c) scanning electron microscopy of cross-sectional view of the DSSC incorporated in the textile (reprinted with permission from [29], Springer Nature, Copyright 2015).

In another study, photoanodes and counter electrodes were introduced into the textile fabric as warp and weft yarns [29]. This fabric was then again sewn into cloth. The photoanodes consisted of dye-loaded, porous TiO_2-coated stainless-steel ribbons with periodic holes, whereas the counter electrode was made of carbon yarn coated with Pt nanoparticles, while nylon filaments were present in between prevented electrical short. The device structure is shown in Figure 15.10. DSSCs have also been manufactured on glass fibre-based fabrics. Such devices showed efficiencies up to 1.5% [30].

15.5 Triboelectric energy harvesters (TENG)

Triboelectric nanogenerators (TENG) work by combining triboelectrification and electrostatic induction. The triboelectric effect arises out of the transfer of electrons or ions or both [31] from one material to another when two materials are rubbed together. The tendency of donating or accepting electrons of a particular material determines its position in the triboelectric series [32]. The triboelectric series is shown in Figure 15.11. The triboelectric property of metal or semiconductors depends on its work function, and the triboelectric phenomenon in insulators can be explained via electron transfer theory, as shown in the case of PTFE and polymethyl methacrylate or in the case of organic insulating surfaces [32] and the electron-transfer mechanism in the contact-electrification effect [33]. Also, mass transfer between the two contacting surfaces has been correlated with triboelectricity, as shown in the case of polyester and PTFE [34, 35].

There are four working modes of TENG, as illustrated in Figure 15.12: vertical contact separation mode, in-plane contact-sliding mode, single-electrode mode, and freestanding triboelectric layer mode. In vertical contact separation mode of TENG, two pieces of dielectric materials with different electron gaining or losing tendencies make periodic contacts with each other. Two electrodes are, respectively, attached to the back of these materials. In contact sliding mode, the arrangement remains the same. But the charge is generated by sliding one surface on top of another. In single-electrode mode, the electrode towards the bottom is grounded. In free-standing mode, two electrodes are placed below a dielectric. The pros and cons of each mode are illustrated in Figure 15.12.

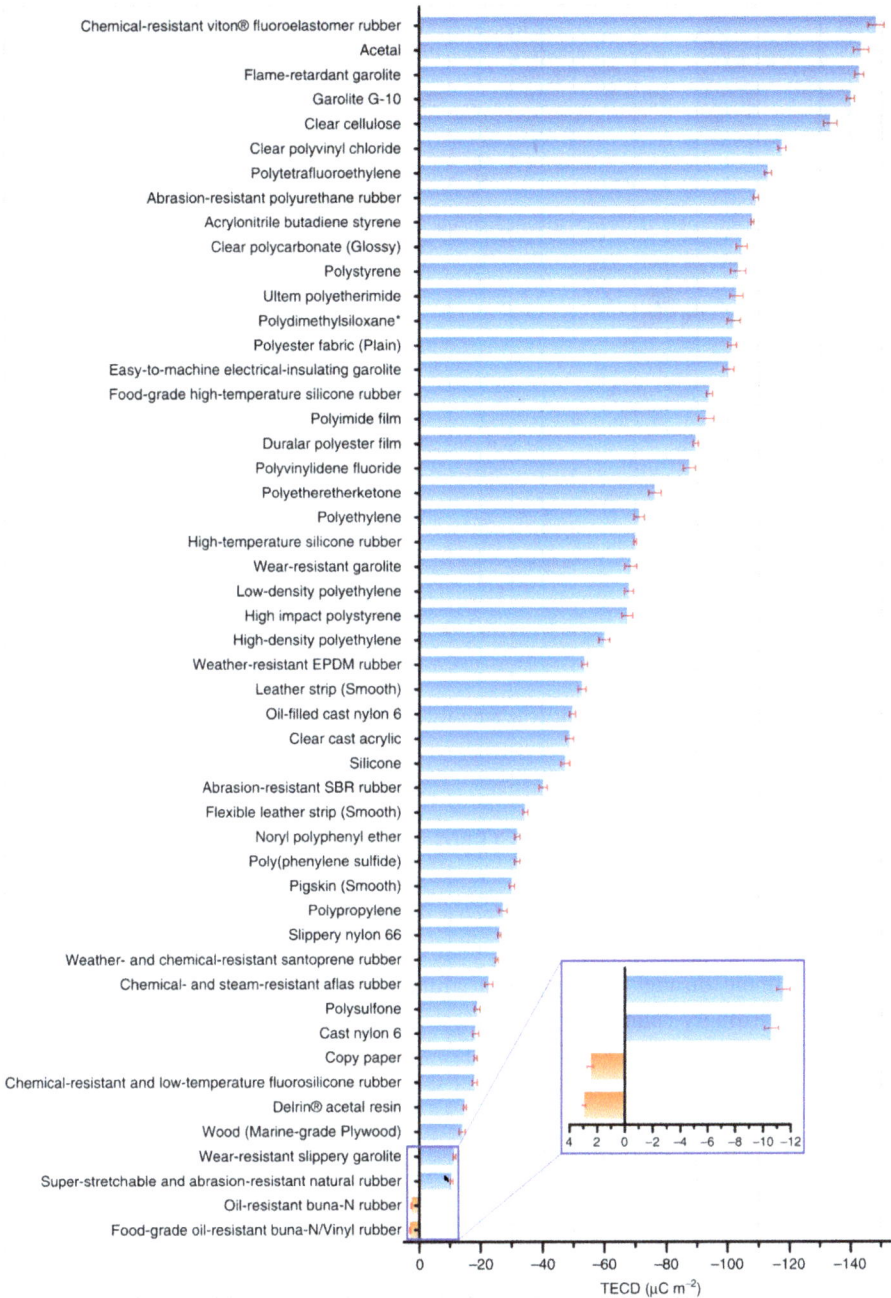

Figure 15.11: The triboelectric series of some known materials (reprinted with permission from [32], Springer Nature, Copyright 2019).

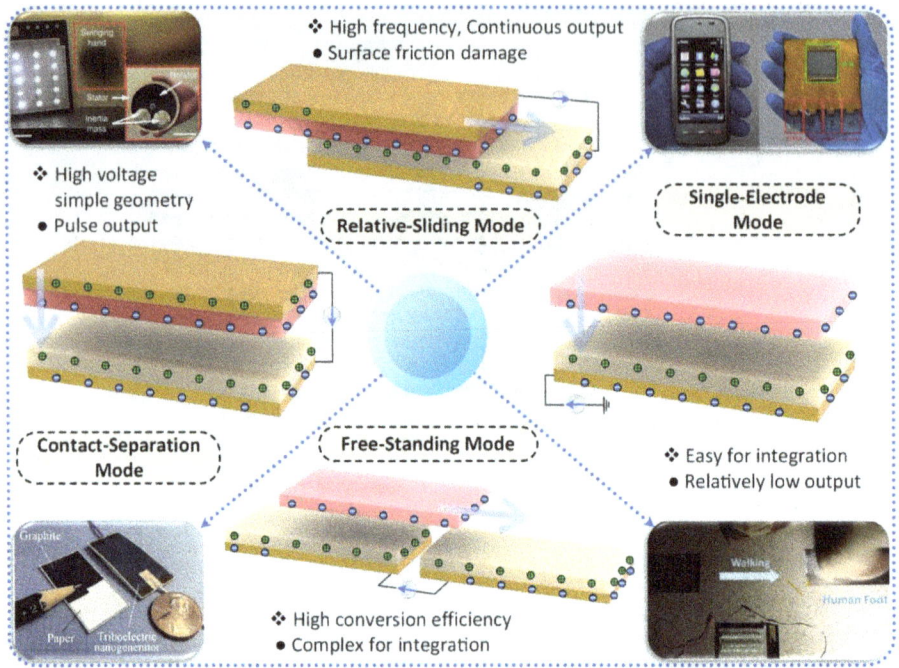

Figure 15.12: Modes of triboelectric nanogenerators with their pros and cons (reprinted with permission from [5], Elsevier, Copyright 2018).

15.5.1 Device architecture of TENG

The majority of the textile-based triboelectric device structures are based on two different fabrics with opposing triboelectric characteristics in a face-to-face arrangement [36]. For example, through effective surface functionalization with a polymeric fluoroalkylsiloxane, ordinary cotton fabrics were converted into negative triboelectric materials, and such functionalized cotton textiles were used as the negative triboelectric material, while nylon cloths were employed as the positive triboelectric material to construct a TENG.

Another method of producing TENGs is to use two distinct types of conductive threads and weave or knit them together to make a swatch of two-in-one triboelectric textiles. Triboelectric textiles were manufactured by both weaving and knitting. Potentials of plain, double, and ribbed fabric architectures for textile-based energy harvesting were investigated [37]. Double-knitted fabric and rib-knitted fabrics showed better performance.

15.5.2 Cellulose-based TENG

Cellulose as natural and abundantly available material has been used extensively for producing TENGs. A flame-retardant conductive cotton fabric and PTFE-coated cotton fabric were used to construct flame-retardant textile-based TENG [38]. The flame retardancy of the fabric was achieved using polyethyleneimine and melamine and phytic acid (Figure 15.13). A maximum peak power density of 343.19 mW/m^2 could be obtained under a tapping frequency of 3 Hz.

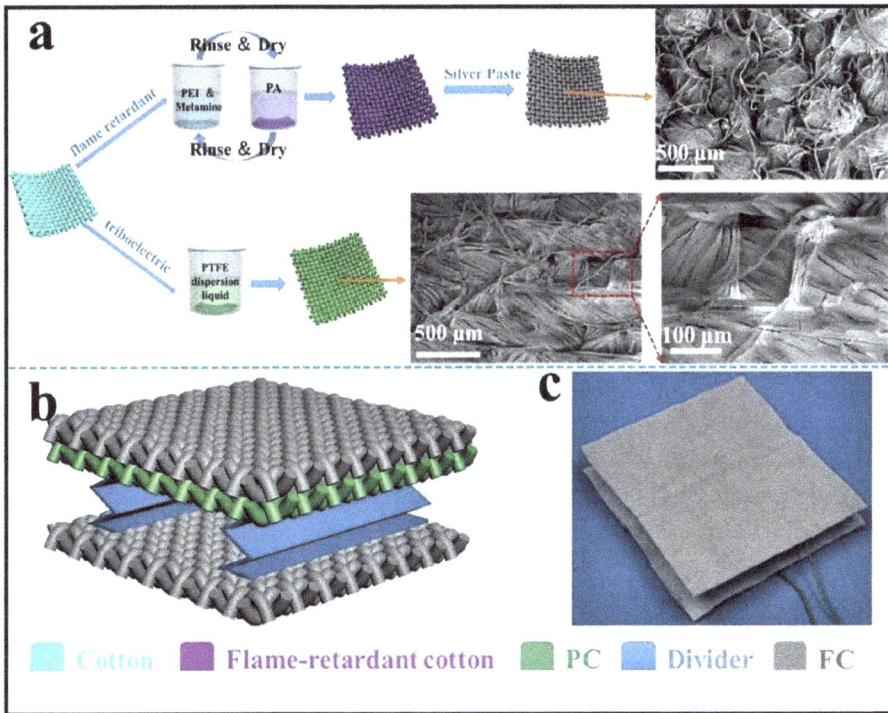

Figure 15.13: (a) Schematic of the preparation of the flame retardant along with SEM of the PTFE-coated and flame-retardant conductive cotton fabrics; (b) schematic diagram; and (c) photographic image of the flame-retardant TENG (reprinted with permission from [38], ACS, Copyright 2020).

The positive and negative layers in TENGs, cotton, and silk were dipped in cyanoalkyl silane and fluoroalkyl silane in a different study [39]. The TENG that was created reached a maximum output current of 50.3 A and a maximum output voltage of 216.8 V. Recently, a turntable-shaped cotton-based TENG for harnessing wind and water flow energy was described [40]. The device rotates to improve movement between the cotton sheet and the fluorinated ethylene–propylene film used to create the device, which is originally propelled by the wind or water flow. About 782 V and 8.9 A, respectively, are the open-circuit voltage and short-circuit current values.

Conjugated conductive polymers have also been used for developing TENGs. Polyaniline/cotton-based flexible and wearable TENGs were developed using low-temperature in situ polymerization [41]. The textured cotton textile with deposition provided a rough surface, which improved the output performance of TENG. The greatest V_{OC}, I_{SC}, and power density values of the TENG device were 350 V, 45 A, and 11.25 W/m^2, at the compression force of 5 N, respectively.

Other than cotton fabric, cellulose in different forms, for example bacterial cellulose [39, 42–44] and cellulose paper, has also been used in constructing TENGs.

15.5.3 Silk-based TENG

Other than cellulose-based fabrics, silk has also been used as a raw material for producing TENGs. A silk fibroin bio-triboelectric generator based on an electrospun nanofibre-networked film [45] was reported. Compared to a smooth cast silk film, the electrospun silk had a substantially greater surface-to-volume ratio and much rougher surfaces, resulting in increased power generation. At the same time, crystalline silk microparticles were found to increase the surface charge density in materials such as polyvinyl alcohol [46]. Such particles could be recovered from discarded *Bombyx mori* silkworm cocoons using a simple alkaline-hydrolysis process.

By microstructuring the surface of PDMS (polydimethylsiloxane), screen-printing technique was effectively refined to produce graphite interdigital electrodes on PDMS. Sequentially, silk fibroin was used to cover the electrodes for developing TENG [47]. A silk nanoribbon-based TENG was also created using a silk nanoribbon film and a regenerated silk fibroin film [48]. To preserve the native meso/nanoscale structure of silk, 0.38-nm-thick silk nanoribbons were directly exfoliated from real silk. The maximum voltage, current, and power density of the bio-output TENG's performance were 41.64 V, 0.5 A, and 86.7 mW/m^2, respectively.

15.5.4 Nylon- and polyester-based TENG

A polyester/nylon fabric with silver electrodes based on TENG was proposed by Zhong and coworkers [49]. The fabric was flexible, washable, and breathable and was shown to be able to harvest energy from a variety of human motions. To improve hydrophobicity, PTFE or SiO$_2$ modified with trichloro(octadecyl)silane was sprayed into the surface of a polyester fabric. With a high water contact angle of 144° and 153°, respectively, the PTFE- and SiO$_2$/OTS-coated polyester textiles displayed good water repellency.

15.6 Piezoelectric energy harvesters

Piezoelectric energy harvesters are designed to convert mechanical energy to electrical energy (named as piezoelectric effect) or vice versa (named as converse piezoelectric effect). The textile industry has moved forwards from its conventional areas and then towards the segment of intelligent and multifunctional textiles acting as sensors and actuators. Therefore, one can make use of an impetus from the piezoelectric materials which on application of stress converts into electrical energy. The harvested electric energy can be used to power different electronic gadgets possessing low power requirement. The piezoelectric materials can be in the form of films, fibres including nanofibres, and mono- and multi-filaments, which can be further integrated into textiles.

15.6.1 Theory of piezoelectricity

The piezoelectric materials show superior electromechanical coupling. The electromechanical coupling is the virtue by which the piezoelectric materials produce electrical displacement on application of mechanical stress and the strain on the application of electric field. The mechanical to electrical coupling is termed as direct piezoelectric effect and the electrical to mechanical coupling is termed as indirect piezoelectric effect.

In the linear elastic region, stress is proportional to strain. The stress–strain relationship is

$$S = T/Y \tag{15.10}$$

$$S = sT \tag{15.11}$$

where s is the compliance, S is stress, Y is Young's modulus, and T is strain.

Within a certain limit of stress, electric displacement is proportional to electric displacement. The slope of the curve is piezoelectric strain coefficient "d" which is a measure of piezoelectricity in ferroelectric materials. The piezoelectric charge coefficient d_{33} quantifies a volume change in a piezoelectric material when subjected to electric field or polarization produced on application of stress (where polarization and stress is in the same direction):

$$D = dT \tag{15.12}$$

The relationship between *electric displacement* (*D*) and *strain* (*S*) as a function of *electric field* (*E*), *stress* (*T*), and ε (dielectric permittivity) is given by

$$\left\{ \begin{array}{c} S \\ D \end{array} \right\} = \begin{bmatrix} s & d \\ d & \varepsilon \end{bmatrix} \left\{ \begin{array}{c} T \\ E \end{array} \right\}$$

(15.13)

15.6.2 Device architecture of piezoelectric energy harvesters

Textile-based woven piezoelectric energy harvesters consist generally of piezoelectric yarn and conductive yarns in order to harvest the energy. It should be borne in mind that the instances of two conductive yarns touching each other should be avoided at any circumstances. However, there can be innumerable configurations in the fabric to augment the generated energy from strain developed in the fabric during its usage. Different textile forms like woven, knitted, electrospun, meltspun, and bicomponent fibres are promising piezoelectric textiles. There are several reports on the growth of piezoelectric materials like ZnO and different metal-doped ZnO on textile substrates.

15.6.3 Single fibre-based piezoelectric nanogenerators

Textile fibres are typically made using three processes: sol–gel extrusion, viscous solution spinning technique, and thermoplastic extrusion [50]. Ceramic fillers are combined with thermoplastic matrix and extruded at temperatures below and above the polymer's melting point in thermoplastic extrusion. The polymer is subsequently removed, and inorganic grains are sintered to ensure that ferroelectric domains are ordered uniformly. For the extrusion of fibre, it is essential to establish a proper balance between the percentage of filler loading and melt viscosity. Piezoelectric powder is combined with polymer solution in the viscous solution spinning process, and the mixture is subsequently coagulated in a non-solvent. Different agents are introduced to regulate the coagulation rate and spin capability. Sol is created with the necessary quantity of precursors prior to the sol–gel extrusion process. The rate of hydrolysis, condensation, and viscosity are factors that affect the spinnability of the sol solution. For the correct viscosity ratio, a long curing time is necessary. However, it should not be left to cure for an extended period of time as this could result in the creation of non-spinnable gel. The sol is extruded in liquid form through a spinneret, where gel fibres are created and gathered on a take-up roller. High temperatures are used to sinter the amorphous gel fibres. The different active fibre composites are depicted in Figure 15.14.

Nylons or polyamide fibres are one of the most extensively used synthetic fibres in the textile industry. Odd-numbered nylons are the most promising piezoelectric materials. Nylon 11 has different crystalline phases namely α, α', Y, and δ', of which the δ'-phase shows the strong piezoelectric activity. This phase has a highly polar smectic pseudohexagonal chain arrangement with disordered H-bonding arrangement between the chains

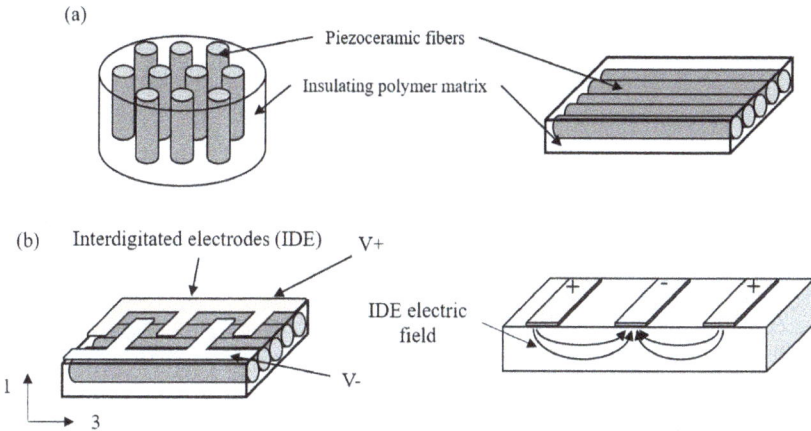

Figure 15.14: (a) Active fibre composites with piezoelectric fibres in polymer matrix; (b) active fibre composites with interdigitated electrodes and fibres; corona poled along the long axis so that the piezoelectric coefficient is d_{33} (reprinted with permission from [50], ACS, Copyright 2022).

[51]. Anwar et al. have studied the piezoelectric effect in electrospun Nylon 11 nanofibres. From X-ray diffraction studies, it was seen that δ'-phase had a largest lateral spacing of 0.424 nm between the sheets. Fast solvent evaporation is necessary for the favourable piezoelectric phase in Nylon 11. Under periodic impact at a frequency of 8 Hz, nanofibres generated maximum open-circuit voltage of 6 V. Figure 15.15 shows the piezoelectric output from the electrospun nanogenerator.

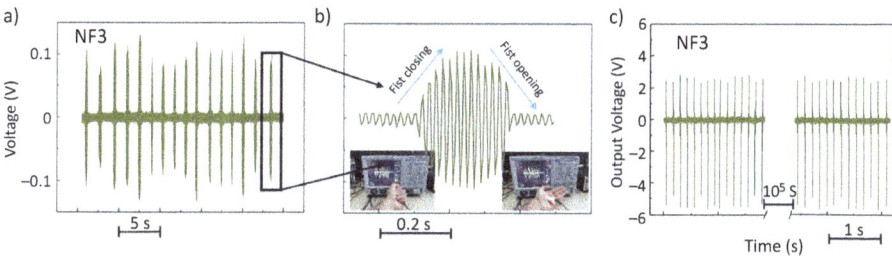

Figure 15.15: (a) Real-time body movement detection by Nylon 11 piezoelectric nanogenerator; (b) zoomed view of (a); and (c) output voltage before and after continuous impact for 10^5 s at 8 Hz (reprinted with permission from [51], Wiley, Copyright 2021).

Park et al. fabricated a piezoelectric sensor by wrapping a poly(vinylidene fluoride) (PVDF) film with a helical structure on a thread core [52]. A customized flexible circuit board was attached on PVDF film as an electrode for harvesting charges. Sinusoidal flexural loading was applied on piezoelectric textile for studying electrical characteristics of the sensor. Different voltages were measured at varied flexural loadings.

Martins et al. have studied the piezoelectric effect of three-layered piezoelectric monofilaments co-extruded using PVDF and two different polypropylene-based conductive polymers [53]. The filaments were corona poled in order to augment the piezoelectric effect. The peak-to-peak voltages of the filaments was reported to be 0.97 and 1.44 V at load values of 11.78 and 19.41 N.

Poly(L-lactic acid) (PLLA) is one of the promising bio-based piezoelectric materials. It can act as a good substitute for piezoelectric polymers like PVDF. The synergistic effect of ceramic filler and biopolymer is advantageous in terms of piezoelectric energy harvesting. Ju Oh and his team of researchers have reported on the piezoelectric properties of melt-extruded PLLA/barium titanate [54]. The post-processing conditions were optimized to obtain favourable β-phase. The structural orientation in the melt-extruded fibres took along the fibre axis along with the transformation to β-phase during drawing and heat-setting processes. The fabric integrated with these fibres generated output voltage of 0.5 V with palm tapping and 0.62 V with the tapping of the side of the hand. The short-circuit current values were in the range of 557–911 nA.

Lund et al. have reported on the bicomponent fibre consisting of PVDF as a sheath material and carbon black-loaded HDPE as a core material[55]. An array of such fibres generated output (peak-to-peak) voltage of ~ 40 mV. These types of fibres present a plethora of opportunities in the field of flexible sensors. There are also reports on multifunctional piezoelectric nanogenerators. Shape-memory polymer (PU) and piezoelectric PVDF are blended to yield a meltspun extruded fibre [56]. It was observed that addition of PU resulted in favourable β-conformations, and addition of reduced graphene oxide augmented the flow of charges towards the surface of the fibre for enhanced energy harvesting capabilities. The nanogenerator generated average open-circuit voltage of ~ 349 mV and recovery ratios up to 100%. Hierarchical structures on nanofibre web or any textile fibre are good options towards increased piezoelectric energy harvesting. Feng et al. reported on the PVDF nanofibre web being sandwiched by conductive substrates with hierarchical ZnO nanostructures [57]. The nanogenerator generated an output voltage of 1.60 ± 0.08 V at 150 kPa and has a linear response in the range of 3–150 kPa. The durability of device was also tested as it could succumb to 3,600 cycles of pressing. Figure 15.16 represents the visuals of the device and the generated output voltage during rain.

Different coating approaches are also employed to fabricate piezoelectric sensors. In one of the studies by Mazbah Uddin et al., different coaxial yarn-based piezoelectric nanogenerators were explored. This technique resulted in in-situ β-formation. The inner electrode was coated with PVDF solution in one of the instances, and touchspun nanofibre-coated inner electrode on the other. Solution-coated coaxial yarn-based nanogenerator generated peak-to-peak open-circuit voltage of 5.12 V and 41.25 nA of peak short-circuit current, while nanofibre-coated coaxial yarn generated peak-to-peak open-circuit voltage of 5.08 V and 29.1 nA of short-circuit current, which could power several LEDs [58].

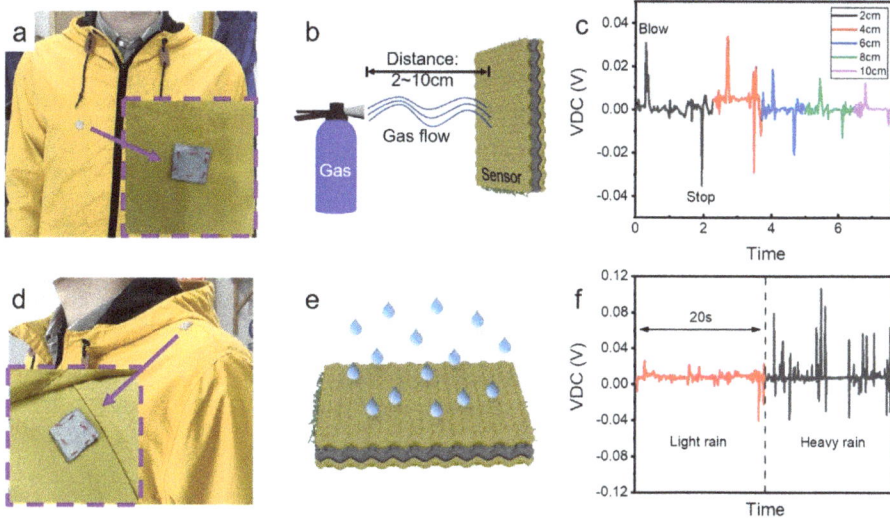

Figure 15.16: Application of the sensor measuring airflow and rainfall: (a) Photograph of the sensor placed on the outdoor jacket; (b) schematic diagram of the device under test; (c) the output voltage generated by the sensor under the flow rate of 0.1 L/s at different locations in the airflow monitoring test; (d) photograph of the sensor on the outdoor jacket; (e) schematic diagram of the device under test; and (f) the output voltage generated by the sensor under different intensities of rainfall in the rainfall monitoring test (reprinted with permission from [57], Elsevier, Copyright 2022).

15.6.4 Fabric-based piezoelectric nanogenerators

The growth of different nanostructures on textile substrate can augment the energy harvesting largely. ZnO nanorods were grown on silver-coated cotton textiles [59]. The uniform and spatial distribution of nanorods led to generation of voltage of ~9.5 mV.

Healthcare monitoring is one of the key areas in the segment of stretchable sensors. Muscle fibre-inspired electrospun/nonwoven textile-based piezoelectric nanogenerator was developed for physiological monitoring. In order to mimic muscle fibres, polydopamine was mixed with $PVDF/BaTiO_3$ during nanofibre preparation [60]. The developed nanofibre-based nonwoven fabric had outstanding sensitivity (3.95 V/N) and displayed long-term stability (<3% reduction in performance after 7,400 cycles).

Knitted fabric has an inherent stretchability due to its inherent fabric structure. Therefore, this unique characteristic of the knitted textile can be explored for piezoelectric energy harvesting. In one of the studies by Anand et al., 3D knitted spacer technology was employed to fabricate a piezoelectric-based knitted textile [61]. The knitted textile consisted of PVDF filaments having electroactive phase of 80%. The textile generated power density in the range of 1.10–5.10 $\mu W/cm^2$ at input pressures of 0.02–0.10 MPa.

A hybrid piezoelectric fibre was fabricated using $BaTiO_3$ nanoparticle and PVDF in the mass ratio of 1:10 [62]. These fibres were integrated in knitted garment and generated maximum output voltage of 4 V and power density of 87 $\mu W/cm^3$. The wearable nanogenerator was able to charge a 10 μF capacitor in 20 s. In Figure 15.17, voltage profiles of different fabric structures are given.

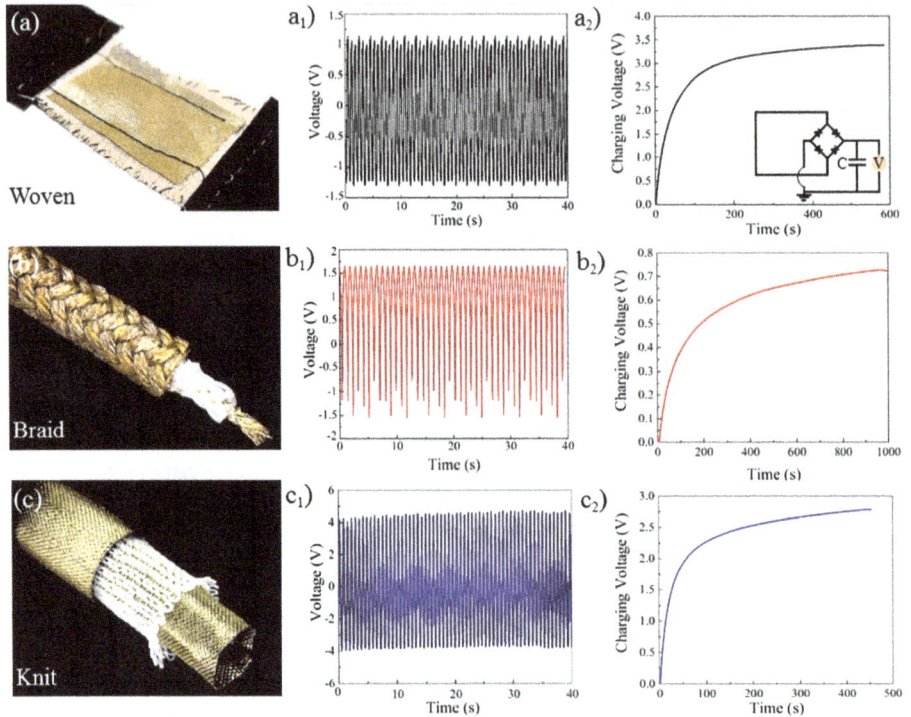

Figure 15.17: Output voltage profiles of the charged capacitor (10 μF) using as-prepared energy harvester based on PVDF/BT10 fibres: (a) representative picture of woven fabric and its (a₁) output voltage versus time of the fabric (a₂) capacitor charging profile; (b) braided fabric and its (b₁) output voltage versus time curve (b₂) capacitor charging profile; (c) circular knitted fabric and its (c₁) output voltage versus time curve and (c₂) capacitor charging profile (reprinted with permission from [62], Wiley, Copyright 2020).

15.7 Conclusions and future outlook

The science and technology of textile-based energy harvesters have greatly improved over the years, and many devices based on a multitude of structures and materials have been reported till date. However, challenges remain for each specific type of energy harvester which still need to be overcome. Mostly, there has been a recent thrust on usage of organic materials like conductive polymers and carbon-based materials to replace the traditionally used ceramics and metals. Incorporation of such organic materials come with their own problems of reduced power output. However, they are perhaps more lucrative in terms of biodegradability and recyclability. There has also been a trend of combining organics with inorganic materials, to complement each other and provide flexibility and increased output together. Further, for realizing truly textile-based self-powered systems that are able to operate under demanding conditions and for a long period of time, tests need to be done to check their durability and stability. Achieving fully biodegradable, scalable, high-performance devices that can function at the same level over extended periods of time is still a challenge.

References

[1]   Wang ZL, Lin L, Chen J, Niu S, Zi Y. Triboelectric Nanogenerators 2016.https://doi.org/10.1007/978-3-319-40039-6.

[2]   Zhang X. Overview of Triboelectric Nanogenerators. Flex Stretchable Triboelectric Nanogenerator Devices Toward Self-Powered Syst 2019:1–18. https://doi.org/10.1002/9783527820153.CH1.

[3]   Li P, Guo Y, Mu J, Wang H, Zhang Q, Li Y. Single-walled carbon nanotubes/polyaniline-coated polyester thermoelectric textile with good interface stability prepared by ultrasonic induction. RSC Adv 2016;6:90347–53. https://doi.org/10.1039/C6RA16532J.

[4]   Dabrowska A, Greszta A. Analysis of the Possibility of Using Energy Harvesters to Power Wearable Electronics in Clothing. Adv Mater Sci Eng 2019;2019. https://doi.org/10.1155/2019/9057293.

[5]   Zhang XS, Han M, Kim B, Bao JF, Brugger J, Zhang H. All-in-one self-powered flexible microsystems based on triboelectric nanogenerators. Nano Energy 2018;47:410–26. https://doi.org/10.1016/J.NANOEN.2018.02.046.

[6]   Gao Y, Xie C, Zheng Z. Textile Composite Electrodes for Flexible Batteries and Supercapacitors: Opportunities and Challenges. Adv Energy Mater 2021;11:2002838. https://doi.org/10.1002/AENM.202002838.

[7]   Ghouri AS, Aslam R, Siddiqui MS, Sami SK. Recent Progress in Textile-Based Flexible Supercapacitor. Front Mater 2020;7:58. https://doi.org/10.3389/FMATS.2020.00058/BIBTEX.

[8]   Elmoughni HM, Menon AK, Wolfe RMW, Yee SK. A Textile-Integrated Polymer Thermoelectric Generator for Body Heat Harvesting. Adv Mater Technol 2019;4:1800708. https://doi.org/10.1002/ADMT.201800708.

[9]   Beeby SP, Cao Z, Almussallam A. Kinetic, thermoelectric and solar energy harvesting technologies for smart textiles. Multidiscip Know-How Smart-Textiles Dev 2013:306–28. https://doi.org/10.1533/9780857093530.2.306.

[10] Chatterjee K, Ghosh TK. Thermoelectric Materials for Textile Applications. Molecules 2021;26:3154. https://doi.org/10.3390/MOLECULES26113154.

[11] Wang L, Zhang K. Textile-Based Thermoelectric Generators and Their Applications. Energy Environ Mater 2020;3:67–79. https://doi.org/10.1002/EEM2.12045.

[12] Enescu D. Thermoelectric Energy Harvesting: Basic Principles and Applications. Green Energy Adv 2019. https://doi.org/10.5772/INTECHOPEN.83495.

[13] Jangra V, Maity S, Vishnoi P. A review on the development of conjugated polymer-based textile thermoelectric generator 2022;51:181–214. https://doi.org/10.1177/1528083721996732.

[14] Schmidl G, Jia G, Gawlik A, Andrä G, Richter K, Plentz J. Aluminum-doped zinc oxide–coated 3D spacer fabrics with electroless plated copper contacts for textile thermoelectric generators. Mater Today Energy 2021;21:100811. https://doi.org/10.1016/J.MTENER.2021.100811.

[15] Klochko NP, Barbash VA, Petrushenko SI, Kopach VR, Klepikova KS, Zhadan DO, Yashchenko OV, Dukarov SV, Sukhov VM, Khrypunova AL. Thermoelectric textile devices with thin films of nanocellulose and copper iodide. Journal of Materials Science: Materials in Electronics. 2021 Sep; 32(18):23246–65.

[16] Klochko NP, Klepikova KS, Zhadan DO, Kopach VR, Chernyavskaya SM, Petrushenko SI, et al. Thermoelectric textile with fibers coated by copper iodide thin films. Thin Solid Films 2020;704:138026. https://doi.org/10.1016/J.TSF.2020.138026.

[17] Zhou J, Wang H, He D, Zhou Y, Peng W, Fan F, et al. Transferable and flexible thermoelectric thin films based on elemental tellurium with a large power factor. Appl Phys Lett 2018;112:243904. https://doi.org/10.1063/1.5034001.

[18] Lu Z, Zhang H, Mao C, Li CM. Silk fabric-based wearable thermoelectric generator for energy harvesting from the human body. Appl Energy 2016;164:57–63. https://doi.org/10.1016/J. APENERGY.2015.11.038.

[19] He X, Shi J, Hao Y, He M, Cai J, Qin X, et al. Highly stretchable, durable, and breathable thermoelectric fabrics for human body energy harvesting and sensing. Carbon Energy 2022;4:621–32. https://doi.org/10.1002/CEY2.186.

[20] Komatsu N, Ichinose Y, Dewey OS, Taylor LW, Trafford MA, Yomogida Y, et al. Macroscopic weavable fibers of carbon nanotubes with giant thermoelectric power factor. Nat Commun 2021 121 2021;12:1–8. https://doi.org/10.1038/s41467-021-25208-z.

[21] Amirabad R, Ramazani Saadatabadi A, Pourjahanbakhsh M, Siadati MH. Enhancing Seebeck coefficient and electrical conductivity of polyaniline/carbon nanotube–coated thermoelectric fabric. J Ind Text 2022;51:3297S–3308S. https://doi.org/10.1177/15280837211050516/ASSET/IMAGES/LARGE/ 10.1177_15280837211050516-FIG2.JPEG.

[22] Ruiz-Calleja T, Calderón-Villajos R, Bonet-Aracil M, Bou-Belda E, Gisbert-Payá J, Jiménez-Suárez A, et al. Thermoelectrical properties of graphene knife-coated cellulosic fabrics for defect monitoring in Joule-heated textiles. J Ind Text 2022;51:8884S–8905S. https://doi.org/10.1177/ 15280837211056986.

[23] Marin G, Funahashi R, Karppinen M. Textile-Integrated ZnO-Based Thermoelectric Device Using Atomic Layer Deposition. Adv Eng Mater 2020;22:2000535. https://doi.org/10.1002/ ADEM.202000535.

[24] Borazan İ, Bedeloglu AC, Demir A. A photovoltaic textile design with a stainless steel mesh fabric. J Ind Text 2022;51:1527–38. https://doi.org/10.1177/1528083720904053/ASSET/IMAGES/LARGE/ 10.1177_1528083720904053-FIG2.JPEG.

[25] Satharasinghe A, Hughes-Riley T, Dias T. A Review of Solar Energy Harvesting Electronic Textiles. Sensors 2020;20:5938. https://doi.org/10.3390/S20205938.

[26] Wilson JIB, Mather RR. Photovoltaic energy harvesting for intelligent textiles. Electron Text Smart Fabr Wearable Technol 2015:155–71. https://doi.org/10.1016/B978-0-08-100201-8.00009-6.

[27] Sharma K, Sharma V, Sharma SS. Dye-Sensitized Solar Cells: Fundamentals and Current Status. Nanoscale Res Lett 2018 131 2018;13:1–46. https://doi.org/10.1186/S11671-018-2760-6.

[28] Yun MJ, Cha SI, Seo SH, Lee DY. Highly Flexible Dye-sensitized Solar Cells Produced by Sewing Textile Electrodes on Cloth. Sci Reports 2014 41 2014;4:1–6. https://doi.org/10.1038/srep05322.

[29] Yun MJ, Cha SI, Seo SH, Kim HS, Lee DY. Insertion of Dye-Sensitized Solar Cells in Textiles using a Conventional Weaving Process. Sci Reports 2015 51 2015;5:1–8. https://doi.org/10.1038/srep11022.

[30] Opwis K, Gutmann JS, Lagunas Alonso AR, Rodriguez Henche MJ, Ezquer Mayo M, Breuil F, et al. Preparation of a Textile-Based Dye-Sensitized Solar Cell. Int J Photoenergy 2016;2016. https://doi. org/10.1155/2016/3796074.

[31] Galembeck F, Burgo TAL, Balestrin LBS, Gouveia RF, Silva CA, Galembeck A. Friction, tribochemistry and triboelectricity: recent progress and perspectives. RSC Adv 2014;4:64280–98. https://doi.org/10.1039/C4RA09604E.

[32] Zou H, Zhang Y, Guo L, Wang P, He X, Dai G, et al. Quantifying the triboelectric series. Nat Commun 2019 101 2019;10:1–9. https://doi.org/10.1038/s41467-019-09461-x.

[33] Xu C, Zi Y, Wang AC, Zou H, Dai Y, He X, et al. On the Electron-Transfer Mechanism in the Contact-Electrification Effect. Adv Mater 2018;30:1706790. https://doi.org/10.1002/ADMA.201706790.

[34] Dharmasena RDIG, Silva SRP. Towards optimized triboelectric nanogenerators. Nano Energy 2019;62:530–49. https://doi.org/10.1016/J.NANOEN.2019.05.057.

[35] Liu C, Bard AJ. Electrostatic electrochemistry at insulators. Nat Mater 2008 76 2008;7:505–9. https://doi.org/10.1038/nmat2160.

[36] Zhang L, Yu Y, Eyer G, . . . GS-AM, 2016 undefined. All-textile triboelectric generator compatible with traditional textile process. Wiley Online Libr 2016;1. https://doi.org/10.1002/admt.201600147.

[37] Kwak SS, Kim H, Seung W, Kim J, Hinchet R, Kim SW. Fully Stretchable Textile Triboelectric Nanogenerator with Knitted Fabric Structures. ACS Nano 2017;11:10733–41. https://doi.org/10.1021/ACSNANO.7B05203.

[38] Cheng R, Dong K, Liu L, Ning C, Chen P, Peng X, et al. Flame-retardant textile-based triboelectric nanogenerators for fire protection applications. ACS Nano 2020;14:15853–63. https://doi.org/10.1021/ACSNANO.0C07148.

[39] Zhang J, Hu S, Shi Z, Wang Y, Lei Y, Han J, et al. Eco-friendly and recyclable all cellulose triboelectric nanogenerator and self-powered interactive interface. Nano Energy 2021;89. https://doi.org/10.1016/J.NANOEN.2021.106354.

[40] Xia R, Zhang R, Jie Y, Zhao W, Cao X, Wang Z. Natural cotton-based triboelectric nanogenerator as a self-powered system for efficient use of water and wind energy. Nano Energy 2022;92:106685. https://doi.org/10.1016/J.NANOEN.2021.106685.

[41] Dudem B, Mule AR, Patnam HR, Yu JS. Wearable and durable triboelectric nanogenerators via polyaniline coated cotton textiles as a movement sensor and self-powered system. Nano Energy 2019;55:305–15. https://doi.org/10.1016/J.NANOEN.2018.10.074.

[42] Jakmuangpak S, Prada T, Mongkolthanaruk W, Harnchana V, Pinitsoontorn S. Engineering bacterial cellulose films by nanocomposite approach and surface modification for biocompatible triboelectric nanogenerator. ACS Publ 2020;2:2498–506. https://doi.org/10.1021/acsaelm.0c00421.

[43] Kim HJ, Yim EC, Kim JH, Kim SJ, Park JY, Oh IK. Bacterial Nano-Cellulose Triboelectric Nanogenerator. Nano Energy 2017;33:130–7. https://doi.org/10.1016/J.NANOEN.2017.01.035.

[44] Shao Y, Feng CP, Deng BW, Yin B, Yang MB. Facile method to enhance output performance of bacterial cellulose nanofiber based triboelectric nanogenerator by controlling micro-nano structure and dielectric constant. Nano Energy. 2019 Aug 1;62:620–7.

[45] Kim H-J, Kim J-H, Jun K-W, Kim J-H, Seung W-C, Kim H-J, et al. Silk nanofiber-networked bio-triboelectric generator: silk bio-TEG. Wiley Online Libr 2016;6. https://doi.org/10.1002/aenm.201502329.

[46] Dudem B, Graham SA, Dharmasena RDIG, Silva SRP, Yu JS. Natural silk-composite enabled versatile robust triboelectric nanogenerators for smart applications. Nano Energy 2021;83. https://doi.org/10.1016/J.NANOEN.2021.105819.

[47] Shao Y, Feng CP, Deng BW, Yin B, Yang MB. Facile method to enhance output performance of bacterial cellulose nanofiber based triboelectric nanogenerator by controlling micro-nano structure and dielectric constant. Nano Energy. 2019 Aug 1;62:620–7.

[48] Niu Q, Huang L, Lv S, Shao H, Fan S, Zhang Y. Pulse-driven bio-triboelectric nanogenerator based on silk nanoribbons. Nano Energy. 2020 Aug 1;74:104837.

[49] Zhou T, Zhang C, Bao Han C, Ru Fan F, Tang W, Lin Wang Z. Woven structured triboelectric nanogenerator for wearable devices. ACS Publ 2014;6:14695–701. https://doi.org/10.1021/am504110u.

[50] Scheffler S, Poulin P. Piezoelectric Fibers: Processing and Challenges. ACS Appl Mater Interfaces 2022;14:16961–82. https://doi.org/10.1021/ACSAMI.1C24611.

[51] Anwar S, Hassanpour Amiri M, Jiang S, Mahdi Abolhasani M, F Rocha PR, Asadi K, et al. Piezoelectric Nylon-11 Fibers for Electronic Textiles, Energy Harvesting and Sensing. Adv Funct Mater 2021;31:2004326. https://doi.org/10.1002/ADFM.202004326.

[52] Park C, Kim H, Cha Y. Piezoelectric sensor with a helical structure on the thread core. Appl Sci 2020;10. https://doi.org/10.3390/APP10155073.

[53] Martins RS, Gonçalves R, Azevedo T, Rocha JG, Nóbrega JM, Carvalho H, et al. Piezoelectric coaxial filaments produced by coextrusion of poly(vinylidene fluoride) and electrically conductive inner and outer layers. J Appl Polym Sci 2014;131:8749–60. https://doi.org/10.1002/APP.40710.

[54] Oh HJ, Kim DK, Choi YC, Lim SJ, Jeong JB, Ko JH, et al. Fabrication of piezoelectric poly(l-lactic acid)/BaTiO3 fibre by the melt-spinning process. Sci Rep 2020;10. https://doi.org/10.1038/S41598-020-73261-3.

[55] Lund A, Jonasson C, Johansson C, Haagensen D, Hagström B. Piezoelectric polymeric bicomponent fibers produced by melt spinning. J Appl Polym Sci 2012;126:490–500. https://doi.org/10.1002/APP.36760.

[56] Chowdhury A, Bairagi S, Ali SW, Kumar B. Leveraging Shape Memory Coupled Piezoelectric Properties in Melt Extruded Composite Filament Based on Polyvinylidene Fluoride and Polyurethane. Macromol Mater Eng 2020. https://doi.org/10.1002/MAME.202000296.

[57] Feng W, Chen Y, Wang W, Yu D. A waterproof and breathable textile pressure sensor with high sensitivity based on PVDF/ZnO hierarchical structure. Colloids Surfaces A Physicochem Eng Asp 2022;633:127890. https://doi.org/10.1016/J.COLSURFA.2021.127890.

[58] Uddin MM, Blevins B, Yadavalli NS, Pham MT, Nguyen TD, Minko S, et al. Highly flexible and conductive stainless-steel thread based piezoelectric coaxial yarn nanogenerators via solution coating and touch-spun nanofibers coating methods. Smart Mater Struct 2022;31:035028. https://doi.org/10.1088/1361-665X/AC5015.

[59] Khan A, Ali Abbasi M, Hussain M, Hussain Ibupoto Z, Wissting J, Nur O, et al. Piezoelectric nanogenerator based on zinc oxide nanorods grown on textile cotton fabric. Appl Phys Lett 2012;101:193506. https://doi.org/10.1063/1.4766921.

[60] Su Y, Chen C, Pan H, Yang Y, Chen G, Zhao X, et al. Muscle Fibers Inspired High-Performance Piezoelectric Textiles for Wearable Physiological Monitoring. Adv Funct Mater 2021;31:2010962. https://doi.org/10.1002/ADFM.202010962.

[61] Anand S, Soin N, Shah TH, Siores E. Energy harvesting "3-D knitted spacer" based piezoelectric textiles. IOP Conf Ser Mater Sci Eng 2016;141:012001. https://doi.org/10.1088/1757-899X/141/1/012001.

[62] Mokhtari F, Spinks GM, Fay C, Cheng Z, Raad R, Xi J, et al. Wearable Electronic Textiles from Nanostructured Piezoelectric Fibers. Adv Mater Technol 2020;5:1900900. https://doi.org/10.1002/ADMT.201900900.

Ajay K. Maddineni and Dipayan Das

16 Fibrous materials for automotive applications

Abstract: The current chapter deals with various kinds of the fibrous materials used for different applications in automotive vehicles. It starts with the fibrous materials used for automotive air filtration. It discusses on typical airborne contaminants and filtration strategies, followed by illustration of basic principles of aerosol filtration. It then proceeds to delineate on the fibrous materials used for thermal and acoustic insulation. It presents the principles of propagation of thermal and noise energies within the fibrous structures. Afterwards, the structural applications of fibrous materials used for safety and reinforcement purposes are included. This chapter also highlights the recent developments that have been taken place on various kinds of fibrous materials for automotive applications. Further, a special emphasis is made to provide the current need of the industry for the fibrous material in the aforementioned applications.

Keywords: Fibrous materials, aerosol filtration, thermal insulation, acoustics insulation, safety restraint, structural reinforcement, automotive

16.1 Introduction

Fibrous materials have gained enormous popularity in the recent years for automotive applications due to their desirable properties in meeting the higher level of air quality, better NVH needs (passenger comfort), fuel efficiency, stringent emission, and safety regulations. They are extensively used in automotive for filtration of engine intake air, filtration for electric vehicles, and filtration for interior cabin. Also, the fibrous materials are widely used for vehicle interior, exterior, and high temperature applications related to thermal insulation, noise reduction, and structural purposes. Nevertheless, the ever-increasing mobility trends and emphasis on light weight to address the environment pollution and safety norms have led to develop the newer functional materials meeting the performance attributes of their intended automotive applications. Research attempts to develop such engineering materials includes composite fibrous structures, incorporating nanotechnology, chemically treating the fibres, and enhanced processing are various trends today. The purpose of this chapter is to review and discuss the engineering and scientific developments of fibrous materials used in aforementioned automotive applications.

https://doi.org/10.1515/9783110759747-016

16.2 Fibrous materials for automotive air filtration

16.2.1 Airborne contaminants and air filtration strategies

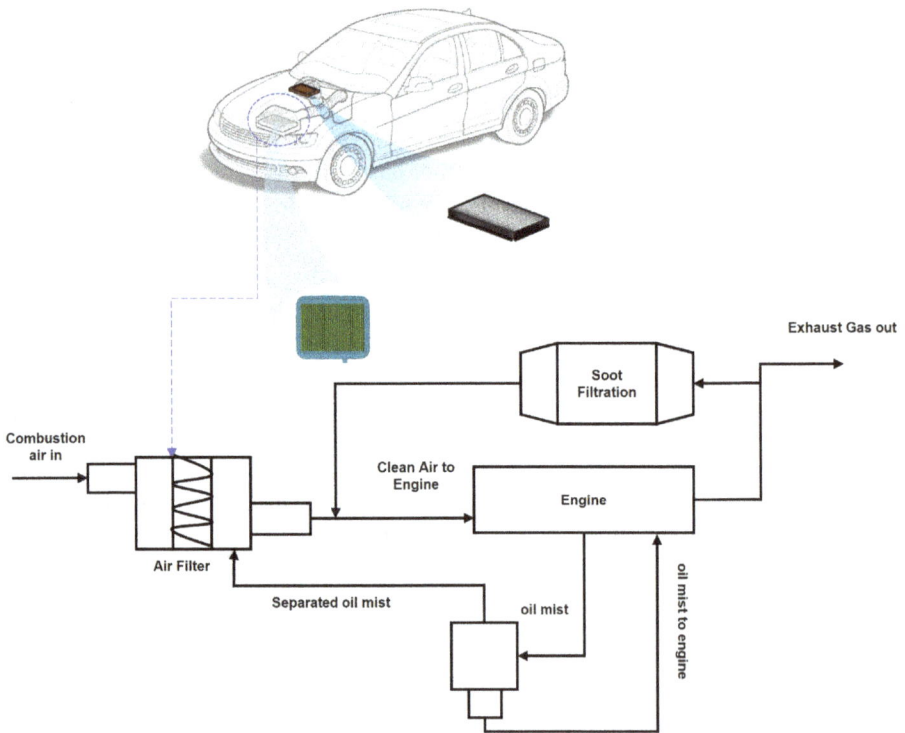

Figure 16.1: Pictorial view of air filtration system in an automotive vehicle.

Filtration plays a vital role in several fluid flow applications, for example, automotive engine filtration to reduce engine wear and protecting electronic sensors, interior cabin air filtration for passenger comfort, particulate air filtration for exhaust emission reduction, respiratory filters for human health, and industrial filters for minimizing ambient air pollution. Among several filtration and separation techniques available, fibrous filter media for filtering the dust particles ranging from as low as nano to submicron to as high as micron level is common nowadays due to its versatile nature in terms of fibre characteristics and media structure. Air filtration plays an important role in deciding the performance of automotive vehicles in terms of fuel consumption and emission reduction and passenger health and comfort. The air filtration system in automotives is quite diverse and complex primarily because of the critical requirements of modern engines, stringent emission regulations, demand for fuel economy, and greater comfort. Figure 16.1 gives a pictorial view of the air filtration system in automotive vehicles. The system includes engine intake air filtration, cabin air filtration, exhaust gas filtration, and oil mist separation. The engine intake air filtration removes the airborne contaminants from entering the engines, which, otherwise, would have caused significant engine wear leading to performance loss, enhanced exhaust emission, high oil consumption, and increased operation cost [1–3]. The cabin air filtration is used to improve the air quality and cleanliness in the vehicle interior as well as to protect the air conditioning system from dust particles such as particulate matter (PM) and other harmful contaminants [2]. The exhaust gas requires filtering the tiny soot particles; otherwise they cause adverse effects on human health and well-being [4]. Also, the oil mist-containing particles ranging from 1 to 10 μm in diameter should be separated before entering the air intake system in order to reduce the oil consumption and also to meet the newer exhaust emission limits [5]. Needless to say that the specific requirements of different kinds of automotive filters vary widely and that all the filters must be specifically designed for their intended purposes because of a wide range of contaminates, operating conditions, and environmental considerations. Today, many functions in automotive filtration can no longer be considered separately; to some extent they are inter-related, forming an integral part of a system and perform complex processes and functions within it, which go well beyond the actual task of filtration. A profound knowledge on automotive air filtration system is therefore necessary to design, develop, and optimize modern automotive vehicles. The automotive engine intake air filtration unit includes air filter housing, pleated filter element, and flat filter media. This is shown in Figure 16.2. The primary function of the filter housing is air filtration and flow management, while the secondary function concerns acoustics, design, water separation, and air mass sensing. The filter housing must be adapted to the limited installation space available. The filter element is the customization of the filter media, which is almost always, due to space constraint, pleated to increase the area of filtration. The filter media may be made from fibrous material or non-fibrous material. Whatsoever be the filter media, the internal combustion engines (ICEs) must receive clean air for combustion of fuels. However, with the advancement of civilization and its growth, the quality of the atmospheric air is deteriorating day by day. This is mainly due to continuous increase in air contaminants such as dust and sand from agricultural fields carried by the wind, particles resulted from the road wear and the motor vehicle component wear, diesel soot from the motor vehicle exhaust, particle sources from industries, organic, and mineral dusts [6]. They do not, however, appear only as solid particles but may be of liquid form, for example, rain droplets, snow, fog, unburnt fuel, and oil mist from blow by gas in crankcase [3]. These air contaminants are influenced by the geographical region, season, urban, or extra urban areas. Figure 16.3 displays the typical contaminants present in air. The most critical factor for filter performance is the wide range of air contaminants at different concentrations, particle size distribution, and chemical composition. As known, the atmospheric contaminants can be usually described by a bimodal distribution [7]. The fine mode contains particles in the range of 0.1 to 2.5 μm, while the coarse mode lies is in the range of 2.5 to 30 μm. The particle sizes ranging from 0.002 to 100 μm are generally found in traffic on highways and dusty environments. On dusty roads, particle sizes larger than 100 μm are found such as mineral and organic particles [3]. The largest particles such as pollen, insects, and some vegetation debris can be easily removed. Particles varying from 2.5 to 15–20 μm in diameter are observed to be the major cause of engine wear. Also, the engine components with large clearances, for say bearings, were more affected by particles of 20–40 μm diameter [8]. The engine intake air is the main source of such contaminants to enter the engine and, to some extent, into the oil. In this way, they penetrate critical areas such as clearance between cylinder liners and piston and between piston rings and connecting rods, where they cause significant wear on the components. Also, these contaminants cause malfunctioning of air flow sensors, which are responsible for metering the fuel in modern engines. These altogether result in loss of power, increased fuel consumption, and higher pollutant emissions.

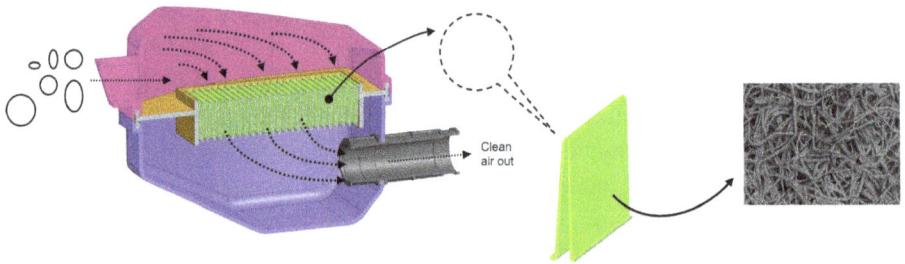

Figure 16.2: Schematic of engine intake air filtration unit in automotive.

Besides particles, ultrafine PM generated from modern engines, harmful gases such as sulphur oxides, nitrogen oxides, volatile organic compounds (VOCs), ozone, and odours would also be getting into the indoor cabins via automotive heating ventilation and air conditioning systems. Also, the suspended and inhalable PM along with the aeroallergens such as pollen, mould spores, dust mites, and pet dander would cause allergic reactions and unwanted symptoms due to sensitive immune response. The size of the black carbon particles that are generated from the diesel engine and the direct injection gasoline engines may vary from 10 to 100 nm, depending on the vehicle operating conditions. The size of the particle in the case of bacteria will range from a few sub-micrometres to 10 μm and are of 8 nm to 0.3 μm in the case of viruses. Further, the black carbons often carry the chemicals and allergens that are adsorbed on its surface. The indoor cabin airborne-contaminant concentration has found to be higher than that found in the ambient atmosphere, and this would put the cabin passengers at high risk upon the exposure. Needless to say, these particles must be removed from the air stream. Therefore, the cabin air filtration is needed to protect the passengers from the harmful contaminants.

There are several strategies available for air filtration depending on the size, type, and concentration of the contaminants to be filtered out. Figure 16.3 displays a schematic

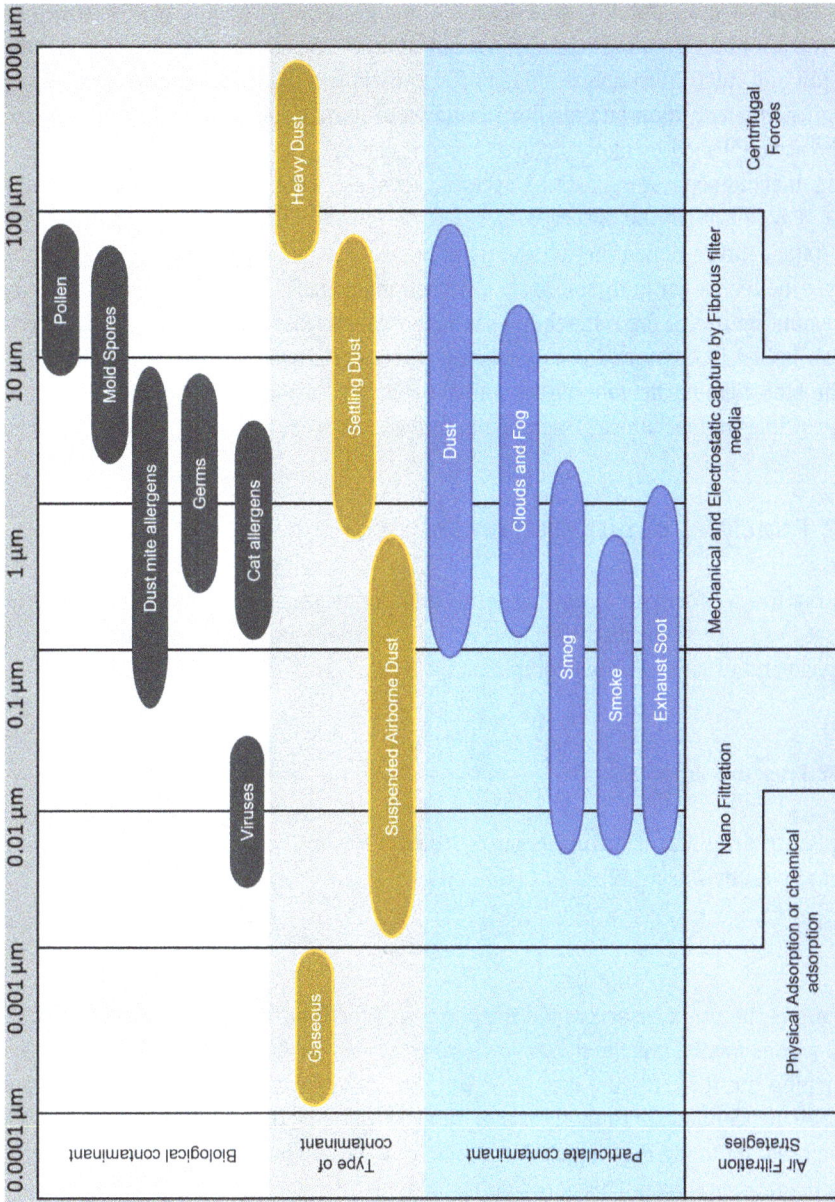

Figure 16.3: Airborne contaminants and air filtration strategies.

of such filtration strategies. It can be observed that particle filtration is an efficient strategy for macro and micro particles. Macro filtration is efficient for particles size greater than 1 µm and micro filtration is efficient for particle less than 1 µm in size. Depending on the dust concentration and air flow, removal of particles of greater than 10 µm size could be accomplished by exposing to external forces such as centrifugal forces. For this purpose, inertial separators such as cyclonic separator and swirl tube separators are used [9]. Further, it can be observed that particle filtration is based on mechanical and electrostatic capture depending on the particle size. It is a known fact that the fibrous filters are ineffective for inorganic gases. Adsorption principles based on activated carbon filters would be useful for removing various inorganic gases and hydrocarbons, which are often linked to unwanted symptoms and discomfort caused by diesel exhaust. Further, the biological particulate matter and the allergens are filtered based on the combined principles of mechanical, electrostatic, and chemical capture.

16.2.2 Principles of filtration and purification

Predicting the performance requirements such as pressure drop and filtration efficiency at the stage of product design is essential in today's shorter product development cycle and greater product complexity.

16.2.2.1 Pressure drop

The pressure drop across a fibrous porous media for an incompressible and laminar flow was given by Darcy [10]:

$$\frac{\Delta P_f}{Z} = \frac{1}{k}\mu U_0 \tag{16.1}$$

where ΔP_f is the pressure across the thickness Z of the filter media, k is the permeability of the filter media, μ is the dynamic viscosity of air, and U_0 is the face velocity of air approaching the filter media. It is observed that this law could be applicable for the flow of air perpendicular to the granular filter media. For fibrous filter media, Davies [11] proposed the following empirical equation that was widely used in practice for estimating the pressure drop. This equation is valid for filter solid volume fraction ranging from 0.006 to 0.3:

$$\Delta P_f = 64 U_0 \frac{Q}{A}\frac{1}{d_f^2}\varphi^{1.5}\left(1 + 56\varphi^3\right)Z \tag{16.2}$$

where φ is the solid volume fraction, d_f is the fibre diameter, Q is the volumetric flow rate, and A is the cross-sectional area of the filter media.

16.2.2.2 Filtration efficiency

The classical filtration theory, based on single fibre filtration efficiency, was primarily developed based on fluid flow field through cell model proposed by Happel [12] and Kuwabara [13]. For studies of filtration in granular filter media, Happel's model is often used, whereas, for studies in fibrous filter media, Kuwabara's model is mostly used [14]. In the following, an analytical expression for predicting the filtration efficiency of fibrous filter media is given:

$$E = 1 - \exp\left(-\frac{4\varphi E_\Sigma Z}{\pi d_\mathrm{f}(1-\varphi)}\right) \tag{16.3}$$

where E indicates the overall filtration efficiency of the filter media and E_Σ denotes the single fibre efficiency.

The basic mechanism by which the airborne particles are captured by a fibre can be divided into five groups. These are known as Brownian diffusion, inertial impaction, direct interception, gravitational settling, and electrostatic attraction. The overall filtration efficiency of the single fibre is assumed to be the summation of filtration efficiencies of all the capture mechanisms [15]:

$$E_\Sigma = E_\mathrm{D} + E_\mathrm{I} + E_\mathrm{R} + E_\mathrm{G} + E_\mathrm{q} \tag{16.4}$$

where E_D denotes single fibre filtration efficiency due to diffusion, E_I indicates single fibre filtration efficiency due to inertial impaction, E_R represents single fibre filtration efficiency due to direct interception, E_G refers to single fibre filtration efficiency due to gravitational settling, and E_q denotes single fibre filtration efficiency due to electrostatic attraction.

It is known that the ability of a filter to remove particles from the air stream is directly related to the size of the particles present in the air stream. The engine intake air filters filter out particulates from airstreams by virtue of four mechanisms: Brownian diffusion, direct interception, inertial impaction, and gravitational settling [15]. These four mechanisms are schematically illustrated in Figure 16.4.

16.2.2.3 Brownian diffusion

The particles of diameter in the range of 0.1 µm and below tend to make random motions due to their interaction with the zigzag gas molecules [17]. As these particles are bumped by the gas molecules; they too begin to bump into other particles and move randomly as well. This phenomenon is identical to Brownian diffusion. This is shown in Figure 16.4(a). The smaller is the particle and the slower is the flow, the more is the time to have zigzag motion, and thereby giving it much better chance to collide with and stick to a fibre surface. The particle collection by diffusion is governed by a

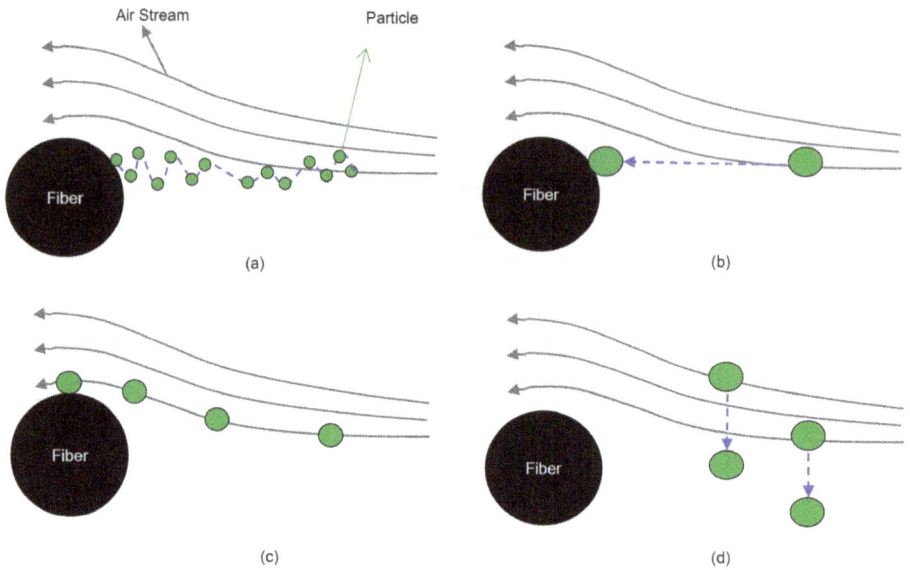

Figure 16.4: Scheme of filtration mechanisms: (a) Brownian diffusion, (b) inertial Impaction, (c) direct interception, and (d) gravitational settling.

characteristic number which is referred to as Peclet number. The Peclet number is defined as follows:

$$P_e = \frac{d_f\, U_0}{D_i} \qquad (16.5)$$

where P_e denotes Peclet number, U_0 refers to face velocity, and D_i indicates diffusion coefficient. The diffusion coefficient is defined as follows:

$$D_i = \frac{C_c K_B T}{3\pi\mu d_p} \qquad (16.6)$$

where C_c denotes Cunningham correction factor, K_B indicates Boltzmann constant, T refers to absolute temperature, μ represents dynamic viscosity of gas (air), and d_p is particle diameter. The Cunningham correction factor is expressed as follows:

$$C_c = 1 + 2.492\frac{\lambda}{d_p} + 0.84\frac{\lambda}{d_p}\exp\left(-0.435\frac{\lambda}{d_p}\right) \qquad (16.7)$$

where λ denotes the free mean path length of gas molecules. In general, as the velocity of approach increases the Peclet number also increases and hence collection efficiency decreases. Lee and Liu [18] derived the following expression to predict the single fibre efficiency to capture the particles by diffusion mechanism

$$E_D = 2.58 \left(\frac{1-\varphi}{\text{Ku}} \right)^{1/3} \text{Pe}^{-2/3} \tag{16.8}$$

where Ku stands for Kuwabara hydrodynamic factor, which is defined as follows:

$$\text{Ku} = -\frac{1}{2}\ln\varphi - \frac{3}{2} + \varphi - \frac{\varphi^2}{4} \tag{16.9}$$

16.2.2.4 Inertial impaction

Inertial impaction occurs when a suspended particle is so large that it is unable to quickly adjust to the abrupt changes in the streamline direction near the boundary of the filter fibre [19]. This is shown in Figure 16.4(b). The particles try to continue their movement in a straight path and hence collide with the fibre where they get trapped and remained by the fibre surface. The particle collection by inertia is governed by a characteristic number which is often referred to as stokes number. Stokes number (Stk) is defined as

$$\text{Stk} = \frac{\rho_p C_c d_p^2 U_0}{18 \mu d_f} \tag{16.10}$$

where d_p represents particle diameter, ρ_p refers to particle density, U_0 denotes approaching velocity, μ indicates dynamic viscosity of fluid, d_f refers to the fibre diameter, and C_c is the Cunningham slip factor. It is reported that if stokes number is greater than one then the particles' ability to follow the fluid decreases, and hence the collection efficiency of the filter increases with increasing Stokes number [20]. According to the model of Lee and Liu [18], the single fibre efficiency due to inertial impaction is expressed as follows:

$$E_I = \frac{(\text{Stk})J}{2(\text{Ku})^2} \tag{16.11}$$

The parameter J, when the interception parameter $Q = d_p/d_f$ is less than 0.4, is defined as follows:

$$J = \left(29.6 - 28\varphi^{0.62} \right)Q^2 - 27.5Q^{2.8} \tag{16.12}$$

The relation shows that if the flow velocity or the particle diameter increases, the single fibre collection efficiency also increases.

16.2.2.5 Direct interception

Interception occurs when particles, which are following a gas streamline, come into close proximity with the filter fibre [19]. Figure 16.4(c) displays the scheme for direct interception. If the particles are not too large enough, the inertial force is not predominant, and they are not even small enough to allow adequate diffusion to happen. For a particle of given size, there are certain streamlines which move close enough to the filter fibre so that particle is trapped in the stagnant zero zone near the fibre surface. Streamlines one particle radius away from the filter does not contribute to the interception mechanism [15]. Lee and Liu [18] derived the following relationship to predict the single fibre efficiency due to interception, which was independent of velocity:

$$E_R = \frac{(1+Q)}{2Ku} \left[2\ln(1+Q) - 1 + \varphi + \left(\frac{1}{1+Q}\right)^2 \left(1 - \frac{\varphi}{2}\right) - \frac{\varphi}{2}(1+Q)^2 \right] \tag{16.13}$$

where $Q = d_p/d_f$ is called interception parameter.

16.2.2.6 Gravitational settling

The penetration of particles may noticeably differ depending on the direction of flow relative to the direction of the gravitational sedimentation of particles. The penetration of particles decreases with decrease in the flow velocity as in both the cases gravity force starts to play. This is shown in Figure 16.4(d). The single fibre gravitation efficiency can be predicted from the following relationship proposed by Thomas et al. [21]:

$$E_G = \frac{\rho_p d_p^2 C_c g}{18\mu U_0} (1+Q) \tag{16.14}$$

where g stands for acceleration due to gravity.

16.2.2.7 Electrostatic attraction

Electrostatic deposition is often neglected as a mechanism for air filtration unless the particles or fibres are charged. Respiratory filters are often electrically charged by means of corona charging process [22]. Particles may be charged triboelectrically to know the charge on the particles and on the fibres to quantify. The single fibre efficiency for electrostatic image forces, E_q, for a neutral fibre and a particle with charge q, is expressed as

$$E_q = 1.5 \left[\frac{(\varepsilon_f - 1)q^2}{(\varepsilon_f + 1)12\pi^2 \mu U_0 \varepsilon_0 d_p d_f^2} \right]^{1/2} \tag{16.15}$$

where ε_f is the relative permittivity (dielectric constant) of the fibre, ε_0 is the permittivity of vacuum, μ is air viscosity, and U_0 is free stream gas velocity. As particle collection by interception and inertial impaction increases, the particle diameter increases, and collection by diffusion increases as particle diameter decreases, and a particle diameter region exists across which no mechanism dominates. This leads to a minimum filtration efficiency for the particles of about 200–500 nm in size. The particle diameter at which the minimum efficiency occurs is termed the most penetrating particle size.

16.2.2.8 Particle deposition

Irrespective of the type of aerosol (solid particle or the liquid droplet), the above-mentioned mechanisms would still be valid for capturing the particles by the fibres. However, both kinds would behave completely different after they capture. In general, filtration process begins with the filter media free from any deposited particle. As time proceeds, accumulation of particles increases and hence pressure drop across the media increases. Most of the published work, mentioned in the preceding section, does not take into account the presence of deposited particles. Consequently, the aforesaid model is used to predict the filtration efficiency of clean filter media. In the case of clogged filter media, a few attempts were made to predict the pressure drop. In addition to the aforesaid five mechanisms, the sieving mechanism and cake filtration also help to capture particles by the fibrous aggregates. The sieve effect stops large particles that are just too big to fit through the open pores of the filter medium. As reported, the smaller was the particle size than the pore size, the less was the particle capture by sieving mechanism. Cakes deposited on the inside of the filter media also act as a new collecting surface for time dependent filtration. This cake filtration mechanism is discussed hereunder.

In the case of clogged filter media, a few attempts were made to predict the pressure drop. Bergman et al. [23] suggested that the total pressure drop across a clogged filter is the summation of pressure drop of clean filter media (ΔP_f) and the pressure drop of the loaded filter (ΔP_g). This is shown as follows:

$$\Delta P = \Delta P_f + \Delta P_g \tag{16.16}$$

Based on this, Davies equation for pressure drop was modified as follows:

$$\Delta P = 64\mu U_0 Z \left(\frac{\alpha}{d_f^2} + \frac{\alpha_p}{d_p^2} \right)^{1/2} \left(\frac{\alpha}{d_f} + \frac{\alpha_p}{d_p} \right) \tag{16.17}$$

where a_p is packing density of collected particles and d_p is particle diameter. Novick et al. [24] proposed a model for pressure drop during loading, similar to Bergman, based on Kozony–Carman approach. It was assumed that the particles are deposited on the surface of the filter. The pressure drop across the cake is expressed as

$$\Delta P_g = \frac{h_k a_g^2 a_{pc} \mu}{C_c (1 - a_{pc})^2 \rho_p} U_0 m \tag{16.18}$$

where h_k is Kozeny constant and it is equal to 5 for spherical particles, a_{pc} is packing density of cake, C_c is Cunningham slip factor, ρ_p is density of particle, and a_g is specific area of the particle.

Thomas et al. [25] did a pioneering work to develop a new semi-empirical model for estimating the pressure drop during clogging process. They divided a filter medium into k number of layers, and the thickness of each layer was considered as Z_k. It was assumed that the particles were loaded uniformly in each layer. The packing density $a_{p,j,t}$ of the deposited particles was estimated by using the following relation:

$$a_{pkt} = \frac{V_{pk}}{V_k} \tag{16.19}$$

where V_{pk} denotes the volume of particles collected in layer k and V_k indicates the volume of layer k. Further, the fraction of flow across the fibres was equal to $1 - a_p/(1 - a)$, and the fraction of flow across the collected particle was equal to $a_p/(1 - a)$. For each time increment and each layer addition, the collection efficiency is estimated using the classical filtration relations discussed earlier. Knowing E_Σ using the existing relations and the particle size distribution upstream of filter media, mass deposited on each layer (due to fibre and particles), packing density, and the diameter of new fibres due the collected particles would be calculated. The total pressure drop during dust loading (ΔP_t) due to fibre and the collected particles is given by

$$\Delta P_t = \sum_{j=0}^{np} \Delta P_{k,t} \tag{16.20}$$

where $\Delta P_{k,t}$ denotes the pressure drop of each layer due to fibres and collected particles and the same was expressed as follows:

$$\Delta P_{k,t} = 64 \mu U_0 Z_j \left(\frac{a_{p,k,t}}{d_{p,k,t}^2} + \frac{a}{d_f^2} \right)^{0.5} \left(\frac{a_{p,k,t}}{d_{p,k,t}} + \frac{a}{d_f} \right) \left[1 + 56 (a + a_{p,k,t})^3 \right] \tag{16.21}$$

where $d_{p,k,t}$ means diameter of the collected particles in the kth slice at time t. One can refer to the work by Thomas et al. [25] for more details. Filtration of liquid aerosols is significantly different from filtration of solid particles [26]. The liquid droplets that are approaching the filter media either penetrate through the fibrous structure without a single collision to capture or collide the fibre and adhere to the fibre to undergo changes depending on the droplet size and how densely the fibrous structure is packed. If the fibrous structure is densely packed, capillary forces will dominate to fill the pore space with the water and possibly clog the filter quickly to increase the restriction of air flow. If the fibrous structure is loosely packed, then the droplets would adhere and stay there to evaporate or combine with the other approaching particle and grow big enough that gravity forces would become dominant. These grown droplets would eventually leave the fibre and get separated. This process is called coalescence. It does not require any filter regeneration.

In an interesting experimental observation, Mueller et al. [27] observed that the fibrous filter media exhibited a drop in filtration efficiency at a higher velocity. A similar observation was also made by Hubbard et al. [28]. This was probably associated with particle bounce and re-entrainment phenomena as shown in Figure 16.5. The occurrence of particle bounce is very common, and it mainly depends on particle size and shape, filtration velocity, impaction surface, and structure of filter media [29–31]. For bigger size particles, the adhesion is generally large enough to stick to the fibre; however, the drag acting on the larger particle is large enough to cause the particle to bounce from the fibre even at a lower velocity. So, their mass inertia leads to high impaction velocities in the fibre matrix and thereby shifting in dust distribution along the depth of the filter media [32], which increases the particle penetration through the media. In case of automotive engine intake application, this leads to an elevated risk to the engine components, associated with exposure to dust particles. It was reported that the spherical particles were more likely to bounce than the non-spherical particles. It was experimentally found by Mullins et al. [33] that the surface area was responsible for adhesion strength of particles to fibres. According to Boskovic et al. [34], on testing the spherical polystyrene latex and the cubic particles of magnesium oxide, it was observed that the cubic particles exhibited lower filtration efficiency than the spherical ones. This was due to the movement of such particles on the fibre, that is, the spherical particles would roll or slide whereas the cubic particles would tumble or slide with less surface area than the spherical ones. When the particle hits the fibre at low velocity the solid particle losses its kinetic energy by deforming itself and the fibre and stick to the fibre. At higher velocities, a part of the kinetic energy is dissipated in the deformation process and the remaining part is converted elastically to kinetic energy of rebound. If the rebound energy exceeds the adhesion energy, the particle will bounce away from the fibre. The total single fibre efficiency E_Σ, as a product of collision efficiency $E_{\Sigma,C}$ and adhesion efficiency $E_{\Sigma,A}$, was defined by Ptak and Jaroszczyk [35] as follows:

$$E_\Sigma = E_{\Sigma,C} E_{\Sigma,A} \tag{16.22}$$

For adhesion efficiency, Ptak and Jaroszczyk [35] proposed an expression as a function particle Reynolds number Re_p and Stokes number Stk of for adhesion efficiency $E_{\Sigma,A}$ based on experiments, as follows:

Particle
Re-entrainment

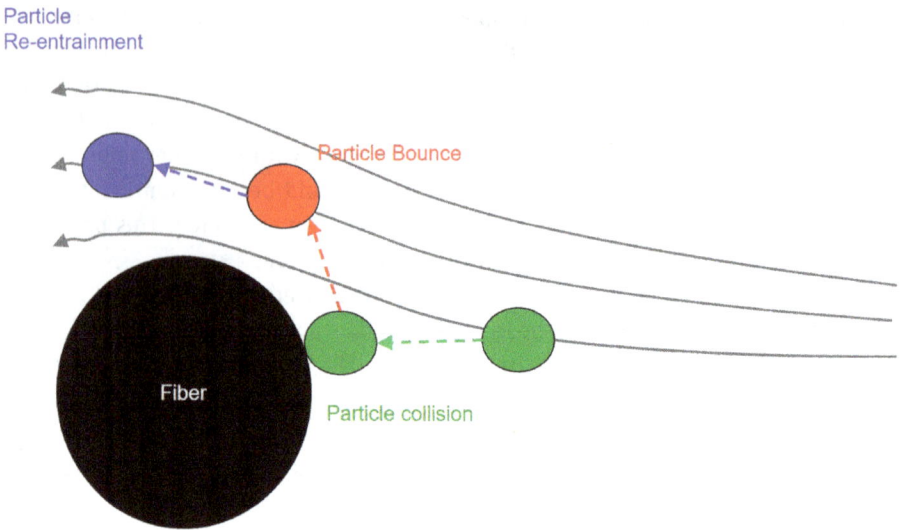

Figure 16.5: Scheme of particle rebound after collision.

$$E_{\Sigma,A} = \frac{190}{\left(Re_p Stk\right)^{0.68} + 190} \tag{16.23}$$

16.2.2.9 Adsorption

Further, the airborne water vapour and gaseous contaminants can be separated out via adsorption. Adsorption is a mass transfer process whereby the contaminant (adsorbate or sorbate) is accumulated on the surface of the solid (adsorbent/sorbent) coated on the fibre. A scheme of such process is shown in Figure 16.6. The driving force of this kind of separation is the residual surface energy of molecules or atoms due to unbalanced forces acting on the adsorbent surface that could form bonds with the adsorbate [36]. Upon collision on the surface of the adsorbent, contaminant would be attracted by these unbalanced forces and adhere to the surface. Such interactions involve direct transfer of electrons between the sorbate and the sorbent to accomplish the separation process. Adsorption process could be physical or chemical based on the kinds of unbalanced forces. Physical adsorption is due to the intermolecular forces such as electrostatic and van der Waals forces. This would generally be occurring at a low temperature with faster rate of adsorption and lower heat for adsorption. The structure of the adsorbate molecules hardly changes as the intermolecular attraction is generally weak, and a little energy is needed to separate the adsorbed

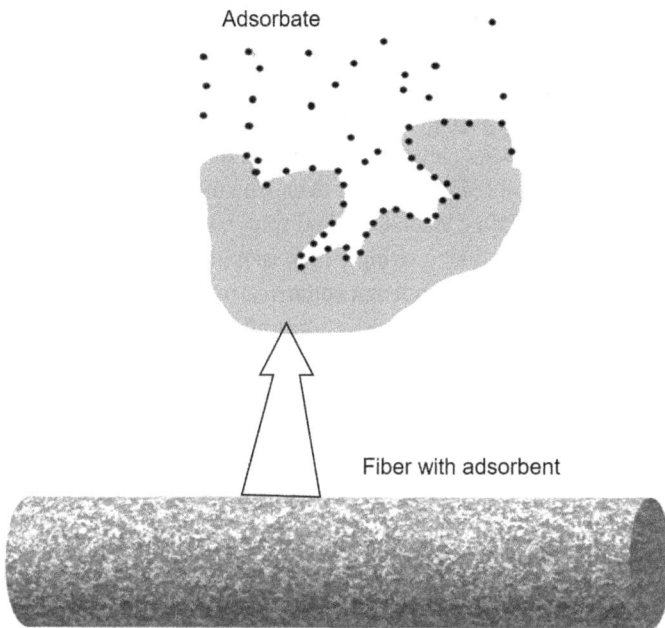

Figure 16.6: Scheme of gas adsorption.

molecules easily. Chemical adsorption is due to the action of formation or destruction of chemical bonds. Often, the physical and chemical adsorption occurs together.

16.2.3 Applications of filter materials

16.2.3.1 Filter materials for combustion engines

The engine intake air filters have evolved significantly from the use of steel mesh and oil bath air filter to sophisticated fibrous filters. Nowadays, a wide variety of filters such as polyurethane foam, cotton gauge, cellulosic papers treated with phenolic or non-phenolic resin, and synthetic non-wovens are used. Each of these filters has its own limitations with respect to the manufacturing processes and structure–property relationship, operating conditions, which are known to affect their filtration performance. As known, filtration efficiency and pressure drop are contradictory, that is, improving one leads to worsen the other, for example, increase in filtration efficiency is possible with increase in thickness of the filter (to increase the collision probability across the filter depth), which increases the pressure drop across the filter [37]. This also limits to dust holding capacity of the filters. Generally, to improve the dust holding capacity of the filter, it is necessary to increase the area of the filtering media, which is why the filters are almost always found to be pleated.

The filter media that are often used for automotive engine intake air filtration are synthetic filter media and cellulose filter media. As synthetic media are manufactured using non-woven technology, they are called non-woven filter media. The non-woven filter media utilize both surface as well as depth loading of dust deposition. Such media are typically designed to be used at face velocities in the range of about 0.25–0.75 m/s [38]. Also, such media exhibit remarkably low pressure drop. The mostly used synthetic media are prepared using needle-punching technology and spun-bond technology. They are very popular in European countries due to stringent emission norms, but less popular in Asian countries as they are expensive. In Asian countries, the most popular filter media which are used for automotive engine intake air filtration are cellulosic paper filter media. They are often called paper filter media [15]. They are thin sheet (single layer) of cellulosic fibres, whose ratio of media surface area to media calliper is very large, which makes paper media dominant in surface filtration, unlike depth filtration in foam or non-woven media. These filter materials are typically designed to be used at face velocities in the range of about 0.05–0.20 m/s [38]. The efficiency of this kind of media is superior to that of the foam filter media. The efficiency increases with the formation of dust cake; however, this may greatly increase the pressure restriction during loading. The dust holding capacity is low due to lower mass deposition per unit area. The resin-treated cellulosic media prepared using wet lay process have the ability to pleat easily with well-defined shape [39]. The cellulosic filter media are comparatively cheaper than the non-woven filter media. The traditional cellulosic paper media and the non-woven synthetic media are generally made up of relatively large fibres whose diameter is usually larger than 20 μm. The filter efficiency of such relatively large fibres is not high for smaller dust particles, which is encountered in on-highway environment. Because of this, the initial efficiency is usually very low with these filter media. The initial condition, that is, the beginning of the filter operation, is considered to be the worst time for the engine because a large amount of dust may reach the combustion chamber. Despite this fact, these media are commonly used in automotive engine intake air filtration. Of the several possibilities available to improve the initial efficiency of the engine intake air filter media, fibre diameter is the dominant factor that has significant effect on filtration efficiency. As the fibre diameter decreases, the net available surface area per unit mass increases. This enormous increase in surface area is known to improve the extent and quality of filtration. But the pressure drop significantly increases with decrease in fibre diameter. Also, improving the thickness of the filter media will improve the filtration efficiency but at the cost of increased pressure drop. Further, the random arrangement of fibres in cellulosic and non-woven filter media results in a wide range of pore size distribution. Large pores offering path of lower restriction to aerosol flow act as a source of pin holes, as the other regions of the filter become increasingly restricted by dust accumulation, the velocity increases through these pores, resulting higher penetration of both uncollected and re-entrained particles. This phenomenon is more dominant at the end of the filter life. Moreover, the trend of downsizing automotive air filter housing and pleated filter element also limit filtration area to accommodate. Reducing filter area results in increased local filtration velocities. For example, reduction of filtration area by 20% results in increased filtration velocity by 25% [40]. Increased filtration velocity leads to high particle kinetic energy that could cause bounce and re-entrainment of dust particles to reduce the collection efficiency. For the space constraint designs, cellulose-based paper filter media may not be beneficial due to its dominant surface filtration than the depth filtration. This greatly impacts the service life by reaching terminating condition much early. To improve the dust capacity, depth filters such as non-woven filter media would be a viable option. However, there exists a limitation with regard to the filtration efficiency beyond the critical velocities. In non-woven filter media, the filtration efficiency drops rapidly than cellulose filter media with increasing velocities. These non-wovens filter materials are still considered to be relatively young than the cellulose-based paper filter media.

To counteract the particle bounce and re-entrainment effect, the fibrous filter media used in automotive engine intake air filtration are sometimes found to be treated with oil [27, 31, 41, 42] as shown in Figure 16.7(a). Also, the oil-treated media have advantage on pressure drop during dust loading which increases dust holding capacity too. The oil film deposited onto the surface of the filter media facilitates the particles to dissipate the kinetic energy on contact, thereby reducing the probability of rebound and re-entrain into the air stream. The oil-treated cellulosic paper filter media displayed higher filtration efficiencies at higher velocities. Owing to oil coating, dust would be deposited in two stages. At first, the particles would be immersed in the oil film, showing practically no change in pressure drop and filtration efficiency. This stage is called the film-loading stage and is responsible for higher dust-loading capacity of oil-treated media. Once the film is saturated with particles, further deposition would lead to the growth of dendrites, leading to steep increase in pressure drop as shown in Figure 16.7(b). Further, the oil impregnation process would suppress the particle rebound to improve the efficiency in the inertial regime of particle capture. Also, the wicking action of oil makes the deposited dust particles more porous than the dust cake formed without oil. Therefore, the overall filtration efficiency of oil-treated cellulosic paper filter media is lower than that of untreated media even though the accumulated dust mass for the oil-treated cellulosic filter media is higher. This is a consequence of extended period of low efficiency during film loading.

An ideal filter should provide higher filtration efficiency, lower pressure drop, and higher dust holding capacity. Despite the difficulties associated with the limited filtration area, balance of these filtration attributes could be achieved with the composite filter media. Recently, such filter media, either prepared from a blend of fine and coarse fibres [43] or layering two different fibrous materials one over another [40, 44] to optimize the filtration performance was studied. Melt-blown layer on synthetic media or cellulosic media is the classical example of this kind of filter media. The finer melt-blown layer helps in achieving higher filtration efficiency, while the coarser cellulosic or synthetic layer helps in lowering pressure drop. Owing to smaller fibre diameter, the melt-blown filters would tend to increase the initial stage of filtration efficiency than the conventional filters such as cellulosic and synthetic filter media [45]. Further, ever-increasing need of filtering the sub-micron-size particles is becoming more challenging nowadays with the advancements in powertrain technology and mobility trends to address the stringent emission and safety regulations [46]. A few attempts were made to address this issue by laminating nanofibrous layer produced by electrospinning process with the cellulosic or non-woven fibrous materials.

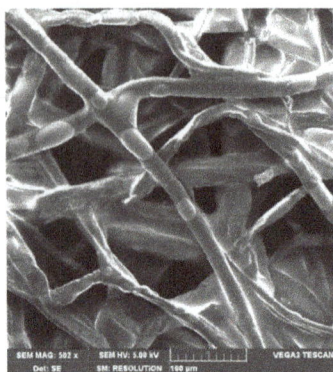

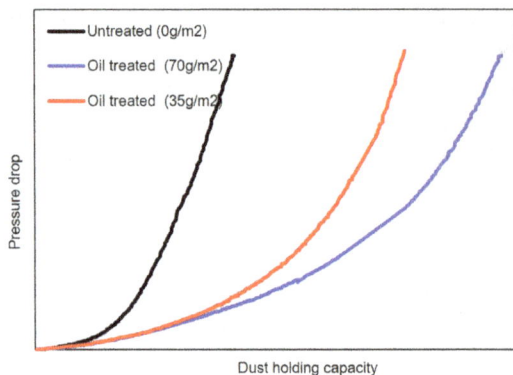

(a)　　　　　　　　　　　　　　　　(b)

Figure 16.7: Oil-treated fibrous filter media (a) with improved dust capacity (b).

Nanofibre filter media exhibit higher surface area to volume ratio due to smaller diameter that are extremely smaller (10–100 times) than those of the melt-blown and spun-bond media. Such fibres would offer very small pore size with interconnected tortuous flow path and have the potential for fibre surface modification that could be suitable for a wide range of applications such as heavy duty and high dusty off-road applications. Such an arrangement of fibres and fibrous layer within the filter media would utilize the advantages of both depth and surface deposition of dust particles. By positioning the melt-blown or nanofibrous layer at the upstream of the filter media, the surface filtration performance can be improved. On the other hand, the cellulosic/needle-felt/spun-bond fibrous layer at the downstream of the arrangement would improve the depth filtration performance. A schematic of such phenomenon is shown in Figure 16.8. The fine fibre media at the upstream of the layering arrangement would lead to better filter regeneration with improved dust cake dislodgement process during vibrations caused by the engine and vehicle excitations. This would achieve a good balance between efficiency and pressure levels during filter lifetime. Also, the dust cleaning process of the cake at regular intervals after reaching the limiting pressure drop across the filter media would be enhanced with excellent surface filtration performance [47].

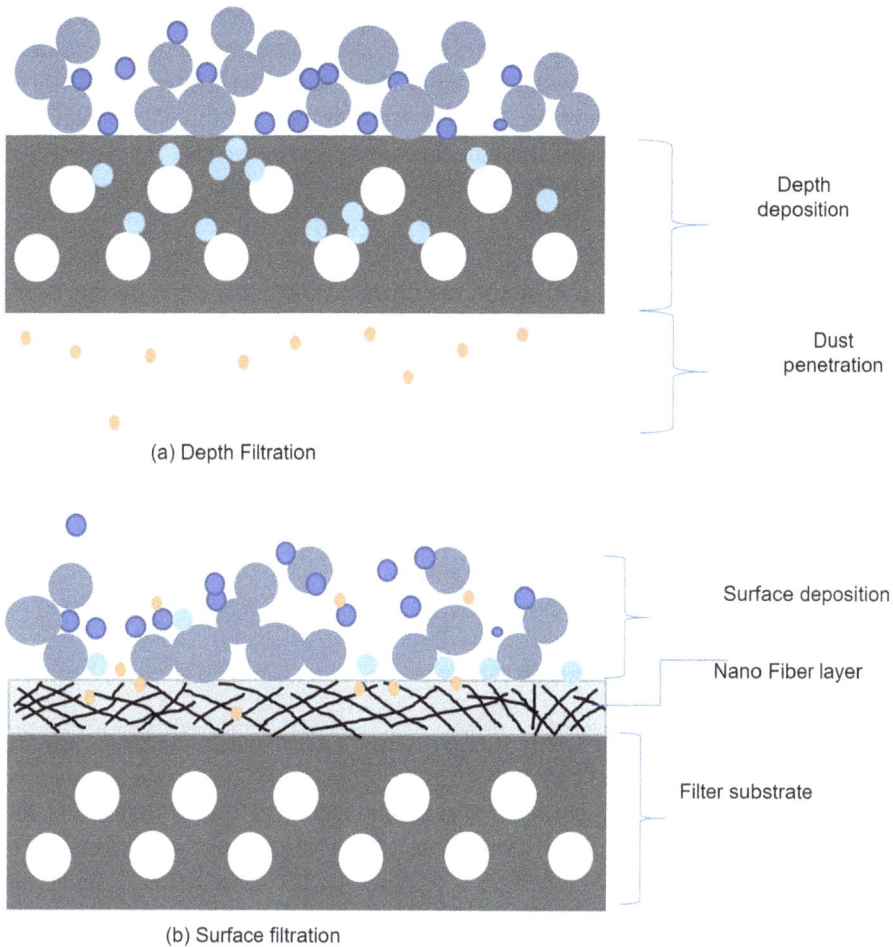

Figure 16.8: Dust deposition for conventional versus nanofibrous air filters.

Often the same filter media would also be exposed to the water droplets generated due to wheel-shredded rain and heavy moisture content in the ambient air and the eventual presence of free water. Presence of water in the filter media would adversely affect the efficiency, resistance, lifetime, or a combination of all three by changing the fibrous structure. This mainly depends on the presence of water content, the type of filter media, structure of filter media, and construction of the filter system. Those media with hydrophilic fibres (such as cellulose) would absorb moisture and swell, causing a decrease in efficiency and lifetime while increasing the resistance [48,49]. Water-repellent property in order to impart lower surface energy and higher contact angle. However, the silicon-based coating is not oil-repellent. Recently, the use of nanocoating has been effectively fulfilling such a requirement. Commercially available nano finishes such as Nano Care®, NanoPel®, and Resist Spills™ is offering the hydrophobic and oleophobic properties of both natural and synthetic fibres as well [50]. Nevertheless, the blend of cellulose and synthetic (usually polyester at 20%–30%) fibres could also be used to improve the mechanical properties of the filter media for limited or seasonal exposure of the moisture levels. However, the synthetic filter media would provide the superior performance of coalescence in the case of aggressive exposure to moisture levels due to inherent hydrophobic nature [40]. Such filter media are depth-loaded and eventually clog the filter media over time and thus cleaning would become very difficult. Nanofibrous media and/or microporous membrane laminated to the cellulosic or non-woven substrates form excellent media for self-cleaning or pulse-cleaning of dust cake formation in the form of mud due to surface filtration performance [47]. Very recently, nanotechnology-based self-cleaning fibrous materials are being explored to reduce the filter maintenance downtime. Such requirement could be accomplished either by modifying the surface topography or by photocatalytic nanocoatings [50]. Similar materials are also seen in the venting of high-pressure gas in the several applications of reservoirs and protective enclosures of hybrid power train and electric vehicles.

16.2.3.2 Filter materials for electric vehicles

Today, there exist two types of propulsion systems for zero emission targets. They are battery electric vehicle (BEV) and fuel-cell electric vehicle (FCEV). In the automotive world, BEV uses lithium-ion batteries to store the electrical energy that drives the electric motor. Protection of highly sensitive electronic components and sensors is really needed whether or not the vehicle is powered by ICE or by battery. Unlike the electronic components used in ICE vehicles, EV batteries have higher power and undergo significant change in the pressure due to temperature fluctuations caused by the electron flow and altitude difference by driving up-hill and/or down-hill [51]. These present a unique challenge and are relatively newer than the ICE vehicles. One of the challenges is the operating temperature difference between the battery pack and the ambient condition. Electro-chemical reactions within the battery cells would generate a heat during the vehicle operation. In a situation where the battery pack encounters with cold spray of water on the road or at car wash, already heated components would be cooled rapidly and generate extreme vacuum within the battery enclosure due to temperature difference. This would pull the ambient air along with the dust, moisture, and oil fumes into the enclosure. These contaminants would corrode the electronics of the EV batteries and shorten the battery life. Therefore, a robust enclosure protection from harsh external conditions, along with adequate gas venting for compensating the pressure and temperature fluctuations, must be needed. This means that the gases that generated from the electro-chemical process of the operating battery should expel from the enclosure to the external atmosphere and the ambient air to flow from outside atmosphere into the enclosure via venting system. This pressure compensation would prevent the battery enclosure from permanent deformation and leaks. Besides the pressure and temperature balance, these venting materials should meet the water tightness (as per IP67, 68, and 69 standards), in case of water wading and high-pressure jet cleaning; otherwise the electronics would be exposed to the corrosive chemicals and salts present in the road water [52]. Filter media used in these applications should be permeable to air or gas but impermeable to liquid and/or oil within the operated pressure difference. These requirements of ventilation process of the battery enclosure could be achieved with the presence of composite materials made membranes such as polytetrafluoroethylene (PTFE) as a barrier and fibrous filter media. PTFE is a synthetic fluoropolymer of tetrafluoroethylene. PTFE material is naturally hydrophobic with excellent chemical and temperature resistant and non-sticky. Also, these materials are highly gas permeable due to their microporous structure and the presence of millions of micropores (0.05 µm) within a reasonably small area [52,53]. However, microporous PTFE filters are typically a few micrometre thin (2 µm) and are weak to sustain the macroscopic forces during the venting process of the gases. It is therefore often laminated to the woven or non-woven fibrous media as a substrate to enhance the structural requirements [54]. Besides the structural aspect, the substrate would enhance the filtration performance and the moisture control as well. PTFE membranes are responsible for separation process by the size exclusion mechanism, unlike the fibrous fitters. The unique pore size distribution of the membranes would lead to the trade-off between airflow and water entry pressure. This causes the membrane fouling and reduces the gas flux during operation. Also, water vapour from the humid air could enter the battery enclosure and forms the liquid due to condensation below the dew point of air. In these cases, the coarser porous substrate material (non-wovens) would be placed at the upstream of the PTFE membrane to collect the larger dust particles to protect the fine membrane from fouling [55–57]. Also, the size of the battery pack that varies with the size of the vehicle would present an interesting challenge to expel the large gas volumes at a high rate. This would open doors to develop newer materials or adopt the fibre-based membranes used for the macro-venting application with the suitable surface-modified coatings (hydrophobic and/or oleophobic). These coatings would allow membrane to repel the water and other liquids up to water entry pressure of 30 mbar but maintain the maximum flow of air/gases [58].

Due to various advantages of hydrogen FCEV over the BEV in terms of high efficiency, fast start up, higher ranges, and short refuelling times have made the proton-exchange membrane (PEM) fuel-cell systems has becoming the key alternative technology for tomorrow's sustainable mobility [59, 60]. Fuel-cell vehicles utilize compressed hydrogen, ambient oxygen, and a PEM fuel-cell stack to produce the electric energy that would be stored in the battery. Like the battery-venting systems discussed in the previous section, venting system is needed for fuel-cell stack enclosure to dissipate the high-pressure hydrogen [61]. Unlike BEVs, FCEVs need intake air filter to clean the ambient air for oxygen to participate in the catalytic reactions. The fibrous air filter media required for cathode side of the fuel cell need to address the unique set of challenges. The presence of the cathode filtration results in the change of oxygen partial pressure that leads to increase the power loss (typically 10%–20%) and reduces the power output of the fuel cell. Further, FCEVs are not only sensitive to particulate contaminants but also to the vapour and gaseous contaminants to meet protection and safety norms. Service life and the performance of fuel cells and the associated component, for example, expensive platinum-coated membranes, may degrade by exposing them to particulates, salts, nitrous oxides, sulphur compounds, ammonia compounds, and volatile organic compounds, which can ingest via cathode side of the air intake system [59, 62, 63]. Like the ICE applications, airborne particulate contaminants separation could be achieved by non-woven filter media with various arrangements, as discussed in above sections that would lead to high efficiency of submicron particles. While the airborne water vapour and gaseous contaminants could be separated via adsorption process.

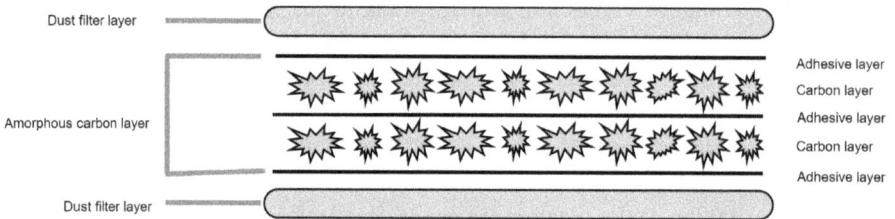

Figure 16.9: Typical filter structure for fuel-cell application.

A wide variety of filtration options are available that allows the cathode air filter to deal with various contaminant types and the concentrations. Two commonly used filter materials namely packed bed materials and microfibrous materials are well suitable for this application to maintain the filtration and power-saving requirements as well. Due to higher sorbent capacity, and lower pressure drop packed bed is the most widely used. The microfibrous materials exhibit a higher single pass removal of contaminants, but they have a lower capacity and higher pressure drop when compared to packed beds. Microfibrous materials also add a new dimension to the problem, through application as a pleated filter media or as a polishing sorbent in a composite bed formation [62]. Activated carbon is a widely used adsorbent material that usually integrated within the filter media as an impregnated form or a separate granular packed bed form. Activated carbon is a microcrystalline material produced by pyrolysis of almost all carbonaceous organic materials such as coal, wood, husks, coconut shells, and walnut shells. Such material exhibits higher microporous surface area with the typical pore size of 10 to 60 Å and excellent adsorption capacity of most gaseous contaminants [36]. A composite bed effectively combines the capacity of a packed bed and the high contacting efficiency of microfibrous materials. An optimized composite bed is capable of higher logs of removal, has a reduced total bed depth, and has a lower pressure drop than a packed bed. A typical example of this kind of a filter media, as shown in Figure 16.9, for fuel-cell application consists of upstream dust holding non-woven layer followed by the adsorption layer and a downstream dust holding layer in the stacked form. In its form, first filtration layer structure is made of short fibres of polyethylene terephthalate material processed using the dry-laid non-woven composed of two layers made of two different diameters. The upper layer and the lower layer are made up of fibres of diameter about 20–150 μm and 5–20 μm, respectively. An amorphous activated carbon layer was bonded between the dust filtering layers using the polyolefin binders for adsorption of gases like sulphur dioxide, toluene, and butane. The second filtration layer is made up of long fibres of polypropylene that was produced by the melt blown process. Fibre diameter of such layer was of 5 μm to filter out the fine dust particles penetrating out of adsorption layer. Such composite filter media exhibit higher dust holding capacity due to upstream filtration layer and filtering fine dust particles via downstream filtration layer. Also, the chemical holding capacity could simultaneously be achieved by the same filter media [64]. Further, the finer the particle size of the adsorbent filter the higher is the single pass filtration efficiency. This is because of the increased surface area of molecular contact and microporous mass transportation elimination. Therefore, it was highly suggested to have a pleated from of the composite filter media that includes a layer of adsorptive filter [62].

Besides the gas venting and cathode air filtration applications, fibrous materials are also widely used as a separator membrane by keeping the electrodes apart from being contact to avoid the electrical short circuits and allowing the ionic charge carriers to pass through to close the electric circuit during electro chemical energy generation. A separator is a thin porous membrane made of several materials such as cellulosic paper, non-woven fabrics, foams, and microporous thin sheets. These materials should exhibit a higher mechanical stability to abusive conditions, electrically stable for higher voltage production, and chemically inert to the other battery cell materials. Recently, the increase in higher power applications and the introduction of new battery cell chemistries have made the separators more complex and sophisticated to meet the trade-off between the mechanical stability and the ion transport properties [65]. Nowadays, electro spun nanofibre-based separators are used due to higher surface-to-volume ratio, higher porosity, and better ion transportation while maintaining thermo-mechanical stability and electro-chemical stability. A few studies on nanofibre composite materials such as PVDF-*co*-chlorotrifluoroethylene, PVDF-*co*-hexafluoropropylene, and SiO_2–Nylon 6,6 were reported as a battery separator. Such materials might replace the microporous membranes. Also, a commonly used separator in the case of proton exchange membrane fuel cell is the electrospun nanofibre layer laminated to high molecular weight non-woven substrate. Nevertheless, the presence of carrier material would minimize the transport hydrogen ions between oppositely charged electrodes. A recent study on nanofibre composites made of sulfonated polyimide reported a remarkably higher proton exchange than its counter one [50].

16.2.3.3 Filter materials for interior air

In the past, it was a common practice for cabin air filtration applications to use conventional fibrous filters that were used to filter the road dust and the PM (fine and coarse) [66]. High efficiency fibrous air filter media that was discussed in the earlier sections would serve its purpose here as well. Nevertheless, there exists the most particle penetrating size of fibrous filter media in the particle size range of 0.03–0.05 μm and that cause the passengers and drivers to higher concentration exposure of submicron to ultra-fine PM. The application of electrostatic forces could augment the lower filtration efficiency of these particles using fibrous filters that operate under the mechanical and thermal forces [67]. Such electrostatic treatment of the fibrous filter media could be achieved using several processes including corona charging, triboelectric charging, and induction charging. Nevertheless, the use of such materials is limited due to the charge decay process with the loaded particles on the fibre, and thus the filtration efficiency decreases for the sub-micron particles. A detailed study on explaining the charge decay process of corona charged non-wovens made of polypropene melt-blown process was carried out [68]. A similar study was conducted on the triboelectric charged non-wovens made of needle-punched process using wool and polypropylene at various mass proportions [69]. However, an increasing need for high filtration efficiency of sub-micron particles and remarkably low pressure drop, the air filters still require significant research to be carried out. Further, the increase in the ambient air pollution levels has also exposed the cabin air filters to the higher dust loads. This also enables the use of low packing density fibrous media made of needle punching process as pre-filter to increase the dust loading capacity and to increase the life of electret filters [70].

Besides the airborne particulates, the protection and safety of the passengers and the driver from other harmful gases and allergens are becoming more important day by day. Separate air filters for each of the mentioned purpose of this kind were being used in the automotive applications. Nevertheless, the recent developments of this kind of filter media motivated the manufacturers to use the multi-layer structure for combined functions that each layer would exhibit its own function. Such developments made to use of the existing dust and high efficiency filter media to incorporate the additional filter layers of adsorption and anti-microbial coatings [71]. Example of such an embodiment consists of multi-layered filter media having particulate non-woven layer, followed by an activated carbon layer arranged in the downstream side of the particulate filter layer to perform adsorption of odour [72]. Also, the arrangement of these non-woven layers with progressively reduced fibre diameter in the air flow direction could effectively filter the particulates, odours, and contaminants such as vapour and gaseous contaminants. The amount of activated carbon substance used for this purpose would depend on the level of contaminant concentration and the filtering capacity needed. Further, another layer of finer melt-blown non-woven adjoined against the particulate and adsorption layer is coated with the anti-microbial and/or biocidal substances including octaisothiazolone, plant extracts, and metal compounds of copper to inhibit the growth and/or kills bacteria and allergens. Several such commercial filter media options are available nowadays. Recent advantages of nanotechnology have allowed to develop the nanoscale anti-microbial coatings [50]. These coating would increase the higher surface area for the collection of allergens and odours. Several metal and metal oxide nanocoating materials such as silver, copper, titanium, zinc, calcium, and magnesium are available today.

16.3 Fibrous materials for insulation purpose

16.3.1 Thermal and acoustic noise sources

Fibrous structures, because of their complex porous structure, are well-known materials for thermal insulation as well as acoustic noise reduction. The need for higher fuel efficiency and lower exhaust emission causes the engine operating temperature to be higher. Further, the demand for higher utilization of thermal energy, light weight, comfort of cabin members, driving experience, and energy savings from air-conditioning systems are driving the use thermal insulation materials in automotive applications. The major thermal source of the automotive vehicles is the vehicular engine itself. Engine compartment insulation prevents the transfer of heat and noise into the vehicle interior cabin. Thermal insulation reduces the thermal load of the cabin enclosures from exposing them to engine temperature and extreme temperatures of various seasons and help economically operate the HVAC systems [73–75]. Recently, cold start emission control strategies put forth the stringent requirement to insulate the exhaust after treatment systems for quicker light off temperature and efficient catalytic reactions [76, 77]. Underbody and luggage trunk insulation is also be needed to block the heat transfer from the exhaust system into the passenger cabin.

Acoustic noise that generated by the vehicle due to engine combustion and gas exchange process, tyre/road interaction, and wind flow at higher vehicle speeds can be transmitted into the vehicle interior in the form of sound waves. This acoustic wave energy based on the pressure levels would be transmitted by the air and could be heard by the driver and passengers of the vehicle cabin and the passers-by. This noise is unpleasant and has a substantial effect on the driving experience and the comfort levels of the cabin members. Also, the noise that is generated within the cabin due to heating, ventilation, and air conditioning system would generate the noise and would be reflected many times to cause cabin discomfort. It is therefore necessary to reduce the sound pressure levels to meet the regulatory requirements and to improve the passenger comfort in the vehicle cabin [78, 79]. The passive method of reducing the noise is by using the sound absorption materials such as fibrous or porous materials placed in the path of sound transmission.

16.3.2 Principles of thermal insulation

Thermal insulation is the process of blocking the flow of heat energy between the objects that exists at two different temperatures. Conduction can be blocked by eliminating the physical contact between the objects. Convection can be eliminated by supressing the fluid motion due to significant friction caused by the fibres against the flow. Radiation can be minimized by minimizing the view factor between the fibres. Heat transfer in the fibrous materials occurs in general via conduction, convection, and radiation. Nevertheless, the flow of heat energy in the fibrous structures was challenging due to the co-existence of coupled heat transfer modes (possibly all three modes) because of its geometrical complexities. Generally, the fibrous thermal insulation materials can be treated as the continuum of two phases namely the solid material that constitutes the fibres and the saturated air present within the pores. Conductive heat transfer can take place in the fibrous material through both the fluid within the interstitial pores, the solid fibre itself, and the fibre contact materials. Also, the movement of the interstitial fluid would exist due to buoyancy forces, usually the convection process. Radiation would also be the dominant mode of heat transfer at the higher temperatures (>573 K) [80, 81].

The effectiveness of thermal insulating material would be indicated by the effective thermal conductivity that depends on the microstructural aspects of fibrous assemblies. Under the assumptions of local thermal equilibrium of two-phase fibrous material with the presence of stagnant saturated fluid within the pores, the effective thermal conductivity could be estimated based on the principle of super position of solid (K_{cond}), gas (K_{conv}), and radiation (K_{rad}). This is stated as follows:

$$K_{eff} = K_{cond} + K_{conv} + K_{rad} \tag{16.24}$$

The fibrous assemblies with lower solidity and having a pore size greater than 1 mm would experience convective mode of heat transfer. The existence of the convective heat transfer could be estimated using the modified Rayleigh number Ra*. Convective heat transfer could be neglected if the Rayleigh number is less than 40 [82, 83]:

$$Ra^* = \left(\frac{g\chi\rho_{air}^2 C_p}{\mu} \right) \left(\frac{k}{K_{eff}} \right) (\nabla T) Z \tag{16.25}$$

where g is the acceleration due to gravity, χ is the volumetric thermal expansion coefficient, ρ_{air} is the mass density of the air, C_p is the specific heat of air at constant pressure, μ is the dynamic viscosity of air, k is the permeability of fibrous assembly, Z is the thickness, and ∇T is the temperature difference across the fibrous assembly. Convection mode of heat transfer might be neglected at higher temperatures, at atmospheric pressure the pore size of less than 1 mm which is usually the case of typical fibrous insulation material [81]. The effective thermal conductivity K_{eff} could be estimated as proposed by Bankvall [84]:

$$K_{\text{eff}} = K_{\text{fibre}} + K_{\text{air}} + K_{\text{rad}} \tag{16.26}$$

Further, thermal conductivity due to conduction of the fibrous material could be treated as the combination of solid fibre conduction K_{fibre} and the air conduction K_{air}. Many studies were conducted in the past to evaluate the thermal conductivity due to fibre, air, and radiation:

$$K_{\text{fibre}} = a\varphi k_{\text{fibre}} \tag{16.27}$$

$$K_{\text{air}} = a(1-\varphi)k_{\text{air}} + (1-a)\frac{k_{\text{fibre}}k_{\text{air}}}{\varphi k_{\text{fibre}} + (1-\varphi)k_{\text{air}}} \tag{16.28}$$

where a denotes the fraction of the fibrous material that is parallel to the heat flow and other fraction would be perpendicular to heat flow. At moderate temperatures, light weight fibrous structure has displayed the thermal conductivity due to radiation accounting for 40%–50% of the total heat transfer [81]. Radiation plays an important role for low density fibrous materials and at even moderate temperatures. Thermal conductivity due to radiation [84] can be best calculated using the following equation:

$$K_{\text{rad}} = \frac{4\sigma d_f T_{\text{m}}^3}{\left[\frac{1}{\beta_R} + \frac{d_f}{Z}\left(\frac{2}{\sum_0} - 1\right)\right]} \tag{16.29}$$

where T_{m} is the mean temperature, $\sum_0$ is the emissivity of the fibre surface, σ is the Stefan–Boltzmann constant, and β_R is a radiation coefficient describing the radiation properties of the fibres and fibrous layers. In an optically thick fibrous medium relative to fibre diameter, radiation traverses only across the short mean free path before interacting with other parts of the fibre matrix, radiative heat transfer could be lumped as an equivalent diffusion process [85]:

$$K_{\text{rad}} = \frac{16\sigma T_{\text{m}}^3}{3\beta_R} \tag{16.30}$$

The heat loss across the thermal insulating material of thickness l could be calculated according to Fourier's law:

$$Q = K_{\text{eff}}A\left(\frac{\Delta T}{Z}\right) \tag{16.31}$$

16.3.3 Principles of acoustic noise propagation

Noise reduction by controlling the sound during its propagation is a widely and effectively used technique using fibrous assemblies. Fibrous materials used in automotive application for controlling the noise during its propagation include sound absorption,

E_r

E_t

E_a

E_i

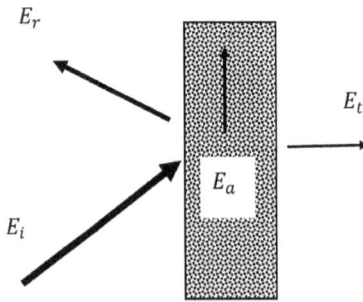

Figure 16.10: Schematic of sound interaction within the fibrous material.

insulation, and damping. When the sound waves incident on the fibrous assemblies, parts of its energy are reflected, absorbed, and transmitted. Absorption of acoustic energy is the result of pore structure that happens at the microscale where the viscous effect is very much dominant. This effect would reduce the speed of the acoustic wave propagation and converts the acoustic energy into the heat energy due to viscous boundary layers between the sound waves and the fibrous assemblies [86]. Total acoustic energy balance of the incident wave (E_i) can be stated as per eq. (16.32) (see Figure 16.10). The ratios of sound energy absorbed (E_a), reflected (E_r), and transmitted (E_t) to the total incident energy are defined as the acoustic absorption coefficient $\varsigma = E_a/E_i$, reflection coefficient $\xi = E_r/E_i$, and transmission coefficient $\tau = E_t/E_i$, respectively, and are the most important attributes of evaluating the acoustic materials

$$E_i = E_r + E_a + E_t \tag{16.32}$$

Zwikker and Kosten [87] proposed an equation to calculate the sound absorption coefficient ς as a function of reflection coefficient ξ for porous materials using eq. (16.33). It is generally assumed that the fibre, a fundamental entity of the fibrous materials, is acoustically rigid. Thus, the elastic wave on the solid part could be ignored, and the fibrous assembly could be treated as the equivalent fluid model characterized by two complex quantities namely characteristic impedance (M_0) and propagation constant (γ) of fibrous material as follows [88]:

$$\varsigma = 1 - |\xi|^2 \tag{16.33}$$

$$\xi = \frac{M - \rho_0 C_0}{M + \rho_0 C_0} \tag{16.34}$$

$$M = M_0 \coth \gamma M \tag{16.35}$$

where M is the impedance of the fibrous material, ρ_0 is the density of air, and C_0 is the speed of sound in air. The prediction of the sound absorption coefficient is possible if the specific acoustic impedance of the fibrous material is known. The available methods to predict the surface acoustic impedance was found in the literature based on the empirical and theoretical considerations that deal with the first principle of sound propagation.

However, the theoretical methods are derived based on the complexity variables such as tortuosity and the porosity, which are difficult to measure or model. This limits the practical use of physics-based models in the engineering and research fields. The empirical models are created based on the pre-supposed knowledge and easy to measure properties such as flow resistivity and the thickness of the fibrous assemblies. Delany and Bazley [89] developed an empirical equation for fibrous materials based on the measurements conducted for sound propagation through several wool materials (glass and rock). The sound propagation in an isotropic homogeneous fibrous assembly was determined as follows:

$$M_0 = \xi + jX \tag{16.36}$$

$$\gamma = \varsigma + j\beta \tag{16.37}$$

The following empirical equations were obtained as a function of the ratio between acoustic frequency f and flow resistivity σ:

$$M_0 = \rho_0 C_0 \left[1 + 0.057 \left(\frac{f}{\sigma} \right)^{0.754} \right] + j\rho_0 C_0 \left[0.087 \left(\frac{f}{\sigma} \right)^{0.732} \right] \tag{16.38}$$

$$\gamma = 0.189 \frac{2\pi f}{C_0} \left(\frac{f}{\sigma} \right)^{0.595} + \frac{2\pi f}{\sigma_0} \left[1 + 0.098 \left(\frac{f}{\sigma} \right)^{0.7} \right] \tag{16.39}$$

Delany and Bazley's [89] method is valid in a frequency range, where $0.01 \leq f/\sigma \leq 1$. The flow resistivity of the fibrous assembly is defined as follows:

$$\sigma = \frac{\nabla P}{U} \frac{1}{Z} \tag{16.40}$$

$$\sigma = \gamma \left(\varphi^\beta \right) \tag{16.41}$$

where γ is the proportionality constant and β is the exponent. The flow resistivity of the fibrous model depends on the microstructural details such as fibre diameter, packing fraction, fibre density, fibre orientation, and binder agents. Further, using eq. (16.32–16.41), one can find the acoustic barrier property that would be used for evaluation of blocking the sound propagation performance through the fibrous material. Such a property is called the transmission loss (STL) as follows [90]:

$$STL = 10 \left[\log_{10} \left(\frac{1}{\tau} \right) \right] \tag{16.42}$$

16.3.4 Applications of insulation materials

The emphasis on thermally and acoustically efficient vehicles (passenger cars, trucks, farm tractors, construction equipment's, etc.) has increased significantly in the past decade due to stringent government regulations across the world [91]. This would likely to with the recent emphasis on light weight and ever-changing mobility trends. Alternative power trains (electrification) to combustion vehicles have created the radical NVH spectrum, which are completely different from traditional vehicles, and also changed the weight reduction strategies. It is therefore essential to better understand the noise sources and noise control materials for better vehicular design [92]. Nowadays, most of the insulating functions are no longer considered to be isolated but many other requirements are equally important to meet. Depending on the applications these may vary and a few of the requirements to be considered are light weight, premium space, easy to install (mouldable/warp around), mechanical durability over lifetime, fire protection (incombustible), water resistant, safety, effective thermal insulation, acoustic performance, and cost-effectiveness [93–95]. A few of most widely used materials for this purpose could be classified as high temperature fibres and low temperature fibres. Examples for high temperature fibre materials include glass, basalt, ceramic, rock wool, silica, carbon, and aramid. Examples for lower temperature materials include cellulose-based non-wovens such as wood pulp, jute, and kemp and synthetic-based non-wovens such as polypropylene, polyester, polyethylene, glass, nylon, carbon, and other suitable ones. The fibres could be pre-processed in the form of non-woven, weaving, knitted, braiding, stitching, and so on are extensively used in automotive applications [93].

16.3.4.1 Vehicular interior and exterior applications

Currently, most of the non-woven materials used in the interior and exterior applications of automotive vehicles are made up of synthetic materials. Among them, polypropylene is widely used due to its easy to production, low cost, light weight, acid and alkali resistance, and stable chemical performance. Nevertheless, the poor melting point (~161 °C) and mechanical of these materials could be reinforced with other fibres such as polyester, glass fibre, nylon, polyamide, and others. The non-woven composites for automotive interior applications however do not meet the growing requirements of mechanical properties, loss of weight, degradability, and cost. The non-woven composites developed with terelyne fibres and polypropylene fibres have improving characteristics but poor bending strength, dimensional stability, and difficult for non-planar moulding aspects. Also, the composites with polypropylene fibres and fibre glass materials improved the bending properties but with increased weight. Further, the vehicle interior applications are complex that should meet various stringent industry standards and legislations. These materials should meet flame resistance for fire safety, abrasion,

wear and impact resistance for occupant injury safety, fibre dust and chemical generation resistance for occupant health, higher strength and flexural rigidity to vibrations
for lifetime endurance, soil and moisture resistance, thermal and acoustic noise insulation for passenger comfort, and increased harness. Recently, on the other hand, concerns
with non-biodegradability of these materials and the environmental consciousness are
increasing [96]. Automotive manufacturers are continuously searching for environmentally sustainable materials to improve the end-of-life recyclability of newly developed
products. A cellulose-based non-woven could serve as better alternative to develop the
environmentally sustainable products. These materials exhibit higher specific strength
and relatively low elongation at break. However, these materials are limited to vehicle
interior applications due to flammability, high moisture absorption, and subsequent
decay of mechanical strength. Nowadays, non-woven composites made up of cellulosic
and synthetic fibres are becoming popular and commercially viable [97, 98].

A cost-effective Source–Path–Receiver model of noise control in the automotive vehicles often blocks the transfer path of the noise source. This could be met by the addition of sound insulation and/or sound-absorbing materials at the vicinity of the noise
source. The absorptive material used for this purpose is usually of porous structure
with less dense and is made of fibre, felt, or glass wools. Also, the fibrous materials
made of densely packed and/or compressed fibrous assemblies act as an insulating material. Sound absorption characteristics of the fibrous media vary with frequency, and
it depends on the fibrous structures and their properties. Also, in general, the fibrous
materials are effective for higher frequency sound waves because of shorter wavelengths
comparable with fibre diameter and pore size. Nevertheless, the increasing expectations
of customer driving comfort as well the meeting the low frequency absorption of sound
need thicker fibrous material, which adds weight to the vehicle. Depending on the application of these materials could be used with backing materials such as films, foils, and
scrims to increase the absorption capacity and thermal insulation [54, 99, 100]. Sound insulating materials are characterized by the transmission loss, and the sound absorption
materials are characterized by the absorption coefficient. Nevertheless, an ideal acoustic
material should have better performance in terms of both absorption and transmission
loss. Usually, the sound absorptive fibrous material is more effective as the thickness corresponds to one-fourth of the wavelength of the frequency of the sound. However, sound
absorbers such as needled cotton fibres (shoddy) and felt applied in the interior applications are limited with thickness of the fibrous material and is typically found to be
25 mm. It is therefore a practice to increase the thickness of the material using another
high dense material as a sound insulator for overall noise reduction [101]. Further, the
acoustic noise generated from structural vibrations can be damped by placing the fibrous
material adjacent to the near field region of the vibrating component [102]. This is however increasing the additional weight of the vehicle. This could be overcome using the
non-woven composite materials.

Fibrous materials for interior noise reduction as an acoustic panel and padding
for automotive application have been widely used in dash panel liners or mats,

engine side fire wall insulation, engine side hood insulation, interior wheel well insulation, trunk compartment trim insulation, flooring underlayment, package trays, and door panels. Several such packaging options of non-woven composite materials have been claimed in Reference [101] to form a composite structure of high-density scrim and loosely packed acoustic fibrous layers. Absorptive non-woven pads might be made from cellulose-based natural fibres and synthetic fibres such as PET or polyester, mineral fibre, glass fibre, and other suitable materials. One of such embodiments consists of polypropylene non-woven with spun-bond and melt-blown layers (SMS or SMMS). The absorptive material of 100 g/m^2 weight felt is formed by air-laid process on the top of the scrim using bi-component (bico) fibres at a mass proportion of 80:20. One of the bi-component fibres includes cellulose fluff pulp and the other one is polyethylene bi-component binder fibre with two denier fineness and 6 mm length. Different layers of air laid non-woven were bounded using thermal bonding technology or by latex binders sprayed during the air laid process and before the heat treatment and the compression process. It was found that the lower was the density and higher was the thickness of air-laid composite the higher was the sound reduction due to sound absorption rather than sound blocking. This was due to the large number of short fibres with a higher fraction in a given volume. Also, the type of fibre material (virgin or recyclable or shoddy) selection was found to be not playing any role, and this is purely dependant on the microstructural details. Interestingly, constant thickness of the fibrous assembly has no effect on the transmission loss. Nevertheless, the sound absorption per unit of basis weight increases with lowering basis weight at 25 mm thickness. Figure 16.11 displays the sound absorption coefficient of commercially availed shoddy, cellulose-based non-woven, and PET non-woven with 25 mm thickness and basis weights of 2,000, 1,000, and 350 gsm, respectively. However, these materials are not so effective at a frequency below 1,000 Hz. The latex treatment added on both sides of the mat would enhance the sound absorption below 1,000 Hz by formation of an additional discrete layer. This latex binder in addition to these composite structures would also enhance the mouldability. For applications that need to handle the lower frequencies, sound absorption materials (cellulose-based and PET) with thickness above 25 mm and higher basis weight are suitable. This might allow the sound waves with higher number of fibres to support viscous dissipation. This indicates that the non-woven fibrous structure would be employed with lower weight but with the same packaging that the conventional materials were employed. Also, by treating the fibre with diammonium phosphate-based flame retardant (SPARTAN™ AR 295, e.g., used here) to meet flammability of interior standard FMVSS-302. Such fibrous materials would reduce deaths and injuries to the vehicle passengers caused by vehicle fires. Further, treating with water proofing agents such as silicone based (MAGNASOFTR® extra emulsion by GE Silicones, e.g., used here) coatings to make the fibrous assembly repellent to water. These materials found their applications in door panels, wheel wells, and engine compartment for example. This would enhance the water repellence by 97%. Another packaging option includes the scrim of

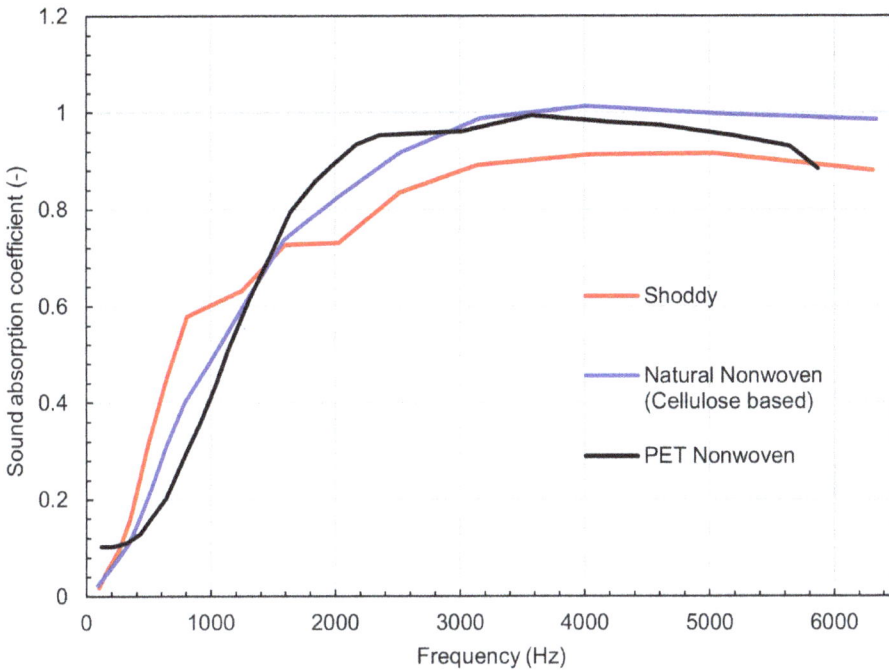

Figure 16.11: Typical woven construction of seat belt application.
Adapted from Cai et al. [74] and Gross et al. [101].

the composite structure to provide higher tensile strength to the material from sagging due to weight added by sound absorptive material and decorative trim.

The use of cellulosic-based non-woven composites made of kenaf, ramie, cotton, jute, flax bagasse, and recycles polyester/substandard polypropylene was investigated for the vehicle interior applications such as headliners, trunk liners, and wall panels [103]. Non-woven composites, suitable for automotive head liner, made of carded and needle-punched process (70:30 kenaf and ramie blended) exhibited significantly lower thermal conductivity than the non-woven composites made by using air-laid process (100% kenaf and 70:30 kenaf and ramie-blended). Also, the non-woven composites with PET scrim, suitable for trunk interior applications, made from various cellulosic fibres in two vegetable and synthetic fibre ratio (35/35/30 with kenaf/jute/cotton + recycled PE + PP using carded, and needle-punched (four times) exhibited lower thermal conductivity by 13%–18% than the non-woven composites made up of recycled PE and PP (70:30). A series of another composition (50/50 with kenaf/jute/flax + PP-blended) made using carded and needle-punched (two times) non-wovens exhibited more thermal conductivity than the aforesaid materials. They concluded that such difference was the result of not using the PE scrim and needle-punched materials. Another set of cellulosic non-woven composites made from bagasse and cotton (70:30) using carded with 6 and 12 layers and needle-punching bonding was sandwiched with eastar biocopolymer melt-blown PE in a 70:30 mass proportions. Here, the six-layered material exhibited less thermal conductivity than the 12-layered one. The higher conductivity of the 12-layered material was due to convective heat transfer mechanism owing to having double thickness. All these materials were diminishing their mechanical properties after 1–6 weeks in soil burial.

A recent study from Cai et al. [74] explored the cost-effectiveness of waste wool (28 μm fibre diameter) materials and virgin wool (11.9 μm fibre diameter) materials, made of needle punched non-woven process, for thermal and acoustic applications for automotive vehicles. They investigated and found the importance of the effect of fibre diameter, non-woven surface, layer structure, thickness, and basis weight on sound absorption and thermal resistance capabilities. It was found that these recyclable fibrous materials exhibited comparable thermal and acoustic properties with traditional synthetic materials like polyester and cellulose/polyester. They suggested materials with basis weight ranging from 460 to 790 g/m^2 and thickness varying from 18 to 29 mm have excellent sound absorption and would be suited for varied sound frequencies within the automotive interior applications. Further, this wool material exhibited excellent soil biodegradability against weight loss with seven times higher than the PET fibres. These recyclable materials had identical anti-fungal and anti-bacterial properties as compared to synthetic fibres.

Nowadays, nanofibrous materials are becoming a potential solution for thin and light weight sound-absorbing materials for vehicle interior applications. Compared to the conventional microfibrous materials, nanofibrous materials exhibit excellent sound absorption capacity due to smaller fibre diameter and higher porosity. A higher surface to volume ratio of these materials would cause higher viscous losses due to friction between fibres and air. In addition to the viscous losses, a part of the energy is absorbed by the vibration caused in the individual fibres and/or fibrous layers [104]. Nanofibrous layers are flexible enough to resonate at its natural frequency to absorb the low frequency sound energy when sound propagation is perpendicular to the fibre arrangement [105]. Two of such commercially available material technologies such as Fredenberg's Evolon® and Elmarco NanoSpider™ were discussed in Reference [50]. Fredenberg's Evolon® was produced from the nano-modified microfilaments with 5 to 10 times thinner than the many other microfibres. Such fibrous structure was used in vehicle interior applications (head liners, doors, carpet backing, dashboard, and under body) with the reduction in thickness and weight by 10–30-fold and with better noise absorption at equal or greater than 2,000 Hz.

16.3.4.2 Vehicular high temperature applications

As various thermal-related issues have made to search for cost-effective solutions, the earliest forms of thermal insulation are still viewed today as a key approach for solving these problems. However, as vehicles have transformed, so have the thermal requirements, materials, and constructions used in textile materials. Recent changes of automotive exhaust emission norms across the globe have emphasized the need of cold start emissions to reduce the permissible limits of NO_x and PM pollutants from the exhaust gases. The after-treatment devices such as selective catalytic converter and particulate filters are imperative to meet the standards. These devices in the diesel and gasoline-operated engines display the working temperature of 250–400 °C. These devices could not realize the catalytic conversion of pollutant gases correctly at the lower exhaust temperatures, which usually called light-off temperatures. Therefore, effective thermal insulation is needed to heat up the aforementioned components and this has become increasingly challenging to the automotive manufacturers because of space constraint [106]. Often these insulating materials are generally sandwiched with the heat shield material for strength and to isolate from external exposure and safety purposes. The skin temperature for such heat shield should be substantially lower than the temperature of the exhaust gases, usually exceeds 600 °C, to mitigate the risk of injury. Insulation mats made of silicate fibres and E-glass fibres are widely used for this purpose. However, the silicate fibre mats and the glass fibre mats have limited resistance at higher temperatures. This would often lead to break down or shrink in the material volume which in turn increases the packing density of the mat and thus the thermal conductivity increases. Such limitations would be countered using the fibre reinforcement with less shrink materials either in the mixed fibre form or as an additional layer. Further, these materials are not bio-persistent and cause serious health issues upon exposure. One such embodiment was claimed to use the mixture of inorganic fibres comprise bio-soluble and refractory ceramic fibres with 80% weight proportion and 20% weight proportion of silica or glass fibres for reinforcement purpose. With the recent developments [107], application of basalt fibres has also becoming a suitable substitute over glass fibres for increasing needs of thermal insulation and human health concerns.

In the high-temperature application of vehicle exhaust, the conduits must be insulated with higher temperature resistant materials to improve the catalytic reaction efficiencies of exhaust after treatment systems [77]. Textile materials such as fibre glass and basalt fibres under various braided forms are widely used nowadays. These fibrous liners in the braided forms that cover or coat the exhaust pipes would reduce the heat loss and thereby aid in better utilization of catalytic processes. In a recent study, basalt fibres in the form of sleeve, winding, and felt and glass fibre in the form of mat with thickness of 5 mm and basis weight of 120 g/m^2 was tested for diesel engine application for the end-use temperature of 780 and 400 °C, respectively. They concluded that the basalt fibre material exhibited better thermal performance than the glass fibre. This was due to larger interior amorphous region that results in the lower thermal conductivity. During the cold stage of the emission duty cycle, the gas temperatures of the exhaust conduit that was covered with basalt fibre material with felt cover attains the best thermal performance with 2.6, 2.9, and 0.5 °C lower than the sleeve type, winding type, and glass fibre mat, respectively. Compared to the felt type, other types of weaving methods exhibit higher pore size and result in higher heat loss and poor thermal insulation.

The noise radiated from the exhaust pipe of the vehicles in general can be suppressed by the principles of reflection and absorption. Generally, automotive muffler geometry for this purpose could be designed to have both reflective and absorptive properties for the entire frequency spectrum. Reflective properties could be used to handle the low frequency and/or specific frequencies coincide with the engine firing frequencies. However, the absorptive mufflers are widely used to handle the frequencies from mid-range to higher range of the spectrum. Staple fibre glass wool and basalt wool made of needle punched non-wovens are widely used materials to attenuate the sound wave generated from the engine gas exchange process. However, the durability of these materials in the muffler applications is challenging due to the exposure of harsh environment during their lifetime. Absorptive wool will be clogged with unwanted matter (PM and water), and this would loosen the packing of wool and lead to blow out into the exhaust system. Further, space constraints, uneven filling, and additional compression from the forces exerted by the exhaust gases would deteriorate the acoustic performance. To counter one such effect, a continuous strand of glass fibre was fluffed up by the roving process along with the binders into a defined shape of the muffler [108].

16.4 Fibrous materials for safety and structural purposes

Safety of the passengers in a vehicle can be achieved by the passive and active safety systems. The passive safety systems include the seat belt, air bag, bumpers, and other structural components of a vehicle. During a vehicle crash, kinetic energy of the car is released into the passenger bodies because of inertia and the passengers would accelerate forward in relation to the vehicle, prior to the impact on the car interior (dashboard and steering wheel) and windshield. The severity of the crash impact would depend on the speed of the vehicle. Such crash impact on the passengers could be prevented using the motion retention system such as seat belt as a shock-absorbing device [109]. In conjunction with seat belts, the driver and passenger air bags reduce the peak acceleration and distribute the restraint loads on the upper part of the body. To utilize the advantages of the coupled actions, these passive restraint systems are nowadays developed for integration.

16.4.1 Seat belts

Currently, the widely used three-point automotive seat belts are made of multi-layer-woven fabric material with higher tenacity polyester and nylon (polyamide 66) [110]. Nylon has higher strength, higher elongation before break, and strain recovery than polyester. However, it is prone to abrasion and offers poor UV resistance. Polyester is popularly used to produce automotive seat belts, owing to its higher stiffness and lower extensibility as compared to nylon. Typically, these materials are used to resist the dynamic load up to 16.7 kN, as per European Directive 77/541/EEC. Such fabric must also meet the abrasion resistance, UV, and heat resistance to meet the reliability requirements [111]. It must be soft and flexible in the longitudinal direction and rigid in the transverse direction on order to have high-dimensional stability. This is because the change in the width or thickness of the fabric would cause web-bunching effect to hinder the sliding efficiency [112]. To meet such requirements, often the fabric was made with warp yarns and weft yarns which are woven in 2/2 or 4/4 twill weaves [113]. In general, the weft yarns have fineness of about 1,000 denier to 1,500 denier and extensibility of 10–20%, while the weft yarns have fineness of 500–700 denier and extensibility of 15–25%. The conventional fabrics are however failed to exhibit the initial restraining ability by elongating over 15% when a collision occurs. A typical force of about 1,100 kg is applied on the passenger at the onset of collision causing the belt to elongate significantly that a passenger might hit the windshield or dash panel. Therefore, the ideal seat belt fabric must exhibit considerable stiffness for abrasion resistant and initial restraining ability at the onset of a collision, and at the same time, it should absorb the shock by elongating during the collision process. The behaviour of the seat belt during the crash events would mainly depend on the interactions between the warp and wept yarns of the fabric. One of such embodiments has been developed by adding an extra yarn in the base-woven structure and extending straight along the length of the seat belt as shown in Figure 16.12. This would give combined properties of higher tensile strength due to extra yarn and lower extensibility than base-woven structure.

Often, to increase the sliding efficiency, the high strength polyester fibre is used. This increases the stiffness of the fabric. This may cause injury to a passenger upon a car crash due to stiffness of the seat belt itself. Nevertheless, by changing the weaving within the webbing from regular to variable herringbone weaving has made the fabric more flexible and thus having low friction to pull the seat belt. This improvement has made 10% lower seat belt pull out force. Also, it has reduced the pressure acting on the body during use [114]. To assess the reliability of such fabric webbings is to understand the failure modes under the typical operational environments [111]. One of such failure modes of the fabric is the decrease in tensile strength caused by the sunlight within the cabin and the abrasion due to the repeated fastening of the seat belt. The fabric would fail if the tensile strength of the material would decrease below 16.7 kN.

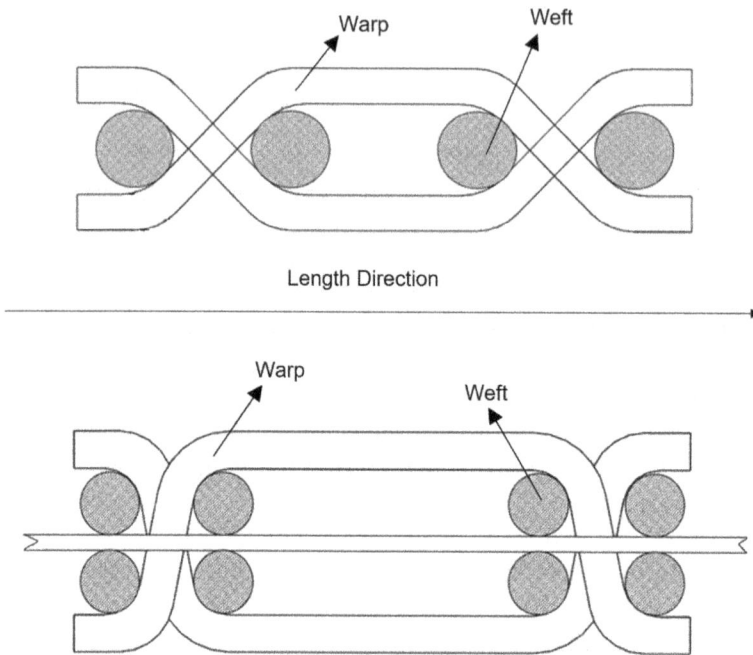

Figure 16.12: Typical woven construction of seat belt application.
Adapted from Koseki [113].

16.4.2 Airbags

Airbags are in general controlled by a central unit which monitors a series of sensors that includes accelerometers to detect the collision events at a speed higher than 35 km/h and sets off the explosive chemical to inflate the fabric. Due to the increasing demand of inflation rate (<50 ms, currently via chemical reaction generating nitrogen gas) and higher impact force involved during collision, air bags are manufactured with high strength-woven fabrics made of polyamide 6, polyamide 6.6, and polyester (PET) [110]. Also, the alternative materials that could provide light weight, high strength (ability of the fabric to resist tearing and bursting), and excellent toughness that are required to withstand the dynamic environments of vehicle crash events are also under use, for example, poly paraphenylene terephthalamide (KEVLAR®) [115]. The linear density and weaving intensity are the two important attributes that could impart the required strength after the weaving process. Typically, these fabrics are made of plain weave or jacquard weave pattern with single layer to avoid the crimping of the fabric [116]. Polyester is not commonly used for this kind of application due to its lower thermal properties than nylon 6,6 [117]. Although, polyamide is a preferable material for air bag fabrics because of its compatibility with inflator gas temperature and relatively high toughness. However, the lower scale mass production due to tighter supply chain has made the air bag cushion manufacturers to prefer PET fibres recently. Further, emerging new applications of air bags using cold gas inflators for roll over curtains has made a significant technical progress in polyester fibres.

Ideal air bag cushion must exhibit higher tenacity, higher toughness, and both lighter and thinner construction to foldup into a compact pack. Higher tenacity is needed to meet the dynamic loads (impact and thermal) that are encountered during air bag inflation operating at a temperature range of −45 to −85 °C. Whereas, a higher toughness is needed to provide the load distribution in high stress areas and avoid fabric bursting and passengers getting hit against the hard surfaces of the vehicle interior. Recent developments of polyester fibre have resulted the higher tenacity and elongation to meet the requirements of airbag applications. Such comparison of the fibre and the yarn properties for the aging conditions of 1,000 h and 110 °C are discussed by Orme et al. [115]. However, the material properties such as density, stiffness, and melting point are still finding the gap. Lower density is preferred for lower weight and high cover factor (percentage area covered by the woven fibres). For fabrics with comparable cover factor, polyester fibre results in heavier weight (by 21%) fabric than nylon (PA 6,6) [115]. Besides the above structural aspects, airbag materials are required to meet the lower initial modulus and higher thermal resistance; otherwise, the airbag fabric would be ripped and teared due to hot gases. This would increase the passenger risk getting burnt. Often the durability requirements of these fabrics are significantly met using silicon coating on the fabric surface. Typically, 25–30 g of silicone would be applied on 1 m² fabric area in case of frontal crash events (impact occurs at short time) and 60–100 g of silicone in case of roll over crash events (impact occurs at longer time) [118]. This silicone coating would also enhance the air permeability property of the fabric by reducing the fabric porosity but increase the weight and cost. Nevertheless, uncoated fabrics exist that make use of fabric construction to achieve the desired air permeability [117]. Regular weaving construction has substantially increased the air permeability when the fabric is subjected to tension during inflation of the airbag. Example of such weaving structure with multiple warp harness arrangements in combination of twill and basket weaves has avoided the increase in permeability during airbag inflation [119].

16.4.3 Structural reinforcement

The automotive industry continues to push their boundaries towards the lower weight for IC engine applications and the electric vehicles as well. Interest in composite materials by the addition of fibres to the polymer resins to increase the strength to weight ratio of the resulting materials has been increasing amongst the automotive manufacturers. Carbon fibre and basalt fibre-reinforced polymer composites have widely been used in automotive application such as body panels, floor panels, roof, and under body structures [120, 121]. However, recent emphasis on recyclability to reduce the carbon footprint has made the automotive industry to look for biodegradable materials. Several biofibres such as jute, flax, hemp, ramie, and kenaf are often used as a reinforcement with polyester, polypropylene, and polylactic acid matrices. This field is still young, with rapid developments associated with the mobility trends for the automotive industry, and it needs a considerable amount of research to be carried out to develop the light-weight products.

Kumar and Das [122] have investigated the poly(lactic acid) biocomposites reinforced with nettle fibres for automotive dashboard panel application. In this study, nettle fibres with fineness of about 25 μm and 55 mm length and PLA fibres of 6 denier and 55 mm length were homogeneously mixed and five different blend proportions by weight (10/90, 25/75, 50/50, 75/25, 90/10) were prepared. Prior to this, nettle fibres were alkali treated to increase the mechanical properties. The composites were developed by employing carding and compression moulding technologies. These composites were evaluated for static and dynamic mechanical properties, thermogravimetric behaviour, and biodegradability. The optimum tensile, bending, and impact properties were obtained at the weight proportion of 50/50 nettle and poly(lactic acid) fibres. The thermal properties of these biocomposites were increased with increasing proportions of nettle fibres. The biocomposites were found to be extremely good in terms of dynamic mechanical properties against temperature, time, and frequency. Further, the biodegradability of these biocomposites was increased with the increase of nettle fibre content. The biocomposite specimens with higher nettle fibre proportion exhibited loss in mechanical properties during the 20 days of soil burial.

Rwawiire et al. [123] have developed a biocomposite based on green epoxy polymer and natural cellulose fabric that was claimed to be suitable for automotive instrumental panel application. In this study, alkali-treated black cloth was reinforced with green epoxy matrix using hand lay-up method. Four black cloth plies with ply angles of 90°, 0°, –45°, and 45° were utilized to fabricate the biocomposite. It was concluded that the alkali-treated biocomposite exhibited tensile strength, Young's modulus, and flexural strength as 33 MPa, 5 GPa, and 207 MPa, respectively. Overall, the biocomposite was found to offer superior strength (33 MPa) to that (25 MPa) typically required for automotive interior applications.

In the recent times, polymer nanocomposites are becoming more popular than the conventional composites for automotive applications. The conventional composite materials require a high loading level of fillers to achieve the desired stiffness and dimensional stability. Nevertheless, the higher filler loading results in higher weight, lower toughness, and poor surface quality. In contrast to the conventional composites, the polymeric nanocomposites have shown improved physico-chemical property at a low level of filler loading, that is, less than 5% [124]. The presence of filler material at nanoscale results in a higher surface to volume ratio and an exponential increase of interfacial interaction between the polymer matrix and the nanofiller [50]. Structurally noncritical parts of automotive such as fascia, cowl, grills, and valve/belt timing covers made of polymer nanocomposites could save nearly 25% of weight. Also, replacement of talc-reinforced materials with highly exfoliated ultrapure clay-PO nanocomposites has resulted in 10% of weight reduction and improved stiffness and toughness [124]. Nowadays, several other nanocomposites are being widely used in the applications such as engine and power train components, automotive body and chassis components, vehicle exhaust and after treatment systems, tyre and tyre chord, protective paints, and coatings [50].

16.5 Concluding remarks

Fibrous materials for automotive applications are always emerging to meet the constantly evolving mobility trends. These materials have gained enormous popularity in the recent years due to their suitable structures and appropriate properties. Their applications in automotive air filtration, thermal insulation, acoustic noise reduction, safety, and structural purposes are discussed. The key design requirements that are driving the development of fibrous materials for automotive applications are identified. Today, the properties of the fibrous materials that involve fluid and thermal flows, acoustic wave propagation, and structural needs are no longer isolated but are designed to meet the combined requirements. Further developments towards fibrous materials, fibre and (or) yarn arrangements, manufacturing processes, nanotechnology, and their combinations to achieve the required properties of next generation fibrous materials will continue to play an important role in deciding the performance of automotive vehicles.

References

[1]   Paril, A. S., Halbe V. G., and Vora, K. C. A system approach to automotive air intake system development. SAE Technical Paper 2006; 2006-26-011. doi:10.4271/2005-26-011.
[2]   Reinhardt, H. and Stahl, U. Recent developments in cabin air filtration. SAE Technical Paper 2006; 2006–01-0270, doi: 10.4271/2006-01-0270.
[3]   Ptak, T. J., Richberg, P., and Vasseur, T. Discriminating tests for automotive engine air filters. SAE Technical Paper 2002; 2001-01-037, doi:10.4271/2001-01-0370.
[4]   Konstandopoulos A. G., Kostoglou, M., Vlachos, N., Kladopoulou, E. Advances in the science and technology of diesel particulate filter simulation. Advances in Chemical Engineering 2007; 33: 213–275.
[5]   Kolhe, V., Sharma, M., Veeramani, K., and Kulkarni, M. Development of advanced oil separator to give uniform oil separation efficiency across engine speed and load conditions. SAE Technical Paper 2012; 2012-01-0179. doi: 10.4271/2012-01-0179
[6]   Jaroszczyk, T., Wake, J., and Conner, M.J. Factors affecting the performance of engine air filters. Journal of engineering for gas turbine and power 2009a; 115: 693–699.

[7] Whitby, K.T., Husar, R. B., and Liu, B. Y. H. The aerosol size distribution of the Los Angeles smog. Journal of Colloid Interface Science 1972; 29: 177–204.

[8] Watson, C. E., Hamgley, F. J., and Burhell, R. W. Abrasive wear of piston rings. SAE Technical Paper 1955; 1955-55-0289. doi:10.4271/550289.

[9] Mund, M.G. and Guhne, H. Gasturbines – Dust – Air cleaners: Experience and trends. ASME publication 1970; 104: 1–27.

[10] Darcy, H. Les fontaines publiques de la ville de Dijon (in French). France: Victor almont; 1856.

[11] Davies, C.N. The separation of airborne dust and particles. Proceedings of Institute of Mechanical Engineers 1952; B1: 185–213. London.

[12] Happel J. Viscous flow regime relative to arrays of cylinders. AIChE Journal 1959; 5: 527–532.

[13] Kuwabara S. The forces experienced by randomly distributed parallel circular cylinders of spheres in a viscous flow at small Reynolds numbers. Journal of Physical Society of Japan 1959; 145: 27–532.

[14] Tien, C. and Rao, B. V. R. Granular Filtration of Aerosols and Hydrosols. UK: Elsevier's Science and Technology; 2007.

[15] Brown, R. C. and Wake, D. Air filtration by interception: Theory and experiment. Journal of Aerosol Science 1991; 22-2: 181–186.

[16] Hutten, I. Handbook of Nonwoven Filter Media, UK: Elsevier's Science and Technology; 2007.

[17] Kim, H. T., Kwon, S. B., Park, Y. O., and Lee, K. w. Diffusional filtration of poly dispersed aerosol particles by fibrous and packed-bed filters. Filtration and Separation 2000; July/August: 37–42.

[18] Lee, K. W. and Liu, B. Y. H. Theoretical study of aerosol filtration by fibrous filters. Aerosol Science and Technology 1982; 1: 147–161.

[19] Davies, C. N. Filtration of aerosols. Journal of Aerosol Science 1983; 14: 147–161.

[20] Otani, Y., Eryu, K., Furuuchi, M., Tajima, N., and Tekasakul, P. Inertial classification of nanoparticles with fibrous filters. Aerosol and Air Quality Research 2007; 7: 343–352.

[21] Thomas, J. W., Rimberg, D., and Miller, T. J. Gravity effect in air filtration. Journal of Aerosol Science 1971; 2: 31–38.

[22] Brown, R. C. Air filtration: an integrated approach to the theory and applications of fibrous filters. Oxford: Pergoman press; 1993a.

[23] Bergman. W., Taylor, R. D., and Muller, H. H. (1978), Enhanced filtration programme at LLL – A progress report, 15th DOE Nuclear Air Cleaner Conference, Boston.

[24] Novick, V. J., Higgins, P. J., Dierkschiede, B., Abrahamson, C., Richardson, W. B., Monson, P. R., and Ellison, P. G. Efficiency and mass loading characteristics of a typical HPEA filter media material. Proceedings of the 21st DOE/NRC Nuclear Air Cleaning Conference, San Diego, 1991; 782–798.

[25] Thomas, D., Penicot, P., Contal, P., Leclerc, D., and Vendel, J. Clogging of fibrous filters by solid aerosol particles experimental and modelling study. Chemical Engineering Science 2001; 56: 3549–3561.

[26] Charvet, A., Gonthier, Y., Gonze, E., Bernis, A. Experimental and modelled efficiencies during the filtration of a liquid aerosol with a fibrous medium. Chemical Engineering Science 2010; 65-5: 1875–1886.

[27] Mueller, T. K., Meyer, J., Thebault, E., Kasper, G. Impact of an oil coating on particle deposition and dust holding capacity of fibrous filters. Powder Technol. 2014; 253: 247–255.

[28] Hubbard, J. A., Salazar, K. C., Crown K. K., and Servantes, B. L. High-volume aerosol filtration and mitigation of inertial particle rebound. Aerosol Science and technology 2014; 48-5: 530–540.

[29] Maddineni, A. K., Das, D., and Damodaran, R. Experimental and Numerical Study on Automotive Pleated Air Filters. SAE Technical Paper 2016; 2016-28-0100. doi: 10.4271/2016-28-0100.

[30] Maddineni, A. K., Das, D., Damodaran, R. M. Air-borne particle capture by fibrous filter media under collision effect: A CFD based approach. Sep. Purif. Technol. 2018; 193: 1–10.

[31] Maddineni, A. K., Das, D., Damodaran, R. M. Inhibition of particle bounce and re-entrainment using oil-treated filter media for automotive engine intake air filtration. Powder Technol. 2017; 322: 369–377.

[32] Hammen, A., Hausle, T., and Sauter, H. Experimental investigation on the particle distribution and rearrangement in filter media. Proceedings of FILTECH Conference 2006, Germany.

[33] Mullins, M. E., Michaels, L. P., Menon, V., Locke, B., and Ranade, M. B. Effect of Geometry on Particle Adhesion. Aerosol Science and Technology 1992; 17: 105–118.

[34] Boskovic, L., Altman, I. S., Agranovski, I. E., Braddock, R. D., Myojo, T., and Choi, M. Influence of particle shape on filtration process, Aerosol Science and Technology 2005; 39: 1184–1190.

[35] Ptak, T. and Jaroszcsyk, T. Theoretical-Experimental aerosol filtration model for fibrous filters at intermediate Reynolds number. Proceedings of the Fifth World Filtration Congress, France, 1990; 566-572.

[36] Christie, J. G. Transport processes and separation process principles. Oxford: Pearson press; 2015.

[37] Payen, J., Vroman, P., Lewandowski, M., Perwuelz, A., Calle-Chazelet, S., and Thomas, D. Influence of fibre diameter, fibre combinations and solid volume fraction on air filter properties of nonwoven. Textile Research Journal 2012; 82: 1948–1959

[38] Bugli, N.J. Automotive Engine Air Cleaners – Performance Trends. SAE Technical paper 2001; 2001-01-1356. https://doi.org/10.4271/2001-01-1356.

[39] Poon, W. S. and Liu, B. Y. H., Fractional efficiency and particle mass loading characteristics of engine air filters. SAE Technical Paper 1997; 970673, doi:10.4271/2001-01-0370.

[40] Mueller, T., Batt, T., Heim, M., Pelz, A., Klein, G.M. Technology change towards fully synthetic air filter elements in engine air filtration. International stuttgarter symposium proceedings; Wiesbaden: Springer; 2016.

[41] Mullins, B. J., Agaranovski, L. E., Braddock, R. D. Particle bounce during filtration of particles on wet and dry filters. Aerosol Sci. Technol. 2003; 37: 587–600.

[42] Yadav, S, and Das, D. Liquid-mediated particle capture by nonwoven filter media for automotive engine intake air filtration. Journal of industrial textiles 2022.

[43] Pradhan A.K., Das D., Chattopadhyay R., Singh S.N. Effect of 3D fiber orientation distribution on particle capture efficiency of anisotropic fiber networks. Powder Technology 2013; 249: 205–207.

[44] Mohamaddi, M. and Banks-Lee, P. Air permeability of multilayered non-woven fabrics: Comparison of experimental and theoretical results. Textile Research Journal 2002; 72: 613–617.

[45] Hollingsworth & Vose 2021. https://www.hollingsworth-vose.com/products/meltblown/ (accessed December 15, 2021).

[46] Malviya, R. Nano-Fiber Filters for Automotive Applications. SAE Technical Paper 2018; 2018-28-0041. https://doi.org/10.4271/2018-28-0041.

[47] Donaldson Filtration Solution 2021a. https://www.donaldson.com/en-in/industrial-dust-fume-mist/technical-articles/ultra-web-media-technology/ (accessed December 15, 2021).

[48] Xu, B., Wu, Y., Lin, Z., Chen, Z. Investigation of air humidity affecting efficiency and pressure drop of vehicle air cabin air filters. Aerosol and air quality research 2014; 14: 1066–1073.

[49] Joubert, A., Laborde, J. C., Bouilloux, L., Chazelet, S., and Thomas, D. Modelling the pressure drop across HEPA filters during cake filtration in the presence of humidity. Chemical engineering journal 2011; 166-2: 616–623.

[50] Joshi, M., Adak, B., Nano-technology based textiles: A solution for emerging auto sector. In Banerjee, B. (Ed.) Rubber nano composites and Nano textiles: perspective in automotive technologies. Walter de Gruyter GmbH & Co KG 2019; 179–23

[51] Freudenberg sealing technologies 2021. https://www.fst.com/corporate/newsroom/press-releases/2021/quick-release/ (accessed December 15, 2021).

[52] Sanders, J. Venting for EV battery packs. SAE Tech Briefs 2020; 2020-09-03 https://www.sae.org/news/2020/09/venting-for-ev-battery-packs (accessed December 15, 2021)

[53]   Kent, M. Venting under pressure. Charged 2014; 14: 30–33.

[54]   Mukhopadhyay, A. Composite nonwovens in filters: applications. 164-210. in Das, D., Poudeyhimi, B., (Eds.) Composite nonwoven materials. London: Woodhead publications; 2014.

[55]   Sascha. B., Stepan, M., Zbiral, R., Beylich, M. Venting device for housing and method of producing same. US patent US20120247338A1: Mann and Hummel GmbH; 2012.

[56]   Sanders, J. Venting assembly and microporous membrane composite. US patent US9317068B2: Donaldson company Inc.; 2016.

[57]   Basham, D. E., Lombardi, N. (2019), Vent for pressure equivalization, W.L.Gore associates, Inc., Newark, Del., US patent 0226574A1.

[58]   Oxyphen Filtration group 2021. https://www.oxyphen.com/technologies/fiber-based-membranes/ (accessed December 15, 2021)

[59]   Harenbrock, M., Korn, A., Weber, A., and Hallbauer, E. Cost-Efficient Cathode Air Path for PEM Fuel Cell Systems. SAE Technical Paper 2020. 2020-01-1176. https://doi.org/10.4271/2020-01-1176.

[60]   Freudenberg 2021. https://www.freudenberg-filter.com/en/world-of-automotive/products/fuel-cell-solutions/ (accessed December 15, 2021).

[61]   Cusumano, T.J., Pittman, D.D., Alessi, Jr. D.P. Ventilation systems for an automotive fuel stack enclosure. US patent US0191805A1: Ford motor company; 2009.

[62]   Kennedy, D.M., Cahela, D.R., Zhu, W.H., Westrom, K.C., Nelms, R.M., Tatarchuk, B.J. Fuel cell cathode air filters: Methodologies for design and optimization, Journal of power source 2007; 168-2: 391–399.

[63]   Özyalcin, C.; Mauermann, P.; Dirkes, S.; Thiele, P.; Sterlepper, S.; Pischinger, S. Investigation of Filtration Phenomena of Air Pollutants on Cathode Air Filters for PEM Fuel Cells. Catalysts 2021; 11: 1339. https://doi.org/10.3390/catal11111339

[64]   Baek, S., Jung, C. Air filter for fuel cell vehicle. US patent US10220343B2: Hyundai Motor co; 2016.

[65]   Orendorf, C.J. The role of separators in lithium-ion cell safety. The electrochemical society Interface 2012; Summer: 61–65

[66]   Vande Hey, J.D., Sonderfeld, H., Jeanjean, A.P.R., Panchal, R., Leigh, R.J., Allen, M.A., Dawson, M., Monks, P.S. Experimental and modelling assessment of a novel automotive cabin PM2.5 removal system. Aerosol Science and Technology 2018; 52-11: 1249–1265.

[67]   Thakur, R., Das, D., Das, A. Electret Air Filters. Separation & Purification Reviews 2013; 42-2: 87–129

[68]   Thakur, R., Das, D., Das, A. Study of charge decay in corona-charged fibrous electrets. Fibers and polymers 2014; 15: 1436–1443.

[69]   Das, D., Waychal, A. On the triboelectrically charged nonwoven electrets for air filtration. Journal of electrostatics 2016; 83: 73–77.

[70]   Walsh, D. C. The Behaviour of Electrically Active and Prefilter Fibrous Filters under Solid Aerosol Load. Ph.D. thesis. Loughborough University of Technology, 1995.

[71]   Keerl, D. Keeping pollen out. Sonderprojekte ATZ/MTZ worldwide 2016; 21:30–33.

[72]   Piry, A., Blum, R., Buchta, E. Antimicrobial filter medium and cabin air filter. US patent US20180085697A1: Mann and Hummel Gmbh and Neenah Gessner Gmbh; 2018.

[73]   Marshall, G.J., Mahony, C, P., Rhodes, M, J., Daniewicz, S, R., Tsolas, N., Thompson, S,M. Thermal management of vehicle cabins, external surfaces, and onboard electronics: An overview. Engineering 2019; 5-5: 954–969.

[74]   Cai, Z. Al Faruque, M.A. Kiziltas, A. Mielewski, D. Naebe. M. Sustainable lightweight insulation materials from textile-based waste for the automobile industry. Materials 2021; 14: 1241. https://doi.org/10.3390/ma14051241.

[75]   Isover technical insulation 2017. https://www.isover-technical-insulation.com/download-documents/brochures/automotive.pdf (accessed December 15, 2021).

[76]   Chauhan D.S., Kumar, A., Lacki, T.S. High temperature resistant insulation mat, US patent US9452719B2: Unifrax 1 LLC; 2016.

[77] Zhao, K.; Lou, D.; Zhang, Y.; Fang, L., Tang, Y. Experimental study on diesel engine emission characteristics based on different exhaust pipe coating schemes. Micromachines 2021; 12: 1155. https://doi.org/10.3390/mi12101155

[78] Mohanty, A.R., Fatima, S. An overview of automobile noise and vibration control. Noise notes 2014; 13-1: 43–56

[79] Recticel Flexible Foams, Belgium. Material developments in car interior and engine compartment, white paper, June 2017; 11–12.

[80] Arambakam, R., Tafreshi, H.V., Pourdeyhimi, B. A simple simulation method for designing fibrous insulation materials. Materials and Design 2013; 44: 99–106.

[81] Can, H., Yue, Z. Calculation of high-temperature insulation parameters and heat transfer behaviors of multilayer insulation by inverse problems method. Chinese journal of aeronautics 2014; 27-4: 791–796.

[82] Arambakam, R., Tafreshi, H.V., Pourdeyhimi, B. Modelling performance of multi-component fibrous insulations against conductive and radiative heat transfer, International journal of heat and mass transfer 2014; 71: 341–348.

[83] Tilioua, A., Libessart, L., Lassue, S. Characterization of the thermal properties of fibrous insulation materials made from recycled textile fibers for building applications: Theoretical and experimental analyses. Applied thermal engineering 2018; 142: 56–67.

[84] Bankvall, C. Heat transfer in fibrous materials, Journal of testing and evaluation 1973; 1(3): 235–243.

[85] Yang, J., Wu, H., Wang, M., Liang, Y. Prediction and optimization of radiative thermal properties of nano TiO2 assembled fibrous insulations. International journal of heat and mass transfer 2018; 117: 729–739.

[86] Sujon, Md. A. S., Islam, A., Nadimpalli, V.K. Damping and sound absorption properties of polymer matrix composites: A review. Polymer testing 2021; 104: 107388.

[87] Zwikker C, Kosten CW. Sound absorbing materials. Elsevier; 1949.

[88] Tang, X., Yan, X. Acoustic energy absorption properties of fibrous materials: A review. Composites: Part A 2017; 101: 360–380.

[89] Delany, M.E. Bazley, E.N. Acoustical properties of fibrous absorbent materials. Applied Acoustics; 1970 3: 105–116.

[90] Zhang, C., Gong, J., Li, H., Zhang, J. Fiber-based composite with dual gradient structure for sound insulation. Composites part B: Engineering 2021; 198-1: 108166.

[91] Owens corning, India, 2021. https://www.owenscorningindia.com/OCIndia/Composites/Silentex/PDF/Silentex_Product_Brochure.pdf (accessed December 15, 2021)

[92] Pluymers, B., Haider, M. Noise, vibration and harshness research needs, priorities & challenges for road transport in horizon 2020. ERPA position paper 2016; 1–5.

[93] Enkler, M.F., Bopp, M. Heat and sound insulating shroud for the engine compartment of motor vehicles. European Patent, EP1104497A1: HP Chemie Pelzer Research and Development Ltd.; 1998.

[94] Daniel, P. General design principles for an automotive muffler. Proceedings of acoustics 2005; 153–158. Busselton: western Australia.

[95] Copley, D.C., Callas, J., Martin, K.L. Sound suppression device for internal combustion engine system. US patent US7635048B2: Caterpillar Inc.; 2006.

[96] Tao, J., Bao, Z., LI, H., Qian, D., Zhao, Y., Yang, Z. Nonwoven composite material for automobile interior trim and preparation method thereof. China Patent CN102173141A Jiangyin xietong automobile accessory Co. Ltd., & Nanjing University of Aeronautics and Astronautics; 2010.

[97] Kamath, M.G., Bhat, G.S., Parikh, D.V., Mueller, D. Cotton fiber nonwovens for automotive components. Journal of engineered fibers and fabrics 2005; 14-1: 34–40

[98] Yachmanev, V., Negulescu, I., Yan, C. Thermal Insulation properties of cellulosic-based nonwoven composites. Journal of industrial textiles 2006; 36-1: 73–87.

[99]   Calçada, M. and Parrett, A. Enhanced Acoustic Performance using Key Design Parameters of Headliners. SAE Technical Paper 2015; 2015-01-2339. https://doi.org/10.4271/2015-01-2339.

[100]   Prashsarn, C., Klinsukhon, W., Suwannamek, N., Wannid, P., Padee, S. Sound absorption performance of needle-punched nonwovens and their composites with perforated rubber. Spinger nature applied sciences 2020; 2:559.

[101]   Gross, J. R., Hurley, J.S., Boehmer, B.E. Moose, R.T. Nonwoven material for acoustic insulation and process for manufacturing. US patent, US0121461A1: Baker Botts LLP, NY; 2008.

[102]   Kim, N.N., Lee, S., Bolton, J.S., Hollands, S., Yoo, T. structural damping by the use of fibrous materials. SAE Technical Paper 2015; 2015-01-2239. https://doi.org/10.4271/2015-01-2239.

[103]   Val. Y., Ioan. N., and Chen. Y. Thermal inculation properties of cellulose based nonwoven composites. Journal of industrial textiles 2016; 36: 73–87.

[104]   Ulrich, T., Arenas, J.P., Sound absorption of sustainable polymer nanofibrous thin membranes bonded to a bulk porous material. Sustainability 2020; 12-6: 2361. https://doi.org/10.3390/su12062361

[105]   Dahl, M.D., Rice, E.J., Groesbeck, E. Effects of fiber motion on the acoustic behavior of an anisotropic, flexible fibrous material. The Journal of the Acoustical Society of America 1990; 87: 54.

[106]   Fernando, J. A., et al. Insulating material for automotive exhaust line tubing and manifolds. US patent US8627853B1: Unifrax 1 LLC; 2007.

[107]   Final Advanced Materials 2021. https://www.final-materials.com/gb/297-basalt-fibre (accessed December 15, 2021).

[108]   Knutsson, G. Preformed sound-absorbing material for engine exhaust muffler. European patent EP0692616A1: Owens-Corning Sweden AB; 1995.

[109]   Koch, G. Stretchable belt and process for the production thereof. US patent US4662487A: Ieperband NV; 1987.

[110]   Richaud, M., Vermeersch, O., Dolez, P.I. Specific testing of textiles for transportation. In Dolez, P.I., Vermeersch, O., Izquierdo, V., (Eds.) Advanced Characterization and Testing of Textiles. the textile institute book series 2018; 399–432.

[111]   Koo, H., Kim, Y. Reliability assessment of seat belt webbings through accelerated life testing. Polymer testing 2005; 24-3: 309–315.

[112]   Dubois, D., Silverthrone, P., Markiewicz, E. Assessment of seat belt webbing bunching phenomena, International journal of impact engineering 2011; 38-5: 339–357.

[113]   Koseki, T. Woven fabric for seat belt. US patent US5376440A: Ikeda Bussan Co Ltd, Hamamatsu Industry Co Ltd; 1993.

[114]   Low friction belt. Nissan Motor Corporation 2021. https://www.nissan-global.com/EN/TECHNOLOGY/OVERVIEW/lfs.html (accessed December 15, 2021)

[115]   Orme, B., Walsh, R., and Westoby, S. Equivalency or Compromise? A Comparative Study of the Use of Nylon 6,6 and Polyester Fiber in Automotive Airbag Cushions. SAE Technical Paper 2014; 2014-01-0509. https://doi.org/10.4271/2014-01-0509

[116]   Lewis, K.R., Quentin Whitfield, G. St. Fabric for airbag. US patent US6022417A: Black and Decker Inc, Invista, North America LLC; 1997.

[117]   Fung, W., Hardcastle J.M. Textiles in automotive engineering. Cambridge: England; Woodhead publishing limited; 2000.

[118]   Elkem silicones 2021. www.elkem.com/silicones/blog/textile-leather/silicone-coatings-keeping-performance-in-check-so-airbags-can-save-lives/ (accessed December 15, 2021).

[119]   Bowen, D.L., Bower, C.L., Sollars, Jr, J. A. Air bag fabric with specific weave construction, US patent US5921287A: Milliken Research Corp; 1996.

[120]   Hiremath, N., Young, S., Ghossein, H., Penumadu, D., Vaishya, U., Theodore, M. Low cost textile grade carbon-fiber epoxy composites for automotive and wind energy applications. Composites part B: Engineering 2022; 198-1: 108–156

[121] Fiore, V., Scalici, T., Di Bella, G., Valenza, A. A review on basalt fibre and its composites, Composite Part B: Engineering 2015; 74: 74–94.
[122] Kumar, N., Das, D. Fibrous biocomposites from nettle (*Girardinia diversifolia*) and poly(lactic acid) fibers for automotive dashboard panel application. Composite Part B: Engineering 2017; 130: 54–63.
[123] Rwawiire, S., et al. Development of a biocomposite based on green epoxy polymer and natural cellulose fabric (bark cloth) for automotive instrument panel applications. Composite Part B: Engineering 2015; 81: 149–157.
[124] Garces, J.M., Moll, D.J., Bicerano, J., Fibiger, R., McLeod, D.G. Polymeric nanocomposites for automotive applications. Advanced Materials 2000; 12-23: 1835–1839.

Bapan Adak, Samrat Mukhopadhyay, Shanmugam Kumar

17 Smart/functional textiles and fashion products enabled by 3D and 4D printing

Abstract: Over the last two decades, researchers, technologists, designers, and manufacturers have made enormous efforts to commercialize additive manufacturing (AM) or 3D printing technology in an array of fields including textile, apparel, and fashion industries. Recently, a great advancement in AM of complex architectures, which are impossible or difficult to produce otherwise, has been reported. Following the success of making metal/polymer-based 3D printed stiff structures, researchers have also explored the potential of this technique for creating flexible materials such as smart textiles. This chapter presents 3D printing as a novel method for the manufacturing of more flexible, cost-effective, and functional textiles via techniques such as screen printing and ink-jet printing and for fabricating smart textile structures which are slightly different from the conventional knitted or woven fabrics but possessing intelligent properties. Specifically, in this chapter, an overview of 3D printing technology, different 3D printing techniques, material selection, and properties of 3D printed objects in the context of manufacturing of smart and functional textiles is discussed. Emerging smart textiles enabled by 4D printing have also been explored which can exhibit transformation in their structure or colour as a function of time in the presence of an external stimulus. Therefore, the transition from 3D to 4D printing, the basic aspects of 4D printing, and materials selection for 4D printing of smart textiles and fashion products are presented here. The subsequent section discusses the potential applications of textiles enabled by 3D and 4D printing. Finally, current challenges and future perspectives of 3D and 4D printing of smart textiles are summarized.

Keywords: Additive manufacturing, 3D printing, 4D printing, smart textiles, shape memory, fashion trends

17.1 Introduction

Textiles with integrated functions of controlling or adjusting their properties based on their applications are called as functional textiles. Smart textile can detect and process the wearer's physiological condition as well as perceive and communicate the environmental conditions [1, 2]. Smart functionalities can be incorporated in textiles through many techniques such as polymeric coating and lamination, functional finishing, and printing [3]. Among different printing technologies, inkjet printing and screen printing are two extensively used techniques for creating functional and smart textiles. For the fabrication of smart wearable electronics, screen printing shows good potential through

https://doi.org/10.1515/9783110759747-017

layer-by-layer printing with various functionalities on the top of textile fabric [4]. The main advantages of this printing method are low-cost patterning, high-volume batch fabrication, processability at room temperature (but need curing step), applicability on any irregular textile surface, and elimination of any photolithographic and chemical etching processes which are required in traditional subtractive microfabrication processes [5, 6]. In comparison with screen-printing techniques, inkjet printing is known for high precision, thin-layer deposition capability, short run length, and tailored/integrated production processes. Although the printing is possible on the rough textile surfaces, multiple printing passes are required to incorporate desired functionality in textiles by inkjet printing, and hence, it increases process cost and manufacturing time. Sometimes, a thin base coat is applied on the textile surface or screen printing are applied is employed to create a better interface and reduce the surface roughness prior to inkjet printing [7, 8]. In addition to different advantages, there are many issues with these conventional printing technologies; for example, inkjet printers are expensive, the inkjet-printed layers on fabric are resistant to stretching and bending, but most fabrics cannot withstand temperatures above 150 °C required for the curing of inkjet-printed layers on fabric. On the other hand, in screen printing, extra spaces are needed for screen warehouse, and the downtime reduces production and increases the manufacturing cost, difficulty in reprocessing of imperfect product as well as generation of pollutants and waste during screen-printing process [9]. Therefore, researchers are in search of new printing technologies to elude these limitations.

Emerging three-dimensional (3D)-printing technology enables the systems to be more easily integrated and tailored, where 3D objects are manufactured directly from digital computer-aided design (CAD) files [10]. The technology of 3D printing, also known as additive manufacturing, has been developed for more than 30 years. In recent years, 3D printing has been recognized as a disruptive technology for future advanced manufacturing systems including textiles for smart applications by depositing polymers on the fabrics. Printing technologies have a potential for integrating or tailoring production process for functional and smart textiles which evade inessential use of energy, water, and chemicals and reduce the waste for improving ecological footprint and productivity. There have been significant advances in materials, hardware, and processes related to 3D printing technology, which has a great potential to transform everything from our everyday lives to the global economy.

There are different 3D printing techniques for the fabrication of smart and functional textiles. These techniques are based mainly on two strategies: (i) deposition process, that is, 3D printing of polymer/metal/ceramic on the surface of the textile fabric for incorporating smart and/or functional properties and (ii) binding process, that is, 3D printing of polymers for making textile structures with some smart and/or functional properties [11]. With the advancement in technology, researchers have developed textile structures enabled by four-dimensional (4D) printing. In 4D, the structure is made by 3D changes (*x*, *y*, *z*) over time (*t*) as indicated by arrow in Figure 17.1. In response to stimuli such as temperature, water, ultraviolet (UV) rays, magnetic energy, and pH, the 4D printed structures can change their shape, colour, function, or other characteristics [12, 13]. This chapter highlights different 3D printing techniques, their features, challenges, and potential smart applications of 3D and 4D printed textiles.

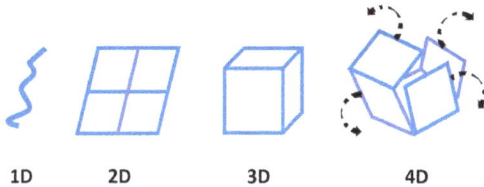

| 1D | 2D | 3D | 4D |

Figure 17.1: Schematic of 1D, 2D, 3D, and 4D concepts. The arrows indicate the direction of change of a 3D structure with respect to time.

17.2 Brief overview of 3D printing techniques

3D printing is a computer-assisted manufacturing (CAM) technology that uses liquids or rigid materials to build a component or the whole product layer-by-layer based on a 3D digital model [14–16]. In contrast to traditional manufacturing processes, 3D printing allows for more complex designs to be designed and printed making the technology more cost-effective and sustainable [17–19]. 3D printing consists of three major stages: (i) pre-processing, (ii) printing and fusing, and (iii) post-processing [11, 20]. The pre-process of this production begins with a model design with an appropriate software, for example, Rhino, AutoCAD, Repetier-Host, Materialise Magics, and 3Ds Max. By adjusting the algorithms, any measurement can be corrected, or any size parameter can be added for individual customers [20, 21]. This will be followed by dividing the model into horizontal layers, followed by converting each layer into *x* and *y* dimensions. To set up a communication between CAM and solid models, the digital file data is transferred to the printer. As soon as the printer receives the design instructions, the CAM enables printing of layers and provides a shape of a particular object by fusing the layers. The post-processing step comprises polishing, sanding, dyeing, or painting for enhancing the surface finish of the object, followed by eliminating unwanted print edges [20, 22]. There are some printers that cannot print entire garments at once. In that case, the entire garment is printed from multiple parts, which are then joined together to form a complete product [22].

17.3 Additive manufacturing-enabled materials for textile/fashion industry

In conventional textiles, a wide range of natural fibres (cotton, linen, wool, silk, etc.) and man-made fibres (viscose, polyester, nylon, acrylic, cellulose acetate, Lycra®/Spadex®, etc.) and fibre blends (polyester/cotton, polyester/viscose, cotton/Lycra®, etc.) are used for making yarns, followed by making of fabric. These fibres are made of different polymers such as polysaccharides (mainly cellulose, hemicellulose, and lignin), proteins (keratin, sericin, fibroin, etc.), polyamide, polyester, and polyurethane (PU). Few of these polymers can be processed by melting and some cannot be melt-processed as they degrade before melting. Recently, by virtue of 3D printing, different polymers [such as polylactic acid (PLA), PU, polyamide, polyethylene terephthalate (PET, polyester), polyvinyl alcohol (PVA) etc.], photopolymer resins, ceramics, and metal powders can be used for producing fabric of various textures, garment components, long dresses, footwear, and metal accessories. Moreover, they can also be 3D printed on the surface of conventional textiles by deposition technique to impart smart or functional properties to the textiles. TamiCare has developed a 3D printing technology called CosyflexTM, which uses liquid polymers, including silicon, natural latex, Teflon and PU, as well as textile fibres such as rayon, cotton, and polyamide [22, 23].

17.4 Different additive manufacturing techniques for making textiles and fashion products

Among different 3D printing techniques, only a few can be used in textile and fashion industry. The choice of the appropriate method or process is crucial for determining whether the design and desired outcome are feasible or not for making smart textiles or apparels. Two different strategies used for producing smart/functional textiles enabled by 3D printing are:

i) **Deposition process:** In this process, a thermoplastic polymer or a photo/UV-curable resin or a metal or a ceramic is printed on the surface of textile fabrics. Generally, fused deposition modelling (FDM) [24, 25], stereolithography (SLA) [16], and 3D microfiber extrusion technique are used in this process.

ii) **Binding process:** This process is used to fabricate and design structures directly using 3D printing techniques, making long dress, apparel components, and jewelleries. Generally, selective laser sintering (SLS) and inkjet-printing techniques are used in this process.

Researchers have found that not all 3D printing techniques available in the market are cost-effective and applicable for producing textiles and fashion items. Therefore, textile and fashion technologists and manufacturers need to choose the most appropriate and cost-effective ones. The different 3D printing techniques which have the potential for producing smart and functional textiles are discussed further.

17.4.1 Material extrusion

It is the most widely used 3D printing technique in which materials (mainly filament or ink) are pushed through a programmable nozzle to create 3D pattern [26]. This process can be further subdivided into three categories: (i) FDM or fused filament fabrication (FFF), (ii) direct ink writing (DIW) or robocasting, and (iii) droplet-based printing.

17.4.1.1 FDM or FFF 3D printing

Based on the availability of a wide range of materials, simplicity of the mechanism, and low initial investment, FDM has become one of the most popular techniques for developing textiles and fashion products. In this technique, thermoplastic filament is used as feedstock. The filament is passed through a heated nozzle where it is melted, extruded, and subsequently deposited layer-by-layer (l-b-l) on a substrate to create a 3D object upon solidification [27] as shown in Figure 17.2(a). A software program initially processes an STL or CAD file, mathematically slicing and orienting it for the building process. Each time a layer is printed, the platform can be lowered according to the selected layer thickness and height. In FDM technique, a porous or cellular structure can be prepared using as movable nozzle with a typical feature resolution of 100–150 µm. FDM printing has a potential for replacing other technologies such as screen printing, gravure, flexography, and inkjet printing for functionalizing and altering different types of textiles [7]. Recently, FDM technique has received increased attention among the material, polymer, and textile researchers as well manufacturers towards enabling functional and smart textiles via deposition of functional polymers or mixtures of functional polymers and polymer nanocomposites on textile fabrics [20, 28].

17.4.1.2 Robocasting or DIW 3D printing

Robocasting or DIW 3D printing is a very useful material extrusion technique for processing 3D printed textiles as it allows for the creation of complex 3D structures in a rapid and facile manner. This technique involves robotic l-b-l deposition of inks or filament-based suspensions, which solidify because of gelation of the paste or because of a rheological transition from pseudoplastic to dilatant state, forming complex structures with programmed gaps [29], as shown in Figure 17.2(e). For the creation of such

complex structures and for smooth running of DIW printing there are a few basic requirements for the inks:

i)   The inks should have controllable viscoelasticity. An optimal balance between the
     shear stress and storage shear modulus (G') is required for printing.
ii) For printing fibre-like structures, shear-thinning property is very important to
     create fibre without breakage or clogging of nozzle.
iii) They should be able to hold their shape immediately after printing that means a
     fast-curing rate is required.
iv) There should be minimal shrinkage of the printed object after drying.

DIW is a vital technique for printing soft matters with curvilinear shapes required for
making textiles with similar or better physical properties compared to the conventional textiles. There are various types of inks such as viscous organic fugitive inks,
hydrogel inks, colloidal gels, and concentrated polyelectrolyte complexes [30]. The
DIW 3D printing can be classified further in two categories as follows:

i)   Ultraviolet-assisted direct write (UV-DIW): For creating free-standing, curvilinear
     structures made of thermoset polymers [31].
ii) Laser-assisted direct-write (L-DIW): For creating curvilinear, free-form metal
     structures [32].

17.4.1.3 Material jetting or droplet-based inkjet printing

In this technique, low viscosity liquids are deposited in the form of droplets, causing alternate jetting and solidification cycles to create a final object (Figure 17.2(d)). There are
many critical factors that affect drop formation including surface tension, density, viscosity, velocity, drop length, and nozzle diameter. During this process, droplets are deposited
on their substrates, where they interact and solidify. Inkjet printing generates droplets
(diameter 10–150 μm) through two methods:

(i) Continuous inkjet printing, which produces droplets even when printing is not
     required.
(ii) Drop-on-demand (DOD) printing, which produces drops when they are required.

Another droplet-based printing technique is hot-melt printing, in which, waxes with
low molecular weights are inkjet-printed, and they solidify upon impact with the substrate, avoiding the need for ink-drying.

In polyJet technique, it is possible to deposit multiple materials in one layer, in
which connective joints and rigid parts can be printed together. With polyJet, beads of
liquid photopolymer resin are selectively dropped from inkjet print heads onto a
building platform using DOD inkjet-printing technique. Afterwards, a roller is used to
even out the surface of the layer. In order to harden the liquid resin, two UV lights
are passed over it multiple times, one after the other. The building platform lowers as

each layer is added and continues until the print is completed [33–35]. Additional print heads containing the separate materials are installed to print single layers with multiple materials. This printer can print multiple material products, which creates a product that has more movement, flexibility, and texture. It is also one of the fastest 3D printing methods and provides a high-quality surface finish. However, polyJet requires mechanical removal of support rafts and moreover, temperature changes, humidity, and UVs can change the product's dimensions.

Nowadays, fashion designers and manufacturers are using polyJet techniques in designing smart garments of different geometries. PolyJet printers from Objet Geometries (OG), called Objet Connex, come in three different sizes. In addition to being able to print items with multiple materials, the Connex printers are excellent for highly detailed garments and accessories. For instance, Iris van Herpen's VOLTAGE collection uses the Objet Connex multi-material technology. It was a collaborative project with architect, designer, and Professor Neri Oxman from Massachusetts Institute of Technology's Media Lab designed 11 ensembles for her collection, two of which were 3D printed by OG using Objet Connex [36].

17.4.2 Binder jetting

To create a 3D object, binder jetting uses glue to bond successive layers of powder material [40]. In this technique, the powder materials are first spread onto a platform in a thin layer. Then, glue is deposited on top of the powder layer by an inkjet print head. Subsequently, as the building platform lowers, powder materials are placed on top to make the next layer. The unused powder is removed with each new bonding layer. Until the product is finalized, the process continues (Figure 17.2(f)). The main advantages of binder jetting technique are: (i) fastest bonding – binds single layers within seconds, (ii) no requirement of support rafts, and (iii) it can print multiple colours simultaneously as single monochrome inkjet head is the substitute for four- or five-colour heads. However, binder jetting generally results in weaker products with uneven surfaces [41, 42]. 3D Systems' Spectrum Z510 binder jetting printer is one example of a prototype printer used in the fashion industry. With 24-bit colour and 600 dots per inch resolution, it is the only binder jet 3D printer in the market that can print multiple colour objects [43].

17.4.3 Vat photopolymerization or stereolithography

In the presence of UV light or visible light (in some cases) certain liquids or soft materials undergo solidification due to free radical photopolymerization or cationic photopolymerization, which are called as photopolymers [44]. The process of Vat photopolymerization or SLA involves solidifying molecular chains of such soft materials by using light in order to link them together to form polymers. In this technique, a macromolecular

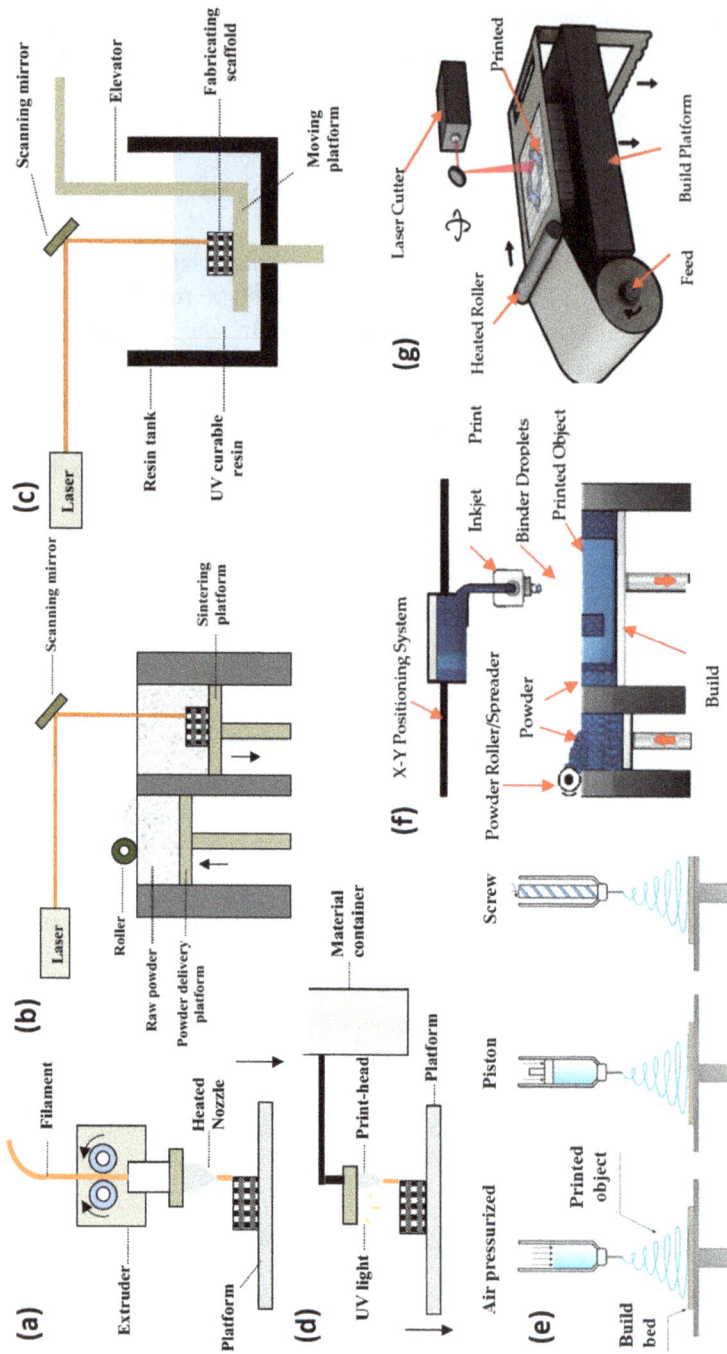

Figure 17.2: Schematic representations of different additive manufacturing techniques: (a) FDM, (b) SLS, (c) Vat photopolymerization, (d) material jetting [37], (e) DIW or robocasting [38], (f) binder jetting, and (g) laminated object manufacturing (LOM) [39].

structure can be built from molecular raw materials, allowing a bottom-up production. Peterson et al. [45] observed that the photopolymers cured with high intensity light had higher overall storage modulus and glass transition temperature in contrast to materials cured at low intensity. Therefore, a greater level of control can therefore be achieved over the final design and properties [45]. The set-up and steps of SLA techniques can be described as follows:

i)   A platform is submerged in a Vat, which contains the photopolymer. At the surface of this platform the photopolymer is exposed to the light, causing it to solidify.
ii) In the subsequent immersion of the platform into the pool, uncured liquid resin of deposited first layer is cleared and coated it with fresh material.
iii) The second layer is then polymerized again by UV light and adheres to the previous layer.
iv) The process is repeated until the product is finished (Figure 17.2(c)).

SLA technique facilitates the relatively fast printing process (product printing completes within few hours depending on the size and complexity of the structure); it is user-friendly and provides high-quality surface finish. However, the main disadvantages of the SLA technique are dependency on the special photopolymers, high cost of photopolymers, and uncured photo-initiator causing cytotoxicity and requiring support rafts.

17.4.4 Selective laser sintering (SLS)

The SLS technique involves fusing tiny particles or powders of polymers (such as nylon, TPU, PE, and PEEK), glass, metals, ceramics, and so on using high-powered lasers. A thin layer of material is first applied to the building platform. As the layer is traced with a laser, the powder is heated just below its melting point so that it becomes a solid object, a process known as sintering [15, 46]. Upon creating the first layer, the building platform lowers, revealing the next layer of powder to be traced and fused. Until the product is ready, the process continues (Figure 17.2(b)). The benefits of SLS 3D printing include: (i) offering designers a wide variety of materials for creating delicate and yet highly functional and durable products, (ii) no use of support rafts, and (iii) requirement of less sanding of the printed object. A disadvantage of SLS is that its surface quality is inferior to that of SLA [33, 46].

Since last decade, SLS is getting immense exposure for making smart textile and fashion products. For example, during the 2013 Paris Fashion Week, Dutch fashion designer Iris van Herpen featured one of the dresses from her haute couture collection, VOLTAGE, based on SLS [47]. It features a lace-like texture made from a flexible fabric-like material (TPU) with a soft hand. It consisted of numerous layers of woven lines created using precision lasers, which mimicked the appearance of yarn. Multiple pieces of the dress have been fused together and dyed black since TPU is originally white powder. With its material and construction, the dress possessed the illusion of movement and organic appearance.

17.4.5 Other 3D printing techniques

Other two additive manufacturing techniques which are getting importance in making smart textile or apparel products are directed energy deposition (DED) and LOM [Figure 17.2(g)]. With DED, large structures can be built from powdered substrates similar to powder bed fusion. However, fabrication speed of DED technique is low, and it possesses inherent surface roughness [48–50]. LOM utilizes l-b-l fabrication of an adhesive-coated roll of fabric, paper, metal, or plastic [51, 52]. However, LOM is limited to materials which can be obtained only in solid sheet form [51, 52]. Table 17.1 summarizes the materials and products of few 3D printing machines including their benefits and challenges.

17.5 3D printing on the fabric surface

In this strategy, different functional polymer or polymer nanocomposites are used as raw material/feedstocks for 3D printing on textile (mainly fabric) surface for improving aesthetic or functional properties.

17.5.1 Aspects to be considered during 3D printing on fabric surfaces

For effective deposition of polymers onto fabrics by deposition techniques (mainly FDM and SLA) the following points need to be considered:
(i)   Good bonding/adhesion of polymer onto the fabrics
(ii)   Drapability for free movement of the printed fabric
(iii) Deformability and recovery when the printed fabric is subjected to daily wear forces
(iv) Colour, design, and other aesthetic properties as per the requirements
(v)   Special functionalities as per the requirements
(vi) Breathability
(vii) Washability

Table 17.1: Comparison of a few additive manufacturing (AM) machines used in the textile and fashion industry [20, 53, 54].

AM method	Printer Company	Maximum printing size	Materials	Benefits	Challenges	Product categories	Brands or designers
FDM	Replicator Desktop MakerBot	200 mm × 250 mm × 150 mm	PLA filament; ABS filament	Flexible materials; compact size; user-friendly; multiple products at once; various quality levels	Support rafts; slower lead time; limited printing size	Dresses; accessories; garment components; prototypes	Francis Bitonti
SLA	Mammoth Materialise	2 m	Photopolymer resins	Large objects; detailed objects; fast lead time. User-friendly; high-quality surface finish	Support rafts; large space needed	Long dresses. Detailed component	Iris van Herpen; Lady Gaga
SLS	PRECIOUS M 080 EOS	80 mm × 95 mm	Metal powder	No support rafts; compact size; fast lead time; high-quality surface finish	Limited printing size; limited end-uses	Jewellery; watches; metal accessories	Dr Richard Hoptroff
Polyjet	Objet Connex Stratasys	490 mm × 390 mm × 200 mm	Liquid photopolymer	Complex and delicate features and small cavities, complex and delicate features and small cavities	Support rafts; lower surface quality	Highly textured dresses and separates	Iris van Herpen
Inkjet	Spectrum Z510 3D Systems	254 mm × 356 mm × 203 mm	Powdered metal and ceramic filaments	Inexpensive; prints in colour; fastest lead time; high-quality surface finish	Weaker products	Shoes; accessories; prototypes	Timberland

17.5.2 Strategies for improving adhesion between 3D printed objects and fabric

Deposition of polymer on fabrics by 3D printing, textile-polymer adhesion, and final properties of the 3D printed textiles strongly depend on the types of polymers and textile substrates as well as method of 3D printing employed. This area requires more research on the potential materials (new or existing) for 3D printing, textile-polymer adhesion, and extrusion/deposition technology. Among different requirements for 3D printing on fabric surfaces (as summarized in previous section), adhesion between printed substrate and fabrics is considered the most important parameter as the other parameters are directly or indirectly depended on this. On the other hand, adhesion between fabric and 3D printed objects depends on few more parameters as summarized below.

i) 3D printing process parameters

Several 3D printing process parameters have significant effect on the adhesion between polymer and textile substrate. The most important 3D printing parameters are extruder temperature, platform temperature, extrusion rate, or printing speed and z-distances between nozzle and printing bed [55].

Diffusion of polymer has a significant effect on the properties of the polymer across polymer-fabric interface and their adhesion strength, which are the functions of polymer-fabric compatibility, temperature, composition, molecular weight of the polymer, chemical structure of the polymer, and orientation of polymer chains. Diffusion of the polymer through the fabric structure improves polymer-fabric adhesion strength and makes the interface strong. When the extruder temperature of the polymer is close to melting point of the base fabric, polymer diffuses into the fabric structure leading to improved adhesion strength between fabric and polymer [56]. Adhesion strength between 3D printed polymer and textiles also depends on the thickness of the deposited polymer which is controlled by the printing speed.

Sanatgar et al. [57] studied the effect of different 3D printing process parameters on the adhesion properties between extruded nylon copolymer and nylon 6,6 fabric. Adhesion strength of the polymer and fabric increased from 5 to 11 N, when the extruder temperature increased from 190 to 230 °C, which can be explained by diffusion theory as discussed above. Moreover, at higher extruder temperature, the depositing polymer softens the base fabric, making a good contact and strong interface. The phase state of the polymer of the fabric is dependent on the platform temperature. The platform temperature has no significant effect on the adhesion strength when the temperature is less than the glass transition temperature (T_g) of the fabric. However, it can play a vital role in controlling adhesion property above the T_g of the polymer. With increasing printing speed up to an optimum level of 50 mm/min, the thickness of the printed layer reduces gradually, which helps in increasing bond strength as the stress dissipates through the finer layer easily. However, with further increase in printing speed, penetration of polymeric macromolecules reduces, resulting in significant reduction in adhesion strength. Grimmelsmann et al. [58] investigated the effect of z-distances between nozzle and printing bed, showing strong impact of pressing the molten polymer between the yarns or even fibres of the fabric substrate on which it was printed, effecting the adhesion property.

ii) **Parameters related to fibres, yarns, and fabrics**

There are several parameters related to fabric such as physical and chemical properties of fibre, yarn structure and count, fabric type (woven, nonwoven, braided, and knitted), weave structure (plain, twill, satin, sateen, leno, etc.), thickness, thread density, and aerial density of fabric and many others which play a great role in controlling the adhesion between fabric and deposited polymers. The roughness of the weave and topography of the textile surface, as well as the level of hairiness on the textile surface, which together form a locking connection, are major influencing factors for controlling adhesion strength. 3D printing on knitwear also relies on these adhesion mechanisms. When yarn inter-sections are increased in weave units, the yarns are brought closer together in contact points, and this results in a reduction in roughness values affecting the adhesion with 3D printed objects.

Meyer et al. [59] observed that polyester fabric made of filament yarn shows less adhesion strength than staple fibre yarn fabrics (linen and cotton). The inferior adhesion strength in case of polyester fabric was a result of low friction between fibre–fibre and fibre-printed polymer. On the other hand, protruding fibres from the linen and cotton fabric surface was advantageous for embedding 3D printed polymer easily which resulted in higher fabric-polymer adhesion strength. They also studied the effect of staple fibre length on the adhesion strength of the printed polymer and found that a higher staple fibre length in case of linen fibre assisted in obtaining higher adhesion strength than cotton and silk fabric made of shorter fibres.

3D printing of polymer on a very tightly constructed fabric may often result in lack of adhesion strength because of lower penetration of the polymer. Mpofu et al. [60] studied the effect of different fabric parameters on the adhesion of the 3D printed PLA polymer onto woven fabric. As per their observation adhesion strength reduced with increasing both the warp and weft densities. While the adhesion strength increased for the fabric made of coarser yarns, higher fabric handle, and also for the fabric having higher thickness and aerial density (Figure 17.3), these changes in adhesion strength with fabric parameters can be explained by the diffusion theory of the molten polymer. They also studied the effect of different fibres, cotton, polyester, polyester/cotton blend, and acrylic on the adhesion properties. Acrylic fabric exhibited highest adhesion strength with 3D printed PLA polymer, while polyester showed the lowest adhesion strength (Figure 17.3).

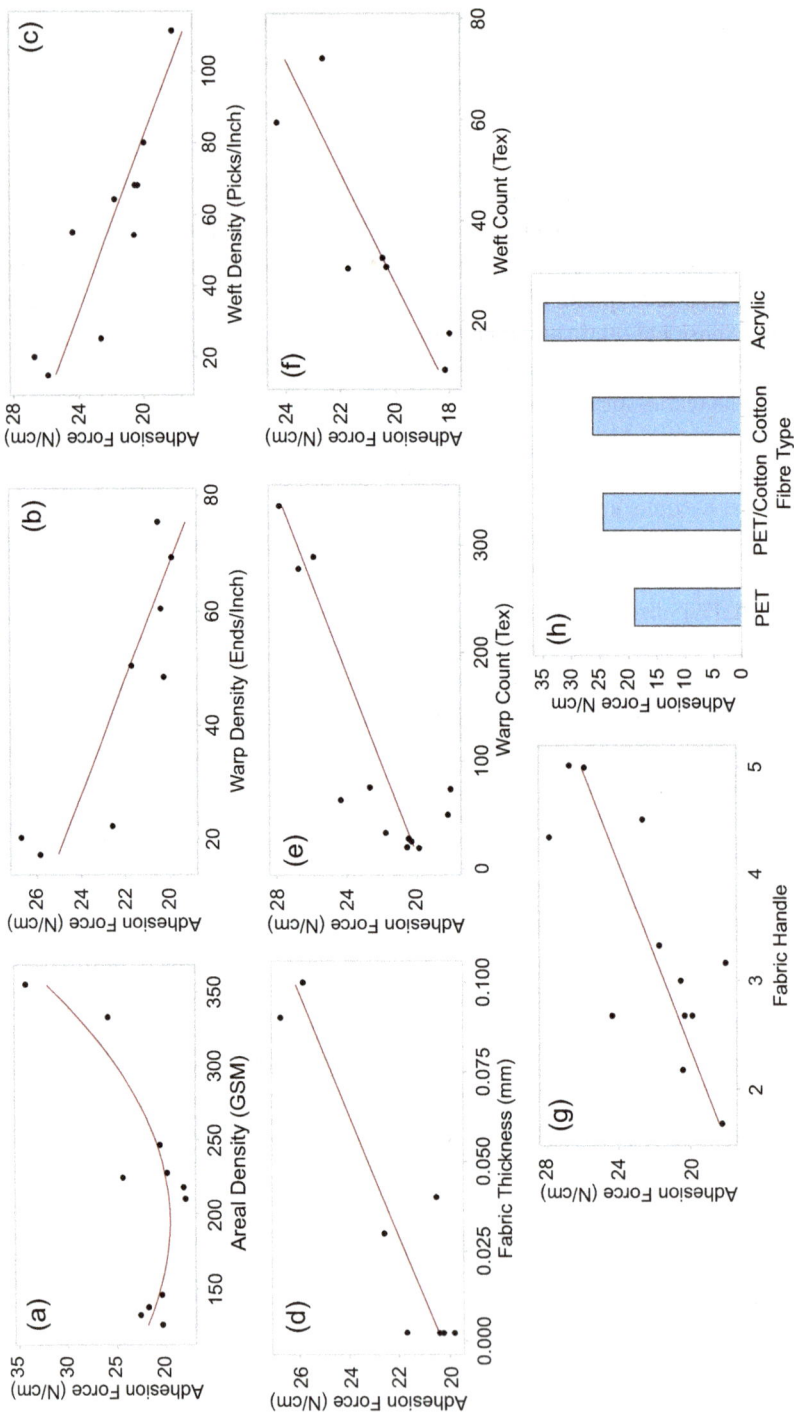

Figure 17.3: Effect of various parameters such as (a) areal density of fabric, (b) warp density in fabric, (c) weft density in fabric, (d) fabric thickness, (e) warp count, (f) weft count, (g) fabric handle, and (h) fibre type on the adhesion strength between fabric and 3D printed PLA polymer [60].

iii) **Surface finish of base fabric**

The wettability of the textile surface by the molten polymer is determined by the textile surface energy and can be controlled using specific treatments, such as desizing, scouring, finishing (surface functionalization), corona discharge, or plasma treatment of the textile before 3D printing.

When the two polymers (used for 3D printing and the fabric substrate) are not chemically compatible or when the fabric is chemically inert (such as polypropylene fabric), the fabrics is treated with some chemicals before printing to improve adhesion strength between 3D printed polymer and the base fabric. Researchers have also explored the possibility of coating of fabric with a suitable polymer to improve adhesion strength as well as to reduce surface roughness before 3D printing. However, for creation of suitable surface system via polymeric coating, a thorough understanding of surface chemistry and wettability is required as they are directly related to the adhesion property. Unger et al. [61] coated the surface of a cotton fabric with different polymers (PLA, ABS, PMMA, and PA) before 3D printing with PLA and ABS. Polymeric coating before 3D printing increased the hydrophobicity of the cotton fabric which subsequently helped to improve adhesion with 3D printed polymers being hydrophobic themselves. Especially, the coating with ABS and PLA improved the adhesion when 3D printing was done with ABS. While the PLA was used in 3D printing, a coating of PMMA or PLA performed excellently for improving adhesion strength. In a similar study, Meyer et al. [59] coated the textile fabrics (made of cotton, linen, silk, and polyester) with PMMA before 3D printing with rigid and soft PLA, while they observed improved adhesion strength for PMMA-coated fabric with 3D printed stiff PLA, PPMA coating did not improve adhesion strength for soft PLA. Researchers are working on factors related to fabric surface for improving adhesion strength. There is a potential for enhancing the adhesion strength by surface functionalization of fabric by treating it with plasma, corona discharge, and so on, or by surface finishing by various methods such as laser micromachining and sanding before 3D printing [62].

iv) **Type of polymer**

Adhesion strength strongly depends on the type of polymer used for 3D printing. For proper adhesion of printed polymer with fabric, some chemical interactions such as polar–polar attraction, hydrogen bonding, or at least van der Waals forces are advantageous with mechanical anchoring. The chemistry of the polymer used for 3D printing as well as the chemistry of the polymer of base fabric determines the chemical interaction. On the other hand, mechanical anchoring is controlled by the diffusion of polymer which finally depends on the molecular weight, orientation of polymer chains, and viscosity of the molten polymer [57].

Among different polymers, till now mostly PLA [28, 57, 63–66] has been explored for making FDM-based 3D printed textiles, while some researchers also used nylon [57], ABS [61], and so on. However, with the advancement of technology and research in the field of material science, more polymers will be explored in future for making 3D printed smart textiles by FDM technique.

v) Filler incorporated polymers

Many times, different fillers are incorporated in polymer matrices for improving the functional properties of polymer to be used for 3D printing. Recently, in the era of nanotechnology, researchers have synthesized many polymer nanocomposites by incorporating different nanomaterials in polymer matrices. These materials, especially conductive polymer nanocomposites, have a huge potential for improving the performance of 3D printed smart and functional textiles. Generally, carbon black (CB), carbon nanotubes (CNTs), graphene, and metallic particles are incorporated in polymer matrices for synthesizing conductive polymer nanocomposites. These fillers play a significant role in controlling not only the conductivity and other functional properties but also adhesion property between polymer and fabric.

Sanatgar et al. [57] studied the effect of nanosized CB and multi-walled carbon nanotubes (MW-CNT) on the adhesion property of PLA/CB and PLA/MW-CNT nanocomposites with a PLA fabric.

17.6 Polymer composites and polymer nanocomposites-based 3D printed textiles

Many literatures reported the 3D printing of polymeric composites containing long and short fibres for enhancing mechanical properties as well as those containing micro or nanoparticles for imparting various functional properties. 3D printing techniques can be used to fabricate complex composite structures with heterogeneous properties in a facile, inexpensive, and rapid manner without the need for moulds. Fibres/particles can be used as reinforcement to improve the functional performance and structural properties of polymer composites. They are often made anisotropic when reinforcing species are oriented within the matrix to enhance their properties or they can be made functional by adding electrically and/or thermally conductive particles such as CB and CNTs, silver, and graphene [67–73].

3D printing can be used to create fibres or yarn-like structures with conductive polymer composite or nanocomposites inks, especially for making wearable sensor and energy harvesting and energy storing systems. Gao et al. [74] used wet spinning with 3D printing to generate a thermally conductive fibrous structure from poly(vinyl alcohol) (PVA) with boron nitride nanosheets (BNNSs) and used it for a thermally regulating smart textile, as shown in Figure 17.4. A BN/PVA dispersion was used as ink in a 3D printer, followed by extrusion into a cooled methanol bath, followed by drying and then hot drawing to align the BNNS. Despite the fact that the fibres were collected in a methanol coagulation bath in a similar manner to wet spinning, the method of extrusion was still called as 3D printing. Essentially, this process combines wet spinning and material extrusion. A subsequent hot-drawing procedure further increased the in-plane thermal conductivity of the composite fibres due to the flow-induced alignment of the BNNSs, as shown in Figure 17.4h. Taking advantage of these fibres, knitted and woven fabrics were made, as shown in Figure 17.4(f, g), demonstrating their scalability and personal comfort-enhancing properties using as weatherable thermal regulation fabric [74].

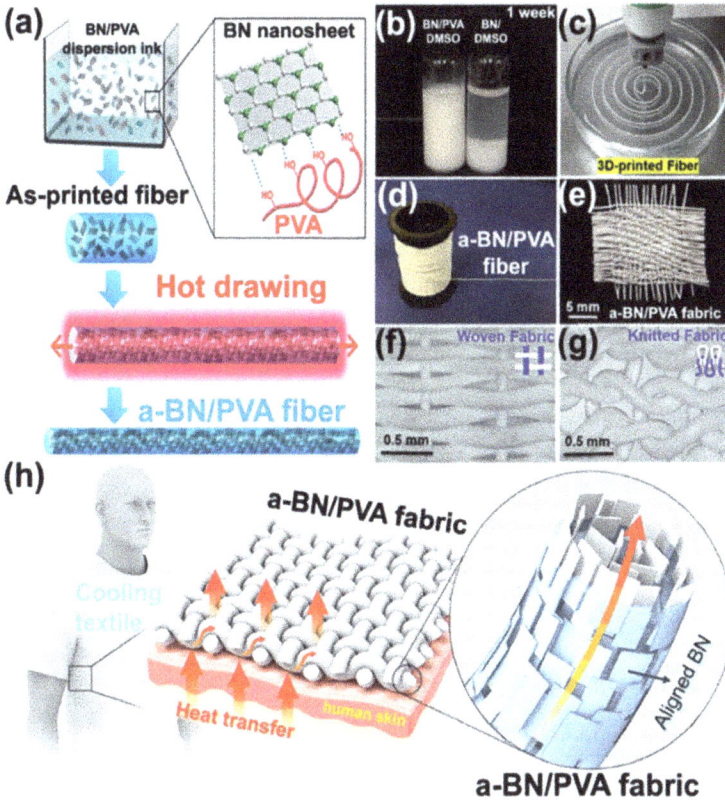

Figure 17.4: 3D printed thermal regulation textiles made from aligned-BN (a-BN)/PVA nanocomposite fibres: (a) process for preparing PVA/DMSO/BNNS dispersion ink for fibre fabrication; (b) as shown by the two solutions, one with and one without PVA after one week, the PVA in the dispersion allows for homogenization of the dispersion; (c) ink is printed into cooled methanol solution using a 3D printing machine; (d) printing a-BN/PVA yarn from fibre on a bobbin was enabled via the scalable production method; (e) fabric woven from a-BN/PVA yarns. a-BN/PVA yarn-based (f) woven and (g) knitted textile structures, with insets illustrating a representative unit from each type; (h) with high and anisotropic thermal conductivity, the aligned BNNS provide pathways for heat transfer from the aligned BN/PVA fabric to the human body (reproduced with kind permission from [74] Copyright 2017, American Chemical Society).

17.7 4D Textiles enabled by 4D printing

Technological advancements in 3D printing along with advanced functional shape-memory polymers (SMPs) have opened an opportunity for researchers to explore "4D Textiles", a hybrid 3D printed textile that can change structural shape over time. First time, Skylar Tibbits introduced the term "4D printing" at a TED conference talk [75]. In addition to 3D space coordinates of 3D printing, the 4D printing technology involves the fourth dimension of time. Therefore, 4D printing can be ascribed to the ability of the printed structure to change shape or function over time (t) in response to external stimuli such as temperature, pressure, water, wind, or light.

17.7.1 Potential applications of 4D printed textiles

Unlike conventional 3D printed objects, 4D printed objects are affected by external stimuli and can change colour, shape, or size significantly over time. 4D materials that are highly accurate and calibrated can be combined to make robust objects which are in high demand for several applications, including healthcare, space suits, military applications, apparel, sports, robotics, and more.

4D printing technology has huge potential in various fields, such as simple changes to bioprinting of organisms, which utilizes smart materials, designs, and printing [76]. This technology has already been applied by the U.S. Army to produce camouflage textiles that can bend light reflected from the clothing so soldiers can hide in certain environments [77].

17.8 Developments in this field of 3D- and 4D printed textile structures and their applications

This technology has a great potential for creating wearables with sensors and thermal heating properties, sportswear that manages body temperature, virtual reality gloves, smart bandages, defence safety equipment, medical equipment, footwear and fashion products, automotive accessories, aerospace accessories, and many more. Few specific applications of 3D printed smart textiles with recent advancements in those fields are described further.

17.8.1 Biodegradable polymer-based 3D printed textiles for various applications

Sabantina et al. [64] examined the mechanical properties of different textile fabric/mesh structures with FDM technique-based 3D printed PLA matrix combinations (Figure 17.5). They realized that threads of base fabric and printed material form a sufficient connection which can be potentially utilized in garments and technical textiles.

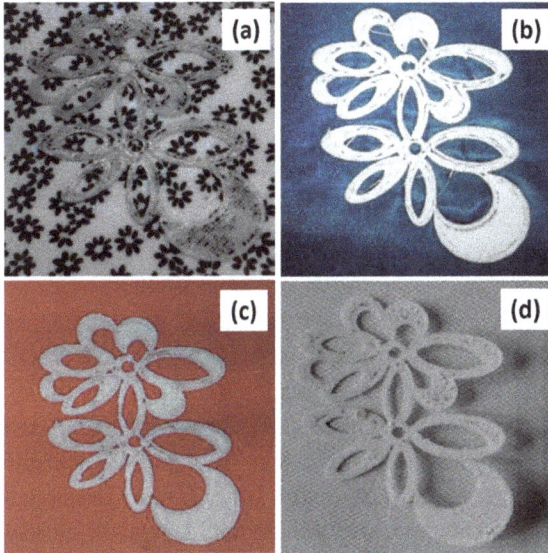

Figure 17.5: Floral pattern printed on (a) cotton fabric, (b) viscose fabric, (c) wool fabric, and (d) polyester net (from left to right) [64].

Pei et al. [63] also examined the adhesion of functional and decorative components printed directly onto fabric via FDM 3D printing. For instance, they printed orthotics braces onto textile, where textile provided comfort and support as they are flexible and breathable, and the polymer structures provided good rigidity and support. PLA was found to have good adhesion and good flexural strength.

A bio-based polymer like PLA that is doped with carbon could be deposited onto textiles to realize inexpensive and sustainable conducting textiles. A new path towards conductive grids would be opened by controlling the bonding mechanism between doped PLA and textiles. Therefore, in the study of Brinks et al. [28], molten PLA was deposited on PET bundles with application of pressure, causing some penetration into the PET bundle as well as forming bonds between molten PLA and the PET fibres making 3D printed conductive textiles. In a similar study, Sanatgar et al. [57] fabricated conductive textiles by 3D printing of MWCNT/PLA nanocomposite on a PLA fabric.

Several researchers have investigated the use of PLA filament as raw material for FDM technique-based 4D printing and the possibilities of combining PLA with nylon fabric to create smart textiles [78]. Nylon fabric that can be thermo-mechanically trained into temporary shapes and returned to permanent forms when heated, while PLA's thermal shape-memory properties will maintain these abilities. Smart textiles can be heat-treated to form custom shapes and then returned to their original flat forms upon being heated above the glass transition temperature (T_g) of the polymer. By using this concept, clothing can be designed in custom shapes and aesthetics, or materials can be encapsulated and released when activated by their environments. Thus, these developments can be applied to 4D printing to further accelerate the development by targeting multidirectional applications such as biomedical devices, smart textiles, and advanced manufacturing.

17.8.2 Chainmail-structured 3D or 4D fabrics for load-bearing applications

Nowadays, polymer/textile researchers and designers are focusing on development of 3D-structured fabric with tuneable mechanical properties, which is inspired by metallic micromachined chain mail fabrics [79] and topologically interlocked elements [80]. In a recent study, Wang et al. [81] constructed interlocked granular particle-based two-layered structured fabric from connecting trusses where each particle is a specially designed hollow 3D structure. The interlock chainmail lattice is prepared by SLS method forming each piece without extra support. Two chain mail layers are stacked together to increase the number of contacts, which can bend/fold over curved objects like chainmail armour showing high flexibility.

Interestingly, the demonstrated chainmail structure has a tuneable bending modulus. Under application of load/pressure, the flexible chainmail structure is changed to a complex jammed structure because of the interlocking of the particles. With the application of a small external pressure (about 93 kPa), the sheets become more than 25 times stiffer in comparison to their relaxed state. This remarkable increase in the bending resistance was resulted from high tensile resistance of the interlocked structure. This work shows a way to manufacture lightweight, additive fabric with tuneable mechanical stiffness, showing a potential to be used in haptic architectures for transportation, wearable exoskeletons, and reconfigurable medical supports.

In a similar study, Ploszajski et al. [82] manufactured magnetically functionalized, lightweight, 4D printed actuating chainmail fabric. At first, a nylon-based cellular structure made of cube-shaped unit particle was fabricated by SLS 3D printing method, which was successively padded in a bath containing a commercial ferrofluid (oil-based magnetic liquid), followed by drying under heat. With the application of a magnetic field to the end of the chainmail structure, the chainmail actuates into a stiff structure resisting the pull of gravity and remain horizontal. This mechanism can be useful for making thin, lightweight, and comfortable wearable assistive devices (exoskeletons).

17.8.3 Biomedical applications of 3D printed textiles

Textile scaffolds are one of the most common types of porous scaffolds that are used in tissue engineering. In medicine, there is a paradigm shift from synthetic implants and tissue grafts to a tissue engineering approach that uses porous scaffold materials integrated with cells and molecules to regenerate tissues. For biological delivery and tissue regeneration, scaffolds must balance temporary mechanical functionality and mass transport [83], which can be obtained by using 3D printed cellular structures. It has been demonstrated that two phases of polylactic acid (PLGA)/polyglycolic acid (PLGA) scaffolds can be fabricated by 3D printing, resulting in the second phase of a L-PLGA/tri-calcium phosphate mixture [84]. In another study, polycaprolactone scaffolds were produced using FDM with porosities ranging from 48% to 77%, yield strengths and compressive moduli ranging from 2.58 to 3.32 MPa and 4 to 77 MPa, respectively [84]. Along with the development of hydrogel systems, 3D printing of electrically conductive hydrogels is one of the most advanced approaches towards rapid fabrication of future biomedical implants and devices with versatile designs and tuneable functions such as wearable heating and sensing systems [85].

Textile structures have been used in surgeries for decades, but the increasing complexity of devices due to technological advances is enabling them to serve a broader range of purposes. Recently, researchers are addressing the shortage of organs for transplantation and therapeutic applications by 3D printing biomaterials that restore defective or diseased tissue. A variety of challenges are encountered in developing novel bio-inks for 3D printing, including their rheological, chemical, physical, and biological properties as well as their cytotoxicity, immune response, and regeneration rate. Recently, chitosan and its composites have been studied as bio-inks for 3D bioprinting to create artificial organs. 3D printed chitosan composites have varying regenerative capacities depending on their size, porosity, stimulation effect, cell interaction, cell adhesion, and differentiation potential of stem cells [86].

17.8.4 3D printed textiles for thermoregulation

As one of the most promising technologies for saving cooling energy and costs, personal cooling technology directs local heat to a thermally regulated environment to provide thermal comfort. To improve the thermal transport properties of textiles for personal cooling, Gao et al. [74] demonstrated the use of thermally conductive, highly aligned BN/PVA fibres (abbreviated to a-BN/PVA) to fabricate textiles using a fast and scalable 3D printing technique. In the a-BN/PVA textile, better cooling properties (55% improvement over commercial cotton fibre) can be achieved due to the improved thermal properties imparted by the thermally conductive and highly aligned BNNSs. 3D printed a-BN/PVA fibres in wearable a-BN/PVA textiles may offer a promising choice for meeting the personal cooling requirement, which could significantly reduce cost and energy demands for cooling entire buildings.

17.8.5 3D printed textiles for energy harvesting and storage applications

With the development of wearable electronics and flexible electronics, electronic textile (E-textile) has attracted tremendous attention. Zhang et al. [87] reported the direct printing of E-textiles comprising core-sheath fibres using a 3D printer equipped with a coaxial spinneret (Figure 17.6). It is possible to print coresheath fibre-based patterns on textile for various purposes using customer-designed coresheath fibre patterns. To demonstrate this, a CNT core-sheath fibre-based smart pattern was fabricated using CNTs as a conductive core and silk fibroin as a dielectric sheath, which was further used as a triboelectric nanogenerator textile. Biomechanical energy could be harvested from human motion through the smart textile, which would have a power density of 18 mW/m^2. A textile supercapacitor was also printed for energy storage. By this way, a self-sustaining E-textile with integrated electronics can be produced on a large scale by printing smart patterns directly on textile.

Wearable applications require flexible and breathable power systems that should not be bulky and rigid. In spite of the tremendous efforts put into developing various 1D energy storage devices with sufficient flexibility, cost, fabrication scalability, and efficiency remain challenging. Wang et al. [88] fabricated a flexible all-fibre lithium-ion battery (LIB) using scalable, high-efficiency, and low-cost 3D printing. To print lithium iron phosphate (LFP) fibre cathodes and lithium titanium oxide (LTO) fibre anodes, highly viscous polymer inks containing CNTs were used. Half-cell configurations demonstrate good high electrochemical performance and flexibility for both fibre electrodes. A quasi-solid electrolyte of gel polymer can be used to assemble an all-fibre LIB using as-printed LFP and LTO fibres. In addition to having a high specific capacity of 110 mAh/g at a current density of 50 mA/g, the all-fibre device is flexible and exhibits a long battery life, which is suitable for future wearable electronic applications that can be embedded in textile fabrics.

17.8.6 Shape-memory or stimuli-responsive applications

Developing new 3D-printable materials will enable us to create new materials with volumetric and shape-shifting properties as they are exposed to different stimuli (4D printing).

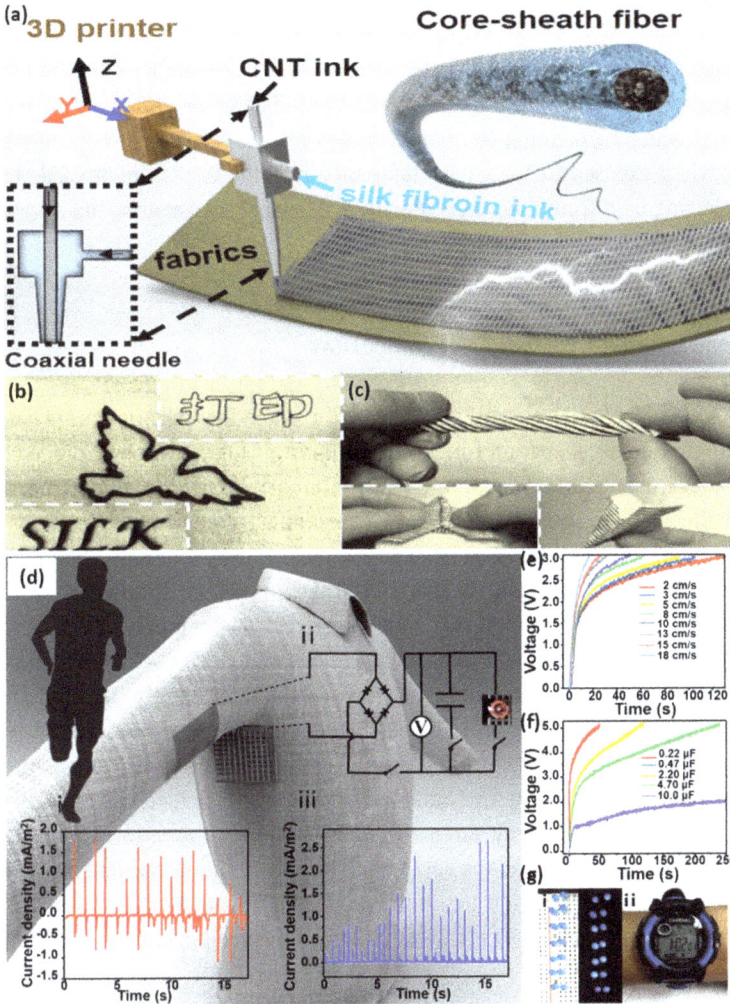

Figure 17.6: An energy-management smart textile pattern made with core-sheath fibres is directly printed on fabrics: (a) 3D printing schematic using a coaxial spinneret, (b) a photograph showing an English lettering pattern for SILK, Chinese characters for PRINTING, and a pattern of a pigeon are some of the patterns, (c) smart textile showing high flexibility when twisted and folded; the application of 3D printed E-textiles in energy management, (d) an illustration showing the performance of smart clothes for energy management. Inset (i) shows a smart gridline pattern generated by an arm moving on an underarm sleeve of a shirt. Inset (ii) shows the power system's rectifying circuit diagram. Inset (iii) shows a rectified output Isc density of the smart pattern, (e) an example of charging curves for a capacitor (3.3 mF) displaced at different speeds using the smart pattern, (f) the smart pattern charges different capacitances at a displacement speed of 13 cm/s to produce charging curves of different capacitances, (g) images of LED lights and an electrical watch powered by 3D printed E-textiles (reproduced with kind permission from [87], Copyright 2018, Elsevier).

17.8.6.1 Space fabric

NASA Jet Propulsion Laboratory proposed an application of shape-memory alloys when they designed "Space Fabric", which was primarily composed of silver squares. This material was prepared using 4D printing technology and is capable of changing its shape and functionality when exposed to heat or UV light. By absorbing light from one side of the fabric and reflecting light from the other side, the space fabric can control the microenvironment thermally [89].

17.8.6.2 4D printed braided structures and composites

Multi-directionally reinforced 4D printed preforms with shape-memory polymers can be used for developing next-generation functional composites taking the advantage of 3D printing technology, 3D textile design, and shape-memory behaviour of polymers. A 4D printed circular-braided tube preform and its silicone elastomer matrix composite exhibit shape-memory behaviour and recovery forces. A shape-memory polymer (SMP), PLA was used to print the preforms using FDM technique and studied how different braiding angles, tube wall thicknesses, and shape recovery temperatures affect the shape-memory behaviour of 4D printed tube preforms and silicone elastomer matrix composites. Shape-memory behaviour of preforms is significantly affected by braided microstructural parameters and shape recovery temperatures, as found in Figure 17.7. By adding a silicone elastomer to the 4D printed tube preform/silicone elastomer matrix composite, not only the radial compressive failure load considerably increased but also the shape recovery force enhanced notably at a wider operating temperature range, resulting in full shape recovery. The stunning mechanical properties, shape-memory behaviour, and the flexibility of microstructural design of textile composites indicate the potential of 4D printed textile composites for many functional applications [90].

17.8.6.3 4D printed laminated Miura-origami structures for actuator applications

The advantages of 3D printing technology and SMPs make it possible to print origami structures in 4D for actuators and reconfigurable devices. Liu et al. [91] investigated the shape recovery progression of Miura-origami tessellations and tubes after folding and unfolding under compressive loads. The shape recovery temperature and loading pattern significantly affected the shape recovery behaviour and recovery force. Shape recovery ratios of over 94% as well as volume changes of up to 289% demonstrated the high shape recovery capability of the specimens showing a potential for actuator application of the 4D printed laminated Miura-origami structure.

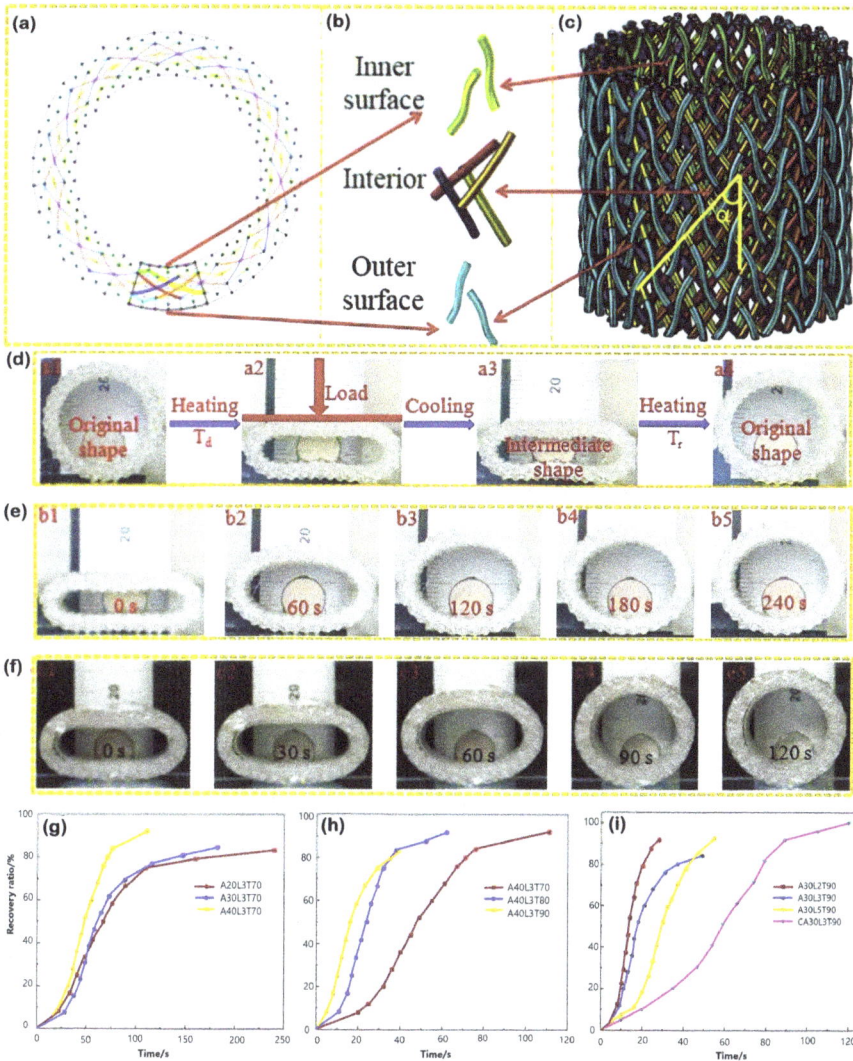

Figure 17.7: 3D-circular braided tube model and specimens with three braiding layers and braiding angle of 30°: (a) yarn contour projection; (b) unit cell model; (c) tube geometrical model; (d) shape-memory cycle of 4D printed circular-braided tube; (e) shape recovery of the specimen A30L3T70; (f) shape recovery of specimen CA30L3T90; (g) shape recovery ratio versus time curves of specimens L3T70 with three braiding angles of 20°, 30°, and 40°; (h) shape recovery ratio versus time curves of specimens A40L3 at three shape recovery temperatures of 70, 80, and 90 °C; as well as (i) shape recovery ratio versus time curves of specimens A30T90 with two braiding layers, three braiding layers, and five braiding layers and the specimen CA30L3T90. (Notes: the legends A, L, and T denote the braiding angle, braiding layer, and recovery temperature, respectively. C denotes composite specimen, and the legend without C means the preform. For instance, A20L3T70 denotes the preform with braiding angle of 20° and three braiding layers at the recovery temperature of 70 °C; reprinted with kind permission [90], Copyright 2022, Elsevier).

17.8.7 3D and 4D printed textiles as fashion products

Several industries, including the fashion industry, are increasingly utilizing 3D printing. From clothing to footwear and accessories, 3D printing offers endless possibilities for the fashion industry. Initially, 3D printing was used only for artistic purposes, but now fashion designers are using 3D printing to make wearable garments. Comparing it with traditional manufacturing processes, using 3D printing technology the design process for fashion industry can be accelerated, production time can be reduced, and inventory, warehousing, packaging, and transportation costs can be lowered.

17.8.7.1 Modeclix

Invention of modeclix has set a benchmark in additively manufactured products, which is scalable, replicable, sustainable, having potential for providing many cost-effective technical solutions. Modeclix is a 3D printed flexible textile, a sheet-like material with many connected links or panels. It has a great potential to overcome the challenges of conventional textile manufacturing techniques. Generally, CAD software is used to design links with spiral arms and to connect the links making flexible sheet-like modeclix structures via SLS 3D printing technique as shown in Figure 17.8(a and b).

In an interesting study, Bloomfield and Borstrock [92] utilized white polyamide PA12-powder (nylon-12) as raw material for SLS 3D printing, while the powder layer supported each new build layer to form sintered nylon link which were linked using an array modifier in the CAD software to give different configurations. The prepared modeclix structures can be post-processed and finished very easily. Moreover, they can be dyed using different dyes (acid, metal complex, etc.) to develop colour as per the customer's requirement. The prepared linked structure can be used for making garments, bags, jewellery, interior screening, toys, and so on (as shown in Figure 17.8(c–i)) to satisfy the market demands in terms of versatile design as well as price points.

17.8.7.2 4D printed shape-changing dress and jewellery

Nervous system, a Massachusetts design studio, has developed a way to 3D-print jewellery and garments with articulated joints so that when they are removed from the printer, they automatically transform from one shape to another shape. They worked on the development of a dress that can be 3D printed as one piece even though they are much larger than the space inside the printer. Furthermore, they develop jewellery with articulating joints that automatically adapt to the shape of the body despite being printed on flat sheets. Hence, this technique is advantageous to print in a particular shape for several reasons: it may be faster, cheaper, and allow for printing larger objects in a smaller volume [93].

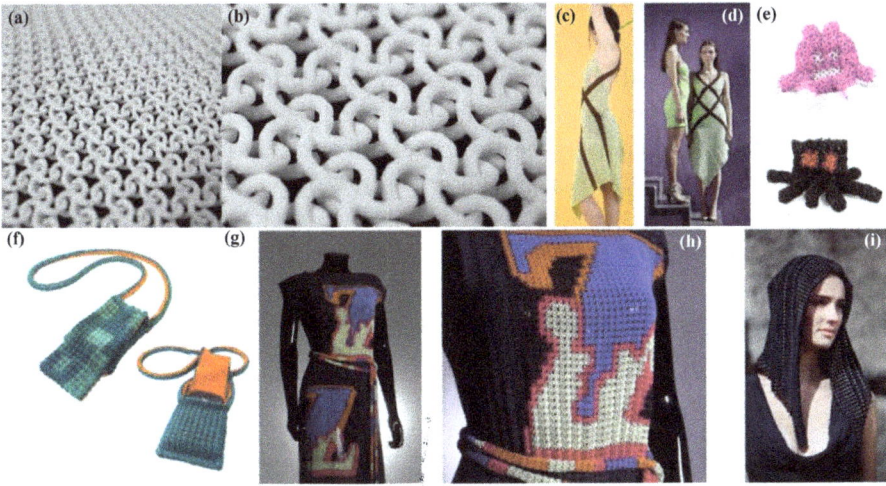

Figure 17.8: (a and b) Additively manufactured modeclix as a sheet with connected links; (c–i) use of modiclix in garments, toys, and accessories (reprinted with kind permission from [92], Copyright 2018, Elsevier).

In their "Kinematics" design system-hinged panels are combined with folds and compression simulations to produce customized designs that are readily printed in 3D (Figure 17.9(a)). As a prototyping system, it is an example of the rapidly developing field of 4D printing, which uses 3D printing in order to create objects that change in shape. The components which worked together to advance the core concept were (i) a hinge mechanism for making complex 3D printed structure as a single part, (ii) a folding simulation to compress the structure into a smaller space, and (iii) a web-based software which allowed the people to create their own design [94].

Using a similar concept, another 3D printing company Shapeways (New York) developed compressed fabrics that unfold into their intended shape after 3D printing and have a good drapability on wearer's body. The patterned structure is composed of 2,279 triangular-shaped panels connected by 3,316 hinges, which are all 3D printed in nylon using SLS technique [95].

17.8.7.3 Petal dress

Using 4D printing technology, the Massachusetts design studio (Nervous System) has created a dress made of 1,600 unique and rigid pieces of nylon (Figure 17.9(b)). The rigid pieces were of small-sized and rounded shapes which were inspired by feathers, petals, and scales. A SLS printing technique was used to convert the nylon power into rigid pieces (solid or porous mass). The pieces are attached to a framework of triangular panels with moving joints and although each interlocking panel is rigid, when put together they behave as a continuous textile. Moreover, these unique rigid pieces were intercommoned by more than 2,600 hinges and assembled to give a shape of ready-to-wear garment. 4D printing made it possible to print an articulated garment in one piece and then modify its shape once removed from the printer. A 3D scan can be used to create a custom fit for the dress, and each element can be adjusted individually for length, direction, and shape. This petals dress was commissioned for an exhibition (Techstyle) which is now on view at the Museum of Fine Arts (Boston) showing the impact of technological advancements in fashion [96, 97].

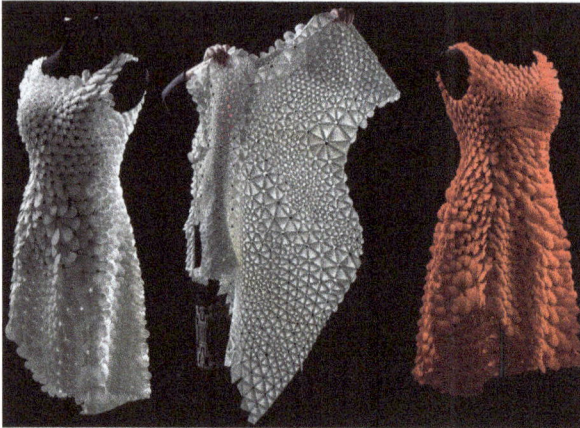

Figure 17.9: 4D printed petal dress made of 1,600 rigid pieces of nylon (photo courtesy: Nervous system and dezeen [96]).

17.8.7.4 Wearable 3D printed structures for interplanetary voyages

The MIT Media Lab team of Neri Oxman has created four 3D printed "wearable skins" that facilitate synthetic biological processes that may one day be used to help humans survive on other planets. In collaboration with a 3D printing company Stratasys, Neri Oxman, and members of the Mediated Matter group at MIT Media Lab designed structures with varying rigidity, transparency, and colour. An Objet500 Connex3 colour, multi-material 3D Production System was used for 3D printing the skins in a range of plastics with different densities, each suited to a different planet in the solar system [98].

Wanderers: An Astrobiological Exploration, which is a project that imagines that living matter, will be embedded into the four pieces (Figure 17.10), as follows:
a) Mushtari: Designed to interact with Jupiter's atmosphere, it is made up of layers that resemble animal intestines and is made up of a continuous translucent strand. It is proposed to place the device around the lower abdomen and consume

and digest biomass, absorb nutrients, generate energy from the sucrose that accumulates in the side pockets, and expel waste.

b) Otaared: In order to survive on Mercury, it builds a protective exoskeleton that is custom fit to the wearer.

c) Zuhal: It is designed to adapt to the vortex storms on Saturn. The swirling textured surface of the bodice would have bacteria that would turn the planet's hydrocarbons into edible matter.

d) Al-Qamar: It is designed as a "wearable biodome", fitting around the neck and over the shoulders, with a shell of algae-based air purification and biofuel collection to produce and store oxygen.

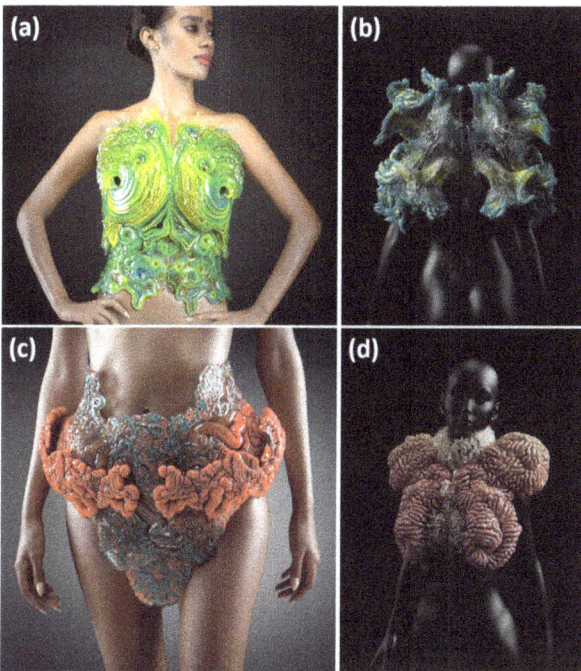

Figure 17.10: Wearable 3D printed structures for interplanetary voyages: (a) Mushtari, (b) Otaared, (c) Zuhal, and (d) Al-Qamar (picture courtesy: Neri Oxman's team at MIT Media Lab and dezeen [98]).

17.8.7.5 Footwears and watches

In addition to garments, 3D printing is revolutionizing the footwear industry as well. Among different 3D printing technologies, SLS and Polyjet techniques are getting huge interest for making 3D printed smart watches and footwears. Additive manufacturing is being used by some important brands, such as Adidas, to create impressive products.

With recycled plastic from the ocean, Adidas created a 3D printed midsole for one of its sneakers. A new process called Digital Light Synthesis (formerly known as CLIP) allows Adidas to create shoes that feature a midsole created in partnership with Carbon 3D [99].

In the footwear industry, 3D printing is mostly used for 3D printing midsoles, which allows for custom-made products that can be adapted to any morphology. In order to create unique, lightweight, highly durable, and perfectly fit shoes, designer Olivier Van Herpt uses 3D scanning and 3D printing [100].

Few companies are also utilizing 3D printing technology for the manufacture of watches. For example, s French brand ABL "Atelier le Brézéguet" is based in Toulouse. The black rings on top and bottom of this watch were made using 3D printing technique by a company specializing in watches. Polyamide material is used to 3D-print these parts [101].

17.9 Challenges and future scope

Though the textile/polymer researchers and designers have proposed many applications for 3D printed textiles, especially made via FDM and SLS, they fall short for many reasons. In fact, printing textiles using any of the current 3D printing methods poses huge challenges due to numerous factors, such as the tiny fibre dimensions, the high porosity of the textile structure, and the discontinuities required within the textile.

In case of FDM, the size of the textile structures is limited by the need for support structure. Commercial FDM is limited to a few polymeric materials, including PLA and ABS, which are not appropriate for textiles. Moreover, the FDM process produces rough surfaces with poor interlayer bonding. In order to produce high-resolution, smooth surfaces with good mechanical properties, emerging light-based 3D printing techniques, are used which require materials that can be photopolymerized, so they are primarily made with polyacrylates or epoxy macromers that have a low molecular weight. Due to resin cure rates and viscosity dependence, this results in limited mechanical properties and increased costs. SLA-printing technique has a speed about 10 times faster than conventional SLS technique. However, to meet the textile material requirements, resin chemistry must be developed further. Additionally, the speed of most of the 3D printers is much lower in comparison with different traditional textile manufacturing processes. For example, yarn spinning speed varies from 1,000 to 2,000 m/min, while the printing speed of FDM is only 0.5–1 mm/h. Thus, 3D printing techniques such as FDM, SLA, and SLS may not be appropriate due to their limitations in extending to a large number of materials, temperature requirements, and simplicity of designs.

In melt-spinning process, extrudate polymer melts are subsequently stretched which causes improvement in orientation and crystallinity, resulting in huge improvement in mechanical properties. However, in the absence of molecular orientation the strength of the fibre is low and recovery is poor, which can also be reflected in the 3D printed textiles. Designers have already created footwear, dresses, or other garments that are impossible to produce with ordinary methods in the textile industry. Although 3D printing is becoming more common in textile and clothing production, it is still a long way from being the norm. 3D printed products, especially those made of pure plastic, lack adequate mechanical properties, especially lack tensile strength, precluding their replacement of common technologies such as knitting and weaving.

SLS technique is used to prepare modeclix or chainmail structures, and this technique has many limitations such as (i) many times, mixing of powder is not uniform/even which causes inconsistent heat distribution throughout the build chamber, corruption in the 3D data file, and generation of soft/brittle-built parts or parts with anomalies; (ii) once the product/structure is made it is difficult to check the part consistency throughout [92]; and (iii) many times, modeclix shows limited stretchability. A connected modeclix panel matrix of 22 × 22 links made of nylon-12 shows a breaking strength of 7 kg across, but if it is pulled with high force, also the links will stretch with the structure or even the links can break. Hence, it cannot be used for textile-based applications where the textile material is put under stress [92].

Moreover, fibre extrusion method requires a polymer having low melt viscosity (high melt flow index) and generally produces weaker fibres. However, the concept remains mostly unexplored and should be explored in future, based on the on-going interest in 3D printing textiles. Thus, the deposition processes combining 3D prints with textile fabrics can be a better way to incorporate them into garments [64].

In spite of many limitations for 3D printing to be used for making textiles, it has many advantages as well, which are (i) potential benefits such as more flexible and cost-effective production of high end products, (ii) minimization of textile waste combined with reduced consumption of energy, water, and chemicals can increase the productivity and improve the ecological footprint, and (iii) 3D printed structures lack residual stresses, which are present in fabrics made from conventional processing techniques and contribute to low strain deformation.

This technology may be used in developing a variety of textile products. However, 3D printing of textiles has not yet reached to industrial scale to a great extent. However, extensive research and development is going on in this field to commercialize various 3D printed products in near future for many smart applications. As per the published articles, the number of polymers explored till date by different researchers/manufacturers for making 3D printed textiles is much less than the polymers used in conventional textile. However, with the advancement of technology and material science, this number will increase hopefully in the near future. In todays' world of nanotechnology, nanocomposites are finding applications in all diverse fields including 3D printing. It is expected that the potential of polymer nanocomposites will greatly be used in making smart textiles with specific functionalities in near future. Textile manufacturing will also benefit from current 3D printing techniques through multidirectional preforms and fibre integration into composites.

17.10 Summary

Manufacturing textiles was probably the first craft invented before farming. Despite the fact that the processes of assembling fibre into textiles have become more precise, faster, and automated over the years, the fundamental nature of textiles has not changed. Recently, fashion and clothes manufacturing have evolved in the way we think about them. Designers are now considering new aspects and the reasons why they choose 3D printing where sustainability and environmentally friendliness is becoming more important. In today's world, textiles are produced with a variety of technologies and are used for a wide range of smart/functional applications, including chemical/thermal protection garments and fashion products. This chapter has presented a broad overview of how process and material parameters interact in fibrous structures and in that context reviewed the current 3D printing technologies, which have both advantages and disadvantages.

A detailed description of different 3D printing techniques, including materials and machines, has been presented in this chapter. The rapid development of 3D-printable materials, printers, and designs led to the creation of 4D printing. Currently, 4D printing has also received increased attention due to its ability to change the form and/or function of structures through temperature, light, pressure, wind, or water. As 3D and 4D printing is an emerging topic for discussion in the textile, garment, and fashion industry, this chapter will provide guidance on how these printing techniques can be used to realize textile architectures, providing information on the benefits it offers as well as the challenges it poses.

References

[1]   Wang Y, Li L, Hofmann D, Andrade JE, Daraio C. Structured fabrics with tunable mechanical properties. Nature 2021;596:238–43. https://doi.org/10.1038/s41586-021-03698-7.

[2]   Ahmed A, Hossain MM, Adak B, Mukhopadhyay S. Recent Advances in 2D MXene Integrated Smart-Textile Interfaces for Multifunctional Applications. Chem Mater 2020;32:10296–320. https://doi.org/10.1021/acs.chemmater.0c03392.

[3]   Joshi M, Adak B. Advances in Nanotechnology Based Functional, Smart and Intelligent Textiles: A Review. Compr. Nanosci. Nanotechnol., Elsevier; 2019, p. 253–90. https://doi.org/10.1016/B978-0-12-803581-8.10471-0.

[4]   Kao H, Chuang C-H, Chang L-C, Cho C-L, Chiu H-C. Inkjet-printed silver films on textiles for wearable electronics applications. Surf Coatings Technol 2019;362:328–32. https://doi.org/10.1016/j.surfcoat.2019.01.076.

[5]   Zeng W, Shu L, Li Q, Chen S, Wang F, Tao X-M. Fiber-Based Wearable Electronics: A Review of Materials, Fabrication, Devices, and Applications. Adv Mater 2014;26:5310–36. https://doi.org/10.1002/adma.201400633.

[6]   Wei Y, Torah R, Yang K, Beeby S, Tudor J. A novel fabrication process to realize a valveless micropump on a flexible substrate. Smart Mater Struct 2014;23:025034. https://doi.org/10.1088/0964-1726/23/2/025034.

[7]   Chauraya A, Whittow WG, Vardaxoglou JC, Li Y, Torah R, Yang K, et al. Inkjet printed dipole antennas on textiles for wearable communications. IET Microwaves, Antennas Propag 2013;7:760–7. https://doi.org/10.1049/iet-map.2013.0076.

[8]   Whittow WG, Chauraya A, Vardaxoglou JC, Yi Li, Torah R, Kai Yang, et al. Inkjet-Printed Microstrip Patch Antennas Realized on Textile for Wearable Applications. IEEE Antennas Wirel Propag Lett 2014;13:71–4. https://doi.org/10.1109/LAWP.2013.2295942.

[9]   Tai Y-L, Yang Z-G, Li Z-D. A promising approach to conductive patterns with high efficiency for flexible electronics. Appl Surf Sci 2011;257:7096–100. https://doi.org/10.1016/j.apsusc.2011.03.056.

[10]   Hashemi Sanatgar R, Campagne C, Nierstrasz V. Investigation of the adhesion properties of direct 3D printing of polymers and nanocomposites on textiles: Effect of FDM printing process parameters. Appl Surf Sci 2017;403:551–63. https://doi.org/10.1016/j.apsusc.2017.01.112.

[11]   Biswas MC. Fused Deposition Modeling 3D Printing Technology in Textile and Fashion Industry: Materials and Innovation. Mod Concepts Mater Sci 2019;2:1–5. https://doi.org/10.33552/mcms.2019.02.000529.

[12]   Choi J, Kwon O-C, Jo W, Lee HJ, Moon M-W. 4D Printing Technology: A Review. 3D Print Addit Manuf 2015;2:159–67. https://doi.org/10.1089/3dp.2015.0039.

[13]   Rajkumar AR, Shanmugam K. Additive manufacturing-enabled shape transformations via FFF 4D printing. J Mater Res 2018;33:4362–76. https://doi.org/10.1557/jmr.2018.397.

[14]   Andrew JJ, Verma P, Kumar S. Impact behavior of nanoengineered, 3D printed plate-lattices. Mater Des 2021;202:109516. https://doi.org/10.1016/j.matdes.2021.109516.

[15]   Schneider J, Kumar S. Multiscale characterization and constitutive parameters identification of polyamide (PA12) processed via selective laser sintering. Polym Test 2020;86:106357. https://doi.org/10.1016/j.polymertesting.2020.106357.

[16]   Kumar S, Wardle BL, Arif MF. Strength and Performance Enhancement of Bonded Joints by Spatial Tailoring of Adhesive Compliance via 3D Printing. ACS Appl Mater Interfaces 2017;9:884–91. https://doi.org/10.1021/acsami.6b13038.

[17]   Ubaid J, Wardle BL, Kumar S. Bioinspired Compliance Grading Motif of Mortar in Nacreous Materials. ACS Appl Mater Interfaces 2020;12:33256–66. https://doi.org/10.1021/acsami.0c08181.

[18]   Kumar S, Wardle BL, Arif MF, Ubaid J. Stress Reduction of 3D Printed Compliance-Tailored Multilayers. Adv Eng Mater 2018;20:1700883. https://doi.org/10.1002/adem.201700883.

[19]   Alam F, Varadarajan KM, Kumar S. 3D printed polylactic acid nanocomposite scaffolds for tissue engineering applications. Polym Test 2020;81:106203. https://doi.org/10.1016/j.polymertesting.2019.106203.

[20]   Vanderploeg A, Lee SE, Mamp M. The application of 3D printing technology in the fashion industry. Int J Fash Des Technol Educ 2017;10:170–9. https://doi.org/10.1080/17543266.2016.1223355.

[21]   Ahn D, Kweon J-H, Kwon S, Song J, Lee S. Representation of surface roughness in fused deposition modeling. J Mater Process Technol 2009;209:5593–600. https://doi.org/10.1016/j.jmatprotec.2009.05.016.

[22]   Yap YL, Yeong WY. Additive manufacture of fashion and jewellery products: a mini review. Virtual Phys Prototyp 2014;9:195–201. https://doi.org/10.1080/17452759.2014.938993.

[23]   TamiCare. Welcome to a new generation of fabric 2013. https://doi.org/http://www.tamicare.com.

[24]   Dugbenoo E, Arif MF, Wardle BL, Kumar S. Enhanced Bonding via Additive Manufacturing-Enabled Surface Tailoring of 3D Printed Continuous-Fiber Composites. Adv Eng Mater 2018;20:1800691. https://doi.org/10.1002/adem.201800691.

[25]   Andrew JJ, Schneider J, Ubaid J, Velmurugan R, Gupta NK, Kumar S. Energy absorption characteristics of additively manufactured plate-lattices under low- velocity impact loading. Int J Impact Eng 2021;149:103768. https://doi.org/10.1016/j.ijimpeng.2020.103768.

[26]   N. Turner B, Strong R, A. Gold S. A review of melt extrusion additive manufacturing processes: I. Process design and modeling. Rapid Prototyp J 2014;20:192–204. https://doi.org/10.1108/RPJ-01-2013-0012.

[27]   Korpela J, Kokkari A, Korhonen H, Malin M, Närhi T, Seppälä J. Biodegradable and bioactive porous scaffold structures prepared using fused deposition modeling. J Biomed Mater Res Part B Appl Biomater 2013;101B:610–9. https://doi.org/10.1002/jbm.b.32863.

[28]   Brinks GJ, Warmoeskerken MMC., Akkerman R, Zweers W. The Added Value of 3D Polymer Deposition on Textiles. 13th AUTEX World Text Conf 2013:1–6.

[29]   Stuecker JN, Cesarano J, Hirschfeld DA. Control of the viscous behavior of highly concentrated mullite suspensions for robocasting. J Mater Process Technol 2003;142:318–25. https://doi.org/10.1016/S0924-0136(03)00586-7.

[30]   Chatterjee K, Ghosh TK. 3D Printing of Textiles: Potential Roadmap to Printing with Fibers. Adv Mater 2020;32:1902086. https://doi.org/10.1002/adma.201902086.

[31]   Lebel LL, Aissa B, Khakani MA El, Therriault D. Ultraviolet-Assisted Direct-Write Fabrication of Carbon Nanotube/Polymer Nanocomposite Microcoils. Adv Mater 2010;22:592–6. https://doi.org/10.1002/adma.200902192.

[32]   Skylar-Scott MA, Gunasekaran S, Lewis JA. Laser-assisted direct ink writing of planar and 3D metal architectures. Proc Natl Acad Sci U S A 2016;113:6137–42. https://doi.org/10.1073/pnas.1525131113.

[33]   Sclater N. Mechanisms and mechanical devices sourcebook. New York: McGraw-Hill Education; 2011.

[34]   Kumar S, Ubaid J, Abishera R, Schiffer A, Deshpande VS. Tunable Energy Absorption Characteristics of Architected Honeycombs Enabled via Additive Manufacturing. ACS Appl Mater Interfaces 2019;11:42549–60. https://doi.org/10.1021/acsami.9b12880.

[35]   Liljenhjerte J, Upadhyaya P, Kumar S. Hyperelastic strain measurements and constitutive parameters identification of 3D printed soft polymers by image processing. Addit Manuf 2016;11:40–8. https://doi.org/10.1016/j.addma.2016.03.005.

[36]   Celaschi F, Celi M. Advanced design as reframing practice: Ethical challenges and anticipation in design issues. Futures 2015;71:159–67. https://doi.org/10.1016/j.futures.2014.12.010.

[37]   Bahraminasab M. Challenges on optimization of 3D printed bone scaffolds. Biomed Eng Online 2020;19:69. https://doi.org/10.1186/s12938-020-00810-2.

[38]   Rafiee M, Farahani RD, Therriault D. Multi-Material 3D and 4D Printing: A Survey. Adv Sci 2020;7:1902307. https://doi.org/10.1002/advs.201902307.

[39]   Vafadar A, Guzzomi F, Rassau A, Hayward K. Advances in Metal Additive Manufacturing: A Review of Common Processes, Industrial Applications, and Current Challenges. Appl Sci 2021;11:1213. https://doi.org/10.3390/app11031213.

[40]   Mostafaei A, Elliott AM, Barnes JE, Li F, Tan W, Cramer CL, et al. Binder jet 3D printing – Process parameters, materials, properties, modeling, and challenges. Prog Mater Sci 2021;119:100707. https://doi.org/10.1016/j.pmatsci.2020.100707.

[41]   Hoskins S. 3D printing for artists, designers and makers. 2nd ed. Bloomsbury Publishing; 2018.

[42]   Stucker BE. Additive Manufacturing Technologies: Technology Introduction and Business Implications. Semant Sch 2011.

[43]   Spectrum Z510 3D Printer n.d. https://scientificservices.eu/item/spectrum-z510-3d-printer/819.

[44]   Edgar J, Tint S. "Additive Manufacturing Technologies: 3D Printing, Rapid Prototyping, and Direct Digital Manufacturing", 2nd Edition. Johnson Matthey Technol. Rev., vol. 59, 2015, p. 193–8. https://doi.org/10.1595/205651315X688406.

[45]   Peterson GI, Schwartz JJ, Zhang D, Weiss BM, Ganter MA, Storti DW, et al. Production of Materials with Spatially-Controlled Cross-Link Density via Vat Photopolymerization. ACS Appl Mater Interfaces 2016;8:29037–43. https://doi.org/10.1021/acsami.6b09768.

[46]   Huang SH, Liu P, Mokasdar A, Hou L. Additive manufacturing and its societal impact: a literature review. Int J Adv Manuf Technol 2013;67:1191–203. https://doi.org/10.1007/s00170-012-4558-5.

[47] Mendoza HR. 3D Printed Fashion Presents Dresses Which Are Actually Comfortable n.d.
      https://3dprint.com/16385/3d-printed-fashion/%0A.
[48] Frazier WE. Metal Additive Manufacturing: A Review. J Mater Eng Perform 2014;23:1917–28. https://
      doi.org/10.1007/s11665-014-0958-z.
[49] Carroll BE, Palmer TA, Beese AM. Anisotropic tensile behavior of Ti–6Al–4V components fabricated
      with directed energy deposition additive manufacturing. Acta Mater 2015;87:309–20. https://doi.
      org/10.1016/j.actamat.2014.12.054.
[50] Gibson I, Rosen DW, Stucker B, Khorasani M, Rosen D, Stucker B, et al. Additive manufacturing
      technologies. Vol 17. Switzerland: Springer.; 2021.
[51] Park J, Tari MJ, Hahn HT. Characterization of the laminated object manufacturing (LOM) process.
      Rapid Prototyp J 2000;6:36–50. https://doi.org/10.1108/13552540010309868.
[52] Feygin M, Pak SS. Laminated object manufacturing apparatus and method. US5876550, 1995.
[53] Z Corporation Ships Spectrum Z510 3D Printing Systems n.d. https://www.3dsystems.com/press-
      releases/z-corporation-ships-spectrum-z510-3d-printing-systems.
[54] Objet350 and Objet500 Connex3 n.d. https://www.stratasys.com/siteassets/3d-printers/printer-
      catalog/objet-350-500-connex3/pss_pj_objet350objet500connex3_0319a.pdf.
[55] Sun Q, Rizvi GM, Bellehumeur CT, Gu P. Effect of processing conditions on the bonding quality of
      FDM polymer filaments. Rapid Prototyp J 2008;14:72–80. https://doi.org/10.1108/
      13552540810862028.
[56] Brown HR. Adhesion between Polymers and Other Substances – A Review of Bonding Mechanisms,
      Systems and Testing. Mater Forum 2000;24:49–58.
[57] Hashemi Sanatgar R, Campagne C, Nierstrasz V. Investigation of the adhesion properties of direct
      3D printing of polymers and nanocomposites on textiles: Effect of FDM printing process
      parameters. Appl Surf Sci 2017;403:551–63. https://doi.org/10.1016/j.apsusc.2017.01.112.
[58] Grimmelsmann N, Kreuziger M, Korger M, Meissner H, Ehrmann A. Adhesion of 3D printed material
      on textile substrates. Rapid Prototyp J 2018;24:166–70. https://doi.org/10.1108/RPJ-05-2016-0086.
[59] Meyer P, Döpke C, Ehrmann A. Improving adhesion of three-dimensional printed objects on textile
      fabrics by polymer coating. J Eng Fiber Fabr 2019;14:155892501989525. https://doi.org/10.1177/
      1558925019895257.
[60] Mpofu NS, Mwasiagi JI, Nkiwane LC, Njuguna D. Use of regression to study the effect of fabric
      parameters on the adhesion of 3D printed PLA polymer onto woven fabrics. Fash Text 2019;6:24.
      https://doi.org/10.1186/s40691-019-0180-6.
[61] Unger L, Scheideler M, Meyer P, Harland J, Gorzen A, Wortmann M, et al. Increasing adhesion of 3D
      printing on textile fabrics by polymer coating. TEKSTILEC 2018;61:265–71. https://doi.org/10.14502/
      Tekstilec2018.61.265-271.
[62] Sauerbier P, Köhler R, Renner G, Militz H. Surface Activation of Polylactic Acid-Based Wood-Plastic
      Composite by Atmospheric Pressure Plasma Treatment. Materials (Basel) 2020;13:4673. https://doi.
      org/10.3390/ma13204673.
[63] Pei E, Shen J, Watling J. Direct 3D printing of polymers onto textiles: experimental studies and
      applications. Rapid Prototyp J 2015;21:556–71. https://doi.org/10.1108/RPJ-09-2014-0126.
[64] Sabantina L, Kinzel F, Ehrmann A, Finsterbusch K. Combining 3D printed forms with textile
      structures – mechanical and geometrical properties of multi-material systems. IOP Conf Ser Mater
      Sci Eng 2015;87:012005. https://doi.org/10.1088/1757-899X/87/1/012005.
[65] Korger M, Bergschneider J, Lutz M, Mahltig B, Finsterbusch K, Rabe M. Possible Applications of 3D
      Printing Technology on Textile Substrates. IOP Conf Ser Mater Sci Eng 2016;141:012011. https://doi.
      org/10.1088/1757-899X/141/1/012011.
[66] Narula A, Pastore CM, Schmelzeisen D, El Basri S, Schenk J, Shajoo S. Effect of knit and print
      parameters on peel strength of hybrid 3-D printed textiles. J Text Fibrous Mater
      2018;1:251522111774925. https://doi.org/10.1177/2515221117749251.

[67] Ahmed A, Jalil MA, Hossain MM, Moniruzzaman M, Adak B, Islam MT, et al. A PEDOT:PSS and graphene-clad smart textile-based wearable electronic Joule heater with high thermal stability. J Mater Chem C 2020;8:16204–15. https://doi.org/10.1039/D0TC03368E.

[68] Faruk MO, Ahmed A, Jalil MA, Islam MT, Shamim AM, Adak B, et al. Functional textiles and composite based wearable thermal devices for Joule heating: progress and perspectives. Appl Mater Today 2021;23:101025. https://doi.org/10.1016/j.apmt.2021.101025.

[69] Adak B. Utilization of Nanomaterials in Conductive Smart- Textiles: A Review. J Text Sci Fash Technol 2021;8. https://doi.org/10.33552/JTSFT.2021.08.000678.

[70] Joshi M, Adak B. 6. Nanotechnology-based Textiles: A Solution for the Emerging Automotive Sector. Rubber Nanocomposites and Nanotextiles, De Gruyter; 2019. p. 179–230. https://doi.org/10.1515/9783110643879-006.

[71] AlMahri S, Schneider J, Schiffer A, Kumar S. Piezoresistive sensing performance of multifunctional MWCNT/HDPE auxetic structures enabled by additive manufacturing. Polym Test 2022;114:107687. https://doi.org/10.1016/j.polymertesting.2022.107687.

[72] Ubaid J, Schneider J, Deshpande VS, Wardle BL, Kumar S. Multifunctionality of Nanoengineered Self-Sensing Lattices Enabled by Additive Manufacturing. Adv Eng Mater 2022;24:2200194. https://doi.org/10.1002/adem.202200194.

[73] Verma P, Ubaid J, Varadarajan KM, Wardle BL, Kumar S. Synthesis and Characterization of Carbon Nanotube-Doped Thermoplastic Nanocomposites for the Additive Manufacturing of Self-Sensing Piezoresistive Materials. ACS Appl Mater Interfaces 2022;14:8361–72. https://doi.org/10.1021/acsami.1c20491.

[74] Gao T, Yang Z, Chen C, Li Y, Fu K, Dai J, et al. Three-Dimensional Printed Thermal Regulation Textiles. ACS Nano 2017;11:11513–20. https://doi.org/10.1021/acsnano.7b06295.

[75] Tibbits S. The emergence of "'4D printing'" 2013.www.ted.com/talks/skylar_tibbits_the_emergence_of_4d_printing.

[76] Murphy S V, Atala A. 3D bioprinting of tissues and organs. Nat Biotechnol 2014;32:773–85. https://doi.org/10.1038/nbt.2958.

[77] Rubežiene V, Padleckiene I, Baltušnikaite J, Varnaite S. Evaluation of camouflage effectiveness of printed fabrics in visible and near infrared radiation spectral ranges. Medziagotyra 2008;14:361–5.

[78] Leist SK, Gao D, Chiou R, Zhou J. Investigating the shape memory properties of 4D printed polylactic acid (PLA) and the concept of 4D printing onto nylon fabrics for the creation of smart textiles. Virtual Phys Prototyp 2017;12:290–300. https://doi.org/10.1080/17452759.2017.1341815.

[79] Engel J, Liu C. Creation of a metallic micromachined chain mail fabric. J Micromechanics Microengineering 2007;17:551–6. https://doi.org/10.1088/0960-1317/17/3/018.

[80] Dyskin A V., Pasternak E, Estrin Y. Mortarless structures based on topological interlocking. Front Struct Civ Eng 2012. https://doi.org/10.1007/s11709-012-0156-8.

[81] Wang Y, Li L, Hofmann D, Andrade JE, Daraio C. Structured fabrics with tunable mechanical properties. Nature 2021;596:238–43. https://doi.org/10.1038/s41586-021-03698-7.

[82] Ploszajski AR, Jackson R, Ransley M, Miodownik M. 4D Printing of Magnetically Functionalized Chainmail for Exoskeletal Biomedical Applications. MRS Adv 2019;4:1361–6. https://doi.org/10.1557/adv.2019.154.

[83] Hollister SJ. Porous scaffold design for tissue engineering. Nat Mater 2005;4:518–24. https://doi.org/10.1038/nmat1421.

[84] Sherwood JK, Riley SL, Palazzolo R, Brown SC, Monkhouse DC, Coates M, et al. A three-dimensional osteochondral composite scaffold for articular cartilage repair. Biomaterials 2002;23:4739–51. https://doi.org/10.1016/S0142-9612(02)00223-5.

[85] Athukorala SS, Tran TS, Balu R, Truong VK, Chapman J, Dutta NK, et al. 3D Printable Electrically Conductive Hydrogel Scaffolds for Biomedical Applications: A Review. Polymers (Basel) 2021;13:474. https://doi.org/10.3390/polym13030474.

[86]   Murugan SS, Anil S, Sivakumar P, Shim MS, Venkatesan J. 3D-Printed Chitosan Composites for Biomedical Applications, 2021, p. 87–116. https://doi.org/10.1007/12_2021_101.

[87]   Zhang M, Zhao M, Jian M, Wang C, Yu A, Yin Z, et al. Printable Smart Pattern for Multifunctional Energy-Management E-Textile. Matter 2019;1:168–79. https://doi.org/10.1016/j.matt.2019.02.003.

[88]   Wang Y, Chen C, Xie H, Gao T, Yao Y, Pastel G, et al. 3D-Printed All-Fiber Li-Ion Battery toward Wearable Energy Storage. Adv Funct Mater 2017;27:1703140. https://doi.org/10.1002/adfm.201703140.

[89]   NASA's Jet Propulsion Laboratory, "Space Fabric" Links Fashion and Engineering n.d. https://www.jpl.nasa.gov/news/news.php?feature=6816%0A.

[90]   Zhang W, Zhang F, Lan X, Leng J, Wu AS, Bryson TM, et al. Shape memory behavior and recovery force of 4D printed textile functional composites. Compos Sci Technol 2018;160:224–30. https://doi.org/10.1016/j.compscitech.2018.03.037.

[91]   Liu Y, Zhang W, Zhang F, Lan X, Leng J, Liu S, et al. Shape memory behavior and recovery force of 4D printed laminated Miura-origami structures subjected to compressive loading. Compos Part B Eng 2018;153:233–42. https://doi.org/10.1016/j.compositesb.2018.07.053.

[92]   Bloomfield M, Borstrock S. Modeclix. The additively manufactured adaptable textile. Mater Today Commun 2018;16:212–6. https://doi.org/10.1016/j.mtcomm.2018.04.002.

[93]   Griffiths A. "4D-printed" shape-changing dress and jewellery by Nervous System. Zeen 2013. https://www.dezeen.com/2013/12/03/kinematics-4d-printed-shape-changing-jewellery-by-nervous-system/.

[94]   Kinematics. Nerv Syst 2013. https://n-e-r-v-o-u-s.com/projects/albums/kinematics-concept/.

[95]   Howarth D. MoMA acquires "4D-printed" dress. Zeen 2014. https://www.dezeen.com/2014/12/09/moma-acquires-first-4d-printed-dress-nervous-system-kinematics/.

[96]   McKnight J. Nervous System creates "4D-printed" dress made of nylon petals and scales 2016. https://www.dezeen.com/2016/03/08/nervous-system-4d-3d-printed-kinematic-nylon-petals-dress-fashion/.

[97]   Biswas MC, Chakraborty S, Bhattacharjee A, Mohammed Z. 4D Printing of Shape Memory Materials for Textiles: Mechanism, Mathematical Modeling, and Challenges. Adv Funct Mater 2021;31:2100257. https://doi.org/10.1002/adfm.202100257.

[98]   Howarth D. Neri Oxman creates wearable 3D printed structures for interplanetary voyages. Zeen 2014. https://www.dezeen.com/2014/11/25/neri-oxman-mit-media-lab-stratasys-wearable-3d-printed-structures-interplanetary-voyages-synthetic-biology/.

[99]   3D printed clothes in 2021: What are the best projects? Sculpteo n.d. https://www.sculpteo.com/en/3d-learning-hub/applications-of-3d-printing/3d-printed-clothes/.

[100] 3D Printed Shoes. Oliv van Herpt Logo 2012. https://oliviervanherpt.com/3d-printed-shoes/.

[101] A.L.B and 3D Printing. Horol Work n.d. https://www.alb-watches.com/en-at-05-3d-printing-watch-alb.php.

Andrew J. Hebden, Parikshit Goswami

18 Functional and smart textiles in care, treatment, and diagnosis of COVID-19

Abstract: The COVID-19 pandemic has provided a great challenge with unprecedented demand for personal protective equipment (PPE) during the initial stages of the outbreak. However, it has also been a tremendous opportunity for the textile industry to innovate within the sphere of PPE as textile products formed the first line of defence against this novel coronavirus giving time to the scientists to develop a vaccine. This chapter provides an overview of the challenges presented by COVID-19 and the key constituent parts of PPE for medical personnel during this time. The construction and features of these items, in addition to the regulations governing these important items of PPE, are discussed. Additionally, the future direction of PPE, particularly with regard to single-use items and the sustainability of PPE supplies, is considered.

Keywords: COVID-19, PPE, masks, healthcare, polymers, airborne virus

18.1 Introduction

Since 2020 a global spotlight has been shone on the area of personal protective equipment (PPE) unparalleled in modern times and has seen a new addition to the urban litter landscape – the single-use mask. COVID-19 has impinged on every aspect of modern life and placed immense strain on healthcare infrastructure and supply chains globally, whilst restricting social and societal freedoms in many countries [1]. The vaccinations played a significant role in controlling the spreading of COVID-19 in later stage, and it has received much attention, but during those early stages of the pandemic it was PPE and thus the textiles industry that led the response while vaccines was under development stage. This chapter looks at the role of functional textiles which have played an important role in the fight against COVID-19 and also the innovation within the textile field can lead to develop smart textiles that will not only speed up the diagnosis of COVID-19 but perhaps find usage within a wider healthcare setting. The COVID-19 pandemic comes against a backdrop of European consumer unease towards single-use plastic, with nations adopting a variety of initiatives to reduce their use. Thus, against this climate it is right to objectively consider if there are other alternatives within the textile and non-woven sphere which can offer a sustainable PPE solution primarily for COVID-19 protection but also in the wider PPE marketplace.

COVID-19 is a respiratory virus which was initially characterized by symptoms including a cough and fever, meaning it exhibits clinical similarities to other illnesses such as seasonal flu. Other symptoms, such as anosmia, ageusia, tiredness, muscle

https://doi.org/10.1515/9783110759747-018

pain, sore throat, headache, diarrhoea, skin rash, or irritated eyes, are all recognized symptoms, with different variants exhibiting different incidence profiles. A COVID-19 diagnosis can be confirmed on the basis of a reverse transcription polymerase chain reaction test or an antigen lateral flow test [2].

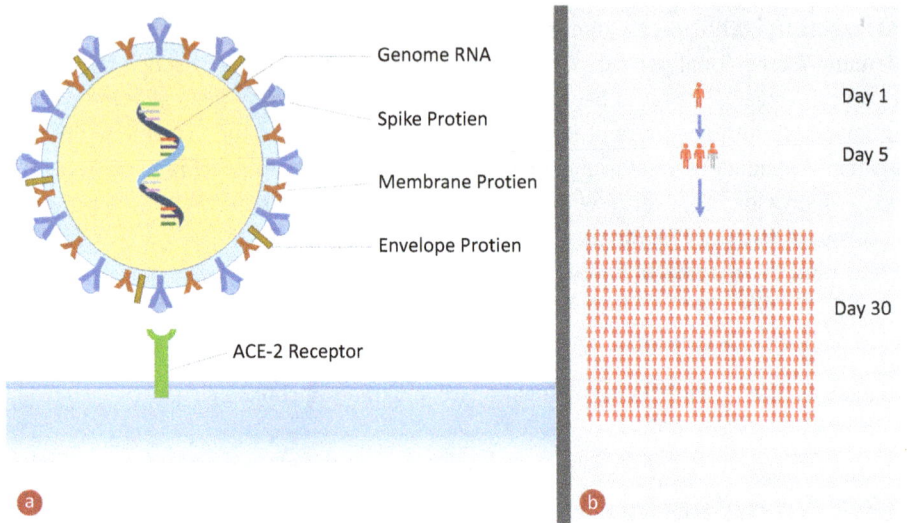

Figure 18.1: (a) COVID-19 structure and (b) mechanistic action and spreading of COVID-19.

The COVID-19 virus is isostructural to other coronaviruses in that it is composed of four structural proteins known as envelope, membrane, nucleocapsid, and spike [3]. It is the spike protein that allows the COVID-19 virus to enter lung cells by attaching to the angio-tensin converting enzyme 2 receptor found on the surface of human cells. Proteolytic cleavage of the spike protein will then lead to membrane fusion and subsequently re-lease of viral RNA into the host cells cytoplasm from where it can be replicated using the host cell's machinery before being released via exocytosis. Figure 18.1 shows the struc-ture and mechanistic action of COVID-19, in addition to its propensity to spread.

COVID-19 is primarily transmitted through contact with an infected person via droplets or aerosols which may be generated through activities such as coughing or talking [4]. Thus, societal restrictions that limit individual's contacts offer one mecha-nism to reduce spread by reducing the probability of encountering an infected per-son. Social distancing then offers a second line of defence, breaking the chains of transmission, by reducing the chance that droplets or aerosols produced will travel from an infected person to another. A third level of protection for the individual in society may be some form of face covering (though not formerly PPE) to contain drop-lets or aerosols. In a medical setting, particularly where COVID-19 patients are been treated, PPE requirements to protect the clinician and break chains of transmission are more stringent. With PPE shortages prevalent at the start of the pandemic it puts clinicians in an unenviable position of putting themselves at additional risk in order to perform life-saving treatment. Enhanced levels of cleaning within clinical and workplace settings and the encouraging of additional hand washing through government campaigns have also played a role in virus control. The viral envelope of the virus is destroyed by standard soap when it comes in contact outside the body [5].

Viruses are well known for their ability to mutate, hence the need for many governments to run annual flu vaccine campaigns to mitigate seasonal flu. Coronavirus is no different with five variants of concern identified (alpha, beta, gamma, delta, and omicron) in the last two years as the virus adapts to increase transmissibility or virulence or indeed to reduce the effectiveness of vaccines. It should also be noted that different governments globally have taken a wide variety of approaches to tackling the pandemic based on their own unique social and economic priorities. These include Sweden, who pursued a strategy without national lockdown or curfew, in contrast to the majority of Europe [6], New Zealand, and Australia, where strict travel conditions were imposed for those outside the country, and when cases were detected, strict state-wide lockdowns were imposed [7] and China who have maintained a zero COVID-19 policy, locking down and isolating entire cities [8]. Only in the fullness of time will it be determined which approach is most appropriate for this type of virus.

18.2 PPE

Throughout the pandemic, the World Health Organization (WHO) released guidance to mitigate the severe PPE shortages encountered globally in an attempt to ensure those at the greatest risk, particularly in healthcare settings, were able to access the most appropriate PPE as just-in-time manufacturing struggled to meet the upsurge in demand [9]. It should also be noted that as scientists and healthcare professionals gained a better understanding of the virus and its modes of transmission, PPE guidance could be adapted to ensure appropriateness. At the height of the pandemic, monthly global face mask usage was estimated at 129 billion, equivalent to 3 million in a minute [10]. Within a clinical setting at the start of the pandemic, healthcare workers (HCWs) were routinely wearing various permutations of face masks, face shields, eye protection, aprons, coveralls, and powered air purifying respirator hoods. Often this depended on what was available at the time and whilst a number of the items are deemed single use, supply shortages often meant extended wearing sessions were necessary beyond the recommended usage times.

18.2.1 PPE constituent parts

It is important that those HCWs required to wear PPE in order to protect from COVID-19 infections have appropriately sized items to wear and have received appropriate training enabling them to don and remove the items sequentially, in a way that prevents infection spread and ensures that the wearer is receiving the full protection afforded by the item. It should also be noted that whilst PPE provides a barrier wearing for an extended period may cause heat stress or a loss of dexterity within the HCW impinging their ability to perform their duties [11]. As shown in Figure 18.2, a typical HCW's PPE will ideally consist of a face shield, or goggles, to protect from droplets, a fluid-resistant mask, or respirator to protect from COVID-19-containing aerosols, a gown, or coverall, which will be fluid-resistant and maybe worn in conjunction with a disposable apron, and single-use gloves are used on the hands. These PPE elements will be discussed in more detail in subsequent sections of this chapter. Where supply issues existed during the pandemic, there were countless examples of HCWs improvising PPE from items such as bin liners in order to be able to deliver life-saving care [12]. Thus whilst this represents the idealized picture for PPE, this was not the reality experienced by many HCWs worldwide throughout the pandemic.

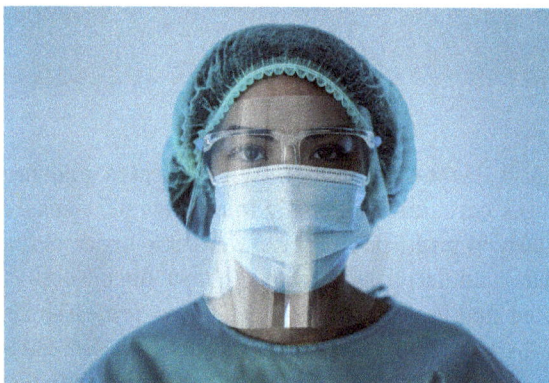

Figure 18.2: Typical PPE for a healthcare worker (reproduced with kind permission, ©2022, iStock).

18.2.2 Raw materials and construction

At the start of the pandemic China was responsible for the largest share of PPE production in the world with the small margins on such items leading to a "just-in-time" supply chain model around the world with healthcare providers holding small inventories of PPE [13]. Globally there was a swift change, particularly with regard to masks that saw them go from a niche product primarily used in clinical settings to a product required by a country's entire population. This put pressure not only on manufacturers and supply chains but also global transportation routes which were also severely disrupted due to COVID-19 restrictions. Using masks as an example, the majority use melt-blown polypropylene in order to install and set up such a production line will take many months, require significant capital investment, and thus it is not something which can be quickly achieved.

One of the great scientific advances of the twentieth century was the development of the polymer industry; plastics were revolutionary as they meant for the first time that the constraints of nature did not limit manufacturing. Swiftly these new materials came to dominate the PPE market, offering sterility and their single-use nature presenting a cost-saving compared to natural alternatives which required laundering. As a result, polymers such as polypropylene, polyester, and polyethylene are typically used in PPE garments such as disposable isolation gowns.

Polyethylene (PE) is thermoplastic and constitutes around a third of annual global plastic production with the majority produced either high-density polyethylene (HDPE) or low-density polyethylene (LDPE). LDPE is formed by the addition of short-chain alpha olefins such as hex-1-ene leading to chain branching, reducing the ability of the molecular chains to tightly pack, thus reducing crystallinity and enabling flexibility. Polypropylene is an isomeric polymer with atactic, isotactic, and syndiotactic forms possible depending on the relative orientation of methyl side groups within the polymer chain. Isotactic polypropylene, with the methyl groups arranged on the same side of the polymer backbone, causes a helical shape to be formed leading to a semi-crystalline polymer that is widely used for the production of non-wovens. It is over 80 years since polyester, the fourth most produced polymer globally, synthesized from two monomers, ethylene glycol and terephthalic acid, was first produced. The impact of an increase in demand for polymers such as polypropylene in the context of PPE should also not be underestimated on the supply of this raw material to competing but equally important sectors.

In addition to polymers used in item construction, laminated polymer films based on PE or thermoplastic polyurethane can also be used within the garment to provide barrier properties. During COVID-19 pandemic, to fulfil high demand of PPE water-repellent and waterproof-coated fabrics got huge importance having option of breathability and multiple time use. Water repellency is imparted to the textile by generally treating with fluorocarbon chemicals. However, as the environmental impacts of fluorocarbon finishes have become apparent in recent years, other alternatives have been investigated including silicones [14]. The water-repellent finished fabrics are generally coated with polyurethane for achieving high waterproofness, which is a potential material for PPE coverall. Sometimes, anti-viral chemicals also are incorporated in finishing and coating formulations for anti-viral property in the coverall fabrics.

18.2.3 Usage/features

Current COVID-19 PPE guidance for the UK which is atypical of the guidance currently in place within Western healthcare was released by the Department of Health and Social Care and the devolved public health agencies in March 2022 for those providing care in health and care settings, a summary of this guidance is shown in Table 18.1 [15]. Under this guidance in the case of droplet creating procedures, fluid-resistant surgical masks and eye/face protection can be worn seasonally. The same applies to respiratory protecting equipment in the case of aerosol-generating procedures. All other PPE must be changed between patients or after completing a procedure in line with best practice.

Table 18.1: UK PPE guidance effective from March 2022 for those providing direct care to a patient with confirmed or suspected COVID-19 [15].

PPE required by type of transmission/exposure	Disposable gloves	Disposable/reusable fluid-resistant apron/gown	Fluid-resistant surgical mask (FRSM)/ respiratory protective equipment (RPE)	Eye/face protection (goggles or visor)
Droplet	Single use	Single-use apron or fluid-resistant gown if risk of extensive spraying/ splashing	Single-use FRSM Type IIR for direct patient care	Single use or reusable
Airborne (when undertaking or if aerosol generating procedures are likely)	Single use	Single-use fluid-resistant gown	Single-use FFP3 or reusable respirator/ powered respirator hood	Single use or reusable

As scientific understanding of COVID-19 has matured during the pandemic, it has enabled PPE usage to be tailored based on risk. Thus, for example, where an aerosol generating procedure is to be undertaken, a higher level of respiratory protection can be employed by utilizing an FFP3 respirator rather than relying on a fluid-resistant surgical mask. This helps to ensure that PPE supplies are utilized to their full potential.

The wearing of PPE can lead to changes in the behaviour of the wearer, making them partake in riskier behaviour or indeed reduce their levels of compliance with other COVID-19 preventative measures, for example, good hand hygiene. The correct wearing of masks can also reduce levels of anxiety experienced by the general public [16]. Whilst we have seen the removal of governmental restrictions in many parts of the world at the time of writing, that does not mean that all individuals have gone back to pre-pandemic behaviours, with significant proportions of the elderly and clinically vulnerable still maintaining some aspects of caution be that wearing PPE, limiting social contacts or not entering crowded spaces.

18.3 Face mask: developments, usage, and efficacy

There are various types of disposable masks in common use: single use, reusable, washable, surgical, and respirator. WHO guidelines state that in order to be classified as a mask, the chin, mouth, and nose must be covered. The relevant standards and efficacy for each of these masks are shown in Table 18.2. As a general trend, assuming no supply shortages, as the level of protection increases, so does the price point, and thus surgical masks are much cheaper to purchase than FFP3 respirators. It should be noted that filtering efficiency is dependent upon particle size and thus the best point at which to test performance is at the most penetrating particle size, in the knowledge that if it performs well at this point it will perform even better in other particle ranges.

Where respirators are used such as those which conform to FFP3, it is important that the wearer is fit-tested and clean-shaven to ensure that the mask will indeed offer the level of protection intended.

Table 18.2: Filtration efficiency of different mask types and relevant standards. NR means no requirement for particle size in relevant standard.

Mask type	Filtration efficiency			Relevant standard
Community face covering	3 μm: ≥70%			UK – BSI Flex 5555
Single-use face mask	3 μm: ≥95% 0.1 μm: NR			China – YY/T0969
Surgical mask	3 μm: ≥95%			China – YY0469
	Level 1	Level 2	Level 3	Europe – EN 14683
	3 μm: ≥95% 0.1 μm: ≥95%	3 μm: ≥98% 0.1 μm: ≥98%	3 μm: ≥98% 0.1 μm: ≥98%	
	Type 1	Type 2	Type 3	USA – ASTM F2100
	3 μm: ≥95% 0.1 μm: NR	3 μm: ≥98% 0.1 μm: NR	3 μm: ≥98% 0.1 μm: NR	
Respirator mask	N95/KN95	N99/KN99	N100/KN100	China – GB2626 USA – NIOSH (42 CFR 84)
	0.3 μm: ≥95%	0.3 μm: ≥99%	0.3 μm: ≥99.97%	
	FFP1	FFP2	FFP3	Europe – EN149:2001
	0.3 μm: ≥80%	0.3 μm: ≥94%	0.3 μm: ≥99%	

A key element of the UK government's COVID-19 strategy, which was mirrored in numerous other countries globally, was to instruct the public to wear face coverings in community settings. Such face coverings are intended to be worn by those not displaying any clinical signs of bacterial or viral infection. These face coverings protect others

that the wearer may come into contact with by containing respiratory droplets or secretions when the wearer is talking, coughing, or sneezing. To this end BSI Flex 5555
was rapidly developed by an advisory group of academics, healthcare providers, and
governmental bodies to provide a standard for such coverings to comply to which is
less stringent than those standards in place for medical masks [17]. Community face
coverings can be either single-use or reusable and formed from multiple fabric layers
be they knitted, woven, or non-woven, they must not contain a valve. The material
used in manufacture should be appropriate for the masks designed period of use, thus
if a reusable mask, it must be capable of tolerating standard domestic laundering.

Surgical masks were first introduced in the 1890s and were used by surgeons in
the operating theatre as a strategy for infection control, with early iterations formed
of multiple layers of cotton gauze or silk [18]. Indeed testing from the time of these
natural fibre masks indicated that bacterial filtering performance increased after
laundering, perhaps as a result of the fibres tightening and shrinkage [19]. By the
1930s, these had been superseded by paper masks and 30 years later in the 1960s, the
single-use mask made from synthetic materials that we know today had emerged.
This was part of a wider drive within western healthcare at this time to single-use
disposable items such as syringes, whilst offering sterility, such items reduced labour
costs in terms of laundering or sterilization.

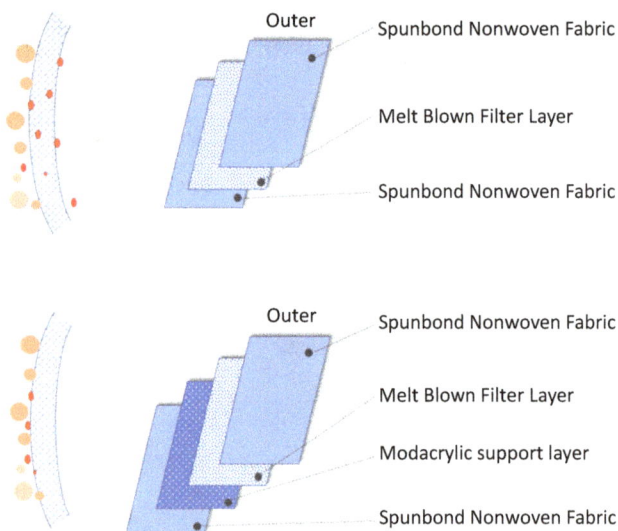

Figure 18.3: Layered structure of both surgical masks (top) and N95 respirators (bottom).

Modern surgical masks are fluid-resistant type IIR, these are 3-ply formed from a central layer of melt-blown non-woven polypropylene, sandwiched between two layers
of spun-bonded non-woven polypropylene as shown in Figure 18.3. The outermost

layer is droplet and stain-resistant, the middle layer is the filtering layer, and the innermost layer is designed for comfort as in contact with the skin.

The masks are pleated, such that one size fits all, have elastic ear loops to secure the mask to the face and in some cases a metal or plastic nosepiece to increase the level of fit in this area. Single-use or general medical masks generally lack the filtration efficiency of surgical masks, but they have a role to play in terms of general protection particularly with regards to containment of the user's aerosols.

The key difference between a respirator and a mask is that a respirator will form an airtight seal with the wearer ensuring that all inhaled air must pass through the filter material. However, it wasn't the medical field which saw the first use of respirators; significant efforts were made to develop respirators, in the form of gas masks during World War I and subsequently World War II. The first N95 respirator, similar to those on sale today was introduced by 3M in 1972 but it was not until the 1990s and the rise of drug-resistant tuberculosis that they saw routine use of respirators in hospitals. COVID-19 has once more necessitated the routine use of respirators in clinical settings, particularly where patients are symptomatic or aerosol-generating procedures are undertaken. In response to the pandemic, 3M upscaled their N95 mask making capacity from 630 million in 2019 to 2.5 billion in 2021 as demand for the masks surged.

In contrast to surgical masks discussed earlier, respirators have additional filtration layers formed using a range of manufacturing techniques which allow higher levels of filtration. The outer and inner layers perform similar functions as surgical masks, namely fluid resistance and wearer comfort. Due to the need to provide a good seal between the edge of the respirator and the wearer, foam strips are often included around the edge of the mask. The inner filtration layers consist of a thick pre-filtration layer and a thin electret non-woven melt-blown material responsible for filtration. KN95 respirators comply with Chinese standard GB2626 and are composed of four layers: outer, filter, cotton, and inner [20]. In comparison with an N95 respirator, the filter layer of a typical KN95 is around eight times thinner, with the cotton layer occupying a considerable amount of the cross-sectional area.

It must be remembered that there are three methods by which particles are filtered within an N95 mask as shown in Figure 18.4. Larger particles (>1 µm) and with sufficient inertia will collide with the fibrous mask material rather than follow the airstream around fibres and thus be filtered. Those particles smaller than 1 µm will generally undergo diffusion and as they are bombarded by air molecules, Brownian motion will occur causing them to become attached to the filter's fibrous layers. Lastly, and a key feature of N95 masks is electrostatic interaction, this occurs between oppositely charged particles and fibres regardless of size. Once particles interact with the mask fibres, they are held in place by strong molecular forces, making them difficult to remove once collected.

The inclusion of coloured elastic bands allows a visual indication to those observing the wearer, the type of mask been worn, and whether it is the right type for the

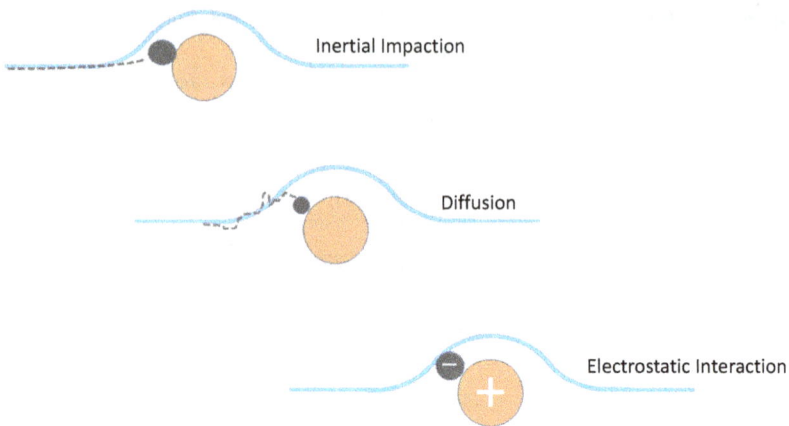

Figure 18.4: Three filtration mechanisms employed by N95-type masks, fibres shown in orange, and particles in grey.

atmosphere they are in. To this end FFP1 masks generally have yellow bands, FFP2 blue or white bands, and FFP3 red.

Some N95 respirators have exhalation valves which makes it easier for the wearer to exhale, and these are not appropriate for COVID-19 applications as they leave the expelled air from the wearer unfiltered and can thus spread virus-containing aerosol droplets. It should be noted that in addition to N95 class respirators, there are also R95 and P95 classes with the difference been the resistance of the respirator to oil, which may be relevant depending on the environment they are used in. N class are not oil-resistant, R class are oil-resistant to some extent, whereas P class are oil proof, but will only last for around 40 h use before they require replacement.

The importance of PPE appropriateness is highlighted by a study by Ferris et al. [21] at Cambridge University Hospitals NHS Foundation Trust. They found that when a switch was made from fluid-resistant surgical masks to FFP3 respirators for those HCWs employed on red wards (COVID-19-positive patients), COVID-19 infections among HCWs which could be attributed to direct contact with COVID-19 patients was dramatically reduced [21]. What should be remembered in this context is that the illness of an HCW not only leads to an increase in the COVID-19 rate but also reduces the pool of trained people available to care for those suffering from the illness.

18.4 Gloves

Gloves form an important defensive barrier in the fight against COVID-19 (Figure 18.5) and should be used where there is a chance that a healthcare professional may come in contact with bodily fluids of which, in the case of COVID-19, respiratory secretions are of particular note. A range of materials are available for medical gloves, including natural options such as latex and synthetic options such as nitrile and vinyl [22]. Care must be taken when wearing gloves to avoid transfer to fomites such as computer keyboards. Gloves should also not be seen as an alternative to good hand hygiene.

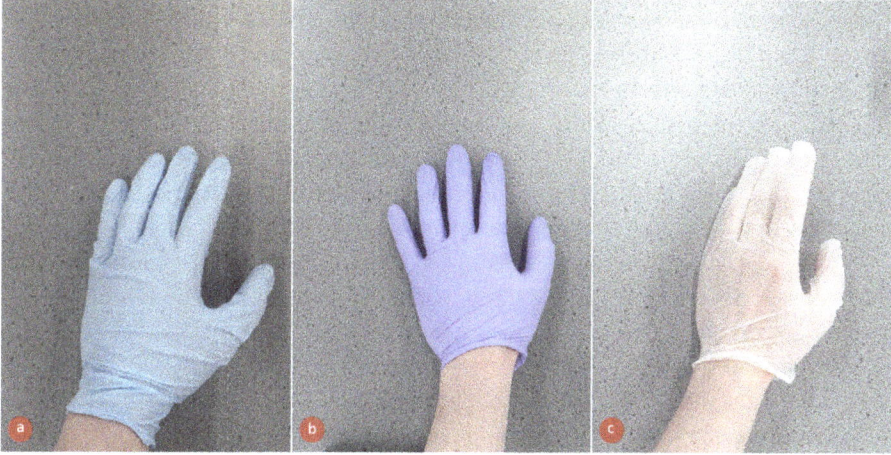

Figure 18.5: Different types of medical gloves: (a) vinyl, (b) nitrile, and (c) latex.

Latex gloves are made from natural rubber extracted via careful tapping from the *Hevea brasiliensis* tree, with the elastic component constituting between 25% and 45% of the total extract. Nitrile gloves gained popularity in the 1990s due to their durability and with the increasing incidence of latex allergies leading to the restriction of the usage of latex gloves in some clinical settings. It should be noted that latex allergies within medical professionals are at a higher rate than the general population [23]. Vinyl gloves formed from polyvinyl chloride are often used as a cheaper alternative to nitrile gloves, but they do not offer the same level of protection meaning they are more appropriate for lower risk tasks. Unpowdered gloves are generally preferred by HCWs compared to the powdered equivalents, but these attract a price premium. Within the EU, the standard for medical gloves is EN 455, composed of four parts; the United States is having a different standard for each material as ASTM D3578 (rubber), D5250 (PVC), and D6319 for nitrile gloves.

18.5 PPE garment fabrics: developments, usage, and efficacy

The concept of PPE stretches all the way back to Roman times, when Pliny the Elder who realized the risks of mining zinc and sulphur developed a mask from an animal bladder to protect workers from the fumes [24]. Later in the seventeenth century during the Black Death outbreak, plague doctors would wear a long beaked mask, black hat, and a waxed gown in an attempt to protect themselves when treating patients afflicted with the plague, which had an 80% mortality rate [25]. In more modern times, tightly woven cotton fabric coated with a waterproof finish began to appear in operating theatres during the 1960s to combat the transfer of bacteria from the surgeon through the conventional surgical gowns in use at the time [26].

From a medical point of view coveralls, or indeed gowns and aprons, need to provide a barrier between the HCW and the patient but also need to be comfortable to wear for an extended period. As a result, they need to have sufficient density to absorb liquid materials such as blood, secretions, and sputum which will be encountered during the normal working day. A denser fabric will provide a higher resistance to liquid penetration, but as density is increased breathability will be reduced. Where HCW are required to wear PPE for extended periods and in some cases entire shifts, it is important that moisture management and thermoregulation are considered.

Single-use gowns are common in the United States, whereas European healthcare have traditionally used a greater number of reusable ones. Disposable gowns are typically formed from synthetic fibres such as polypropylene or polyester formed into non-woven materials which can be bonded to plastic films or undergo chemical or physical treatments to enhance the liquid barrier properties of the resultant material. In contrast reusable gowns tend to be 100% cotton, 100% polyester, or a blend of the two. By forming the fibre into a tightly woven structure and then the addition of water-repellent finish, and polyurethane coating, liquid penetration can be reduced even after a significant number of laundry cycles.

Thus, water absorbency can be reduced in such protective clothing by the addition of coatings or water-repellent finishes. Work by Verbeek et al. [27] evaluated a number of controlled studies involving HCW in full PPE and were able to conclude that the combination of powered, air-purifying respirator, and coverall was able to offer a greater level of protection than an N95 mask and gown [27]. Additionally evidence pointed to those wearing long gowns exhibited lower levels of contamination than those wearing coveralls due to the difficulty in donning and doffing this form of PPE without contamination. Thus, there is a need for training to minimize contamination particularly where HCWs are not familiar with wearing PPE.

The importance of water repellency in PPE is highlighted by the potential of garments worn by infectious inpatients to act as fomites and cause cross-contamination either between patients or onto other surfaces within the healthcare setting leading to viral spread. In a COVID-19 context, where highly absorbent fibres will absorb the droplet liquid and isolate the virus within the fabric structure, natural fibres such as cotton, silk, and wool exhibit higher absorption capacity when compared with synthetic fibres such as polypropylene and polyester [28]. In these cases where low absorbency fibres are used in garments or textile items such as curtains, there is potential for the liquid to wick along the fibre or be maintained on the surface of the item.

18.6 Potential use of electrospun nanofibres to combat COVID-19

Electrospinning is a technology that has a history stretching back over 120 years [29] but it is only in the last 25 years that the technique has seen widespread use, with research papers on the area increasing since the mid-1990s with commercially relevant machines arriving in the market more recently. Application of a high voltage to a liquid droplet during the electrospinning process, such that the body of the liquid becomes charged, leads to a change in the surface tension of the droplet which is counteracted by the resultant electrostatic repulsion. A critical point is reached at which point a jet of liquid is ejected from the surface. This jet, known as a Taylor cone, flies towards the grounded target whilst been elongated by a whipping process in addition to undergoing solvent evaporation. Where conditions are appropriately optimized, this leads to thin uniform fibres been deposited on the collector with tuneable diameters typically in the micron range but can be tuned to the nanometre range. There are many parameters which can be altered to control the nature of the fibres produced including distance between the spinneret and collector, motion and size of collector, electric potential, needle gauge and shape, solution properties including solvent, flow rate, conductivity, viscosity and surface tension, polymer concentration, polymer molecular weight, molecular-weight distribution and architecture (branched, linear, etc.), ambient conditions including temperature, humidity, and air velocity in the chamber. The small diameter fibres created during the electrospinning process offer a high surface area to volume ratio ideal for the creation of active fibres with embedded functionality and are thus particularly applicable to the area of filtration. However mechanical strength of the randomly oriented electrospun fibres can be an issue due to weak fibre–fibre interactions. This can be improved by forming aligned nanofibrous mats by utilizing a rapidly rotating collector [30].

Work by Wang et al. [31] has utilized silk cocoons as a biocompatible and renewable material in the manufacture of nanofibrous air filters via electrospinning from formic acid solutions. Membranes produced showed filtration efficiency of 96.2% for 0.3 μm particles and a weight of 3.4 g m^{-2} around 2% of the basis weight of the equivalent layer in typical commercial respirator with similar filtration levels. Silver-doped fibres were also spun by the addition of AgNO$_3$ to the spinning solution, which was reduced to silver nanoparticles in situ by formic acid and anti-bacterial performance evaluated against *Escherichia coli* and *Staphylococcus aureus.*

The solubility of cellulose in aqueous media and common organic solvents provides a barrier to its usage in electrospinning, meaning that alternatives such as cellulose acetate have been explored. De Almeida et al. [32] used cellulose acetate in conjunction with cetylpyridinium bromide, a cationic surfactant to produce uniform nanofibrous mat with fibre diameters in the range of 200–300 nm. Filtration efficiency was calculated at 99.9% for particles sized 0.07–0.3 μm which would include the COVID-19 virus [32]. Polyacrylonitrile can easily be electrospun into fibres from solvents such as *N,N*-dimethylformamide and *N,N*-dimethylacetamide, forming fibres with high mechanical properties and good thermal stability and lend themselves to usage in filtration applications [33, 34]. Indeed Bortolassi et al. [35] deposited novel silver/polyacrylonitrile electrospun fibres onto a non-woven substrate and were able to show anti-bacterial activity against *E. coli* in addition to filtration of sodium chloride aerosol particles between 9 and 300 nm diameter, with filter efficiency as high as 100% in some cases.

One of the issues with facemasks during the pandemic is they shroud the face making it difficult for observers to read facial expressions and for those with hearing difficulties it prevents lip reading. To this end He et al. [36] have utilized polyvinyl alcohol electrospun fibres, supported on a 3D printed framework for stability to produce a transparent facemask with filtering capabilities.

The formation of composites scaffolds can be achieved via electrospinning allowing tailored structures to be produced for such end uses as biosensing, drug delivery, tissue repair, or regeneration and wound healing, specifically within a COVID-19 context such structures lend themselves to viral filtration or detection [37]. The architecture of the functional scaffold can be controlled by controlling the fibre alignment (aligned, cross-aligned, or random) and architecture or indeed the geometrical characteristics of the fibre.

Work by McCarthy et al. [38] involving the usage of aligned, layered, 3D gelatine-coated PCL nanofibre matrices for COVID-19 swabbing focused on improving sample collection and test sensitivity as shown in Figure 18.6 [38]. This led to a reduction in the false-positive rate and a 10-fold increase in detection when using the nanofibrous swabs compared to standard cotton swabs. Such swabs have potential for a wide range of diagnostic uses. Electrospun biosensors for detection of pathogens such as *E. coli* and bovine viral diarrhoea virus have already been developed, with the high surface area offered by electrospun membranes enhancing the signal sensor to noise ratio [39, 40] and utilizing a fluorescent output upon detection, this offers a potential pathway for rapid detection of COVID-19.

Patil et al. realized a biodegradable electrospun nanofibrous mask with encapsulated phytochemicals exhibited excellent bacterial filtration efficiency (97.9%). [41] In the PLA electrospinning solution, the traditional Indian herbal extracts from *Azadirachta indica* and *Eucalyptus citriodora* were added. The mask was then composed of a PLA needleless electrospun layer sandwiched between two cotton fabric layers.

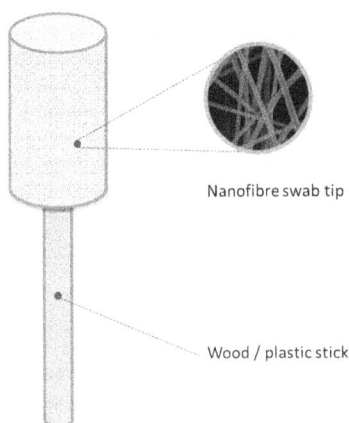

Nanofibre swab tip

Wood / plastic stick

Figure 18.6: Structure of 3D gelatine nanofibrous COVID-19 swabs.

18.7 The role of durable nanofinishes: making breathable and washable fabric-based PPE coverall fabric

Richard Feynman first introduced the idea of nanotechnology in 1959 during a speech entitled "There's Plenty of Room at the Bottom" but it was 15 years later when the Japanese scientist Norio Taniguchi coined the word "nanotechnology" [42]. In a textile context, nanofinishing offers a wealth of opportunities including introducing properties to textiles which conventional finishing cannot whilst not effecting the handle of the fabric or breathability, whilst also using a smaller amount of active material relative to that used in the equivalent bulk finishing. Within healthcare and the COVID-19 pandemic finishes around anti-viral/anti-bacterial, anti-stain, and self-cleaning have particular pertinence. Such properties can be imparted using a range of techniques such as chemical vapour deposition, electrospraying, nanocoating, and sol–gel deposition. In general, such nano-treatments can be classified in one of three ways: nanolayers, nanoroughness, or nanostructures.

Nanoparticles of metals such as copper, gold, silver, titanium, and zinc all exhibit anti-viral activity [43, 44], even at low dosing levels, but clearly cost can be prohibitive depending on the element used. There is the potential for the creation of a variety of morphologies at the nanoscale, be they nanorods, nanocubes, nanospheres, or indeed a host of other shapes. In each case they will have differing surface areas and potentially surface area to volume ratios leading to differing chemical and physical properties. Care also needs to be taken where there is a need to launder garments to ensure that leaching of metals into waste water is not an issue [45]. In addition to metal particles, metal oxides such as TiO_2, ZnO, MgO, Mn_3O_4, copper oxides, and iron oxides can also be used to antimicrobial effect [46–49]. Ibrahim [47] utilized silver and zinc oxide nanoparticles, applying them to wool/cotton and wool/viscose blends via a pad, dry microwave fixation process, imparting anti-bacterial, UV protection, and anti-crease properties to the material which showed good resistance to washing [47].

In situ simultaneous formation and deposition of nanoparticles lead to a time-saving as well as reduction in energy and chemical consumption whilst still offering excellent durability. However extreme pHs that can be required during in situ synthesis can lead to fibre damage and control of nanoparticle shape can be difficult during this method. Radetić and Marković [49] described the formation of copper oxide nanoparticles on a cellulosic material which was first oxidized and then coated with poly-carboxylic acids. These negatively charged groups allowed the sorption of positive Cu^{2+} ions, which were then reduced in situ under alkaline conditions to form copper-based nanoparticles. Upon drying and exposure to air these were further oxidized to form cupric and cuprous oxide nanoparticles which exhibited anti-microbial activity.

Control of surface roughness by the introduction of nanoroughness will also influence the surface energy and thus adhesion and wettability of the material surface. Plasma treatment, either environmental or low pressure, is a well-known method of introducing surface roughness to textiles whilst not affecting the bulk properties of the material and can be carried out in a dry, environmentally sympathetic manner [50]. Plasma technology can also be used for surface activation, increasing the interaction between nanoparticles and fibres and thus reducing the leaching potential during washing [51].

The application of nanocoatings, which are typically less than 100 nm in thickness, generally does not affect the breathability and handle of the underlying fabric [52]. Fluorocarbon finishes provide durable water and oil repellence but have received much attention due to bioaccumulation within nature particularly with C_8-based polymers where perfluorooctanoic acid is a known pollutant [53]. Alternative fluoropolymer synthetic routes have thus been developed and a general trend towards shorter length C_4 or C_6 fluorocarbons is been observed. Work by Jongprateep et al. [54] applied a uniform aqueous coating of TiO_2 nanoparticles via immersion to a polycotton (65:35 polyester:cotton), with four successive immersions to build up the coating. A fluoropolymer-containing solution was then applied to the material several times, with air-drying between applications. The resultant textile showed good water repellency, maintaining a contact angle of around 130° after 10 washes [54].

Often science tries to imitate nature and this is seen in the case of water repellence; lotus plants exhibit super hydrophobic surfaces with contact angles of water droplets greater than 150°. Such droplets will also roll off the surface of the leaf at very shallow angles (<10°), removing dust from the surface as they go, thus keeping the leaf surface clean. There are a number or companies involved in this area including Nano-Tex, who incorporate nano-whiskers into fabric producing lightweight water, and stain, repellent material which retains its breathability [55]. Work by Chen et al. [56] produced a biomimetic structure via electrospinning of polyvinylidene fluoride with ZnO nanoparticles followed by hydrothermal synthesis of ZnO nanowire arrays which were then coated with oleic acid. The resultant membranes were superhydrophobic, breathable, and self-cleaning which could make them suitable for wearable textiles.

18.8 Coating/lamination options for making breathable and washable fabric-based medical gowns or PPE coverall fabric

Within a medical context and particularly when in the midst of a pandemic, health-care staff need to wear additional PPE for extended periods of time, often whilst doing physically demanding tasks, for example, cardiopulmonary resuscitation. Thus, it is important to be able to maintain a level of protection whilst delivering wearer comfort in order to ensure PPE compliance, often whilst working within cost constraints. Behind gloves, gowns are the second most used piece of PPE in a healthcare setting [57]. Modern healthcare relies on petrochemical-based polymers such as polyester, polyethylene, polypropylene, and nylon, in addition to natural fibres, such as cotton, for coverall fabrics. Such materials offer differing levels of protection from infectious agents, oil and/or water-repellent properties, and breathability depending on their nature of construction. In general, the greater the overall weight of the material, the greater the level of protection and material strength but the lower the level of breathability. Protective gowns are either destined to be single use and thus disposable or reusable and thus subject to laundering. Reusable gown fabrics are often woven or knitted, whereas in comparison non-woven fabrics are typically utilized for single-use items. The properties of the gown are often intrinsically linked to the physical and chemical properties of the fibres from which they are constructed.

18.8.1 Coated washable/reusable woven fabrics

The passage of a microorganism through a fabric depends on four key factors: (i) the size, shape, and surface characteristics of that organism; (ii) the properties of the carrier for that organism; (iii) the physical and chemical nature of the fibres forming the fabric; and (iv) type of chemical finish, coating, and film (for lamination) used. As COVID-19 is an airborne virus, the organism involved is around 0.3 µm in diameter. It is transmitted via droplets and small particles; thus the ability of a fabric to repel such water-based droplets or to isolate them within the structure via wicking is extremely important.

Reusable woven fabrics which can be laundered are typically formed from cotton, polyester, or a blend of the two fibres. A tightly woven plain weave is utilized and may incorporate a chemical finish onto the surface to further enhance the barrier to liquid penetration [57]. Maqsood et al. [58] investigated the effect of weave structure on the barrier effectiveness of surgical gowns and were able to conclude that a more compact structure achieved by using higher thread densities and a shorter weave float enhanced barrier properties but resulted in lower levels of air permeability. An alternatively reusable fabric construction method involves the sandwiching of a membrane layer between knitted and woven layers of polyester. The application of coatings to the woven fabric results in the reduction of the fabric's surface energy and thus enhances the repellent properties. Coatings such as acrylics, polyurethane or polytetrafluoroethylene enhance the water-repellent properties of the fabric [59–61]. In combination of fluorocarbon-based water repellent (WR) finish and polyurethane/acrylic-based coating, the fabric become waterproof, which can resist even penetration of any pathogens through blood. The performance coated fabric may become wash-durable by optimizing the WR finish recipe and coating add-on. By choosing a suitable breathable grade of polymer for coating, this coated fabric can also become breathable, improving its comfort for prolonged use. Moreover, before the coating, the fabric can be treated with different antimicrobial/antiviral chemicals/nanoparticles (in combination of WR finish or separately) to generate antimicrobial/antiviral property on the fabric surface.

Fluorocarbon finishes offer a reliable method to create a water/oil-repellent surface, but perfluorinated alkyl substances have received much negative press in recent years leading to enhanced environmental regulations, which has seen industry move away from C_8 chain fluorocarbons in favour of C_6 [62]. Alternatives to fluorocarbons are receiving considerable attention such as the utilization of siloxanes [63] and the work by Mullangi et al. [64] to develop organic frameworks. Where garments are suitable for repeated use and are thus laundered, a form of tracking system must be in place to monitor usage, this could be in the form of a bar code or radio frequency chip as a result when items reach their usage limit they can be withdrawn from circulation.

18.8.2 Film-laminated non-woven fabrics

The most common surgical gowns in use consist of a spun-bonded/melt-blown/spun-bonded (SMS) non-woven composite formed from polypropylene as shown in Figure 18.7. The outer layers provide strength to the composite, whilst the melt-blown core layer acts as the filter layer composed of much finer fibres than the outer layers typically around 1–4 μm. In order to enhance the filtration efficiency or to enhance barrier protection, additional melt-blown layers can be incorporated to form SMMS or SMMMS composites [65].

The properties of such non-woven composites can be further enhanced by the addition of fluorochemical or anti-microbial finishes. Spun-bonded polypropylene can be

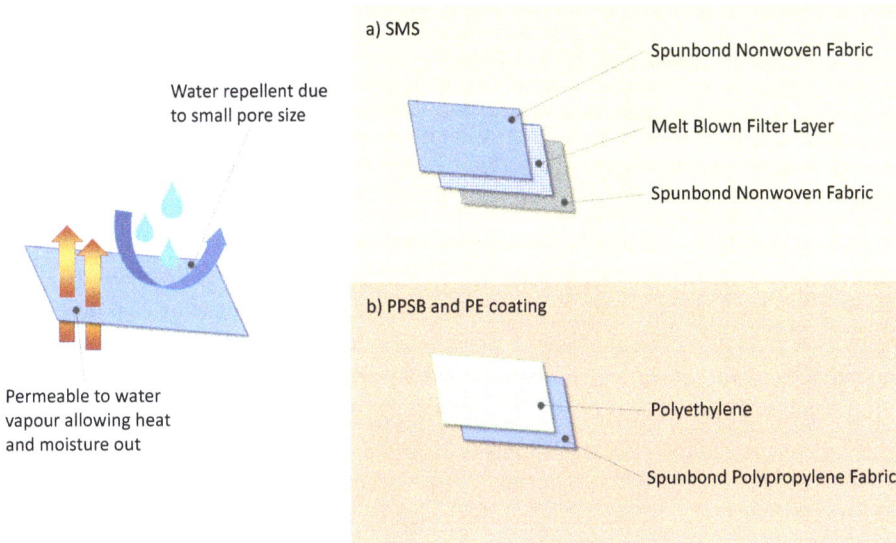

Figure 18.7: Typical features of multi-layer PPE structures and individual layers of (a) spun-bond melt-blown spun-bond (SMS) and (b) spun-bond polypropylene (PPSB) with polyethylene (PE) lamination.

laminated with a layer of polyethylene [66]. Due to its impervious nature, such material provides excellent protection from biological hazards making it an ideal material for surgical or isolation gowns but does not allow the passage of moisture through the film unlike microporous polyethylene film.

Careful control of the pore size within the film allows the passage of water vapour (around 40×10^{-6} µm in size) through the film transporting both heat and moisture from the wearer [67]. This allows the wearer to experience a degree of comfort whilst simultaneously preventing the passage of water droplets, the smallest of which are approximately 10 µm in diameter, through the material in the reverse direction, which in turn could allow the passage of pathogens. The key challenge of wearer comfort has been addressed directly by healthcare professionals during the pandemic. One such innovation developed by an Indian navy surgeon to cope with the hot and humid conditions is known as NavRakshak PPE and comfortably exceeded the minimum 3/6 required by the Indian government in the synthetic blood penetration test as outlined in ISO 16603 [68].

Such non-woven garments are single use and delivered in a sterile state having a positive impact on infection control. In addition, these garments can be removed post-procedure or when contaminated and disposed of as clinical waste. However, this approach offers environmental concerns particularly during times of pandemic when PPE usage is heightened [69].

18.9 Additional textiles/wearable technologies to combat COVID-19

In due course we will see a new generation of PPE that not only protects but also detects providing environmental and/or user feedback to the wearer. COVID-19 has exhibited the greatest mortality rate on those with underlying health conditions and the desire to monitor these patients remotely without the need to take them into a clinical setting, so-called telemedicine, is likely to increase. The proliferation of wearable non-textile products such as wristbands and smartwatches already provide real-time data on the user's activity. As the field of fibretronics matures, smart or e-textiles which include sensors for temperature or breathing rate will start to be realized, key parameters relatable to the current COVID-19 pandemic [70]. Substantial interest is been garnered in wearable biosensors which have the potential to provide continuous real-time monitoring via non-invasive monitoring of appropriate biomarkers in bodily fluids such as sweat or saliva [71]. Key to realizing this potential will be sensor miniaturization, accurate measurement against a background of potential contaminants and correlation of analytes in fluids such as sweat compared to concentrations in the wearer's blood.

Whilst the incorporation of contact tracing capabilities within textiles may be possible in due course, highlighting close contacts during the course of the day, it remains to be seen whether that is the most appropriate direction of travel, given that so many of the population, particularly from a European point of view, have a smartphone, and the technology has been shown to already work in that area, with some reservations around individual's privacy [72].

18.10 COVID-19 relevant testing/evaluation of textiles

Such was the demand for PPE at the height of the pandemic that some standards were relaxed or additional COVID-19-specific guidance was released around certain items of PPE to relieve issues in certain countries. This is exemplified by the USA, who made use of a range of emergency use authorizations primarily for surgical masks but also for respirators, face shields, and gowns during the initial phase of the pandemic [73]. As supply pressures have reduced due to increased production, some of these relaxations have been removed. Table 18.3 shows the current testing required for a surgical gown in Europe with two different performance regimes, with US standards shown in Table 18.4. It should be noted that in the US level 3 and 4 gowns are considered to be a Class II medical device and thus require approval by the FDA.

EN ISO 22612 is concerned with the penetration of bacteria on particles in the size range of typical human skin scales or indeed organic/inorganic particulates, which

Table 18.3: Selected characteristics to be evaluated and standard performance parameters for surgical gowns within a European context [74].

Characteristic	Test method	Requirement	
		Standard performance	
		Critical product area	**Less critical product area**
Microbial penetration – dry	EN ISO 22612	N/A	≤300 CFU
Microbial penetration – wet	EN ISO 22610	$\geq 2.8\ I_B$	N/A
Cleanliness microbial/bioburden	EN ISO 11737-1	≤300 CFU/100 cm^2	
Liquid penetration	EN ISO 811	≥20 cm H$_2$O	≥10 cm H$_2$O
Bursting strength – dry	EN ISO 13938-1	≥40 kPa	
Tensile strength – dry	EN 29073-3	≥15 N	

Table 18.4: PPE protection levels and standards in the United States [65].

Protection level	Relevant standard	Impact penetration	Hydrostatic pressure	Applications
1. Minimal risk	ANSI/AAMI PB 70:12 AATCC 42	(water) ≤ 4.5 g	N/A	Basic care, standard hospital medical unit. Not suitable for use in operating theatres.
2. Low risk	ANSI/AAMI PB 70:12 AATCC 42 and 127	(spray) ≤ 1.0 g	≥20 cm	Suitable for minimally invasive surgical procedures including those that present a slight risk of fluid exposure such as blood drawn from vein, suturing, intensive care unit, and pathology lab.
3. Medium risk	ANSI/AAMI PB 70:12 AATCC 42 & 127	(spray) ≤ 1.0 g	≥50 cm	Supplied in sterilized packaging, suitable for arterial blood draw, inserting an IV, emergency room, trauma
4. High risk	ANSI/AAMI PB 70:12 ASTM F1670 and F1671 Phi-X174	Synthetic blood used (for surgical drapes)	Viral Penetration Test (for surgical and isolation gowns): no penetration at 2 psi	Suitable for procedures involving a high risk of fluid exposure and viral penetration up to 60 min. Suitable for infectious disease (non-airborne) and large amounts of fluid exposure over long periods.

may penetrate a barrier material in the dry state. To determine penetration, talc contaminated with *Bacillus subtilis* is placed on the test piece suspended above a sedimentation plate [75]. The apparatus is then vibrated and the talc that penetrates captured in the sedimentation plate which is then incubated at 35 °C for 24 h and the number of colonies produced determined.

Of particular note in the case of COVID-19 given its route of transmission is penetration through protective garments, for example, gowns or coveralls when wet, EN ISO 22610 deals with microbial penetration carried by a liquid [76]. The test specimen is placed on an agar plate, onto which the bacteria-containing donor material is placed and then covered with a sheet of HDPE. The application of two steel rings holds the three sheets together whilst an abrasion-resistant finger moves over the surface during a period of 15 min. The bacterial plates are incubated, allowing the barrier capability and penetration over time to be calculated.

In terms of waterproofness for gowns a hydrostatic pressure test is conducted according to EN ISO 811, with a 100 cm^2 area of the test fabric been clamped into the tester and subjected to an increasing hydrostatic pressure at a rate of 10 cm H$_2$O/min $\pm$ 0.5 cm H$_2$O/min until the point of penetration [77]. For high performance the fabric must withstand 1,000 mm/H$_2$O in critical areas, it is up to the manufacturer to identify critical areas; otherwise all areas are required to meet the critical specification.

Bursting strength is calculated by securing the test fabric above a highly expansive diaphragm and increasing the fluid pressure beneath the diaphragm causing distension until the point of bursting. When calculating bursting strength according to EN ISO 13938-1 the diaphragm pressure must be subtracted from the mean bursting pressure [78]. The diaphragm pressure is the pressure required to distend the diaphragm to the mean bursting distension without a test specimen in place. In this case a bursting strength of 40 kPa is required to pass the standard in all critically product areas both under wet and dry conditions.

The penetration of bodily fluids through protective equipment is a potential transmission pathway for some diseases, for example, hepatitis and acquired immune deficiency syndrome can be transmitted via blood. In its current form, COVID-19 is an airborne transmitter, spread by aerosols, or droplets. However, protective equipment must resist a variety of transmission pathways particularly in a medical environment. ISO 22609:2004 is concerned with the penetration of synthetic blood through a medical face mask and is designed to mimic the puncture of a small blood vessel under typical physiological blood pressures [79]. In the test procedure, the mask is secured on a specimen holding fixture, 300 mm from the tip of the canula that will dispense the synthetic blood, in a horizontal arrangement. A plate with a 5 mm aperture is placed in front of the mask, 10 mm from the mask surface, and this prevents the initial high velocity blood from impacting the mask. Following conditioning of the mask for at least 4 h at (85 ± 5)% relative humidity at (21 ± 5) °C to simulate the conditions created by the wearer when worn. At three different test pressures (10.6, 16.0, and 21.3 kPa), 2 mL of synthetic blood is dispensed from a canula with internal diameter of 0.84 mm, giving velocities of 450, 550, and 635 m/s, respectively. The outcome of the test is either pass/fail, depending if the blood has penetrated, the experimental set-up is shown in Figure 18.8. A similar experimental set-up can be utilized to test fabric to be used in coveralls and similar protective garments as shown in Figure 18.9.

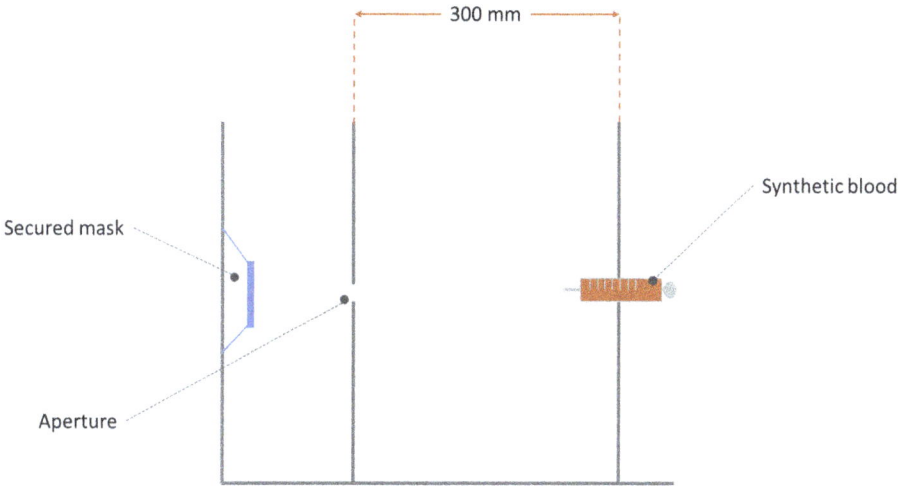

Figure 18.8: Blood penetration testing set-up for masks in accordance with ISO 22609:2004.

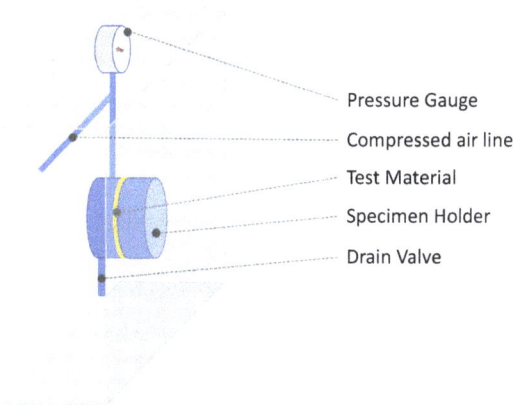

Figure 18.9: Experimental set-up for blood penetration of protective garments.

Bacterial filtration efficiency (BFE) for those surgical masks covered by BS EN 14683 or ASTM F2100 can be calculated by using *S. aureus* [80, 81]. The mask specimen under test is clamped into an aerosol chamber with an aerosol of *S. aureus* drawn through the mask under vacuum. BFE is calculated as a percentage based on the number of colony forming units (CFUs) that pass through the mask compared to the number present in the challenge aerosol.

Table 18.5: Penetration of filter material as defined in BS EN 149-2001.

Classification	Maximum penetration of test aerosol	
	Sodium chloride test 95 L/min (% max)	Paraffin oil test 95 L/min (% max)
FFP1	20	20
FFP2	6	6
FFP3	1	1

Filtration efficiency for respirator style masks can be tested using the sodium chloride and paraffin oil penetration test as defined in BS EN 13274-7:2019. The maximum penetration of test aerosol for each mask classification is shown in Table 18.5. In terms of the sodium chloride test, an aerosol of sodium chloride particles is generated by atomizing an aqueous solution of the salt such that the median particle size distribution is between a diameter of 0.06 and 0.10 μm. The concentration of the salt is measured both pre and post-filter by way of flame photometry at 589 nm, the characteristic emission for sodium. A similar procedure can be used for atomized paraffin oil with detection achieved via a light scattering aerosol photometer.

18.11 A sustainable future for PPE

As highlighted earlier in this chapter, the majority of PPE is single use and derived from petrochemicals. In addition, the contaminated nature of the waste after use adds an extra dimension of complexity when it comes to recycling or re-use particularly in the early days of the pandemic when society is dealing with a new unknown pathogen. The persistence of waste plastic within the environment is a hot topic currently in the Western world and the disruptive impact that microplastics can have on ecosystems is only beginning to be understood [82]. What is clear is that single-use masks are a new menace within the urban-litter landscape, finding their way into streams and rivers and in due course our oceans [83]. During 2021, French company Plaxtill have taken 30,000 masks collected via recycling bins in local shops from Locminé. After exposing the shredded masks to ultraviolet light designed to destroy any virus (or bacterial) contamination, the masks are melted down and reformed into school geometry sets consisting of a ruler, square, and protractor and distributed to school children [84]. However at the peak of the pandemic over 30,000 masks were used globally every second, thus whilst this proves a concept, solutions need to be on a much larger scale. This is an approach that has been used in other sectors of the textile industry such as Adidas' Futurecraft Loop shoes, where end of life and disposal is as important as the item's primary function [85].

During the pandemic in the face of limited supplies, people began to look for ways to safely reuse PPE supplies including N95 masks, which under normal circumstances have a limited usage lifetime [86, 87]. Indeed Juang and Tsai [88], the latter who invented and patented the N95 filter material were quick to publish a simple method for N95 mask sterilization which retained 92.4–98.5% of the filtering efficiency. Their methods recommended either using four masks on a 4-day rotation based on a 3-day survival time for the COVID-19 virus – this resulted in no change in the mask's properties. Alternatively heating the mask to 70 °C in an oven on a wooden dowel for 1 h retained 98.5% of the filtering efficiency. Whilst boiling for 5 min led to insignificant charge loss but saw filtering efficiency dip to 92.4%.

18.12 Conclusion

It should be noted that whilst COVID-19 has been the most severe and challenging new disease encountered in living memory, it is not the first time that novel coronaviruses have disrupted lives in recent years and posed a global public health issue. The severe acute respiratory syndrome coronavirus between 2002 and 2004 infected over 8,000 people of which around 9% died [89]. Middle East respiratory syndrome coronavirus (MERS-CoV) followed in 2012, whilst the number of cases was smaller (2,500), the mortality rate was much greater, in the region of 35%. Additionally, the fact that R_0 is close to 1 for MERS has played an important role in its control compared to 4 for SARS and 2.5–3 for the original variant of COVID-19 [90]. As a society and a textile industry we must be careful that all lessons are learned from the past, but that they do not cloud our judgement of the future as we begin to move into a new phase of living with COVID-19 and responding to it more like seasonal influenza. The next pandemic to affect society may not be coronavirus-based, thus whilst there are textile innovations that can be made, perhaps there is as much to be gained from appropriate usage, stockpile requirements, and industry's obsession with just-in-time manufacturing.

The role of PPE in Western society is complex, and it is interesting that having moved from reusable PPE items such as cotton masks in the early nineteenth century, societal pressures find industry and academia once more searching for PPE that aligns with one of the 3Rs: reduce, reuse, and recycle. Moving forward, it is evident that healthcare systems' resilience levels must be raised before the world faces another pandemic of this magnitude, and textiles can play a key role in this effort.

References

[1]   Coccia M. The impact of first and second wave of the COVID-19 pandemic in society: comparative analysis to support control measures to cope with negative effects of future infectious diseases. Environmental Research. 2021;197:111099.

[2]   Jarrom D, Elston L, Washington J, Prettyjohns M, Cann K, Myles S, et al. Effectiveness of tests to detect the presence of SARS-CoV-2 virus, and antibodies to SARS-CoV-2, to inform COVID-19 diagnosis: a rapid systematic review. 2022;27:33–45.

[3]   Huang Y, Yang C, Xu X-f, Xu W, Liu S-w. Structural and functional properties of SARS-CoV-2 spike protein: potential antivirus drug development for COVID-19. Acta Pharmacologica Sinica. 2020;41:1141–9.

[4]   Jayaweera M, Perera H, Gunawardana B, Manatunge J. Transmission of COVID-19 virus by droplets and aerosols: A critical review on the unresolved dichotomy. Environmental research. 2020;188:109819–.

[5]   Ijaz MK, Nims RW, de Szalay S, Rubino JR. Soap, water, and severe acute respiratory syndrome coronavirus 2 (SARS-CoV-2): an ancient handwashing strategy for preventing dissemination of a novel virus. PeerJ. 2021;9:e12041–e.

[6]   Claeson M, Hanson S. COVID-19 and the Swedish enigma. The Lancet. 2021;397:259–61.

[7]   Blair A, de Pasquale M, Gabeff V, Rufi M, Flahault A. The End of the Elimination Strategy: Decisive Factors towards Sustainable Management of COVID-19 in New Zealand. 2022;3:135–47.

[8]   Lu G, Razum O, Jahn A, Zhang Y, Sutton B, Sridhar D, et al. COVID-19 in Germany and China: mitigation versus elimination strategy. Global Health Action. 2021;14:1875601.

[9]   WHO. Q&A on coronaviruses (COVID-19). 2020.

[10] Prata JC, Silva ALP, Walker TR, Duarte AC, Rocha-Santos T. COVID-19 Pandemic Repercussions on the Use and Management of Plastics. Environmental Science & Technology. 2020;54:7760–5.

[11] Lisa M. Casanova, William A. Rutala, David J. Weber, Mark D. Sobsey. Effect of single- versus double-gloving on virus transfer to health care workers' skin and clothing during removal of personal protective equipment. American Journal of Infection Control. 2012;40:369–74.

[12] Graham LD. The right to clothing and personal protective equipment in the context of COVID-19. The International Journal of Human Rights. 2022;26:30–49.

[13] Lagu T, Artenstein AW, Werner RM. Fool Me Twice: The Role for Hospitals and Health Systems in Fixing the Broken PPE Supply Chain. J Hosp Med. 2020;15:570–1.

[14] Sharif R, Mohsin M, Ramzan N, Sardar S, Anam W. Synthesis of Bio-Based Non-Fluorinated Oil and Water Repellent Finishes for Cotton Fabric by Using Palmitic Acid, Succinic Acid, and Maleic Acid. Journal of Natural Fibers. 2022:1–12.

[15] Government U. Infection prevention and control for seasonal respiratory infections in health and care settings (including SARS-CoV-2) for winter 2021 to 2022. In: Agency UHS, editor. 15 March 2022 ed2022.

[16] Xu Q, Mao Z, Wei D, Fan K, Liu P, Wang J, et al. Association between mask wearing and anxiety symptoms during the outbreak of COVID 19: A large survey among 386,432 junior and senior high school students in China. Journal of Psychosomatic Research. 2022;153:110709.

[17] BSI Flex 5555 Community face coverings V2.1. London: British Standards Institution; 2021.

[18] Strasser BJ, Schlich T. A history of the medical mask and the rise of throwaway culture. Lancet. 2020;396:19–20.

[19] Quesnel LB. The efficiency of surgical masks of varying design and composition. 1975;62:936–40.

[20] Yim W, Cheng D, Patel SH, Kou R, Meng YS, Jokerst JV. KN95 and N95 Respirators Retain Filtration Efficiency despite a Loss of Dipole Charge during Decontamination. ACS Applied Materials & Interfaces. 2020;12:54473–80.

[21] Mark Ferris, Rebecca Ferris, Workman C. FFP3 respirators protect healthcare workers against infection with SARS-CoV-2. Authorea. 2021.

[22] Anedda J, Ferreli C, Rongioletti F, Atzori L. Changing gears: Medical gloves in the era of coronavirus disease 2019 pandemic. Clin Dermatol. 2020;38:734–6.

[23] Yip E, Cacioli P. The manufacture of gloves from natural rubber latex. Journal of Allergy and Clinical Immunology. 2002;110:S3–S14.

[24] Rice P. Industrial Hygiene for the Construction Industry. ASSE Professional Development Conference and Exposition 2013. p. ASSE-13-702.

[25] Mussap CJ. The Plague Doctor of Venice. Internal Medicine Journal. 2019;49:671–6.

[26] Bernard HR, Cole WR, Gravens DL. Reduction of iatrogenic bacterial contamination in operating rooms. Ann Surg. 1967;165:609–13.

[27] Verbeek JH, Rajamaki B, Ijaz S, Sauni R, Toomey E, Blackwood B, et al. Personal protective equipment for preventing highly infectious diseases due to exposure to contaminated body fluids in healthcare staff. Cochrane Database of Systematic Reviews. 2020.

[28] Owen L, Shivkumar M, Cross RBM, Laird K. Porous surfaces: stability and recovery of coronaviruses. 2022;12:20210039.

[29] Boys CV. On the Production, Properties, and some suggested Uses of the Finest Threads. Proceedings of the Physical Society of London. 1887;9:8.

[30] Andersson RL, Ström V, Gedde UW, Mallon PE, Hedenqvist MS, Olsson RT. Micromechanics of ultra-toughened electrospun PMMA/PEO fibres as revealed by in-situ tensile testing in an electron microscope. Scientific Reports. 2014;4:6335.

[31] Wang C, Wu S, Jian M, Xie J, Xu L, Yang X, et al. Silk nanofibers as high efficient and lightweight air filter. Nano Research. 2016;9:2590–7.

[32] de Almeida DS, Martins LD, Muniz EC, Rudke AP, Squizzato R, Beal A, et al. Biodegradable CA/CPB electrospun nanofibers for efficient retention of airborne nanoparticles. Process Safety and Environmental Protection. 2020;144:177–85.

[33] Du Z, Cheng J, Huang Q, Chen M, Xiao C. Electrospinning organic solvent resistant preoxidized poly (acrylonitrile) nanofiber membrane and its properties. Chinese Journal of Chemical Engineering. 2022.

[34] Heikkila P, Harlin A. Electrospinning of polyacrylonitrile (PAN) solution: Effect of conductive additive and filler on the process. Express Polym Lett. 2009;3:437–45.

[35] Bortolassi ACC, Nagarajan S, de Araújo Lima B, Guerra VG, Aguiar ML, Huon V, et al. Efficient nanoparticles removal and bactericidal action of electrospun nanofibers membranes for air filtration. Materials Science and Engineering: C. 2019;102:718–29.

[36] He H, Gao M, Illés B, Molnar K. 3D Printed and Electrospun, Transparent, Hierarchical Polylactic Acid Mask Nanoporous Filter. Int J Bioprint. 2020;6:278–.

[37] Zhou Y, Liu Y, Zhang M, Feng Z, Yu D-G, Wang K. Electrospun Nanofiber Membranes for Air Filtration: A Review. 2022;12:1077.

[38] McCarthy A, Saldana L, Ackerman DN, Su Y, John JV, Chen S, et al. Ultra-absorptive Nanofiber Swabs for Improved Collection and Test Sensitivity of SARS-CoV-2 and other Biological Specimens. Nano Letters. 2021;21:1508–16.

[39] Luo Y, Nartker S, Miller H, Hochhalter D, Wiederoder M, Wiederoder S, et al. Surface functionalization of electrospun nanofibers for detecting *E. coli* O157: H7and BVDV cells in a direct-charge transfer biosensor. Biosensors and Bioelectronics. 2010;26:1612–7.

[40] Quirós J, Amaral AJR, Pasparakis G, Williams GR, Rosal R. Electrospun boronic acid-containing polymer membranes as fluorescent sensors for bacteria detection. Reactive and Functional Polymers. 2017;121:23–31.

[41] Patil NA, Gore PM, Jaya Prakash N, Govindaraj P, Yadav R, Verma V, et al. Needleless electrospun phytochemicals encapsulated nanofibre based 3-ply biodegradable mask for combating COVID-19 pandemic. Chemical engineering journal (Lausanne, Switzerland : 1996). 2021;416:129152.

[42] Taniguchi N. On the Basic concept of Nanotechnology. Proceeding of the ICPE. 1974.

[43] Galdiero S, Falanga A, Vitiello M, Cantisani M, Marra V, Galdiero M. Silver Nanoparticles as Potential Antiviral Agents. 2011;16:8894–918.
[44] Ibrahim Fouad G. A proposed insight into the anti-viral potential of metallic nanoparticles against novel coronavirus disease-19 (COVID-19). Bulletin of the National Research Centre. 2021;45:36.
[45] Nawaz T, Sengupta S. Silver Recovery from Laundry Washwater: The Role of Detergent Chemistry. ACS Sustainable Chemistry & Engineering. 2018;6:600–8.
[46] Abbas M, Iftikhar H, Malik MH, Nazir A. Surface Coatings of TiO2 Nanoparticles onto the Designed Fabrics for Enhanced Self-Cleaning Properties. 2018;8:35.
[47] Ibrahim NA, Emam E-AM, Eid BM, Tawfik TM. An Eco-Friendly Multifunctional Nano-Finishing of Cellulose/Wool Blends. Fibers and Polymers. 2018;19:797–804.
[48] Faisal S, Naqvi S, Ali M, Lin L. Comparative study of multifunctional properties of synthesised ZnO and MgO NPs for textiles applications. Pigment & Resin Technology. 2022;51:301–8.
[49] Radetić M, Marković D. Nano-finishing of cellulose textile materials with copper and copper oxide nanoparticles. Cellulose. 2019;26:8971–91.
[50] Zille A. 6 – Plasma technology in fashion and textiles. In: Nayak R, editor. Sustainable Technologies for Fashion and Textiles: Woodhead Publishing; 2020. p. 117–42.
[51] Radetić M, Marković D. A review on the role of plasma technology in the nano-finishing of textile materials with metal and metal oxide nanoparticles. Plasma Processes and Polymers. 2022;19:2100197.
[52] Temesgen AG, Turşucular ÖF, Eren R, Ulcay YJIJAMR. Novel applications of nanotechnology in modification of textile fabrics properties and apparel. 2018;5:49–58.
[53] Kredel J, Gallei M. Ozone-Degradable Fluoropolymers on Textile Surfaces for Water and Oil Repellency. ACS Applied Polymer Materials. 2020;2:2867–79.
[54] Jongprateep O, Mani-lata C, Sakunrak Y, Audcharuk K, Narapong T, Janbooranapinij K, et al. Titanium dioxide and fluoropolymer-based coating for smart fabrics with antimicrobial and water-repellent properties. RSC Advances. 2022;12:588–94.
[55] Paramsothy M. Nanotechnology in Clothing and Fabrics. 2022;12:67.
[56] Chen R, Wan Y, Wu W, Yang C, He J-H, Cheng J, et al. A lotus effect-inspired flexible and breathable membrane with hierarchical electrospinning micro/nanofibers and ZnO nanowires. Materials & Design. 2019;162:246–8.
[57] Kilinc FS. A Review of Isolation Gowns in Healthcare: Fabric and Gown Properties. Journal of Engineered Fibers and Fabrics. 2015;10:155892501501000313.
[58] Maqsood M, Nawab Y, Hamdani STA, Shaker K, Umair M, Ashraf W. Modeling the effect of weave structure and fabric thread density on the barrier effectiveness of woven surgical gowns. The Journal of The Textile Institute. 2016;107:873–8.
[59] Jassal M, Khungar A, Bajaj P, Sinha TJM. Waterproof Breathable Polymeric Coatings Based on Polyurethanes. Journal of Industrial Textiles. 2004;33:269–80.
[60] Mukhopadhyay A, Midha VK. A Review on Designing the Waterproof Breathable Fabrics Part I: Fundamental Principles and Designing Aspects of Breathable Fabrics. 2008;37:225–62.
[61] Ghezal I, Jaouachi B, Sakli F. Investigating the Performances of a Coated Plain Weave Fabric Designed for Producing Protective Gowns. Advances in Materials Science and Engineering. 2021;2021:4260411.
[62] Jung H, Kwon J, Jung H, Cho KM, Yu SJ, Lee SM, et al. Short-chain fluorocarbon-based polymeric coating with excellent nonwetting ability against chemical warfare agents. RSC Advances. 2022;12:7773–9.
[63] Rutkevičius M, Pirzada T, Geiger M, Khan SA. Creating superhydrophobic, abrasion-resistant and breathable coatings from water-borne polydimethylsiloxane-polyurethane Co-polymer and fumed silica. Journal of Colloid and Interface Science. 2021;596:479–92.

[64]   Mullangi D, Shalini S, Nandi S, Choksi B, Vaidhyanathan R. Super-hydrophobic covalent organic
       frameworks for chemical resistant coatings and hydrophobic paper and textile composites. Journal
       of Materials Chemistry A. 2017;5:8376–84.
[65]   Karim N, Afroj S, Lloyd K, Oaten LC, Andreeva DV, Carr C, et al. Sustainable Personal Protective
       Clothing for Healthcare Applications: A Review. ACS Nano. 2020;14:12313–40.
[66]   Midha VK, Joshi S, Dakuri A. Surgical gown fabrics in infection control and comfort measures at
       hospitals. Indian Journal of Fibre & Textile Research 2022;47.
[67]   Baker B, Mo Q, Lawson RP, O'Connor D, Korolev A. Drop Size Distributions and the Lack of Small
       Drops in RICO Rain Shafts. Journal of Applied Meteorology and Climatology. 2009;48:616–23.
[68]   Kumar S, Fatima U, Shah R, Gupta U. International Journal of Scientific Research & Engineering
       Trends. 2021;7.
[69]   Haque MS, Sharif S, Masnoon A, Rashid E. SARS-CoV-2 pandemic-induced PPE and single-use plastic
       waste generation scenario. Waste Management & Research. 2021;39:3–17.
[70]   Xu L, Liu Z, Zhai H, Chen X, Sun R, Lyu S, et al. Moisture-Resilient Graphene-Dyed Wool Fabric for
       Strain Sensing. ACS Applied Materials & Interfaces. 2020;12:13265–74.
[71]   Kim J, Campbell AS, de Ávila BE, Wang J. Wearable biosensors for healthcare monitoring. Nature
       biotechnology. 2019;37:389–406.
[72]   Jalabneh R, Syed HZ, Pillai S, Apu EH, Hussein MR, Kabir R, et al. Use of Mobile Phone Apps for
       Contact Tracing to Control the COVID-19 Pandemic: A Literature Review. In: Nandan Mohanty S,
       Saxena SK, Satpathy S, Chatterjee JM, editors. Applications of Artificial Intelligence in COVID-19.
       Singapore: Springer Singapore; 2021. p. 389–404.
[73]   Rowan NJ, Laffey JG. Unlocking the surge in demand for personal and protective equipment (PPE)
       and improvised face coverings arising from coronavirus disease (COVID-19) pandemic –
       Implications for efficacy, re-use and sustainable waste management. Science of The Total
       Environment. 2021;752:142259.
[74]   13795-1:2019 Surgical clothing and drapes. Requirements and test methods Surgical drapes and
       gowns. British Standards Institution; 2019.
[75]   22612:2005 Clothing for protection against infectious agents – Test method for resistance to dry
       microbial penetration. International Organization for Standardization; 2005.
[76]   22610:2018 Surgical drapes, gowns and clean air suits, used as medical devices, for patients, clinical
       staff and equipment – Test method to determine the resistance to wet bacterial penetration.
       International Organization for Standardization; 2018.
[77]   811:2018 Textiles – Determination of resistance to water penetration – Hydrostatic pressure test.
       International Organization for Standardization; 2018.
[78]   13938-1:2019 Textiles – Bursting properties of fabrics – Part 1: Hydraulic method for determination
       of bursting strength and bursting distension. International Organization for Standardization; 2019.
[79]   22609:2004 Clothing for protection against infectious agents – Medical face masks – Test method
       for resistance against penetration by synthetic blood (fixed volume, horizontally projected).
       International Organization for Standardization; 2004.
80]   14683:2019 Medical face masks. Requirements and test methods. British Standards Institution; 2019.
[81]   F2100-21 Standard Specification for Performance of Materials Used in Medical Face Masks. American
       Society for Testing and Materials; 2021.
[82]   Everaert G, Van Cauwenberghe L, De Rijcke M, Koelmans AA, Mees J, Vandegehuchte M, et al. Risk
       assessment of microplastics in the ocean: Modelling approach and first conclusions. Environmental
       Pollution. 2018;242:1930–8.
[83]   Aragaw TA. Surgical face masks as a potential source for microplastic pollution in the COVID-19
       scenario. Marine Pollution Bulletin. 2020;159:111517.
[84]   Plaxtil. Recycling Masks, https://www.plaxtil.com/recyclagemasques Accessed on 2nd April 2022.
[85]   The bigger picture: Adidas Futurecraft loop. Engineering & Technology. 2019;14:58–9.

[86] Czubryt MP, Stecy T, Popke E, Aitken R, Jabusch K, Pound R, et al. N95 mask reuse in a major urban hospital: COVID-19 response process and procedure. Journal of Hospital Infection. 2020;106:277–82.

[87] Steinberg BE, Aoyama K, McVey M, Levin D, Siddiqui A, Munshey F, et al. Efficacy and safety of decontamination for N95 respirator reuse: a systematic literature search and narrative synthesis. Canadian Journal of Anesthesia/Journal canadien d'anesthésie. 2020;67:1814–23.

[88] Juang PSC, Tsai P. N95 Respirator Cleaning and Reuse Methods Proposed by the Inventor of the N95 Mask Material. Journal of Emergency Medicine. 2020;58:817–20.

[89] da Costa VG, Moreli ML, Saivish MV. The emergence of SARS, MERS and novel SARS-2 coronaviruses in the twenty-first century. Archives of Virology. 2020;165:1517–26.

[90] Guarner J. Three Emerging Coronaviruses in Two Decades: The Story of SARS, MERS, and Now COVID-19. American Journal of Clinical Pathology. 2020;153:420–1.

Index

3D printed E-textile 35
3D printing 37, 47–49, 684–689, 691–692,
    694–695, 697–704, 706, 708–714
3D spacer fabrics 612
4D printed textile 700, 706
4D printing 683, 684, 700, 702, 706, 709, 714

absorbable sutures 560
acoustic barrier 663
acoustic frequency 663
acoustic impedance 662
acoustic noise 659
acoustic wave energy 659
activated carbon 640, 656
active coatings 98, 129
active heating and cooling 335
active safety 670
active smart coatings 166
active smart textiles 3
active thermoregulation 23
actuator 193, 446–447, 470
adaptive camouflage 405
adhesive 132, 136
adsorptive filter 657
advanced textiles 330, 332, 345, 351
aesthetic 63–64, 82, 89–90
air contaminants 637
air filter 378
air filtration 636
air intake system 656
air laid 666
air permeability 673
airbags 672
alginate fibres 544
alkaloids 556
allergens 658
amicrobial textile 2
antibacterial 124, 551–556, 576, 581
anti-biofouling coating 163
anti-fungal 600, 668
anti-inflammatory 576
antimicrobial 8, 16, 43, 69, 72, 79–82, 86–88, 98,
    125, 261, 268, 552, 555, 557, 658
antimicrobial sutures 574
anti-static 2, 19, 261
antiviral 79, 82, 725

artificial cells 129
artificial muscle 578
artificial organs 535
asymmetric membrane 596
atomic layer deposition 13, 15, 615
automotive interior 664
automotive textiles 8, 47

bacterial resistance 537
ballistic 375, 377, 379
band structure theory 269
bandages 534, 537–542, 545–546, 569,
    571–572, 581
basalt fibres 669
battery electric vehicle 654
bi-component fibres 666
binder jetting 689
bioactive surfaces 591
biochemical process 553
biocidal 659
biocompatible 565
biocomposites 674
biodegradable 674
biodegradable polymer 535, 701
biofuel cells 34
biological particulate 640
biomaterials 170, 569
biomechanical energy 34–35, 41
biomimetic coating 171
bio-mimicking 346, 350
biopolymer coatings 157
biosensor 13, 734, 740
bio-soluble 669
body armour 4, 375, 379
boron nitride nanosheets 698
braided 538–539, 577, 669, 706
braiding 664
breathable lamination 173
Brownian diffusion 641
brush coating 111, 281, 283, 316
bullet proof vest 43

cabin air filtration 636, 658
cake filtration 645
calender coating 119
Camo net 413

https://doi.org/10.1515/9783110759747-019

camouflage 26, 43, 402, 403, 407, 413, 423
capacitors 268, 315
carbon black 17–18, 222
carbon nanotubes 14–15, 17–18, 222, 236, 380, 386
carbon-based nanomaterials 13
cardiac support device 535–536
carpet backing 668
carsolchromic 25
catalytic converter 669
cathode air filtration 657
ceramic fibres 669
chain mail fabrics 702
characteristic impedance 662
chemical exfoliation 275–276, 280, 316
chemical polymerization 231
chemical synthesis 275–276, 278, 280, 316
chemical vapor deposition 13, 97, 126, 128, 158, 275–277, 316
chemical warfare 375, 377–378, 384
chitosan 535, 538–539, 556, 568–569, 580
chromic textiles 458
chronic venous ulcers 541
clogging process 646
CNTs 187, 388
coated textiles 151–152, 162, 165, 178
coating 16, 18, 20, 26, 29, 43–44, 50, 111, 113–114, 121, 123–124, 127, 152, 238, 725, 732, 736–737
collection efficiency 646
comfort 542, 562, 567, 571–572, 574, 581
compression bandages 534, 542, 571–572, 581
compression garments 533, 541, 571
computer-aided design 47, 684
computer-aided manufacturing 47
computer-assisted manufacturing 685
conducting polymer 141
conduction 660
conductive fabric 142, 565
conductive fibres 141
conductive heating 11
conductive polymers 17, 20, 29, 159, 624, 628, 630
conductive textiles 17–18, 20, 98, 111, 141
conductivity 543, 563–564, 580
contact angle 653
contact-electrification effect 620
continuous inkjet 688
corona charging 658
cotton 538–540, 546, 548, 560, 562

coverall 724–725, 732, 737
COVID-19 721–727, 729–730, 733–735, 740, 742, 744–745
crash impact 670
curtain coating 120
CVD 126–128
cytotoxicity 576–577, 580

damping 662
dashboard 668, 670
decorative lamination 176
degumming 560
depth filtration 650
dip coating 112, 281, 285, 288, 293, 312, 315–316
direct coating 106–107
direct ink writing 687
direct interception 641
directed energy deposition 692
disposable masks 536
doctor blade 106, 114
door panels 666
drop-on-demand 688
Dry heat lamination 134
DSSC 618
dust holding capacity 649
dust loading 646
dyeability 184, 188, 198
dye-sensitized solar cells 37

E. coli 305
E. faecalis 305
eco-friendly 118, 132, 134
ECWCS 43, 380–382
elastomeric yarns 542
electret filters 658
electric double-layer capacitors 40
electrical conductivity 125, 142, 221, 239, 268–270, 272–273, 286–287, 289, 298, 300, 313, 315–316, 566
electrical heating 268
electrically conductive coating 165
electrocardiogram 564
electrochemical polymerization 231
electrochromic materials 29
electrochromic textiles 361
electroconductive textiles 17, 222, 234
electrodeposition 15, 20
electroless method 165
electroless plating 20, 103

electromagnetic interference 126, 165, 253
electromagnetic shielding 12
electronic sutures 578
electronic textiles 17
electron-transfer mechanism 620
electrospinning 175, 238, 577, 652, 733
electrospraying 128
electrospun 733–734
electrospun nanofibre 128, 175, 451
electrostatic attraction 641
electrosynthesis 231
electrothermal conversion 335
EMI shielding 126, 140–141, 240, 268, 311–313,
    453–454
emissivity 333, 336–337, 341
emotional wounds 568
energy consumption 329, 337
energy harvesting 622, 698, 704
energy storage 268, 312, 314, 609
energy-harvesting textiles 8, 42, 50
engraved roll 135
enzyme treatment 104
epitaxial growth 275–276, 279, 316
E-textile 4, 17, 22, 50, 268, 285, 300, 311, 376
ethylene-vinyl acetate 172
exfoliated nanocomposites 185
exhaust emission 669
exhaust gas filtration 636
extended cold weather clothing system 380,
    398, 418
extreme environment 330–331, 350

fabric webbings 671
face mask 723, 742
faradaic capacitance 42
felt 669
fibrous material 635
filter 383
filtration 535, 550
filtration efficiency 640
filtration velocity 651
finish coat 121
fire retardant 8
Flame lamination 132
flame retardancy 98, 125, 375, 377, 385–386, 388
flame-retardant 73, 75, 78, 88, 623
flame-retardant textile 2
flammability 666
flexible electronics 704

flow resistivity 663
fluoropolymer 153, 736
fluoropolymer coating 153
foam coating 118
fuel-cell electric vehicle 654
functional clothing 8, 44
functional textiles 1–2, 8, 11, 14, 17, 20, 24, 49–50,
    375–377
fused deposition modelling 47, 687
fused filament fabrication 687

gas barrier 184, 188, 206–207, 209
gas sensor 251
gas venting 654
gaseous contaminants 648
Gauge factor 306
glass fibre 669
glide ratio 438–439
glycosaminoglycan 556
gown 724, 726, 732, 740
graft copolymer 100
grafting 100–101
gram-negative bacteria 553
graphene 17–19, 24, 40, 234, 288–289, 293,
    296–297, 300, 312, 316, 380, 385–386, 388–389
graphene oxide 19
graphene quantum dot 19, 474
graphene woven fabrics 310
graphite 17–18, 24
gravure coating 114

HAPO chamber 430, 432
HDPE 14, 135, 725
headliners 667
healthcare device 562
heat dissipation 330, 337, 343, 345
heat transfer 330, 332–333, 343, 345–346, 659
hexagonal boron nitride 207
high performance fibre 379, 399
hollow polypropylene 535
hospital bed linen 536
hot melt laminating adhesive 139
Hot melt lamination 133
hybrid capacitors 40
hybrid generators 34
hydroentangled nonwoven 549
hydrogel 568–569, 571
hygrothermal resistance 173
hypsochromic shift 361

ibuprofen 577
immobilization 600
impact resistance 375, 377, 379, 390
in situ chemical polymerization 237
in situ vapour phase polymerization 237
in vivo 569, 572, 574, 576–577
induction charging 658
inertial impaction 641, 647
infection-resistant dressings 596
inherently conductive polymers 457
initial efficiency 650
inkjet printing 683, 686–688
intercalated nanocomposites 185
interfacial interaction 674
interfacial pressure 542
intermolecular forces 648
interplanetary voyages 710–711
ion implantation 126
ionization 553
IR protection 12
IR radiation 336–337, 339
IR transmittance 331, 334, 339, 348

Jacquard weave 672
Joule heating 24
Joule's effect 240, 245

Kevlar 379, 381
kinetic energy 647
Kiss roll coating 115
KN95 727, 729
knee braces 542
knife coating 106
knife-over-air 106
knife-over-roller 107
knitted 540, 542, 544, 610, 664
knitting 18, 26, 32, 45
knot security 580
knotting strength 560

laminated fabric 385–386
laminated object manufacturing 692
laminated textiles 152
lamination 13, 171, 654, 737, 739
Laser ablation 102
latex treatment 666
laundry 567, 572–573
layer-by-layer assembly 11
layer-by-layer printing 684

layered double hydroxides 204
L-b-L 199
LDPE 14, 725
light-emitting fabrics 44
limiting oxygen index 203
lithium-ion batteries 654
lithography 20
Lotus Effect 10
LTA aircrafts 445–446

material jetting 688, 690
mechanical exfoliation 275, 279, 316
mechanochromic textile 366
medical textile 8, 46, 533–534, 536, 538, 540, 542,
     547, 549, 551, 553, 581
Melt intercalation 188
meltblown 550, 651, 666, 668, 725, 728–729
memory polymer coating 166
metal nanoparticles 13
metal oxide 10, 553
microcomposite 185
microorganisms 550–552, 557
microporous film 173
microporous membrane 162, 654
microwave absorbent textiles 402
microwave absorption 12, 403, 405
microwave heating 13
military and defence 1, 42
military parachutes 43
military surveillance 445
Miura-origami structures 706
modeclix 708
moisture absorption 665
moisture management 12
molecular layer deposition 615
monofilaments 546, 560, 577
montmorillonite 188, 207
mothproofing 125
MSCN 412–413
multifunctional 64, 86–89
multi-walled carbon nanotube 236
muscle activity 564, 572
muscle damages 574
MWCNTs 187
MXene 17, 21, 223, 235, 239

N95 respirator 729
nano-Ag 337–338
nanocellulose 13

nanocoating 184, 653, 735
Nanocomposite coatings 194
nanocomposite membranes 190, 193
nanocomposites 14, 183–185, 210, 385–388, 390, 674
nanoencapsulation 552
nanofibre 14, 23, 380, 652, 668
nanofibrous membrane 565
nanofiller 674
nanofinishing 64–65, 68, 83, 184
nanoparticles 186, 202
nanoporous polyethylene 337, 342
nanosilver 375, 383
NanoSphere® 10
nanotechnology 2, 8, 10–11, 29, 43, 45, 49–50, 63–64, 89, 375, 377, 381, 383–384, 386, 389–390
nanoweb 176
natural fibres 666
NBC 43, 378, 383, 385–386
n-doping 20
near-field communication 563
needle-punching 650
nettle fibres 674
N-halamines 555
non-allergic 536
non-biodegradability 665
non-elastic sutures 578
Non-intumescent 161
non-Newtonian fluid 376, 379
nonwoven 46, 538–540, 544, 549–551, 565, 649
nonwoven composite 664, 666
nuclear biological and chemical 398

oil coating 651
oil film 651
oil mist 637
optical fibres 569
organic solar cells 37
orthopaedic implants 535
outdoor jacket 616
oxidative catalysis 10

P. aeruginosa 305
paper filter 650
parachute 439–440, 442, 445
paragliding 438, 441, 444
particle bounce 647, 651

particle filtration 640
particulate filters 669
passenger air bags 670
passive heating 329, 331, 333–334, 336
passive safety 670
passive smart 375–376
passive smart textiles 3
passive thermoregulation 23
PCM 23–24
Peclet number 642
PEEK 691
peelable coating 158
Peltier effect 24, 142, 257
PENG 34, 36, 196
perfluorinated alkyl substances 738
perovskite solar cells 37
personal protective cloths 536
personal protective equipment 378, 383
personal thermal comfort 329
petal dress 709
pH sensor 13, 252
Phase change material 368, 376, 383
photocatalytic 72, 87, 268, 309–310
photocatalytic nanocoatings 654
photochromic 25–26, 28
photochromic textile 25, 27, 356–357
photocurrent 37
photovoltaic cells 615
photovoltaic effect 37
photovoltaic fibres 616
photovoltaic textiles 616
pH-responsive textiles 363
physical vapour deposition 13, 126
piezochromic 25
piezochromic textiles 362
piezoelectric fibre 630
piezoelectric nanogenerator 33, 629
piezoelectric-based knitted 629
piezoresistive sensors 268, 306
plasma 11, 67, 87, 101, 128, 554, 574
pleated filter 656
PN junction 615
polyacetylene 226–227
polyacrylic 122
polyaniline 228, 555
polycationic 555
polydiacetylens 233
poly(dimethylsiloxane) 565

polyester grafts 548
polyethylene 725, 737–739, 742
polyglactin 560
polyglycolic acids 560
poly(hexamethylene biguanide) 555
polyJet technique 688
poly(L-lactic acid) 628
polymer nanocomposites 16, 49, 188
polymeric coating 153
polymerization 122
poly(m-phenylene) 228
polyolefins 14
polyoligomeric silsesquioxanes 204
poly(phenylene vinylene) 232
polyphenylenes 227
(poly(p-phenylene sulphide) 228
polypropylene 14, 725, 728, 732–733, 737
polypyrrole 231–232
polysaccharides 591
polytetrafluoroethylene 162
polythiophene 230
polyurethane 14, 123, 125, 156, 725, 732, 738
Polyurethane ureas 156
polyvinyl chloride 14, 122
polyvinyl fluoride 445
polyvinylidene chloride 14
polyvinylidene fluoride 445
porosity 132
porous scaffolds 703
Powder adhesives 134
Powder coating 119
PPE 721–726, 730, 732, 735, 737, 740–740, 745
pressure drop 640
pressure garments 542, 545
pressure sensitivity 306–307
pressure sensors 306–307
primer coat 121
programmable fibres 408
proliferation 547
protective clothing 14
protective textiles 8, 42, 375, 377
proton-exchange membrane 655
pseudocapacitors 40
PTFE membrane 655
PVDF 196, 627–628

quaternary ammonium compounds 552, 555
quaternary ammonium salt 201
quinones 557

radar absorbing material 411
radar absorbing textiles 405
radiative cooling 339, 341–343
radiative heating 334–337
reactive coating 156
reactive oxygen species 202, 553
recyclability 674
reduced graphene oxide 19, 238, 401, 404
respirator 723–724, 726–727, 729–730,
    732–733, 744
respiratory filters 644
responsive textiles 8, 22, 29
reusable gowns 732
reusable mask 728
reusable PPE 745
reverse roll coating 115
robocasting 687
roll coating 113
rotary screen coating 116

S. aureus 305
sanitary napkins 538, 551
scatter coating 119
screen printing 683, 687
scrim 666
seat belt 670
Seebeck coefficient 259, 610
Seebeck effect 609
selective laser sintering 47, 690, 691
self-cleaning 3, 10, 14, 16, 64, 68–72, 86–87,
    102, 125
self-healing 166, 567–568, 581
self-tightening capacity 579
semiconductor 10, 15
sensor 397–399, 409, 446–449, 455–457, 461, 470,
    608, 734, 740
separation 648
service life 656
shape memory polymers 166
shape-changing dress 708
shape-memory 704, 706–707
shape-memory alloys 706
shape-memory textiles 8, 24
shear thickening fluids 43, 399
sheet lamination 171
shielding effectiveness 165
sieving 645
silicate fibre 669
silicon coating 673

silicon dioxide 161
silicon fibres 535
silicone 666
silk fibres 536
silver-coated yarns 563
single-use gowns 732
single-use mask 721, 728
single-walled carbon nanotubes 236
sleeping bag 424–426, 429
slot-die coating 109
smart body armour 398
smart coatings 98, 125
smart skin material 454
smart suture 578
smart textiles 2, 5, 375, 378, 389
smart watches 711
SMS 738–739
sniper suit 414
soft body armour 43
soft tissue implants 535
soil biodegradability 668
solar cells 25, 34, 37, 50
solar radiation 335, 342
solar reflectance 334, 342
solar reflectivity 334–335, 342
solid-phase micro-extraction 129
solvatechromic 25
solvatochromic dyes 360
solvatochromic textiles 360
solvatochromism 361
Solvent-borne laminating adhesive 138
Solvent-less laminating adhesive 138
sol–gel 11, 98, 102, 129, 131, 735
sound absorption 661
sound pressure level 659
sound propagation 662
soybean 542, 546
space fabric 706
space suits 43
spacer fabrics 542–544
speciality coatings 153
spin coating 108
spinning 17–18, 20, 23, 32
spun-bond 550, 652, 666, 739
spunlaced 549
sputter coating 127
sputtering 20, 99
stab resistant body armour 43
stain repellent 10

Staphylococcus aureus 576
stealth-coated textiles 410
stereolithography 47, 689
STF 379, 381
stimuli 2–3, 8, 16, 24–25, 29, 31
stimuli-responsive materials 355
stimuli-responsive textiles 356, 367
stimuli-responsive wound dressing 567
Stokes number 643
strain sensor 242, 248
stress relaxation 542
stress-induced orientation 32
stretchable heaters 302
structural health monitoring 455–456
superabsorbent 44, 85
supercapacitors 38, 40–42, 50
supercooling 23
superhydrophilicity 125
surface filtration 650
surgical mask 726–730, 740, 743
surgical wounds 543
sutures 533, 535, 537–539, 559–561, 574, 576–581
SWCNTs 187
synthetic blood penetration test 739

tannins 557
technical textiles 16, 42, 144, 378
tendons 538, 577
TENG 196, 620
terahertz technology 416
textile antenna 563
textile batteries 609
textile scaffolds 703
textile-based supercapacitors 609
thermal bonding 666
thermal comfort 22–24, 329–332, 336, 343,
    348–349, 351
thermal conductivity 330–332, 343, 345, 347–348,
    351, 544, 661
thermal insulation 332–334, 343, 345–346, 377,
    382, 659
thermal lamination 172
thermal radiation 332, 335, 342–343
thermal resistance 572–573
thermal well-being 330
thermochromic 25, 28
thermochromic textiles 26, 358
thermoelectric device 609
thermo-electric effect 257

thermoelectric energy generators 609
thermoelectric figure-of-merit 257
thermoelectric generators 34
thermoelectric nanogenerator 612
thermoregulating 12, 22
thermoregulation 330, 333, 336, 339–340, 343,
    351, 703
thermoregulatory textiles 22
thin-film deposition 126
transfer coating 115
transition metal oxides 29, 40
transmission loss 663, 665
transparent lamination 174–175
triboelectric charging 658
triboelectric nanogenerator 33, 399, 451, 619
triboelectric sensors 565
triboelectric-piezoelectric nanogenerators 37
triboelectrification 620
triclosan 552, 555, 576

ultra-high molecular weight polyethylene 43, 379
Ultrasonic lamination 134
UPF 76–77, 86, 88
UV protecting 268
UV protection 12, 99, 102
UV resistance 671
UV-curable laminating adhesive 140

vacuum filtration 281–282, 316
van der Waals forces 648
vascular grafts 538, 545
vascular implants 545

Vat photopolymerization 689–690
viscose fibres 535
volatile organic contents 137

warp knitting 541–542
washing durability 133
water repellency 725
water resistant 162, 664
water vapour permeability 543–544, 546, 572
waterborne laminating adhesive 137
waterborne polyurethane 289
waterproof 99, 116
waterproof fabric 2, 162
waterproof garment 152
wearable electronics 683, 704
wearable motherboard 448, 450
wearable nanogenerator 630
wearable thermoelectric generators 609
weather resistant textiles 43
weaving 17–18, 26, 32, 34, 664
weft knit structures 541
wet spinning 238
wicking 536, 651
wind chill 419
wireless connection 571
World Health Organization 723
wound dressing 533, 538, 543–544, 547, 550,
    567–569, 571, 576, 581
wounds 559–560, 567, 569, 571, 577
woven bandages 571
woven gauze 544
woven structures 547–548

www.ingramcontent.com/pod-product-compliance
Lightning Source LLC
Chambersburg PA
CBHW080337220326
41598CB00030B/4533